ISBN 978-1-330-73431-5
PIBN 10098501

1 MONTH OF
FREE
READING

at

www.ForgottenBooks.com

By purchasing this book you are eligible for one month membership to ForgottenBooks.com, giving you unlimited access to our entire collection of over 1,000,000 titles via our web site and mobile apps.

To claim your free month visit:

www.forgottenbooks.com/free98501

English
Français
Deutsche
Italiano
Español
Português

www.forgottenbooks.com

Mythology Photography **Fiction**
Fishing Christianity **Art** Cooking
Essays Buddhism Freemasonry
Medicine **Biology** Music **Ancient**
Egypt Evolution Carpentry Physics
Dance Geology **Mathematics** Fitness
Shakespeare **Folklore** Yoga Marketing
Confidence Immortality Biographies
Poetry **Psychology** Witchcraft
Electronics Chemistry History **Law**
Accounting **Philosophy** Anthropology
Alchemy Drama Quantum Mechanics
Atheism Sexual Health **Ancient History**
Entrepreneurship Languages Sport
Paleontology Needlework Islam
Metaphysics Investment Archaeology
Parenting Statistics Criminology
Motivational

THE DEPOSITS

OF THE

USEFUL MINERALS AND ROCKS

MACMILLAN AND CO., Limited
LONDON · BOMBAY · CALCUTTA
MELBOURNE

THE MACMILLAN COMPANY
NEW YORK · BOSTON · CHICAGO
DALLAS · SAN FRANCISCO

THE MACMILLAN CO. OF CANADA, Ltd.
TORONTO

THE DEPOSITS

OF THE

USEFUL MINERALS & ROCKS

THEIR ORIGIN, FORM, AND CONTENT

BY

PROF. DR. F. BEYSCHLAG
GEH. BERGRAT, DIREKTOR DER KGL. GEOLOG.
LANDESANSTALT, BERLIN

PROF. J. H. L. VOGT
AN DER UNIVERSITÄT, KRISTIANIA

PROF. DR. P. KRUSCH
ABTEILUNGSDIRIGENT A. D. KGL. GEOLOG. LANDESANSTALT U. DOZENT A. D. KGL.
BERGAKADEMIE, BERLIN

TRANSLATED BY

S. J. TRUSCOTT
ASSOCIATE ROYAL SCHOOL OF MINES, LONDON

VOL. II

LODES—METASOMATIC DEPOSITS—ORE-BEDS—
GRAVEL DEPOSITS

WITH 176 ILLUSTRATIONS

MACMILLAN AND CO., LIMITED
ST. MARTIN'S STREET, LONDON
1916

TRANSLATOR'S PREFACE

I HAVE gratefully to acknowledge the several helpful reviews upon Vol. I. of this translation.

It has been suggested [1] that it would have been better not to have used eruptive as synonymous with igneous, but rather with volcanic. That is a suggestion which many would urge. The authors, however, whose views I am representing, actually use eruptive. Moreover, this term, in association with sedimentary as counterpart, is common among both British and American authorities. Geikie, for instance, in his *Text-Book of Geology* gives eruptive undoubted preference over igneous, and, according to him,[2] the eruptive rocks include both the plutonic and the volcanic.

It appeared to me also that in discussing ore-deposits the term eruptive assisted in conveying the idea of the part played by material coming upwards through the crust. The term sedimentary, its counterpart, similarly conveyed the idea of settlement upon the crust. In these two words we therefore have the magmatic and meteoric sources of ore-deposits suggested. The term igneous, properly speaking, should have aqueous for counterpart, these two terms suggesting the elements fire and water respectively, a suggestion less pertinent than the one above.

F. L. Ransome regrets the confusion in the English and American terms for the principal oxidized zinc ores. To avoid this confusion I have adopted a suggestion by Prof. Cullis, and described these ores as zinc carbonate and zinc hydrosilicate respectively, and the mixture of the two as zinc oxidized ore.

I have at times been doubtful whether the expression 'payable' in connection with ore-deposits should be continued, or whether profitable or workable should not be substituted. Of these two alternatives, how-

---

[1] *Mining Magazine*, Vol. XII. p. 114.   [2] Page 719.

ever, the former appeared to me more applicable to an enterprise than to an ore-deposit. Moreover, profit is an indefinite term and one which Rickard felt compelled to eliminate from his definition of ore.[1] I therefore have not used its derivative.

Workable, similarly, seemed to raise the question as to whether the dimensions of the deposit, physically speaking, allowed it to be worked. It appeared more applicable therefore to beds of coal or ironstone, where, the whole material of the deposit being the valuable commodity, size was the primary factor. Where, however, as with most metalliferous deposits, content is the factor first to be determined, it becomes pertinent to use a term suggestive of relative content. Pay and its derivatives have been used in this connection for generations, not only colloquially but also in monographs and technical papers. Pay-streak, pay-gravel, pay-shoot, etc., are expressions which have received the sanction both of long usage and authority; so also is payable; while payability conveniently expresses the ability to pay the cost of working, at least.

All these terms are found in Murray's *Oxford Dictionary*, payable being defined as follows :

1. Of a sum of money, a bill, etc. . . .
2. Mining (in active sense); of a mine, a bed of ore, a vein of metal, etc.: That can be made to pay, or yield adequate return for the cost of working; capable of being profitably worked.

Rickard, who otherwise discountenances the use of payable, says of this dictionary :[2] "The *Oxford Dictionary* is the ultimate authority in our language. It is the function of a dictionary . . . to record the words that have, after probation, found a place in our language."

Accepting a suggestion of Prof. Henry Louis that flucan was a doubtful rendering of *Gangtonschiefer*, I have in this present volume translated that word as lode-slate. This material is the altered, crushed, and sometimes ore-impregnated slaty material occasionally found in the lode-filling. Speaking generally, it might be considered as included in the more frequent term fault-rock.

Flucan I have taken to be an occurrence rather than a material, and to be the equivalent of *Lettenkluft*, which literally means clay-fissure. The term flucan formerly covered two things — namely, the clayey material found in fissures, and the clay-filled fissure itself. In this

---

[1] *Mining Magazine*, Vol. X. p. 257. [2] *Trans. I.M.M.* Vol. XIX. p. 589.

work the clayey material I have described as gouge, and the clay-filled fissure as a flucan.

The expression clay-parting I have used for a clay-filled fissure parallel to the bedding or to the walls of a deposit.

Sahlband I have translated as lode-wall or, more simply, wall when speaking of a lode.

In Vol. I. I translated *Graben* and *Horst* as tectonic depression and tectonic elevation respectively. In this volume I have used subsidence and uplift, though perhaps trough-subsidence and block-uplift would be more expressive.

These two volumes, I. and II., form the complete work on ore-deposits. The third volume necessary to conform to the title, " The Deposits of the Useful Minerals and Rocks," has, so far as is known, not yet appeared.

I take pleasure in acknowledging my indebtedness to Miss M. B. Handy for many suggestions and for relieving me of countless details in this translation.

S. J. TRUSCOTT.

LONDON, *Sept.* 1915.

CONTENTS

ORE-DEPOSITS

[1] Touching the points mentioned under the young gold-silver lodes.

[1] Touching the points mentioned under the young gold-silver lodes.

[2] Partly in Austria.

[3] Touching the points mentioned under the young gold-silver lodes.

LIST OF ILLUSTRATIONS

THE YOUNG GOLD-SILVER LODES

THE lodes belonging to this group for the greater part carry both gold and silver; more seldom they carry either gold or silver; while sometimes they carry silver and lead. It is characteristic of them, and especially of the largest and richest, that they occur in geologically young and chiefly Tertiary country, in association with numerous intrusions of eruptive rock, between which rock and the deposits both the closest connection and the most obvious dependence exist.

In Europe, lodes of this character are met in the Carpathians and along a mountain range near Cartagena in south-eastern Spain. The lead-silver deposit of Pontgibaud in France, on the western side of the large Tertiary eruptive area in Auvergne, may likewise be considered as belonging to this group. It is however in the extensive Andes of Chili, Bolivia, and Peru, in the mountain ranges of Mexico, in the Great Basin of the United States and, continuing farther north, in the Sierra Nevada and Rocky Mountains, that they have their greatest development. There they do not end but are found to the north again, in Alaska.

Lodes of similar character are next met in Japan and then, farther to the south, along the east coast of Asia, and in Sumatra, Borneo, Celebes, and the Philippines.

This disposition of these deposits indicates a distribution coincident with the geologically young mountain chains which, to the east and west, border the Pacific. The Tertiary area of Hauraki in New Zealand, where among others the famous Waihi gold-silver deposit occurs, is, so far as is known, a disconnected and isolated occurrence.

It is of particular interest that lodes of this group do not occur in those Tertiary ranges which have not to any extent suffered intrusion by young eruptive rocks. The Alps and the Pyrenees, for instance, contain no such lodes.

As already mentioned,[1] the young gold-, gold-silver-, silver-, and silver-lead lodes are distinguished from the old gold-silver and silver-lead lodes not only by their geological age and the association with

[1] *Ante*, p. 185.

young eruptive rocks, but also by their association with certain alteration zones of those rocks, the propylitization zone for instance, with which in general they appear to be connected. It must be remarked, however, that where no index to age is forthcoming the difference between the two groups is not always pronounced, though mineralogically a preponderance of sulpho-salts is in many cases characteristic of the younger group. When therefore determining whether any particular lode should be placed in the one group or in the other, no single criterion may be taken as decisive, but the sum of all.

This division into a young and an old group, first proposed by F. v. Richthofen,[1] has since been adopted by many other authorities, such for instance as Suess,[2] Vogt,[3] Lindgren, Ransome, and Spurr.

The Relation of the Young Silver-Gold Lodes to the Young Eruptive Rocks.—The spacial and genetic connection between these lodes and young eruptive rocks is observable in every district where such lodes occur, whether the particular eruptives be of Miocene age, as they chiefly are ; of Lower Tertiary, as they occasionally are ; or of Late Cretaceous, as they are in isolated cases. The lodes occur preferably in volcanic chimneys, the so-called ' necks,' but also in the country-rock immediately adjacent. Less frequently they occur in portions of the eruptive farther removed from the centre of extrusion.

The eruptive rocks concerned are in most cases andesite or dacite, are often also rhyolite, are at times trachyte or even phonolite, but very seldom are of basalt. In districts where erosion has cut deep or where mining operations have penetrated to greater depths than usual, these rocks here and there show a normal granitic structure, indicating a consolidation under conditions productive of plutonic rocks ; such occurrences have been described in the Hodritz valley near Schemnitz, at the Comstock, at Cripple Creek, etc.

While in the case of the tin- and the apatite lodes constant association with acid and basic rocks respectively has been established, the young gold-silver lodes show no such settled dependence, but maintain a close connection with an eruption as a whole, whereby they often occur associated with rocks of varied petrographical character, all of which however must have been derived from one and the same stock magma.[4] Furthermore, although W. Moericke [5] with reference to the gold-, silver-, and

[1] See Literature of Hungary and the Comstock.

[2] *Zukunft des Goldes*, 1877, and *des Silbers*, 1892.

[3] *Zeit. f. prakt. Geol.*, 1895, p. 485 ; 1898, p. 388.

[4] See the descriptions of Schemnitz-Kremnitz, Transylvania, Cartagena, Cripple Creek, Goldfield, Tonopah, etc., which follow.

[5] W. Moericke, *Die Gold-, Silber-, und Kupfererzlagerstätten in Chile und ihre Abhängigkeit von Eruptivgesteinen*, Freiburg i. B. 1897.

copper deposits of Chili found that the silver lodes as a rule were more often connected with the basic rocks and the gold lodes with such as were acid, no confirmation of this preference has elsewhere been observed.

The young silver- and gold lodes usually cross all the rocks belonging to ne particular eruptive epoch, and their formation must consequently be considered as belonging to a very late phase of the eruptivity. Here and there a lode is found cut by still younger eruptives ; sometimes also, as for instance at Schemnitz and at Pachuca, the lodes do not occur in the youngest of the Tertiary flows present, these flows doubtless belonging to a later eruptive epoch ; while in exceptional cases, as for instance at Tonopah in Nevada, the different eruptive rocks contain different lode-systems.

Hot springs or gas exhalations of varied description are frequently found in the neighbourhood of these lodes ; these are to be regarded as the last efforts of an expiring eruptive activity. In isolated cases such springs even occur in the lodes themselves, this occurrence at Comstock being well known. In that mine at a depth of 900 m. such large amounts of water having a temperature of 75° C., were met, that further mining operations were suspended. A similar hot spring broke into the Smuggler Union Mine at Telluride, Colorado, at a depth of 600 metres. At Cripple Creek and Tonopah, as well as at Mazarron and Pontgibaud, the miner has had at times to combat carbonic acid exhalations, though in the case of Mazarron it is questionable whether these were of volcanic origin.

In some cases the close relationship between the ore occurrence and volcanic phenomena is also suggested by a striking increase of the temperature in depth, such increase being far above the normal. This was the case at Comstock and at Tonopah.

Deposition of the Ore in Relation to the Surface.—As stated already, most of the deposits belonging to this group were formed in Middle Tertiary time. Since then erosion has lowered the surface, though naturally not to the same extent as would have been the case if the deposits had been formed at some more ancient period. In 1909 Ransome estimated the depth eroded at Goldfield and, together with Lindgren in 1906, that at Cripple Creek, to be at most 300 metres. Although every district will of necessity have its own figure, it may be said that the above relatively low figure affords some idea of such erosion in general.

Mines belonging to this group only in the rarest cases reach to depths greater than 750 metres. Adding this figure to the depth eroded, it may be said that these lodes are known to a depth which at the most is not more than 1·25 km. below the surface existent at the time of their formation. When it is realized, for instance, that the silver mines at Kongsberg in Norway have been exploited to an equivalent absolute depth of some

3–5 km., it is evident that the mineralogical character of lodes of different age may only be properly compared in relation to such absolute depths, and not to present depths.

In many Tertiary lode districts, as for instance Goldfield in Nevada,[1] Cripple Creek in Colorado,[2] Potosi in Bolivia,[3] and Mazarron in Spain,[4] it has been particularly remarked that the number of lodes at the surface is much greater than at a depth of 400–500 metres. In the neighbourhood of the present surface—that is, some hundred metres below that which existed when the lode became formed—the ore in some districts occurs chiefly in contraction fissures, while in greater depth proper tectonic fissures are the rule. Moreover, even in districts where tectonic fissures alone occur the same numerical decrease in depth may be observed. This phenomenon is probably due to the greater resistance which in greater depth the rock opposes to the fracturing forces.

Propylitization.—The country-rock of these lodes is almost invariably more or less altered,[5] such alteration, as indicated in Figs. 299 and 300, often continuing for a considerable width. This is all the more striking in that it is repeated faithfully in situations of the most varied geographical distribution. It is an interesting fact also that this alteration is not always accompanied by ore but, as illustrated in Fig. 305, it is also met where the fissure is either entirely, or almost free from ore. Not only has this alteration taken place in andesite, dacite, rhyolite, trachyte, phonolite, syenite, and diorite, the rocks associated with the lodes, but it has also been found in such rocks as have become involved by reason of their accidental proximity to them. The Jurassic melaphyre at Boicza in Transylvania has, according to Semper, suffered in this manner ; and, according to Lindgren and Ransome, so also has the pre-Cambrian granite of Cripple Creek.

As is well known, the andesite of the Carpathians, altered in this manner, was in the 'sixties regarded by F. v. Richthofen as an independent eruptive rock which in his opinion was the oldest or first member of the Tertiary eruptives and by him accordingly named propylite, a name since retained to designate this altered rock, though it has long been established that the rock originally so termed was no new primary species but a secondary product.

Propylitization is not uniform in all districts. Propylite proper, —which according to Rosenbusch is a pathogenetic alteration particularly of andesite or andesite-dacite—is widely distributed. In its formation the original rock became bleached, friable, and impregnated more or

[1] Ransome, 1909.   [2] Lindgren and Ransome, 1906.
[3] Steinmann, 1910.   [4] Pilz, 1905, 1906.
[5] *Ante*, p. 134.

less with pyrite. In addition, new minerals were formed, chlorite, sericite, calcite, and epidote, particularly; quartz, adularia, etc., often; and kaolin occasionally. In this connection it is interesting to note that many of the white and clayey occurrences formerly regarded as kaolin, are, according to more recent investigation, in reality sericite.

In process of propylitization the ferro-magnesian silicates, augite, hornblende, biotite, etc., are decomposed earlier than the felspars. At decomposition they provide the material for the secondary formation of chlorite more particularly, but also of calcite, epidote, quartz, etc. Pyrite occurs enveloping or enclosed within the ferro-magnesian silicates, but often also in the place of original magnetite, the iron for its formation having been derived from original minerals, while the sulphur, as sulphuretted hydrogen or as an alkaline sulphide, entered with the mineralizing solution. The felspars are altered chiefly to sericite, though calcite, epidote, quartz, kaolin, etc., are formed to a less extent. In the case of dacite the quartz dihexahedra, in consequence of their greater resistance, are occasionally found unaltered in the resultant propylite.

The following analyses serve to emphasize the difference between propylitized and fresh rocks. Silica and alumina, where decomposition is not extreme, are not greatly displaced. Some magnesia always, and some phosphoric acid and titanic acid often, are removed. Lime occasionally becomes diminished, though often, on the other hand, in consequence of the secondary presence of calcite, it is increased. Soda suffers invariably. Potash, in consequence of the formation of sericite or adularia, is generally increased, the increase being at times considerable. Although propylite without exception contains a good deal of pyrite, the total quantity of iron present is nevertheless usually less than in the original rock.

The silica content, owing to the removal of various bases, is occasionally found to be substantially higher. At times, not by the addition of silica but by the removal of bases, a porous rock remains, consisting chiefly of quartz. On the other hand, a considerable addition of silica resulting in the silicification of the rock sometimes takes place, as for instance at the Csetatye near Verespatak in Transylvania. At Tonopah also, the country-rock in the neighbourhood of the lodes is in places altered to a rock consisting chiefly of quartz with some sericite and adularia, while, farther away, calcite and some sericite are the principal secondarily-formed minerals.[1]

[1] Spurr, *loc. cit.*

[TABLE

| | Hornblende-Andesite. | | | Hornblende-Dacite. | | |
| | Fresh. | Altered. | | Fresh. | Altered. | |
	1a.	1b.	1c.	2a.	2b.	2c.
SiO_2 . . .	57·42	57·99	58·98	63·45	61·78	76·61
Al_2O_3 .	17·61	17·59	11·21	15·26	14·89	8·31
Fe_2O_3 . .	2·34	1·56	1·45	2·28	2·08	1·08
FeO . .	3·77	2·37	2·42	3·01	2·51	0·59
FeS_2 . .	...	1·42	3·13	...	0·65	3·59
MnO . . .	0·43	0·21	0·11	0·36	0·28	0·11
MgO . . .	2·19	2·01	1·43	1·29	1·08	0·51
CaO . . .	5·69	5·45	8·11	3·44	3·16	3·61
Na_2O . .	3·22	1·98	0·61	2·21	2·18	0·29
K_2O . .	1·94	1·65	3·93	1·78	3·68	1·98
H_2O . .	3·47	3·45	3·69	4·00	4·94	1·51
TiO_2 . .	0·68	0·51	0·11	0·75	0·69	0·28
P_2O_5 . .	0·31	0·35	0·06	0·29	0·30	0·11
CO_2 . . .	0·95	3·89	4·69	1·08	2·01	1·87
Totals . .	100·02	100·43	99·93	99·20	100·23	100·45

The above analyses are of material from the Hauriki - Goldfield, New Zealand: No. 1 from Thames; No. 2 from Waihi; 1a, 2a are of fresh rock; 1b, 2b are of partly altered rock; 1c, 2c are of much altered rock.[1]

| | Latite-Phonolite. | | Granite. | |
	Fresh.	Altered.	Fresh.	Altered.
SiO_2 . . .	59·38	56·74	66·20	59·58
Al_2O_3 . . .	19·47	20·30	14·33	16·00
Fe_2O_3 . . .	1·60	1·06	2·09	0·30
FeO . . .	1·19	...	1·93	0·65
FeS_2 . . .	...	4·65	0·12	4·78
MgO . . .	0·36	0·23	0·89	0·03
CaO . . .	1·96	0·57	1·39	2·03
Na_2O . . .	7·80	0·62	2·58	0·98
K_2O . . .	5·83	13·36	7·31	11·93
H_2O . . .	0·80	1·48	1·31	1·13
TiO_2 . . .	0·58	0·58	0·65	0·75
P_2O_5 . . .	0·08	0·25	0·25	0·32
SO_3 . . .	0·37	...	...	...
Cl . . .	0·22	...	...	...
CO_2 . . .	...	...	0·36	0·26
Totals . .	100·05	100·10	99·74	99·66

The above analyses are of material from Cripple Creek; the altered rocks here contain much sericite in addition to adularia.[2]

Propylitization was first closely investigated by G. F. Becker in 1883

---

[1] Finlayson, *Econ. Geol.* IV., 1909.
[2] Lindgren and Ransome, 1906.

in connection with the Comstock Lode, and by Béla v. Inkey in 1885 in connection with the occurrence at Nagyag. Since then W. Lindgren in a paper entitled *Metasomatic Processes in Fissure Veins*,[1] has discussed this subject, and descriptions of separate occurrences have appeared in recent years by Lindgren, Ransome, Spurr, etc., in connection with the Cripple Creek, Goldfield, Tonopah, and other gold-silver deposits in North America. Finlayson also has contributed to the subject in descriptions of Hauraki, New Zealand.

Besides propylite in the narrow sense of the term, there are in some cases other similarly formed alteration products. Sometimes, for instance, the decomposed rock on either side of a lode or fissure is remarkable for its richness in sericite and calcite. In other cases the sericite is accompanied by kaolin; in others again kaolin and alunite, $K_2O.Al_2O_3.4SO_3.6H_2O$, have become abundantly formed in the country-rock. A recent paper by Ransome[2] and a special report upon the Goldfield district[3] by the same authority, both deal fully with the formation of this last-named mineral. From the analyses Ia and IIa given below, and from determinations of specific gravity and of pore

		Dacite.		Content: Grammes per 100 cc. of Rock.	
		Fresh, Ia.	Altered, IIa.	Fresh, Ib.	Altered, IIb.
SiO_2	. . . .	59·95	60·53	160·07	151·27
Al_2O_3	. . . .	15·77	15·32	42·11	38·27
Fe_2O_3	. . . .	3·34	0·20	8·91	0·50
FeO	. . . .	2·34	0·14	6·23	0·35
FeS_2	. . . .	0·00	7·20	0·00	18·00
MgO	. . . .	2·73	0·06	7·38	0·15
CaO	. . . .	5·84	0·41	15·57	1·02
Na_2O	. . . .	3·07	0·84	8·21	2·09
K_2O	. . . .	2·52	1·06	6·73	2·64
H_2O {Below 110° C. .		0·95	1·33	2·55	3·31
{Above 110° C.	.	2·00	6·60	5·34	16·49
TiO_2	. . . .	0·82	0·80	...	...
ZrO_2	. . . .	0·02	0·01	...	...
CO_2	. . . .	0·00	0·00	0·00	0·00
P_2O_5	. . . .	0·26	0·27	...	...
SO_3	. . . .	0·00	5·97	0 00	14 91
F	. . . .	0·00	Trace	...	...
MnO	. . . .	0·09	Trace	...	...
BaO	. . . .	0·11	0·06	...	...
SiO	. . . .	0·13	...	...	...
Total	. .	99·94	100·80	263·00	249 00

[1] 'Genesis of Ore Deposits,' *Amer. Inst. Min. Eng.* XXX. and XXXI., 1902.
[2] F. L. Ransome, 'The Association of Alunite with Gold in the Goldfield District, Nevada,' *Econ. Geol.* II., 1907.
[3] 1909, *loc. cit.*

volume, it is calculated that 100 cc. of fresh rock on the one hand, and of altered rock on the other, contain the weights of different minerals given in columns I*b* and II*b* respectively.

The altered rock of which the chemical composition is given in columns II*a* and II*b*, is calculated to consist mineralogically of :

49·38 per cent Quartz.	7·20 per cent Pyrite.
23·99 ,, Kaolin.	2·53 ,, Water.
15·73 ,, Alunite.	1·17 ,, Undetermined.

In some samples the presence of diaspore was established.

The quartz-alunite rock forms at times pseudomorphs after primary felspar. During the process of alteration some silica and alumina, very much alkali, almost the whole of the lime and magnesia, and some iron, are removed. The greater part of the iron however goes to the formation of pyrite.[1]

Propylitization—not only in the narrow sense of the term but including also the alteration to sericite-calcite, quartz-kaolin-alunite, etc.—differs, as Lindgren conclusively demonstrated, essentially from surface-weathering, and is of such a character as may only be satisfactorily explained by the action of heated waters ascending in fissures. Such waters saturating the‖shattered and often somewhat porous country-rock were at times able to effect an alteration for distances as great as one kilometre or more on either side of the fissure. With flat-lying lodes, as illustrated in Figs. 299 and 300, the hanging-wall is usually more highly altered than the foot-wall, a result which expresses the endeavour of the solutions to take the shortest route to the surface.

From the iron content of the original rock, and by the action of sulphur contained in the heated waters either as sulphuretted hydrogen or as an alkaline sulphide, practically without exception pyrite became formed. The frequently observed abundant formation of calcite and carbonates postulates a content in carbonic acid which in many cases must have been high, though in other cases this acid was absent or nearly so. The increment of potassium, which occasionally is considerable, indicates at times an equally considerable amount of that element in these heavy-metal solutions. The secondary formation of alunite and selenite is due to the action of sulphuric acid, this acid, according to Ransome, having presumably resulted from the oxidation of the sulphuretted hydrogen in the solutions, by descending air. Under these conditions sulphurous acid may also sometimes have been formed.

In many instances payable ore has been found in the propylite.

The Gangue-Minerals.—The most widely distributed gangue-mineral in the young gold-, gold-silver-, silver-, and silver-lead lodes is quartz,

[1] *Postea*, p. 549.

which now and then is accompanied by chalcedony, and exceptionally by opal; calcite and dolomite generally occur to a less extent; siderite is found plentifully in some lodes in southern Spain, as for instance at Mazarron; barite is fairly common, though generally in but small amount; rhodochrosite is not uncommon, its occurrence having been particularly remarked in Hungary, North America, Mexico, and Japan; rhodonite has similarly deserved particular mention at Kapnik, Verespatak, at several places in Mexico, and in the Tertiary silver lodes at Butte, Montana; the occurrence of alabandite has been established at Nagyag, Kapnik, etc.; adularia occurs now and then; [1] while zeolites seldom occur.

Fluorite in most districts is completely absent, both from the lodes as well as from the propylitized country-rock, this being also the case with the other compounds of fluorine and of chlorine, except such minerals as cerargyrite, etc., the occurrence of which is limited to the oxidation zone. This absence shows that, like boron, fluorine and chlorine played little part in the formation of the lodes here being described. The exceptional occurrence of fluorite in the very rich telluride-gold district of Cripple Creek, where the gangue consists approximately of 60 per cent quartz, 20 per cent fluorite, and 20 per cent dolomite, is therefore all the more striking. Some fluorite with native gold and a little gold telluride is also found in the quartz lodes of the Judith Mountains, Montana.[2]

Against this exceptional occurrence however, is the fact that fluorite is absent both from the well-known Tertiary telluride gold mines of Transylvania as well as from the important telluride gold deposits of Western Australia; there can therefore be no regular association of fluorine and tellurium in the group of lodes being described. Further, the almost complete absence of fluorine is a feature common to both the young and the old gold lodes. In the case of the silver lodes, on the other hand, it is worthy of remark that the occurrence of fluorite in the old lodes is frequent, while in the young lodes it is either absent or present only as an exception.

The Ores and the Relation between the Gold and the Silver.—In the old gold lodes and the old silver- and silver-lead lodes the two precious metals usually occur apart. Although silver in small amount is almost always present in the old gold lodes, it is seldom that this amount is more than one-fifth to one-third of the gold present. Similarly, any gold content with the old silver- or silver-lead lodes is extremely small; for instance, at Kongsberg one part of gold is present for approximately 10,000 parts of silver, and at Freiberg for 5000–10,000 parts.[3]

[1] Lindgren, 'Orthoclase a Gangue Mineral,' *Amer. Journ. Sc.* V., 1898; *U.S. Geol. Survey*, 20th Ann. Rep., 1900.
[2] Weed and Pirsson, *U.S. Geol. Survey*, 18th Ann. Rep. III., 1896–1897, pp. 445-614.
[3] Vogt, *Zeit. f. prakt. Geol.*, 1896, p. 389.

With the young gold lodes, on the other hand, it is often the case that the two metals occur in such measure that both are of economic importance. This is a feature of the occurrences in Hungary, of many places in the Great Basin of the United States, in Mexico, South America, Japan, Sumatra, and elsewhere. The relation between the two metals in some of the Hungarian mines may be gathered from the following figures : [1]

		One Part of Gold to Parts of Silver.
Transylvania	Nagyag, Muczari, Verespatak . .	about 1·0
	Boicza	,, 1·5–2·0
	Ruda	,, 2·0–3·0
	Kajanel, Main Lode	,, 10·0
Carpathians	Kreuzberg	,, 10·0
	Borsabanya	,, 10·0–12·0
	Nagybanya	,, 2·0–2·5
	Veresvir	,, 25·0–30·0
	Felsöbanya	,, 50·0–60·0
	Kapnik	,, 100·0
	Alt-Rodna	,, 150·0
	Kremnitz	,, 2·5–5·0
	Schemnitz, average	,, 50·0
	Schemnitz, some lodes . . .	,, 6·8

For occurrences outside of Europe the following figures are illustrative :

		One Part of Gold to Parts of Silver.
Nevada	Comstock	22·5
	Tonopah	100·0
Colorado	Custer Co.	45·0
	Clear Creek	80·0
	Rico Mountains	125·0
Idaho	Owyhee	35·0
Japan	Numerous lodes	5·0–100·0
Sumatra	Redjang Lebong	10·0

In view of the variety of the relation between the two metals the lodes of the young gold-silver group may for convenience of description be divided as follows :

1. Gold lodes proper, containing little silver ; as for instance those of Cripple Creek with 1 of gold to 0·1 of silver ; those of Goldfield with 0·15 of silver ; many other lodes in North, South, and Central America, and in New Zealand, etc.
2. Gold-silver lodes, these being exceedingly well represented.
3. Silver lodes proper, these being well represented in Mexico, the United States, South America, etc.

With some of the young gold lodes the gold in its primary condition occurs either entirely, or to a large extent in the form of different compounds of tellurium and gold. In this respect Nagyag in Transylvania, with gold telluride exclusively and no primary native gold, is famous. On the

---

[1] Vogt, *Zeit. f. prakt. Geol.*, 1898, p. 388, and supplement.

other hand, many of the lodes in Transylvania carry gold telluride together with native gold, while most carry native gold with no gold telluride. Still more important are the rich telluride lodes of Cripple Creek, Colorado, including those of the Boulder district to the north and the Telluride district to the south-west. Nevertheless, even in Colorado, in most of the mining districts the lodes carry native gold only. Gold telluride occurs also in the lodes of the Black Hills, Dakota. Although therefore in many places gold telluride plays a prominent part in the young gold lodes, in the majority of cases the gold is either associated with pyrite or it occurs native.

Auriferous telluride also occurs plentifully in Western Australia, where however it has not yet been possible to state with certainty the age of the accompanying eruptive rocks.

Auriferous gravels in association with the young gold lodes, as will be further discussed when describing the occurrences in the United States, play but a very small part.

Selenium occurs fairly plentifully at Tonopah in Nevada, where one part of gold occurs for every 100 of silver; and at Redjang Lebong in Sumatra, where the silver content is ten times that of the gold; while it has also been observed in smaller amounts elsewhere. As already stated,[1] selenium and tellurium in this connection appear to mutually replace one another. Selenium also occurs occasionally in the old gold-silver lodes, as for instance in the gold-bismuth-selenium lodes at Fahlun.[2]

The silver occurs chiefly in the form of sulpho-salts, such as pyrargyrite, proustite, stephanite, polybasite, tetrahedrite, etc.; or as argentite. Native silver as a primary deposit also occurs here and there.

The gold- and silver minerals in these lodes are accompanied by pyrite, chalcopyrite, sphalerite, galena, and often also by arsenic-, antimony-, and bismuth minerals. In the young silver-lead lodes, auriferous galena is naturally well represented. Nickel- and cobalt minerals are extremely rare. The occurrence of tin in this connection is mentioned when describing the lodes of Bolivia.

The minerals present with the young gold-silver lodes are in general distinguished from those of the old lodes in that generally they include proportionally more sulpho-salts and other arsenic- and antimony minerals, and, exceptionally, even tin minerals.

Division into Sub-Groups.—According to the nature of the minerals present the young gold-silver lodes may be divided into a number of sub-groups. The following may thus be differentiated:

    1. Telluride gold lodes with quartz, calcite, or dolomite, and much
        fluorite; as at Cripple Creek.

[1] *Ante*, p. 83.   [2] *Ante*, pp. 166, 314.

2. Telluride gold lodes with calcite, rhodochrosite, quartz, etc., but with no fluorite; as at Nagyag.

3. Lodes with both telluride and native gold in the same quartz gangue; as at Offenbanya.

4. Lodes with native gold but without telluride, yet generally with some silver and small amounts of other ores. These are chiefly quartz lodes characterized by ordinary propylitization. Examples of such occur at Kremnitz, many in Transylvania, in the United States, in Mexico, in South America, Japan, Sumatra, etc. Some of these lodes carry selenium, when also traces of tellurium are generally found.

5. Lodes with native gold but without telluride, and with but little silver or other ores; quartz gangue; characterized by the alteration of the walls to alunite and kaolin; as at Goldfield in Nevada.

6. Gold-silver lodes carrying the two metals in such proportion that their values are roughly equal; with but little galena, though often some arsenic, antimony, etc.; and with quartz as chief gangue-mineral. The most famous of such lodes is the Comstock Lode of Nevada. At Tonopah, comparatively speaking, much selenium occurs.

7. Gold-silver-lead lodes with the two precious metals having approximately equal value; with much galena and other ores·; and with quartz as the principal gangue; as some lodes at Schemnitz and elsewhere.

8. Silver lodes with little gold or galena; with quartz again as principal gangue-mineral; as numerous lodes in Mexico and elsewhere.

9. Lead-silver lodes with little gold but with much galena; with quartz as chief gangue - mineral; as several lodes in the Schemnitz district and numerous occurrences in Mexico and South America.

10. Galena lodes with a relatively low silver content—1–3 kg. per ton—and practically without gold; as at Mazarron in Spain, where siderite is the characteristic gangue-mineral.

11. Silver lodes with relatively much copper, as at El Pasco in Peru.

12. Silver lodes with tin ores, as in Bolivia.

If in addition the relative abundance of calcite and barite were taken into account, a still greater sub-division might be made.

Mineralogically and metallurgically all these sub-groups differ markedly from one another in the relative proportions of gold, silver, lead, as well as copper, zinc, tin, etc. Too much importance should however not be attached to this difference, because often in one and the same limited district the individual lodes may show considerable fluctuations in this respect, and also because different sub-groups may occur quite close together. In this connection the Schemnitz district is particularly interesting, though in spite of all the variety in the sub-groups there present, it is probable that all in their main features were formed by the same chemical-geological processes.

It must further be added that the well-defined and geologically allied groups occurring within one metal province often differ widely from one another in their mineralogical and metallurgical relations. For instance, the lodes in Transylvania with gold telluride, those with telluride and native gold, and those with native gold without telluride but with some silver, are quite distinct from one another. In the Schemnitz-Kremnitz district where the two towns so named are situated but 25 km. apart, the sub-groups Nos. 4, 7, and 8 are all represented. Similarly in Colorado representatives of Nos. 1, 3, 4, 6, and 7 are found. The two districts of Goldfield and Tonopah, though geologically so similar and but 45 km. apart, nevertheless differ materially in the proportion of gold to silver, and in the alteration of the country-rock ; at Goldfield, No. 5 only, with 1 of gold to 0·15 of silver, is represented ; at Tonopah, No. 6, with 1 part of gold to 100 parts of silver.

Instances of geologically, mineralogically, and metallurgically differing sub-groups occurring at small distance from one another or even in close spacial connection, are numerous.

For these reasons, and unlike the Freiberg school and the text-books of Beck and Stelzner-Bergeat, we have not based our classification upon the above-mentioned sub-groups, but have taken the view that geological uniformity and relationship are more important factors, though naturally at the same time the metal- and mineral combinations have been given due consideration.

Primary and Secondary Depth-Zones.—Many of the Tertiary gold- silver lodes are remarkable for their great richness, this having been strikingly the case in North, Central, and South America. In the upper levels particularly, enormous bodies of silver-, silver-gold-, and gold ores have sometimes been found. With the silver lodes these bonanzas in the great majority of cases represented local enrichment in the oxidation zone, when to the happy circumstance of their occurrence was added the fact that such silver ores were amenable to treatment by the simplest of metallurgical processes. From such occurrences the fabulous amounts of silver produced by Mexico, Bolivia, and Peru, in the sixteenth, seventeenth, and eighteenth centuries, were derived. Enrichment in the cementation zone, otherwise so frequent, appears with these lodes to play but a small part. According to Lindgren and Ransome, both of whom were well acquainted with the sulphide enrichment in the Butte district, a similar zone of enrichment is wanting at Cripple Creek and at Goldfield ; while Steinmann, in reference to the silver lodes at Potosi, Oruro, and El Pasco in South America, remarks that these either have no cementation zone or one of but little importance.

A large concentration of metal in the cementation zone demands the presence of considerable quantities of sulphide ores to provide by oxidation the ferric sulphate necessary to dissolve the gold, silver, copper, etc. Such conditions are found with many deposits, but do not exist at Cripple Creek, Goldfield, and other occurrences belonging to the young gold-silver group.

Krusch, with respect to the lodes of Western Australia, observed that, just as at Cripple Creek, the oxidation zone there ends immediately above the primary zone, and he suspected that the absence of a cementation zone in this case might be connected with the presence of tellurium.

With many of those Tertiary gold- and silver lodes which were extraordinarily rich at first, it was the subsequent experience that in the primary zone and in depth the metal content diminished. This was remarkably the case at Potosi in Bolivia,[1] and with many other deposits in South America and Mexico. At Cripple Creek also and at Goldfield, an impoverishment of the gold began in depth as the primary zone was entered. In several districts—Potosi, Cripple Creek, Mazarron, for instance—it has also been particularly remarked that not only does the number of lodes decrease in depth but also the amount of metal contained per square metre upon the lode plane. In the very richest districts particularly, the bulk of the metal appears to have been deposited at depths less than 500 m. below the surface. Much of this afterwards became removed by erosion, though at the same time a considerable portion migrated downwards and was there retained. It is from such enrichments in the oxidation zone that the tremendous quantities of silver were obtained.

Primary variation in depth has received expression in the case of some of the richest of these lodes by obvious impoverishment as depth was attained ; while in other cases it has been marked by the occurrence of other ores. For instance at the El Pasco silver mine in Peru the copper content materially increases in depth ; with many silver mines in Mexico, etc., sphalerite, or sphalerite and galena increase in depth, while the silver diminishes.

The valuable contents of these lodes are also often found concentrated in ore-shoots, this having been particularly the case at Comstock.[2] Such

---

[1] *Postea,* p. 578.   [2] *Postea,* p. 529.

accumulations of ore appear however to be primary; whether secondary processes contributed to their formation has not yet been established.

The Economic Importance of the Young Gold-Silver Lodes.—As discussed in a later section, these lodes have latterly yielded more than one-half of the annual silver production of the world and roughly one-fourth that of the gold. Figures of total production in the case of the best known of these deposits are given below, in addition to which other similar figures have already been given.[1]

Potosi, Bolivia, since 1545, about 30,000 tons of silver, worth about 300 million sterling.

Guanajuato, Mexico, since 1558, about 15,000 tons of silver, worth about 160 million sterling.

Zacatecas, Mexico, 1548–1832, about 14,000 tons of silver, worth about 150 million sterling.

Comstock, Nevada, 1859–1902, silver and gold, worth about 77 million sterling.

Cripple Creek, Colorado, 1891–1910, about 330 tons of gold, worth about 46 million sterling.

Pachuca, Mexico, 1522–1901, more than 3500 tons of silver, worth more than 31 million sterling.

Chañarcillo, Chili, since 1832, silver worth about 22 million sterling, or according to other data about 60 million sterling.

St. Eulalia, Mexico, 1703–1890, silver, etc., worth some 27 million sterling.

Fresnillo, Mexico, from 1833 to 1863 only, 902 tons of silver, worth some 8 million sterling.

Cripple Creek has of late years produced about 22 tons of gold, worth about 3 million sterling, annually; and Goldfield in Nevada, gold worth about 1·5 million sterling. Tonopah in 1908 produced 223 tons of silver valued at £900,000, and gold to the value of £340,000, making a total of about £1,240,000. In 1877, when work on the Comstock Lode was at its zenith, the total value of the gold and silver produced was about 7·5 million sterling.

The Relation to other Groups of Deposits.—The young gold-silver lodes occur chiefly within volcanic vents and in the immediately surrounding country-rock. They are characterized by an intense metamorphism of that rock, and were deposited at no great depth below the surface. Such is their geological circumstance.

The old gold-, silver-, and silver-lead lodes, on the other hand, represent fissure-fillings unaccompanied as a rule by any such alteration of the country-rock, though now and then a secondary formation of calcite and

[1] *Ante*, pp. 202, 478.

sericite quite different from propylitization may be noticed. Further, the considerable continuance of the gold- and silver content in depth indicates that deposition took place at considerable depths below the surface. The young and the old lodes therefore differ from each other much as do the intrusive and the extrusive rocks, though no sharp demarcation between the two exists.

With the young quicksilver deposits the young gold-silver lodes have this in common, that both groups are connected with young and chiefly Tertiary eruptions. In some districts the two groups occur in close spacial connection; the Mexican and Peruvian gold-silver- and quicksilver deposits, for instance, are found in the same mountain ranges; while the Comstock Lode with its recent hot springs lies only some 9–10 km. distant from the quicksilver deposits of Steamboat Springs.[1] The young quicksilver deposits not infrequently possess a small silver- or gold content,[2] and cinnabar has here and there been found in the young gold-silver deposits, as, for instance, at Schemnitz, Nagyag, Cripple Creek, Goldfield, etc. No deposits have however yet been found in which both gold-silver ore and quicksilver ore occur together and of like importance, and the two groups of deposit therefore, in spite of their common association with young eruptive rocks, are quite distinct.

The relation of the young gold-silver lodes to the copper lodes is explained when describing the Butte district. There, in Tertiary granite or quartz-monzonite, rich copper lodes with one part of silver to 400 of copper occur associated with silver lodes, these latter having the same character as those associated with flows and dykes.

The relations with, and differences between the silver-tin lodes of Bolivia and the ordinary tin deposits characterized by fluorine minerals, are dealt with when describing the Bolivian lodes.

The same Tertiary eruptive rocks as those with which the young gold-silver lodes are associated, have here and there also formed contact-deposits, wherein as a rule auriferous and argentiferous copper- and lead-zinc ores occur together with garnet and wollastonite. A very instructive example of this is found at Offenbanya in Transylvania, this occurrence being illustrated in Fig. 49. In this connection the silver-gold deposit of Elkhorn in Montana, and others in Mexico, are deserving of mention. Sometimes also metasomatic deposits occur in the neighbourhood of these young eruptive rocks, such deposits being more particularly of lead, zinc, and silver. Examples of these are found at Mazarron-Cartagena in Spain,[3] and at several places in the United States.

Genesis of the Young Gold-Silver Lodes.—The spacial connection with

---

[1] *Ante*, p. 467. [2] *Ante*, p. 458.

[3] *Postea*, p. 547.

Tertiary or sometimes late Cretaceous eruptives, the occurrence of hot springs and gas exhalations at or near the lodes, and the propylitization of the country-rock, justify, as has already been stated, the conclusion that the deposition of the ore and gangue of these lodes took place from heated waters closely associated with actual eruption. The considerable width of rock observed at times to have suffered propylitization indicates that these waters were present in considerable volume; doubtless therefore they contained the precious metals in great dilution. The minerals present, and particularly the calcite, sericite, chlorite, epidote, kaolin, and alunite of the propylitized rock, show that the solutions were chiefly of an aqueous nature, and consequently that at the time of deposition the critical temperature of water, 365° C., was not passed. It is interesting to note in this connection that Lindgren and Ransome in 1906 estimated the temperature of the waters from which the Cripple Creek ores were deposited at 100°–200°, and the pressure at about 100 atmospheres.

The secondary formation of pyrite observable everywhere in propylite indicates in the solutions a sulphur content in the form of sulphuretted hydrogen, alkaline sulphides, etc., while the frequent though not invariable occurrence of secondary calcite and other carbonates indicates that carbonic acid must in many cases also have been present. That proportionally much more lime and magnesia than alkali became removed during propylitization may be because the solutions usually carried alkaline salts. In many cases a considerable addition of potassium has been demonstrated to have resulted, indicating a not inconsiderable potassium content for the particular solutions.

The formation of alunite, that is alunitization, is due to the action of sulphuric acid, such acid having resulted from the oxidation of sulphuretted hydrogen or alkaline sulphides;[1] the presence of secondary selenite may be similarly explained. On the other hand, the generally very sparing occurrence of barite and celestine probably indicates that the solutions originally, that is before such oxidation, usually contained but few SO_4-ions.

The generally complete absence of fluorite and other primary fluorides and chlorides indicates either that the -ions of these halogens were absent from the solutions, or that they were present in but small amount. The exceptional occurrence of fluorite at Cripple Creek is explained by Lindgren and Ransome not by the presence of free hydrofluoric acid, but by a small amount of alkaline fluoride in the solutions; sodium fluoride and potassium fluoride attack lime silicates with the formation of fluorite which is extremely difficult of solution in water.

By far the most important gangue-mineral is quartz, and the solutions

---

[1] *Ante*, p. 522.

accordingly invariably carried silica, though it is not yet known in what combination ; probably it occurred as hydrated silicic acid, H_4SiO_4, or a soluble silicate such as K_4SiO_4. With few exceptions the quartz is exclusively deposited in the fissures and not in the metasomatically altered country-rock, this fact indicating that the amount of silica in the solutions was generally comparatively low. Several authorities— Becker in his quicksilver monograph [1] and Lindgren and Ransome in their paper upon Cripple Creek,[2]—suggest that the silica was present in colloidal form, the rock walls then acting towards it as a semi-permeable membrane, allowing it to diffuse with difficulty into the country-rock, while the ordinarily -ionized salts were allowed a free entry. This view however appears very questionable.

Some pyrite is always found in these gold-silver lodes though the amount is generally low. It is sometimes accompanied by marcasite. The iron content of the pyrite in the propylite is derived chiefly, if not exclusively, from the iron originally present in the rock ; [3] the amount of iron contained in the solutions, in spite of the wide distribution of the pyrite, was therefore probably quite low. The same may be said of the manganese.

With many of these lodes the sulpho-salts, such as tetrahedrite, pyrargyrite, proustite, stephanite, etc., occur in relatively large amount, which may be because arsenic and antimony were present in the solutions as alkaline sulpho-salts.

Résumé of Foregoing Considerations.—Propylitization and the nature of the lode material together indicate that the solutions from which they resulted originally carried, little silica and iron ; generally some carbonic acid ; invariably much sulphur, either as sulphuretted hydrogen or alkaline sulphide ; often much potassium ; and frequently sulpho-salts in appreciable amount. It follows therefore that they were either neutral or alkaline, but not acid. The silica accordingly probably entered the fissures as an alkaline salt such as K_4SiO_4 and not in the colloidal form. The solutions naturally were very dilute and of different composition in different fissures.

While the old lead-zinc-silver lodes often exhibit a crusted or combed structure indicating that the solutions rising at different times were of different composition, this structure is either absent from the young lodes or is but seldom developed ; the constant or almost constant composition of the solutions in any one fissure thereby suggested, is in harmony with the idea of a large volume.

[1] Becker, ' Quicksilver,' *U.S. Geol. Survey, Min. Resources,* 1892, p. 156.
[2] Lindgren and Ransome, Cripple Creek, 1906.
[3] *Ante,* p. 522.

Among other properties silver is soluble as nitrate, as double thio-sulphate with alkali, and as double cyanide with alkali, though such solutions probably play little part in ore-deposition. Chlorides convert silver salts into silver chloride, which is easily soluble in an excess of the particular chloride, sodium chloride for instance; this property in particular cases may have been of great importance in the redistribution of the silver in the oxidation and cementation zones. It is probable, however, that at the primary deposition of the silver in these lodes some other silver compound was present, as the absence of chlorine minerals indicates that chlorides were not well represented in the original solutions. Silver is further soluble as sulphate, Ag_2SO_4; as $AgHCO_3$, in carbonic acid water;[1] and in an excess of ferrous salts, etc.

Gold is soluble in liquids containing chlorine; in hydrochloric acid when either chromic acid, selenic acid, antimonic acid, or arsenic acid is present; in bromine, alkaline cyanides, ferric sulphate, etc. Ferrous sulphate on the other hand is a well-known precipitant of gold. As already mentioned, the iron sulphates are of the greatest importance in the development of the phenomena characteristic of the oxidation and cementation zones. The solubility of gold in alkaline sulphides as a double sulpho-salt is interesting.[2] Under the application of high pressure gold is also soluble in sodium- and potassium silicates; and at about 200° C., in moderately strong sodium carbonate solution.[3]

In the solutions from which these lodes were formed the gold was probably in very great dilution. The content in the lode material is only in exceptional cases more than 50 grm. per ton, equivalent to 0·005 per cent. If it be reckoned that the lode material amounts to 1 per cent of the weight of the depositing solution, which is placing it high, then the gold content in this solution would be 0·00005 per cent; in all probability it was even considerably less.

Concerning the nature of the solutions these were generally complex, with many cathions and anions, while possibly colloids also were present. In solutions of bicarbonates—and for gold, in those of alkaline silicates and alkaline sulpho-salts also—gold and silver may be present in appreciable amount. From these solutions the silver would be precipitated chiefly as sulphide, but also as a sulpho-salt; the sulphide might be independent silver sulphide or this sulphide contained in galena. The gold would be extremely readily precipitated when the solutions came in contact with pyrite or other sulphide ore,[4] which is doubtless the reason that pyrite

---

[1] *Ante*, pp. 136, 219.
[2] G. F. Becker, *Amer. Journ. Sc.*, 1887, XXXIII. p. 199; Liversidge, *loc. cit.*
[3] Bischof, *Lehrbuch der chem. phys. Geol.* 2nd edit. III. pp. 838, 843; Liversidge, *Roy. Soc.*, New South Wales; Dölter, *Monatsbericht*, II. p. 149, etc.
[4] *Ante*, p. 139.

is so often auriferous. The precipitation of the gold as telluride may result from reactions similar to those which precipitate silver, lead, etc., as sulphides.

Though the above considerations relative to the deposition of the ore in these lodes are well based, further synthetical experiments are highly desirable in order to obtain a better idea of the conditions under which such deposition took place.

The ultimate source of the gold and silver has been the subject of discussion for many years. The theories of ascension, descension, and lateral secretion, have already been discussed,[1] and are further discussed in a later chapter. With regard to the young gold-silver lodes it must be emphasized that these were formed by the heated waters circulating towards the close of that period of eruption which, generally speaking, took place in Tertiary time. It is therefore to a certain extent justifiable to consider that these waters together with their metal content were derived directly from eruptive magma. With the advancing cooling and crystallization of that magma the residual magmatic aqueous solutions continued to become more and more charged with CO_2, etc., till as the final stage of the eruption they issued as hot springs. These in their course through the already consolidated crust would supposedly take up certain constituents, which would become more and more concentrated in them. According to Lindgren and Ransome,[2] the hot springs of Cripple Creek on their way through granite existing in depth, took up alkaline fluoride from fluorite occurring within that granite, and deposited this again as fluorite in the lodes above. In this way is explained the unique position among the young gold-silver lodes, occupied by the gold occurrences of Cripple Creek by reason of the fluorite they contain. It is not necessary to assume that all the minerals found deposited in these lodes were derived from the eruptive magma ; some material may more conceivably have been derived by lateral secretion in the widest sense of that term ; and some again by the leaching action of the waters along their course.

Hungary

LITERATURE

B. v. Cotta. Über Erzlagerstätten Ungarns und Siebenbürgens, Gangstudien, IV. 1862.—F. v. Richthofen. ' Studien aus den ungarisch-siebenbürgischen Trachytgebirgen,' Jahrb. d. k. k. geol. Reichsanst. XI., 1860 ; ' Principles of the Natural System of Volcanic Rocks,' Mem. Calif. Acad. of Sc. I., 1868, Pt. 1, reviewed in Neues Jahrb. f. Min. Geol. Pal., 1868, pp. 852-854.—L. Litschauer. ' Über die Verteilung der Erze in den (ungarischen) Lagerstätten,' Zeit. f. pr. Geol., 1893.—Exhibition brochures, Budapest, 1896 ; Paris, 1900.
Schemnitz: M. V. Lipold. ' Der Bergbau von Schemnitz,' Jahrb. d. k. k. geol. Reichs-

---

[1] *Ante*, pp. 189, 190.   [2] *Loc. cit.* 531.

anst. XVII., 1867 ; Souvenir publication at centenary of the Mining and Forestry Academy in Schemnitz, containing geological section by Faller, 1871.—ZEILER ET HENRY. ' Les Roches éruptives et les filons métallifères du district de Schemnitz,' Ann. d. Mines, III. 7, Paris, 1873.—J. W. JUND. ' On the Ancient Volcano of the District of Schemnitz,' Quart. Journ., 1876.—G. VOM RATH. Sitz.-Ber. Niederrh. Ges., 1877, pp. 291-324.—E. HUSSAK. ' Über die Eruptivgesteine bei Schemnitz,' Sitz.-Ber. d. Akad. d. Wiss. LXXXII. Vienna, 1880.—HUGO BÖCKH. ' Über das Altersverhältnis der in der Umgebung von Selmecz-banya vorkommenden Eruptivgesteine,' in Földtani Közlony, XXXI., 1901.—Geological Map of the Schemnitz Lode District by J. v. Pettkó, 1853, and Josef Szabó, 1883.

Kremnitz : ' Windakiewicz,' Jahrb. d. k. k. geol. Reichsanst. XVI., 1866.—ALEX. GESELL. Mitt. Jahrb. ungar. geol. Anst. XI., 1897-1898.

Nagybanya, with Felsöbanya, Kapnik, etc. : ' Géza Szellemy,' Zeit. f. pr. Geol., 1894.—ALEX. GESELL. ' Montangeol. Aufnahme über Nagybanya,' Jahrb. d. ungar. geol. Anst., 1891, 1892 ; ' Felsöbanya,' 1893.—GÉZA SZELLEMY, Erzlagerstätten des Vihorlat-Gutin-Trachytgebirges, Montangeol. Millenniumkongress, Budapest, 1896.

Transylvania : v. HAUER UND STACHE. Geologie Siebenbürgens, 1885. — SEMPER. ' Beitrag zur Kennt. der Goldlagerstätten des siebenbürgischen Erzegebirges,' Abhd. d. preuss. geol. Landesanst. Part 33. Berlin, 1900. Chiefly petrographical : H. HÖFER. Jahrb. d. k. k. geol. Reichsanst. XVI., 1866.—G. VOM RATH. Niederrh. Sitz.-Ber., 1874, 1876, 1879.—C. DOELTER. Jahrb. d. k. k. geol. Reichsanst. XXIX., 1878 ; Tschermaks Min. Mitt., 1874, 1880. Chiefly mineralogical : A. SCHRAUF. ' Über die Telluerze Siebenbürgens,' Zeit. f. Krist. Min. II., 1878.—A. KOCH, reviews in Zeit. f. Krist. Min. X., 1885, p. 96 ; XI., 1886, p. 262 ; XIII., 1887, ·p. 65, 607 ; XVII., 1890, p. 505.—H. B. v. FOULLON. ' Gediegen Tellur von Faczebaja,' Verh. d. k. k. geol. Reichsanst., 1884. Chiefly geological : SEMPER, see above. BÉLA v. INKEY, Nagyag, Budapest, 1885.—F. POŠEPNÝ. Jahrb. d. k. k. geol. Reichsanst. XVIII., 1868 ; XXV., 1875 ; Österr. Zeit. f. Berg- u. Hüttenw., 1894 ; Genesis der Ezlagerstätten, Vienna, 1895, p. 114-120.—G. PRIMICS. The South-Western Portion of Transylvania (in Hungarian), 1896.

Verespatak : HAUER. Jahrb. d. k. k. geol. Reichsanst. II., 1851.—GRIMM. ibid. III., 1892. — ALEX. GESELL. Jahresber. d. k. ung. geol. Landesanst., 1898 (1901). — P. T. WEISS. Jahrb. d. ung. geol. Landesanst. IX., 1891 ; upon Ruda-Brad, Berg- u. Hüttenm. Ztg. LIII., 1894 ; upon Nagy-Almás, ibid. LIV., 1895. STEINHAUSZ. ' Über Nagyag,' Österr. Zeit. f. Berg- u. Hüttenwesen, LII., 1904.—BAUER. ' Der Goldbergbau der Rudaer 12 Apostelgewerkschaft bei Brad,' Leobener Jarb. LIII., 1905.—K. v. PAPP. ' Über Karács-Czebe,' Zeit. f. prakt. Geol., 1906.—M. V. PÁLFY. ' Das Goldvorkommen in siebenbürgischen Erzegebirge und sein Verhältnis zum Nebengestein der Gänge,' ibid., 1907.

Reconnaissance Notes on Transylvania by Beyschlag and Vogt, 1896 ; on Schemnitz-Kremnitz by Vogt, 1896. Written communications from the Hungarian Geological Department and from Prof. Böckh to Vogt, 1910, with numerous statistics.

The Carpathians form that circle of hills which, incomplete towards the south-west, connects the Alps with the mountain land of the Balkan Peninsula. These hills, formed in Tertiary times, are arranged in a large number of scattered groups in broken connection with one another.

The geological structure is as follows : the oldest crystalline rocks form a number of isolated complexes surrounded to a large extent by Triassic, Rhaetic, Jurassic, and Cretaceous beds, and to a less extent by Devonian, Carboniferous, and Permian. Enveloping all of these is an upland of undisturbed Tertiary beds with which, particularly within the Carpathian circle, extensive young eruptive areas are associated. The most important of these Tertiary eruptive areas are :

1. The district of Schemnitz-Kremnitz in the western portion of the Carpathians.

2. That in the vicinity of Kaschau.
3. That in the neighbourhood of Ungvár.
4. That around Nagybanya, Felsöbanya, and Kapnik, in the central portion of the mountain chain.
5. The eastern portion of Transylvania near the Roumanian frontier.
6. The Transylvanian Erzgebirge.

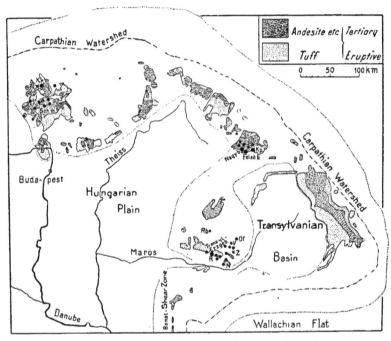

FIG. 292.—Map of the more important Tertiary eruptive areas and gold-silver lodes of Hungary and Transylvania.

Schm, Schemnitz ; *Krm*, Kremnitz ; *Nagyb*, Nagybanya ; *Felsöb*, Felsöbanya ; *Kp*, Kapnik ; *Of*, Offenbanya ; *V*, Verespatak ; *Z*, Zalatna ; *N*, Nagyag ; *R*, Ruda ; *Rb*, Rezbanya. The dotted line represents the boundary between mountain and plain.

Among the Tertiary eruptive rocks, andesite and dacite have the widest distribution, after which come rhyolite, obsidian, and basalt, while trachyte in the usual sense of this term, and phonolite, are absent. In the deep-cut Hodritz valley near Schemnitz the Tertiary magma is seen consolidated as an hyp-abyssal plutonic rock, diorite or grano-diorite.

For the district of Schemnitz-Kremnitz, H. v. Böckh [1] gives the following eruptive sequence beginning with the oldest : pyroxene-andesite, diorite

[1] *Loc. cit.*

with grano-diorite, biotite-amphibole-andesite, and finally rhyolite. These
rocks formed together one connected outpouring of which the time of
maximum outbreak was Lower and Middle Miocene. Much younger in

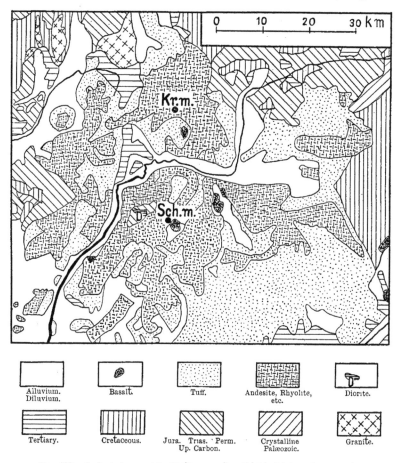

Alluvium. Diluvium.	Basalt.	Tuff.	Andesite, Rhyolite, etc.	Diorite.
Tertiary.	Cretaceous.	Jura. Trias. Perm. Up. Carbon.	Crystalline Palæozoic.	Granite.

FIG. 293.—Geological map of the Tertiary eruptive district of Schemnitz-Kremnitz.
Hungarian Geological Survey.
Schm, Schemnitz ; *Krm*, Kremnitz.

age was the extrusion of basalt, the youngest rock in the vicinity of
Schemnitz and there but poorly represented. The eruption of the
andesite and dacite at Nagybanya continued from Middle Miocene to
the Sarmatian horizon of the Upper Miocene, or, at some places, even to

the Pliocene ; and in Transylvania, from the Upper Cretaceous right through to the Sarmatian.

The Hungarian gold-silver lodes are, spacially as well as genetically, most closely associated with these young eruptives. The most impoitant mining districts are those of Schemnitz-Kremnitz, Nagybanya-Felsöbanya-

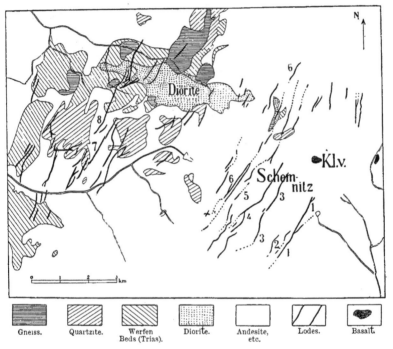

| Gneiss. | Quartzite. | Werfen Beds (Trias). | Diorite. | Andesite, etc. | Lodes. | Basalt. |

FIG. 294.—Map of the Schemnitz district. J. Szábo, 1883.   The white aɾeas consist preponderatingly of different andesites.

1, Gruner lode ; 2, Stephan lode ; 3, Johann lode ; 4, Spital lode ; 5, Biber lode ; 6, Theresia lode ; 7, Brenner lode ; 8, Elisabeth lode ; K.l.v. basalt sheet on the Kalvarienberg.

Kapnik, and the Transylvanian Erzgebirge. The total gold and silver production of Austria-Hungary since 1493, statistics of which year are the earliest available, has been as follows :

	Silver.	Gold.	Authority.
1493–1875	7770 tons	460,600 kg.	Soetbeer
1875-1900	1247 ,,	54,558 ,,	Neumann

Of these amounts somewhat more than one-half of the silver and the greater part of the gold were derived from the above-mentioned Hungarian districts, from whence in addition during the eight years 1901–1908 further amounts of 137,161 kg. of silver and 27,930 kg. of gold were obtained, of which latter amount 20,213 kg. came from Transylvania. These Hungarian mines consequently during the period 1493–1908 produced some 5000 tons of silver and roughly 500 tons of gold, beside which many of them were worked at an earlier period, long before statistics were instituted.

The silver production of Hungary decreased from 23,636 kg. in 1901 to 12,612 kg. in 1908. The gold production on the other hand has kept fairly regularly between 3250 and 3500 kg. per year, some 2250–2700 kg. of which come from Transylvania.

The Hungarian production during the year 1907 was contributed to by the various districts as follows :

	Gold.	Silver.	Lead.
	Kg.	Kg.	Kg.
Schemnitz . . .	114	4541	351
Kremnitz	27	78	...
Nagybanya . . .	687	1712	8
Felsöbanya . . .	37	1857	769
Kapnik	19	1651	206
Transylvania, including :	2537	...	...
Nagyag	84	195	...

Of the above figures for Transylvania, 1713 kg. of gold were returned by the Ruda-12-Apostles' mine.

Schemnitz and Kremnitz about 25 km. farther north, belong to the same eruptive area. In the middle of the Tertiary period there was here a large volcano of approximately the present size of Etna, which volcano subsequently suffered erosion to such an extent that in the Hodritz valley, as mentioned before, diorite and grano-diorite representing that portion of the magma which consolidated in depth, are now exposed. In the vicinity of Schemnitz there is according to H. v. Böckh the following sequence of formations :

Triassic : Werfen beds ; limestone ; and quartzite.

Eocene : Nummulitic beds.

Miocene : Pyroxene-andesite tuff, the oldest ; pyroxene-andesite, 56 per cent SiO_2 ; augite-diorite, 60 per cent SiO_2, with grano-diorite, 67 per cent SiO_2, and aplite dykes ; biotite-amphibole-andesite tuff ; biotite-amphibole andesite, 56 per cent SiO_2 ; rhyolite tuff ; and finally, rhyolite with 77·5 per cent SiO_2.

Pliocene : Basalt.
Diluvium and Alluvium.

In this neighbourhood post-volcanic effects are still represented by solfataras, mofettes, fumaroles, and hot springs, these last often being used as medicinal baths. The lodes traverse the older beds and the Miocene eruptives, but not the basalt, than which consequently they are older. They are roughly parallel to one another and have a north-north-east strike, while the most important occur within an area about 12 km. long and 10–11 km. wide. By far the greater number are found in andesite, only a few being found in the Hodritz diorite and other rocks. In the adjacent sandy and slaty sediments the fissures quickly split up and disappear.

The lode material consists in greater part of quartz, in which numerous highly propylitized fragments of the country-rock are scattered. Composite lodes appear frequently, indicating that the lode fissure had in places been re-opened and re-filled repeatedly, the so-called Schemnitz *Blätter* or layers being thus formed. Of these in any lode-system the oldest are the ore-bearing quartzose layers, and the youngest the clayey layers consisting of triturated and decomposed material and carrying but little ore. In this manner large lode widths reaching as much as 15–20 m. or more, were formed ; the payable width however is usually much smaller, being generally between 1·5 m. and 5 metres.

The andesite on both walls is generally propylitized for a width of 10–20 m. and occasionally for as much as 100 m., though in the Hodritz diorite this alteration continues for but 5–10 metres. In addition, the rocks in the neighbourhood of the lodes are much decomposed and here and there even altered to a soft kaolin-like mass. Sometimes also silicification of the country-rock may be observed.

As already mentioned, quartz is the most important gangue-mineral ; it often occurs as the variety amethyst. Calcite, mangano - calcite $CaMn(CO_3)_2$, and other carbonates ; some barite ; and some zeolites, are also found. Fluorite is practically absent, this being also the case with most of the other Hungarian young gold-silver lodes with the exception of those at Kapnik-Felsöbanya. The *Zinopel* of the Schemnitz lodes is an intimate mixture of quartz and specularite.

The most important ore-minerals are argentite, stephanite, galena, polybasite, pyrargyrite, proustite, tetrahedrite, etc. ; less important are sphalerite, pyrite, marcasite, and chalcopyrite. Native silver is occasionally found as secondary moss or wire upon argentite, just as at Kongsberg. Native gold is sometimes visible to the naked eye, though generally it is too finely distributed to be seen. Cinnabar, though seldom, does occasionally occur. Stibnite, the occurrence of which at Kremnitz is so pronounced, is at Schemnitz practically absent ; while tetrahedrite, which at Kapnik plays so important a part, is insignificant at Schemnitz.   Secondary minerals such as pyromorphite, cerussite, etc., are found in large amount.

The ore is practically free from nickel, cobalt, bismuth, and tin ; arsenic is but poorly represented, though antimony, from the occurrence of the sulpho-salts, is more abundant.   A small amount of copper is won as a by-product in treatment.   Sphalerite occurs in such small amount that it cannot be separated by concentration.

The variability of the relation between gold, silver, and lead, is striking. In this district no fewer than four sub-groups may be differentiated, these being as follows :

1. Silver lodes containing stephanite, pyrargyrite, proustite, argentite, polybasite, etc., with a relatively low gold content and almost free from galena.   The gold amounts to about 10–14 parts per 1000 parts of silver. Such lodes, as for instance the Alt Antoniostollen and the Einigkeit lodes, occur in the Hodritz diorite, from which they continue into the adjacent andesite.   They carry quartz and some calcite and are mostly from 0·25 m. to 3 m. in width.

2. The Grüner lode of quartz carrying silver ore fairly rich in gold, but with little galena.   In the upper levels the proportion of gold was only some 12 parts per 1000 of silver, a proportion which in greater depth rose to 140, the value of the gold being then substantially higher than that of the silver.   Although this lode is several kilometres long, the rich ore is limited to certain shoots.   At a depth of about 400 m. one such ore-shoot was 200 m. long and 1–2 m. wide.

3. The Johann lode containing galena and stephanite chiefly and characterized by the presence of *Zinopel*.

4. The Spitaler lode carrying galena chiefly, some sphalerite, and but little silver mineral, etc.   Owing to the number of layers, the width of this lode rises in places to 20 m., and exceptionally to 40 m., though seldom more than 5 m. is payable.   This lode in relation to the mineraliza-tion, the minerals present, and its brecciated structure, presents a striking resemblance to some of the Clausthal lodes.

The lodes mentioned under groups 2, 3, and 4 occur chiefly in andesite.

At Schemnitz there are accordingly represented : lodes with silver minerals but almost without galena ; lodes with auriferous silver minerals and some galena ; lodes intermediate between the two foregoing groups ; and finally, lodes in which galena preponderates and but little silver or gold is contained.   These are all more or less of the same age ; in any case no such sequence of relative age as is possible for instance at Freiberg, can be established.   Similarly, no rules concerning ore-shoots or enrichment at junctions or intersections have been possible of formulation.

The average ratio of the metals to one another in the Schemnitz mines belonging to the Hungarian Crown, has of late been one part of gold to 40–50 of silver, and one of silver to 80 parts of lead, these figures pertaining to the ore won after the drop in the price of silver had caused the lodes relatively rich in gold to receive greater attention.

Mining operations at Schemnitz are extremely old having presumably been begun some time before the year 750. About the year 1600 there were already more than 400 mines in operation. The Joseph II. Adit, 16,538 m. long, was begun in 1782 and finished in 1878. The mines have now reached a maximum depth of some 700 metres. They employ 2500 miners. Since the fall in the price of silver at the commencement of the 'nineties they have been worked at considerable loss, the average yearly loss during the period 1903–1907 having been about £46,000.

The auriferous quartz lodes of Kremnitz likewise occur in propylitized andesite. They too are often developed as composite lodes, reaching then a width of 10–15 m., while the numerous simple lodes are generally but 1–2 m. wide. The gold is seldom visible to the naked eye. Among the ore-minerals, pyrite and stibnite are common, while galena is almost completely absent. Operations in this district, which began in the twelfth century or perhaps even before, have latterly declined to a considerably smaller scale, recent figures of production having been only 27–46 kg. of gold and 76–141 kg. of silver annually, that is, one part of gold to 2·5–3 parts of silver.

Nagybanya, Felsöbanya, and *Kapnik* belong to one and the same eruptive area, which, as illustrated in Fig. 292, consists chiefly of andesite and to a less extent of rhyolite. Felsöbanya lies about 9 km. and Kapnik about 35 km. to the east of Nagybanya. In spite of these relatively small distances from one another the nature of the ore at these three places and the character of the lodes, differ considerably. At Nagybanya the lodes consist chiefly of auriferous quartz with some silver minerals, pyrargyrite particularly ; the production of late years has varied between 600–687 kg. of gold, 1219–1712 kg. of silver, and but 6–10 tons of lead per year ; the relation between gold and silver is as 1 of gold to 2–2·5 of silver. At Felsöbanya silver-lead lodes carrying some gold occur ; the production of late has been 37–51 kg. of gold, 1857–2579 kg. of silver, and 768–1064 tons of lead per year ; or 1 part of gold to 50–60 parts of silver. Kapnik is characterized by silver-lead-zinc lodes containing comparatively little gold ; recent figures of production have been 15–24 kg. of gold, 1651–2405 kg. of silver, and 206–296 tons of lead per year ; or 1 part of gold to 90–120 parts of silver.

The lodes of any one district are usually found to be more or less parallel, and all have quartz as principal lode-filling ; at Kapnik in addition there is much fluorite. These lodes generally are known for the variety of minerals contained, many of which are often beautifully crystallized. Prominent among these minerals are, rhodochrosite, barite, rhodonite, tetrahedrite, stibnite, bournonite, jamesonite, freieslebenite, etc., and sometimes, though seldom, wolframite.

Mining here also is likewise extremely old, having been begun about the end of the eleventh century, or even before. Most of the mines belong to the State. The distribution of other gold-silver lodes in the Carpathians, such as are no longer worked, is indicated in Fig. 292.

The Transylvanian Erzgebirge contains the most important gold occurrence in Europe. The auriferous area, with Offenbanya, Zalatna, Nagyag, and Karacs at its corners,[1] has a length of about 55 km. and a maximum width of about 33 km. It consists of a core of Archaean rocks over which are spread : melaphyre and Jurassic limestone ; Carpathian sandstone of Cretaceous, and perhaps also in part of early Eocene age ; and finally, Tertiary sediments and tuffs, chiefly Miocene. With these last are associated Tertiary eruptives, chiefly andesite and dacite and to a less extent rhyolite, these forming a number of detached areas, most of which are small.

Mineralogically, the occurrence at Nagyag is particularly characterized by the occurrence of gold in combination with tellurium [2] in the form of sylvanite, nagyagite, some petzite and krennerite. At Offenbanya [3] some lodes carry their gold exclusively as tellurides, sylvanite chiefly but also nagyagite ; others contain both tellurides and native gold ; while others again contain native gold only. In other Transylvanian mines gold tellurides, as well as the other tellurides, hessite, tetradymite, tellurite, and native tellurium, occur only as curiosities ; in fact in most of the lodes, including those which are most productive, gold telluride is completely absent, and free gold only occurs. About nineteen-twentieths of the present gold production of Transylvania is derived from free gold, and only the one remaining part from tellurides, this coming almost entirely from Nagyag.

The gold is generally accompanied by a number of silver minerals, such as argentite, pyrargyrite, proustite, stephanite, native silver, etc. ; by pyrite, galena, and sphalerite ; by antimony minerals, such as stibnite, bournonite, tetrahedrite, etc. ; and finally, by some arsenic minerals, arsenopyrite especially. The native gold of most of the lodes contains a relatively high proportion of silver, the well-known gold crystals of Verespatak, for instance, containing 30–35 per cent. When to this is added the silver contained in the silver minerals proper, the amount of this metal present is often many times that of the gold.[4]

---

[1] Map by v. Papp, *Zeit. f. prakt. Geol.*, 1906, p. 306.
[2] *Ante*, p. 80.   [3] *Ante*, p. 526.   [4] *Ante*, p. 524.

Among the gangue-minerals quartz usually occupies the first place. Calcite is often abundant sometimes even equalling or exceeding the quartz. The frequent occurrence of manganese minerals is characteristic; rhodochrosite and mangano-calcite are often well represented, as for instance at Nagyag and Verespatak; rhodonite is often found at Verespatak; while alabandite occurs here and there at Nagyag.

The lodes, which as a rule are small, in general follow tectonic, and not contraction fissures. The immediate country-rock in the different districts is as follows:

At Nagyag: dacite rich in hornblende.

At Offenbanya: dacite rich in hornblende, merging into hornblende-andesite.

At Hondal, Troicza-Tresztya, Nagy-Almas, and Korabia-Vulkoj: hornblende-andesite.

At Muszari: andesite and melaphyre or melaphyre tuff.

At Verespatak: dacite poor in hornblende, and rhyolite; Carpathian sandstone and a rock known as *Lokalsediment*.

At Boicza: melaphyre with quartz-porphyry; and a limestone patch in melaphyre, near hornblende-andesite.

In Transylvania, far back in ancient times, gold was won at several places, partly by washing the rather poor gravels. Some idea of the production in more recent times may be obtained from the following figures:

1770 roughly	300 kg. gold.	1890 roughly	1570 kg. gold.
1787 ,,	700 ,,	1895 ,,	2274 ,,
1810 ,,	210 ,,	1900 ,,	2260 ,,
1842 ,,	1140 ,,	1905 ,,	2725 ,,
1858 ,,	700 ,,	1908 ,,	2311 ,,

The increase of late has been chiefly due to rich pockets discovered in the Ruda-12-Apostles' mine, which itself has latterly produced some 1600–1900 kg. yearly. Nagyag produces yearly about 80–100 kg. of gold and 200–250 kg. of silver. Nagyag and Verespatak are in part State mines, the others are in private possession. The limit of payability in Transylvania is considered to be as follows:

Amount of Ore per Unit of Lode Plane.	Yield of Gold.
0·5 ton per sq. m.	15 grm. per ton.
1·0 ,, ,,	12 ,, ,,
1·5 ,, ,,	10 ,, ,,
2·0 ,, ,,	8 ,, ,,

At Nagyag the lodes occur within an eruptive throat or chimney consisting of propylitized dacite, though andesite also appears. Within an area about 1000 m. long and 950 m. wide an extraordinary number of steeply inclined and approximately north-south lodes are found, these being generally but 10 cm. wide and seldom as much as 30 cm. The occurrence is illustrated in Fig. 295. In addition to gold telluride,[1] native gold, though seldom, is also found. Its derivation from the telluride may be explained in the same way as the silver horns upon argentite at Kongsberg.[2]

The lodes themselves may be divided into three divisions :

1. The quartz-telluride lodes, containing quartz with sylvanite and more seldom with nagyagite ; and pyrite and tetrahedrite.

2. The pink manganese-telluride lodes, with rhodochrosite ; some quartz, alabandite, and nagyagite ; and tetrahedrite, pyrite, bournonite, as primary minerals ; and with arsenic, sulphur, etc., as secondary minerals.

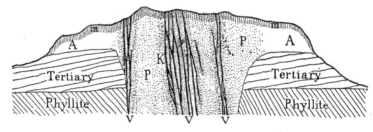

FIG. 295.—Idealized section at Nagyag. B. v. Inkey, 1885.

A, dacite ; P, propylite ; K, kaolinized dacite immediately along the lodes ; V, lodes ; m, surface weathering.

3. The base-metal lodes, containing galena, sphalerite, and pyrite, with calcite and dolomite.

With many of the lodes a breccia occurs [3] which doubtless, both in regard to its fragments and matrix, is the result of friction and trituration along the fissure walls. The most productive ore-bodies, some being occasionally very rich, are found where many lodes intersect to form a chimney or shoot.[4] In 1883 or 1884, for instance, one such body in three days yielded gold to the value of £2300 from an area of two square metres on the lode plane and a thickness of about 0·20 metre.

Mining at Nagyag began in the year 1747. The element tellurium was discovered at this place. The mines at present are worked from the Franz-Joseph adit which has a length of 5012 metres. According to B. v. Inkey, from the commencement of mining in 1748 to the year 1882, Nagyag produced 39,995·6 kg. of gold-silver bullion worth £2,193,000, equivalent to about 18,000 kg. of gold and 22,000 kg. of silver ; after the payment

---

[1] *Ante*, p. 543.   [2] *Ante*, p. 131.
[3] *Glauch.*   [4] *Erzstock.*

of all costs the net profit in some years amounted to £420,000. By 1902 the total production had increased to 46,335 kg. of such bullion, while in the six years 1903–1908 some 556 kg. of gold and 1387 kg. of silver were produced, bringing the total production of gold to about 22,000 kg. or 22 tons.

The country around Verespatak — so well known historically and for its gold crystals — consists of : dacite and rhyolite, these rocks forming the peaks known as Kirnik and Boj, the latter containing the celebrated Csetatye deposit ; Carpathian sandstone ; and *Lokalsediment,* a rock regarded by Pošepný as a sediment but which according to Semper is probably an outpouring of volcanic mud. In the dacite, strikingly large dihexahedra of quartz are found. A little distance away andesite also occurs.

The lodes, which contain gold crystals, free gold, auriferous pyrite, quartz, rhodochrosite, etc., occur in the dacite, rhyolite, and sediment, while in the porous Carpathian sandstone most of them quickly disappear. The area in which they occur has a length of about 2·5 km. and a breadth of about 1·5 km. Individually they are generally of little width and but short extent along both strike and dip. Sometimes however they occur in such considerable number and so close together that the whole rock mass has to be mined. In this way no doubt were formed, the large old Roman excavation, the ' Csetatye mare ' [1] near the top of the Boj, and the so-called Katroncza chimney, 100 m. deep and 20–40 m. wide. In this district during the two years 1823 and 1824 alone, gold to the value of £80,000 was recovered. The metasomatic features of this occurrence are mentioned later.[2]

Mining at Verespatak began in the time of the Romans, A.D. 106–276. Following an ancient local mining law, extremely small areas of cubical or spherical shape were formerly granted to a large number of small corporations or to individual persons. Such grants on account of their shape have no extension in depth. Every such corporation or person has its own small and primitive battery, so that within a fairly small district there are no less than about 6000 stamps. In these batteries apparently but 40 per cent of the gold content is recovered. In addition, the Hungarian Mining Department is working here, as well as a large French company. Much gold was won formerly though now the output is small.

The Muczari and Ruda mines adjacent to one another and now belonging to one company, have produced of late the greater portion, some two-thirds of the total production of Transylvania. In their neighbourhood a melaphyre tuff of great extent is broken by several eruptions of andesite. In the principal mine the northern portion of this andesite is crossed by a fault striking N.W.-S.E. which carries ore and is known as the Klara lode, one of the principal lodes. A second fault, striking approximately north-south, cuts through the eastern portion of the andesite and forms the Carpin lode. Where these two cross the ore was richest, a shoot there, about 50 m. long and 80 m. deep, having yielded some thousands of kilo-grammes of gold. The other lodes likewise are tectonic fissures ; the amount of ore they carry generally diminishes as the distance from the centre of the andesite increases.

The occurrence at Boicza is singular in so far that the lodes occur in the Jurassic melaphyre and quartz-porphyry; the distance from the nearest outcrop of Tertiary hornblende-andesite is, however, but 3 kilometres.

CARTAGENA AND MAZARRON IN SOUTH-EASTERN SPAIN

LITERATURE

A. OSANN. ' Über die Eruptivgesteine und über den geologischen Bau des Cabo de Gata,' Zeit. d. d. geol. Ges. Vol. XLI., 1889, and Vol. XLIII., 1891.—R. PILZ. ' Die Bleiglanzlagerstätten von Mazarron,' Zeit. f. prakt. Geol., 1905 ; ' Die Erzlagerstätten von Cartagena,' ibid., 1908 ; Dissertation über Mazarron. Dresden and Freiberg, 1906. In these works the pertinent literature, and especially that in Spanish, is given.

The narrow stretch along the coast from Cabo de Gata to Cabo de Palos near Cartagena was in Tertiary times the scene of great volcanic activity, which included extrusions of liparite, dacite, andesite, and basalt. Accord-ing to Osann the centres of eruption appear to be arranged in three parallel lines, as illustrated in Fig. 296. In close association with these eruptives, and particularly with the andesite and dacite, ore-deposits are found in the neighbourhood of Mazarron and Cartagena.

At Mazarron the lodes occur within an area about 8 km. long and 3 to 4 km. wide ; partly in dacite ; partly in mica-schist, amphibolite, dolomite, and quartzite, surrounding the dacite ; and partly along the contact between the two. In the upper levels the lodes within the dacite appear to be chiefly contraction fissures. At depths of 400 and 500 m., on the other hand, it is probable that the majority are tectonic fissures, these being frequently of composite character. As illustrated in Fig. 297, the number of lodes diminishes rapidly in depth, while of those which continue, most become impoverished at a depth of 400–500 metres. The most important ore-minerals are galena, sphalerite, pyrite, marcasite, some chalcopyrite, etc. The galena generally carries 1·5 kg. of silver per ton, and exceptionally as much as 3–6 kg. The gangue consists of siderite, calcite, dolomite, some barite, and quartz. Selenite occurs as a secondary mineral. Magnetite and specularite are found in very small amount, while the presence of quicksilver as a mineralogical curiosity has been established.

In the oxidation zone much limonite, cerussite, zinc carbonate, pyromorphite, mimetesite, etc., and some native silver, are found.

The lode district around Cartagena has a length of 10 km. and a width of 5 km. Like the occurrence at Mazarron, the lodes occur partly within the eruptive rock ; partly between this and the Triassic dolomite or slate which surrounds it ; partly as cross-courses and bedded lodes in the slates ; and finally as chimneys, pipes, and bedded deposits in the dolomite. The eruptive rock is chiefly andesite. The most im-

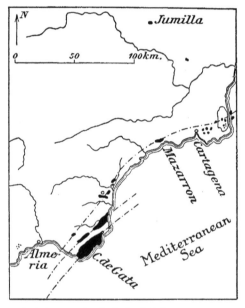

Fig. 296.—Map of the Tertiary eruptive belts (black) between Almeria and Cartagena, Spain.
Osann, 1891.

portant primary ore is galena, which contains on an average 1–1·5 kg. of silver per ton. Sphalerite, copper sulphides, much pyrite, and ferro-manganese minerals, also occur. The oxidation zone, in addition to those ores mentioned as occurring in that zone at Mazarron, contains a good deal of cerargyrite. At the outcrop of one of the principal mines, the Monto de los Azules, wood-tin was found, a fact which recalls the Potosi type of deposit. This wood-tin was considered to be of secondary formation.

The country-rock of the lodes is highly altered to a propylitized rock carrying much sericite and kaolin. In this connection, both at Mazarron and Cartagena alunite occurs in such amount as to have been formerly worked for alum, 7·5 tons of the raw material having been required to produce one ton of alum. Pilz expresses the opinion that the formation of the alunite was not the direct consequence of the volcanic activity but rather that of the decomposition of the pyrite in the lodes. It nevertheless appears possible and even probable that, as at Goldfield,[1] the alunite is in greater part, if not exclusively, a direct fumarolic product due, conjointly with the lodes, to the action of heated waters.

Exhalations of carbonic acid are not infrequently encountered in the mines at Mazarron, the issuing gas consisting by volume of 93·5 per cent CO_2, 5·6 per cent nitrogen, 0·9 per cent oxygen, and traces of water. The amount given off is sometimes so great as to seriously disturb the work ; on February 16, 1893, twenty-eight workmen and officials lost their lives by such an issue of gas. In one place mofettes were still in active operation ten years after their first appearance as water springs containing gas. At times the gas is found under pressure in fissures, and frequently when breaking through from eruptive rock to slate, violent issues of gas have occurred. Pilz in 1905 regarded such gaseous accumulations under pressure as closely associated with the actual eruption of the dacite, though in 1908, on the other hand, he indicated the possibility that they arose by the action of sulphuric-acid waters upon carbonates in the lodes and country-rock.

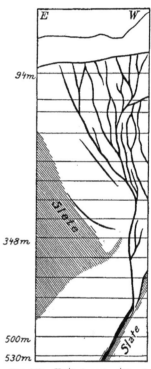

FIG. 297.—Vertical section through the Santa Ana mine at Mazarron, Spain. Pilz, *Zeit. f. prakt. Geol.*, 1905.

The white areas next to the slate represent dacite ; on top is a sand- and clay covering.

These deposits were worked by the Phoenicians, the Carthaginians, and the Romans, the works of the latter reaching to a depth of 360 metres. Subsequently, for more than a thousand years they lay untouched, until those at Cartagena were opened again in 1839, and those at Mazarron in 1870. Latterly operations at both places have been quite brisk ; thus, in 1904 the mines at Mazarron produced more than 30,000 tons of lead ore with 58 per cent of lead, and 1694 tons of zinc ore with 30–40 per cent of zinc, while 5000 tons of iron ore were also won as a by-product. At Cartagena the production

[1] *Ante*, p. 522.

in the same year was still larger, namely, 80,000 tons of lead ore and some 85,000 tons of zinc ore.

In the neighbourhood of Cabo de Gata also, the position of which is given in Fig. 296, there are occurrences of andesite, dacite, and liparite, associated with and carrying a large number of deposits. At Pinar in the Sierra de Bedar, especially, such deposits are worked for galena and to a less extent for copper ore. The occurrences of iron ore at Serena and Tres Amigos, from which about 100,000 tons of limonite are produced yearly, are regarded as metasomatic replacements of siderite ; they occur chiefly at the contact between limestone and slate. In the Sierra Almagrera, near the coast and approximately midway between Cabo de Gata and Mazarron, auriferous galena mixed with siderite was formerly worked under great difficulty with water, the lodes apparently permitting the entry of sea-water. At Herrerias silver- and iron ores are worked.[1]

PONTGIBAUD IN FRANCE

LITERATURE

LODIN. ' Étude sur les gîtes métallifères de Pontgibaud,' Ann. d. Mines, Sér. 9, I., 1892, pp. 389-505; extracted in Zeit. f. prakt. Geol., 1893, pp. 310-319.—The geological maps of Moulins, Gannat, and Clermont.

The lode district of Pontgibaud, 14 km. long and about 4·5 km. wide, in the department Puy-de-Dôme, lies on the west side of the Tertiary eruptive region of Auvergne.[2] In the middle of this district, at Chalusset, a small extinct volcanic cone rises. The lodes occur in gneiss and mica - schist, along fissures following old granitic dykes. Quartz is the principal gangue, though barite also occurs. The principal ore is galena, which occurs with some pyrite and sphalerite ; bournonite, tetrahedrite, etc., are more seldom. The silver content of the galena diminishes in depth. It is considered that the fissures, along which, as before mentioned, carbonic acid exhalations were remarked, are of Miocene age. Mining operations, which at the latest began in the year 1554, were from the 'sixties to the 'eighties of last century quite important, while as late as 1890 there were still 489 persons employed ; now they are stopped.

THE UNITED STATES

LITERATURE

WALDEMAR LINDGREN. ' The Geological Features of the Gold Production of North America,' Trans. Amer. Inst. Min. Eng. XXXIII., for 1902 ; ' A Geological Analysis of the

---

[1] F. Fircks, ' Über einige Erzlagerstätten der Provinz Almeria in Spanien,' Zeit. f. prakt. Geol., 1906.

[2] Michel-Lévy, 'Massif du Mont-Dore,' etc., Guide géol. des excursions du VIII. Congr. géol. intern., 1900, XIV.

Silver Production of the United States in 1906,' U.S. Geol. Survey, Bull. 340, 1908.—
F. V. RICHTHOFEN. Works cited in the description of the Comstock Lode.—EDWARD
SUESS. Zukunft des Goldes, 1877 ; Zukunft des Silbers, 1892.—S. F. EMMONS AND G. F.
BECKER. Geological Sketches of the Precious Metal Deposits of the Western United States,
Washington, 1885.—J. F. KEMP. Ore Deposits of the United States and Canada.—
C. R. VAN HISE. 'A Treatise on Metamorphism,' U.S. Geol. Survey, Mon. XLVII., 1904.
For the different mining districts reference is recommended to the comprehensive
descriptions published by the U.S. Geol. Survey, these in former years being particularly
by G. F. Becker, S. F. Emmons, A. Hague, Whitman Cross, and of late years particularly by
J. S. Diller, J. D. Irving, W. Lindgren, R. L. Ransome, and J. E. Spurr. In these monographs detailed and exhaustive bibliographies are collected ; Lindgren and Ransome, for
instance, give in their work upon Cripple Creek, published in 1906, a list of thirty-eight
previous papers on that district ; and Ransome in 1909 a list of seventeen works upon
Goldfield. In the U.S. Geol. Survey, Bull. 340, 1908, there is on pp. 153-156 a list of the
Survey publications upon gold and silver. Only the most important can be mentioned here.
    Comstock in Nevada.—Principal Work : G. F. BECKER. ' Geology of the Comstock
Lode and the Washoe District, with Atlas,' U.S. Geol. Survey, Mon. III., 1892. In addition,
F. v. RICHTHOFEN. The Comstock Lode : Its Character and probable Mode of Continuance
in Depth, San Francisco, 1866 ; ' The Natural System of the Volcanic Rocks,' Cal. Acad. Sc.,
1867 ; Zeit. d. d. geol. Ges., 1868, p. 663.—CLARENCE KING. Geology in Exploration of the
40th Parallel, U.S. III., 1870, pp. 1-96.—F. ZIRKEL. Microscopical Petrology in Explora-
tion of the 40th Parallel, IV., 1876.—J. A. CHURCH. .The Comstock Lode. New York, 1879.
—E. LORD. ' History of Comstock,' U.S. Geol. Survey, Mon. IV., 1883.—A. HAGUE and
J. P. IDDINGS. ' On the Development of Crystallisation in the Igneous Rocks of Washoe,'
U.S. Geol. Survey, Bull. 17, 1885.—Answered by BECKER. ' The Washoe Rocks,' Cal.
Acad. Sc. II. Bull. 6, 1886 ; Amer. Journ. Sc. XXXIII., 1887.
    Elsewhere in Nevada. *Goldfield.*—Principal Work : F. L. RANSOME. ' The Geology
and Ore Deposits of Goldfield,' U.S. Geol. Survey, Professional Paper 66, 1909 ; extract by
Ransome in Econ. Geol. V., 1910.—J. E. SPURR. ' Geology of the Tonopah Mining District,'
U.S. Geol. Survey, P.P. 42, 1905 ; ' Ore Deposits of the Silver Peak Quadrangle,' U.S.
Geol. Survey, P.P. 55, 1906.—J. S. CURTIS. ' Silver-lead Deposits of Eureka,' U.S. Geol.
Survey, Mon. VII., 1884.—A. HAGUE. ' Geology of the Eureka District,' U.S. Geol. Survey,
Mon. XX., 1892.
    Cripple Creek in Colorado.—Principal Work : W. LINDGREN and F. L. RANSOME.
' Geology and Gold Deposits of the Cripple Creek District,' U.S. Geol. Survey, P.P. 54, 1906.
—W. CROSS and R. A. F. PENROSE. ' The Geology and Mining Industries of the Cripple
Creek District,' U.S. Geol. Survey, 16th Ann. Rep. II., 1895.
    Elsewhere in Colorado.—W. CROSS and S. F. EMMONS. ' Geology of Silver Cliff and the
Rosita Hills,' U.S. Geol. Survey, 17th Ann. Rep. II., 1896.—J. E. SPURR, G. H. GARREY,
and S. H. BALL. ' Economic Geology of the Georgetown Quadrangle,' U.S. Geol. Survey,
P.P. 63, 1908.—N. M. FENNEMAN. ' Geology of the Boulder District,' U.S. Geol. Survey,
Bull. 265, 1905.—F. RICKARD. ' Gilpin County,' Trans. Amer. Inst. Min. Eng. XXVIII.
1899—S. F. EMMONS. ' The Mines of Custer Co.,' U.S. Geol. Survey, 17th Ann. Rep. II.,
1896.—W. CROSS. ' Geology of the Rico Mountains,' U.S. Geol. Survey, 21st Ann. Rep.
II., 1900.—F. L. RANSOME. ' The Ore Deposits of the Rico Mountains,' U.S. Geol. Survey,
22nd Ann. Rep., 1902 ; ' Report on the Economic Geology of the Silverton Quadrangle,'
U.S. Geol. Survey, Bull. 182, 1901.—J. D. IRVING. ' Ore Deposits of the Ouray District,'
U.S. Geol. Survey, Bull. 260, 1905.—J. A. PORTER. ' The Smuggler Union Mines,
Telluride,' Trans. Amer. Inst. Min. Eng., 1896.—C. W. PURINGTON. ' On the Mining
Industries of the Telluride Quadrangle,' U.S. Geol. Survey, 18th Ann. Rep. III., 1898.—
S. H. BALL. ' Southern Nevada,' U.S. Geol. Survey, Bull. 308, 1907.—S. F. EMMONS.
' Eureka,' U.S. Geol. Survey, Mon. XII., 1886.—S. F. EMMONS and J. D. IRVING.
' Eureka,' U.S. Geol. Survey, Bull. 320, 1907.—J. E. SPURR. ' Aspen,' U.S. Geol.
Survey, Mon. XXXI., 1898 ; Econ. Geol. IV., 1909.
    Elsewhere in other States.—W. LINDGREN. ' The Gold and Silver Veins of Silver City,
De Lamar, and Other Mining Districts in Idaho,' U.S. Geol. Survey, 20th Ann. Rep. III.,
1900.—J. M. BOUTWELL, A. KEITH, and S. F. EMMONS. ' Economic Geology of the Bingham
Mining District, Utah,' P.P. 38, 1905; Bull. 213, 225, 260.—S. F. EMMONS and J. E.
SPURR, Economic Geology of the Mercur Mining District, Utah. U.S. Geol. Survey, 16th

Ann. Rep. II., 1895.—S. F. Emmons. 'The De Lamar and the Hornsilver Mines; two types of Ore-Deposits in the Deserts of Nevada and Utah,' Amer. Inst. Min. Eng., 1901.— G. W. Tower and G. O. Smith. 'Geology and Mining Industry of the Tintic District, Utah,' U.S. Geol. Survey, 19th Ann. Rep. III., 1899.—W. H. Weed and L. V. Pirsson. 'Geology and Mineral Resources of the Judith Mountains of Montana,' U.S. Geol. Survey, 18th Ann. Rep. III., 1898.—In addition, 'Annual Contributions to Economic Geology,' U.S. Geol. Survey.

It is fitting to begin this description with the following figures of the production of the United States.[1]

	Gold.	Silver.
	Tons.	Tons.
1800–1848 yearly average about .	1·5	0·0
1851–1855 ,, ,, .	88·8	9·3
1856–1860 ,, ,, .	77·1	6·2
1861–1865 ,, ,, .	66·7	174·0
1866–1870 ,, ,, .	76·0	301·0
1871–1875 ,, ,, .	59·5	564·8
1880	54·2	943·0
1885	47·8	1124·6
1890	49·4	1695·5
1895	70·5	1441·1
1900	117·6	1793·4
1905	132·7	1745·3
1910	144·5	1755·4

The discovery of alluvial gold in California in the year 1848 was, as is well known, followed by an intense working of these gravels, the zenith of production having been reached in 1853. With the exhaustion of these deposits the gold production of the entire country sank to a minimum in the years 1882–1890. Then work upon the many gold lodes began. Now, within the last twenty years, the production has again increased, not to any great extent by reason of the discovery of gravels in Alaska, but owing chiefly to the fruition of such mining districts as Cripple Creek, Goldfield, etc., most of which belong to the Tertiary group. Outside of Alaska there is now but little gold won from gravels. Thus, in 1901 gold to the value of 66 million dollars was won from lodes, and to the value of 12·2 millions from gravels; of these amounts no less than 8·2 millions came from Alaska, practically all of this being from gravels.

In the United States the first silver mine of any importance began work in the year 1859, on the Comstock Lode. Since the fall in price in 1892–1894 the production of silver has remained fairly constant.

W. Lindgren [2] divides the gold deposits in the United States, Mexico, and Canada, into the following groups :

[1] B. Neumann, Die Metalle, Halle, 1904 ; The Mineral Industry ; Die statistischen Tabellen der Frankfurter Metalgesellschaft.
[2] Loc. cit., Trans. Amer. Inst. Min. Eng. XXXIII., 1903.

1. *Contact - Deposits.* — These have little importance in the United States but attain somewhat greater significance in Mexico.[1]

2. *Pre - Cambrian Lodes.* — These in the United States are found particularly in the Appalachian Mountains of Georgia, North and South Carolina, Tennessee, Maryland, Virginia, and in the Black Hills of South Dakota.

3. *Cretaceous Lodes.*—These occur on the Pacific Coast within a long belt extending from Mexico through the central portion of California, where they are largely developed, on to Northern California, Oregon, and Idaho, and finally to British Columbia and Alaska. These lodes are quartz lodes containing free gold and auriferous sulphides. They occur in connection with granite and diorite, and, on account of the great denudation they have suffered as well as the coarse character of the gold, they are accompanied by important auriferous gravels. This older gold-quartz belt extends along the east side of the Sacramento Valley and the Sierra Nevada, the quicksilver zone of the Coast Ranges keeping nearer the Pacific sea-board.

4. *Late Cretaceous Lodes, in part Early Tertiary.*—Such occur in the central belt of the Central and Eastern Cordilleras, particularly at different places in Arizona, Nevada, Utah, Colorado, Idaho, Montana, etc.

5. *Tertiary Lodes, chiefly Post-Miocene.*—These occur in association with Tertiary eruptive rocks, particularly andesite and dacite ; more seldom rhyolite and basalt ; and exceptionally phonolite ; all of which in this connection are characterized by propylitization.

Occasionally these lodes carry either silver or gold exclusively, but generally both metals occur in approximately equal-value amounts. Many of these deposits are characterized by extraordinarily rich but limited bonanzas, while in many cases a decrease in value in depth has been established. The gold in these lodes is mostly finely distributed throughout the quartz, on account of which, and also because of the comparatively limited amount of denudation which from their lower age they have experienced, these Tertiary occurrences are accompanied by auriferous gravels to a less extent than are the older Californian lodes.

The following statistical-geological statement of the gold production of Mexico, the United States, and Canada, is based upon figures from Lindgren's work, to which data for the year 1908 have been added. The figures given are in million dollars, roughly equivalent to 1·5 tons of gold. The distribution among the different lode-groups can naturally not be exact. The gold won from gravels has been reckoned with those lodes from which such gravels have presumably arisen.

[1] W. H. Weed, ' Elkhorn Mining District, Montana,' *U.S. Geol. Survey*, 22nd Ann. Rep. W. Lindgren, ' The Character and Genesis of Certain Contact-Deposits,' *Trans. Amer. Inst. Min. Eng.*, February 1901 ; W. H. Weed, ' Ore-Deposits near Igneous Contacts,' *ibid.* October 1902.

GOLD PRODUCTION OF NORTH AMERICA IN MILLION DOLLARS

	Total from Discovery to 1908.	Geological Distribution. Pre-Cambrian.	Mesozoic (Pacific Coast).	Late Cretaceous (Central).	Tertiary.	1900. Total.	1900. Tertiary included.	Total 1908.
UNITED STATES								
Alaska	30·7	...	29·7	...	1·0	8·2	0·4	19·9
Washington	21·4 ?	...	10·0	...	11·4	0·7	0·5	0·3
Oregon	54·5	...	54·0	...	0·5 ?	1·7	...	0·9
C....fila	1380·0 ?	...	1350·0	...	30·0	15·8	1·0	19·3
Idaho	112·8	...	90·0	...	22·8	1·7	1·0	1·4
Montana	203·5 ?	...	...	200·0	3·5 ?	4·7	...	3·2
South Dakota	90·0	74·0	...	...	16·0	6·2	2·4	7·7
Wyoming	1·0 ?	1·0	...	...	...	1·0	...	...
Colorado	251·1 ?	...	...	34·0	217·1	28·8	26·1	22·9
Utah	27·0	...	...	25·0	2·0 ?	4·0	?	3·9
Nevada	250·0 ?	...	...	20·0	230·0	2·0	2·0	11·7
Arizona	42·1	...	22·1	...	20·0	4·2	2·2	2·5
Nw Mexico	176	...	...	...	10·0 ?	0·8	0·4	0·3
Appalachian States	47·0	47·0	...	7·6 ?	...	0·3	...	0·3
Total for United States	2528·7	122·0	1555·8	286·6	564·3	79·2	36·0	94·2
BRITISH NORTH AMERICA								
Nova Scotia	13·7	13·7	...	...	...	0·6	...	...
Quebec	2·0	2·0 ?	...	...	...	...	...	...
Ontario	1·2 ?	1·2 ?	...	...	...	0·3	...	...
British	70·7	...	70·7	...	...	4·7	...	...
N.W. Territory	52·6	...	52·6	...	...	22·3	...	...
Total	140·2	16·9	123·3	...	...	27·9	...	...
Mexico	200·0 ?	...	40·0 ?	...	160·0 ?	9·0	7·0 ?	...
Total for North America	2868·9	138·9	1719·1	286·6	724·3	116·1	43·0	...

The Tertiary gold lodes and the closely allied silver lodes occur over a very large metal province, which, associated with many and large Tertiary extrusions, stretches from Mexico in the south, to the Great Basin lying between the Sierra Nevada on the west and the Rocky Mountains on the east. Along this extent the lodes occur in greatest number in Colorado, Utah and Nevada, Arizona and New Mexico ; farther to the north in California, Oregon, Washington, Idaho, Wyoming, and Montana, they are not so numerous.

Lindgren [1] made also a similar statistical-geological statement for the silver lodes of the United States, which he divided into the following groups:

1. *Old Silver Lodes.*—These occur in Montana, Idaho, and elsewhere. They are found in granite or are associated with porphyry. Quartz is the usual gangue. The lodes are often rich near the surface, where secondary sulphides and sulph-antimonites have been formed. In the primary zone below water-level impoverishment often sets in. Occasionally the lodes carry comparatively much galena.

2. *Lodes in Tertiary Eruptives, Rhyolite, Dacite, Andesite, etc.*— These consist preponderatingly of quartz, occasionally with some chalcedony, and in many cases with adularia. The primary ore is chiefly argentite, this being accompanied by relatively small amounts of lead-, zinc-, and copper sulphides. In dry climates the ore in the upper levels has in places been greatly enriched by oxidation and the consequent formation of sulph-antimonites. Typical examples of such enrichment are found in the Comstock Lode, at Tonopah in Nevada, Mogollon in New Mexico, and Silver City in Idaho. The lodes belonging to this group, in addition to silver, usually contain a valuable amount of gold.

3. *Deposits in Limestone.*—These in general are associated with intrusions of granite, diorite, monzonite, or porphyry. Most are rich in lead, some also in copper and zinc. Quartz and calcite are the important gangue-minerals. Secondary silver sulphides and sulph-antimonites are more seldom seen in the alteration zones of these deposits, though, on the other hand, much native silver and cerargyrite have sometimes been found, as for instance at Leadville in Colorado, and Lake Valley in New Mexico. Several of these occurrences belong to the metasomatic and contact-metamorphic deposits, as such deposits are defined in this work.

These three groups are probably the products of the same mineralizing processes, their differences being due to different depths below the surface at the time of original deposition, to varying physical conditions, and to the influence of the country-rock, whether for instance that were limestone, andesite, or granite, etc. The second group probably belongs to the late Tertiary, the third to the earliest Tertiary.

[1] Lindgren, *U.S. Geol. Survey*, Bull. 340, 1908.

A good deal of silver is also won as a by-product when treating copper ores, this being particularly the case at Butte, Montana.

From a total of 57·4 million ounces of silver produced in 1906, 40·4 millions came from lead-, copper-, and zinc ores, and only 16·8 millions from silver-quartz lodes proper. Of this last figure, lodes within Tertiary eruptives yielded 10·3 million ounces, of which 7·5 millions came from ores containing both gold and silver. Many of the occurrences found in the neighbourhood of, but not actually within such Tertiary eruptives, must also be counted as belonging to that group.

THE SILVER PRODUCTION OF THE UNITED STATES

	Units of 1000 oz. = 31·1 kg.		
	1890.	1900.	1908.
Washington . . .	28	225	87
Oregon	18	115	56
California	1,063	941	1,704
Idaho	3,138	6,429	7,558
Montana	13,511	14,195	10,356
South Dakota . . .	105	536	197
Colorado . . .	18,376	20,484	10,150
Utah	7,005	9,268	8,451
Nevada	4,697	1,359	9,509
Arizona	1,813	2,996	2,900
New Mexico . . .	1,251	434	401
Michigan	15	102	294
Missouri	...	...	49
Tennessee	...	...	61
Texas	323	477	447
Alaska	9	73	205
Total . . .	51,355	57,647	52,441

From 1859 to about 1880 the State of Nevada, including the Comstock Lode, ranked first among the silver-producing states. In Colorado and Nevada the Tertiary gold-silver deposits especially are widely distributed, while in Utah this is the case with the closely allied Tertiary silver lodes.

In *Colorado*, at present the most important producer of gold and silver in the United States, if Leadville and some other supposedly metasomatic occurrences be excluded only Tertiary deposits connected with eruptives occur. The disposition of the different mining districts is shown in Fig. 298. That of Cripple Creek in Tellur County, where the deposits are remarkable for the considerable amounts of gold telluride they contain, has of late yielded more than one-half of the total gold production of Colorado. North of Cripple Creek occur the deposits of Clear Creek, Gilpin and Boulder counties. These are chiefly connected with andesite dykes. They carry both gold and silver, though latterly the gold recovered has exceeded the silver in value. In Gilpin County the lodes contain sulphide ore and free gold; Boulder County yields chiefly telluride gold ore; Clear Creek County produces smelting ore with much silver. In the last-named county the production during the period 1859–1904, according to Spurr and Garrey, was gold to the value of 16·1 million dollars, silver to 63·6 million, lead to 3·8 million, copper to 0·5 million, and zinc to 0·04 million dollars.

Sixty-five kilometres south of Cripple Creek, in Custer County, the

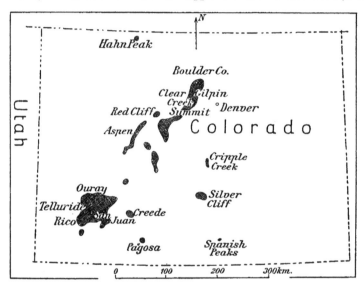

FIG. 298.—Map of the more important gold-, silver-, and lead districts in Colorado. Spurr and Garrey, *U.S. Geol. Survey*, P.P. No. 63, 1908.

Silver Cliff, and Rosita Hills mines, described by Whitmann Cross and S. F. Emmons,[1] occur. These together from 1880 to 1894 produced gold to the value of $1,822,327, and silver to $4,055,625.

In south-western Colorado, in the San Juan, San Michel, and Ouray counties, the Juan goldfield occurs. There some of the lodes carry silver only, others gold and silver, while others again carry gold only. In isolated cases telluride-gold and silver ores are found in considerable amount; hence the name Telluride for one of the districts. Most of these occurrences lie in thick andesite- or rhyolite flows. According to Ransome,[2] the mines of the Rico Mountains during the period 1879–1900 yielded about

[1] *Loc. cit.*, 1896.   [2] *Loc. cit.*, 1902.

73,000 oz. of gold and 9 million oz. of silver, or approximately 1 part of gold to 125 of silver.

In *Utah,* the Tintic and Horn Silver districts, among others, belong to the Tertiary group. Both carry silver principally and gold subordinately. At the Horn Silver mine the lodes occur at the contact of rhyolite with limestone ; they carry a lead-silver ore with a little gold. The Cornonabe mine, which also carries lead-silver ore, occurs in andesite.

Of the occurrences in *Nevada,* the Comstock Lode, the district of Goldfield containing gold chiefly, and that of Tonopah, chiefly silver, are described more closely below. In addition, the Eureka district presumably associated with rhyolite, the Tuscarora district in young eruptives, and the De Lamar district, are deserving of mention. Eureka yields gold to the extent of one-third the value of its total production, silver and lead to two-thirds. The two other districts yield both silver and gold.

The occurrences in the other states are illustrated by the following brief mention of representative deposits. *Arizona.*—At the Commonwealth mine in Cochise County, where the lodes occur in rhyolite and are very productive, gold forms one-third and silver two-thirds of the value of the production. *California.*—Many important mines working silver lodes in rhyolite occur in Bernardino County, such lodes presumably being connected with the Tertiary gold lodes. In addition, many young Tertiary lodes are found in the eastern foot-hills of the Sierra Nevada ; the Bodie mine in andesite worked one such lode containing much gold and silver. *Idaho.*—The Owyhee gold-silver lodes in basalt and rhyolite near the Nevada boundary are worthy of mention. From these, according to Lindgren, during the period 1880–1893 some 313,448 oz. of gold and 10,540,870 oz. of silver were produced. Farther to the north is the Custer mine. The Rocky Bar, Atlanta, and the recently discovered Thunder Mountain lodes, which apparently occur in rhyolite, probably also belong to this group. *Oregon, Washington, Alaska.* — Some Tertiary precious - metal lodes occur in these states, among them being those in andesite at the Apollo mine. *Montana.*—A considerable portion of the silver production of Montana comes from the copper district of Butte, where during the period 1892–1900 copper to the value of 331 million dollars, silver to 86 million, and gold to 14·5 million dollars, were produced.

THE COMSTOCK LODE

The outcrop of this lode in the Washoe district of Nevada, near the Californian boundary, is situated about 1970 m. above sea-level on the east slope of the Virginia Mountains, one of the north-eastern spurs of the Sierra Nevada, in latitude N. 39° 20' ; the distance from Steamboat

Springs, where the interesting recent deposits of quicksilver occur, is but
9–10 km.[1] The lode itself occurs within a large Tertiary eruptive mass
consisting in greater part of andesite. Investigation of the more detailed geological position of this occurrence was considerably facilitated
by the rock exposures in the Sutro Tunnel, made during the period

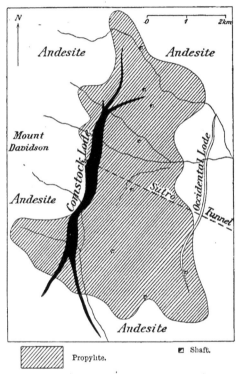

FIG. 299.—Plan of the Comstock Lode (black), showing the extension of the extreme
propylitization. Becker.

1868–1878, which reached the deposit at a depth of 500 m. after having
been driven 6·4 kilometres.

Becker,[2] in addition to granite occurring some little distance away,
differentiated the following eruptive sequence beginning at the oldest :
granular diorite, porphyritic diorite, quartz-porphyry, older diabase,
younger diabase—the so-called black dyke—older hornblende-andesite,
augite-andesite, younger hornblende-andesite, and finally basalt, the
youngest rock in the sequence. According to later investigation by

[1] *Ante*, pp. 461, 467.   [2] *Loc. cit.*, 1882.

Hague and Iddings,[1] many of these chemically so closely related rocks merge gradually into one another, the textural differences depending upon the less or greater depth at which they became consolidated. Rocks which consolidated near the surface are more glassy, while those which consolidated in depth where cooling took longer are holocrystalline and granular. The augite-andesite is therefore but a facies of the granular diorite and of the older diabase ; the hornblende-andesite stands in similar relation to the porphyritic diorite ; the quartz-porphyry is partly a dacite, partly a rhyolite ; while the younger diabase dyke must be considered a basalt dyke. According to these authorities also, augite-andesite is the

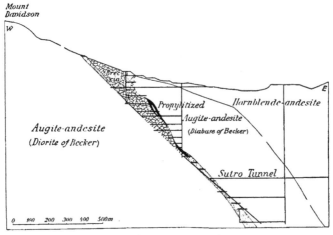

FIG. 300.—Transverse section across the Comstock Lode. Becker.
The hatching represents the lode mass, the black the worked portion, the remainder being lode breccia.

oldest rock, then hornblende-andesite, and finally mica-hornblende-andesite, dacite, rhyolite, and basalt, the last named having but little extent.

This powerful lode, according to Becker, occupies a large fissure along which a correspondingly large movement has taken place. Following the nomenclature of this authority diabase forms the foot-wall and diorite the hanging-wall. According to Hague and Iddings however, one and the same andesite occurs on both walls. All authorities nevertheless are now probably agreed that the whole sequence of eruptives at Comstock are of Tertiary age, and that the textural differences are referable to different conditions of consolidation.

The principal lode, about 4·5 km. long, strikes N. 15° E. and dips about 45° E. To the north as well as to the south, as illustrated in Fig. 299,

---

[1] *Loc. cit.*, 1885.

it splits into branches. Including these the total length is almost 7 km. The deposit itself is a wide quartz-breccia lode containing a series of separate and enormously rich bonanzas. The lode material, consisting of quartz with highly propylitized and often quite clayey fragments of the country-rock, is generally more than 100 feet in width, though in places, as illustrated in Fig. 300, it may be more than 100 metres. In depth also it splits into branches. In addition to quartz ; calcite, selenite, and the zeolites chabasite and stilbite, occur, but only to a small extent.

In spite of its large width there is at most places along its extent so little ore that the lode generally is not payable ; the rich ore is concentrated in a series of bonanzas of relatively huge dimension. These bonanzas, which altogether occupy but one six-hundredth part of the lode plane, lie irregularly along that plane, and, as seen from Fig. 301, hold somewhat better in depth than along the strike. Their width is occasionally as much as 15 m. or more. Some of them come right to the surface, though most were first met underground.

The most important silver minerals of these bonanzas are argentite, stephanite, and argentiferous galena ; less important are pyrargyrite, proustite, polybasite, native silver, and, near the surface, cerargyrite. Gold occurs chiefly as free gold finely distributed. In addition, sphalerite, pyrite, and chalcopyrite occur. The composition of the ore may be gathered from the following analyses :

	California Mine.	Ophir Mine.		Savage Mine.	Kentuck Mine.
SiO_2 . . .	67·50	63·40	SiO_2	83·90	91·50
S	8·75	7·92	Fe_2O_3	1·95	0·83
Au . . .	0·079	0·059	Al_2O_3	1·25	1·13
Ag . . .	1·75	2·79	Mn_2O_3	0·64	...
Fe . . .	2·25	5·46	MgO	2·82	1·37
Cu . . .	1·30	1·60	CaO	0·85	1·42
Zn . . .	12·85	14·46	Ag_2S	1·08	0·12
Pb . . .	5·75	4·15	Au	0·02	0·0017
Sb . . .		0·09	ZnS	1·75	0·13
			CuS	0·30	0·41
			PbS	0·36	0·02
			FeS_2	1·80	0·92
			Alkali	1·28	1·05
			H_2O	2·33	0·59

The Occidental lode, the position of which 2·3 km. east of the principal lode is indicated in Fig. 299, is without economic significance.

The far-reaching propylitization of the country-rock, which as indicated in Fig. 300 extends chiefly in the hanging-wall of the lode, has already been mentioned.[1]

[1] *Ante*, p. 518.

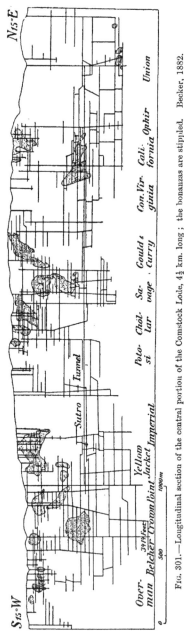

Fig. 301.—Longitudinal section of the central portion of the Comstock Lode, 4½ km. long ; the bonanzas are stippled. Becker, 1882.

The Comstock Lode, apart from the enormous richness of its bonanzas, is also widely known because of the rapid rise of temperature and the inrush of hot water in depth. The temperature of the solid rock rises on an average about 1° F. for every 33 feet, that is 1° C. for every 19·5 m., this being considerably more rapid than would be expected from the ordinary disposition of geotherms in depth. This hot water which first became noticeable at 600–800 m. was at 900 m. during the winter of 1880–1881 so powerful as to necessitate pumping at the rate of 7 million cubic metres of water yearly. The water had a temperature of 75° C. and in spite of good ventilation the water-saturated air left the mine at a temperature of 35° C.

As pointed out by Becker, this high temperature does not arise from the kaolinization or oxidation of sulphides in the country-rock, but is a direct result of eruptive activity. The large quantity of these hot waters is all the more striking in that the district is a dry one, very little rain falling. The high temperature on the 900 m. level rendered further sinking impossible, and since at the same time the higher-lying bonanzas were becoming gradually exhausted, mining operations dwindled, till in 1892 they were finally suspended. Since then, except for a little sporadic work, the mines have either lain idle, or the dumps and residues from earlier times have been re-worked.

The Comstock Lode from the

commencement of operations, in 1859, to 1891 yielded 4820 tons of silver and 214 tons of gold,[1] having together a value of 351·2 million dollars or about £73,000,000. If the further results up to 1902 be added these figures become respectively 369·5 million dollars or £77,000,000. It is worthy of remark that this production was from a lode about 4·5 km. in length worked down to a depth of 900 m., and that consequently the Comstock Lode represents the richest concentration of precious metals yet encountered. The zenith of production was reached in the year 1877 when gold to the value of 14·5 million dollars and silver to the value of 21·8 millions were obtained, figures which represented almost one-third of the gold production of the United States at that time, and almost one-half of the silver. Up to December 31, 1880, from a gross revenue of 306 million dollars or £63,750,000, dividends amounting to 118 million dollars or £24,500,000 were distributed.

The weight relation of gold to silver in the total production was as 1 : 22·5, a relation from which, as might be expected, variations occurred, both in the different bonanzas as well as in the mass of each individual bonanza. The following table of the value relations appertaining to the different groups of mines up to the year 1882, formulated by Becker, is of interest in this connection.

	Percentages of Value.	
	Gold.	Silver.
	Per cent.	Per cent.
Gold Hill Group	47	53
Central Group	36	64
Bonanza Group	47	53
All together, to end of 1865 . . .	32	68
,, ,, ,, 1882 . . .	42·5	57·5

The discoverer, or one of the discoverers of this world-famous lode, a Canadian, Henry Comstock by name, after making much money in the beginning, died in 1870, a beggar.

CRIPPLE CREEK, COLORADO

Gold telluride in quartz-fluorite-dolomite lodes and in the neighbourhood of a phonolitic intrusion

In the Cripple Creek district, about 2900 m. above sea-level, a Tertiary phonolite intrusion occurs in pre-Cambrian granite and slate. According to Lindgren and Ransome the eruptive sequence was probably as

[1] *Ante*, p. 202·

follows : first latite-phonolite and syenite ; then phonolite again, constituting the principal eruption, and trachyte-dolerite ; and later, basic dykes

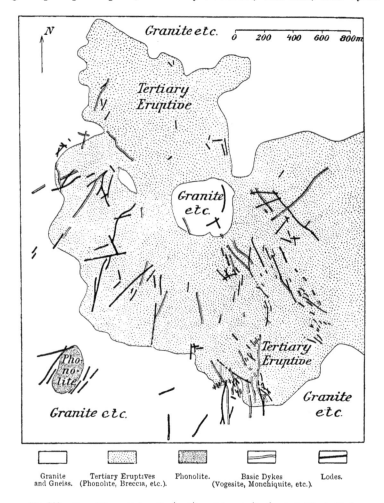

FIG. 302.—Map of the more important portion of the eruptive trunk at Cripple Creek,
with the accompanying lodes and basic dykes.   Lindgren and Ransome, 1906.

of vogesite, monchiquite, and trachy-dolerite, all of which belong to the
same petrographical province.   In addition, breccias are common, while
finally, in the neighbourhood of Cripple Creek, dykes of rhyolite occur,

though these have little extent. There are no eruptive flows or sheets at Cripple Creek. The eruptive area has a length of about 5 miles and a width of 3 miles, embracing therefore 12·7 square miles or approximately 33 sq. km. Leaving the breccias out of consideration, phonolite occupies 73·5 per cent of this extent, latite-phonolite 23·9 per cent, and the other rocks but 2·6 per cent. This phonolite is a nepheline-phonolite with some sodalite and nosean.

The lodes are found concentrated within the eruptive chimney, principally in breccia and phonolite, but to a less extent also in the surround-

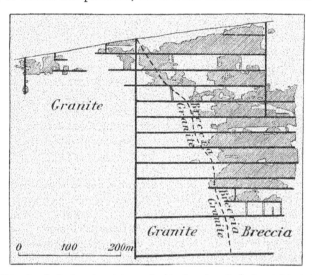

Fig. 303.—Longitudinal section of the ore-shoot in the Independent mine at Cripple Creek. Lindgren and Ransome, 1906.

ing older rocks, granite, etc. The most important lodes, yielding together about £2,500,000 yearly, are found within a circle having a radius of but 2·5 km. Broadly speaking, the steep lodes show a radial grouping around a centre situated in the northern portion of the occurrence; often they follow the well-defined walls of the younger basic dykes. Although in general the individual lodes are not wide, as indicated in Fig. 302 they are often found in connected series.

The gold occurs in the form of tellurides, calaverite chiefly, sylvanite subordinately, and other tellurides of gold, silver, and lead, to a still smaller extent. Native gold occurs secondary but not primary; pyrite is common. In addition, tetrahedrite, stibnite, and small amounts of galena, sphalerite, molybdenite, etc., occur. The weight relation of gold to silver is approximately as 1 : 10. In the rich oxidation zone, which generally extends to a depth of about 70 m., the gold tellurides are in greater part decomposed and the gold is free. Emmonsite[1] and tellurite[2] are also found secondary in this zone.

Of the gangue, quartz with some chalcedony and opal, forms about 60 per cent, while fluorite and dolomite equally divide the remainder. Roscoelite,[3] rhodochrosite, celestine, etc., are also found, though in very small amount. Adularia is not uncommon. On account of the small width of the lode fissures, gangue-minerals are not present in great amount. A considerable portion of the ore consists of the propylitized and metasomatically altered country-rock.[4] Mineralization probably took place directly after the intrusion of the youngest basic dykes.

Mining at Cripple Creek began in 1891. Since 1898, gold to the value of about 15 million dollars or £3,100,000 has been won yearly. The total production to the end of 1905 amounted to 232,750 kg. of gold, equivalent to 154·6 million dollars or £32,200,000 ; to the end of 1910 it probably amounted to about 330 tons of gold, equivalent to 220 million dollars or £46,000,000. Cripple Creek in not quite twenty years will accordingly have produced from telluride ores about fifteen times as much as Nagyag in 160 years.

A few years ago there were more than twenty shafts deeper than 300 m. at work at Cripple Creek. The ore on an average carries about 50 grm. of gold per ton. Rich shoots occur not only in the Tertiary eruptives and breccias but also in the adjacent granite. No influence of the country-rock upon the gold content of the lodes has been established. The amount of gold appears to diminish below the 300 m. level. The El Paso tunnel, which cuts the lodes in depth, was completed in 1903.

GOLDFIELD, NEVADA

Gold-quartz lodes with unimportant bismuth, etc. ; lodes characterized by alunitization of the country-rock

The rich district of Goldfield, first discovered in 1902, is situated in Western Nevada, about 1600 m. above sea-level, near a desert land, and in the neighbourhood of several other Tertiary eruptive districts, that of Montezuma lying 11 km. to the south, and Tonopah 45 km. to the north. The geological structure of the district is illustrated in Fig. 304, which is taken from Ransome's work.

The oldest known beds consist of what is considered to be altered

---

[1] TeO_2 with some Fe_2O_3.
[2] TeO_2.
[3] Vanadium-mica.
[4] *Ante*, p. 521.

Cambrian and granite, which are intruded and covered by eruptive rocks of Eocene to Pliocene age. These eruptives occur as flows and intrusions, and consist chiefly of rhyolite, latite, dacite, andesite, and basalt, this last occurring only in flows. In addition various tuffs and breccias occur. The older of these eruptives were covered by the lacustrine Siebert

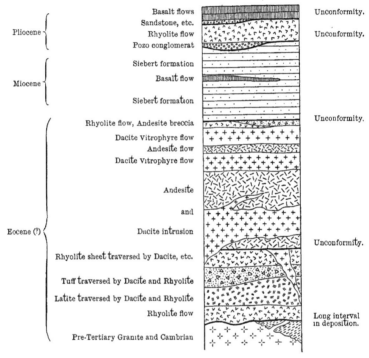

FIG. 304.—Diagrammatic section of the Goldfield district. Ransome, 1909.

formation, 300 m. thick, after the deposition of which a not inconsiderable denudation took place before the basalt was outpoured.

The alunitization already described[1] extends over a large irregular-shaped area, often 1 to 2 km. wide, illustrated in Fig. 305. Within this, and especially in the neighbourhood of the town of Goldfield, the ore fissures are very numerous. Those there present account for 95 per cent of the total gold production, although they are contained within an area barely 1·5 km. long by 1 km. wide. Most of the fissures occur in intrusive dacite and but few in andesite or other rock.

[1] *Ante*, p. 522.

The ore-minerals are native gold with some pyrite, bismuthinite, famatinite,[1] and small amounts of enargite, goldfieldite [2] with 17 per cent tellurium, chalcopyrite, galena, sphalerite, pyrargyrite, proustite, etc. At least 95 per cent of the gold is native, only a small portion being in combination. Silver is very subordinate, but 1 part occurring for every 7·5 parts of gold. The most important gangue-mineral is quartz, this being accompanied by kaolin, alunite, barite, selenite, and other secondary sulphates. Calcite does not occur in the lode material.

After tremendous eruptive activity followed undoubtedly an extensive

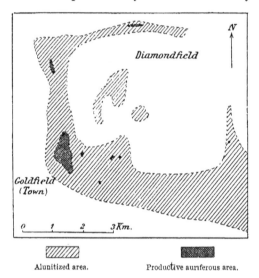

Alunitized area. Productive auriferous area.

Fig. 305.—Plan of the Goldfield district. The white areas are chiefly andesite, dacite, and vitrophyre. Ransome, 1909.

period of thermal activity, from which the alunitization resulted. Later still, probably in late Miocene or early Pliocene time, a second more limited thermal period began, to which the introduction of the gold was due. Since the period of ore-deposition the surface has at the most been lowered about 300 m. by erosion. The ore now occurs in irregular fissures, which, being limited along the strike, Ransome did not regard as lodes [3] but as veins.[4]

The production, which in 1903 was small, rose in 1904–1905, and

[1] Copper-antimony-arsenic sulphosalt.
[2] $5CuS (Sb, Bi, As)_2 (S, Te)_3$.
[3] Ransome, 'lodes or veins.'
[4] Ransome, 'ledges,' see Preface to Vol. I.

still more rapidly in 1906–1907, amounting in the latter year to 406,756 oz. or 12,876 kg. of gold, equivalent to \$8,455,725 or £1,750,000 when the small amount of silver present is included. Up to the end of

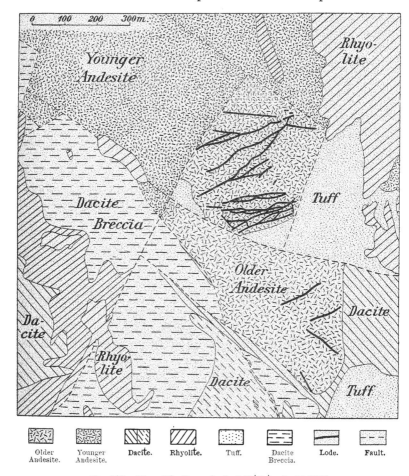

Fig. 306.—Map of the Tonopah silver district. Spurr, 1905.

that year 954,466 oz. or 26,684 kg. of gold, and 116,188 oz. or 3612 kg. of silver, had been produced, equivalent to a total value of 19·8 million dollars or £4,125,000. The present production is about £1,500,000 per year. The ore contains on an average about 50 grm. of gold per ton.

The oxidation zone though very irregular extends generally to a depth of about 50 metres.　A few years ago most of the shafts were only 250–300 m. deep.　The value of the ore appears to diminish in depth.

TONOPAH, NEVADA

The Tonopah silver-field was first discovered in 1900.　It lies about

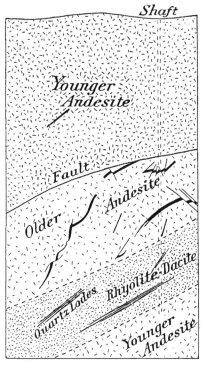

FIG. 307.—Section through the Montana-Tonopah mine.
Older andesite with lodes ; younger andesite ; intrusive rhyolite-dacite with lodes.

45 km. north of Goldfield and about 1800 m. above sea-level.　Both districts are on the western margin of the Great Basin.

The sequence of the Tertiary eruptives at Tonopah is as follows : an older andesite of a hornblende-biotite variety ; a younger andesite of a biotite-augite variety ; and later a rhyolite-dacite.　Basalt also occurs though only to an insignificant extent.　The rocks at this place also, as at Goldfield, are overlaid by the lacustrine Siebert formation.　The whole district is traversed by many faults, some of which have considerable throw. These are indicated in Fig. 306.

The more important lodes, carrying 1 part of gold to approximately 100 parts of silver, are found in the older andesite, but not in the younger eruptives. The formation of these lodes consequently took place immediately after the extrusion of the older andesite and before that of the younger rocks. The deposition of the ore probably took place fairly near the surface. These lodes contain the silver minerals, argentite, polybasite, stephanite, etc., with some chalcopyrite, pyrite, a little galena, and sphalerite. Selenium is also present, probably in association with the silver minerals. Quartz is the principal gangue ; with it calcite, sericite, and adularia occur to a less extent. In the upper levels secondary cerargyrite, pyrargyrite, argentite, and native silver, are found.

In addition to these older lodes there are younger lodes associated with the rhyolite-dacite eruption. These also contain silver minerals in a quartz gangue. The gold content is relatively higher than with the older lodes, in spite of which, however, these lodes have not the same economic importance.

The country-rock in the neighbourhood of the lodes is greatly propylitized. The temperature rises in depth almost as rapidly as at Comstock.[1]

Of the 9,508,464 oz. or 295·7 tons of silver produced by Nevada in 1908, no less than 7,172,396 oz. or 223 tons came from Tonopah, in addition to which Tonopah in that year also yielded gold to the value of $1,624,475 or £338,475.

MEXICO

LITERATURE

Vol. XXXII., 1902, Trans. Amer. Inst. Min. Eng. is devoted to Mexico. In it, among others, the following papers deserve mention.—J. G. AGUILERA. 'The Geographical and Geological Distribution of the Mineral Deposits of Mexico.'—E. HALSE. 'On the Structure of Ore-Bearing Veins in Mexico.'—W. H. WEED. 'Notes on Certain Mines in the State of Chihuahua, Sinoloa, and Sonora'; and 'Notes on a Section across the Sierra Madre of Chihuahua and Sinoloa.'—E. ORDONEZ. 'The Mining District of Pachuca.'—W. P. BLAKE. 'Notes on the Mines and Minerals of Guanajuato.'—J. W. MALCOLMSON. 'The Sierra Mojada and its Ore-Deposits.'—J. P. MANZANO. 'The Mineral Zone of Santa Maria del Rio, San Luis Potosi.'

Among older works are the following : ALEX. V. HUMBOLDT. Essai politique sur le Royaume de la Nouvelle-Espagne, III., 1811.—P. LAUR. Ann. d. Mines, 6 Sér. XX., 1871.— S. RAMINEZ. Noticia histórica de la riqueza minera de Mexico, etc., Mexico, 1884.—E. HALSE. Articles in Eng. and Min. Jour., 1894, 1895 ; and papers in Trans. Amer. Inst. Min. Eng. XVIII., XXI., XXIII., XXIV.—Special descriptions by AGUILERA, ORDONEZ, SANCHEZ, RANGEL, GONZÁLES Y CASTRO, upon Pachuca, 1897, and Real del Monte, 1899, etc.—E. FUCHS and L. DE LAUNAY. Traité des gîtes minéraux, 1893.—J. D. VILLARELLO, T. FLORES, and R. ROBLES, upon Guanajuato, guide to the Internat. Geological Congress, Mexico, 1906.—R. ROBLES, upon Hidalgo del Parral, ibid.—ANTONIO DEL CASTILLO. Geological and Ore-Deposit Map of Mexico, 1889.—Written communications from Ordonez to Vogt.

---

[1] *Ante*, pp. 517, 562.

In Mexico, the country richest in silver and at present responsible for the greatest production of that metal of any country of the world, the Tertiary eruptives, chiefly Miocene and post-Miocene, have a tremendous distribution, especially in the Sierra Madre the immediate continuation of the Rocky Mountains, and in the hill ranges and enclosed plateau near

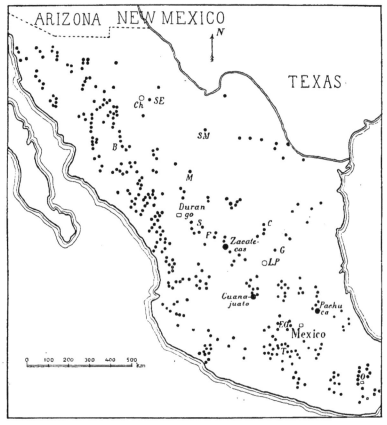

FIG. 308.—Map of the silver- and gold deposits of Mexico. A. del Castillo, 1889.

Ch, Chihuahua ; SE, Santa Eulalia ; SM, Sierra Majoda ; M, Mapimi ; B, Batopilas ; S, Sombrerete ; F, Fresnillo ; C, Catorce ; G, Guadalcazar ; LP, San Luis Potosi ; EO, El Oro ; T, Tasco ; O, Oaxaca.

Mexico City. These Tertiary rocks include andesite, dacite, rhyolite, obsidian, perlite, trachyte, phonolite, basalt, etc., with attendant plutonic rocks, agglomerates, and tuffs. According to the geological map of Mexico, these rocks, chiefly as eruptive flows, occupy about half the surface of an area 750 km. long and 200–300 km. wide in the above region.

Of the sedimentary formations present the Cretaceous has the largest extent.

In Mexico too, most of the lodes, carrying silver chiefly but also some gold, are connected both spacially and genetically with young eruptive epochs ; indeed the majority of these lodes occur actually within Tertiary eruptives, chiefly in andesite, more seldom in rhyolite, etc., but also in granite and diorite. In addition, there are many which though found in sedimentaries are in the vicinity of eruptives. Of these Aguilera says,[1] " It is evident that they are related to and dependent upon andesitic Tertiary eruptive rocks." The silver belt, as indicated in Fig. 308, first extends as the continuation of the metal province of Arizona, Nevada, etc., along the Sierra Madre, especially its western slope, and then farther south occupies the central plateau mentioned above. Still farther east-south-east many similar lodes are found in the State of Oaxaca, so that the length of this belt within Mexico reaches the astounding figure of 2200 kilometres.

In these Tertiary silver lodes gold is always present, though generally in such small amount as 1 part of gold to 140–400 parts of silver. The principal lode of the Promontario mine in Durango for instance, occurring in rhyolite-porphyry, produced from December 5, 1896, to August 18, 1906, 179·1 tons of silver and 493·2 kg. of gold, or 1 part of gold to about 360 parts of silver.[2] Beside the silver lodes there are in many districts gold- or gold-silver lodes of the same age and genesis, for instance, that important producer of recent years, the El Oro, situated about 100 km. west-north-west of Mexico City ; the lodes at and near this mine occur in andesite and Mesozoic sediments. In addition, the following gold lodes which occur entirely in andesite may be mentioned : Taviche in Oaxaca, Ixtlan in Tepic ; Cerro Colorado in Chihuahua ; and the lodes at Guadelupe-y-Calvo, formerly so famous, likewise in Chihuahua, with 1 part of gold to some 10 parts of silver.

In addition to these Tertiary lodes, Aguilera, Ardonez, and Lindgren [3] mention two other classes of gold deposit, namely :

(a) Contact occurrences of gold ore with copper ore in diorite and limestone of late Cretaceous or Tertiary age. Such are found more particularly on the eastern slope of the Sierra Madre, well-known instances occurring at Encarnacion and San José del Oro in Tamanlipas, Mazapie in Zacatecas, and farther south at Santa Fé in Chiapas. As described later, a considerable proportion of the copper deposits of Mexico are of contact character, such deposits standing in genetic association with fairly young eruptives.

[1] *Loc. cit.*, 1902.   [2] Church Lincoln, *Trans. Amer. Inst. Min. Eng.*, 1907.
[3] *Ante*, p. 553.

(b) Gold lodes with little silver, in granite and other old rocks. These occur chiefly on the west coast of Mexico, in Sonora, Sinaloa, Tepic, Guerrero, and Oaxaca. To these belong among others the lodes of the second most important gold district of Mexico, Minas Prietas in Sonora. These lodes are comparable with those of California.

According to an estimate by W. Lindgren [1] the gold production of Mexico is approximately distributed, as to some 20 per cent from the lodes just mentioned, which are probably Mesozoic ; and as to 80 per cent from the Tertiary and chiefly post-Miocene gold- and gold-silver lodes.

Galena and sphalerite, etc., are found in most of the silver lodes of Mexico, especially in depth. In addition, both these sulphides occur particularly in metasomatic deposits in Cretaceous limestone and slate, usually in connection with eruptive rocks,[2] andesite and rhyolite particularly ; from these deposits the bulk of the expanding lead production of Mexico is derived ; the galena strangely enough is rather poor in silver. Occurrences of this kind are found at : Santa Rosa de Muspuiz, Sierra Mojada, and Mula in Coahuila ; Naica and Los Adargas in Chihuahua ; La Velardena and Mapimi in Durango ; Cerralvo in Nuevo León ; Zimapán, Pechuga, Cardonal, and Lomo de Toro in Hidalgo ; Caltepec, Santa Ana, and Tehuacán in Pueblo ; Bramador in Jalisco ; Sombrerete, Mazapie, and Noria de Angeles in Zacatecas ; and Huetamo in Michoacán.

The following statement of the production of silver, gold, and lead in Mexico from ages past to present time will give an idea of the position.

		Tons of 1000 Kg.		
		Silver.	Gold.	Lead.
Yearly Average	1521–1544	3·4	0·2	Nothing or but little
	1545–1560	15·0	0·2	
	1561–1580	50·2	0·3	
	1581–1660	84·7	0·4	
	1661–1700	106·2	0·4	
	1701–1740	197·3	0·6	
	1741–1780	333·7	1·1	
	1781–1810	559·6	1·4	
	1811–1840	302·6	1·0	
	1841–1870	458·1	1·8	
	1871–1875	601·8	2·0	
	1880	701·0	1·4	
	1885	772·7	1·5	17,500
	1890	1211·6	1·7	22,300
	1895	1582·3	8·7	68,000
	1900	1786·9	13·5	84,700
	1905	1700·2	24·2	75,000
	1910	2291·3	33·7	110,000

[1] Ante, p. 554.   [2] Aguilera, 1902, p. 572.

The total production of Mexico from 1851 to 1909 may be estimated at 122,500 tons of silver, worth some 925 million sterling ; and about 450 tons of gold, worth about 75 million sterling. Elisée Reclus,[1] basing himself to some extent upon the same data, gives the following figures : from 1521 to 1890 silver to the value of about 800 million sterling, and gold 36 million sterling. The data available for the earlier years are however quite unreliable.

As far back as 1519 when Cortes arrived in Mexico, the Aztecs were found to possess enormous treasure of precious metal, and particularly of gold. Soon afterward several mines were started, Pachuca for instance in 1522, Zacatecas in 1546, Durango somewhat later, and Guanajuato in 1558, while the patio process was introduced in 1557. Under Spanish rule gold and silver mining flourished exceedingly ; during, and for some time after the War of Independence it fell ; while now, again, within the last twenty years, favoured by the construction of many railways, it is particularly active.

Of the famous silver-mining districts, Pachuca and Real del Monte lie about 90 km. to the north-east of Mexico City ; Guanajuato and Veta Madre about 275 km., and Zacatecas and Veta Grande 525 km. to the north-west. Others worthy of mention are Villanueva, Fresnillo, etc., in Zacatecas ; Guadalcázar, Catorce, San Pedro near San Luis, etc., in San Luis Potosi ; Parral, Santa Eulalia, and Batopilas in Chihuahua ; Chipioneña and Carmen in Sonora. Beside these there are a considerable number of other mines, so that the present production of the country is derived from many lodes and is distributed among all the states with the exception of Yucatan. The Tertiary silver lodes generally occur high up in mountainous country ; Tasco for instance is 1600 m. above sea-level, Pachuca 2460 m., Real del Monte 2765 m., Guanajuato 2000 m., and Zacatecas 2500 metres.

These Tertiary lodes have quartz generally—often with amethyst and chalcedony—as principal gangue-mineral; in addition calcite, and some-times also barite. Rhodochrosite, rhodonite, and apophyllite are common. Fluorite on the other hand is absent from most lodes, or only occurs here and there and in small amount. The most common primary silver minerals are argentite, pyrargyrite, proustite, stephanite, polybasite, tetrahedrite, etc. These are accompanied by pyrite, galena, sphalerite, etc.

In the oxidation zone—which with the more important lodes extends occasionally to a depth of 100–150 m.—cerargyrite, bromargyrite, and native silver are found in addition to the usual iron- and manganese oxides, while gold is also often present in considerable amount. These easily amalgamable ores rendered possible the large early production of

[1] *Géographie universelle*, Paris, 1891, Vol. XVII. p. 294.

precious metal. Below this zone masses of silver minerals, chiefly concentrated in bonanzas, often follow. Deeper still the proportion of galena and sphalerite, etc., increases, and most of the lodes so rich above become impoverished. Finally, in depth it is often enough the case that a non-argentiferous lead-, lead-zinc-, or lead-antimonial ore-body is found.

The Fresnillo mines illustrate this impoverishment in depth. These mines, opened in 1824, had in 1863 reached a depth of 405 m. The total production from 1833 to 1863 was 902,268 kg., during which period, according to Laur, the average silver content was as follows :

1835	.	.	. 0·225 per cent.	1854	.	.	. 0·063 per cent.
1839	.	.	. 0·146 „	1859	.	.	. 0·062 „
1844	.	.	. 0·115 „	1863	.	.	. 0·056 „
1849	.	.	. 0·078 „				

In accordance with this decrease of value in depth the Tertiary silver mines in Mexico are usually not particularly deep, and though the Valenciana mine on the Veta Madre near Guanajuato some years ago reached a depth of 622 m., a depth of 500 m. is rarely attained even in the most famous mines ; generally it fluctuates between 400 and 500 metres.

Many of the Mexican silver lodes attain a considerable length along the strike; the Vizcaina, Analco, and San Cristóbal at Pachuca, for instance, have lengths of 16 km., 6 km., and 4 km. respectively, though the width is seldom more than 7 m. The exposed length of the Veta Cantera at Zacatecas is more than 12 km., the width being 12–15 m. on an average, though occasionally more than 30 m. The neighbouring Veta Grande has a similar or perhaps even greater length. The famous Veta Madre at Guanajuato is likewise many kilometres long and occasionally even more than 150 m. in width, so that in mass it is comparable to the Comstock ore-body.

These powerful lodes of Mexico often exhibit a brecciated structure and it is probable that they invariably represent faults. Composite lodes are common. At Pachuca and Real del Monte, which are but 5 km. apart, considerable outbreaks of andesite took place in Miocene times ; later, rhyolite followed, with dacite, obsidian, pitchstone, and tuffs ; and finally basalt. The lodes though chiefly parallel shew many bifurcations and linked veins. At Pachuca, for instance, four principal lodes are worked, namely, the Vizcaina, El Cristo, San Juan Analco, and Santa Gertrudis, besides the neighbouring lodes of Real del Monte. These lodes, which in greater part occur in andesite, are younger than the rhyolite but older than the basalt. Those in most of the other districts, Zacatecas for example, have approximately the same geological position. In others, as that of Guanajuato, they occur in sedimentary formations, the Cretaceous, Triassic, etc., though in close proximity to Tertiary eruptives.

The following figures relative to the production of individual districts will be of interest. The Santa Eulalia district 25 km. east of the town of Chihuahua has since 1703, or roughly during the course of 200 years, produced silver to the value of 28 million sterling ; and the Batopilas district about 12 million sterling.[1] Chihuahua, when copper and lead also are considered, is now the most important mining district of Mexico. Pachuca [2] from its discovery, in 1522, to 1901 yielded more than 3500 tons of silver worth more than 31·5 million sterling. One single bonanza of elliptical outline, having the dimensions $1000 \times 400 \times 2\frac{1}{2}$ m., yielded in the course of ten years a value of close upon £3,000,000 ; another from 1853 to 1883, close upon £6,000,000 ; while an earlier bonanza is stated to have been richer still. Zacatecas, including Veta Grande, is stated from 1548 to 1832 to have produced silver to the value of almost 150 million sterling, equivalent to some 14,000 tons of silver, though according to other data this figure is too high. Guanajuato [3] in silver and gold has produced as follows : 1701 to 1800, some 279·7 million dollars ; 1801 to 1829, some 85·8 million dollars ; 1830 to 1887, some 277·6 ; equivalent to a total of 643·1 million dollars or about 134 million sterling. To this must be added the very considerable production from 1558 to 1700, and that since 1887. Some idea of this latter may be gathered from the fact that during the period 1900–1903 the value produced was 6·2 million dollars. The total production of the Veta Madre at Guanajuato, Humboldt estimated at 80 million sterling. The total production hitherto from Guanajuato may probably also be put down at some 160 million sterling, equivalent to some 17,000 tons of silver. The most imposing impression of the silver production of this country is obtained when it is considered that during the period 1899–1908 almost 2000 tons of silver were produced annually, the actual average having been 1890 tons. Even after the fall in the price of silver at the beginning of the 'nineties, the silver production of Mexico still continued to rise.

In the small Republics of Central America also, several Tertiary silver-gold deposits occur, one of the better known being that of San Juancito in Honduras,[4] which in 1903 produced 21,266 kg. of silver and 113 kg. of gold. The output of gold was greater formerly.

[1] Fuchs and De Launay, 1893.
[2] Ordonez, loc. cit., 1902.
[3] Trans. Amer. Inst. Vol. XXXII., 1902, p. clxxxix.
[4] Leggett, Trans. Amer. Inst. Min. Eng. XVII., 1889.

LODES OF THE SOUTH AMERICAN CORDILLERAS AND THE BOLIVIAN
SILVER-TIN LODES

LITERATURE

G. STEINMANN. Über gebundene Erzgänge in den Cordilleren Südamerikas. Intei-
national Congress, Düsseldorf, 1910 ; ' Gebirgsbildung und Massengesteine in den Cordilleren
Südamerikas,' Geol. Runds. I., 1910, Pts. I.-III. ; ' Über die Zinnerzlagerstätten Bolivias,'
Zeit. d. d. geol. Ges., Jan. 1907 ; ' Observaciones geologicas effectuadas desde Lima hasta
Chanchamayo,' Bol. Cuerpo, Ing. Min. Peru, Lima, 1904 ; Die Entstehung der Kupfererz-
lagerstätte von Corocoro und verwandten Vorkommnisse in Bolivia. Rosenbusch Celebra-
tion, Stuttgart, 1906.—A. W. STELZNER. ' Die Silber-Zinnerzlagerstätten Bolivias,' Zeit.
d. d. geol. Ges. II., 1897, wherein the works of A. v. Humboldt, A. d'Orbigny, D. Forbes,
H. Reck, A. Gmehling, etc., are mentioned. The following works deal especially with Potosi :
A. F. WENDT. ' The Potosi Bolivia Silver District,' Trans Amer. Inst. Min. Eng. XIX.,
1891.—WIENER. ' Oruro,' Ann. d. Mines, Paris, Sér. 9, V., 1894.—W. R. RUMBOLD.
' The Origin of the Bolivian Tin Deposits,' Econ. Geol. IV., 1909.—EVERDING. ' Unter.
lagen zu einer bergmännischen Lagerstättenbegutachtung im bolivianischen Zinnerz-
distrikt,' Glückauf, 1909, p. 1325. The works of Domeyko, Möricke, etc., upon Chili are
cited when describing the copper lodes of Chili.

By far the greater number of the metalliferous lodes of the Cordilleras
of South America are of Tertiary age. They are always associated with
eruptive rocks upon which, both in their occurrence and extension, they
are manifestly dependent. Along the 6000 km. length of these Cordilleras the eruptives, according to Steinmann, appear in three forms.
The volcanoes which were active in late Tertiary and Diluvial time have
long been known. These in their extension coincide essentially with the
principal mountain range ; they however carry no ore ; no lodes are found
either in those which are active or those which are extinct, while even the
necks of those which have been eroded appear to be equally free.

In these Cordilleras those eruptive rocks which probably belong to
early Tertiary must be regarded as the vehicles of the ore. These consist
partly of granular plutonic rocks of granitic or dioritic character, constituting the second form of eruptive occurrence ; and partly of porphyritic rocks of liparite-trachyte or andesite-dacite nature, constituting
the third form. These older eruptives have a much larger distribution
than the younger rocks mentioned above, which in addition are generally
more basic. Many not unimportant deposits, especially of gold- and
copper ore, occur in connection with the early Tertiary grano-diorites in
the Andes ; but more important still are the lead, silver, copper, zinc,
tin, and gold occurrences regularly associated with the andesitic and
allied rocks which may be observed everywhere in Peru, Bolivia, and
farther south in Chili and Argentina.

The intrusions of andesite or andesite-liparite, and the lodes associated
with them, are found concentrated in a wide belt embracing the principal
mountain range. In the north of Chili and Argentina, and in Bolivia,

this belt is in places 500 km. wide, a width which northwards and southwards diminishes to 250 km. and even to 100–150 km.   To the east, along the ranges which descend on the one side to the lowlands of Brazil, Bolivia, and Argentina and on the other towards the Pacific, the andesites and liparites, and with them the lodes, are less extensive.   In detail the association between these rocks and the lodes is more evident still. The andesite and allied rocks generally appear as dykes, lenses, or bosses, which vary from those of small dimension to such as are 10–20 km. across. From their geological situation these may most fittingly be regarded as inclined or vertical laccoliths, which ended blindly without reaching the original surface, and which consequently were not generally accompanied by craters or tuffs.

The lodes with silver sulphide minerals, etc., and locally with tin and gold content, exhibit generally the most intimate connection with these deep early-Tertiary laccoliths.   Ordinarily they occur in the eruptive itself or in the closest proximity thereto.   This is the case for instance at Potosi, Oruro, Huanchaca, and other Bolivian deposits ; and at Cerro de Pasco, Huallanca, Ticapanupa, Tarica, Morococha, Hualgayoc, and other places in Peru.   Limited occurrences consisting of one or two small lodes of little extent are numerous in the Cordilleras, these being associated with small dyke-like eruptive masses.   All the larger and more productive districts, on the other hand, are associated with extensive masses which either consist of large single peaks, as for instance at Cerro de Potosi and Chorolque, or form composite massives as at Cerro de Pasco, Morococha, Oruro, etc.   Steinmann from this draws the conclusion that in the Cordilleras there exists a quantitative relation between the bulk of the ore vehicle and the number and content of the lodes produced by it.   Similar quantitative relations have already been noticed in connection with some magmatic eruptive deposits.[1]

The lodes of the Cordilleras are in many places found concentrated in the boundary region between the eruptive and the surrounding sediments, as illustrated by the diagrammatic representation of the occurrence at Cerro de Pasco in Fig. 309, where the laccolith has been freed from its mantle of sediments.   The uncommonly numerous lodes traversing the marginal portions of the eruptive and adjacent sediments at this place, were remarkable for abnormal richness in silver. In depth these lodes decreased both in number and content.   The Socavon Real adit, put in at great expense at the foot of the hill and 680 m. below the summit, disappointed the hopes upon which it was started, in that it encountered but few lodes, and these relatively poor.

The Tertiary gold-silver lodes of the Cordilleras usually contain quartz

[1] *Ante*, pp. 247, 288, 295.

as the principal gangue - mineral. Fluorite, zeolites, carbonates, and barite are absent from most, though the two last-named appear abundantly in some. With these lodes also, gold and silver are closely associated, these two metals either occurring together in the same lode or in separate though neighbouring lodes. For instance, in the important silver district of Hualgayoc the gold is practically limited to one single lode. Humboldt [1] estimated the average annual precious-metal production of Potosi up to the commencement of the nineteenth century at 481,830 marks [2] of silver and 2200 marks of gold, or 1 part of gold to 200–250 parts of silver.

Many of the lodes are characterized by well-defined primary depth-zones. Those of Cerro de Pasco. for instance, which in times past were responsible for the greatest silver production of Peru and for centuries were worked almost exclusively for silver, in depth passed over in part to

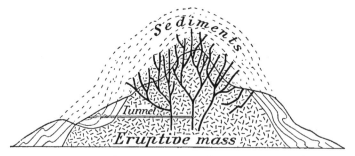

Fig. 309.—Diagrammatic section across Cerro de Potosi, showing the collection of lodes at the contact of the eruptive with the sediments, and the diminution of the number of lodes in depth. The dotted lines represent the beds removed by erosion. Steinmann, 1910.

become copper lodes. In the case of the silver-tin lodes of Bolivia, the tin according to Steinmann is generally found concentrated in the upper levels, while the silver ore is found below. Many of the silver lodes proper contained quantities of secondary silver minerals in the oxidation zone, such for instance as native silver and cerargyrite, which are easily amalgamable ; it is nevertheless a striking fact that secondary enrichment such as would constitute a cementation zone is, according to Steinmann, either entirely absent or extremely infrequent.

The silver-tin deposits of Bolivia, which have been more particularly investigated by Stelzner, are of especial interest. They occur in the Eastern Cordilleras or Cordillera Real where active volcanoes are absent, and in the high plateau confined between these and the Western or Coast Cordilleras. In the northern portion of this extent is situated the Titicaca Lake at an altitude of 3854 m. above sea-level. The highest point of these

---

[1] Citation by Soetbeer.   [2] 1 mark = 8 oz.

Eastern Cordilleras, which have an average height of 4700 m., is the Illampu with a height of 7513 metres. The average altitude of the Western Cordilleras is 4550 metres. La Paz and Oruro, together with the places lying along the line Oruro-Ujuni to the south, are the delivery stations for the tin ore, the mines lying almost exclusively east of this line. The deposits of the north-western portion of the Eastern Cordilleras yield tin ore almost exclusively, with quite subordinate silver, bismuth, wolfram, and antimony ores. On the other hand, in the adjoining portion of the broad lode belt to the south-east, silver ores play a prominent part. While formerly silver-mining conducted in the rich ores of the upper levels was alone of economic importance, tin-mining has latterly become more and more prominent. The districts of Potosi and Huanchaca are to-day the most important of the silver-mining districts.

The Western Cordilleras consist of Mesozoic strata, chiefly Jurassic or Cretaceous, which have been repeatedly intruded by young eruptives. Their extent is marked by a long row of volcanoes, some of which are still active. The high tableland between the two Cordilleras is a desolate sandy steppe almost without vegetation and often assuming the character of a salt desert. The Eastern Cordilleras consist of Palæozoic slates, quartzites, and grauwackes, chiefly Silurian and Devonian, which sediments in the most highly contorted portion of their occurrence are seen to be underlaid by granite. All these rocks are traversed by an abundance of Tertiary eruptives.

Those unique lodes which contain both silver and tin extend from the 16th parallel in the southern portion of Peru, to the 22nd parallel and perhaps even still farther south. The length of the belt in which they are contained is accordingly about 800 km. It is 300 km. wide. The most important districts are Carabuco, Avicaya, Milluni and Huayna-Potosi, Monte Blanco in the Quinza-Cruz mountains, Colguiri, Oruro, Morococha, and Huanuhi, Llallagua, Colquechaca, Potosi, Porco, Pulacayo, Huanchaca, Chocaya, Tasna, Chorolque, etc. In this last district the mines are from 3500 to 5200 m. above sea-level. The lodes usually carry tin as well as silver in one and the same lode, the intergrowth of the two ores being generally so intimate that the ore is first chloridized and then amalgamated or cyanided for silver, and afterwards dressed for tin. Several lodes in part carry silver ore without tin, or tin ore without silver.

The primary silver minerals are principally sulpho-salts, antimonial tetrahedrite in the first place, then pyrargyrite, proustite, and stephanite, etc. Argentite, the new mineral sundtite, etc., also occur. These minerals, which are here regarded by different authorities as primary, are those which in other districts are found in the cementation zone. Other undoubtedly primary minerals present are pyrite, arsenopyrite, pyrrhotite, chalcopyrite, stibnite, galena, sphalerite, ullmannite, and bournonite, and occasionally abundant bismuth ores. The most important tin ore is cassiterite. The sulpho-stannates stannite,[1] plumbostannite, canfieldite, franckeite, and cylindrite also occur, the first-named being found in some lodes in notable amount. The three last-named contain germanium. The silver-germanium sulpho-salt, argyrodite, which contains 6·5 per cent of germanium, also occurs. The Bolivian silver-tin lodes are relatively the richest in germanium of any hitherto investigated. Tin and germanium belong, as is well known, to the same periodic system. Wolfram, elsewhere the constant associate of cassiterite, is here represented only in some lodes.

The most important gangue-mineral is quartz, which is occasionally accompanied by some calcite and barite. The characteristic minerals of the typical tin lodes,[2] fluorite, tourmaline, lithia-mica, topaz, apatite, and other combinations rich in fluorine and boron are, on the other hand, either completely absent or have only been established as mineralogical rarities. Fluorite is extremely uncommon, while tourmaline, so characteristic of many of the Chilian copper lodes, occurs only sporadically in the Bolivian silver-tin lodes.

Along the lodes a kaolinization of the country-rock is often found, and sometimes a silicification, while greisen formation, otherwise so characteristic of tin, receives no mention. These Bolivian silver-tin lodes, rich in tin, differ essentially therefore in this respect from the ordinary tin lodes, though certain resemblances remain.[3] On the other hand, mineralogically, chemically, and geologically, they agree in their broad lines with the normal Tertiary silver lodes, though naturally with the difference that the Bolivian lodes carry cassiterite and other tin minerals which the normal silver lodes, with but few exceptions, do not.[4]

Stelzner put forward the Bolivian lodes as representative of what he termed the Potosi type in contradistinction to the Schemnitz type of Groddeck. These lodes belong none the less to the Tertiary silver lodes which are distributed along the entire length of the Cordilleras, from Ecuador or Colombia in the north, to Chili and Argentina in the south. It is nevertheless striking that, occurring over a length 800 km. along this lode belt, they should be characterized by richness in tin, while the silver lodes to the north and south, similarly situated geologically, contain none of that metal.

The Bolivian lodes at the outcrop have a stanniferous gossan, in which, according to Stelzner, wood-tin is present as a secondary mineral

---

[1] Cu_2FeSnS_4 with 27·5 per cent Sn.
[2] *Ante*, p. 413.   [3] *Ante*, p. 423.   [4] *Ante*, pp. 423, 548.

derived from primary sulphide tin ores. The ores at the outcrop, containing native silver, cerargyrite, pyrargyrite, proustite, etc., are locally termed *Pacos*, the undecomposed ores in depth are *Negrillos* or black ores, while those between the two are termed *Mulattos*.[1]

In addition to the above-described lodes characterized by the common occurrence of silver and tin and by the comparative absence of the usual tin minerals, there are also in the Eastern Cordilleras, according to Rumbold, a number of tin lodes which carry, in addition to quartz, a considerable amount of tourmaline, and which mineralogically and geologically closely resemble the ordinary tin type.[2] These appear to be associated with a quartz-porphyry which, according to the above-mentioned authority, is older than the Tertiary eruptives. This however requires confirmation. Such tin lodes poor in sulphides are found more particularly at Oruro and in the neighbourhood of Tres Cruces, 90 miles to the north. The greater part of the Bolivian tin ore produced in recent years has probably been derived rather from these more characteristic tin lodes than from the combined tin-silver occurrences.

The economic importance of the silver lodes of the South American Cordilleras may be gathered from the following figures of production, of which the earlier are, however, somewhat uncertain :

		Colombia.	Bolivia.	Peru.	Chili.	Argentina.	Ecuador.
				Tons of Silver.			
Yearly Average	1545–1560	...	183	48	...	...	...
	1561–1580	...	152	46	...	...	...
	1601–1620	...	206	103	...	...	...
	1641–1660	...	139	103	...	...	...
	1681–1700	...	93	103	...	...	...
	1721–1740	...	43	103	1	...	...
	1761–1780	...	84	122	2	...	...
	1801–1810	...	97	151	7	...	...
	1821–1830	...	42	58	6	...	...
	1841–1850	...	66	108	45	...	...
	1861–1870	...	81	72	57	...	...
	1880	...	265	158	122	...	...
	1885	10	245	49	210	11	...
	1890	20	301	66	124	15	...
	1895	54	643	115	150	36	...
	1900	87	325	204	178	12	...
	1905	31	205	156	26	2	1
	1910	43	218	202	44	4	2

The total silver outputs from the three principal countries of South America up to 1910 were as follows :

Peru since 1533   .   .   .   35,000 tons of silver.
Bolivia „ 1545   .   .   .   48,000 „ „
Chili   „ 1545   .   .   .   6,600 „ „

[1] *Ante*, p. 219.   [2] *Ante*, p. 413.

From these figures it appears that Bolivia with its stanniferous lodes takes first place. The richness of the silver deposits of Potosi discovered in 1545 was enormous,[1] and the total production of this district alone is given as some 30,000 tons. According to Soetbeer the production during the period 1545–1600 amounted to more than one-half the world's production at that time. In depth the lodes became poorer and the district consequently declined. In addition to Potosi other rich lodes have been worked in Bolivia; the *Compagnie Huanchaca de Bolivia* for instance, from 1873 to 1888 produced silver to the value of 50·6 million dollars, of which amount 19·5 million were distributed to the shareholders.

The history of the Bolivian and Peruvian silver mines is briefly as follows : After the conquest of the country by Pizarro in 1533 the output of silver, particularly from the district around Potosi, was very considerable. In the eighteenth century the easily treated ore from near the surface being in greater part exhausted, a decline followed, which during the War of Independence in the early part of the nineteenth century, 1809–1825, became more and more pronounced. The building of railways, however, to remote mining districts in the 'seventies brought about a revival which reached its zenith in the 'eighties, to be followed in turn by a decline consequent upon the fall in the price of silver during the years 1892–1894.

The Bolivian tin ores were formerly either not worked or only inadequately so, for lack of communication. Since the completion of the railways however this particular mining has developed considerably. It is especially the north-western portion of the Eastern Cordilleras which is stanniferous. There, on the southern slope of the Illampu-Illimani mountains at a height of about 5000 m., the tin mines of Huayna-Potosi and Mullini are situated. To the south-east, separated by the valley of the La Place, is the Quinza-Cruz massive, which is reckoned to be particularly rich in tin. The more important mines, most of which however are only in process of development, are the Monte Blanco, Huanchaca de Inguisiri, Concordia, Santa Rosa, and the Capacabana, all about 5000 m. or more above sea-level. South-east of the Quinza-Cruz mountains lie the mines of Colquiri, and isolated on the west slope, that of Araca. The most important tin occurrences at present are El Balcon and Penny Duncan at Huanuni, 30 km. from Machacamarca ; and Avicaya, Totoral, and Antequera, 10–25 km. from Paznia. The deposits at Patino and Illalayua in the environs of Unicia are also deserving of mention.

The country-rock on either side of the lode is generally silicified to a light-grey quartzitic rock. The lode itself consists chiefly of quartz,

[1] *Ante*, p. 580.

cassiterite, and pyrite, while arsenopyrite, chalcopyrite, sphalerite, galena, and bismuthinite, occur to a less extent.

In the north-western part of the Eastern Cordilleras the characteristic ore is a silicified rock traversed by a network of small veins and fissures filled with cassiterite and pyrite, these minerals sometimes also forming lenses or nests. On the wall a compact layer of mineral 2–5 cm. thick often occurs. As a rule the tin content of the pyrite cannot be distinguished macroscopically, though occasionally crystal individuals may be seen. According to Everding it would appear that cassiterite with quartz occurs more plentifully as an upper primary depth-zone, while compact pyrite containing tin forms a lower zone. The payable portions contain 3–6 per cent of tin on an average, though occasionally the content is as high as 15 per cent. The oxidation zone—which generally extends to a depth of 60 m. below the surface and sometimes as much as 300 m.— consists of limonite and brown-stained rock fragments with unaltered cassiterite.

According to the statistics of the *Metallurgische Gesellschaft*, Frankfort, the weight of metallic tin in the Bolivian output of ore has at different periods been as follows :

1885	.	.	.	225 tons of tin.	1900	.	.	.	6,950 tons of tin.
1890	.	.	.	1660 ,, ,,	1905	.	.	.	13,000 ,, ,,
1895	.	.	.	4100 ,, ,,	1910	.	.	.	23,000 ,, ,,

Bolivia therefore now produces about one-fifth of the world's production of tin.[1] It is responsible at the same time for a material portion of the world's small bismuth production, the districts of Tasna and Chlorolque in the south being the contributors.

JAPAN

LITERATURE

'Geology of Japan,' The Imperial Geol. Survey of Japan, Tokio, 1902 ; reviewed in Spurr's previously cited work upon Tonopah.—Mining in Japan, Past and Present. Bureau of Mines, Department of Agriculture and Commerce. Japan, 1909.

The gold and silver production of Japan, as may be seen from the following table, has of late years considerably increased ; the gold output is to the extent of about one-twentieth derived from gravels :

1875	.	.	.	191 kg. gold	7,630 kg. silver.
1885	.	.	.	294 ,,	26,150 ,,
1895	.	.	.	983 ,,	79,280 ,,
1905	.	.	.	5078 ,,	91,390 ,,
1908	.	.	.	5762 ,,	136,240 ,,
1909	.	.	.	3922 ,,	127,916 ,,
1910	.	.	.	4284 ,,	143,597 ,,

[1] *Ante,* p. 424.

Of the gold- and silver lodes of Japan those of Tertiary age only are
concerned in this present brief description ; at all events the largest proportion of the mines working to-day, some of which are producing considerably,
exploit Tertiary lodes.   These occur both in sedimentary formations and
in eruptive rocks.   In many cases it is not a question of a simple fissure
but of a shattered zone associated with intense impregnation in Tertiary
tuff, schist, or liparite.   Often with these deposits, the bodies of which
so far appear to increase in depth, copper ore also occurs.   To such as
these belong the important gold lodes at Poropets in Hokkaido, and Washi-
nosu in Rikuchu, as well as the silver deposits of Fukuishi near Omori in
Iwami, and Matsuoka and Hata in Ugo.   Tertiary eruptives, such for
instance as andesite, dacite, liparite, and basalt, are strongly represented
in Northern Japan, and particularly along the central range.   In these
eruptives, and especially in andesite, most of the gold- and silver occurrences
are found.

The principal deposits are : Hoshino in the province of Chikugo, where
quartz lodes with pyrite, sphalerite, gold, and silver, occur in andesite ;

Gold   Silver   Quicksilver   Copper

Fig. 310.—Map of the

Lead Tin Antimony Iron Manganese Graphite Sulphur Coal Petroleum

mineral deposits of Japan.

Serigano in Satsuma, quartz lodes in andesite with pyrite, chalcopyrite, gold and silver, 78 kg. of gold having been produced in 1908 ; the Yamagano district in Satsuma, many productive quartz lodes in andesite with calcite, pyrite, gold, argentite, etc., 1 part of silver occurring with 5 parts of gold, 370 kg. of gold having been produced during 1908 ; Sado or Aikwa in Sado, quartz lodes in andesite and tuff, with calcite, dolomite, selenite, native gold and silver, argentite, chalcopyrite, pyrite, galena, sphalerite, and less often stephanite, pyrargyrite, etc., 427 kg. of gold having been produced in 1908 ; Zuiho in Formosa, where in 1908 about 280 kg. of gold were produced from lodes in Tertiary sediments ; Kago in Satsuma, with gold lodes in andesite and liparite ; Otani in Satsuma, with lodes in liparite and Tertiary sediments ; Ushio and Okuchi in Satsuma, where from gold lodes in andesite 435 kg. of gold were produced in 1908 ; and finally, Poropets in Hokkaido, where from lodes in liparites and Tertiary sediments 216 kg. of gold were produced in 1908.

The following important mines produce silver chiefly : Kanagase and Tasei, Ikuno in Tajima, with silver- and copper ore in liparite, 6590 kg. of silver having been produced in 1908 ; Innai in Ugo, with silver ore in andesite, liparite, and Tertiary sediments, production in 1908, 2950 kg. of silver ; and Tsubaki in Ugo, with silver- and lead ores in Tertiary sediments and andesite, 38,700 kg. of silver having been produced in 1908.

In addition, many other similar occurrences have been recorded in the provinces Iwami, Ugo, Rikuchu, Kaga, Juwashiro, Iwaki, Mino, Bizen, etc. These are found in andesite and liparite and occasionally also in Tertiary tuff. With some of them silver predominates, with others gold. An intimate admixture of sphalerite, galena, chalcopyrite, pyrite, with gold and silver in varying amount, is found widely distributed in Japan. Such is known locally as *Kuromoro* or black ore. Probably not less than two-thirds of the silver production of Japan in 1908 was derived from this fine-grained mixture. Some silver is also won as a by-product in treating copper ores ; from lead ores on the other hand a surprisingly small amount, only one-twentieth of the total production, is recovered.

SUMATRA

LITERATURE

S. J. TRUSCOTT. Trans. Inst. Min. Met. X. pp. 52-73 ; reviewed in Spurr's Tonopah work previously cited.—W. LIEBENAM. 'Review of Truscott's Paper,' Zeit. f. prakt. Geol., 1902, p. 225.—P. KRUSCH. Untersuchung und Bewertung von Erzlagerstätten, II. Edit. p. 188.—Written communications from Müller-Herrings to Krusch.

In south-western Sumatra [1] is situated a lode district consisting of a disturbed zone 30 km. in length, along which seven hot springs occur.

[1] *Zeit. f. prakt. Geol.*, 1902, p. 227.

The lodes of this district are in part large gold-silver lodes with quartz and chalcedony gangue ; they have hypersthene-andesite in the hanging-wall and rhyolite of Miocene age in the foot-wall. The former in the vicinity of the lode is propylitized.

The lodes worked in the Redjang Lebong and Lebong Soelit mines in another district, carry gold partly free and partly combined with selenium. The peculiar composition of the| gold ore, which has not yet been definitely determined, has all along excited the interest of those geologists who have visited these mines. The gold- and silver minerals are very finely distributed throughout a quartz and chalcedony gangue which, near the walls especially, exhibits crusted structure. The chalcedony is more plentiful in the neighbourhood of the walls than towards the middle of the lode. The rule, demonstrated in the laboratory, that solutions with equal silica content when cooler deposit chalcedony and when hotter tend to the formation of quartz, apparently here finds confirmation.

The lode at Redjang Lebong reaches up to 22 m. in width, while it has so far been developed for a length of 300 m. along the strike. To the north and south at either end it pinches out. The ore treated per year amounts to about 100,000 tons containing 30 grm. of gold and 250 grm. of silver per ton, or 1 of gold to 8 of silver. Selenium occurs to an extent equal to 2–5 per cent of the bullion recovered. It appears to be associated with the silver rather than with the gold. The highly seleniferous slags from smelting have been treated for selenium. Spurr [1] recognizes an analogy between this deposit and those at Tonopah in Nevada.[2] In the year 1906, 1426 kg. of fine gold and 7600 kg. of fine silver were produced, while in the same year from Lebong Soelit, about 18 km. farther to the west, the production was 463 kg. of gold and 645 kg. of silver.

The Hauraki Goldfield, New Zealand

LITERATURE

J. Park, F. Rutley, Ph. Holland. 'Notes on the Rhyolites of the Hauraki Goldfield,' Quart. Jour. LV., London, 1899.—A. M. Finlayson. 'Geol. of the Hauraki Goldfield,' Econ. Geol. IV., 1909, wherein many publications in New Zealand are cited.—W. Lindgren. Eng. Min. Jour. Vol. LXXIX., 1905, p. 218.—J. R. Don. Trans. Amer. Inst. Min. Eng. XXVII., 1898.—Schmeiser and Vogelsang. Die Goldfelder Australiens, 1897, pp. 92-98.

This goldfield which is situated in the Cape Colville peninsula and neighbourhood, North Island, New Zealand, is connected with a Tertiary eruptive area 125 km. long and 15 to 30 km. wide, consisting chiefly of andesite,

---

[1] *Loc. cit.* [2] *Ante*, pp. 525, 570.

dacite, and rhyolite.[1]   The lodes found there carry quartz with free gold principally, but occasionally also gold- and silver tellurides.   They contain but little pyrite, chalcopyrite, sphalerite, arsenopyrite, stibnite, pyrargyrite, proustite, etc.   Some are gold-silver lodes, as for instance those in the Waihi district, where 1 part of gold is produced to every 30 parts of silver.   The deposits occur chiefly in andesite and dacite.   The propylitization in connection with these lodes has been exhaustively studied by Finlayson.[2]

In the Thames district, from the discovery of gold in 1867 to the year 1897, gold and silver to the value of about 7·5 millions sterling were won, of which 6 millions were obtained from an area of but 3 sq. km.   These lodes were generally poor when their whole length and breadth were considered ; they contained however some especially rich shoots or bonanzas.   According to Finlayson these shoots were primary and not secondary in character. In depth they became impoverished.   Not far away is the Coromandel district and, somewhat to the south, the Karangahake and Waihi districts with gold-silver lodes.   The Waihi mine in this latter district from 1890 to 1907 yielded gold to the value of 6·25 million sterling.   At other places in New Zealand gold gravels occur, some of which are worked.   As will be found stated later,[3] during the last decade the gold production of New Zealand has risen.

WESTERN AUSTRALIA

LITERATURE

SCHMEISSER and VOGELSANG.   Die Goldfelder Australasiens, Dietrich Reimer, Berlin, 1897 ; English Translation by H. Louis, Macmillan & Co., 1898.—A. GIBB MAITLAND. Bibliography of the Geology of Western Australia, Perth, 1898.—HALSE.   'Observations on some Gold-Bearing Veins of the Coolgardie, Yilgarn, and Murchison Goldfields. W.A.,' Trans. Fed. Inst. Min. Eng. XIV., 1897–1898 ; ibid. XX., 1900–1901.—H. C. HOOVER. 'The Superficial Alteration of Western Australian Ore Deposits,' Trans. Amer. Inst. Min. Eng. XXVIII., 1898 ; reviewed Zeit. f. prakt. Geol., 1899, p. 87.—BANCROFT.   'Kalgoorlie, W.A., and its Surroundings,' Trans. Amer. Inst. Min. Eng. XXVIII., 1899.—T. BLATCH-FORD.   'The Geology of the Coolgardie Goldfield,' Geol. Survey of W.A. Bull. 3, 1899 ; reviewed Neues Jahrb., 1901, II.—A. GIBB MAITLAND.   'The Mineral Wealth of Western Australia,' Geol. Survey, Bull. No. 4, Perth, 1900.—P. KRUSCH.   Über einige Tellur-Gold-Silberverbindungen von den westaustralischen Goldgängen. Zentralblad f. Min. etc., 1901, No. 7.—T. A. RICKARD.   'The Telluride Ores of Cripple Creek and Kalgoorlie,' Trans. Amer. Inst. Min. Eng. XXX., 1901 ; 'The Veins of Boulder and Kalgoorlie,' ibid. XXXIII., 1903.—KUSS.   'L'Industrie minière de l'Australie occidentale,' Ann. d. Mines, 9, XIX., 1901.—General Ground-Plan of Boulder Group of Mines, Kalgoorlie, W.A., published by Cie. Belge des Mines d'Or Australiennes, Ltd., Liège.—Geological Map of Kalgoorlie by A. GIBB MAITLAND and W. D. CAMPBELL.   Geol. Survey of W.A., 1902.—E. S. SIMPSON. 'Notes from the Departmental Laboratory,' Geol. Survey of W.A., Bull. No. 6, 1902.— P. KRUSCH.   'Beitrag zur Kenntnis der nutzbaren Lagerstätten Westaustraliens,' Zeit. f. prakt. Geol., 1903.

---

[1] Geological Map in *Quart. Journ.*, cited in the Literature.
[2] *Ante*, p. 518.   [3] *Postea*, p. 598.

Western Australia forms approximately the western third of the Australian continent. Its principal goldfields lie to the east and south-east of the town of Kalgoorlie, near the eastern boundary.

According to Woodward, the rocks of this country have been folded into a number of north-south anticlines and synclines, and have been intruded by numerous eruptive rocks. The ordinary section shows granite, gneiss, and schist, which in their disposition form six zones or belts. Reckoning from west to east, only the fourth belt in which the Southern Cross mine occurs, and the sixth belt with the two principal districts of Kalgoorlie and Coolgardie, are auriferous. In the immediate neighbourhood of the two latter districts, Gibb Maitland and Campbell, beside such surface formations as laterite, record the presence of slate, quartzite, quartzite-schist, felsite, amphibolite, porphyrite, mica-schist — all of which probably represent dynamically metamorphosed eruptive rocks—as well as peridotite and its varieties.

The age of these rocks is not clear ; the tectonics likewise have been but little investigated, the geological mapping of one section of an unexplored whole being connected with almost insurmountable difficulty. At present, only the disposition of the rocks is known ; it is realized that in the gold district not only amphibolites occur, as was formerly supposed, but that slates, etc., are also present. According to Krusch, the rock designated amphibolite is in no sense a single rock but rather a number of different rocks. Two groups of amphibolites may be distinguished, namely, the schistose amphibolite described by Schmeisser and Vogelsang, on the one hand, and the granular hornblende rocks which occur within this thinly bedded variety and show no concern for its bedding, on the other. Gibb Maitland and Campbell recognized this difference and, going farther, divided the massive group into hornblende-, chlorite-, and actinolite-amphibolites. According to Krusch the schistose group is likewise not simple. Concerning the age of these metamorphosed eruptives, it is probable that not only older rocks but younger also suffered deformation.

The gold lodes of Kalgoorlie are intimately associated with the amphibolites. Although the age of these rocks at present cannot be definitely settled, nevertheless, according to Krusch, from the occurrence of the deposits and the nature of their filling, it may be fairly safely concluded that the Western Australian gold lodes belong to the young gold-silver group. Only exceptionally are they simple fissure-fillings ; more usually they are composite lodes,[1] that is, they are veined zones consisting of a large number of small fissure-fillings of fairly parallel strike, from which an intense impregnation and replacement of the country-rock proceeded. Quartz and metalliferous minerals were thus introduced, often in such

[1] *Ante*, p. 40.

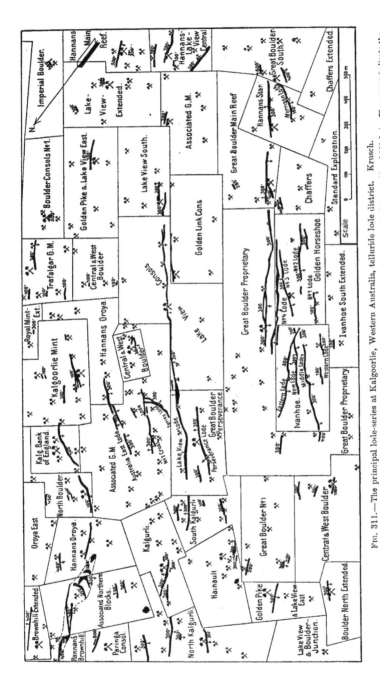

FIG. 311.—The principal lode-series at Kalgoorlie, Western Australia, telluride lode district. Krusch.

X = shafts ; the figures along the lodes indicate the depth of the level in feet, though those marked 100' include developments between 50 and 100 feet. The arrows indicate the dip usually between the level indicated and the next lowest.

amount that the country-rock became entirely replaced. In this manner compact bodies of quartz with disseminated mineral often arose such as might at first sight give the impression of being simple lodes, while in reality they were formed chiefly by replacement. In such cases it could be seen upon closer investigation, however, that there existed no sharp boundary with the country-rock but that on either wall both silicification and mineralization gradually diminished. Such lodes are illustrated in Figs. 53, 54, and 55. In width they sometimes attain several metres.

The lodes in general strike north-west, and in spite of repeated junctions are broadly speaking parallel. According to Krusch, they may in the central district be divided into three groups. Of these the first and western group includes the lodes in the Ivanhoe, Golden Horseshoe, Great Boulder Proprietary, Great Boulder Main Reef, Hannan's Star, and Great Boulder South mines. The second group lies to the north-east and includes the Great Boulder Perseverance, Lake View Consols, Golden Link Consolidated, Central and West Boulder Associated, South Kalgurli, Hainault, North Kalgurli, and the Kalgurli lodes. East of this again the lodes of the third group extend through the Kalgoorlie Mint, Kalgoorlie Bank of England, North Boulder, Hannan's Oroya, Associated Northern Blocks, Paringa Consolidated, Brownhill Extended, and Hannan's Brownhill.

A certain discontinuity expressed by branching or by an apparent or actual disappearance in depth, is characteristic of individual lodes. The veined zones likewise split up arbitrarily, so that the number of veins in a particular area may occasionally be doubled. Not infrequently on the other hand such a composite lode pinches out completely.

The many lodes lie close together within an area some 3 sq. km. in extent, the so-called ' Golden Mile.' In this small space over a hundred mines work and the individual properties are consequently small, relatively few having attained any considerable production. The best known are the Great Boulder Proprietary, Ivanhoe, Golden Horseshoe, Great Boulder Perseverance, Lake View Consols, etc.

The lode material consists chiefly of quartz, carbonates being but sparingly represented. The quartz contains auriferous pyrite with gold- and other tellurides. These sulphides and tellurides are intergrown in the most intimate manner, and though with the miner it is customary to speak of the mass as sulphide ore, this intimate mixture is meant. In the primary zone the tellurides are particularly characteristic, the light conchoidal calaverite, the dark conchoidal petzite, and the quicksilver telluride, coloradoite, being the most frequent, while the other tellurides, krennerite, hessite, altaite, etc., are more uncommon.

Analyses of tellurides or telluride ores have shown that tellurium in part may be replaced or represented by selenium, this latter element indeed being found in light telluride ores to the extent of several per cent. The intergrowth of telluride gold with free gold is particularly interesting and often seen, such gold without doubt being primary in character. Sulphides such as galena, sphalerite, enargite, etc., are less noticeable. The last-named mineral has this claim to attention, that for some time it was taken for gold telluride. These minerals are accompanied by tourmaline, the presence of which was first remarked by Mariansky, the discoverer of the hitherto unrecognized and consequently unappreciated tellurides.

The gold is not uniformly distributed along the strike but concentrated in shoots which either descend fairly vertically into depth, as was the case at the Ivanhoe, illustrated in Fig. 79, or pitch at an inclined angle, as at the Associated Northern Blocks, illustrated in Fig. 80.

In the case of every goldfield the behaviour of the gold content in relation to depth is most important. Western Australian mining tells the story of decrease of value in depth, though such may be slow. Whether this decrease is due to the disappearance of gold telluride, or whether both the telluride and the auriferous pyrite decrease, are questions which have not been closely investigated. All the lodes near the surface are more or less decomposed by the action of meteoric waters. While those auriferous deposits where the gold is chiefly or exclusively associated with sulphides often show two very characteristic depth-zones—an oxidation zone with its gold in greater part leached, and beneath this an abnormally rich cementation zone [1]—the Western Australian telluride lodes display oxidation and primary zones only, no cementation zone exists. As the oxidation zone carries free gold exclusively, the recovery of the gold from the ore in that zone is simple; the extraction of the gold from the telluride ore, on the other hand, is more difficult. The boundary between these two zones is therefore not only of interest in the study of ore-deposits, but also of importance to the miner and metallurgist; it is consequently most carefully entered upon the mine plans, such an entry being illustrated in Fig. 82. Its course generally proves to be very irregular; while in one mine it may be found but 20 m. below the surface, in the next it may be found at many times that depth.

The free gold resulting from the decomposition of the primary tellurides in the oxidation zone occurs in four forms so characteristic that from hand specimens it may be said whether a particular piece of ore showing visible gold came from the oxidation zone of a telluride deposit, or not. These forms are as follows :

    1. Fairly lustreless, mustard-coloured, earthy, loose cavity-fillings, in the form of blotches and coatings, such gold being known as ' mustard ' gold ; illustrated in Fig. 88.

2. Filmy coatings of very fine crystals in cracks and crevices, such gold being known as ' flake ' gold.

3. Crystals in cavities, matted to larger aggregates, some of which have weighed several kilogrammes. On account of its spongy appearance such gold, which according to Simpson contains but 0·09 per cent of silver, is known as ' sponge ' gold.

4. Spots, stars, and small irregular splashes, occurring as thin coatings in fissures or cracks ; illustrated in Fig. 89.

While with other gold deposits a cementation zone occurring directly above the primary zone may be so rich as to necessitate the greatest care in appraising such deposits, the oxidation zone which occurs immediately above the primary zone of these auriferous telluride deposits proves to be poorer than this latter. The following figures pertaining to one of the principal mines afford a comparison between the ores of the two zones in respect to their value. It is seen that the ore of the secondary or oxidation zone is in this case but half as rich as that of the primary zone :

AVERAGE OF THE ENTIRE LODE MASS

	Oxide Ore.	Sulphide Ore.
West Lode about	6 dwt.	about 12 dwt.
No. 2 ,,	,, 9 ,,	,, 13 ,,
No. 3 ,,	,, 14 ,,	,, 20 ,,
No. 4 ,,	,, 19 ,,	,, 44 ,,

From these considerations it follows that in the formation of the oxidation zone some removal of the gold must have taken place. Perhaps in this removal, abnormal as it is with gold deposits, the presence of tellurium and selenium may be of significance ; every metallurgist knows for instance, that gold is soluble in a solution of selenic acid.

Relation between Gold and Silver.—Gold and silver, as the analyses of the tellurides indicate, replace and .represent one another in all possible proportions. While calaverite may contain from traces up to 4·8 per cent of silver, and krennerite from 3 to 4 per cent, sylvanite contains 9–10 per cent, and petzite 40–43 per cent. All these contain their proper proportion of gold. More argentiferous still is hessite, the pure telluride of silver, which may be regarded as the extreme member of this sequence.

Concerning frequency of occurrence, calaverite poor in silver is the most common, this being followed by petzite which· is rich in that metal. Unfortunately, no results of investigation are available to show how much of the gold in these deposits is associated with the tellurides and how much with the sulphides. No exact analytical data therefore exist from which the relation of gold to silver may be calculated. Some idea may however be obtained from other figures. The value of the bullion from each separate mine, that is the gold-silver alloy recovered by treatment and afterwards parted in the refineries, remains fairly constant. The three most important mines, the Great Boulder Proprietary, the Ivanhoe, and the Golden Horseshoe, produce gold from 79 to 87 per cent fine. Weighting the fineness at each mine by the production, an average of 83 per cent of gold and 17 per cent of silver is obtained, that is to say, the bullion of the most important series of lodes in the Kalgoorlie district consists as to five-sixths gold and as to one-sixth silver.

Still more interesting are the figures when the recovery by amalgamation in batteries and pans is kept separate from that by the cyanide process. The battery gold is then seen to contain 91·1–94·3 per cent of gold and 5·7–8·9 per cent of silver, and the cyanide gold 66·6–78·1 per cent of gold and 21·9–33·3 per cent of silver. The battery gold, which is chiefly free gold, is therefore poor in silver when compared with the argentiferous gold obtained by cyanidation, which in greater part is derived from auriferous pyrite and tellurides. The primary free gold of the Kalgoorlie district consists accordingly of twelve parts of gold to one of silver, while the mineralized gold contains 3 parts of gold to one of silver. The explanation of this striking difference must be sought in the genesis of the deposit. In the endeavour to judge of the phenomena which could so result the following possibilities deserve consideration :

1. A change in the composition of the solutions may have taken place whereby the relation of gold to silver may have altered during deposition.
2. The mineral solutions may have remained unaltered but different precipitants may have become active, one after another.
3. The mineral solutions may have remained unchanged but two differing precipitants may have been operative at one and the same time, both of which were effective for gold, and one, in addition, strongly effective for silver.
4. The mineral solutions may have remained in general unaltered, and but one precipitant| may have been active, which however possessed the property of precipitating native gold comparatively pure, and at the same time mineralized gold alloyed with silver.

The last case appears the most simple ; it explains quite well the structure of the ore and the simultaneous formation of non-argentiferous and argentiferous gold.

It is interesting to consider the limit of payability on this field, and its fluctuations. In the year 1903 this was 15 grm., of gold per ton ; to-day the reduction processes have so improved that in spite of extremely high wages the cost is covered by 8–10 grm., any content above this figure being profit. The richest mine of this goldfield, the Great Boulder Proprietary, in its yearly report for 1910 gave the following average figures of cost for different years :

In 1907 24s. 9d. equivalent to 9·1 grm. per ton.
,, 1908 25s. 8d.   ,,   9·5   ,,   ,,
,, 1909 26s. 0d.   ,,   9·6   ,,   ,,
,, 1910 26s. 2d.   ,,   9·7   ,,   ,,

The figures in the second column obviously express the limit of payability of the ore in this mine as a whole. For comparison, the value of the ore-reserves at different levels in the Main Shaft and on the Main Lode are as follows :

Feet.				Dwt. per Ton.	
400–500	.	.	.	30·06	
500–600			.	8·60	
600–700			.	18·51	
700–800			.	28·52	Average value
800–900			.	32·07	to 1300 ft. =
900–1000			.	20·95	22·74 dwt.
1000–1100	,		.	26·52	
1100–1200			.	19·12	
1200–1300			.	20·33	
1300–1400			.	14·17	
1400–1500			.	13·52	
1500–1600			.	13·45	Average value
1600–1750	.		.	10·21	1300 to 2350 ft.
1750–1900			.	11·18	= 13·8 dwt.
1900–2050			.	11·12	
2050–2200			.	17·30	
2200–2350			.	19·48	

These figures taken in their entirety indicate that in the case of the richest mine at Kalgoorlie there is also a decrease in gold content in depth.

The Kalgoorlie goldfield was discovered in the beginning of the 'nineties. In 1896 the telluride ores were recognized. The economic conditions under which these mines worked at the time of their discovery were very unfavourable. The plateau on which they occur is without water, for which, according |to Gmehling, twopence-halfpenny was paid per gallon. To-day Kalgoorlie is connected with Perth by railway, and the goldfield may now be reached from the coast after fifteen to twenty hours in a comfortable express train. The water question has also been solved in a generous manner by the government, which in 1903 laid a pipe-line capable of delivering 5 million gallons per day all the way from Perth, a distance of 325 miles, selling the water at the rate of six shillings per 1000 gallons, whereas previously the price for the same quantity had been about forty-six shillings. As with the water, all fuel and mine stores must be transported all the way from the coast.

The present importance of this field, compared with which the other goldfields of Western Australia are of little account, may be gathered from the following statistics :

GOLD PRODUCTION IN OUNCES OF FINE GOLD

Year.	Western Australia.	Australasia.	World's Production.
1902	1,819,308	3,989,083	...
1903	2,064,801	4,315,759	...
1904	1,983,230	4,220,690	...
1905	1,955,316	4,156,194	...
1906	1,794,547	3,984,538	...
1907	1,697,554	3,659,693	20,121,423
1908	1,647,911	3,546,912	21,448,554
1909	1,595,263	3,447,227	22,230,116

According to these figures the production of Western Australia approximates one-fourteenth of the world's production and not quite one-half of the total production of Australasia, towards which in 1909 Victoria contributed 654,222 oz., Queensland 455,577 oz., New South Wales 204,709 oz., Tasmania 44,777 oz., South Australia 7500 oz., and New Zealand 485,179 oz. Western Australia therefore in respect to its gold production easily takes first place among the countries of Australasia, a prominence which makes it all the more regrettable that since 1903 there has been a continuous decline, such as must be connected with decrease of value in depth, particularly below about 700 metres.

ALTENBERG NEAR SEITENDORF

LITERATURE

G. GÜRICH. 'Beiträge zur Kenntnis der niederschlesischen Tonschieferformation,' Zeit. d. d. geol. Ges. Vol. XXXIV.—J. ROTH. Explanatory Text with the Geological Map of Lower Silesia. Berlin, 1867.—B. KOSMANN. 'Der Metallbergbau im Schmiedeberger und Katzbach-Gebirge,' Breslauer Gewerbeblatt, Vol. XXXIII.—v. FESTENBERG-PACKISCH. Der metallische Bergbau Niederschlesiens, 1881.—v. ROSENBERG-LIPINSKY. 'Beiträge zur Kenntnis des Altenberger Erzbergbaus,' Jahrb. d. k. preuss. geo. Landesanstalt für 1894, p. 161.—A. SACHS. Die Bodenschätze Schlesiens, Erze, Kohlen, nutzbare Gesteine. Leipzig, 1906. Veit & Co.

Altenberg in Silesia, so long as only the Bergmannstrost lode with its arsenic-lead-silver content was worked, was known as a lead-silver mine ; the subsequent discovery of copper-bearing lodes caused it next to be placed among the copper deposits ; while quite recently attention was called to the high gold-silver content of newly discovered lodes, and now, according. to Beyschlag and Krusch, Altenberg is more properly classed with the gold deposits.

The district consists of highly metamorphosed, dark grey, bluish-black, and dark green slates, the age of which, though not yet definitely established,

may with some assurance be considered as Silurian. These sediments
alternate with sheets of prevailingly brick-red or speckled porphyry, diabase,
and schalstein, and are crossed by younger porphyry dykes radiating
from neighbouring porphyry peaks. In relation to the ore-deposits
two eruptive rocks having little observed extension underground, are of
especial importance. Of these the first has long been known as olivine-
kersantite,[1] while the second, according to microscopic examination, may
be regarded as a propylite the parent rock of which has not yet been
determined. The olivine-kersantite maintains the closest connection with
the Bergmannstrost lode, in that it often occurs within the actual fissure
of that lode, thicknesses of it alternating with ore. From the exposures
underground this eruptive is undoubtedly older than the lode. Apparently
a re-opening of the fissure along the eruptive took place, and a veined
zone following the kersantite was formed, which subsequently became
filled by metalliferous material deposited from ascending solutions.

The lodes of Altenberg are simple lodes. That best known is the
above-mentioned Bergmannstrost lode which has a characteristic filling
of arsenopyrite with galena, sphalerite, tetrahedrite, etc., together with
considerable silver. Further work prosecuted underground to prove the
many lines of ancient workings to the north, has revealed the presence
of eight other lodes. Most of these, like the Bergmannstrost, strike
a little south of east and north of west, though the Wandas-Hoffnung
and the Hermanns-Glück lodes extend in a north-east direction. Almost all,
including the Bergmannstrost, dip 60°–75° to the north; occasionally they
are flatter or steeper, but seldom do they dip the other way. Along the
drives the change from one country-rock to another is abrupt. The ore in
the slates usually carries more precious metal than that in the porphyry.
The lode material has an irregular-coarse structure, the barren gangue
occurring in smaller quantity than the ore. The entire width sometimes
consists of solid chalcopyrite or arsenopyrite, these two minerals apparently
replacing and representing one another.

The character of the lode-filling in the northern lodes differs from
that of the Bergmannstrost in so far that with them copper plays an
essential part. In addition, the high gold content of the Mariä-Förderung
and the Wandas-Hoffnung is particularly noteworthy.

Poorer and richer parts alternate in both strike and dip. Towards
the east the lodes appear to become impoverished ; the lode-filling then
consists chiefly of quartz and subordinately of siderite, with ore sparingly
distributed. Towards the west the lodes are affected by disturbances
which have locally robbed the lode of its ore.

Seeing that the district is not yet widely recognized as auriferous, some

[1] After Krusch.

figures indicative of the precious-metal content will be of interest. The
following were those obtained by Beyschlag and Vogt in their examination

			Grammes per Ton.	
Lüschwitzgrund Lode	.	.	.	Trace gold	46·0 silver.
Mariä-Förderung Lode	.	.	.	16·5 „	170·6 „
Olgas-Wunsch Lode	.	.	.	3·0 „	72·0 „
Wandas-Hoffnung Lode	.	.	.	26·6 „	221·6 „
Bergmannstrost Lode	.	.	.	Trace „	146·0 „

If the gold content be compared with that of the silver, it is seen that
in the relation between these two metals the occurrence at Altenberg
possesses great similarity to the lodes of the young gold group. It is
true that on surface no young eruptive rock to which the gold and silver
content might be referred is known, and that the Bergmannstrost, in
addition, is in all probability a lead-silver lode ; there is however a very
good suspicion that the gold present in the northern lodes is considerably
younger than the filling of the Bergmannstrost. Although the question
has not definitely been settled, present information indicates that Altenberg
should be classed under the young gold-silver group.

THE OLD GOLD LODES

WHILE the young gold-silver lodes without exception maintain the closest association with young Tertiary eruptives, the old lodes now to be described exhibit in general no connection with such ore-bringing rocks, and in those exceptional cases where such may be observed, the particular eruptives are old. Apparently therefore the connection between these lodes and eruptive magmas is not so close as is the case with the young gold-silver lodes.

Concerning the form of the deposit a difference may likewise often be remarked. While, for instance, in the case of the young gold-silver lodes composite lodes, in addition to simple lodes, often play an important part, those of the old group, so far as yet known, are exclusively simple lodes, the filling being usually sharply separate from the country-rock. In addition, with the older lodes the length along the strike is more considerable ; the one which has been followed for the greatest length is the Mother Lode of California, this indeed being among the most important fissure-fillings known. Naturally with such an extension as this lode has, it is not a question of one and the same simple fissure throughout, but a series of fissures so arranged that when one pinches out a new one sets in, a little to the side, to maintain the continuity. In dip also these lodes often have a considerable extension, some of the Californian mines, for instance, are at present working at depths from 700 to 1000 metres. The width reaches some few metres at the most, this being materially less than the maximum width of the young gold-silver lodes. The nature of the separation of the lode from the country-rock is always an important factor, and in this connection also there is a difference ; with the young gold-silver lodes an impregnation of the country-rock often plays an essential part, while with the old gold lodes this is seldom the case ; the occurrence at Roudny in Bohemia is however an exception.

The filling of the old gold lodes is simple. Usually quartz is by far the most abundant gangue-mineral present, so that the lodes strictly speaking may be regarded as a special form of quartz lode. The most frequent ore is auriferous pyrite, in which, when in the primary zone, the gold content is not usually to be seen, free gold being present only to a subordinate extent. It is consequently advisable to assay for gold all quartz lodes which carry pyrite. Only in exceptional cases does the amount of pyrite so increase that the lodes become pyrite lodes, with quartz subordinate. The other sulphides, such as galena, sphalerite, chalcopyrite, and the sulpho-compounds, play usually no important part.

To the same extent that the gold content is difficult to recognize in the primary zone, so does it become prominent as the lode decomposes under the action of meteoric waters. It was indeed from experience with these lodes that the importance of the migration of the gold content as a factor in reckoning possible ore-reserves was first appreciated. As previously stated, in the decomposition of pyrite by meteoric waters, ferric sulphate is formed which dissolves gold; in the presence of pyrite therefore gold continues to be dissolved so long as oxygen is present. After the consumption of the oxygen the undecomposed sulphides of greater depth act reducingly upon the descending solution, precipitating the gold as free gold, the affinity of this for oxygen being very small. The gold of the cementation zone may always be recognized by the fact that it either coats primary sulphides or fills cracks within them, or, as is more frequently the case, it occupies fissures and cavities within the quartz, such quartz being generally light brown in colour. Wetting the quartz is of assistance in discerning such gold. In the oxidation zone the precious metal occurs very sparingly, and when occurring is generally found in close association with a gelatinous limonite resulting from pyrite.

The recognition and determination of the secondary and primary zones [1] of these lodes is of the greatest importance to all concerned, to the economic geologist as well as to the miner, on account of the large differences in the metal content of the different zones.

Since the old gold lodes consist chiefly of quartz, by weathering they often become freed from the less resistant country-rock upon their flanks, when they protrude at the surface as walls, ledges, or reefs; such an outcrop is illustrated in Fig. 312. In this they further differ from most of the young gold-silver lodes. As a rule the greater portion of the oxidation zone has long surrendered to erosion, and the enriched cementation zone often appears at the surface. The relative ease with which such lodes are discovered, free gold being readily detected by simple crushing and washing, has often given untrained prospectors a reputation for a knowledge of ore-deposits which they were far from deserving.

Chemically, the gold of these lodes is usually substantially purer

[1] *Ante*, pp. 211, 212.

than that of the young gold-silver lodes. As previously stated, these latter carry gold and silver in variable proportion so that all gradations exist between gold lodes with little silver and silver lodes with little gold. In the case of the group now being described however, the lodes are gold lodes pure and simple. The consequence is that the bullion recovered from the young gold-silver lodes usually has a different composition from that of the old lodes ; while with the former there may be 60 per cent of gold and 40 per cent of silver, with the latter there is at least 90 per cent of gold and at most 10 per cent silver.

Concerning primary depth-zones, experience with the old gold lodes appears to have been more favourable than with those of the young gold-silver group. Mining operations have nevertheless shown that a

FIG. 312.—Outcrop of a gold lode with granite blocks in the background, Iramba plateau, German East Africa. Scheffler.

large percentage of the old lodes also have only proved profitable in the cementation zone, the primary ore having been unpayable. Where however the primary ore is rich enough to work, the gold content may continue to depths of 1000 m. or more. With these lodes therefore the ore is more persistent in depth than is the case with the young gold-silver lodes.

Finally, mineralogically it is interesting to note that tourmaline is more often found with the old lodes than with the young, while gold telluride, which occurs plentifully in some young gold-silver districts, is a rare occurrence in the old lodes. As already mentioned, auriferous pyrite is the most common ore-mineral of these lodes, the other sulphides occurring subordinately. There are however gold lodes in which the gold is accompanied particularly by arsenopyrite, stibnite, and sometimes also by bismuthinite. Mineralogically therefore several classes may be differentiated; of these however, the first-named, that associated with pyrite, is by far the most important.

The gold-quartz lodes with their rich cementation zone naturally tend to the formation of auriferous gravels. It has therefore almost invariably been the case that before the lodes were more closely investigated attention was first devoted to such gravels. Since also such gravel-mining was very cheap and the yield often very considerable, lode-mining was only undertaken when the decline in the productiveness of the gravels compelled such new endeavour. Consequently in new districts of this character, production generally rises very rapidly at first and large profits are returned; and then, with the approaching exhaustion of the gravels a decline gradually becomes established which as a rule it is not possible to stay, even though with all energy it is sought to make good the deficit from the gravels by development upon the lodes.

THE GOLD DEPOSITS OF CALIFORNIA

LITERATURE

J. D. WHITNEY. Geol. Survey of California, I., 1865.—v. RICHTHOFEN. 'Reisebericht aus Californien,' Zeit. d. d. geol. Ges., 1864, XVI., p. 331.—E. REYER. 'Über die Gold-gewinnung in Californien,' Zeit. f. Berg- Hutten- u. Salinenwesen, 1886, XXXIV.—H. W. FAIRBANKS. 'Geology of the Mother Lode Region,' 10th Ann. Rep. State Mineralogist of California.—H. W. TURNER. 'Notes on the Gold Ores of California,' Amer. Jour. Sc. LVII., 1894, p. 467; reviewed Zeit. f. prakt. Geol., 1896, p. 275.—W. LINDGREN. 'The Gold-Silver Veins of Ophir, California,' 14th Ann. Rep. U.S. Geol. Survey, 1894, p. 249; 'Characteristic Features of California Gold-Quartz Veins,' Bull. Geol. Soc. Amer., 1895, Vol. VI. p. 221; reviewed Zeit. f. prakt. Geol., 1895, p. 423; ' The Gold-Quartz Veins of Nevada City and Grass Valley, California,' 17th Ann. Rep. U.S. Geol. Survey, Pt. II., 1896; 'The Geological Features of the Gold Production of North America,' Trans. Am. Inst. Min. Eng., 1903, XXXIII. p. 790; ' Ore Deposition and Deep Mining,' Econ. Geol., 1905, Vol. I. No. 1, —B. KNOCHENHAUER. ' Der Goldbergbau Kaliforniens und sein Ertrag in Vergangenheit, Gegenwart, und Zukunft,' Berg- u. H.-Ztg. LVI., 1897, California-Paris Exposition Commission, 1900.—PRICHARD. ' Observations on Mother Lode Gold Deposits,' Trans. Amer. Inst. Min. Eng. XXXIV., 1904, p. 454.—Lists of literature in Kemp's Ore-Deposits of the U.S.; by H. Ries in Econ. Geol.; and by Lindgren.

The following exposition is based chiefly upon the above-cited excellent work by Lindgren, the greatest authority upon the Californian gold deposits, and especially upon the resumé beginning on page 49 of that work. Three groups of deposits occur : firstly, those on the Mother Lode between Mount Ophir and Placerville; secondly, those to the north near Grass Valley and around Nevada City; and thirdly, those to the south-east of Placerville near Grizzly Flat. The relation of these places to one another is seen from Fig. 313.

Historical.—Gold was first found in the Sierra Nevada in January 1848 at Coloma, Eldorado Co.; in 1850 several thousand men were working in the Nevada district; and in 1856 Nevada City was already a town of close upon a thousand houses. In 1880 the number of inhabitants of that city was 21,000, a number which afterwards diminished when, in consequence of the introduction of hydraulic methods, less hand labour was required. An equally rapid development followed in the other districts of this gold belt. At first, production increased by bounds, only however to decrease as the gravels became exhausted and the miner was forced to work the primary deposits. In this regard Californian experience may be taken to represent the normal sequence of events which may be expected to develop in every newly discovered gold-quartz field.

Geological Circumstance.—According to Lindgren, the Sierra Nevada consists in greater part of members of the Calaveras formation, this formation embracing the Palæozoic beds of this mountain chain. Of these beds those of Carboniferous age predominate, though the right of such to be considered Carboniferous can be verified in but few places. Concerning the disturbances to which this region was subjected, the first plication took place at the end of the Palæozoic or at the beginning of the Mesozoic period; this was accompanied by eruptive outbreaks which began in part during the Carboniferous. In Jurassic-Triassic time the greater portion of the Sierra Nevada was dry land; in such relatively small areas as were occupied by lakes, the Mariposa beds consisting of dark carbonaceous slates and of volcanic tuffs, were laid down. Then followed another period of intense orogenics, to which enormous masses of eruptive material of varying character and structure owe their existence. During this period both the recent and the older rocks were intensely plicated, while diabase and porphyrite with their tuffs appeared as flows and dykes. Finally, as the mightiest phase of the volcanicity, came the eruption of the grano-diorite magma. Under such intense tectonic conditions numerous fissures became formed which in part were filled with gold ore. The formation of the gold lodes in the Sierra Nevada may therefore to a certain extent be regarded as the last phase of the Mesozoic crustal revolution in that region. Since that time these mountains have been dry land. In the latter portion of the Neogene or latest Tertiary period, volcanic eruption and plication began anew, rhyolite and andesite being extruded to form those gigantic lava masses which give to the Sierra Nevada their present configuration.

Since according to Lindgren the gold lodes are of Cretaceous or in part of pre-Cretaceous age, the relation they maintain to the older north-north-west mountain folds is striking. These lodes form many systems, and often appear on the surface in long lines of quartz outcrops.

Previous to the exhaustive investigation of Lindgren in 1895, the numerous works upon Californian lodes and mining districts had not led to any general conclusions, as the individual authors had only investigated

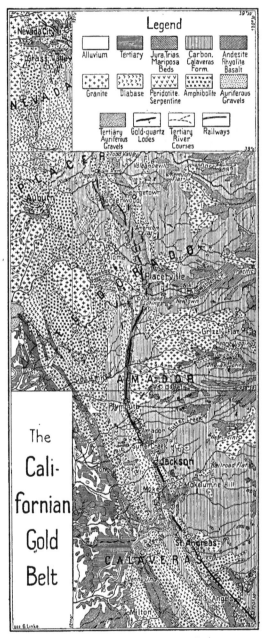

FIG. 313.—The Californian gold belt.

separate districts and a general review had not fallen to the lot of any single authority. The dependence of the lodes upon the belt of metamorphic slates was already recognized by Whitney. While in the granite areas of southern California but few lodes occur—and these only close to the slate boundary—in the slate region of middle and northern California many are found, all the different rocks within the metamorphic belt participating fairly equally, ore-deposits being found in granite, diorite, gabbro, serpentine, quartz-porphyrite, augite-porphyrite, hornblende-porphyrite, diabase, amphibolite, as well as in more or less altered clay-slate, sandstone, and limestone.

Since gold-quartz conglomerates are found in the foot-wall of the metamorphic series, some of the lodes most certainly were of very great age; as pointed out by Whitney and von Richthofen however, the majority and the richest are late Jurassic or early Cretaceous, and to be regarded as the thermal after-effects of the eruptive phenomena of that time.

In detail, both the strike and dip of the lodes are very variable, owing to numerous more recent mountain movements, and to the fact that the rocks in which the fissures were originally formed offered varying degrees of resistance to fracture. In the massive rocks, for instance, the fissures are more or less sharply defined and in good line; in the slate they have followed different planes and become indefinite; while in rocks of medium solidity they are developed as networks of fissures. The dip varies between 20° and 70°. The width is extremely variable, reaching in isolated cases 5 m. though generally much less. The length is also very variable, though in most cases it is short; seldom may a lode be followed for more than one or two kilometres, the gigantic Mother Lode being however a unique exception.

The lode-filling consists chiefly of milk-white quartz of irregular structure generally and but seldom crusted. Other gangue-minerals occur locally and only in relatively small amount; thus calcite and dolomite, chiefly on the walls; some whitish and greenish micas, albite, titanite, ilmenite, and anatase. The gold likewise is irregularly distributed in the mass and is usually of microscopic fineness; occasionally it is visible as flakes, threads, or small irregular blotches; more rarely it forms larger masses, some having been found up to 25 kg. in weight. Generally it contains but little silver, this metal only very exceptionally reaching 30 per cent. The precious metal is accompanied by auriferous pyrite, pyrrhotite, chalcopyrite, sphalerite, and galena; less frequently by arsenides, especially arsenopyrite; more seldom still by antimonides or tellurides; while marcasite is hardly ever present.

The lodes in grano-diorite are almost invariably richer in sulphides than those in other rocks; pyrrhotite for instance appears to be limited to lodes in grano-diorite. In dark clay-slate the sulphides are represented almost entirely by pyrite with occasionally some arsenopyrite. Lodes in gabbro often contain copper. Exceptions to these generalizations are however so numerous that the influence of the country-rock upon the ore present, cannot be regarded as well defined.

With the larger lodes the precious metal is more often found concentrated on the walls, while the middle part of the lode is unpayable; the gold content may however be uniformly distributed throughout the whole section of the lode. Often irregular bodies of payable ore occur within extensive unpayable areas. Sometimes elongated lenticular ore-shoots exist the pitch of which, independent of the dip of the lode, is generally steep and seldom below 45°. The width of such shoots varies between

1 and 100 m., while the length may be 600 m. or more. If in depth a shoot gives out, another is often found at greater depth along the same line. Many of the lodes contain the gold in small nests or bunches. With an increase in the gold content an increase in the quantity of sulphides present is always observed.

The alteration of the country-rock along the lodes is particularly characteristic. In intensity this generally varies with the width of the lode; when the width is great the alteration may extend as much as 10 metres. Only very acid massive rocks and certain carbonaceous slates appear to have been but little affected. Serpentine appears to have suffered most, with increase of lime and decrease of magnesia. In the immediate neighbourhood of the lodes the country-rock not infrequently exhibits parallel jointing, pressure schistosity, or brecciated structure.

Individual Occurrences.—(1) At Nevada City and Grass Valley are found a large number of lodes situated roughly 20 miles north of the last outlying representatives of the Mother Lode, from which lode they also differ in their general character. Their width is small but the gold content is comparatively high. Free gold occurs both near the surface and in depth, while in addition there is a variable gold-silver content in the sulphides present. The strike is extremely variable, though speaking generally it may be said that the lodes are arranged in two main systems, a north-south and an east-west. The dip is low and irregular, nor can any general rule concerning it be formulated. In relation to production the Grass Valley district comes first, then follows that of Nevada City, while the Banner Hill district ranks third. A considerable portion of the output comes from the Eureka, Idaho, Rocky Bar, North Star, Empire, and Providence mines.

From the ordinary type of gold-quartz lode some lodes of less importance in Grass Valley deviate, in that chalcopyrite and bornite are the valuable minerals, and calcite, quartz, and felspar, the principal gangue-minerals.

(2) The Mother Lode of California, forming as it does with its 150 km. of length the most important payable lode in the world, is widely known. It consists of a large number of linked fissures which together form a gigantic veined zone at the foot of the Sierra Nevada, this position being indicated in Fig. 313. This zone generally dips 50°–70° to the east, as do the slates in which it occurs. The width of the separate lodes may reach as much as 10 m., but usually it is less than 1 m. The frequent connection of the fissures with eruptive rocks is remarkable. The quartz, which is usually milk-white, tends to break into parallel flakes, such quartz being known as ribbed quartz. The gold is either free and finely distributed, or associated with pyrite. As with all the gold lodes in California, the

concentration of the precious metal in ore-shoots is characteristic. To the north and south, at both ends, the Mother Lode splits up.

(3) The deposits at Grizzly Flat to the south-east of Placerville are less important. They occur at the contact of the Calaveras slates with granite, and likewise carry gold in a quartz matrix.

With all the Californian gold lodes the relation of the gold content to depth is interesting. Lindgren in his work, *Ore Deposition and Deep Mining*, reports the experiences of some of the mines working on the Mother Lode at depths which in 1905 varied between 1766 and 2863 feet. The Kennedy at 2700 feet, Central Eureka at 1900–2200 feet, Oneida at 1900 feet, and Gwin at 2000 feet, all worked at a profit the ore at these respective depths, this ore containing free gold and being of similar character to that near the surface. A difference was only in so far observed that pocket-like enrichments were less frequent in the deep workings than they had been above. According to Lindgren, in California the hope to find payable ore in still greater depths is justified, though he was of opinion that in any case at a depth of 5000 feet temperature would put a stop to mining. Assuming that a thickness of 3000 feet had been removed by erosion, he came to the conclusion that in California free gold was originally deposited down to a depth of 6000 feet below the then surface. In 1909 some mines in Amador Co. were working upon the Mother Lode at a depth of 3400 feet, at which depth the character of the ore showed no alteration.

Knochenhauer gives 15 to 20 grm. per ton as the average content of the payable ore of California, though the content of the cementation ore has not infrequently reached 160 grammes.

Concerning genesis, Lindgren concluded that the aqueous solutions from which the gold-quartz lodes were formed, in addition to silica contained large amounts of carbonic acid, calcium carbonate, and sulphur, the last as sulphuretted hydrogen or sulpho-salts. Such waters however are in nature only found ascending, and generally they are hot springs. The ultimate source of the gold must therefore be sought deeper, and perhaps in granitic rocks or magmas.

The earlier gold production of California may be gathered from the table on p. 554, in considering which it must be remembered that at the beginning gravel-mining played the greater part. The present annual production is 18–21 million dollars, that of the United States as a whole being about 100 millions. In the year 1908 gold to the value of 19,329,700 dollars was won, in addition to not quite 2,000,000 dollars of silver. In 1909 the gold production was estimated at 21 million dollars, with practically the same silver production as before. In the gold production the proceeds from both primary and gravel-deposits are included. Of the gold produced from quartz lodes the Grass Valley district in Nevada Co. yields the largest quantity, its output exceeding by far that of individual counties upon the Mother Lode. All the gold-quartz mines together produce yearly about two and a half million tons of ore, of which two millions are milling ore with a gold content of 5–5·75 dwt. per ton, while the remainder is copper ore containing gold and silver. The quartz mines produce annually some million more dollars than the gravel-deposits. Those of the Mother Lode in the Amador, Calaveras, El Dorado, Mariposa, and Tuolumne counties, yield three-quarters of the total milling ore, the content of which at 4 dwt. per ton is nevertheless less than that of the remaining counties, where the lodes though smaller are richer.

THE TREADWELL DEPOSIT, ALASKA

LITERATURE

G. F. BECKER. 'Reconnaissance of the Gold Fields of Southern Alaska,' 18th Ann. Rep. U.S. Geol. Survey, 1898.—A. C. SPENCER. 'The Geology of the Treadwell Ore-Deposit,' Trans. Amer. Inst. Min. Eng., Oct. 1904; 'The Juneau Gold Belt, Alaska,' U.S. Geol. Survey, Bull. 225, 1903, p. 28.

This deposit, occurring on Douglas Island opposite Juneau City, Alaska, is remarkable in that primary gold ore of a remarkably low content is worked at a profit. An albite-hornblende rock, sometimes described as a soda-syenite and sometimes as albite-diorite, is traversed by a number of quartz stringers carrying both free and mineralized gold. From these the country-rock involved is more or less impregnated. In addition to pyrite; chalcopyrite, arsenopyrite, pyrrhotite, sphalerite, and galena also occur. Quartz is uncommon. Since the main ore-body is an intrusive zone up to 150 m. in width, vast quantities may be mined at an extremely low cost, and although the average content of the ore varies in general between 3 and 5 grm. per ton, a net profit is obtained. It must be expressly understood however, that such a low content does not in general suffice for profitable mining, and accordingly the abnormally low costs obtaining in the Treadwell mines must not be taken as being applicable elsewhere.

THE GOLD-QUARTZ LODES OF AUSTRALASIA

LITERATURE

E. LIDGEY. 'Report on the Ballarat East Goldfield,' Spec. Rep. Dep. of Mines Victoria, 1864.—W. B. WITHERS. History of Ballarat, 1870.—MURRAY. Report on the Geology and Mineral Resources of Ballarat, 1874.—WOLFF. 'Das australische Gold, sein Lagerstätten und seine Associationen,' Zeit. d. d. geol. Ges., 1877, XXIX.—R. L. JACK Report on the Geology and Mineral Resources of the District between Charters Tower

Goldfields and the Coast, 1879.—T. A. RICKARD. 'The Bendigo Goldfield,' I.-III., Trans. Amer. Inst. Min. Eng., Oct. 1891, Oct. 1892, Aug. 1893.—E. F. PITTMANN. 'On the Geological Structure of the Wyalong Goldfields,' Record. Geol. Survey. N.S.W., 1894.—E. J. DUNN. 'Report on the Bendigo Goldfields,' Dep. of Mines. Melbourne, 1893.—BABU. 'Les Mines d'or de l'Australie,' Ann. d. Mines, 1896, IX. 315-395.—R. ALLEN. 'Report in Connection with the Underground Plans of the Ballarat West Mine,' Dep. of Mines, Vict., 1897.—SCHMEISSER and VOGELSANG. Die Goldfelder Australasiens. Berlin, 1897.— W. LINDGREN. 'Occurrence of Albite in the Bendigo Veins,' Econ. Geol. Vol. I., No. 2, 1905.—H. P WOODWARD. 'The Auriferous Deposits and Mines of Menzies, North Calgoorlie Goldfield,' Geol. Survey, Bull. 22, Perth, 1906.—J. W. GREGORY. 'The Indicators of Ballarat,' The Mining Journal, Jan. 20, 1906 ; 'The Ballarat East Goldfield,' Mem. Geol. Survey, Victoria, No. 4, 1907.

In addition to the telluride gold deposits of Western Australia and the Tertiary goldfields of New Zealand there are in Australasia many important gold-quartz districts of greater geological age. In Western Australia, as a matter of fact, mining did not begin upon the telluride lodes but upon quartz lodes, though these were not of any great importance. Of greater significance are the occurrences in Victoria, Queensland, and New South Wales, the positions of which, under the names of the different goldfields, are indicated in Fig. 314. These have been studied more particularly by Woodward, Rickard, Dunn, Gregory, Schmeisser, and Lindgren.

In Victoria, the goldfields of Bendigo and Ballarat, Ararat, Maryborough, and Castlemaine, to the west, are particularly noteworthy, as are those of Beechworth and Gippsland, to the east. Among these the Bendigo district takes first place. The lodes there occur in a highly plicated Ordovician series of dark slates alternating with fine-grained sandstones, which series somewhat south of Bendigo is intruded by granite or quartz-monzonite. The deposits have the peculiar form of 'saddle reefs,' illustrated in Figs. 315 and 316. These are the fillings of bedded cavities which, as the strata became folded, were formed at the crests of the anticlines and to a less extent in the troughs of the synclines. They extend unbroken for many kilometres along the anticlinal axes, generally agreeing in strike and dip with the country-rock, and forming lode-like ore-bodies extending down the anticlinal limbs, such bodies being known as the saddle 'legs.' Ten and more of such saddle lodes have been found in one vertical plane. In the Bendigo district within a strip about 1400 m. wide three main series of such lodes are found running parallel and striking north. These, as indicated in Fig. 316, are known respectively as the New Chum, Garden Gully, and Hustler series. Besides these saddle lodes the slates contain leaders and more or less irregular ore-bodies which in part are of considerable dimension and more or less directly connected with the main lodes ; these are locally termed 'spurs' and 'makes'; in the north of the district particularly they are very rich.

The auriferous quartz of the Bendigo district separates cleanly from the black slate of the country-rock which, practically speaking, it does not replace. In appearance it is milk-white and almost glassy, and shows but little sign of pressure. Quartz masses with intercalated lamellæ of slate are not infrequent, a fact which moved Dunn to put forward the theory, which we cannot endorse, that the force of crystallization had contributed to the enlargement of the cavities in which the ore is now found, that is to say, he conceived these lodes to have been formed in a manner similar to that suggested by Bornhardt for the siderite lodes of Siegerland. If however the force of crystallization, which doubt-less exists, were powerful enough to produce such effects, the consequences of such colossal pressure as must be postulated, would have been evident first of all upon the country-rock directly in touch with the quartz.

Carbonates of lime, magnesia, and iron are sometimes mixed with the quartz, being often found along the walls. Chlorite in small bent, though definite crystals can also at times be observed.

The sulphides are limited almost exclusively to pyrite and arsenopyrite with a small amount of galena, these minerals occurring particularly in the slate directly adjacent to the quartz; crystals of pyrite indeed are not infrequently found embedded one-half in quartz and the other half in black slate. Lindgren, whose description is here chiefly followed, found albite in the auriferous quartz of the South New Moon, one of the most productive mines around Eagle Hawk. In that mine the ore does not occur in a saddle lode but in the form of a spur or irregular quartz mass in the black slate.

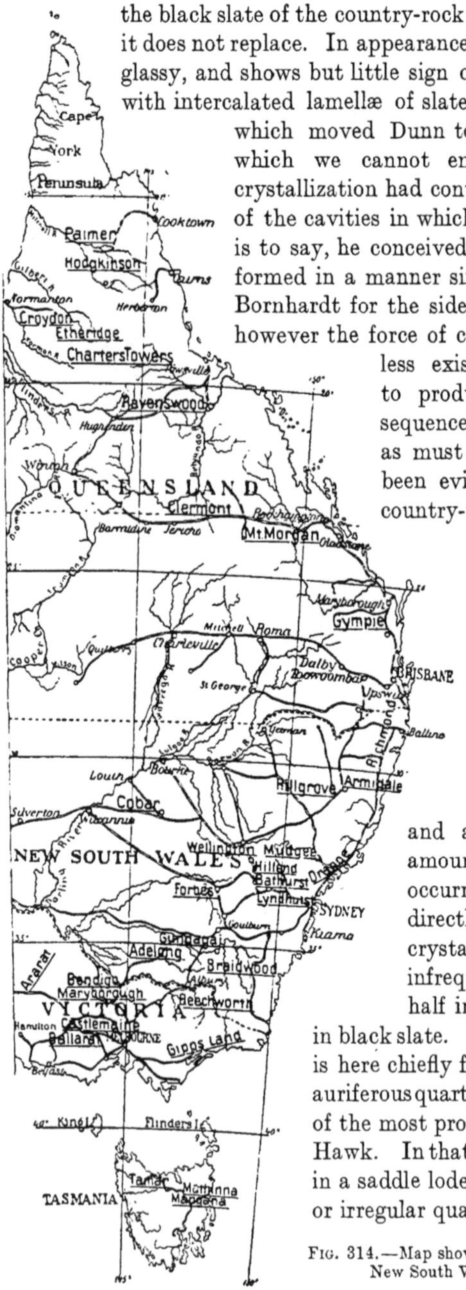

Fig. 314.—Map showing the goldfields of Victoria. New South Wales, and Queensland.

The gold in general is coarse-grained, and in massive, white, almost glassy quartz, may often be observed in large pieces.

With regard to the maintenance of gold content in depth, many of the mines are already more than 1000 m. deep ; indeed the New Chum Railway mine in the year 1906 cut an apparently still payable saddle lode at a depth of almost 1400 metres. The limit of payability is put at 7 dwt., or about 10 grm. of gold per ton. Generally speaking, experience has shown that the ore above 2500 feet was considerably richer than that obtained below that depth.

Up to 1906 the district had altogether produced approximately 20·5 million oz. of gold, of which 6 millions were derived exclusively from primary ore, while the remaining 14·5 millions came partly from primary ore and partly from gravels. In the year 1904 the production amounted to 206,000 oz., and the average value of the ore was 10 dwt. or 15·5 grm. per ton.

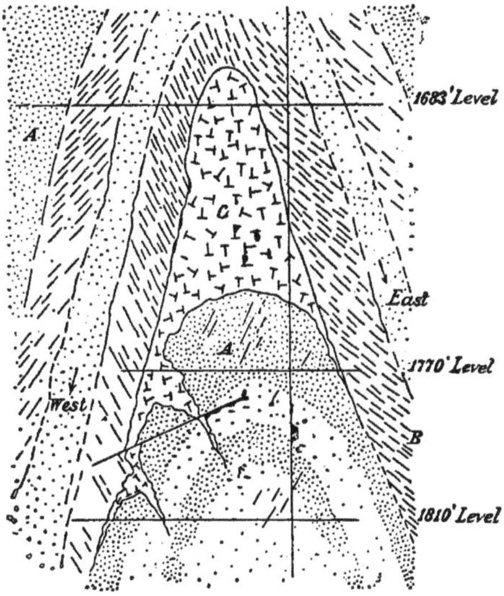

FIG. 315.—Section of an anticlinal lode or saddle reef in the New Chum Cons. mine. Schmeisser, *Zeit. f. prakt. Geol.*, 1898, p. 100.

A, sandstone ; *B*, slaty sandstone with quartz stringers ; *C*, quartz.

In the Ballarat goldfield lying to the south of Bendigo, the country-rock consists of alternating layers, from 2 inches to 1 foot in thickness, of slate and sandstone, the latter being in part quartzitic. All are fine-grained and free from conglomerate, ripple marks indicating that they are of shallow-water formation.

Lindgren called attention to the fact that in spite of their small thickness some of these layers could be followed for great distances, though in general, according to Gregory, petrographical change may be sudden. No fossils have yet been found. The beds were formerly described as Lower Silurian. According to Gregory however, in the eastern portion of the Ballarat plateau they are Lower Ordovician, while at Ballarat they must even be regarded as pre-Ordovician. They strike north-north-west and dip generally to the west, though as the result of comparatively recent movement they now occur in steep anticlines and synclines. East of Ballarat granitic rocks appear, which have effected contact-metamorphism of the slates. Finally, the Carboniferous glacial deposits in the east of the plateau are noteworthy.

The gold-quartz lodes of the district may be divided into three groups, those respectively of Little Bendigo, Ballarat East, and Ballarat West, the general trend of these districts being approximately parallel, and north-south.

Little Bendigo, the most north-easterly district, embraces, over a length of two miles, several parallel lode-series, the more important of which are known as Monte Christo, Band of Hope, and Black Hill. At the last-

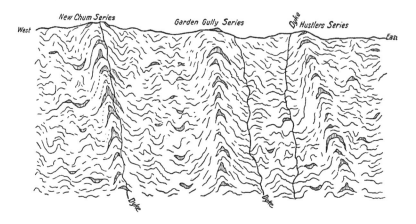

FIG. 316.—Diagrammatic section through the Bendigo goldfield. Rickard.

named locality the Ballarat East series begins, this extending some seven or eight miles southwards from Black Hill. About one mile to the west, the Ballarat West series extends over a length of three miles so disposed that the most northerly mines lie opposite the more important of the most southerly mines of the Ballarat East series.

The occurrences at Little Bendigo are cut off to the north by a large fault. The most important of these is the Monte Christo series, this series, as illustrated in Fig. 317, being confined between two parallel faults some 80 feet apart. The slate and sandstone layers within this space are traversed by a large number of flat quartz veins. Near the surface, where the gold was concentrated in well-defined shoots, these veins were rich ; below 300 feet however they have often proved to be unpayable.

An important feature of the Ballarat district are the so-called 'indicators' which are definite slate layers capable of being followed for great distances. With the Monte Christo series for instance, in the Metropolitan mine, as illustrated in Fig. 317, a slate band known as the Jarvis indicator occurs in the middle of the so-called lode.

The Ballarat West goldfield is the most south-westerly of the district, where, the surface being covered by a basalt flow, the lodes do not outcrop. The auriferous quartz occurs in large irregular masses roughly lenticular in form, from both sides of which and from below, a large number of quartz leaders shoot out. The lodes of this district are arranged in two series known respectively as the Consols Lode and the Star Lode.

The Ballarat East goldfield is the most important of the district. In it those large nuggets were found which, reaching up to 90 kg. in weight, made this goldfield famous. The district contains in addition a large number of well-defined lodes or lenticular quartz masses, which in many of their features recall the celebrated quartz lodes of California. Here however the distribution of the gold is not so simple ; instead of well-defined vertical lodes there are many narrow layers running approximately horizontally, and since much of the quartz is barren the work of mining is difficult. The distribution of the gold however is in some respects regular.

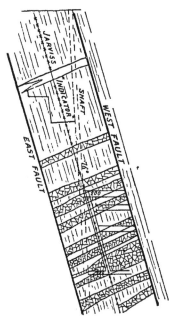

FIG. 317.—Section of the Metropolitan lode, Monte Christo series, showing the Jarvis Indicator. Gregory.

Although the greater portion of the quartz is barren it is traversed by richer bands and patches which always occur where the flat quartz layers or veins cross narrow vertical beds of a dark slate. Such slate beds are the indicators. The quartz veins where they cross such an indicator may be extremely rich, though eighteen inches away they may, practically speaking, contain no gold. The miners therefore follow these indicators, whereby they arrive more easily at the richer patches and save themselves much useless development work. It is interesting to observe that north of this district and of Black Hill many other quartz lodes occur, but without gold.

Among the mines of this goldfield deserving of mention are the Black

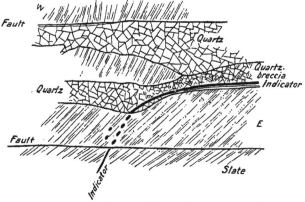

FIG. 318.—Course of the Britannia United Indicator on the 987-foot level of the Victoria United mine.   Gregory.

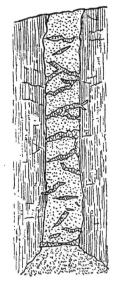

FIG. 319.—Section across a ladder lode at Waverley, Victoria. Phillips and Louis.

Hill, the oldest gold-quartz mine in Victoria ; the Victoria United ; and the East Change.   The gold won in the Ballarat East district is very pure, being about 995·5 fine.

The 'ladder lodes' of Waverley, Victoria, illustrated in Fig. 319, are of similar character to the deposits at Beresowsk.   With them the quartz occurs in an almost vertical, partly decomposed greenstone dyke, which can be followed with almost parallel walls for roughly two and a half miles in the same direction as the slates in which it is found. This decomposed eruptive rock is traversed by horizontal leaders, the thickness of which varies from one inch to two feet.   According to Louis some of these leaders yielded ore of exceptional richness, while the average content was always high.   These dykes, which are known as 'mullocky reefs,' pass, at a depth of 70–200 feet, into undecomposed crystalline rocks where their exploitation is considerably more difficult.   It is stated that the decomposed greenstone itself possesses a low gold content.   The metal won from these lodes is remarkable for its great purity.

The Gold Deposits of Brazil

LITERATURE

v. Eschwege. Beiträge zur Gebirgskunde Brasiliens, 1819.—Ad. Metzger. Report on the Mines of Passagem, Raposos, and Espiritu Santo. Paris, 1885.—M. P. Ferrand. L'Or à Minas Geraes, Ouro Preto, 1894 ; reviewed Zeit. f. prakt. Geol., 1896, p. 123.— E. Hussak. 'Der goldführende Kies-Quarz-Lagergang von Passagem in Minas Geraes, Brasilien,' Zeit. f. prakt. Geol., 1898, p. 345 ; 'Report upon the Morro Velho Mine,' Eng. Min. Jour., Oct. 1901, p. 485.—G. Berg. 'Beiträge zur Kenntnis der Goldlägerstätten von Raposos in Brasilien,' Zeit. f. prakt. Geol., 1902, pp. 81-84.—Scott. 'The Goldfield of the State of Minas Geraes, Brazil,' Trans. Amer. Inst. Min. Eng. Vol. XXXIII., 1902, p. 406.— O. A. Derby. 'Notes on Brazilian Gold-Ores,' Eng. Min. Jour. LXXIV., 1902, p. 142.

The auriferous belt of Brazil lies in greater part along the range of hills, 1000-1713 m. in height, known as the Serra de Espinhaco, which traverses the central portion of Minas Geraes and forms the watershed between the rivers Doce and São Francisco. The boundaries of this goldfield may be taken to pass through the towns of Santa Lucia to the north, Brumado to the south, Ponte Nova to the east, and Paraopeba to the west.

The geological age of the rocks composing this district has not yet been determined. According to Derby they are Cambrian and Lower Silurian, the sequence in the Ouro Preto district from top to bottom being : upper mica-schist ; limestone ; itabirite with jacutinga, a sandy micaceous iron ore ; clay-slate ; schistose quartzite ; mica- and talc-schists ; and finally, gneiss and granite. As elsewhere, mining here first commenced upon the gravel-deposits. According to whether they are at the present river-level or upon higher terraces, the Brazilian miner divides these deposits into veias, taboleiros, and grupiaras. The lodes may be divided into : (a) bedded lodes, which here are wrongly termed contact lodes ; (b) lodes in slate and quartzite ; and (c) the so-called jacutinga lines in itabirite.

The bedded lodes, almost coinciding in strike and dip with the bedding, extend along the Ouro Preto range in a strip beginning roughly 2 km. west of the town of that name and reaching to a point about 4 km. north-east of Marianna. In part they are probably of eruptive origin. The principal mines exploiting such lodes are the Passagem and the Morro Santa Anna, which work lenticular quartz bodies 1-15 m. in width, 10-100 m. in length, and generally of greater extent in depth. The occurrence at Passagem near Ouro Preto, which may be regarded as typical of this class of deposit, is described later in greater detail.

The so-called cross lodes in slate are much more numerous. These consist of lenticular quartz bodies with dimensions subject to great

variation; upon one of the largest the celebrated Morro Velho mine is
working; other occurrences are limited to narrow leaders of auriferous

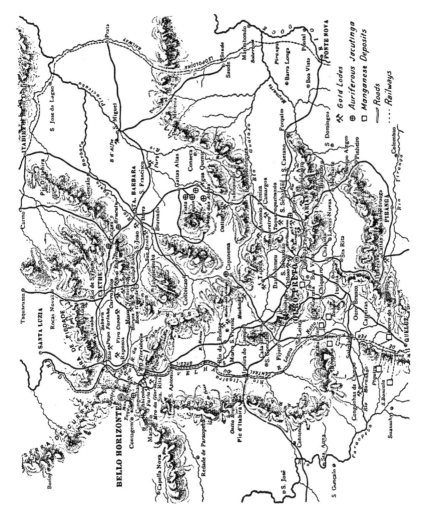

quartz. These lodes likewise follow the bedding. The quartz is often
accompanied by arsenopyrite, pyrrhotite, and limonite, and more seldom,
as for instance at Morro Velho, by carbonates. Occasionally, large amounts
of metallic gold are found in small veins, though such rich zones are

generally of little extent. As far as present experience goes the gold content remains fairly constant in depth.[1]

The lodes in quartzite are not important nor are any at present being worked. The last mine of this class which continued to work, Catta Branco by name, shut down more than fifty years ago.

The so-called jacutinga lines in itabirite are usually not more than one centimetre in thickness. In these the gold is coarse-grained. In the Gongo-Socco mine, once famous, and in others of less importance,

∴∵∴	{ Gneiss or Granite.
≋	Mica- and Talc-schists.
℥℟	Schistose Quartzite.
≋	Clay-slate.
⚏	Itabirite.
≋	Auriferous Jacutinga in Itabirite.
⊞	Limestone.
∷∷∷	Canga or Ferruginous Conglomerate.

Fig. 321.—Section through the Gongo Socco mountains.

masses of gold mixed with jacutinga have been found weighing several kilogrammes. Numerous traces of previous mining operations upon auriferous jacutinga in itabirite are encountered between Ouro Preto and Marianna.

The Passagem Deposit

Seven kilometres to the east of Ouro Preto, the principal town of the Province Minas Geraes, one of the most productive gold mines of Brazil exploits the bedded quartz lode of Passagem, which, striking north-east and dipping 18°–20° to the south-east, is roughly conformable with the country-rock. The more detailed circumstances of its bedding may be gathered from Fig. 322. Quartziferous mica-schist forms the lowest member of the rock-sequence present, and this in contact with the lode

[1] Scott, 1903.

passes over to the so-called contact quartzite ; then comes the quartz lode
which has a cryptocrystalline schist for its immediate hanging-wall ;
while above this again lies the itabirite, with its zone of decomposition
and lateritization known locally as *Canga*.

Of these rocks, the so-called contact quartzite and the quartz lode
claim special attention. The first, according to Hussak, is greenish-white in
colour and distinctly schistose in structure, the mica being sericitic. It is
most closely connected with the quartz lode, in which it forms lenticular
masses which in some places take up the whole lode width, while in other
places they are completely absent.

The hanging-wall rock, the so-called cryptocrystalline schist, consisting of thin layers of minute aggregates of quartz grains, these layers

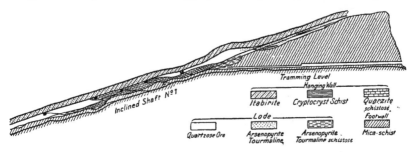

FIG. 322.—Section of a bedded quartz lode at Passagem near Ouro Preto.
M. P. Ferrand, *Zeit. f. prakt. Geol.*, 1898, p. 346.

being separated by narrow beds of a straw-yellow to light-brown amphibole
often resembling asbestos, is very thinly bedded. The quartz layers
contain abundant metalliferous particles, and, though seldom, extremely
finely divided gold.

This lode, regular in dip, deviating but little in strike though
varying considerably in thickness, consists of an auriferous pyritic quartz,
or in greater detail, of a milk-white quartz with some tourmaline and
arsenopyrite, and to a less extent some pyrite and pyrrhotite. What here
is termed a lode is more probably a series of lenses sometimes constricted,
sometimes expanded, at times auriferous, and at times barren. It
is generally the thicker lenses which consist of schistose quartzite and
barren milk-white quartz. Unlike the hanging-wall schist, the quartzite,
which as stated above forms the foot-wall, contains no precious metal. The
richest portions are those containing fine crystals of arsenopyrite and black
tourmaline. The gold content may then be 150–200 grm. per ton, but when
more quartz is present it is lower. Clean quartz from the lode contains
but 2–3 grm. per ton, though when bands of tourmaline render this quartz

schistose the gold content rises to 10–15 grm. The pyrite also, contrary to usual experience, only carries gold when it is intergrown with tourmaline or when it has started to decompose ; in such cases a content of 20–30 grm. of gold per ton is often found. With the gold some bismuth and silver are associated. Such lode-filling has been proved for 700 m. along the strike, and 450 m. in depth. Concerning genesis, Hussak, as the result of observation on the spot and of microscopic investigation, has come to the conclusion that the lode is of intrusive origin, being in fact an ultra-acid granitic apophysis. According to him it broke through the quartz-schist which it fractured and in part absorbed, forming a distinct contact zone on both walls. It is younger than the rocks now found in its hanging-wall.

During the period 1864–1873 from 104,000 tons of ore treated 753·5 kg. of gold were recovered ; and during 1884–1893 from 257,626 tons 2375 kg. In addition about 36 kg. of metallic bismuth per year was separated from the gold.

THE RAPOSOS DEPOSIT

According to Berg,[1] although of no great economic importance this deposit, situated in lat. 19° 58′ S. and long. 43° 49′ W., represents geologically a remarkable formation, in that pyritic sulphides are concentrated in the form of chimneys pitching at an angle into depth. The country around consists chiefly of pre-Cambrian clay-slate and phyllite, which are occasionally represented by chlorite- and sericite-schist, and which alternate with itabirite rich in magnetite, and with different calcareous quartzites ; these schistose north-north-east striking rocks, dipping at an angle of 35° to the east, are pierced by two bosses of diabase. The geological structure is therefore quite similar to that of the important gold deposit of Morro Velho next described.

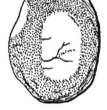

FIG. 323. — Section through a gold ore-chimney 20 cm. diameter, at Raposos, Brazil. Berg, *Zeit. f. prakt. Geol.*, 1902, p. 82.

The rocks in detail are in different places highly contorted, compressed, and elongated. It is particularly where elongation is evident that the ore occurs in regularly arranged chimneys. These are generally 3–6 m., but may occasionally be as much as 12 m. in diameter, though on the other hand they may be as little as 20 cm. ; a chimney of the latter dimension is illustrated in Fig. 323. They lie regularly arranged in the plane of the schist, though not always following the line of the dip. Surrounded by schist, the substance of the chimney consists of quartz and pyrite.

[1] *Zeit. f. prakt. Geol.*, 1902.

The middle is occupied by minutely though distinctly brecciated quartz in which some individual pyrite stringers and veinlets are found, while pyrite and quartz, intimately mixed, occur more particularly around the|periphery. Arsenopyrite remarkable for a high gold content is not infrequently found with the pyrite. Chalcopyrite, pyrrhotite, and sphalerite are very subordinate, while magnetite, presumably remaining from the original itabirite, is common. Not infrequently the chimneys show a stemmed structure, each then consisting of a number of parallel and close-lying smaller individuals. Berg explains this formation by assuming that the crystalline schist under lateral pressure experienced an elongation producing a columnar structure,[1] along which later, and probably in connection with the diabase, the gold solutions arose. The irregularity of the rock and the repeated change from brittle quartzite and itabirite to elastic flexible schist, prevented the formation of regular lode fissures. The solutions had therefore to find their way between the crushed quartzite and the itabirite, the calcite and magnetite constituents of which latter favoured the metasomatic replacement of the country-rock, with the result that where under conditions of undisturbed bedding epigenetic bedded deposits would have arisen, epigenetic ore-chimneys became formed.

THE MORRO VELHO DEPOSIT

At Morro Velho a lode having the form of a column with an elongated oval cross-section, is worked. In this district gneiss and granite form the basement upon which mica-schist, calc-schist, schistose quartzite, clay-slate, itabirite, and younger mica-schist, have been laid. In so far as the gold is concerned, the grey pre-Cambrian schists and phyllites intercalated with itabirite and quartzite but below the main zone of itabirite, are of importance. According to Derby the lode traverses a bed of calc-schist which, containing mica and chlorite, may in general be described as a mica-schist. It is interesting to note that though it contains calcium carbonate in large amount, the carbonates of magnesia and iron are almost completely absent. The lode strikes east-west and has been worked for a length of roughly 200 metres. It is from 1 to 35 m. in width and has been followed for an inclined depth of about 1800 metres. This mine is one of the deepest in the world, having as far back as 1901 already reached a vertical depth of 1030 metres.

The filling consists of a fine-grained mixture of siderite, dolomite, calcite, quartz, and to all appearances also of albite. The large amount of carbonates in the lode-filling is remarkable. Typical vein-quartz, on the other hand, appears in but subordinate amount and only in some

[1] *Griffelstruktur.*

places. Among sulphides, pyrrhotite, pyrite, arsenopyrite, and chalco-pyrite are found, while sphalerite and galena are accounted rarities. According to Wilder the ore contains 30–40 per cent of sulphides, 30–40 per cent of carbonates, and 20–30 per cent of quartz. Pyrrhotite is the prevailing sulphide, and siderite the prevailing carbonate. An average sample, representative of the ore mined for several months, gave 28·5 per cent of pyrrhotite, to which, according to Wilder, may be added 5·04 per cent arsenopyrite, 2·5 per cent pyrite, and 0·66 per cent of chalco-pyrite, making altogether 36·7 per cent of sulphides.

A small amount of smoky quartz finely distributed throughout, is considered favourable to the gold content. Should on the other hand the quartz resemble typical vein-quartz, or pyrite be abundant, the gold content appears to diminish. This deposit therefore, as is also the case with other Brazilian deposits, deviates from the normal type of gold-quartz lode. With increasing arsenopyrite the gold content of the whole mass increases.

At the east end of the lode and in contact with the country-rock, graphite occurs. Beautiful albite crystals are also not infrequent though generally they are only of small dimension. Other features of interest are the aplitic texture of the ore and the occurrence of fissure water in which, according to analyses, almost all the elements found in the lode are contained. Finally, the absence of typical eruptive rocks from the vicinity is noteworthy.

In the year 1901 this mine produced 152,238 tons of ore, of which 140,855 tons were treated yielding 99,197 oz. or 3085 kg. of fine gold. On an average the ore contains about 0·7 oz. per ton.

THE GONGO SOCCO DEPOSIT

This deposit belongs to the so-called jacutinga lines. Its geological position is illustrated in Fig. 321.

Gneiss and granite again form the basement upon which mica-schist, calc-schist, schistose quartzite, and clay-slate lie one after the other. The last of these rocks forms the immediate foot-wall of the itabirite, the hanging-wall of which is limestone. All these rocks dip approximately 45° to the south. The Gongo Socco mine is situated on the central railway of Brazil, at a point roughly 30 km. east of Sabara and 1000 m. above the sea. The auriferous jacutinga is at most 15 cm. thick. In the centre of its thicker portions it contains pieces of pure gold in the form of irregular aggregates plates and wires, such aggregates weighing from a few grammes up to several kilogrammes. Two-thirds of the gold won is found in this form, and but one-third finely distributed throughout the substance of the itabirite itself. Captain Lyon in 1830 reported that a single native recovered in one day specimens of gold ore which yielded 10 kg. of metal. Other remarkable finds made in the years 1829 and 1830 are reported to have yielded between 47·6 and 193 kg. of gold. The whole itabirite formation indeed carries gold though seldom in such amount as to make it worth working. The composition of jacutinga, which in reality is ferruginous itabirite, may be gathered from the following analyses :

$$
\begin{array}{llll}
Fe_2O_3 & . & . & . & 97\cdot00 \text{ per cent} \\
SiO_2 & . & . & . & 1\cdot60 \text{ ,,} \\
Al_2O_3 & . & . & . & 1\cdot10 \text{ ,,} \\
Mn_2O_3 & . & . & . & 0\cdot60 \text{ ,,}
\end{array}
$$

From 1826 to 1839 about 11,000 kg. of gold were obtained from this mine. Unfortunately in 1840 the water in the deepest level could no longer be mastered, and in 1856 when a depth of 140 m. had been reached operations were discontinued. The highest output of any year was made in 1832 when 1578 kg. were obtained ; in 1856 when work was stopped the output had dropped to 29 kilogrammes.

Gold Production of Minas Geraes.—The development of gold mining in this district may be gathered from the following figures : In 1896, 1963 kg. were produced ; 1897, 2071 kg. ; 1898, 3267 kg. ; 1899, 3974 kg. ; 1900, 4811 kg. ; and in 1901 about 5000 kg. Since then the production has fallen considerably, Brazil in 1910 only producing 2972 kg. The total output from 1820 to 1901 amounted to 85 million sterling.

GOLD DEPOSITS AT SEKENKE IN GERMAN EAST AFRICA

LITERATURE

Reports of J. KUNTZ, SCHLENZIG, and F. SCHEFFLER in manuscript.—K. SCHMEISSER. Die nutzbaren Lagerstätten der deutschen Kolonien. Lecture. Berlin, 1910.

The village of Sekenke is situated about 10 km. west of the steep descent from the Iramba plateau, upon a flat undulation about 15 km. long and 3 km. wide, between the Wembere and Chironda streams. Here many lenticular lodes occur in the contact belts around different eruptives, mostly of dioritic nature. These rocks, as well as the lodes they contain, strike north-south and dip at a high angle. Of fifteen such lenses yet known five are payable, three of these constituting the Dernburg lode. The length of these lenses varies between 50 and 300 m., while the thickness in places reaches 3 metres. For every metre of depth they yield something more than 1000 tons of ore, so that down to a depth of 27 m. about 30,000 tons are available. The secondary variations of metal content in depth are very well developed. The cementation zone comes right to surface, samples from it occasionally assaying several thousand grammes of gold per ton ; the average assay of sixty samples, after rejecting abnormally high results, gave 47 grm. per ton. The average content in shaft No. 1 of the Dernburg lode, down to 26 m. was 60 grm. Since, according to Kuntz, rich ore still continues below water-level, this occurrence at Sekenke presents a case where the cementation zone has by subsidence been depressed below the ground-water level. This phenomenon in our opinion is probably associated with the formation of the steep drop from the Iramba plateau.

East of this occurrence gold lodes occur upon the Iramba plateau itself, where formerly extensive work was carried on. The rocks there are slates, probably Palæozoic, these being intruded by granitic rocks chiefly, but also by diorite. The lode-outcrops on account of their quartzose character stand out boldly upon the ground, and in this case also the cementation zone comes to the surface. Samples of the ore from this zone occasionally assay as high as 4 kg. of gold per ton. Development work however has shown that at but little depth the primary zone commences. In this the quartz is impregnated with pyrrhotite and pyrite, and the gold content quickly decreases till the ore becomes no longer payable. This occurrence of pyrrhotite is particularly interesting.

THE GOLD LODES OF SOUTH AFRICA

LITERATURE

F. W. VOIT. ' Übersicht über die nutzbaren Lagerstätten Südafrikas,' Zeit. f. prakt. Geol., 1908, p. 209.—P. R. KRAUSE. ' Über den Einfluss der Eruptivgesteine auf die Erzführung der Witwatersrand-Konglomerate und der im dolomitischen Kalkgebirge von Lydenburg auftretenden Quarzflöze nebst einer kurzer Schilderung der Grubenbezirke von Pilgrimsrest und De Kaap,' Zeit. f. prakt. Geol., 1897, p. 12.—SAWYER. The Goldfields of Mashonaland. London, 1894. — K. SCHMEISSER. Die nutzbaren Bodenschätze der deutschen Schutzgebiete. Berlin, 1902.

Among the old gold lodes, those of the De Kaap goldfield in the Transvaal and those of Rhodesia claim especial attention. The crystalline schists of South Africa lying upon the fundamental gneiss constitute the principal horizon of these quartz lodes, which occur of all dimensions, from microscopically small quartz veins involving the whole bedded formation in which they occur, to the most extensive of ordinary lodes. Generally only the smaller veins are auriferous, the larger lodes consisting more often of barren quartz. The ore-bodies are usually lenticular and are found sometimes along one plane, sometimes along another ; they are also often associated with dyke-like occurrences of eruptive rock. The minerals include pyrite, chalcopyrite, galena, sphalerite at the Sheba Queen mine, stibnite at La France, arsenopyrite, and other sulphides. The

distribution of these is not uniform, they are more usually found concentrated in ore-shoots. It is characteristic of all these deposits that
they are generally only payable in the cementation zone.

The most important deposits are those of the Barberton district, these
constituting the De Kaap goldfield. According to Krause and Voit, in this
goldfield numerous lodes traverse the steep crystalline schists which lie
upon the granite. Some of these are ordinary cross lodes, while others are
bedded. More generally they strike with the schists in which, accompanied
by diabasic rocks, they occur, the lodes often being especially rich where
they intersect these rocks. Of the many mines working in this district
the Sheba Queen is the most interesting. In this mine talc- and chlorite-
schists overlaid by quartzite are traversed by such a number of small
quartz veins that the whole complex, to a thickness of about 100 feet, can
be worked. This is therefore a deposit closely related to the composite
lodes as defined by Krusch. At Pigg's Peak, where quartzite is traversed
by numerous veins, and at Zwartkopje, the circumstances are similar. At
this latter place the gold is contained in a ferruginous banded silica-schist,
at its contact with a grass-green schist which presumably represents a
deformed eruptive rock. The occurrences at Steynsdorp to the south of
Barberton are noteworthy in that the gold, in addition to being associated
with silica-schist, is also connected with itabirite. It occurs sometimes
intimately associated with stibnite, and sometimes irregularly distributed
in native condition through snow-white quartz containing no sulphides.
Analogous lodes are found in the Vryheid district of Zululand and elsewhere. The economic importance of all these occurrences is but small.

In Rhodesia, the gold lodes generally occur in the neighbourhood of
diabase, in east-west striking crystalline schists forming larger and smaller
patches in granite. Since these deposits are generally poor immediately
at the surface and only become payable at a little depth therefrom, it
would appear as though the partly impoverished oxidation zone were here
still represented ; in addition, both in Mashonaland and Matabeleland
the cementation zone appears to continue to greater depths than in the
Transvaal. Two types of lode may be distinguished, firstly, those which
reach 1 m. in width and form ore-shoots 100–140 m. long ; and secondly,
large impregnation zones with fairly constant content for lengths of almost
1000 metres. The deposit worked at the Wanderers mine in the Selukwe
district has for instance a width of 40–60 feet. The Giant mine works an
auriferous itabirite in which the auriferous zones consist of lenticular
bodies, some of which are connected by narrow cross veins.

The lodes in the Malmani dolomite occur in a substantially younger
country - rock. These at the surface often have a width of 10 m., a
width which in depth invariably contracts, suggesting that it is largely

due to oxidation-metasomatism. Although these deposits are often of considerable extent at the surface, they generally have but a low gold content. The best known is the Mitchell lode. The deposits of this class in the Lydenberg district, described later, are more important. The occurrence of copper with the gold connects those deposits with the auriferous copper deposits, while the pronounced replacement phenomena in the limestone, on the other hand, are indicative of metasomatic processes.

GOLD-COPPER DEPOSITS IN GERMAN WEST AFRICA

LITERATURE

G. GÜRICH. Deutsche-Südwestafrika, Reisebilder aus den Jahren 1888 und 1889.— E. STROMER V. REICHENBACH. Die Geologie der deutschen Schutzgebiete in Afrika. Munich, 1896.—K. SCHMEISSER. 'Die nutzbaren Bodenschätze der deutschen Schutzgebiete,' Lecture, Colonial Congress. Berlin, 1902.—F. W. VOIT. 'Beiträge zur Geologie der Kupfererzgebiete in Deutsch-Südwestafrika,' Jahrb. d. k. pr. geol. Landesanst., 1904.— L. MACCO. Die Aussichten des Bergbaues in Deutsch-Südwestafrika, Berlin, Dietrich Reimer, 1907.

Gold-copper ore is won in the Pot mine on the Swakop river, where the deposit, as described by Voit, is a garnetiferous layer intercalated in gneiss and sparsely impregnated with copper, such impregnation having proceeded from a fissure.

Schmeisser mentions the copper deposit at Hussab also as being auriferous. There, in a large patch of gneiss lying upon granite, an extensive quartzite-schist bed is intercalated, in the hanging-wall of which, zones of mica-schist impregnated with copper ore occur. It must be assumed in this case also that the impregnation proceeded from fissures.

More important are the auriferous copper deposits of the Rehoboth district, which were carefully investigated by the Eichmeyer expedition. The most promising of these occur on the Groot and Klein Spitzkop, some 20 km. to the north-west of Rehoboth. There, highly contorted mica-schists are traversed by a large number of linked veins which in general strike east-west and dip to the north. At the top of these two hills the veins form a regular maze in which five only are sufficiently definite to be followed. The copper ore occurs sometimes as malachite impregnations, sometimes as chalcocite, bornite, chrysocolla, decomposed chalcopyrite, or in fact all those ores characteristic of the oxidation and cementation zones. From the occurrence of remnants of pyrite the conclusion may be drawn that the primary ore consists chiefly of pyrite. The gold occurs either as free gold or associated with pyrite. The wedges of country-rock occurring between converging veins have also assayed 3–4 grm. of gold and 20 grm. of silver per ton. Quartz and calcite are the principal gangue-minerals, these being sometimes accompanied by

siderite. The vein-quartz is compact, resinous in lustre, and milk-white
to red in colour, but strangely enough it never carries finely divided gold
this metal always occurring in grains, up to 4 grm. in weight. In contra
distinction to this, the friable and dull quartz and the whitish-brown

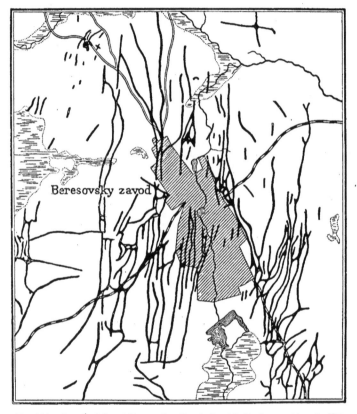

FIG. 324.—Beresite dykes at Beresowsk. Karpinsky, 'Guide des excursions du VII.
Congrès Géolog. Intern. 1897,' *Zeit. f. prakt. Geol.*, 1898, p. 23.

quartz carry gold so fine as to be undiscernible by the naked eye ; analys
give 3·2 grm. of gold together with 28 grm. of silver per ton. In spite
the association of the gold with chalcocite, no gold could be detected
such pieces of chalcocite as were freed from lode material, though su
contained approximately 76 per cent of copper. This must be taken
evidence that the chalcocite was subsequently cemented by gold.

Gold Lodes of Beresowsk in the Urals

LITERATURE

F. Pošepný. ' Die Golddistrikte von Berezov und Mias im Ural,' Arch. f. prakt. Geol.
II., 1895, p. 490.—A. Karpinsky. Guide des excursions du VII. Congrès Géolog. Intern.,
1897, V. p. 42.—R. Beck. ' Die Exkursion des VII. Internationalen Geologenkongresses
nach dem Ural,' Zeit. f. prakt. Geol., 1898, p. 16.

The geological position of these deposits is very interesting. The
district consists of dynamo-metamorphosed talcose, chloritic, and phyllitic
schists, which at the surface are decomposed to a reddish mass. These, as
illustrated in Fig. 324, are traversed by
numerous dykes of a microgranite known
as beresite, which dykes, 2 to 14 m. wide,
strike north-south and dip vertically ; they
likewise are decomposed to a considerable
depth. The primary deposits are associated
with extensive gravel-deposits which often
lie below à peat covering, this being used
for fuel. The gold veins run at right angles
to the beresite dykes, within which they are,
practically speaking, confined, though as
illustrated in Fig. 325, they often continue
a short length in the country-rock beyond.
These veins are usually but a few centimetres in thickness though exceptionally
they may be more than one metre. In their
occurrence they greatly resemble the ladder
lodes of Australia, illustrated in Fig. 319.

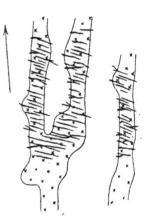

FIG. 325. — Diagrammatic representation of the gold lodes at Beresowsk. Beck.

The lode-filling consists of auriferous quartz, the gold being partly
free and partly associated with pyrite, with which latter lead- and copper
sulphides sometimes occur. Karpinsky states that the gold content
varies between 2·5 and 30 grm. per ton, with an average of probably
13 grm., though in places it may reach as much as 250 grammes.
In connection with the question of the genesis of these deposits
it is significant that the neighbouring Schartasch granitite is stated to
contain up to 1 grm. of gold per ton, though this rock, unlike the beresite,
is quite undecomposed. The decomposition of the beresite, on the other
hand, is so advanced that analyses permit only doubtful comparisons.
If however this granitite contain primary gold, then the beresite dykes,
which represent later pulsations from the same parent magma, must also
originally have contained primary gold, which quite conceivably became
concentrated in the contraction fissures which opened as the rock cooled.

The Hohe Tauern in the Eastern Alps

LITERATURE

B. Cotta. Geologische Briefe aus den Alpen, Leipzig, 1850, p. 146.—A. v. Groddeck. Die Lehre v. d. Lagerstätten der Erze, Leipzig, 1879, p. 206.—F. Pošepný. Arch. f. prakt. Geol., 1880, Vol. I. p. 487.—M. Vacek. Verhandl. d. k. k. geol. Reichsanst., 1893. ' Die Untersuchung des Bergbauterrains in den Hohen Tauern,' Commission's Report, published by the Minister of Agriculture, Vienna, 1895. ' Das Bergbauterrain in den Hohen Tauern,' Jahrb. des naturhistor. Landesmuseums von Kärnten, 1897, Part 24.—P. Krusch. ' Die Goldlagerstätten in den Hohen Tauern,' Zeit. f. prakt. Geol., 1897, p. 77.

According to Pošepný, gold mining in the Hohe Tauern must be accounted as belonging to the oldest mining in Europe. The district traversed by the lodes contains two large gneiss massives known respectively as the Ankogel and the Hochnarr. Both of these form gentle elevations, and to the north and south both are overlaid by crystalline schists, including mica-schist, calc-mica-schist, limestone, phyllite-schist, and chlorite-schist. The beds are horizontal or but flatly inclined, and are faulted along north-south fissures. The area lying between the two massives and consisting likewise of gneiss and mica-schist, is, according to Pošepný, a syncline enclosing younger rocks. More recent investigations by Geyer and Vacek of the Geological Survey are however available. According to the latter, the schistose mantle consisting of members of different age overlays the central gneiss. This gneiss in its upper portion merges into hornblende-gneiss, upon which lie sericitic schist and quartzïte, these in turn being overlaid by the upper hornblende-gneiss.

In this area of crystalline schist, lodes and bedded deposits are found. Cotta and Pošepný considered the lode-like deposits to be true lodes. They are filled with quartz and other gangue and with fragments of country-rock, while in addition to free gold they contain auriferous and argentiferous sulphides. According to Cotta the Tauern lodes are equivalent to the silver-quartz lodes of Freiberg. Pošepný considered them as belonging to the sulphide lead-zinc group of Breithaupt, though at the same time, in their quartz and stibnite content, resembling the silver-quartz lodes. Finally, v. Groddeck regarded the Tauern lodes as belonging to his very comprehensive Australian-Californian type.

Cotta particularly remarked that these lodes, like the Freiberg silver lodes, become poorer or even barren when leaving the gneiss and entering the mica-schist. In the Sieglitz, the gold content in one case increased considerably as the mica-schist was approached, though in that rock itself it ceased. In the Rauris also, enrichment in the neighbourhood of rocks not suited to the formation of fissures has been noticed, and Reissacher mentions that on the Goldberg accumulations of precious metal occur in gneiss near a black slate, while in the slate itself the lode pinches to a barren fissure. Alberti on that hill counted twenty-six lodes striking north-east and in general dipping south-east. For the principal complex of mines at this place Pošepný formulated the following fissure-systems : the Herrnstolln or Fröberling system, the Habersberg fissure, the Haberland and Goldberg system, and the Kirchgang and Bodner fissures. Within the indistinctly bedded gneiss certain schistose beds occur, which, striking towards the position of the sun at nine o'clock, are known as ' nines.' These have an influence upon the gold lodes, in that these latter only maintain a regular strike in the gneiss between these beds, but not within the beds themselves, where their course is indefinite.

The change in the lode-filling on the north side of the Hohe Tauern is interesting. The Sieglitz-Pockhart-Erwies lode-series, which may be followed for about 6400 m., carries, so long as it is in gneiss, decomposed country-rock with quartz, dolomite, arsenopyrite, pyrite, chalcopyrite, galena, argentite, and some light coloured gold. In the limestone overlying the gneiss, on the other hand, the width of the lode, otherwise but small, increases to 20–60 m., and the lode then contains ankerite, siderite, galena, chalcopyrite, sphalerite, and zinc oxidized ore. Gold however is absent, and only occurs again when the gneiss is re-entered. On the Silberpfennig the lodes as soon as they penetrate calc-mica-schist carry pyrite, siderite, and galena, but no longer any gold.

On the south slope of the Hohe Tauern, pyrite deposits occur in the schists above the gneiss, those of Grossfragant at Waschgang and those in the Gössnitz being the best known.

The gold content of the Hohe Tauern lodes is sometimes considerable, and in places as much as 500 grm. or more per ton. A separation of the data available into such as were derived from the secondary and primary zones respectively, can unfortunately no longer be made. On the Rathausberg, in the first half of the seventeenth century an average of 31·7 grm. of gold was maintained, and in the second half 36 grm. ; in the first half of the eighteenth century 22·1 grm., and in the second half 20·5 grm.; in the first half of the nineteenth century 12·6 grm.; or in general 22·7 grm. of gold bullion per ton. Of the Rauris the following figures of content are given : in the second half of the seventeenth century 46·6 grm. ; in the first half of the eighteenth century 33·2 and 26·9 grm. ; in the second half 20·0 grm. and 16·0 grm. ; and in the first half of the nineteenth century 37 grm. and 30·0 grammes. According to Pošepný a general average of 25·5 grm. may be assumed.

New development work is now proceeding upon the Sieglitz lode near Böckstein. In October 1910 Krusch was able to investigate the deposit in the deep winze from the Georg adit. There the well-defined simple lode in the crystalline schists strikes north-north-east and dips comparatively steeply east-south-east. In places, as the result of repeated re-opening of the fissure, false walls are found. From the later fissures an intense impregnation proceeded, which not only involved the old lode material but also the formerly barren country-rock, and this latter to such an extent that much of it is now payable. The width of the lode varies between 70 cm. and 2 metres. The filling consists of quartz and of rock fragments which in places are greatly silicified and almost invariably impregnated with ore. Of the sulphides, arsenopyrite and pyrite are particularly characteristic. While the latter occurs chiefly impregnated and rarely in good crystals, the arsenopyrite is not only found in the impregnated material but also in a continuous streak, which after hugging the hanging-wall, jumps the central portion of the lode to continue along the foot-wall. This streak is a younger formation in a pre-existing quartz-pyrite lode either free from, or containing but very little arsenopyrite. Whether the present gold content belongs partly to the original lode-filling, or is exclusively associated with the younger arsenical filling, cannot be definitely determined. It is certain however that by far the greater part of the gold is associated with the arsenic, and referable therefore to the younger filling. The highest result obtained from the samples taken by Krusch was 276 grm. per ton, though the average is reckoned to be about 45·4 grammes. The silver varies between 15 and 211 grm., being 80 grm. on an average.

The Occurrence of Gold at Schellgaden in the Lungau Tauern

LITERATURE

F. Neugebauer. Lecture, Vienna Mineralogical Society, May 2, 1904. 'Austrian Geological Survey,' Section St. Michael, G. Meyer, K. k. geol. Reichsanst., 1891, 1892, 1893.—F. Beyschlag. 'Der Goldbergbau Schellgaden in den Lungauer Tauern,' Zeit. f. prakt. Geol., 1897, p. 210.

The Lungau is the seat of an ancient and flourishing mining industry, concerning which however reliable data are only available from the fourteenth century. The basement rock of the whole Lungau consists of three occurrences of gneiss; to the powerful gneissic granite of the Ankogel massive belong the mountain masses of the Reisegge group, those of the Hafen Eck, and those of the Hochalpenspitze, all of which strike into this district. According to Geyer these are altered eruptives of obvious eruptive character in the foot-wall but of schistose structure in the hanging-wall. The second gneissic massive overlies peripherally that just mentioned, and according to Vacek and Geyer consists of normal hornblende-gneiss and schist; the Kareg range and the Schlattming mass belong to this occurrence. The third occurrence consists of blocky two-mica gneiss. Over these three gneissic cores the schists of the Lungau Tauern spread as a covering; they consist of garnet-mica schists and of calc-phyllites, these two formations being separated from each other by a plane of considerable disturbance.

For the gold deposits the hornblende-gneiss, which is an alternation of hornblende-gneiss proper, hornblende-free schistose gneiss, and green schist, appears to be the most important.

The Gannthal gold deposit occurs in green mica-schist with fibrous structure. In its upper horizons this mica-schist carries large quartz lenses often containing small amounts of sulphides, which, though occasionally worthless, often constitute, and especially where they are plentiful, rich pyrite deposits. These consist chiefly of pyrite but also contain chalco-pyrite, bornite, some sphalerite, and arsenopyrite. In the quartz fine mica films are intercalated. The pyrite deposits have generally a thickness of 0·25 to 2·00 m., though in the Barbara district 8 m. has been reached.

The strike of the Schellgaden deposits is usually roughly north-south. The gold content of the quartz led in early times to prospecting work. In those days, on account of the pronounced bedding generally seen, the occurrences were regarded as beds conformable in strike and dip with the formation. The bedded nature was however doubted by later authorities—Milichofer and Rosegger—who, assuming the presence of a dislocation zone and remarking the splitting of the deposit into veins, considered the deposits to be lodes. Beyschlag however in 1897 pointed out that the occurrences could not aptly be described as lodes. According to him the ore-bearing lenses are not irregularly distributed in the country but occur in a north-south line which, taken alone or together with other neighbouring parallel occurrences, has the character of a zone of disturbance, with which disturbance the introduction of the ore is probably connected. The so-called beds are therefore narrow zones following the principal fissure and containing quartz lenses in good number and metalliferous. These zones are crossed and dislocated by barren east-west fissures. While therefore Beyschlag inclines to the bedded nature of the quartz lenses and assumes a subsequent intro-duction of the ore, Neugebauer as the result of his investigation comes to the conclusion that the quartz, gold, and sulphides, were simultaneously deposited from one and the same solution and are younger than the country-rock, and that consequently the deposits are entirely epigenetic and represent the filling of cavities which arose as the Alps were folded into position.

The most important deposits are found at Stüblbau in the Gannthal, at Maradlwand, and at Zaneischg. The extension of the occurrence has been proved for more than 2 km., along which length however the continuity of the ore is broken by unpayable and barren patches.

The Gannthal deposits are therefore true bedded lodes which were probably formed by the same happenings as those to which the gold occurrences in the Hohe Tauern owe their existence. The filling consists of quartz, such quartz being of indistinct crystalline habit and exhibiting those evidences of pressure or stress as are characteristic of all quartz deposits occurring in the older geological formations. The presence of scheelite is interesting, this mineral being typical of tin lodes. The sulphides, among which pyrite is most noticeable, rarely form solid masses but are distributed fairly equally throughout. Where however such masses are more plentiful they usually occur parallel with the mica films which divide the quartz into small irregular lenses. The minerals present form an assembly having much in common with the mineral-association of the Hohe Tauern. The gold occurs in small particles in the quartz itself, though it also occurs as an accessory constituent with the sulphides. The gold content of these sulphides, from the results available, varies between 5 grm. and 69 grm. per ton, in addition to which silver is present to the extent of 10–40 grm. per ton. The concentrate recovered in treatment contains 500–600 grm. of gold and 200–300 grm. of silver per ton.

The Gold Occurrence at Roudny in Bohemia

LITERATURE

F. POŠEPNÝ. 'Goldvorkommen in Böhmen,' Arch. f. prakt. Geol., II., 1895.—P. KRUSCH. ' Über die Goldlagerstätten von Roudny in Böhmen,' Zeit. d. d. geol. Ges., 1902.—O. EYPERT, ' Der Golderzbergbau von Roudny in Böhmen,' Österr. Zeit. f. d. Berg- u. Hüttenwesen, 1905.—R. BECK. Lehre von den Erzlagerstätten, 1909.

The gold occurrence at Roudny, about 60 km. south-south-west of Prague and 15 km. east of Woditz in the department of Borkowitz, has of late years become more widely known. According to Beck, the rock which forms there a flat, partly wooded ridge between the Liboun valley and another coming from Ramena, consists chiefly of biotite-gneiss traversed by numerous bosses and dykes of a tourmaline-granite, and much metamorphosed. This granite has often a pegmatitic and sometimes an aplitic character. According to Krusch, on the other hand, the biotite-gneiss is a pressure-deformed granite which has assumed a fibrous, gneiss-like structure; in many places it merges into granite. It contains in addition large and small amphibolite inclusions, which are resorbed along their outlines. The gneiss, gneiss-granite, and amphibolite, are all traversed by aplite.

In this rock complex a number of fissures occur, which generally strike east-west, dip some 60°–70° to the north, and are arranged in systems.

These fissures have generally a width of but a few millimetres, seldom centimetres, and are filled with quartz and pyrite. On either hand an alteration of the granite has taken place, this being chiefly in the form of an impregnation with quartz and pyrite. Though these fissures are in general parallel along the strike they often intersect in dip. In addition, numerous intersecting veins similarly filled proceed from the fissures and traverse

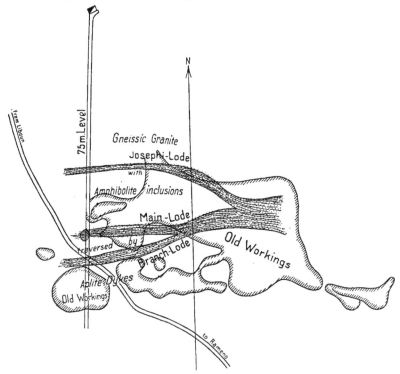

FIG. 326.—The gold lodes at Roudny. Krusch, *Zeit. d. d. geol. Ges.*, 1902.

the granite, the felspar of which is sometimes kaolinized and sometimes, as is also the case with the biotite, replaced by quartz and pyrite.

By this far-reaching silicification and pyritization the granite zones in which the fissures occur have become altered to pyrite- and quartz-impregnated zones, such zones being mined as single occurrences ; they have as a rule no sharp boundaries against the normal gneiss-granite. Three such zones are known, these from north to south being the Josephi lode, the Main lode, and the Branch lode. These dip steeply to the north and vary considerably in width, this dimension sometimes reaching as much as 20 metres. Continued towards the east, as illustrated in Fig. 326, they unite to form one deposit.

The gold content of these veins and impregnation zones is chiefly associated with the pyrite ; gold however occurs free and finely divided in the quartz ; while finally, it is found as flakes and indefinite crystals upon the cleavage-planes of quartz and pyrite. As usual with most gold deposits, here also the content varies greatly, from a few grammes to more than 100 grm. per ton. In general it is found that receding from the fissures the gold regularly decreases ; and that the coarse pyrite crystals contain less gold than the aggregates of fine crystals. It is further found that where the impregnation zones come together the gold content is more than usually high. The included patches of amphibolite contain practically no gold though they are not free from pyrite. The aplite dykes dislocate the gold lodes. The auriferous zones are crossed by north-south fissures which are either barren or carry but little ore. The numerous old workings around Roudny would indicate a large number of lodes.

The intimate association of the quartz, pyrite, and gold, suggests their simultaneous formation. Since, however, irregular intergrowth of pyrite and quartz is more often observed on the walls leaving the centre occupied by quartz alone, it would appear that the deposition of the quartz must have continued longer than that of the pyrite. The age of the impregnation is probably great ; in any case impregnation had already taken place before the aplite dykes were intruded and before the system of north-south fissures was formed.

The Gold Deposits at Hussdorf-Wünschendorf in Silesia

In the Palæozoic slates between Lähn and Greiffenberg, as indicated by old workings upon the surface, many lodes striking roughly north-south, occur. At Hussdorf two such lodes, distinguished respectively as I. and II., dip to the east and are connected by a diagonal lode striking east-west. At Wünschendorf, but one has been explored, this striking north-north-east and dipping east-south-east. At the In-der-Pinge [1] mine a lode, following the contact between Silurian limestone and slate, strikes north-west and dips north-east.

These lodes are fissure-fillings with definite walls. In width they vary from a few centimetres to more than a metre. The lode material consists of quartz, auriferous arsenopyrite, and pyrite. Although when these sulphides are well developed they occupy almost the entire width, their occurrence is more usually limited to a sprinkling in the quartzose material. Fragments of country-rock within the lode are seldom found,

[1] In the old workings.

nor is silicification or impregnation of the walls often observed. The lodes are affected by a large number of disturbances both along the strike and in dip. Along the strike the lode is cut into pieces often not much more than a metre in length, such pieces being generally but little displaced. Where these faults cut the lode at an oblique angle any lateral displacement of the lode takes the form of an overlap. Along the dip, vertical displacements have taken place along clay-partings, dipping flatly in various directions.

These lodes are characterized by well-defined secondary migration of the metal content. The oxidation zone has been eroded to such an extent that the cementation zone almost comes to the surface. It consists of the residual quartzose gangue with coatings and skins of limonite and auriferous pyrite and arsenopyrite. In the primary zone the lode width quickly diminishes and at no great depth only unimportant fissures remain. The gold content is not exclusively connected with the arsenopyrite but also with the pyrite. The ore in the cementation zone sometimes contains as much as 40 grm. per ton, whereas in the primary zone the content is limited to a few grammes, nor have any ore-shoots been established. Mining operations on this occurrence are but on a small scale and work only proceeds occasionally.

NORWAY, SWEDEN, AND FINLAND

Quartz lodes with traces of gold have been established in many places in these countries. Gold lodes proper, however, with a higher gold content have seldom been proved, while rich lodes are unknown.

The lodes at Ädelfors in Småland and at Eidsvold in the Christiania district occur in the fundamental schists. At both places, and especially in the eighteenth century, these lodes were worked, though at great loss. The position of Eidsvold is indicated in Fig. 147. Similar deposits are found in northern Finland.

On the island of Bömmelö, situated south of Bergen off the west coast of Norway, some gold-quartz lodes traverse what are chiefly intensely altered basic rocks belonging to the Cambrian-Silurian formation.[1] These lodes, which occur in the neighbourhood of magmatic-intrusive pyrite deposits, have since 1882 been worked sporadically, producing only about 100 kg. of gold.[2]

At Svartdal in Telemarken, some quartz-tourmaline lodes in quartz-diorite occur carrying native gold with much bismuthinite and some copper ore, etc. Further details of these will be found under the description of

---

[1] Ante, p. 305.
[2] H. Reusch, Bömmelöen og Karmöen, 1888.

the copper ores of Telemarken given later. Reference may also be made to a previous mention of the gold lodes at Fahlun, where the gold is accompanied by bismuth.[1]

THE METASOMATIC GOLD-DEPOSITS

Gold deposits of metasomatic origin are but seldom encountered, though isolated occurrences of such are found both with the young gold-silver- and with the old gold groups. Since all over the world gold ore consists chiefly of quartz, metasomatic processes resulting in the formation of gold deposits must of necessity be identical with the processes of silicification, or impregnation with quartz. Among the occurrences here to be described only that at Lydenburg, where limestone or dolomite has been altered to quartz, presents the true type of metasomatic formation. All the other deposits represent replacements of country rock consisting of sandstone, slate, etc., by silica, and may accordingly be regarded as extreme cases of quartz impregnation; they are therefore closely related to lodes.

LYDENBURG IN THE TRANSVAAL

LITERATURE

J. KUNTZ. ' Über die Goldvorkommen im Lydenburger Distrikt,' Zeit. f. prakt. Geol., 1896, p. 433.—P. R. KRAUSE. ' Über den Einfluss der Eruptivgesteine auf die Erzführung der Witwatersrand-Konglomerate und der im dolomitischen Kalkgebirge von Lydenburg auftretenden Quartzflöze, nebst einer kurzen Schilderung der Grubenbezirke von Pilgrims-rest und De Kaap,' Zeit. f. prakt. Geol., 1897.—F. W. VOIT. ' Übersicht über die nutzbaren Lagerstätten Südafrikas,' Zeit. f. prakt. Geol., 1908.

The dolomite beds of Lydenburg, lying upon clay-slates, rise near Pilgrim's Rest up to as much as 1000 m. above the level of the valley; intercalated with them, even to the top of the range, are numerous sheets of trap.

The primary gold deposits in this district were discovered during the exploitation of gold gravels. Krause describes them as quartz beds, and points out that some of them, especially in the neighbourhood of the trap where that rock is decomposed, contain gold up to 600 grm. per ton. Such a content would probably be due to cementation, since with all the occurrences at Lydenburg, so far as at present known, the secondary processes of decomposition have played an important part. The shallower portions of the quartz beds at Jubilee Hill and the New Clewer Estate have apparently been leached by hot springs, such leaching having given to the quartz a spongy, porous, and pumiceous character, so

---

[1] *Ante,* pp. 314, 315.

that it easily crumbles and covers the floor of the working. The beds are
only in places horizontal; more usually they form flat anticlines and
synclines. In greater detail the circumstances of their bedding are illus-

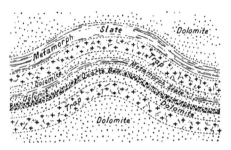

FIG. 327.—Section showing the geological position of the auriferous quartz bed and the trap sheets
in the Malmani dolomite, Lydenburg.   Krause, *Zeit. f. prakt. Geol.*, 1897, Fig. 8.

trated in Figs. 327 and 328. In only one case, that illustrated in the
above-mentioned Fig. 328, did Krause observe them to be intruded by
an eruptive. The dyke in that case dragged the auriferous material right
to surface with it, indicating that it was younger than the deposit. It

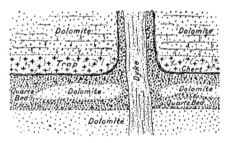

FIG. 328.—Section in the New Clewer Estate mine, Lydenburg.   A diabase dyke crosses the
quartz bed ; both the quartz and trap extend along the dyke.   Krause, *Zeit. f. prakt. Geol.*,
1897, Fig. 9.

is characteristic of these quartz beds that they do not always keep to the
same horizon. The payable thickness is 0·15–0·70 m., or on an average
0·40 m., with an average gold content of 30 grm. per ton. Some copper
accompanies the gold.

Voit considers rightly that these are metasomatic occurrences and
represent certain limestone bands which have become altered to quartz.
They present therefore in all probability the rare case of a typically
metasomatic gold deposit formed from limestone.

The occurrence at the Csetatye near Verespatak, mentioned when describing the young gold-silver lodes, also belongs to the metasomatic occurrences.[1] In view of the former description, only those criteria will here be mentioned which are indicative of the metasomatic character of this deposit. The rock forming the Csetatye hill has been differently described by different authors, a remarkable fact explicable in that, apart from the complicated conditions of bedding, the dacite as well as the accompanying tuffs and sediments have suffered profound alteration. As stated when describing the lodes of the Verespatak district, in addition to lodes, ore-chimneys appear, which column-like and of small section sink into depth. According to Pošepný the intersection of a large number of fractures to form an inextricable maze represents the simplest form of such chimneys. At such intersections the rock is particularly decomposed and impregnated with silica and pyrite. These chimneys are usually connected with eruptive breccias of which the cementing material is porous, pumiceous rhyolite. In this district two deposits particularly were famous for their richness in gold, the Katroncza ore-chimney and the Csetatye. It is the latter only which is of interest here. A vast excavation near the summit of the Boj hill, dating from the time of the Romans and presumed to have been made largely by fire-setting, owes its existence to the presence of an important deposit which occurs partly in the eruptive rock but to a greater extent either in the adjacent *Lokalsediment*, or at the contact of the one rock with the other. Tremendous blocks of Carpathian sandstone were borne upward in the dacite magma. All the rocks exhibit pronounced silicification so that their original petrographical characters are to a great extent obliterated. Breccias of older rhyolite and dacite cemented by younger porous rhyolite traverse the *Lokalsediment* as well as the completely silicified dacite. Originally there occurred here a confused network of irregular fissures ramifying through dacite, *Lokalsediment* and Carpathian sandstone, replacing these rocks to a great extent by quartz and pyrite. Pošepný proposed for such deposits the name of ' ore-typhoons.'[2] The ore-minerals found in this deposit have already been briefly stated.[3]

MOUNT MORGAN IN QUEENSLAND

LITERATURE

J. MACDONALD CAMERON. Managers' Report, Mount Morgan Company, March 26, 1887.—R. L. JACK. Rep. Geol. Survey, Queensland, 1884 ; and Mount Morgan Gold-Deposits, 1892.—T. A. RICKARD. ' The Mount Morgan Mine,' Trans. Amer. Inst. Min. Eng., 1891, Vol. XX.—K. SCHMEISSER. Die Goldfelder Australasiens, Berlin, 1897. Dietrich Reimer.—S. F. EMMONS. ' Structural Relations of Ore Deposits,' Trans. Amer. Inst. Min. Eng. XVI. p. 804.

---

[1] *Ante*, p. 546. [2] *Erztyphone*. [3] *Ante*, p. 546.

This deposit occurring in central Queensland south-west of Rock-hampton, is one of the most interesting and important in the world. Its prominent place in literature is in no small measure owing to the circum-stance that it was formerly regarded as a geyser deposit, an origin which would make it unique among economic gold deposits. This mine works as a quarry the summit of a hill 500 feet above the surrounding level and 1225 feet above the sea, in a district forming part of the fore-ground to that important mountain chain which, known in New South Wales as the Blue Mountains and in Victoria as the Australian Alps, traverses the three colonies parallel to the east and south-east coasts of the Australian continent. The tectonics of the occurrence are simple. The upper

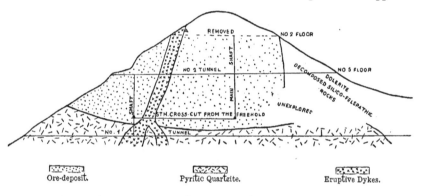

Ore-deposit.   Pyritic Quartzite.   Eruptive Dykes.

Fig. 329.—Section through the Mount Morgan deposit. Rickard, *Trans. Amer. Inst. Min. Eng.*, 1891, Vol. XX. p. 143.

portions of the hill consist of Desert Sandstone presumably belonging to the Cretaceous. Below this comes a sequence of grauwacke, quartzite, slate, and occasionally serpentine. Eruptive dykes belonging to at least two geological epochs traverse these sediments, those of dolerite, in part interbedded, being particularly noteworthy. The ore-deposit constitutes the summit of the hill. Broadly speaking it forms a flat cone, the basin-shaped base of which rests upon pyritiferous quartzite. Both this quartzite and the ore-deposit, as illustrated in Fig. 329, are traversed by dykes.

The ore is very friable so that its comminution presents no difficulty. In petrographical character it varies greatly. In the opencut workings a bluish-grey crushed quartz, greatly resembling ore from the Comstock Lode, is often observed. In other places silicified haematite is found, this being generally regarded as the characteristic type of Mount Morgan ore. These ferruginous masses resemble the outcrops of ordinary gold-quartz lodes as such often appear in California and Victoria. In addition, other dark heavy ironstone, which might very well be taken for the gossan of a sulphide lode, is also met, as well as light-coloured reddish ore such as might arise in the oxidation zone of a copper deposit. One portion of the material has been crushed to a sugar-like, lightly compacted powder, while another forms solid masses. Stalactitic structures, in part filling cavities, are also not infrequent. Some bluish-black iridescent ore was very rich. In the Freehold adit a nest of white, porous, friable rock, considered to be sinter, was encountered which, doubtless because of the numerous air-inclusions, was so light that it floated on water. Generally the ore is dull, though formerly it yielded beautiful specimens of gold quartz some of which may now be seen in the Sydney Museum ; in general these consisted largely of limonite and free gold. As a rule however the gold is finely distributed and difficult to recognize in its environment of iron oxide.

However distinct these different varieties of ore may be, all consist essentially of quartz. The deposit may therefore be regarded as a quartz mass, variable in colour and specific gravity, and traversed by north-west striking dykes. Below a depth of about 90 m. an auriferous and pyritiferous quartzite sets in ; while deeper still the presence of an auriferous copper ore containing 3·5 per cent of copper and 12 grm. of gold was proved by boring. The extension of the occurrence has now been fairly definitely determined ; to the south-west it is bounded by a large felspar dyke, and its greatest extension therefore apparently lies to the north and north-east.

The genesis of this deposit is a much disputed question. R. L. Jack regarded it as a geyser deposit ; J. Macdonald Cameron considered it an auriferous zone traversed by a network of gold-quartz veins ; while some mining engineers expressed the opinion that it was the gossan of a large pyritic lode. Rickard, putting aside all these three theories, considered it in greater part to be a metasomatic occurrence. To him it represented a highly altered part of a shattered country which had been saturated with mineral solutions and in part replaced by auriferous quartz. If this theory be accepted this deposit is another of the rare cases of metasomatic gold deposits. We consider there is much in favour of this view. It explains quite naturally the variable character of the ore by reason of the variable character of the original rock, as well as the great width of the deposit. The grauwacke appears to be the rock which has suffered replacement to the greatest extent, this rock generally overlying the quartzite which forms the base of Mount Morgan. That in any case it is a deposit closely allied to the old gold group is evidenced by the purity of its gold, comparative purity being characteristic of the lodes of that group.

The economic importance of this occurrence may be gathered from the following figures of production for the year 1907 : [1]

District.	Gravel Gold.	Lode Gold.	Total.
	Oz.	Oz.	Oz.
Charters Towers . . .	846	174,706	175,552
Gympie	466	63,716	64,227
Mount Morgan	354	145,420	145,774
Ravenswood	364	34,466	34,830
Croydon	...	13,411	13,411
Clermont	4,364	421	4,785
Etheridge and Woolgar . .	812	7,290	8,102
Other districts	2,127	17,074	19,201

Concerning the gold content of Mount Morgan ore, some idea may be formed from the following data. In 1889 this mine produced 323,542 oz. of gold, having a value of £1,331,484 from 75,415 tons of ore, or 4 oz. 6 dwt. per ton. From this high figure the content slowly but steadily diminished in depth. While formerly 100 grm. per ton was nothing unusual, in 1903 the average recovered was but 15 grammes. In 1906 the depth of the mine was about 250 metres.

UNITED STATES

Deposits very similar to that of Mount Morgan are found at Red Mountain Basin, Colorado, where portions of an andesitic rock occur, the basic constituents of which have been extracted by silicated waters and replaced by quartz. The quartzose masses thus formed, presenting as they did considerable resistance to atmospheric agencies and erosion, remain to-day outstanding as hills. Although these occurrences have no economic significance they are particularly interesting, inasmuch as the geyser theory was also applied in explanation of their genesis.

As pointed out by Rickard in his paper on Mount Morgan, the Yankee Girl and Bassick mines in Colorado to a certain extent belong to this class of occurrence. The first-named lies in the San Juan district and is well known. In that occurrence several crush zones enclose between them a triangular mass which has been so decomposed and replaced by circulating solutions that practically a solid mass of ore has resulted.

[1] Official Year Book of the Commonwealth of Australia, No. 2, 1909.

REVIEW OF THE GOLD-SILVER PRODUCTION, AND ITS DISTRIBUTION
AMONG THE DIFFERENT CLASSES OF DEPOSIT

LITERATURE

A. SOETBEER. 'Edelmetallproduktion und Wertverhältnis zwischen Gold und Silber,'
Petermanns Mitteilung, Supplementary Part 57, Supplementary Vol. XIII., 1879.—
B. NEUMANN. Die Metalle u.s.w. nebst Produktions- und Preisstatistik, Halle, 1904.
The first-mentioned work gives figures up to 1875 ; the second from 1876–1900. Further,
'The Mineral Industry,' 'Mineral Resources of the United States,' Zeit f. prakt. Geol.,
Fortschr. d. prakt. Geol., etc. (ante, pp. 207, 208). For olden times and for the semi-
civilized countries the figures are only approximate.

WORLD'S TOTAL GOLD PRODUCTION

(In Tons)

.	Before 1875, in greater part since 1493.	1876 to 1900.	1901 to 1909.	1910.	Total to 1900.
Austria-Hungary, since 1493 .	460·7	53·6	31·6	3·4	549·3
Sweden	...	1·5	0·4	...	...
Germany	...	...	0·8	...	...
Russian Empire, since 1741 . .	1033 7	965·9	335·9	60·4	2395·9
Canada	...	...	203·1	15·4	218·5
United States, since 1849 . .	2026·1	1582·0	1173·7	144·5	4926·3
Mexico, since 1521 . . .	265·0	99·2	213·6	36·2	614·0
Central America	...	...	24·9	4·1	...
Guiana	...	...	58·5	7·6	...
Colombia	...	...	37·1	4·7	...
Venezuela	...	...	2·4	...	...
New Granada, since 1537 . .	1214·5	116·9			1331·4
Peru, since 1533	163·6	6·5	11·8	0·9	182·8
Bolivia, since 1545 . . .	294·0	6·8 }	10·3	... }	694·9
Chili, since 1545	263·6	118·9 }		1·3 }	
Brazil, since 1691	1037·1	43·2	30·3	3·0	1113·6
British India	...	...	148·7	18·2	...
British Farther India . . .	...	...	17·9	2·2	...
Dutch East India	...	...	18·7	3·3	...
China	...	...	92·1	15·2	...
Korea	...	...	36·0	3·0	...
Japan	...	...	34·7	6·7	...
Africa, since 1493 . . .	731·6	706·8	1413·7	263·1	2115·2
Australia, since 1851 . . .	1812·0	1434·0	1089·1	98·3	4433·4
Other countries . . .	151·6	353·1	15·3	10·5	...
Totals	9453·3	5581·6	5001·1	701·0	20737·1
Value in millions sterling . .	1285	765	687	101	2838

REMARKS ON THE ABOVE TABLE

From 1876–1900 Germany produced altogether 41 tons of gold, almost exclusively
from imported ores. In the figures for Chili during the period 1876–1900, the production
of Guiana and Venezuela from 1876–1895 is also included. Since 1884 the preponderating
part of the African production has come from the Transvaal.

The relation between the production of gold and that of silver in different periods has been as follows :

From 1500–1875 . . . 1 of gold to 19·1 of silver.
„   1876–1900 . . . 1   „   17·6   „
„   1901–1908 . . . 1   „   10·0   „

The relation between the value of equal weights of gold and silver at different periods has been :

1493–1600	.	.	.	1 : 11·80
1601–1700	.	.	.	1 : 14·10
1701–1800	.	.	.	1 : 14·98
1801–1850	.	.	.	1 : 15·68
1851–1875	.	.	.	1 : 15·52
1876–1880	.	.	.	1 : 17·88
1881–1885	.	.	.	1 : 18·64
1886–1890	.	.	.	1 : 21·16
1891–1895	.	.	.	1 : 27·05
1896–1900	.	.	.	1 : 33·29
1901	.	.	.	1 : 34·68
1905	.	.	.	1 : 33·87
1908	.	.	.	1 : 38·67

The average yearly production of gold in different periods has been :

1493–1600	.	.	.	7,100 kg.
1601–1700	.	.	.	9,100
1701–1800	.	.	.	19,000
1801–1820	.	.	.	14,800
1821–1840	.	.	.	17,750
1841–1850	.	.	.	54,760
1851–1860	.	.	.	201,790
1861–1870	.	.	.	188,160
1871–1880	.	.	.	169,880
1881–1890	.	.	.	154,450
1891–1900	.	.	.	319,170
1901	.	.	.	392,705
1903	.	.	.	493,083
1905	.	.	.	568,232
1908	.	.	.	667,071
1910	.	.	.	701,019

With the discovery of gravel gold in California in 1848, and in Australia in 1849, the world's yearly production suddenly increased to ten times its previous amount, that is, from 20,000 to 200,000 kg. These gravels becoming in greatest part exhausted before many decades had passed, a drop in the production followed, this continuing till in the 'eighties a minimum was reached. Then, as in the United States and Australia gold-quartz lodes in number became worked, the production rose again. Such lodes in North America became developed more particularly after the 'nineties. Production from the auriferous conglomerates of the Transvaal began in the middle of the 'eighties after which it quickly reached gigantic proportions ; while the world's production was further increased in the middle of the 'nineties by the output from the important telluride lodes of Western Australia, these two latter increments being responsible for the pronounced upward swing of late years.

The three most important classes of gold deposit are, the ore-beds as represented by the Witwatersrand conglomerate, the young gold lodes, and the old gold lodes, these last being accompanied by important gravel-deposits.

The Witwatersrand conglomerate has latterly produced about 35 per cent of the world's total production.

From the young gold lodes, of late years about 45 per cent of the production of the United States has been derived and approximately 80 per cent of that of Mexico. Lodes belonging to this group have also yielded by far the greater portion of the gold won in Central America, Colombia, Peru, Bolivia, and Chili; a considerable portion of the output of Japan; and in addition the entire production of Hauraki in New Zealand, and of Hungary. The total production of these young gold lodes reaches roughly somewhat more than 100,000 kg. per year; in 1910 it was about 110,000 kg. If to this amount be added that of the telluride lodes of Western Australia then one-quarter of the world's total production is derived from this group.

Still more important are the old gold lodes with their appurtenant placers. These yield approximately one-half of the present production of the United States, and almost the entire production of Canada, the Russian Empire, Queensland, New South Wales, Victoria, South Australia, and many other countries. It may be taken therefore that approximately one-third of the world's production is obtained from the old lodes and the metasomatic occurrences associated with them. Since it is from this group that the extensive gravel-deposits were derived, the proportion of the world's production referable to it was formerly much greater; at the beginning of the 'fifties, for instance, it was responsible for about nine-tenths of the world's production.

A small amount of gold comes from contact-deposits, these being particularly represented in Mexico; while a still smaller amount is obtained from copper- and other deposits.

Altogether therefore, the distribution of the world's present production among the different classes of deposit with their associated gravel-deposits is approximately as follows :

Witwatersrand conglomerate	about 35 per cent.
Old gold lodes	„ 33 „
Young gold lodes	„ 25 „
Contact-deposits	„ 1–5 „
Other deposits	small amounts.

[TABLE

WORLD'S TOTAL SILVER PRODUCTION FROM MINES
(In Tons)

	Before 1875, in greater part since 1493.	1876 to 1900.	1901 to 1909.	Total to 1909.
Germany, since 1493 . . .	7,905	4,568	1,549	14,022
Austria-Hungary, since 1493 . . .	7,770	1,247	488	9,505
Russian Empire, since 1741 . . .	2,429	246	46	2,721
Spain and Portugal	...	1,755	1,150	...
Greece	...	...	259	...
Italy	...	393	220	...
France	...	...	180	...
Norway, since 1642	799	149	59	1,007
Great Britain	...	...	52	...
Sweden, since 1506	252	64	10	...
Turkey	...	...	152	...
Remaining Europe	6,331	2,625	...	...
Chili, since 1721	2,609	3,513 ⎫		
Bolivia, since 1545	37,718	7,927 ⎪	4,478	89,612
Peru, since 1533 . . .	31,222	2,145 ⎭		
Colombia . . .	...	...		
Mexico, since 1521	76,205	28,281	17,112	122,148
United States, since 1851 . . .	5,272	36,181	15,522	56,985
Canada	...	...	3,056	...
Japan	...	980	757	...
East Indies	...	...	71	...
Australia	...	5,662	3,792	9,454
Africa	...	...	115	...
Other countries	2,000	2,646	...	...
Totals . . .	180,511	98,325	49,191	328,532
Value in millions sterling . .	1,595	635	186	2,466

[TABLE

AVERAGE YEARLY SILVER PRODUCTION AT DIFFERENT PERIODS
(In Tons)

	1851 to 1855.	1876 to 1880.	1891 to 1895.	1900.	1905.	1909.
Germany . . .	49	164	176	168	181	166
Austria-Hungary . .	35	48	57	62	58	31
Russia	17	9	12	5	6	4
Spain and Portugal .	...	...	58	99	124	148
Greece	...	...	...	31	26	26
Italy	...	...	24	23	24	25
France	...	...	87	14	9	18
Norway . . .	...	...	5	5	8	7
Great Britain . .	...	...	8	7	5	4
Sweden	...	...	3	2	1	1
Turkey	...	...	...	4	17	24
Remaining Europe .	72	110	113	...	...	0·3
Argentina . . .	...	...	25	12	2	4
Chili	68	110	84	178	126	44
Bolivia	73	252	470	325	205	213
Peru	77	58	85	204	156	195
Ecuador . . .	...	...	...	...	...	2
Colombia . . .	...	...	44	87	31	41
Mexico	466	663	1448	1787	1700	2300
United States . .	8	1176	1693	1793	1745	1702
Canada	...	...	...	138	330	867
Japan	...	16	53	54	75	133
East Indies . . .	...	...	...	3	6	15
Australia . . .	...	...	506	415	391	509
Africa	...	...	...	...	19	34
Other countries . .	4	14	65	...	...	...
Totals . . .	870	2620	4958	5400	5050	6500
Price in shillings per kilogramme . .	182 180	176 172	107 105	83·61 81	82·34 80	70·27 69

These tables refer to the production from mines and not to that from smelting works, these latter being always partly employed in treating imported ores. Exceptionally, the figure for France during the period 1891–1895 probably includes the silver from some imported ore.

To the Tertiary silver lodes belong : the preponderating number of the silver lodes in Mexico ; those in Central America, Ecuador, Peru, and Bolivia ; the greater number of the silver lodes of Chili ; a large number of the lodes of the United States, these approximately providing one-half of the present production of that country ; almost all the lodes of Japan and Hungary ; several of the most important deposits of Spain ; and some deposits in other countries. The total silver production up to and including 1909 may be said to be 330,000 tons valued approximately at 2500 million sterling. Of this total somewhat more than one-third was derived from Mexico, and somewhat less than one-third from Central and South America. If to the preponderating portion from those countries be added one-half of the total production of the United States, together with that of Hungary, the largest part of that of Japan, and a portion of that of Spain, it follows that approximately two-thirds of the total silver production hitherto, has been derived from Tertiary lodes. Considering only recent years, then a proportion of something over one-half, say 55 per cent, is obtained. With respect to silver this geological group is consequently the most important group. Owing to the extremely rich accumulations of ore found with many of these lodes [1] and the consequently low cost of production, these Tertiary lodes have largely determined the price of silver. They were in fact responsible for the drop in price which occurred a few decades ago, except in so far as that drop was referable to the introduction of gold coinage in some countries.

The old silver- or silver-lead lodes, such for instance as those of the Erzgebirge, Příbram, the Harz, Kongsberg, etc., when compared with the Tertiary lodes, are seen to be of less importance. The richest of this old group are those recently discovered at Temiskaming in Canada. Including these, the old lodes are responsible for some 15 per cent of the world's silver production to date, a percentage which of late years has somewhat increased.

An important part of the total silver production, some 10 per cent, comes from the metasomatic deposits, which carry lead-zinc ores chiefly. The contact-deposits yield only about 5 per cent. Among copper lodes also, are some which are quite rich in silver, such for instance as those at Butte, Montana, where latterly about 250 tons of silver have been produced yearly. The production of silver from such lodes may reach some 7·5 per cent of the total production. The silver content of the Mansfeld copper-shale, the yearly silver output from which amounts to about 100 tons or not quite 2 per cent of the world's production, is well known. Altogether therefore, approximately one-tenth of the total production is obtained as a by-product in copper smelting.

Finally, the bed-like pyrite occurrences such as the Rammelsberg deposit, the intrusive pyrite deposits, and others,. yield some silver.

The average proportions of the world's silver production for which of late years these different classes of deposit have respectively been responsible, are therefore approximately as follows :

Young silver-gold group	about 55·0 per cent.
Old silver-lead group	„ 15·0 „
Metasomatic lead-silver-zinc group	„ 10·0 „
Copper lodes	„ 7·5 „
Contact-deposits	„ 5·0 „
Copper-shale group	not quite 2·0 „
Other groups	a small amount.

Although these figures are only approximate they are sufficient to place beyond doubt the great importance of the young Tertiary lodes in the silver production.

[1] *Ante*, p. 529.

THE LEAD-SILVER-ZINC LODES

As with most lodes so also with these, an association between the lode material and eruptive rocks or tectonic disturbances can in many cases be traced, in that the deposits have to all appearances been formed by mineral solutions which themselves were consequent upon the intrusion of old eruptive rocks. Concerning the distribution and importance of this group, it may be said that by far the greater number of all lodes belong to it. With regard to form, both simple and composite lodes—the latter in the sense of Cotta—may be distinguished, the former being narrow while the latter generally are wide, the width in extreme cases exceeding 100 metres. The most important natural concentrations of lead-silver-zinc ore are generally found in such composite lodes. In their extent along the strike these lodes vary greatly ; there are, on the one hand, small fissures of but a few decimetres in length, and, on the other, lodes which continue for several kilometres.

Although, as with all lodes, the mineralization is generally irregular, in this regard the simple lodes differ from the composite, in that these latter, besides being dependent upon the composition of the solution and the possible influence of the country-rock, are, in the distribution of their contained ore, also dependent upon the inclusion of large fragments of country-rock. Since with such inclusions the alternation of rock and ore is subject to no regular law, it follows that in opening up such lodes chance is a greater factor than with the simple lodes. It is often the case therefore, that in mines which exploit composite lodes considerable development work must of necessity be undertaken in barren rock, in order to expose new ore-bodies, or ore-shoots as they are sometimes termed. This primary irregularity in distribution upon the lode plane has sometimes been so pronounced as to have discredited entire districts.

The continuation in depth is likewise very variable, the only permissible statement being the well-known tenet, that those lodes which have considerable extent along the strike usually also exhibit considerable persistence in depth. On the surface the lodes, according to the nature of their filling, are more or less well defined ; where much quartz is present they may form ledges or reefs which subsequently break, scattering fragments to mark their course ; more often however the lode-filling consists chiefly of slaty material in which the ore is either disseminated or confined to veins. In such a case the lode does not differ materially from the country-rock in hardness and its course may often only be distinguishable by a dull reddish-brown colour, while when the surface rubble is thick even this may not be noticeable.

The ores of these lodes are chiefly galena, sphalerite, and pyrite, while the remaining sulphides, and especially the sulpho-salts, though they may often be present, are generally of less importance. The gangue sometimes consists chiefly of quartz, and sometimes of carbonates, barite, etc. Siderite in some districts occurs frequently, sometimes to the extent that it becomes an important saleable constituent, though in most cases it remains but a troublesome factor in concentration and not to be considered as other than gangue. The intimate growth of siderite with sphalerite is particularly unfavourable since these two cannot be separated by gravity concentration, and magnetic separation is necessitated.

With these lodes the enclosed fragments of country-rock play a remarkably important part. From the nature of the composite lode it follows that the greater portion of the width is often taken up by rock which has become more or less characteristically altered by the same mineral solutions which subsequently filled the interstices partly with ore. Lindgren, on such processes of alteration, formulated a classification of lodes which is scientifically of great interest. With sandstone or grauwacke this alteration generally consists in the further introduction and crystallization of silica. With slaty rocks the changes are more complex ; the fragments of slaty material, often quite large, may become so altered by the mineral solutions and by pressure as to retain hardly any resemblance to their original condition.

The form taken by simple fissures is also greatly dependent upon the nature of the country-rock ; while for instance in sandstone and grauwacke the fissures are regular and simple, in slate on the other hand they are much split up. In this latter circumstance a veined zone arises, often causing a bulge in the width of the lode. By mountain movement and mineral solution slate fragments may become so changed that they finally consist of innumerable pressure lenses traversed in all directions by veins filled with ore or gangue. In this extreme stage of alteration they represent what is known as lode-slate ; [1] such material is illustrated in Fig. 114.

The mineral-intergrowth is more various in the group of lodes now

---

[1] *Gangtonschiefer.* See Prefaces to Vols. I. and II.

under consideration than in any other. Sometimes, without apparent reason, the most complete examples of ordinary and concentric crustified structure are found, as in the Oberharz ; sometimes the structure is preponderatingly irregularly coarse, as in the lodes of the Berg Uplands ; while finally, there are many cases where the different constituents of the lode-filling are so intimately intergrown that they can hardly be distinguished by the naked eye. As was particularly mentioned when describing the various forms, structure has a material influence upon the cost of concentration. With the lead-zinc lodes this is so much the case that in neighbourhoods where clean lead- or clean zinc deposits are worked with profit, other lodes with intimate intergrowth of the two ores may often be unpayable. When discussing structure it was also indicated that beside the primary intergrowth, secondary or pseudo-intergrowth might arise, as for instance when a subsequent re-opening of the lode fissure allowed younger heavy-metal solutions to enter. As previously explained, concentric or cocade ore—illustrated in Fig. 126 and particularly characteristic of lead-zinc deposits—arises when layers of different ores have been deposited around a centre, which may be represented by a rock fragment. In such a structure each layer farther from the centre is younger than one inside and nearer to the centre. A singular case of such concentric ore arises when layers of siderite alternating with others of quartz have been metasomatically replaced by galena or sphalerite, the quartz remaining unaltered. In many cases it may be determined whether primary or secondary concentric ore is present, by observing whether the crystal faces occurring on the outline of a sulphide layer are those of the sulphide present or those of a mineral which has been replaced.

From the foregoing it is seen that with these lodes the different age of the separate portions of the lode-filling can play an important part in their structure. Occasionally several stages of lode-filling may be distinguished which in relation to age may be widely separate.

In the investigation of these lead-silver-zinc lodes the primary and secondary depth-zones are of great importance. Experiences in primary ore are far from numerous, yet mining at Freiberg, the Oberharz, and in the Berg district, indicates that there are certain regularities.[1] For instance, in those few cases where the lodes contain a little tin this is concentrated in the uppermost levels, while the tin-free lead- and zinc sulphides occur below. Such lodes may be said to have a tin gossan. With lead and zinc it has often been remarked that lead represents a shallower primary zone than zinc. In such cases galena is found in greatest quantity near the surface, following which comes a zone of galena and sphalerite which continues

[1] P. Krusch, ' Eine neue Systematik primärer Teufenunterschiede,' *Zeit. f. prakt. Geol.*, 1911, p. 129.

into depth, the latter mineral predominating more and more till finally the galena is practically excluded. Such mines in their early days were lead mines, then lead and zinc mines, and finally zinc mines exclusively. This sequence in mineralization should be apparent in the figures of production over a series of years ; not however when the figures of the district as a whole are reviewed, but when each mine is considered separately, and when care is taken that in the case of irregular mining or of the exploitation of different deposits by one management, only those portions of the total output are considered which belong to one particular deposit and to known depths.

The siderite zone in lead-zinc mines is particularly interesting. In some cases this mineral is found below the zinc, in what may justifiably be regarded as a third primary zone or, when the upper zone of tin is reckoned, a fourth. It has however been pointed out by Dr. Schulz [1] that with the lodes of the Berg district siderite also occurs in the upper levels, though its presence there has been obscured by the preponderance of galena and sphalerite, compared with which the siderite is unimportant and of little saleable account. Even however should the amount of siderite not increase in depth, it is nevertheless the case that below the zinc zone may follow a zone distinguishable from those above by containing siderite only.

The nature of the country-rock is also often of material influence upon the primary depth-zones. When describing the form of lodes the difference in sandstone and grauwacke on the one hand, and in slate on the other, was remarked. Such difference in form could not have been without influence upon the circulation of the heavy-metal solutions ; while, for instance, in the open fissure of the grauwacke these moved unhindered and new metal-liferous material presented itself continually, movement in the shattered zone of the slate was more or less impeded. It is indeed remarkably often the case that in sandstone and grauwacke the mineralization is pronounced, while in slate an impoverishment is experienced. Of this, for example, Denckmann was able to produce evidence in the Ramsbeck district. Such a difference has usually nothing to do with the geological age of the beds but rests solely upon mechanical factors. From such observations care must therefore be taken not to draw conclusions too far-reaching. If for instance one bedded complex consist chiefly of grauwacke, and another, younger or older, chiefly of slate, many an observer might be inclined straight away to hold the difference in age responsible for any difference in mineralization, without reflecting that age need have played no part, but that petro-graphical character alone might have produced the effect. In very rare cases moreover the opposite experience is met, the ore being found chiefly in slate and to a less extent in other rocks. The necessity to

[1] *Glückauf*, 1910.

enter cautiously upon generalizations and to treat every district individually is therefore apparent. When forming conclusions relative to the
mineralization in depth or along the continuation of the strike, these
primary depth-zones are very important.

The secondary depth-zones with the lead-silver-zinc lodes, though
these were discussed when describing the particular ores,[1] demand here
some further description. Galena and sphalerite in the oxidation zone
become altered to the well-known and corresponding oxides. In the
cementation zone, however, there is no enrichment either of lead or
zinc, presumably because these two sulphides are already richer in metal
than any other combination met in nature. The cementation zone is
distinguished from the primary zone rather by the amount of precious
metal present as native metal upon the fractures and surfaces of the
sulphides, such precious metal being usually but sparingly distributed
in the primary zone. The depth-zones therefore in so far as they relate
to the lead- and zinc ores themselves, are immaterial, and only merit
consideration in connection with the silver content.

The fact that with these lodes the three metals, lead, silver, and zinc,
are often most closely associated, has a great influence upon the market
in these metals. Should, for instance, silver be in greater demand,
greater production to meet this demand is only possible when lead and
zinc are also produced. Similar circumstances likewise attend the production of the other two metals. While therefore with the more isolated
metals such as platinum, gold, tin, and copper, an effective regulation of
production is possible, in respect of lead, zinc, and silver, the miner and
metallurgist are to some extent helpless.

The silver content in the primary ore varies greatly, though in general
it may be said that it maintains an attachment to lead rather than to zinc.
While galena containing 500 grm. of silver per ton is quite common,
sphalerite usually contains less than 50 grm., though up to these limits all
possible variations are found. As the silver is an admixed constituent
the ores richest in this metal are mostly, though not always, close-
grained or compact, while those which contain little or no silver are often
coarsely crystalline. This difference is more noticeable with galena than
with sphalerite, doubtless because the former is more argentiferous than
the latter. In applying this rule care must be taken to be certain that the
ore is primary, since in the cementation zone the coarsely-crystalline
galena is just as argentiferous as the finely-crystalline, seeing that such
galena when it precipitated the silver was already crystallized. The silver
content of the cementation ore is extremely variable : it may indeed be
represented by all possible amounts up to many kilogrammes per ton.

---

[1] *Ante,* pp. 86-88.

While under the microscope no native silver is usually to be observed in primary galena or sphalerite, with the richly argentiferous cementation ore on the other hand, it is usually seen as a fine coating or film upon the numerous cleavage-planes of both the galena and the sphalerite.

According to the predominant gangue, ore, or ores, the old lead-silver-zinc lodes, like the young gold-silver-lead lodes, may be divided into several subdivisions, thus :

1. Calcite - silver lodes, with calcite predominating but also with quartz, barite, fluorite, etc., and silver minerals. With these lodes galena and sphalerite are either absent or are but sparingly present.

2. Carbonate lead lodes, with calcite or dolomite, and occasionally rhodochrosite, quartz, etc. In these galena and sphalerite occur more particularly, silver minerals being less common.

3. Barite-lead-silver lodes, with barite, quartz, calcite, fluorite, and galena and sphalerite, often also with a small amount of silver mineral.

4. Quartz - silver lodes, with quartz chiefly, and silver minerals, a little galena, sphalerite, etc.

5. Sulphide or sulphide quartz-lead lodes, with quartz as most important gangue, and galena, sphalerite, and different sulphides. With these lodes silver minerals are either absent or but sparingly present.

6. Siderite lodes, with much siderite and quartz as gangue, and with galena, sphalerite, etc.

7. Silver-cobalt lodes, with silver minerals, arsenical cobalt and nickel minerals, galena and sphalerite more seldom. With some of these the occurrence of native bismuth is characteristic, with others that of uranium. These lodes therefore form a link with those next to be defined.

8. Silver-cobalt-uranium and cobalt-uranium lodes.

Recent investigation has shown that the above grouping cannot always be applied, as all possible gradations between the different subdivisions may occur, and the primary depth-zones are sufficiently pronounced that the present surface may coincide with the most varied primary ore. Similar mineralogical composition does not necessarily postulate similar age and genesis.

The best examples of the calcite-silver lodes are those at Kongsberg and St. Andreasberg. The most important representatives of the carbonate lead lodes are probably the dolomite-lead lodes at Freiberg wherein the silver minerals are comparatively speaking strongly represented; to this group belong also many of the Clausthal lodes. The barite-lead-silver lodes and the quartz-silver lodes are also typically developed at Freiberg. Generally speaking however, these old quartz-silver lodes are fairly scarce, while among the young Tertiary lodes such lodes predominate. It is probable that this relation between these two groups is one dependent upon primary variations in depth on a large scale. As was indicated when describing the Tertiary lodes, silver is chiefly deposited in the neighbourhood of the surface, while in greater depth galena and sphalerite often predominate. That being so, and it being also the case that the upper zone of the old lodes has generally been removed by erosion, it happens to-day that the deposit now being worked represents that of original great primary depth. Since such depth in this connection is the province of the sulphides, the probability of discovering old quartz lodes with silver minerals is small.

The sulphide or sulphide quartz-lead lodes are the most common. With these, quartz is the prevailing gangue, while carbonates, barite, fluorite, etc., are either completely absent or but poorly represented. Beside galena and sphalerite, which occur in varying relation to one another, much pyrite and arsenopyrite also are found, some chalcopyrite, and numerous other sulphides. In some cases tetrahedrite, pyrargyrite, proustite, and other silver minerals occur, these being practically absent from other lodes. Among the many sub-groups at Freiberg the sulphide lead lodes are undoubtedly the most important, such being also well represented in the other Saxon mining districts. It may indeed be said that the greater number of the lodes of the world belong to the lead-zinc group, and that among these the sulphide lead lodes constitute the greatest percentage.

In many of the lodes hitherto described siderite is completely absent, while in many others it is present in very small amount; in but few cases is it so abundantly present that the occurrence may be described as a siderite lead-zinc deposit. Such deposits are more particularly found in the Rhenish Schiefergebirge. Siderite as gangue is still more rare with the Tertiary deposits, though cases of its occurrence with these are known, as for example at Mazarron and Cartagena.

With most of these lodes nickel and cobalt are completely absent, a fact which becomes evident upon smelting, since in the furnace products both these metals are either absent or only present as traces. With some of these lodes, on the other hand, cobalt and nickel are more abundant and the passage is thus prepared to the silver-cobalt lodes. With these latter a large number of subdivisions might be formulated. Generally the amount of cobalt is larger than that of nickel, this being also the case with the arsenical cobalt-nickel lodes. With some, as with those at Annaberg in Saxony and Temiskaming in Canada, bismuth is practically absent; with others on the other hand, as with those at Schneeberg, native bismuth and bismuth minerals are so abundant that a separate silver-cobalt-bismuth or cobalt-bismuth subdivision has been set apart for them, this subdivision receiving further mention when describing the Upper Erzgebirge and the Temiskaming districts.

The old silver-lead-zinc lodes are in general characterized by their freedom from gold or by their poverty in that precious metal. Though of little importance, an exception to this is provided by some lodes belonging to the old quartz- and sulphide quartz-lead subdivisions, as for instance the Bergmannstrost lode in Lower Silesia where arsenopyrite and chalcopyrite are abundantly present, and the lodes in the district of Svenningdal in northern Norway which greatly resemble those of the sulphide lead subdivision of Freiberg. The concentrates from the quartz-silver lodes at Freiberg contain 0·5 to 8 grm. of gold per ton.

With some of these lodes quicksilver occurs in minimal amounts, as for instance at Kongsberg, where the native silver is remarkable for a small quicksilver content. Cinnabar also has been reported as a mineralogical curiosity in some lodes ; at Clausthal it occurs in minute amounts together with native quicksilver and seleniferous quicksilver. Lodes wherein the silver minerals are accompanied by quicksilver-tetrahedrite occur at some places,[1] as for instance at Brixlegg and Schwaz in the Tyrol, at places in Bosnia, etc. Such occurrences in general, however, are rare and of little importance, so that it may be said that a sharp line exists between the lead-silver-zinc lodes and the quicksilver lodes proper.

With most of the lead-silver-zinc lodes tin is so completely lacking that its presence may not even be detected in the furnace products from silver works. Exceptionally, it is found in small amounts at Přibram [2] and in the sulphide quartz-lead lodes at Freiberg.[3] The mineral associates of tin—wolframite and scheelite—are found now and then as mineralogical rarities in some lodes.[4] The line between the tin lodes and the lead-silver-zinc lodes is however not only sharp in relation to the minerals contained but also in regard to the country-rock.[5] This sharp separation between the two does not however exclude the possibility that both kinds of lode may occur in the same district. Such is indeed the case not only in Cornwall but also in the Erzgebirge.[6] The tin lodes then are the older, while the various lead-silver-zinc lodes present, became formed during the later periods of a long protracted mineralization.

Pyrite is present with all these lodes. With some, such as those at Kongsberg and St. Andreasberg, pyrrhotite occurs, in part well crystallized. Specularite has been found as a mineralogical curiosity, while magnetite appears to be completely absent. The iron minerals are more particularly well represented in the sulphide lead lodes, though even these usually carry

---

[1] *Ante*, p. 457.   [2] *Postea*, p. 704.   [3] *Postea*, p. 674.
[4] *Ante*, p. 423.   [5] *Ante*, p. 423.   [6] *Ante*, pp. 425, 434.

more lead and zinc than iron. Chalcopyrite, though generally playing quite a subordinate part, is hardly ever absent; its occurrence is more fully discussed when describing the copper lodes.

Concerning the relation of silver to lead, at Kongsberg silver is present to approximately double the amount of lead; with most of these silver lodes however, one part of silver is found to 2–10 parts of lead.[1] From this high ratio all gradations exist to lead lodes with but little silver. According to statistics collected over a period of almost fifty years, in the Freiberg district where the sulphide lead lodes relatively poor in silver are the most common, the ore produced showed an average of one part of silver to 105–175 parts of lead. The sulphide lead lodes by themselves probably carry 250–300 times as much lead as silver, and with these must be reckoned others still poorer in silver, in which for instance the amount of lead reaches as much as 1000 times that of the silver. It is worthy of remark however that the galena is very seldom entirely free from silver, and that the silver content of the lead lodes is generally much higher than that of the metasomatic lead deposits.

In these lodes sphalerite and galena always occur together, the latter being generally more abundant, though [2] this relation is often reversed in depth.

With some of these lodes boron silicates occur to a small extent. In the calcite-silver lodes at Kongsberg, for instance, some axinite is found, and in those at St. Andreasberg, some datolite. Tourmaline on the other hand appears to be absent. Occasionally, chlorite, adularia, albite, epidote, tremolite, and other silicates, occur. Rhodonite and rhodochrosite are at times abundant, the former more particularly with the silver lodes proper. Witherite and strontianite are rare, barite is notoriously common, while the barium zeolite, harmotome, is characteristic of some lodes. Apatite has been found in some of the old lead-silver-zinc lodes in minimal amounts and as a great rarity; with most deposits it is completely absent. In this feature also the difference between the lead-silver-zinc lodes on the one side, and the tin lodes on the other, is marked. The phosphoric acid occurring in the pyromorphite and other phosphates of the oxidation zone, is derived from the country-rock. Compounds of selenium and tellurium occur occasionally as mineralogical curiosities.

The old lead-silver-zinc lodes occur more particularly in the older geological formations up to and including the Culm. In the Upper Carboniferous they are already less frequent, while in the Triassic, Jurassic, and Cretaceous, occurrences are isolated. In the Tertiary their place is taken by the lodes of the young group. The age of the country-rock however puts but the lower limit to the age of the lodes. As previously explained,[3]

---

[1] *Postea*, p. 676.   [2] *Ante*, p. 653.   [3] *Ante*, p. 67.

the determination of the age of the lode-filling needs the most careful investigation. The different subdivisions show themselves in general to be independent of the petrography of the country-rock. All those at Freiberg with their varied filling occur in the same gneiss. Further, the calcite-silver lodes at Kongsberg have for their country-rock mica- and hornblende-schist, those at St. Andreasberg, Devonian silica-schist and grauwacke, while those at Silver Islet have Algonkian schist and gabbro. The barite-lead lodes occur at Freiberg, in gneiss ; at Sarrabus in Sardinia, in clay-slate and granite ; and at Bleiberg, in Triassic limestone. The sulphide quartz-lead lodes at Freiberg and Kuttenberg traverse gneiss ; at Linares and La Touche, granite ; and at Svenningdal,[1] limestone with narrow intercalated thicknesses of mica-schist.

The lodes are often found in areas of considerable tectonic disturbance, and the fissures in many cases are fault planes along which considerable movement has taken place. Most of the mining fields lie in the im- mediate neighbourhood of eruptive rocks or actually within such rocks. Dalmer, for instance, has established the local connection of the lode districts of the Erzgebirge with the granite. Nevertheless, just as with the Tertiary gold-silver lodes, there is no genetic dependence upon any particular eruptive. There is rather a general genetic association between eruptivity and mineralization, the lodes having resulted from metalliferous solutions which appeared as the consequent phenomena of eruptive activity.

Many lodes have been developed to considerable depth below the present surface : Přibram for instance to 1100 m., Kongsberg to 900 m., Freiberg to 700 m., Clausthal to 900 m., and St. Andreasberg to about 820 metres. If to these depths figures be added to represent the erosion such lodes have suffered—which on account of their great geological age is considerable— it may be reckoned that the lead-zinc lodes are known to a depth of some 4 km. below the surface as it existed at the time of their mineralization.

Considering that the lode-filling in general is independent of the mineralogical nature of the country-rock, that the lodes are mostly con- nected with tectonic disturbances, that they occur in the neighbourhood of eruptive rocks, and that they carry ore assuredly to a depth of at least 4 km. from the original surface, the conclusion is justified that the material of the silver-lead-zinc lodes was derived from depth. In this matter, as also in respect to the chemical nature of the solutions, the discussion given in connection with the Tertiary lodes applies here also. It is particularly the case with this group of lodes that ore-shoots are found in the neigh- bourhood of lode intersections. In other cases the ore is more particularly concentrated where the lode fissure crosses certain rocks, the nature of the country-rock having obviously exerted an influence upon the filling.

---

[1] J. H. L. Vogt, *Zeit. f. prakt. Geol.*, 1902, pp. 1-8.

The economic importance of these old lead-silver-zinc lodes is dealt with elsewhere. It is sufficient here to state that of late years they have yielded some 15 per cent of the world's production of silver,[1] some 33 per cent that of lead, and some 11–14 per cent that of zinc.

THE SILVER DEPOSIT AT KONGSBERG, NORWAY

LITERATURE

TH. KJERULF and T. DAHLL. 'Kongsbergs Erzdistrikt,' Nyt Mag. f. Naturw. XI., 1861.—C. F. ANDRESEN. 'Über die Gangformationen zu Kongsberg,' Verhandl. d. 10. Sitz. d. skandin. Naturforscher, 1868.—TH. HIORTDAHL. 'Über güldiges Silber zu Kongsberg.' Nyt Mag. f. Naturw. XVI., 1869.—G. ROLLAND. 'Mémoire sur la géologie de Kongsberg,' Ann. d. Mines, Paris, Sér. 7, XI., 1877.—A. HELLAND. 'Über den Betrieb des Kongsberg-Silberwerkes,' Arch. f. Mathem. und Naturw. X., 1886.—THS. MÜNSTER. 'Über die Kongsberger Mineralien,' Nyt Mag. f. Naturw. XXXII., 1892.—CHR. A. MÜNSTER. 'Über die Zusammensetzung des Kongsberger Silbers und über einen Sekundärprozess bei seiner Bildung,' Nyt Mag. f. Naturw. XXXII., 1892 ; 'Kongsberg-Erzdistrikt,' Ges. d. Wiss. Christiania, 1894, I.—P. KRUSCH. 'Das Kongsberger Erzrevier,' Zeit. f. prakt. Geol., 1896.—J. H. L. VOGT. 'Über die Bildung des gediegenen Silbers, besonders des Kongsberger Silbers durch Sekundärprozesse aus Silberglanz, etc., und ein Versuch zur Erklärung der Edelheit der Kongsberger Gänge an den Fahlbandkreuzen,' Zeit. f. prakt. Geol., 1899. Government Commission Reports for 1835, 1865, 1885, and 1903.—Private information from C. BUGGE, taken from a monograph upon the Kongsberg field to be published later.

Kongsberg, some 80 km. west of Christiania, is situated in an Archaean area which, as illustrated in Fig. 65, consists chiefly of gneiss and granite-gneiss, different gabbro rocks and hornblende-schists, several mica- and chlorite-schists, etc. The rock known formerly as the grey Kongsberg gneiss is in reality a foliated granite, more particularly a soda granite ; at several other places, in addition, there is a red foliated granite rich in microcline. Among the gabbros, olivine - gabbro with ophitic structure—olivine-hyperite—normal gabbro, and norite, are represented. Other varieties worthy of mention are uralite-gabbro, banded-gabbro, banded quartz-gabbro and gabbro-schist, and amphibolite and amphibolite-schist. The mica- and chlorite-schists occur rather sparingly, while of still less extent are the many small diabase- and diabase-porphyrite dykes.

The prevailing strike of the crystalline schists is N. 10° W. with a dip generally 70°–80° to the east. These schists are often remarkable for the occurrence within them of extremely fine layers of pyrrhotite, pyrite, and some chalcopyrite, this being particularly the case with the hornblende-, mica-, and chlorite-schists, but also with the foliated and highly schistose granite. This occurrence of pyritic layers in crystalline schists constitutes the Kongsberg fahlbands. The amount of such sulphides is low, being generally only one or two per cent, though in places it may be somewhat

[1] *Ante*, p. 649

higher. The distribution and extent of these fahlbands may be seen from the afore-mentioned Fig. 65 as well as from Fig. 330, the former however being now somewhat out of date. Sulphides may also occasionally be observed in the relatively but little foliated grey granite, from which circumstance together with the fact that sulphide veins often cross the foliation of the granite, Vogt came to the conclusion that the sulphide layers of the fahlbands are intrusive.[1]

The Kongsberg lodes are in greater part calcite lodes carrying native

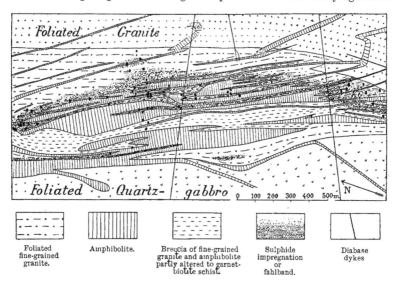

Foliated fine-grained granite.	Amphibolite.	Breccia of fine-grained granite and amphibolite partly altered to garnet-biotite schist.	Sulphide impregnation or fahlband.	Diabase dykes

Fig. 330.—Geological map of a portion of the Overberget. C. Bugge.
KG, Kongens mine; GH, Gottes Hilfe-in-der-Not mine.

silver; they cut across the schist and within the fahlbands are metalliferous. By far the most important ore is native silver, which occurs in the form of wire, moss, or plates, and quite exceptionally in crystals. Large masses have occasionally been found, the largest weighing as much as 500 kg. Sometimes the silver is remarkable for a small but variable quicksilver content, which rarely reaches as much as 2 per cent and generally is below 0·5 per cent. The gold content of this silver is remarkably low, being in the majority of cases only 0·002–0·005 per cent. Golden silver has been found as a mineralogical curiosity in certain separate quartz lodes. With the native silver, argentite occurs, sometimes in masses up to 100 kg. in weight, though altogether responsible for but one or two per cent of the

[1] Ante, p. 340.

total output. It is seldom that stretches of any length are encountered where argentite is preponderatingly present and native silver subordinate, or even sections where both are present in approximately equal amount. Pyrargyrite and proustite are great rarities, and occurrences of other silver minerals such as stephanite are isolated. Sphalerite, galena poor in silver, chalcopyrite, pyrite, and pyrrhotite, occur in very small amount.

The native silver at Kongsberg is in Vogt's opinion in greater part not of primary origin but has arisen secondarily from argentite, or exceptionally from pyrargyrite and proustite. It is often found as wires or horns sprouting from argentite, these wires or horns having sometimes small balls of argentite at the end, as illustrated in Fig. 140. As pointed out by G. Bischof [1] and other authorities, such a manner of occurrence may only be explained by secondary processes, as for instance by the action of oxygen, hydrogen, or water-vapour, etc., upon heated argentite, according to the formula [2] $3Ag_2S + 2H_2O = 6Ag + 2H_2S + SO_2$. In close proximity to the silver it is not uncommon to find some carbon, the presence of which is doubtless due to a reduction by carburetted hydrogen, thus, $2Ag_2S + CH_4 = 4Ag + 2H_2S + C$.[3] Most of the wire silver, it is true, shows no trace of argentite ; its twisted, grooved, and tapering form is however identical with other similar wires where a derivation from argentite can be definitely established. According to Vogt the two habits are of analogous formation, the difference being that the argentite in one case was partly, and in the other case completely decomposed, this decomposition taking place chiefly before, but also during the deposition of the oldest generation of calcite.

The most important gangue-mineral is calcite, after which come fluorite and quartz, the former often occurring in beautiful crystals. Barite, axinite, adularia, albite, chlorite, hornblende-asbestos, and prehnite are also seen, as well as different zeolites, such as apophyllite, desmine, stilbite, harmotome, and laumontite. In addition, anthracite — an analysis of which by Helland showed 95·5 per cent carbon, 1·9 per cent hydrogen and 2·2 per cent oxygen — often occurs, being regarded as favourable to the silver content. The calcite frequently occurs in several generations, fluorite and quartz as a rule only in two generations.

The sequence in age is often as follows : after a first generation of quartz follow most of the sulphides, including the greater portion of the argentite ; with these sulphides the older calcite is approximately contemporaneous ; then come fluorite and adularia, albite and barite, and different carbonates ; and finally the zeolites, the pyrite, and the youngest

[1] *Poggendorfs Ann.* 60, 1843.
[2] *Ante,* p. 131.
[3] Höfer, ' Erdölstudien,' *Wiener Akad. Verh.,* 1902.

calcite generation. The argentite and its alteration to native silver belong to one of the early phases of formation.

The lodes at Kongsberg are generally of little width, this dimension usually varying from the thickness of paper up to 10 cm., and but seldom reaching more than 33 cm. The strike is mostly west-south-west or approximately at right angles to that of the crystalline schists ; often several parallel lodes or veins are bunched together. A few, known locally as bedded lodes, agree in strike with the schists.

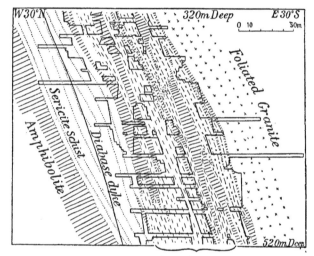

Garnet-biotite schist.

Fig. 331.—Longitudinal section of the richest portion of the Kongens mine, between 320 and 520 m. in depth. C. Bugge.

The rocks, represented as in Fig. 330, are from foot-wall to hanging-wall : amphibolite, sericite-schist, garnet-biotite schist, and fine-grained foliated granite, the last three alternating in certain zones with pyrite impregnations and amphibolite ; foliated granite in the hanging-wall. The areas indicated as having been stoped represent the argentiferous portions of the lode plane.

Concerning the distribution of the silver, the experience of more than a century shows that this is exclusively or almost exclusively confined to those areas on the lode plane where, as illustrated in Fig. 331, that plane intercepts the fahlbands, or in the case of the bedded lodes where such follow the fahlbands. Where the lodes, leaving the fahlbands, continue into the adjacent rock free from sulphides they become poor or even barren, silver in fact does not continue to be found beyond 5 m. or at the most 10 m. from the fahlband zone.

The two principal mining areas are Overberget and Unterberget, these lying respectively immediately west and east of a north-south zone of grey foliated granite 10 km. long and 1–1·5 km. wide.[1]   In Fig. 65 this granite is entered as the ' Middlebergsband.'   At Overberget the fahlband zóne attains a width of 150–300 m., and at Unterberget 100–200 m., this width being made up of an alternation of mica-, chlorite-, and hornblende-schists, with amphibolite and highly foliated granite or gneiss-granite. In this total width, as illustrated in Fig. 330, usually only some of the bands carry sulphides.   The horizontal extension of the silver content along each separate lode is limited ; in the richest portions of the Kongens mine it is 80–95 m. at the most ; while in other cases it is only 20–40 m., or even 10–20 metres.   Across one and the same fahlband the silver content varies considerably.   No definite rule concerning its distribution may be formulated, though, as illustrated in Fig. 331, the largest amounts of silver occur at the intersection with certain of the sulphide beds.   No fixed relation between the sulphide content of a fahlband and the silver content in the lode has been established in general, nor can it be said at Kongsberg that any enrichment occurs at lode intersections or junctions.

According to Bugge, a rapid alternation in the composition of the rocks constituting the fahlband zone is favourable to the silver content. This author calls attention to the fact that the best lodes appear in the neighbourhood of the narrow diabase- and diabase-porphyrite dykes which run approximately parallel to the lodes.   The Kongsberg lodes are younger than the crystalline schists and the above-mentioned dykes. Bugge has recently shown that these latter are closely associated with the Devonian eruptives of the Christiania district, which district, as is well known, is bounded by powerful faults along which in several places ore has been found.[2]   It is therefore possible or even probable that the formation of the Kongsberg lodes is connected with the downthrow of the Christiania district.   The distance from the most southerly mine at Kongsberg to the Silurian beds of the Christiania district does not reach 1 kilometre.

The erosion these lodes have suffered since their formation may, according to Vogt, amount to 3 km. of vertical height.   The deepest mine has reached a depth of 900 metres.   On these figures deposition of ore took place at a depth of some 4 km. below the surface at the time of deposition.

The Kongsberg lodes are normal lodes which, in the occurrence of calcite, fluorite, pyrrhotite, axinite, and the zeolites, and also in the noble character of the ore carried, are closely related to those at St. Andreas-berg in the Oberharz.   The silver ore being accompanied particularly by, and apparently contemporaneous with the calcite, can only have been

[1] Vogt's Map, *Zeit. f. prakt. Geol.*, 1902, p. 6.
[2] J. H. L. Vogt, ' Über die Erzgänge zu Traag in Bamle, Norwegen,' *Zeit. f. prakt. Geol.*, 1907.

deposited from aqueous solution wherein to all appearances carbonic acid was contained. This appearance is strengthened by the fact that silver carbonate is easily soluble in carbonic acid water.[1]

The kindly influence of the fahlbands has been observed not only at Kongsberg but also in the small but analogous occurrence at Hisö near Arendal ;[2] it is therefore of a general character and cannot be explained by lateral secretion. The fahlbands, both rock and sulphide, are completely undecomposed and practically without silver. The sulphides by themselves—generally only amounting to 1–2 per cent of the entire mass— according to Chr. A. Münster, contain but 0·0003–0·00055 and exceptionally 0·002 per cent of silver ; the content of the fahlbands as a whole would therefore be about 0·00025 per cent of silver. The enrichment at the fahlbands is therefore probably due to precipitation brought about by these themselves, from solutions circulating in the fissures. Some authorities [3] advocated the view that the fahlbands, on account of the sulphides they contain, conduct electricity better than the surrounding rocks, and that the precipitation of the silver may consequently have been brought about by terrestrial electric currents. This view however does not explain the fact that the silver is primarily deposited as sulphide. Vogt in 1899 endeavoured to explain the influence of the fahlbands in that carbonic acid solutions circulating in the fissures, attacked the country-rock, causing at the fahlbands the generation of a small amount of sulphuretted hydrogen from the sulphides there present, this gas then bringing about the precipitation of the silver. From a dilute solution containing silver, lead, copper, iron, zinc, etc., silver sulphide would be the first to be precipitated by a small amount of sulphuretted hydrogen.

The silver lodes at Kongsberg are distributed over an area some 30 km. long and 5–10 km. wide, though at places the width may be as much as 15 km. The number of the lodes is very large ; most however have but the thickness of a sheet of paper. Since 1622, when silver was discovered, some 150 mines have at different times been worked. Among these the Kongens mine at Overberget had in 1911 reached a depth of 900 m. ; two other mines, Gottes-Gabe and Gottes-Hilfe-in-der-Not, a depth of 700–730 m.; and some ten others, depths between 300–700 metres. Altogether at Kongsberg, where most of the mines belong to the State and where work has practically continued unbroken since 1624, 561,177 kg. of silver were won from that date to 1815, and 421,399 kg. from 1816 to the commencement of 1909, these totals making together roughly 982,000 kg. with a value of 7·5 million sterling. The net profit received by the State from 1830 to 1890 amounted to 1·1 million sterling.

---

[1] *Ante*, p. 136.   [2] J. H. L. Vogt, *Geol. För. Förh.* VIII., 1886.
[3] Durocher in 1849, Vogt in 1886, Chr. Münster in 1894.

The richest ore-body in the district, illustrated in Fig. 331, occurs in the Kongens mine. From this, during the period 1830–1890 and between depths of 230 m. and 600 m. from the surface, 274,313 kg. of silver were produced from an area of about 29,802 sq. m. on the lode plane, or an average of 9–10 kg. of silver per square metre. The richest portion actually yielded 23,000 kg. of silver from 1200 sq. m. on the lode plane, or roughly 20 kg. per square metre. In depth this rich ore-shoot diminished in size, eventually pinching out at 750–800 metres. Latterly, a new and very rich ore-shoot has been developed in the Samuel mine at Unterberget. At present four mines are being worked at Overberget and one at Unterberget, these having together an annual output of about 8000 kg. of silver, and employing about 300 men. Since the heavy drop in the price of silver in 1892–1893 the production has practically only covered the cost, the losses of some years being made good by the small profits of others. Approximately three-quarters of the production is won from ore containing roughly 70 per cent of silver, such ore going direct to the smelter. The remainder is at present won by the cyanidation of poor concentrate.

THE OCCURRENCE OF SILVER AT TEMISKAMING, CANADA

LITERATURE

WILLET. G. MILLER. ' The Cobalt-Nickel Arsenides and Silver Deposits of Temiskaming,' Report, Bureau of Mines of Ontario, Toronto, 1905, II. ; 1907, II.—Visit to Cobalt and Sudbury of the British Association for the Advancement of Science, August 1909 ; Toronto, 1909.—W. CAMPBELL and C. W. KNIGHT. The Paragenesis of the Cobalt-Nickel Arsenides and Silver Deposits of Temiskaming, 1907.

In this district, lodes carrying silver associated particularly with cobalt were discovered in 1903 during the construction of the Temiskaming and Northern Ontario railroad. These occur on the northern shore of Temiskaming Lake, at the boundary of Ontario and Quebec, about 150 km. north-east of Sudbury.[1] The old rocks of this district form the following sequence : as oldest, the Keewatin beds consisting of diabase, granite-porphyry, granite, etc. ; then at least 500 feet of almost horizontal conglomerates, breccias, grauwackes, and slates of the Lower Huronian ; next the Middle Huronian with its conglomerates, quartzites, etc. ; and finally, the thick diabase flows and intrusions of the Upper Huronian. The lodes are steep in dip and considerable in number. They occur chiefly in the conglomerates and grauwackes of the Lower Huronian, but few being found either in the Keewatin beds beneath, or in the Upper Huronian diabase above. By the year 1909 an area of about

---

[1] *Ante,* p. 289.

20 sq. km. situated or centred around the new town of Cobalt had become recognized as metalliferous. The lodes on an average are but 10–16 cm. wide, the extreme variation being from the thickness of a knife blade to somewhat more than half a metre. The filling consists of rich solid ore with a little calcite and quartz as gangue.

The most important ore-minerals are native silver, with some native bismuth and graphite; the arsenides, smaltite, cobaltite, chloanthite, domeykite, Cu_3As, and to a less extent niccolite; while the antimonide dyscrasite, Ag_6Sb, also is common. The sulpho-salts pyrargyrite, proustite, and tetrahedrite are subordinate, as also are the sulphides argentite, millerite, galena, pyrite, sphalerite, and bornite; the arsenates erythrite and annabergite, on the other hand, near the outcrop occur quite plentifully. Analyses of two parcels of hand-sorted ore, No. I. of 354 tons and No. II. of 537 tons, are as follows :

				I.		II.	
Silver	.	.	.	4·80	per cent.	4·16	per cent.
Cobalt	.	.	.	8·26	,,	6·89	,,
Nickel	.	.	.	4·74	,,	3·09	
Arsenic	.	.	.	34·61	,,	30·91	,,

These lodes are accordingly distinguished by the combination of silver, cobalt, nickel, and arsenic, the cobalt far exceeding the nickel in amount. In addition, some antimony and bismuth are found. Sulphur is present to a less extent than arsenic.

Operations so far have been confined to the neighbourhood of the surface, where native silver is the principal ore and where, in addition, dyscrasite, pyrargyrite, proustite, argentite, etc., are common. Native silver often occurs in large plates or slabs, the heaviest so far discovered weighing 744 kg., of which some 90 per cent was silver. It is probable that the native silver at Temiskaming has been formed by secondary processes.

Campbell and Knight, as the result of microscopic investigation of the fine-grained and compact ore, established the following sequence in age, beginning from the oldest :

1. Cobaltite and chloanthite.
2. Niccolite.
3. Calcite.
4. Argentite.
5. Native silver.
6. Decomposition products, such as erythrite, annabergite, etc.

Although the pre-Glacial weathered zone was in greater part removed during the Glacial period, the lode-outcrops may still be recognized by a considerable amount of bright-coloured cobalt- and nickel arsenates, and by asbolane. In consequence of the afore-mentioned removal by ice and the small amount of sulphides present in the ore, the oxidation zone as found to-day is but a few feet deep.

Concerning the behaviour of these lodes in depth, owing to the short life of the industry but few observations have been possible. No experience is available to show how these lodes, most of which occur in the relatively thin Lower Huronian, will comport themselves in the diabase of the Keewatin below. From the relative ages of the various minerals present and judging from the description of the deposits, it appears highly probable that at present a very rich cementation zone is being worked, and that only with the greatest caution may any conclusions respecting the silver content in depth be hazarded. According to Miller, a genetic connection between the lodes and the late Huronian diabase eruptions exists, in so far that the mineral solutions rising from depth and having the basic magma reservoirs as their ultimate source, are presumably the last echoes of that eruptive activity.

The economic significance of the Temiskaming or Cobalt district may be gathered from the following figures :

	Number of Mines.	Production.				Total Value in Dollars.
		Silver.	Cobalt.	Nickel.	Arsenic.	
		Oz.	Tons.	Tons.	Tons.	
1904	4	206,875	16	14	72	136,217
1905	16	2,451,356	118	75	549	1,473,196
1906	17	5,401,766	321	160	1,440	3,764,113
1907	28	10,023,311	739	370	2,958	6,301,095
1908	30	19,437,875	1224	612	3,672	9,284,869
1909	31	25,897,825	1533	766	4,294	12,617,580
1910	41	30,645,181	1098	604	4,897	15,603,455
Totals	167	94,464,189	5049	2601	17,891	49,180,525

Of the total value given in this table, silver is responsible for $48,368,333, this being chiefly native silver. About 2 per cent of the value is thus left to be accounted for by the remaining metals. These figures indicate how essentially this is a silver field. Seven years after starting it had produced roughly 2900 tons of silver, which is more than Freiberg produced during the whole of the nineteenth century, or more than half the total Freiberg or the total Comstock production. According to an estimate of the United States Mint the world's production of silver in 1910 was 217·8 million ounces. Of this amount Temiskaming produced roughly 14 per cent. The district is without doubt the richest in silver of any yet known belonging to the old group of lodes. The cobalt won as a by-product nearly satisfies in itself the world's consumption. Its entry into the market caused such a considerable drop in the price that the cobalt-blue market was almost completely ruined. The arsenic recovered had a similar effect upon its market.

SILVER ISLET ON THE CANADIAN SHORE OF LAKE SUPERIOR

LITERATURE

E. D. INGALL. Ann. Rep. Geol. Survey. Canada, 1887.—KEMP's Ore-Deposits.

The lodes of Silver Islet and neighbourhood, some 800 km. to the west of Temiskaming and near Port Arthur, are essentially calcite-silver lodes with native silver as principle ore and with subordinate cobalt-arsenic minerals. Less frequent are argentite, some sulpho-minerals, native bismuth, domeykite, etc., while nickel minerals are extremely rare. The abundance of graphite in the lodes is remarkable.

The quantity of silver in these deposits is not to be compared with that at Temiskaming. The Silver Islet mine, begun in 1868 and stopped in the 'eighties, produced silver to the value of about 3·25 million dollars, corresponding to about 100 tons. Including the production of some smaller mines in the neighbourhood, the total value of the silver production is $4,770,000.

The lodes at Silver Islet cross the Animikie stage of the Algonkian slate, in which a wide gabbro- or norite dyke occurs. To the depth of 375 m. reached it was the experience that these lodes carried ore only where they crossed the gabbro. In this there appears to be some analogy to the occurrence at Kongsberg; at both places also the principal ore is native silver, while there is further similarity in the common occurrence of carbon, represented at Silver Islet by graphite and at Kongsberg by anthracite. At the latter place however the cobalt minerals are absent.

The metal-association at Temiskaming on the other hand, consisting as it does of silver, cobalt, arsenic, some nickel and bismuth, most resembles that at Annaberg in Saxony, that at Joachimsthal in Bohemia, or that at Chalanches in the Dauphiné. There exists in addition a certain similarity between Temiskaming and Schneeberg, though at this latter place there is relatively more bismuth and less silver. Silver Islet forms therefore an intermediate stage between Kongsberg on the one hand and Annaberg-Joachimsthal on the other.

THE LODES OF THE ERZGEBIRGE

THE FREIBERG DISTRICT

LITERATURE

J. F. W. v. CHARPENTIER. Mineral. Geographie der chursächsischen Lande, 1778.—
H. MÜLLER. 'Die Erzlagerstätten nördlich und nordwestlich von Freiberg,' Cotta's
Gangstudien, I. p. 101, 1847.—W. VOGELGESANG. 'Die Erzlagerstätten südlich und süd-
östlich von Freiberg,' Cotta's Gangstudien, II. p. 19, 1848.—H. MÜLLER and B. R. FÖRSTER.
Gangstudien aus dem Freiberger Revier. Freiberg, 1869.—H. MÜLLER. Die Freiberger
Erzlagerstätten in Freibergs Berg- und Hüttenwesen, II. Edit., 1893, p. 32; Die Erzgänge
des Freiberger Bergreviers. Monograph of the Geological Survey of Saxony, Leipzig, 1901.
—C. GÄBERT. 'Die geologischen Verhältnisse des Erzgebirges' in 'Das Erzgebirge' by
Prof. Dr. ZEMMRICH und Dr. C. GÄBERT, Landschaftsbilder aus dem Königreich Sachsen,
Vol. II., 1911.

The Freiberg district, as indicated in Figs. 332 and 333, is a large one,
including not only the immediate surroundings of Freiberg and Brand,
but also the occurrences at Oederan, Bräunsdorf, Bieberstein-Nossen
Oberreinsberg, Dittmannsdorf, etc. Tectonically it represents a dome
consisting of the two principal divisions of the Erzgebirge gneiss, these
being the older or grey gneiss, which is a biotite-gneiss, and the younger
or red gneiss, a muscovite-gneiss.

The deepest horizons of the grey gneiss consist of the Freiberg biotite-
gneiss, a coarse-scaled and markedly jointed rock, which occupies the
outlying surroundings of Freiberg and extends over Dippoldiswalde and
Glashütte to Nollendorf and Graupen. It is an eruptive gneiss which
merges into a rock of perfect granitic structure with foreign inclusions.
Towards its upper horizons this old gneiss becomes fine-scaled, while in its
uppermost section the occurrence of bedded dykes of typical augengneiss
is noteworthy.

The younger or red gneiss in its main occurrence is also dome-shaped,
though in addition it forms persistent bed-like or flat lenticular masses
which are found both in the deepest horizons exposed around Freiberg, as
well as in the mica-schist formation, which is materially younger. It is
regarded as a laccolith or bed-like intrusion of eruptive character.

In the upper portions of the Erzgebirge gneiss formation numerous
bed-like intercalations of such sediments as limestone, grauwacke, conglomerate, quartzite, mica-schist, phyllite-like rock, and garnet-mica
rock, occur between the eruptive members. These intercalations are to
be regarded as detached patches of the slate formation which formerly,
as indicated in Fig. 334, completely covered the gneiss.

Against this dome of eruptive gneiss lie sedimentary rocks. These
in the neighbourhood of the granite are highly altered to mica-schist and
garnet-mica rock, into which gneissic material, represented by gneiss and

gneissic mica-schist, was injected. Such rocks constitute the inner contact

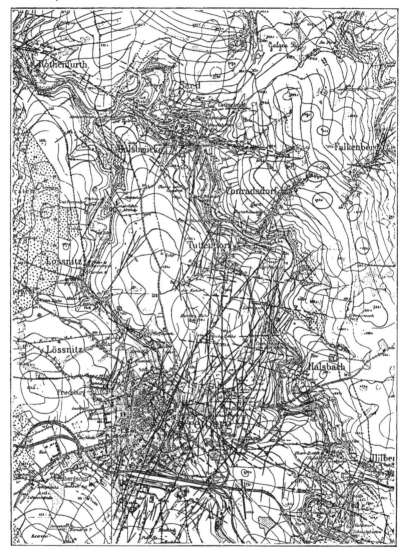

FIG. 332.—Map of the Freiberg lode district, neighbourhood of Freiberg.

zone. Resting on this inner zone comes the outer contact zone of quartz-

and albite-phyllites, these towards the hanging-wall merging into micaceous

FIG. 333.—Map of the Freiberg lode district, neighbourhood of Brand.

phyllite. Finally, come the unaltered rocks, which are but slightly metamorphosed in their bottom layers. These last, according to fossil evidence, belong to the Cambrian.

The Freiberg lodes represent fissure-fillings, most of which dip steeply, and along which the country-rock is more or less highly altered. It was from this district that the scientific investigation of ore-deposits first proceeded. Here Werner formulated his lode theories, and here also the existence of lode-groups in individual districts was first recognized, these groups being based upon difference in geological age, strike, and filling. Such local classification as this, though doubtless apt in the case of Freiberg, has unfortunately become too generalized. Though groups based on age might be formulated for some districts, in other districts variations in strike or filling are no conclusive evidence of difference in age.

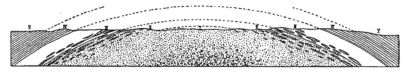

FIG. 334.—Diagrammatic section through a gneiss dome (granite-gneiss laccolith).

I, Central portion of the dome laid bare by denudation, and free from patches of the slate formation. II, Mantle containing highly contact-metamorphosed patches of slate conformably embedded in the gneiss. III, IV, and V, Tilted schists and slates resting upon the gneiss dome, with many intercalations. III, The inner contact-aureole of mica-schist and garnet-mica rock with injected gneissic material. The secondary injection into this aureole of the later or red gneiss, as well as the intrusion of this latter into the gneiss dome itself, are not indicated. IV, Outer contact-aureole or zone of quartz and albite-phyllites and, further in the hanging-wall, micaceous phyllites. V, Unaltered slates, slightly metamorphosed only in the foot-wall, with Cambrian fossil-remains in places.

In the Freiberg district more than 1100 lodes are known, the silver content of which has made the district famous. According to their strike these are locally divided into:

High Lodes, N.-N.E. (*Stehende Gänge*).
Early „ N.E.-E. (*Morgengänge*).
Late „ E.-S.E. (*Spätgänge*).
Low „ S.E.-S. (*Flache Gänge*).

Two systems of lode-groups are known, an older and a younger, these being distinct both in relation to filling as well as to the time of formation. For instance, in the lodes of the younger system arsenopyrite, chlorite, molybdenite, tin ores or stanniferous black sphalerite, have never been observed. These systems and their groups are as follows:

(I.) The older system—

1. The quartz-silver lodes. Ore: argentite, pyrargyrite, proustite, native silver, argentiferous pyrite, and arsenopyrite. Gangue: quartz and hornstone particularly.

2. Sulphide lead-copper lodes. Ore: pyrite and marcasite, galena, sphalerite, arsenopyrite, and chalcopyrite. Gangue: quartz.

3. Silver- or dolomite - lead lodes.   Ore : argentiferous galena, sphalerite, pyrargyrite, proustite, native silver, pyrite, tetrahedrite, and argentite.   Gangue : quartz, dolomite, and rhodochrosite.

4. Copper lodes.   Ore : pyrite, chalcopyrite, chalcocite, and bornite. Gangue : quartz.

To this older system the tin lodes, described previously,[1] belong ; these in the Freiberg district are but few in number and without economic importance.

(II.) The younger system—

1. The barite-lead-silver lodes.   Ore : galena, chalcopyrite, pyrite, marcasite, sphalerite, wurzite, and a little bournonite.   Gangue : barite, fluorite, quartz, and hornstone.

2. Iron-manganese lodes.   Ore : either hæmatite and specularite, or limonite, yellow iron ore and psilomelane.   Gangue : quartz in the former case ; barite and clay in the latter.

Concerning strike, there are broadly speaking two main directions ; namely, a north-north-easterly, maintained by the majority of the sulphide lead- and quartz-silver lodes ; and a west-north-westerly, maintained by most of the barite-lead lodes.   The silver-lead lodes strike indifferently in both these two directions.

It would appear that the silver was associated rather with the grey gneiss than with any other feature, and that the lodes become poorer or impoverished upon entering the red gneiss or mica-schist.

The quartz-silver lodes are found within an area about 22 km. long, extending from Oederan to Nossen.   In width they vary from 0·1 m. to 1 m. ; in strike some of them, as for instance the Neue-Hoffnung-Gottes and Alte-Hoffnung-Gottes lodes, may be followed for several kilometres ; while in depth some have been proved for 460 metres.

The sulphide zinc-lead-copper lodes occur more particularly at Halsbrücke, Berthelsdorf, Brand, and Erbesdorf, in the neighbourhood of Freiberg.   To this class belong the Hohe Birke-Stehende, 4 km. long and developed to a depth of 400 m. ; and the Rotegrube-Stehende, worked in the Himmelfahrt mine to a depth of 400 m. and followed for a length of 5 km.   The galena of these lodes often assays 70–80 per cent of lead with 0·2–0·3 per cent of silver, or 2000–3000 grm. per ton.   Chalcopyrite is usually very poorly represented in the zinc-lead lodes.   The black sphalerite which is particularly characteristic of these lodes carries, according to Stelzner and Schertel, microscopic inclusions of cassiterite crystals ; cassiterite and wolframite have on the rarest occasions even been observed without the aid of the microscope.

[1] *Ante*, p. 425.

The silver- or dolomite-lead lodes are found more especially around Brand and Erbesdorf ; they are generally 600–1000 m. long and have been developed to a maximum depth of 600 metres. The galena of these lodes contains 75–85 per cent of lead, and 0·4–0·6 or more seldom 2 per cent of silver, that is, 4–20 kg. of silver per ton. The pitchblende content of these lodes is noteworthy, though further description of this is deferred.[1] In 1885 the mineral argyrodite was discovered as a rarity in these lodes. It was in this mineral that Cl. Winkler in 1886 discovered the element germanium.

The barite-lead group is represented by about 200 lodes, these occurring in the gneiss and mica-schist around Grosschirma, Halsbrücke, Falkenberg, Hilbersdorf, and Oederan. These lodes vary in width from 0·45 m. to 4 metres. The length may be very considerable, that known as the Halsbrücker-Spat having a length of 8 km. with 400 m. of proved depth. In this group two subdivisions are differentiated, typical barite-lead lodes with chalco-pyrite and pyrite ; and silver-cobalt lodes, distinguished by the occurrence of silver minerals, though such occurrence is but limited. With these lodes seleniferous and vanadiferous pitchblende ores are found, while the presence of marcasite containing 0·5–0·75 per cent of thallium is also worthy of mention.

The iron group has never been of much importance. It is repre-sented for instance by some lodes in the red gneiss at Niederseifenbach in the upper Flöha valley, and by one in granite-porphyry at Holzhau.

The silver- and the lead lodes at Freiberg, with gangue consisting of barite, dolomite, and rhodochrosite particularly, are characterized by their crusted or combed structure, with which numerous fine crystals of the above-mentioned minerals are associated. With the quartz lodes on the other hand, or where quartz is the most important gangue-mineral, this structure is less prominent.

The richest ore-bodies at Freiberg occur chiefly at lode junctions or intersections.

Unfortunately, owing to the discovery of other large silver deposits, especially those in America, and the consequent drop in the price of the metal, mining in this world-famed district where the more important mines have reached depths of over 600 m., has been reduced to a decadent condition, and there are now but few mines working. The total pro-duction has been as follows :

1163–1523	.	.	.	1,958,800 kg.
1524–1835	.	.	.	1,754,983 ,,
1836–1896	.	.	.	1,529,174 ,,
Total 1163–1896	.	.	.	5,242,957 kg. worth £45,400,000

[1] *Postea*, p. 714.

For the five years 1877–1881 when the industry was still at its zenith, the lode area worked, the ore won, and the payment received for ore, were as follows : [1]

1877–1881.	Lode area worked.	Ore produced.	Payment therefor.
	Sq. m.	Tons.	£
Silver-quartz lodes . . .	52,267	25,631	171,000
Sulphide-lead lodes . . .	264,807	183,521	557,000
Silver-dolomite lodes . .	69,090	22,143	169,000
Barite-lead-silver lodes . .	55,269	21,781	110,000
Totals . . .	441,433	253,076	1,007,000

From these figures which represent a yearly average of about £200,000, it is seen that the sulphide-lead lodes were the most important.

The following figures, giving the average production from one square metre of lode area when all the lodes are considered together, will be found interesting :

	Ore won.	Silver contained.	Lead contained.	Payment received.
	Kg.	Kg.	Kg.	Shillings.
1851–1855	178·4	0·256	31·9	34·9
1866–1870	262·3	0·269	41·4	46·8
1877–1881	283·8	0·277	48·9	45·7
1886–1890	253·4	0·286	30·2	35·1
1891–1895	270·5	0·288	40·5	30·9

These figures show how during the period 1886–1890, and still more during 1891–1895, the drop in the price of silver was felt. In the figures of payment the small amounts received for copper, zinc, arsenic, sulphur, and uranium, are included.

The relation between silver and lead naturally varies greatly in the different groups. According to Stelzner [2] the averages per square metre for one or more years from Beschert-Glück representing the silver-lead lodes, and from Himmelfahrt representing both the barite- and the sulphide-lead lodes, were as follows :

	Silver.	Lead.	Nickel-Cobalt.
	Kg.	Kg.	Kg.
Silver-lead lodes	0·386	0·75	0·016
Barite lodes	1·052	2·10	0·089
Sulphide-lead lodes	0·230	61·45	0·001

The present position of mining at Freiberg is discussed later.[3]

[1] Müller, 1901.   [2] *Zeit. f. prakt. Geol.*, pp. 401-402.   [3] *Postea*, p. 683.

THE LODES OF THE UPPER ERZGEBIRGE

LITERATURE

H. MÜLLER. 'Über eine merkwürdige Druse auf einem Schneeberger Kobaltgange,' Zeit. d. d. geol. Ges., 1850, p. 14; 'Der Erzdistrikt von Schneeberg, u.s.w.,' Cotta's Gangstudien, III., 1860; Die Erzgänge des Annaberger Reviers, Explanatory text to the Geological Map, Leipzig, 1894, p. 96.—G. C. LAUBE. Aus der Vergangenheit Joachimsthals, Prague, 1873; Geologie des Böhmischen Erzgebirges, Prague, 1876; Geologischen Exkursionen, 1884.—A. FRENZEL. Mineral Lexikon für das Königreich Sachsen, Leipzig, 1874, p. 167.—M. WEBSKY. 'Silberhornerz im St. Georgs Schacht zu Schneeberg,' Zeit. d. d. geol. Ges., 1881, Vol. XXXIII. p. 703.—F. BABANEK. 'Über die Erzführung der Joachimstaler Gänge,' Oesterr., Zeitschr. f. Bergbau- u. Hüttenwesen, 1884, pp. 1, 21, 61; Geolog. bergmännische Karte mit Profilen und Bildern von den Erzgängen in Joachimsthal. Published by the Minister for Agriculture, 1891.—LOHRMANN. 'Einiges aus der geologischen Vergangenheit des Erzgebirges,' Annaberger Berichte, 1898.—The following explanatory texts with the geological map of Saxony: F. SCHALCH, Johanngeorgenstadt Section; A. SAUER, Wiesental Section; and K. DALMER, E. KÖHLER, and H. MÜLLER, Schneeberg Section.—C. GÄBERT. 'Die geologischen Verhältnisse des Erzgebirges,' in Das Erzgebirge, by Prof. Dr. Zemmrich and Dr. C. Gäbert, 1911.

These lodes, famous by reason of the silver-cobalt mining to which they have given occasion, occur around Annaberg, Buchholz, Marienberg, Scheibenberg, Oberwiesental, Schneeberg, Johanngeorgenstadt, and Joachimsthal, the positions of which, both geologically and geographically, are given in Fig. 335. They lie to the south-west of the Freiberg-Brand district just described.

The silver lodes at Annaberg were discovered in 1492, since when more than 300 lodes have become known in the Annaberg-Buchholz district. In this district the country consists of grey contact gneiss with conformably intercalated layers of quartzite, hornfels, grauwacke, conglomerate, crystalline limestone, and amphibolite, all of which belong to the contact zone of the gneiss, while between Schlettau and Scheibenberg there is a considerable occurrence of augengneiss and coarse gneissic granite. All these rocks are either foliated eruptives or contact-metamorphic rocks. They mantle over the red Erzgebirge gneiss of Reizenstein-Katharinaberg and the grey eruptive gneiss of Freiberg.

The lodes belong partly .to the sulphide-lead- and tin groups of the older system, and partly to the cobalt-silver- and iron-manganese groups of the younger system. The more important have always been the cobalt-silver lodes, which group themselves in definite districts or fields wherein two principal lines of strike prevail, one north-south, the other east-west. Generally speaking these lodes may only be followed for 800 m. or so along the strike, and from 100 m. to 400 m. at the most, in depth. The most important of these fields is that around Annaberg, the centre of the former silver-cobalt mining.

The metalliferous filling consists of different silver minerals such as

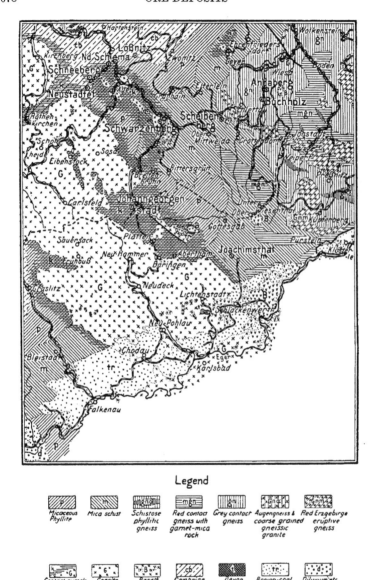

Legend

Micaceous Phyllite	Mica schist	Schistose phyllitic gneiss	Red contact gneiss with garnet-mica rock	Grey contact gneiss	Augengneiss & coarse grained gneissic granite	Red Erzgebirge eruptive gneiss

Contact aureole of the Granite	Granite	Basalt	Cambrian	Devon	Brown-coal formation	Diluvium etc

Fig. 335.—Geological map of the Upper Erzgebirge mining district. Scale, 1 : 500,000.
H. Credner.

dyscrasite, native silver, black silver, argentite, argentiferous marcasite, cerargyrite; of cobalt- and nickel minerals; and of pyrite, sphalerite, chalcopyrite, and scarce bismuth minerals. The gangue consists of quartz, barite, fluorite, and dolomite. Uranium ores represent a special primary zone, which is more closely described in the chapter dealing with uranium lodes. The deposits of this district are particularly rich at lode junctions and at intersection with certain seams known as *Schwebende*, these being bed-like or lode-like, flatly dipping attrition zones of highly decomposed gneiss, mica-schist, or clay, the last-named being sometimes quite sooty-black by reason of contained carbon. The attrition zones themselves contain no ore. We regard them as older than the lodes, towards the parent solutions of which they acted as impervious barriers, so impounding these solutions that great accumulations of ore became precipitated in the immediate neighbourhood.

The production of the Annaberg mines from 1496 to 1600 amounted altogether to 315,500 kg. of silver and 2423 tons of copper, having a total value of about 1·2 million sterling; from 1701 to 1850 about 7855 tons of cobalt ore were produced.

The Marienberg district to the north-east of Buchholz-Annaberg, in addition to the tin lodes described in the section on tin ores,[1] contains others belonging to the silver - cobalt- and sulphide zinc-lead groups. The geological circumstances are the same as those at Annaberg, though Marienberg lies still nearer the red eruptive gneiss of Reitzenstein-Katharinaberg. Two interesting lode - systems embracing together more than 100 lodes occur, which, striking north-east and north-west respectively, are fairly at right angles to each other; these directions correspond respectively to the Erzgebirge and Hercynian folds. The extent of these lodes along the strike, as exemplified in the cases of the Bauer-Morgengang, the Elisabeth-Flachengang, and the Eleonore-Morgengang, reaches at times more than 3 kilometres. The minerals present are galena, sphalerite, pyrite, silver-cobalt minerals, and pitchblende, while quartz, barite, fluorite, and dolomite, occur as gangue. Enrichment in precious metal was found at lode junctions and at intersection with the above-mentioned flat-lying attrition zones, here known as the 'black seams.' According to official figures, from 1775 to 1795 the ore won assayed 78 per cent of silver, though otherwise the general average has been but 0·94 per cent. Compared with Freiberg the lodes of this district are richer, but more broken. From 1520 to 1600 silver- and copper ore to the total value of about £625,000 was produced, of which amount about £220,000 was distributed as dividends. This district to-day is of no importance.

1 *Ante*, p. 425.

The Scheibenberg-Oberwiesental district likewise has no significance.

The lodes of the Schneeberg district, discovered in 1471 at a time when mining at Freiberg was declining, their discovery accordingly imparting new life to Saxon silver mining, are of real importance. Schneeberg, to the west of Annaberg, lies between the Eibenstock granite to the south, the Kirchberg granite to the west, and the granite outlier of Aue and Oberschlema to the east, in an area of Cambrian clay-slate and phyllite, which rocks to a considerable extent have been altered by the granite to spotted schist and andalusite-mica schist. The lodes continue also in considerable number into the granite.

Among the lodes an older system, containing tin, copper, and sulphide zinc-lead ores, may be differentiated from a younger system, with silver ores and quartzose cobalt-bismuth and iron ores. Although of the older system the copper lodes formerly were not without importance, the sulphide-zinc-lead lodes only will be discussed here. These contain abundant arsenopyrite, sphalerite, galena, pyrite, and chalcopyrite, less abundant tetrahedrite and molybdenite, with quartz as the principal gangue.

More important still however are the silver- and cobalt-silver lodes of the younger system which, though small in number, were formerly remarkable for their exceeding richness in silver. These occur in contact-metamorphic schists at Schneeberg and Schlema, as well as at Bockau and Aue, and carry silver minerals with subordinate cobalt, nickel, and bismuth, in a barytic gangue. It is stated that in the year 1477 a mass of mixed argentite and native silver weighing about 20 tons was found in the Fürsten adit of the famous St. Georg mine, at a spot where several lodes came together.

The most important lodes of the Schneeberg district are however those of the cobalt-bismuth group upon which mining is still proceeding. These occur within an area 10 sq. km. in extent, of which Neustädel is the centre. At that place the lodes are in such number that according to Müller the occurrence may be compared to a gigantic stockwork. The lodes strike west-north-west or north-north-west, dip steeply, and are usually 0·5 to 3 m. wide. Some are known for a length of 3 km. along the strike, and for more than 300 m. in depth. Along their extent they often split into veins which afterwards reunite. The lode-filling consists of grey and white smaltite, earthy cobalt, bismuth-linnæite, erythrite, niccolite, chloanthite, frequent native bismuth, bismite, a little silver ore, arsenopyrite, pitchblende, and other rarer uranium ores; together with quartz, hornstone, dolomite, and calcite, as gangue. According to Beck the ore in these cobalt lodes ceases as soon as the fissures penetrate the granite below the schists. An exception to this however is provided in the Weisser-Hirsch mine which is celebrated for its uranium ore; in that mine rich cobalt ores are also found within the granite. In this district also, flat attrition zones with carbonaceous alum-shale material are associated with a betterment in the metalliferous content.

The production of the Schneeberg mines, which belong partly to the State and partly to cobalt-blue works, may be gathered from the following table :

Year.	Argentiferous Co, Ni, and Bi Ore.	Uranium Ore.	Value in Pounds sterling.
1905	239·5	1·5	28,300
1906	235·75	...	26,850
1907	214·5	...	17,500
1908	207·75	...	20,600
1909	235·5	...	20,200

The Johanngeorgenstadt district in Saxony, where mines still continue in operation, lies to the south-south-east of Schneeberg and immediately at the boundary with Bohemia. The geological circumstances of this district are similar to those at Schneeberg ; highly metamorphosed Cambrian slates occur, which in part have been altered to andalusite-mica schist by the Eibenstock granite massive to the west, and by the Plattenberg granite outlier situated somewhat to the east. The lodes belong to the silver- and iron groups. The filling is quite similar to that of the Annaberg district though richer in bismuth- and uranium ores, the latter being particularly found at the Vereinigt mine in the Fastenberg. The importance of this district in bismuth mining may be seen from the following table :

Year.	Bismuth Ore.	Pitchblende.
	Metric Tons.	Metric Tons.
1905	52·3	2·7
1906	42·8	2·5
1907	39·0	0·9
1908	39·9	...
1909	42·7	...

The Joachimsthal district in Bohemia, lies south-east of Johanngeorgenstadt, in the neighbourhood of the great line of disturbance with which the celebrated springs of Carlsbad, etc., are connected.

Geologically, this district consists of mica-schist, which rests upon the Eibenstock granite to the west, and to the north is in turn overlaid by the Cambrian beds of the Johanngeorgenstadt district. This mica-schist also bears evidence of contact with the granite ; it consists petrographically of dark phyllite-like graphitic mica-schist, banded mica-schist, calcite-

and scapolite-mica schist, gneissic mica-schist, and coarsely fibrous garnetiferous mica-schist ; it also contains interbedded layers of limestone and
hornblende rocks. These schists are crossed by dykes of quartz-porphyry
and isolated dykes of basalt and phonolite.

The lodes, which belong to the silver-cobalt group, occur chiefly in
the phyllite-like graphitic mica-schist, which forms several south-east
striking zones dipping north and north-east. The principal directions of
these lodes are north-south to north-north-east, the so-called 'midnight
lodes,' and east-west or east-south-east, the so-called 'early lodes.' The
latter dip 50°–80° always to the north ; the former 40°–80°, sometimes
to the east and sometimes to the west. The width generally varies between
0·15 and 0·6 m., though exceptionally it reaches 1–2 metres.

The ore consists of smaltite, bismuthinite, native bismuth, bismite,
nickel minerals, and pitchblende, the occurrence of this last being discussed
in a later chapter. At the lode junctions, silver enrichments consisting of
native silver, pyrargyrite, proustite, argentite, and black silver, are found.
The lode-filling, particularly that of the early lodes, consists largely of slaty
and clayey material and less frequently of quartz or calcite. In the eastern
extent of the midnight lodes hornstone and dolomite occur plentifully.
The distribution of the ore is irregular and discontinuous. The primary
depth-zones and the part played by the pitchblende are referred to when
dealing with the uranium lodes.

The most important lodes of this still famous district are the Hilde-
brand, Geister, Fluder, and the Edelleutstollen lodes. The important
mines are the State mines of Joachimsthal, the Edelleutstollen, and
the Hilfe-Gottes. The development of the district was begun in the
sixteenth century. In 1520 more than 8000 miners were employed and
hundreds of mines were at work. Rich ore-bodies were found at first
immediately below the surface and some of the mines worked without lamps. It was in this district that silver was first made into coins,
these being known as ' Joachimsthaler ' ; hence the term ' Thaler.' At
present however the district is only of importance in relation to uranium
ores for the extraction of radium, the output of these ores latterly
having been as follows :

1896	.	.	. 30·00 tons.	1903	.	.	.	9·18 tons.
1897	.	.	. 51·00 ,,	1904	.	.	.	8·08 ,,
1898	.	.	. 52·00 ,,	1905	.	.	.	16·35 ,,
1899	.	.	. 46·00 ,,	1906	.	.	.	0·00 ,,
1900	.	.	. 17·00 ,,	1907	.	.	.	0·00 ,,
1901	.	.	. 16·15 ,,	1908	.	.	.	9·18 ,,
1902	.	.	. 11·00 ,,	1909	.	.	.	8·08 ,,

Concerning the geological age of the Erzgebirge lodes, the first rending of
the older fissures and the first filling of the lodes took place presumably in late

Carboniferous-Rotliegendes time ; that this filling itself was subsequently again rent is probable. These old lodes were the source of the amethyst and quartz boulders which are found in the Cretaceous beds in the neighbourhood of Freiberg. The filling of the younger lodes was probably contemporaneous with the main fissures of the Harz, in which case they would be of Miocene age.

The condition of lead-silver mining in Saxony is greatly dependent upon the metal prices. These of late years have been very low. The silver contained in the ore delivered from the Saxon mines to the Freiberg smelting works in the year 1910 amounted to 6421·8 kg. as compared with 7898·8 kg. in the year previous, the corresponding values being £17,000 and £19,600 respectively. Freiberg itself not many decades ago produced 25,000–30,000 kg. of silver yearly. The market conditions for lead also have been unsatisfactory. The lead ore sent from the Saxon mines to the Freiberg works in 1909 contained 1487 tons of lead valued at £12,800, whereas in the year previous these figures were 1493 tons and £13,400 respectively. Owing to the control of the zinc convention the condition of the zinc market in 1909 was more favourable.

According to official statistics the ore produced in the Freiberg district in 1909 was 11,120 tons valued at £32,000. The Marienberg-Scheibenberg district, apart from a few tons of zinc ore, produced nothing. The Johanngeorgenstadt district produced 2662 tons valued at £4240 ; the Schneeberg district, 2746 tons at £23,100. The total production of Saxony in rich silver ore and argentiferous lead-, copper-, arsenic-, zinc-, and sulphur-ores in 1909 amounted to 7617 tons worth £34,300. In addition, 4117 tons of arsenopyrite, pyrite, and chalcopyrite, were produced, having a value of £2450 ; 173 tons of sphalerite worth £245 ; 288 tons of bismuth-, cobalt-, and nickel ores worth £23,100 ; and 0·29 tons of pitchblende worth £37.

THE OBERHARZ

The Harz mountains represent the south-east striking, or Hercynian core of a mountain range consisting of Devonian and Culm beds, these beds themselves being disposed in north-east or Dutch folds. To the north this range is separated from the fore-ground by lines of dislocation, so that a combination of anticline and uplift exists.

It was formerly considered that a sinuous and undulating anticlinal axis of the oldest rocks existed, upon the two lateral flanks of which, and in three synclines occurring between them, the upper beds were laid. Of these anticlinal flanks, that to the north-west would be represented by the Oberharz, and that to the south-east by the Unterharz, while the three synclines would be those at Elbingerode and Selkemulde to the north, and that at Ilfeld to the south. In addition, the geological position is materially conditioned by two granite intrusions of Carboniferous age, the Brocken massive occurring between the Oberharz and the Elbingerode syncline, and the Ramberg massive between the Elbingerode and Selkemulde synclines. Farther to the east the Mansfeld syncline adjoins the Unterharz. Recent investigation however has not supported this widely held tectonic representation, though the surveys are not yet sufficiently advanced to allow a new anticlinal axis to be formulated.

The Oberharz is a high plateau consisting chiefly of Culm beds disposed in conformity with Dutch folds, that is to say, as illustrated in Fig. 35, they strike north-east. Beneath these in the northern portion of the district, Devonian beds appear, which, as illustrated in Fig. 22, with north-east strike and much detailed folding, form roughly an air-anticline normally overlaid on its north-west and south-east flanks by Culm beds, and to the south-west and north-east bounded by fault escarpments. The fore-ground consists of Zechstein, Triassic, Jurassic, and Cretaceous beds.

THE CLAUSTHAL LODES

LITERATURE

C. ZIMMERMANN. Die Wiederausrichtung verworfener Gänge und Flöze, 1828.—A. v. GRODDECK. ' Über die Erzgänge des nordwestl. Oberharzes,' Zeit. d. d. geol. Ges., 1866, p. 693 ; ' Erläuterungen zu den geognostichen Durchschnitten durch den Oberharz,' Zeit. f. d. Berg- Hütten- und Salinenwesen im preussichen Staate, 1873.—v: KOENEN. ' Beitrag zur Kenntnis von Dislokationen,' Jahrb. der preuss. geol. Landesanst. for 1887.—F. KLOCKMANN. Berg- und Hüttenwesen des Oberharzes, Stuttgart, 1895, p. 43.—L. BEUSHAUSEN. ' Das Devon des nördlichen Oberharzes mit besonderer Berücksichtigung der Gegend zwischen Zellerfeld und Goslar,' Abhandl. d. k. p. geol. Landesanst., New Series, Part 30, p. 73.—B. BAUMGÄRTEL. Oberharzer Gangbilder. Leipzig, 1907.—ZIRKLER. ' Über die Gangverh. der Grube Bergmannstrost bei Clausthal,' Glückauf, 1897, 1900.—G. KÖHLER. ' Beitrag zur Kenntnis der Erdbewegungen u.s.w.' Berg- und Hüttenm. Ztg., 1897, p. 343 ; ' Ein weiterer Beitrag zur Kenntnis der Erdbewegungen,' Berg- und Hüttenm. Ztg., 1901, p. 201.

The lodes of the Oberharz extend over the entire Clausthal plateau, around Grund, Wildemann, Clausthal-Zellerfeld, Lautenthal, Bockswiese, and Schulenberg, in a district 18 km. long and 8 km. wide. In greater part and as illustrated in Fig. 56, they are fault fissures filled with ore and gangue. Such other lodes as represent the filling of fissures along which no movement took place, play but a subordinate part. The general strike is east-south-east to south-east, and the dip 70°–80° to the south. Generally the lodes do not occur singly and independently, but several more or less parallel fissures occurring close to one another are linked

together by subsidiary branches; lodes so related to one another form
what is best described as a lode-series. The occurrence is illustrated in
Figs. 3 and 336 representing the different lode-series in the Oberharz,

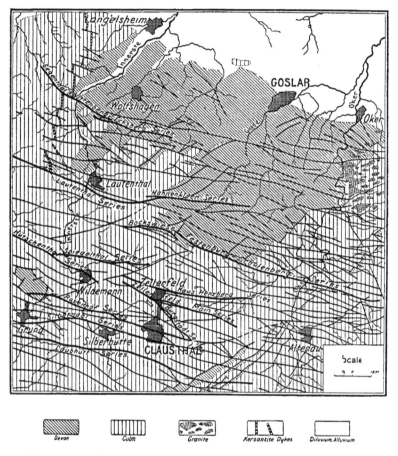

FIG. 336.—The different lode-series of the Oberharz. L. Beushausen, *Kgl. Pr. Geol. Landesanst.*

many of which extend 8–10 km. or more along the strike. These from
north to south are as follows :

    1. The Gegenthal-Wittenberg Series.
    2. The Lautenthal-Hahnenkleer Series.
    3. The Bockswiese-Festenburg-Schulenberg Series.

4. The Hütschenthal-Spiegethal Series.
5. The Haus Herzberg Series.
6. The Zellerfeld Main Series.
7. The Burgstadt Series.
8. The Rosenhof Series.
9. The Silbernaal Series.
10. The Laubhütte Series.

In form, most of the lodes—having generally a distinct clay-parting on the foot-wall while towards the hanging-wall merging gradually into undisturbed country—are composite lodes in the sense of Cotta. The width in consequence is often considerable and may be as much as 40 m., nor to the depth of 910 m. so far reached has any general diminution of width been established. The well-known flucan-faults or *Ruscheln* of the Harz are generally without influence upon the lodes, though occasionally a splitting or deflection takes place where these are crossed ; such splitting for instance occurs with the Rosenhof Series, and such deflection with the Zellerfeld and Burgstadt Series. In the numerous slickensides and in the occurrence of lode-slate there is evidence that earth movements subsequently took place along the lode fissures.

The lode-filling consists of ore, gangue, and country-rock. The most important ore is galena with 0·01–0·3 per cent of silver, that is, 100–3000 grm. per ton ; next in importance comes sphalerite, which at Lautenthal preponderates ; chalcopyrite, tetrahedrite, and bournonite are uncommon ; gersdorffite only occurs locally ; and the selenides of lead and copper only as rarities. Quartz is the prevailing gangue, calcite is less frequent, and the occurrence of barite is limited to the different southern series, and principally to the western portion of these. This occurrence of barite can be demonstrated to be referable to the Zechstein country; indeed throughout Germany the distribution of this mineral often coincides with that of the Zechstein and Bunter formations. The extent to which the country-rock, —grauwacke and clay-slate—participate in the lode-filling, varies.

The lode structure is generally banded or crusted ; less frequently it is irregularly coarse. The zonal incrustation of fragments of country-rock to form concentric ore, illustrated in Fig. 126, is characteristic. Quartz and galena appear to be the oldest minerals, following which come chalcopyrite, sphalerite, and calcite, in varying proportions, while barite, siderite, strontianite, and marcasite, probably represent a younger generation. The ore generally occurs in shoots which, contrary to the steeply dipping lode in which they lie, are flat, having usually a dip of not more than 45°. Where two lodes come together rich ore-bodies are often found.

The first fracturing of the lode fissures was presumably closely connected with the upheaval of the Brocken massive in Upper Carboniferous time, subsequently to which, and in response to later tectonic phenomena, these fissures were again repeatedly opened. The first filling however took place immediately after the first fracturing. Since, though seldom, galena is found in fissures and cavities in the Zechstein, it may safely be assumed that with the later tectonic phenomena metalliferous solutions also ascended.

The importance of the lodes of the Oberharz may be gathered from the following figures. In the year 1908 eight mines together produced 265,000 tons of argentiferous galena and sphalerite, containing on an average 7 per cent of lead and 8·7 per cent of zinc. This ore had a value of about £202,000 at the mine. During the same year the total production of lead-zinc ore in Germany was 2,913,000 tons with 11 per cent of zinc and 3·9 per cent of lead, while the total value of the lead-, silver-, and zinc ores amounted to £1,815,000.

St. Andreasberg

LITERATURE

H. CREDNER. 'Geogn. Beschreibung des Bergwerksdistriktes von St. Andreasberg,' Zeit. d. d. geol. Ges., 1865. — C. BLÖMECKE. Die Erzlagerstätten des Harzes. Vienna, 1885.—F. KLOCKMANN. Berg- und Hüttenwesen des Oberharzes, p. 50, Stuttgart, 1895. —A. BODE. 'Das Nebengestein der St. Andreasberger Silbererzgänge und dessen Beziehungen zur Erzführung,' Zeit. d. d. geol. Ges., 1908, Monatsbericht No. VI. p. 133.— WERNER. 'Die Gangverhältnisse von St. Andreasberg,' Der Bergbau, 1908, No. 47 ; 'Die Silbererzgänge von St. Andreasberg im Harz,' Glückauf, Berg- und Hüttenm. Zeit., 1910, Nos. 29 and 30.

Though a distance of but 16 km. intervenes, the district of St. Andreasberg is quite different from that of Clausthal. The lodes at St. Andreasberg occur in Palæozoic beds bordering the Brocken massive, to the south of the Bruchberg range. The extension of the individual Palæozoic stages is dependent upon considerable dislocations, and particularly upon the two outside faults known respectively as the Neufang flucan to the north, and the Edelleute flucan to the south. These two breaks enclose between them a pronounced wedge of ground, the thin end of which, as illustrated in Figs. 155 and 337, is directed to the west. North of the Neufang flucan the Culm, represented by grauwacke, clay-slate, and silica-schist, occurs, while the wedge itself and the country south, are Devonian. In addition to Lower Devonian beds, which may be correlated with the Upper Coblenz Series, the deep-sea facies of the Middle Devonian are, according to Bode, represented by the Wissenbach slate. Further, according to the most recent correlation, the Lower Devonian, with a considerable shortage of its members, along the Neufang flucan abuts against the Culm to the north. The Lower Devonian beds in this district are therefore especially concerned in the ore-deposits. The northern portion of the district is already in the contact zone of the Brocken granite; the beds there strike north-east, are much contorted and sometimes inverted, and usually dip to the south-east.

The silver lodes are exclusively found in the wedge between the two flucans mentioned, this wedge being some 3 km. long and at its base 1 km. wide. The northern or Neufang flucan is an overthrust which strikes north-east and dips steeply to the south-east. Along it the Lower Devonian beds have been forced up over the Culm. It is some 12 m. wide and cuts the formation at an acute angle. The southern or Edelleute flucan is likewise about 12 m. wide; it strikes east-west and dips steeply to the south. According to Bode, its southern wall in the eastern portion appears to have sunk, while in the western portion on the other hand it appears to have risen, that is, to have been overthrust. The complete part played by these dislocations in the stratigraphy of the district will however only be made clear by still further investigation.

According to all appearances the Neufang flucan represents an overthrust whereof the direction of the movement has not yet been satisfactorily established. Within the wedge two overthrusts of smaller width, known respectively as the Abendrot and Silberburg flucans, occur, though these unfortunately are no longer approachable for further investigation. It is assumed that these three overthrusts, the Neufang, Abendrot, and Silberburg, are contemporaneous among themselves but older than the Edelleute flucan, which is regarded as an ordinary fault.

Although the silver lodes are limited to this wedge between the outside flucans, other lode fissures containing ore are found unfettered by this limitation. Iron lodes, for instance, occur north of the Neufang flucan and also farther to the west, while barite-copper lodes are found to the south of the Edelleute flucan. The north-west-striking Engelsburg lode, situated roughly 1·5 km. south-east of St. Andreasberg, and containing chalcopyrite, silver-free galena, calcite, and quartz, is well known.

Within this wedge the silver lodes occur chiefly to the west, where according to their strike they form two series, one striking north-west and the other striking east. To the former series belong the Wennsglückt, Jakobsglück, Samson, Andreaskreuz, Franz-August, Felizitas, Fünf-Bücher-Moses, and the Prinz Maximilien lodes, while to the latter belong the Neufang, Gnade-Gottes, Julian, Bergmannstrost, and Morgenrot lodes. All dip steeply to the north or north-east, those of the first series at 80°–90°, those of the second at 70°–80°· At the Neufang flucan they split up or are dragged till they disappear. At the Abendrot and Silberburg

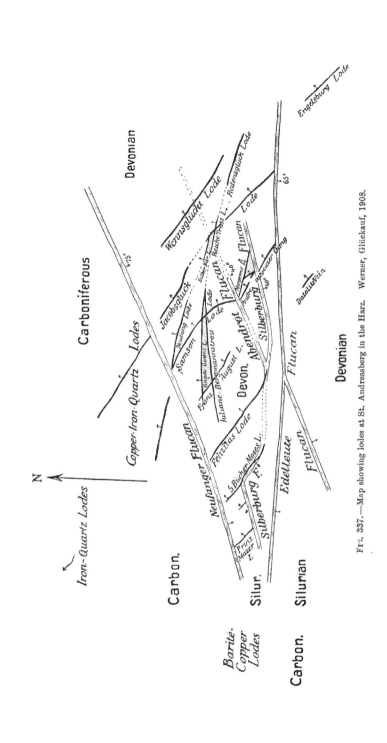

FIG. 337.—Map showing lodes at St. Andreasberg in the Harz. Werner, Glückauf, 1908.

flucans they behave similarly though with this difference, that they reappear and continue on the other side of these flucans. Their relations to the Edelleute flucan can no longer be studied ; Werner however assumed that this fault is younger than the lodes. All the lodes become smaller in soft slaty rock, and particularly in the Wissenbach slate above the diabase. The silver lodes are simple lodes, generally less than 1 m. in width and with sharply defined walls. They are often severed by secondary faults which strike north - east and dip south-east, the eastern portion of the lode at such severance being usually forced over the western ; the lateral displacement does not however amount to more than 1–2 metres. Werner assumed that these secondary faults are younger than the lodes and contemporaneous with the Neufang flucan.

The lode-filling, which is almost invariably fast cemented to the rock, consists chiefly of whitish calcite of an older generation, with impregnations, veins, and nests of quartz, and of galena, sphalerite, native arsenic, pyrargyrite, dyscrasite, arsen-argentite, and native silver. Less important are breithauptite, niccolite, smaltite, and fluorite, while barite is quite an isolated occurrence. The ore- and gangue-minerals in vugs belong to a second generation, one which is excellently developed crystallographically, in which connection Klockmann has called attention to the beautiful and many - faced crystals of calcite with pyrargyrite, and pyrostilpnite. Noteworthy in addition are the numerous zeolites, such as apophyllite, analcime, harmotone, desmine, stilbite, and natrolite ; and also datolite and fluorite. The last-named is infrequent and only of mineralogical interest ; according to Werner it is younger than the quartz but older than the younger calcite and the zeolites.

The Wennsglückt lode to the north-east, exhibits some points of exceptional character. In the upper levels it carried limonite ; below this for some depth it was barren ; while in greater depth it contained sporadic chalcopyrite, a little galena, tetrahedrite, proustite, and pyrargyrite.

Barite as a part of the lode-filling has only been observed in the Prinz Maximilien lode and in a branch of the Samson lode, in both cases close under the surface. It has however also been found together with zeolites and pyrite in the country-rock, at a depth of 750 metres. The distribution of the ore-bodies along the lode plane is usually quite irregular. The primary relations of galena to sphalerite are interesting; the galena in depth appears to give place to the sphalerite.

The difference between these lodes and those at Clausthal is striking. At St. Andreasberg, where silver greatly exceeds the other metals in value, zeolites are common, while fluorite is exceptional, and barite practically absent. Lossen regarded these differences as expressive of different depth-zones, the St. Andreasberg lodes representing a deeper zone nearer the Brocken granite than that which would include the lodes at Clausthal. Klockmann however is of opinion that the diabase so frequently present at St. Andreasberg and absent from Clausthal, materially contributed to these differences. According to Bode, the distribution of the ore-bodies is partly dependent upon the nature of the country-rock, calcareous rocks such as intercalated limestone or calcareous diabase, producing an enrichment.

Werner divided the ore-minerals into three different groups according to the manner of their formation, namely : (1) those deposited from original solution ; (2) those formed in depth by the action of subsequent solutions upon those of the first group ; (3) those formed in the upper levels and gossan by the action of meteoric waters upon the first group ; in this last group are also included those formed in the mines by mine water. According to him the original ores are dyscrasite, arsenic, galena, sphalerite, tetrahedrite, and less frequently, antimony, breithauptite, smaltite, löllingite, chalcopyrite, pyrrhotite, and pyrite. Among these he distinguished three ages, the oldest of which included the native metals, arsenic, and antimony, etc. ; the next, galena, sphalerite, chalcopyrite, and tetrahedrite ; while the youngest included pyrrhotite and pyrite, together with the zeolites extracted from the country-rock. Simultaneously with the formation of the pyrrhotite and pyrite an alteration of the earlier original ores began, whereby native silver resulted in the deeper horizons, and pyrargyrite in the shallower. Then also the bulk of the zeolites became formed. Upon subsequent oxidation by meteoric waters the usual oxidation minerals resulted.

The view of Werner concerning the age and formation of these ores, which ascribes a material part to halurgo-metamorphism, that is, the action of saline solutions, can only be endorsed in part by the authors. In their opinion the features observed at St. Andreasberg do not need the application of this metamorphism for their elucidation. These ores are in no way different from those of other silver districts where it has not been thought necessary to call this metamorphism in aid. In Werner's discussions of the subject the failure to mention cementation ores is striking, since such must have been present, as those solutions which brought about oxidation must lower down also have effected cementation. Again, Werner assumed several periods of mineralization by ascending heavy-metal solutions ; later solutions must therefore have come into contact with sulphides already deposited from the earlier solutions, and cementation ores must have resulted.

The rich silver ores at St. Andreasberg differ from the rich cementation ores of other districts only in that they have been found to continue to an abnormal depth, it being otherwise usual to meet such ores more particularly above ground-water level. In this respect the St. Andreasberg district resembles the silver-copper district of Butte, Montana, where likewise all possible theories were advanced in explanation of the appearance of cementation ores at such relatively great depth; till finally, Emmons explained the circumstance by tectonic causes, namely, a depression of the district, which so far as the deposit was concerned would be equivalent to a rising of the ground-water level.

So long however as the earth movements in the wedge which forms the lode district of St. Andreasberg are not understood, the formation of the rich silver ores there will lack satisfactory explanation. The important question still to be answered is whether this wedge in the course of geological time has, in relation to its surroundings, sunk, or been raised. According to Krusch the nature of the ores suggests a depression of the district, equivalent to a gradual rising of the water-level. In such a case rich cementation ores might very well be found deep below the present water-level. Galena and sphalerite, as before, would be primary ores, galena forming an upper primary zone, and sphalerite a lower. The silver content of the former, probably a high one, would provide the wherewithal to form rich cementation ore, since after having been taken into solution by the descending surface waters, and after the consumption of the oxygen brought in by these waters, it would become precipitated by the reducing action of the galena and sphalerite. The sinking of this wedge would then take such cementation ore into depth with it.

The occurrence of zeolites may perhaps be ascribed to the direct influence of the eruptive rocks found in the neighbourhood. To this influence indeed Krusch ascribed the striking difference between the lodes of this district and those at Clausthal, being moved thereto by the fact that the lodes at Clausthal, towards the south, that is in the direction of St. Andreasberg, possess an increasing silver content, those around Grund having actually the highest silver content of any sulphide lead-zinc lode in Germany, in primary ore.

The lodes at St. Andreasberg, discovered in 1521, reached their zenith in the second half of the sixteenth century, after which followed a period of quiescence. Then, after being taken up again about the middle of the seventeenth century, work proceeded without cessation till 1910, when on the exhaustion of the ore-bodies it again stopped. Of late years the output has been but a few tons of ore annually.

THE LODES OF THE RHENISH SCHIEFERGEBIRGE

1. THE BERG DISTRICT

LITERATURE

E. BUFF. Beschreibung des Bergreviers Deutz. Bonn, 1882.—A. SCHNEIDER. Lagerstättenkarte des Bensberger Gangreviers. Bonn, 1882.—W. PETERSSON. 'Die Blende- und Bleigruben Berzelius und Lüderich im Bergischen Lande,' Berg- u. Hüttenm. Ztg., 1899, p. 601.—L. SOUHEUR. 'Greenockit, Wurzit, und Smithsonit von der Grube Lüderich bei Bensberg,' Zeit. f. Kristall. u. Min., 1884, Vol. XXIII. p. 549.—H. v. DECHEN. Erläuterung der geologischen Karte der Rheinprovinz und Westfalens. Bonn, 1870.— ZELENY. 'Das Unterdevon im Bensberger Erzdistrikt u.s.w.,' Arch. f. Lagenstätt. Geol. Landesanst. Berlin, 1912.

The Berg Hills, consisting of Middle and Lower Devonian slates with subordinate intercalated limestone, rise out of the Rhine valley east of Cologne. The beds strike north-east and include from the oldest Gedinnian with its lower arkose and upper red-coloured slate on the one side, to what appears to be upper Middle Devonian limestone on the other ; since however the boundaries of the individual beds are determined in part by large breaks it cannot yet be said whether the sequence of beds so embraced is complete. At Bensberg for instance, the Gedinnian comes against considerably younger limestones of the Lower and Middle Devonian.

This Devonian formation around Berg-Gladbach, Bensberg, Imme-keppel, and Engelskirchen, contains many lead-zinc lodes, the more important of which have been or are worked in the Lüderich, Bliesenbach, Weiss, Berzelius, Castor, and Pollux mines. The majority of these lodes are found to the south of the Berg-Gladbach limestone syncline, where they extend to the east and south, to the Bröhl river and to the Sieg respectively; few occurrences are known in the Devonian slate north of this syncline. To the north the district is bounded by the Dierscheid and Lennef rivers.

The Lower and Middle Devonian slate, consisting of grauwacke and subordinate clay-slate, rests upon the Gedinnian, with which it forms anticlines and synclines, these in places being steeply inclined. As Zeleny has shown, a relation undoubtedly exists between the tectonics of the district and the occurrence of lodes, these latter occurring along faults and subsidences, a similar relation in the case of Siegerland having been demonstrated by Denckmann. This geological position may be particu-larly well observed in the occurrences at and in the neighbourhood of Lüderich.

While the strike of the lodes varies greatly, the dip, which is generally between 60°–70° and but rarely as low as 45°, is always steeper than that of the country-rock, though in the same direction. In relation to width and

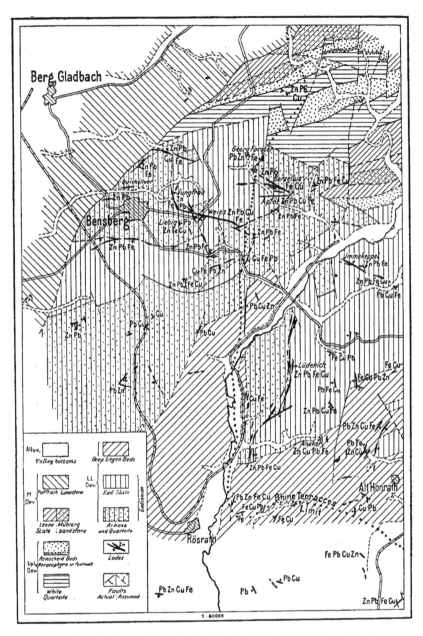

FIG. 338.—Map showing lodes in the Berg district. Zeleny, *Geol. Landesanst.* 1912.

nature the lodes exhibit great differences. The narrow lodes show distinct walls, some of these being clean fissures and others clay-partings. The filling consists of decomposed clay-slate and fragments of grauwacke and arkose, traversed by quartz veins and ore. With large lodes this brecciated structure is less pronounced. In these, large blocks of country-rock, which though wrenched from the parent mass, much folded, broken, contorted, and traversed by fractures, have usually maintained their internal coherence, constitute the principal mass, which consequently often presents a banded appearance. With lodes so filled a distinct separation from the country-rock no longer exists, and often only through the unmistakable occurrence of a lithomarge-like material in cracks in these large blocks, may it be determined whether any particular mass is within the lode or not. The clay-slate within the lode is often transformed to black lode-slate traversed by numerous fracture planes and slickensides. Where many individual veins come together and form what is best described as a lode zone, the foot-wall is in places well marked and definite, while not infrequently, on the other hand, and chiefly in the hanging-wall, there is a gradual passage from shattered material to regular country-rock. These wide occurrences are in fact composite lodes in the sense of v. Cotta.

Along the strike the lodes are of very different length; while most attain but 50 m., the large zones can be followed for some kilometres. The Max lode for instance can be followed for 1 km., and the Lüderich lode-series for about 4 kilometres.

Within the lode the ore is irregularly distributed, ore-bodies being surrounded by barren or poor parts. In such bodies the ore may be either irregularly coarse or disseminated. They may either occupy the entire width, in which case the richest parts are often found at the walls, or they may occur more in the centre of the lode, in which case they sometimes have their own walls, and sometimes are intergrown with the poorer material. They almost always agree in strike with the lode, and usually form a series of lenticular or globular bodies within the lode. In depth they often contract in dimensions like a funnel, and deeper still are succeeded by other bodies similarly disposed.

The ore consists chiefly of compact, fine-grained to coarse galena with a silver content usually between 200 and 500 grm. per ton, but which may rise to as much as 7000 grm. Of similar importance and frequency is the coarsely-crystalline black-brown sphalerite which always carries cadmium. Chalcopyrite is found almost everywhere, though in such small amount as to be without economic importance; pyrite also is not uncommon in fractures and vugs. Siderite is quite common, and cases are known where lodes worked to-day for their contained galena and sphalerite, were originally worked for siderite. As this ore in some mines also occurs plentifully in depth, it does not appear to constitute any particular primary horizon. With sphalerite and galena, however, the case is different. Generally with these lodes galena in depth is replaced by sphalerite, the latter therefore representing a deeper primary zone. The amounts of these ores present, varies in different lodes as well as in one and the same lode. At Lüderich the average relation for twenty-nine years between sphalerite and galena was as 100 : 5·7 ; at Blücher as 100 : 10·9 ; at Berzelius as 100 : 31·5 ; and at Apfel as 100 : 41·2. In several cases developments in depth have been less favourable, massive rich ore has given place to disseminated ore or to ore mixed with country-rock, and in general a considerable impoverishment has become established.

The large old workings which mark the outcrops of the larger lodes are evidences of earlier mining operations, doubtless upon lead and silver enrichments. Concerning such workings no reliable data are however available, though it is certain that they were worked by the Romans. Reliable records pertaining to the Lüderich mine date back to the year 1250, when Archbishop Konrad of Hochstaade is stated to have worked the mine in order to obtain money for the building of the Cologne Cathedral. The present working of this, the most important mine of the Berg district, dates from the late 'fifties, when the recovery of zinc from sphalerite was introduced. The importance of Lüderich may be gathered from its production. This in 1880 amounted to 21,742 tons of zinc ore and 6461 tons of lead ore, having together a total value of about £92,000 ; and in 1911 to 12,600 tons of sphalerite and 1570 tons of galena. This mine is now working to a depth of 80 m. below the level of the valley.

2. The Holzappel Lode-System

LITERATURE

A. BAUER. 'Die Silber- Blei- und Kupfererzgänge von Holzappel an der Lahn, Wellmich und Werlau am Rhein,' Karstens Archiv für Bergbau u.s.w. Vol. XV., 1841.—FR. WENKENBACH. 'Beschreibung der im Herzogtum Nassau an der unteren Lahn und dem Rhein aufsetzenden Erzgänge, sowie eine kurze Übersicht der bergbaulichen Verhältnisse derselben,' Nassauisches naturwissensch. Jahrb., 1861, Vol. XVI.—L. SOUHEUR. 'Die Lagerstätte der Zink- Blei- und Kupfererzgrube Gute Hoffnung bei Werlau am Rhein,' Jahrb. d. k. geol. Landesanst., 1892, p. 96.—Erläuterungen zu Blatt Schaumburg der geologischen Spezialkarte von Preussen 1892, by E. KAYSER, the Ore-Deposits by A. SCHNEIDER.—EINECKE. 'Die südwestliche Fortsetzung des Holzappeller Gangzuges zwischen der Lahn und der Mosel,' Bericht der Senkenbergischen naturforschenden Gesellschaft, 1906.—HOLZAPFEL. 'Das Rheintal von Bingerbrück bis Lahnstein,' Abhandl. d. k. pr. geol. Landesanst., New Series, XV., 1893.—Descriptions of the Mining Districts Wiesbaden-Dietz, 1893, by HOLZAPFEL, ULRICH, KÖRFER, etc., published by the Royal Mining Department at Bonn.—W. SCHÖPPE. 'Der Holzappeler Gangzug,' Kgl. Preuss. Geol. Landesanst., Arch. f. Lagerstättenforschung, Part III., Berlin 1911.

The country of the Lower Lahn consists chiefly of Lower Devonian, which to the east is overlaid by Middle and Upper Devonian. The beds strike north-east and occur in many parallel folds, which being over-turned the beds almost invariably dip south-east. In relation to the lodes, only the Lower Devonian beds come into question, namely, the Hunsrück slate as lowermost bed, and then the Lower and Upper Coblenz,

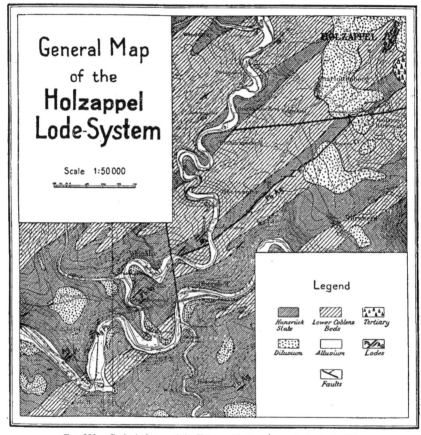

Fig. 339.—Geological map of the Holzappel lode-system. Scale, 1 : 50,000.

these consisting of various clay-slates, grauwackes, and quartzites, with porphyroid slates and diabase. In consequence of the numerous faults and the rarity of horizons with characteristic fossils, the correct correla-tion of any particular bed is not easily determined. The porphyroid slates belong to the deepest horizon of the Lower Coblenz; these rocks are generally regarded as dynamically metamorphosed eruptives and tuffs.

The rock known as white rock deserves special notice. This formerly was regarded either as a bedded occurrence, in which case it would represent an altered slate or porphyroid, or as a dyke, when it would represent a decomposed diabase. Schöppe endorses the view expressed by Rosenbusch that this white rock has resulted chiefly from thermal metamorphism, and only recognizes as white rock that which occurs in dyke form and was originally diabase. According to him, this white rock is older than the present lodes, the solutions which formed these being the last to invade the rock in question.

Both the beds and lodes are affected by a large number of tectonic disturbances, of which the various lateral displacements along the lodes are the oldest, these being presumably of Devonian age. Somewhat younger are the disturbances which brought about a change in level. These strike somewhat more northerly than the beds, and probably represent overthrusts ; they dip at an angle of 10°–30° to the south-east and pinch the lodes both in strike and dip. They are younger than the lodes and probably belong to post-Culm orogenics. True faults are represented by fissures cutting across the formation at an acute angle. Of these, the two most important are known respectively as the ' morning ' and ' evening ' main faults.

According to Kayser, whose view is endorsed by Schöppe, the Holzappel lode-system is a composite strike-fault not invariably accompanied by dislocation. The general coincidence in strike and dip of the lode and country-rock, in spite of small transgressions, is justification for regarding the deposit as a bedded system. This system strikes east-north-east and dips 52° to the south-east. It consists of five lodes, of which one is the main lode, three are branch lodes, and the fifth a transverse lode. The width of the main lode varies between 0·6 and 7 m. ; in depth it appears to be more regular, though contractions in its width are not infrequent. At Holzappel it has been followed for more than 2200 m., and at Leopoldine and Luise for more than 1200 metres. It has been opened to the sixteenth level, that is, to a depth of 342 m. below surface. Of the branch lodes, that in the foot-wall, 0·2–0·3 m. in width, is the most important ; that in the hanging-wall is 0·10–0·15 m. wide ; while that out in the hanging-wall and on the thirteenth level 40 m. distant from the main lode, consists of two veins each 0·15 m. wide. The transverse lode strikes south-south-west and dips 72° to the east. Its width, which is 0·50 m. in the upper levels, diminishes in depth to 0·10–0·12 metres. The influence of the country-rock is evident only in the form of the fissure, which is most regular in grauwacke-slate, splits up in the raw grauwacke, and is still more indefinite in soft clay-slate. Although generally the lodes of this district are composite, simple lodes also occur.

The principal ore-minerals are argentiferous galena, sphalerite, siderite, chalcopyrite, tetrahedrite, and less frequently pyrite, while quartz forms the principal gangue. Calcite and dolomite are found together with sphalerite as a later generation along transverse fractures. The distribution of the ore within the lode is fairly regular since no large barren stretches are met. The richest masses of galena were found in the main lode, from the upper levels down to the third deep level 42 m. below the surface. The silver content is greatest with the fine-grained and compact varieties, wherein it amounts to as much as 0·15 per cent, or 1500 grm. per ton, while the general average at present is but 48–55 grm. per ton. The siderite was found in the upper levels of the main lode. The tetrahedrite gives the impression of being a cementation ore. The amount of chalcopyrite decreases rapidly in depth.

Concerning relative age, the determination by Schöppe that the galena is younger than the sphalerite and the older generation of chalcopyrite, is important. The siderite is in part older than the diabase, though it still continued to be formed after the intrusion of that rock. In Siegerland, not far distant, it is entirely older than the diabase. The following figures from Schöppe indicate the importance of the Holzappel mines :—

Year.	Lead Ore.	Silver (per Ton) in Lead Ore.	Zinc Ore.	Percentage recovered from Raw Ore.	
				Galena.	Sphalerite.
	Tons.	Grm.	Tons.		
1896	3099	57·59	8374	7·760	20·823
1897	3603	58·45	8201	8·306	18·905
1898	3501	55·83	8581	7·697	18·806
1899	3488	54·31	8965	7·254	18·644
1900	3758	68·41	8632	7·607	17·461
1901	3393	76·57	9342	6·952	19·143
1902	3335	66·23	9593	6·546	19·834
1903	4237	69·22	8806	8·201	17·044
1904	4930	58·96	8258	9·240	15·470
				Lead	Zinc
1905	4023	62·51	5766	7·76	11·13
1906	4147	61·69	9572	7·15	16·50
1907	4455	65·52	9196	7·25	14·98
1908	4290	76·37	9817	7·14	16·35
1909	3241	67·94	9965	5·32	16·34
1910	3282	72·73	9650	5·29	15·55

Since from 1905 to 1910 the lead recovered has varied from 5·3 to 7·8 per cent of the ore treated and the zinc from 11·1 to 16·3 per cent, it is evident that the ore from the Holzappel mines is richer than the average of such ores in Germany, which contain but 4 per cent of lead and 11 per cent of zinc.

3. The Ems Lode-System

LITERATURE

Wenkenbach. ' Beschreibung der im Herzogtum Nassau an der unteren Lahn und dem Rhein u.s.w. aufsetzenden Erzgänge, sowie eine kurze Übersicht der bergbaulichen Verhältnisse derselben,' Nassauisches naturwissenschaftliches Jahrb. Wiesbaden, 1861, Vol. XVI.—G. Seligmann. ' Beschreibung der auf der Grube Friedrichsegen vorkommenden Mineralien.,' Verhandl. des naturhistorischen Vereins der preuss. Rheinlande und Westfalens, 1876, Vol. XXXIII.—Geologische Spezialkarte von Ems, by E. Kayser, surveyed 1884 and 1885, with explanatory text.—Beschreibungen der Bergreviere Wiesbaden und Dietz, 1893, by Holzapfel, Ulrich, Körfer, etc., published by the Royal Mining Department, Bonn.

The Ems lode-system occurs in Lower Devonian beds—according to Kayser the Upper Coblenz and the Lower Coblenz quartzite—in a position about 13 km. west of the Holzappel system just described.   These beds, which petrographically consist of clay-slate, grauwacke, and quartzite, occur in north-east striking anticlines and synclines, affected by many disturbances.   The general geology of the district is illustrated in Fig. 340.

In these rocks many lead-zinc lodes are found collected in groups and series.   From north to south the following groups deserve mention, those of Hohe-Buchen, Silberkaute, Silberkäutchen, Kellersberg, Merkur, and to the south of Ems those of Malberg, Bergmannstrost, and Friedrichssegen. The most important of these are Friedrichssegen at Oberlahnstein, Bergmannstrost, and Merkur.   The ore is associated with a zone of soft grauwacke and clay-slate belonging to the Lower Coblenz. This zone strikes north-east, dips 75° to the south-east, and is 120–150 m. wide.   Both walls are marked by clay fissures, known respectively as the main foot-wall and hanging-wall flucans ; at Merkur the hanging-wall flucan contains galena and sphalerite up to a width of 0·5 metre.   The ore-bodies, the lengths of which are dependent upon their angle to these walls, are cut into detached lengths by fissures running more or less parallel to one another. Of these, at Friedrichssegen there are more than twenty-four, seventeen of which, distributed over a length of 1400 m. carry ore.   At Merkur seven ore-bodies are known over a length of 2300 m., the width of these being generally below 10 m., but rising exceptionally to 20 metres.   The structure of the ore, including the gangue, is sometimes irregular and sometimes crusted.   The ore consists of argentiferous galena, sphalerite, siderite, and chalcopyrite, less frequently of millerite, linnaeite, and native silver.   Quartz is the principal gangue, while calcite and dolomite occur but subordinately.   Of these ores, in the main lode at Friedrichssegen galena and sphalerite are in approximately equal amount, some 8–9 per cent ; siderite is estimated at about three times this percentage, while

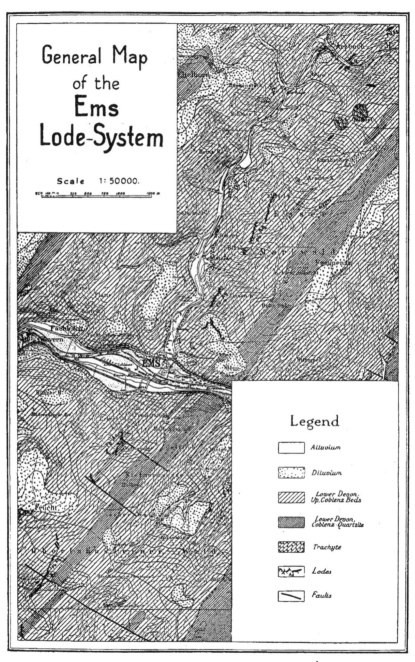

General Map
of the
Ems
Lode-System

Scale 1 : 50000.

Legend

Alluvium

Diluvium

Lower Devon.
Up. Coblenz Beds

Lower Devon.
Coblenz Quarizite

Trachyte

Lodes

Faults

Fig. 340.—Geological map of the Ems lode-system.   Scale 1 : 50,000.

chalcopyrite occurs to the extent of about one-twentieth of the galena. The silver content of the clean galena, containing about 65 per cent of lead, is about 500 grm. per ton.

At Ems the oxidation and cementation ores are well developed. These include pyromorphite, native silver and copper, malachite and azurite, cuprite, bournonite, silver amalgam, tetrahedrite, etc.

Mining in this district began very far back. It is stated that at Friedrichssegen work was already active under the Romans, which is all the more probable in that the ore-bodies outcropped with considerable width and high silver content. Reliable mention however begins with the commencement of the thirteenth century, when Kaiser Friedrich II. granted a loan to the 'Cologne Pits,' as they were then termed. The Merkur and Bergmannstrost mines, dating back to 1158, are also of great age. At present the mines in this district belong to the Stollberger Gesellschaft at Aachen. The production in 1910 amounted to 6447 tons of galena, 7558 tons of sphalerite, 222 tons of copper ore, and 7044 tons of siderite.

4. THE RAMSBECK LODES

LITERATURE

E. HABER. 'Der Blei- und Zinkerzbergbau bei Ramsbeck,' Zeit. f. d. Berg- Hütten- u. Salinenw. im preuss. Staate, 1894, Vol. XLII. p. 77.—E. SCHULZ. Geologische Übersicht der Bergreviere Arnsberg, Brilon und Olpe im Oberbergamtsbezirk Bonn, Bonn, 1877 ; 'Geologische Übersichtskarte der Bergreviere Arnsberg, Brilon, Olpe sowie des Fürstentums Waldeck,' Korrespondenzblatt des naturhistorischen Vereins für Rheinland und Westfalen, 1887, Vol. XLIV.—Investigations by A. DENCKMANN, in manuscript.

In the neighbourhood of Ramsbeck, a village in the mining district of Brilon, several mines, long in operation, are found over a superficies some 14 km. long and 12 km. wide. According to Denckmann the lodes of this district occur in the Ramsbeck beds of the Devonian, which beds consist of an alternation of grauwacke and clay-slate, crossed by numerous faults. In relation to filling and strike, two groups may be differentiated, namely, the unimportant limonite lodes striking north-south and dipping steeply to the east, and the really important lead-zinc lodes striking east-west. These latter occur in great number and, in spite of their patchy character and narrow width, they may be followed for considerable distances along the strike. In the western portion of the district they dip to the south at 12°–15°, and in the eastern portion at 25°–30°, these angles being generally flatter than the country-rock. These east-west lodes cross the country-rock at an acute angle, though occasionally a lode may be found to continue for some distance along the contact between grauwacke and slate. The tendency of the principal lodes to break into parallel veins is noteworthy.

Denckmann has collected evidence showing that the filling of the lodes in the grauwacke is substantially richer than that of those in the slate ; while in the former country the fissures are wide, regular, and carry considerable sulphides, in the latter they split up and become poor. Among the disturbances to which they have been subjected, the flat slides, which chiefly occur near the surface, are noteworthy. In the grauwacke these are definite, while in the slate, on the other hand, they are indefinite. Along them the hanging-wall portion of the formation has generally been thrust in a northerly direction, the extent of this thrust being seldom more than 100 metres. In addition, true faults with steep dip and little throw are present in large number.

The ore-minerals include galena, sphalerite, and subordinate pyrite. and chalcopyrite ; with these a little siderite is associated ; quartz is the principal gangue. The ore occurs very disconnectedly, being limited to small bodies which alternate with more extensive barren parts. The most important mineral is galena, which contains 0·027 to 0·065 per cent of silver and is always intergrown with quartz and sphalerite. The occurrence of roundish inclusions of milky quartz in the fine-grained, almost compact, and often argentiferous galena, is characteristic of the Ramsbeck lodes. It may be that these are the remnants of a quartzose gangue which, prevailing formerly, has since been replaced by galena. Coarse crystals of pure galena are seldom found. The sphalerite, mostly coarsely-crystalline in texture and chestnut-brown in colour, is also intergrown with other minerals and gangue. With the quartz other gangue-minerals occur, such as siderite, calcite, dolomite, and barite, though to a less extent.

The genetic relations at Ramsbeck are by no means easy of determination, the fissures having been repeatedly re-opened ; the different minerals are not contemporaneous. Moreover, further investigation alone will be able to indicate the extent to which subsequent replacement of the earlier fillings proceeded.

These mines, formerly held by many, have since the year 1859 been held by one company.[1] In 1890 the output was 4025 tons of galena and 2924 tons of sphalerite ; and in 1910, 2113 tons of galena and 7252 tons of sphalerite.

5. The Lodes of the Velbert Anticline

LITERATURE

Die Lintorfer Erzbergwerke, published upon the occasion of the Düsseldorf Industrial Exhibition, 1880.—SCHRADER. 'Das Bleierzvorkommen bei Lintorf,' Korrespondenzblatt des naturhistorischen Vereins für Rheinland und Westfalen, 1880, p. 60.—v. GRODDECK. 'Über die Erzgänge bei Lintorf,' Zeit. für Berg- Hütten- und Salinenwesen, 1881, XXIX.

---

[1] Gesellschaft für Bergbau, Blei- und Zinkfabrikation zu Stolberg und in Westfalen.

p. 201.—Schrader. 'Die Selbecker Erzbergwerke,' Korrespondenzblatt des naturhistorischen Vereins für Rheinland und Westfalen, 1884.—v. Schwarze. Zinkblende- und Bleierzvorkommen zu Selbeck, 1886.—Küppers. 'Die Erzlagerstätten im Bergrevier Werden am Rhein,' Mitteilungen aus dem Markscheiderwesen, 1892, VI. p. 28.—H. E. Boeker. 'Die Mineralausfüllung der Querverwerfungsspalten im Bergrevier Werden, u.s.w.,' Glückauf, 1906.—E. Zimmermann II. 'Kohlenkalk und Culm des Velberter Sattels im Süden des westfälischen Karbons,' Jahrb. d. k. pr. geol. Landesanst., 1909, II. p. 369.

South of the Westphalian coalfields, upon the Velbert anticline and another Devonian anticline adjoining to the north, these anticlines pitching east under the Carboniferous, occurs a series of metalliferous mines which unfortunately are no longer in operation. These occur not only in the Devonian beds of the anticlinal core but also in the Carboniferous limestone, the silica-schist and alum-slates of the Culm, and in the Millstone Grit. The situation is shown in Fig. 11. The country-rock on the anticlinal limbs strikes north-east, or roughly at right angles to the lodes. These latter are particularly interesting in that they represent the south-easterly continuation of the transverse faults found in the Rhine-Westphalian coalfields. Between these faults and lodes there exists no difference other than that while the latter are in greater part filled with ore and gangue, the former carry but little ore.

The extension of the lodes along the strike has in some cases, as for instance with the Lintorf main lode, been proved for several kilometres. The width similarly may be several metres. The separation between lode and country-rock is generally ill-defined, particularly at Selbeck.

The ore consists of galena, which is very pure and carries but little silver, of sphalerite to a subordinate extent, and a little chalcopyrite, while marcasite and pyrite are abundantly present; the gangue-minerals are calcite, dolomite, some barite, and quartz, while in addition fragments of the country-rock are fairly common. The presence of barite is interesting in that these lodes are, comparatively speaking, far from the main barite zone of Westphalia, the occurrence of which zone around Gladbeck, etc., is doubtless referable to the presence of Zechstein and Trias in that locality. Genetically, the country-rock, and particularly the alum-slates, probably played a material part in the ore-deposition.

Mining operations in this district have this technical interest, that the transverse faults crossing the Ruhr and the Rhine carried so much water from these rivers that work had to be stopped. At times the pumps had to raise considerably more than 100 cbm. per minute.

Přibram in Bohemia

Literature

W. Vogelgesang. 'Die Přzibramer Erzniederlage,' Cotta's Gangstudien, 1850, I. p. 305.
—E. Kleszczynski. 'Geschichtliche Notizen über den Bergbau um die Stadt Příbram,'

Jahrb. d. k. k. Bergakademien V. für 1855.—Fr. Babánek. ' Zur Kenntnis der Příbramer Erzgänge,' Österr. Zeitschr. f. d. Berg- und Huttenwesen, 1878.—J. Schmid. Bilder von den Erzlagerstätten zu Příbrama. Published by the Minister of Agriculture, Vienna, 1887. Montangeologische Beschreibung des Příbramer Bergbauterrains und der Verhältnisse in der Grube nach dem gegenwärtigen Stande des Aufschlusses in diesem Terrain. Published by the Department of Mines; edited by W. Göbl in 1893.—F. Pošepný. ' Beitrag zur Kenntnis der montangeologischen Verhältnisse von Příbram,' Arch. f. pr. Geol. II., Freiberg, 1895.—Guide to the International Geological Congress, 1903, I.—A. Hofmann. ' Neues über das Příbramer Erzvorkommen,' Österr. Zeitschr. für den Berg- und Hüttenwesen, 1906, No. 10.

This silver-lead district is centred around the towns of Příbram and Birkenberg on the left bank of the river Moldau to the south-west of Prague. The country consists of Lower Silurian grauwacke,[1] known by the geological department as the Příbram slates and sandstones, which about 3·5 km. to the south-east of Příbram give place to granite and phyllite. The lowest member of this grauwacke formation is known as the first slate zone. Upon this lies the first sandstone zone, which exhibits synclinal bedding, in consequence of which its westerly dip changes gradually to a steep dip to the east. Then in upward sequence comes the second slate zone, which is arranged fan-like, its beds dipping first to the east and then to the west. This in turn is overlaid by the uppermost member, the second sandstone zone, which dips gently to the west. Beyond this last zone and still towards the centre of the great Bohemian Silurian syncline, the Jinec beds follow conformably.[2]

The rocks of the slate zones are argillaceous-quartzose, argillaceous-micaceous, very fine-grained, and compact slates, the hardness of which depends upon the proportion of quartz present. The rocks of the sand-stone zones, on the other hand, are generally grauwacke-sandstones, developed sometimes as conglomerates with quartz pebbles and quartzose or quartzose-argillaceous matrix, and sometimes as more or less fine-grained sandstones. The matrix is variously coloured, in consequence of which the individual beds of the sandstone zones present a variety of appearance.

According to Grimm, the deposition of the first sandstone followed immediately after that of the first slate, the two zones at contact merging into one another. On the other hand, the contact of the first sandstone with the second slate above, is marked by a clay-parting, while along the contact above this again, a sulphide seam was encountered. While this latter is of little significance, the clay-parting forms roughly the north-west boundary of the metalliferous district, and is from one to several decimetres in thickness. It strikes north 60° E. and dips 70° to the north-west. The parting itself is filled with a dark grey to deep black stiff clay and fragments of country-rock. The beds of both zones strike with this parting, but dip in the opposite direction.

[1] The Étage B of Barrande.   [2] The Étage C of Barrande.

All these zones are intruded by many dykes and bosses of greenstone, which strike from north 15° W. to north 30° E. and are in two series, the Hatĕ and Birkenberg series respectively. With these dykes the lodes are intimately associated. The greater number and the best, either follow the greenstone in strike and dip, or occur within it or in its immediate vicinity, often in fact at the contact of slate and sandstone.

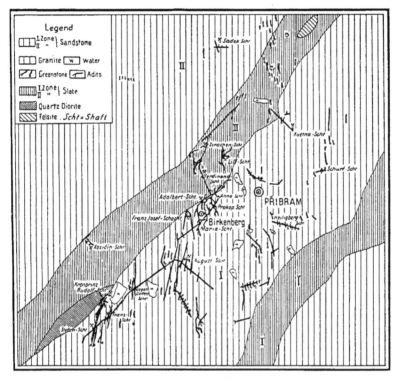

FIG. 341.—Geological map of the Přibram district. Scale, 1 : 100,000. J. Schmid.

For short distances only do these lodes cross the bedded rocks, and then only to return to the greenstone again. In addition to the dykes accompanied by lodes there are others not so accompanied, and others again which are associated with calcite fissures. In no case has any sort of relation been established between the width of the dyke and the mineralization of the lode found accompanying it.

The lodes are either lead-silver lodes or ironstone lodes. While no work is now done upon the latter, the developments upon the former may be said to be excellent. The most important lodes are found in the first sandstone zone, those of ironstone occurring upon the flat synclinal limb to the east, while those of lead are limited to the steeply dipping limb to the west. Along such fissures as cross both limbs the passage from one class of lode-filling to the other is gradual. Among the lead lodes those at Birkenberg are the most important. There within a space of 600 m. fourteen greenstone dykes having a total width of 124 m., and nine lodes, have been exposed, some of these latter having been followed for over 1000 m. along the strike and more than 1100 m. in depth, the width being sometimes as much as 10 metres. In the neighbourhood of the clay-parting the lodes split up.

The principal ore consists of argentiferous galena with sphalerite, siderite, and pyrite; and the gangue of calcite, dolomite, quartz, and barite. Less frequently, tetrahedrite, pyrargyrite, proustite, stephanite, native silver, boulangerite, jamesonite, and bournonite, are found. In the poor zones the lode-filling consists essentially of sphalerite, siderite, and calcite, together with fragments of the country-rock and argillaceous schistose material. Although the ore often displays crustification its structure in general is subject to great variation. The term 'lean ore'[1] is applied to those light to dark grey, fine-grained to compact quartzose masses throughout which galena, pyrargyrite, proustite, native silver, stephanite, tetrahedrite, bournonite, boulangerite, etc., are finely distributed. One analysis showed such ore to contain 17·56 per cent of galena with 0·26 per cent silver, 4·79 per cent sphalerite, 17·11 per cent siderite, and 47·65 per cent of quartz. In the Johanni lode, a north-west lode near the Anna shaft, pitchblende occurs in small aggregates of kidney shape and hazel-nut size, along a 2–5 cm. streak in the foot-wall.

The only observed influence of the country-rock upon the lodes is that solid and tough rocks appear to have resisted the formation of the original fissure. The effect produced on the country-rock by the lodes themselves is a limited bleaching and a slight impregnation with small particles of ore, for a width of 10 cm. at most. The lodes of the first slate zone are of no great importance. Those of the second are interesting in that the form of the fissure is different from that in the first sandstone zone to the east, this difference being probably referable to the different character of the two rocks. The Birkenberg lodes, occurring chiefly in the first sandstone zone, extend to the clay-parting, beyond which, as illustrated in Fig. 341, they continue in the slate in the hanging-wall of that parting. In the second sandstone zone no lodes at present are being worked; such as there are contain poor ore, that is, sphalerite, galena, and a little tetrahedrite, with quartzose gangue, in disconnected ore-bodies.

[1] *Dürrerz.*

Concerning the silver content of the galena, Hofmann has published much detailed information.[1] It was formerly supposed that the silver in the Adalbert main lode between surface and depth increased from 0·07 to 0·7 per cent, equivalent to 0·063 per cent per 100 metres. The more careful determinations of Hofmann upon clean material have however

Section shewing relation of greenstone dykes—hatched—to the lodes.

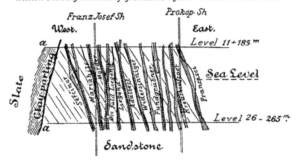

Intersections of the lodes with the clay-parting, projected on the plane of that parting.

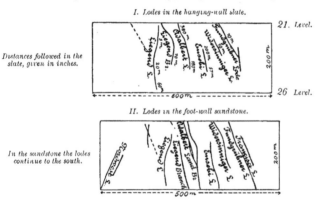

FIG. 342.—Lode sections at Příbram. J. Schmid.

demonstrated that with few exceptions the lead content remains practically constant at 77·5–82·5 per cent for all horizons, while neither the silver content nor the antimony content, which vary between 0·31 and 0·675 per cent, and between 0·32 and 0·86 per cent respectively, show rule or regularity in their variation. The depths over which his investigation relative to the silver was made, extended from 310 m. to 1099 metres.

[1] *Österr. Zeit. für Berg- und Hüttenwesen*, 1906.

Another interesting fact in connection with the galena of the Adalbert main lode is that it contains tin, the amount varying between 0·02 and 0·2 per cent. This tin content, according to Hofmann, is presumably referable to the presence of stannite ; it continues almost to the depth of 1100 metres.

The earliest available records of mining operations at Přibram date from the beginning of the sixteenth century. In the year 1900 the ore won amounted to 300,000 tons, from which a concentrate representing 7·5 per cent of the whole was obtained, this concentrate on treatment having yielded 40,000 kg. of silver and about 5000 tons of lead. In 1910 the output was 47·7 tons of fine silver, 3390 tons of soft lead, 596 tons of antimonial lead, 155 tons of zinc ore, and 50·5 tons of antimony ore. The Adalbert shaft is now approximately 1100 m. deep.

The Lodes at Linares, Spain

LITERATURE

CARON. ' Bericht über eine Instruktionsreise nach Spanien im Jahre 1878,' Zeit. für Berg- Hütten- und Salinenwesen, 1880, XXVIII.—PEDRO DE MESA Y ALVAREZ. ' Memoria sobre la zona minera Linares-La Carolina,' Revista Minera. Madrid, 1889.—A. O. WITTELS-BACH. ' Fragen und Anregungen, die sich an das Auftreten der Erze im Gangrevier La Carolina-Sta Elena (Spanien) knüpfen,' Zeit. f. prakt. Geol., 1897.—PAUL F. CHALON. ' Contribution à l'étude des filons de galène de Linares,' Revue universelle des mines (4), III., 1903, p. 282.—' Der Bleiglanzbergbau bei Linares-La Carolina in Spanien,' Berg- und Hüttenm. Ztg., 1904, Vol. XLIII.

This very important lead district, 35 km. long in an east-west direction and 30 km. wide from north to south, lies to the south of the Sierra Morena. Geologically it consists of several granite plateaus situated north and east of Linares, which rise like islands through Cambrian and Silurian beds, over which in turn others of the Triassic and Miocene form a more recent covering.

In so far as the ore-deposits are concerned the granite comes most into question, the Palæozoic beds having but little importance in this respect. The deposits are most numerous in the granite mass occurring immediately north of Linares, the lodes there being almost exclusively in the granite. The smaller occurrences around La Carolina and St. Elena lie farther to the north-east, around and within another granite mass. Finally, another occurrence is found at Arquillos east of Linares, the lodes there being also found chiefly in granite. Altogether some 1200–1300 deposits are known, 300 of which were being worked in 1903.

The granite, which is the principal country-rock, has the most varied composition and texture. Granulite is common. The Cambrian and Silurian beds resting upon these eruptive rocks consist of clay-slate, quartzite, arkose-sandstone, and grauwacke.

The strike of the powerful lodes occurring in the granite north of Linares is north-east, with a dip of 75°–90° to the north-west. A few, and these of poor content, strike east and dip south. The La Cruz and Alamillos lodes, the positions of which are indicated in Fig. 343, are known

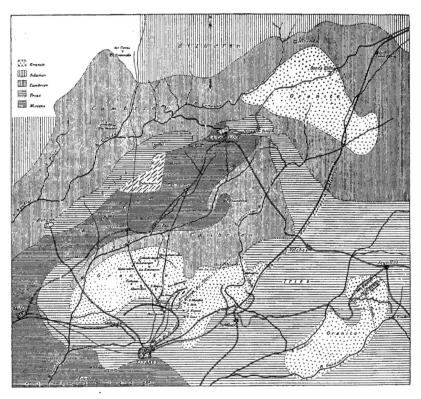

Fig. 343.—Map of the Linares lode district, shewing the positions of Arquillos, La Carolina, and St. Elena. Chalon, *Revue universelle des Mines*.

for 4–6 km. along the strike. The width usually varies up to 2 m., though where a lode is much split it may be as much as 8 metres.

The most important ore is galena, which contains silver up to about 100 grm. per ton ; sphalerite, pyrite, and chalcopyrite are less important. Quartz is the principal gangue, dolomite, barite, and siderite being more uncommon. Fragments of the granite sometimes form a considerable portion of the lode-filling. The copper minerals are somewhat enriched near the surface ; at La Cruz such copper ore was formerly mined. Although in the lode the galena occurs as solid masses, the distribution of these is by no means regular. In the principal mine of Arrayanes near Linares, for instance, ore-bodies of more than 100 m. in length and 1 m. in width alternate with stretches of poorer ore.

At Arquillos, the principal lodes, Las Prolongas and Santa Agueda, likewise strike north-east and dip at a steep angle, to the north or to the south. In width they hardly ever exceed 1·75 metres. The lode-filling is similar to that at Linares, described above.

The lodes at La Carolina and St. Elena, which traverse both slate and granite, generally strike north-west. With these also the principal ore is galena, while sphalerite is subordinate. Chalcopyrite and pyrite are more common. Quartz is the principal gangue-mineral, then barite. A comparison of the silver content with that of the occurrences at Linares is interesting; here it is 600–1000 grm. per ton, or six times as much as at Linares. In depth however it decreases steadily. The average silver content of the entire district is 180 grm. per ton.

This district reached its zenith in 1889, having in that year produced 118,325 tons of lead. In 1909 the production was 78,848 tons, valued at about £400,000, or almost three-fifths of the lead production of Spain.

THE RADIO-ACTIVE URANIUM LODES

On account of the close and particular association of radium with pitch-blende, which mineral is found sometimes with tin ores and sometimes with silver-, silver-gold-, and other ores, the radio-active deposits come within the study of ore-deposits.

Traces of radium are found in the earth's crust, both in the solid rock as well as in the circulating water; a slight radio-activity for instance may be found in almost all household water. As far as our knowledge of radio-active deposits goes, radium is exclusively associated with uranium. All radium lodes are found in granite districts or in slates which have been highly altered by the intrusion of granite. Radium-bearing thor-uranium, including the two minerals bröggerite and cleveite, and other radium and uranium minerals such as fergusonite, are found in the granite-pegmatite dykes of Norway and other countries. Radium and uranium are therefore acid elements.

Fluorite is found not only in uranium-bearing tin lodes but also in lodes, such as those at Joachimsthal, which contain no tin. In all cases however, those in Cornwall included, the close association of uranium ores with sulphide silver ores, less frequently with those of silver-gold,

copper, cobalt, and nickel, is remarkable.   Although it is only in Cornwall
that tin ores occur with uranium, in other uranium districts such tin ores
are found in neighbouring lodes.

From present experience it may therefore be said that those sulphide
ores which appear within or in the neighbourhood of tin-bearing granite
are possible uranium ores.   It is noteworthy that lodes without sulphide
ores but with only uranium-mica, do not appear to be promising ; should
however this mica change in the primary zone to pitchblende, the deposit
becomes promising.

According to the latest researches, and particularly the recently
published quantitative determinations of Mdlle E. Gleditsch in the
laboratory of Madame Curie at Paris, the relation between radium and
uranium with most minerals is practically constant.   Thus, with the
uranium-rich pitchblende of Joachimsthal and of Cornwall, the allied
uranium-rich minerals bröggerite and cleveite of the Archaean pegmatite
dykes of Norway, and the uranium-poor minerals fergusonite, samar-
skite, etc., found also in pegmatite dykes, this relation varies between
1 part of uranium to $3 \cdot 21 \times 10^{-7}$ and $3 \cdot 64 \times 10^{-7}$, or approximately 1 part
of radium to 3 million parts of uranium.   With carnotite the potassium-
uran-vanadate, autunite the potassium-uranium mica, and chalcolite the
copper-uranium mica—these two last being uranium phosphates, while all
are to be regarded as often, if not always of secondary formation—the
proportion of radium is occasionally relatively lower than this.   It is
remarkable that the relation between radium and uranium in the bröggerite,
cleveite, and fergusonite of the Archaean pegmatite dykes of Norway, etc.,
varies but little from that obtaining in the pitchblende of the late
Carboniferous or Permian lodes of Cornwall, or from that of the approximately contemporaneous lodes at Joachimsthal.

Concerning primary and secondary depth-zones, with ordinary
uranium ores it has been observed that uranium ochre and uranium
carbonate are both exclusively secondary, that uranium-micas are sometimes primary and sometimes secondary, while pitchblende is entirely
primary.   At Joachimsthal it may be demonstrated that the pitchblende
zone represents a deeper primary zone than that occupied by the cobalt-
and nickel ores, while it may be expected that in a few years the nature of
primary zone which lies deeper still, will be disclosed in the State mines.

Limited occurrences of uranium, such as are repeatedly found with
the silver-lead- and dolomite-lead lodes at Freiberg, are to be distinguished
from those which contain uranium in such amount as to be worked
especially for that metal.   Of such uranium deposits two classes may be
formulated, the uranium-tin lodes and the uranium-silver lodes, with or
without cobalt and nickel.

1. The Uranium-Tin Lodes of Cornwall

LITERATURE

D. A. MacAlister. ' Geological Aspect of the Lodes of Cornwall,' Econ. Geol. Vol. III., July–Aug. 1908, No. 5.—C. Schiffner. Radioaktive Wässer in Sachsen, Freiberg, Part 1, 1908 ; and Part 2, 1909.—P. Krusch. ' Über die nutzbaren Radiumlagerstätten und die Zukunft des Radiummarktes,' Zeit. f. prakt. Geol., 1911, p. 83.

The uranium-tin lodes of Cornwall and South Devon have been closely studied. In this district, a description of which has already been given,[1] pitchblende has been demonstrated to be present in several tin- or tin-copper lodes, which in general are distinguished by abundant tourmaline and which occur partly within and partly in the vicinity of granite. The uranium mine at Grampound, which is associated with the third granite mass reckoned from the east,[2] is particularly rich in uranium ore, of which it produces on an average some 20–30 tons yearly. From 1896 to 1906 the output, which in 1907 was 72 tons, varied between 6 and 105 tons, this wide range indicating the irregularity of the ore. These figures, it must be noted, are not on a basis for comparison with those of other districts, since the percentage of uranium contained is not given.

2. The Uranium-Silver-Nickel-Cobalt Lodes at Joachimsthal, Bohemia

LITERATURE

Fr. Babánek. Beschreibung der geologisch - bergmännischen Verhältnisse der Joachimsthaler Erzlagerstätten in geologisch-bergmännische Karte mit Profilen von Joachimsthal u.s.w. Vienna, 1891.—J. Step und F. Becke. ' Das Vorkommen des Uranpecherzes zu St. Joachimsthal,' Sitzungsbericht der Akademie der Wissenschaften, Vol. CXIII. Part 1. Vienna, 1904.—P. Krusch. ' Über die nutzbaren Radiumlager-stätten und die Zukunft des Radiummarktes,' Zeit. f. prakt. Geol., 1911, p. 83.

Lodes carrying uranium have long been known in the Bohemian and Saxon Erzgebirge, the number of late years, in consequence of the attention turned to radium, having considerably increased. The occurrence around Joachimsthal, which is the richest and has been most studied, is particularly important. It is worked partly by the State and partly by the Edelleutstollen Gesellschaft. It occurs in a crystalline schistose country closely associated with granite and consisting principally of gneiss, mica-schist, amphibolite, etc. The so-called Joachimsthal slates are regarded as particularly rich in uranium ore.[3] Of these, the relation between the petrographically differing complexes does not appear to have been sufficiently determined to allow any statement of relative age to be formulated.

[1] *Ante*, pp. 431-436. [2] *Ante*, p. 432. [3] *Ante*, p. 682.

The lodes, of which there are a considerable number, form two separate systems, one striking north-south and the other east-west. While the latter generally complete their length without disturbance, the former are in most cases found in displaced sections. The investigation of Krusch upon the property of the Edelleutstollen Gesellschaft showed that lode deflections and not faults were the cause of this discontinuity, and that the east-west lode-system was older than the north-south. The irregularity in the direction in which the sections of the north-south system are displaced is thus explained.

Generally the north-south lodes are richer than the east-west. Silver-, cobalt-, and nickel ores are frequent. With these ores pitchblende occurs, equal in value though less in amount. It was formerly supposed that the pitchblende was exclusively limited to the north-south lodes; recent developments however have established its presence in the east-west lodes also, though these remain substantially poorer in uranium than the north-south.

With the north-south lodes carbonates form the greater portion of the gangue; with the east-west lodes quartz occurs in addition. According to Step and Becke the sequence in age is: quartz, the oldest, uranium ore and dolomite, the youngest. It is interesting and characteristic that near the aggregates of pitchblende both the dolomite and calcite become reddish brown in colour, a fact of use as an indicator to the occurrences of pitchblende, these often being quite isolated.

Pitchblende is often more abundant at the intersections of the east-west with the north-south lodes. Frequently the uranium ore is found in close association with slate; in such cases it occurs either as the oldest crust enveloping a fragment of slate, or separated from the slate only by an earlier crust of quartz, an observation which appears to have been first made by Step.

In general the sulphide and sulph-arsenide silver-, cobalt-, and nickel ores represent a higher primary zone than the uranium ores. As before mentioned, it is probable that within the next few years it will be ascertained, at least at places in the State mines, what primary zone follows the uranium ores in still greater depth.

The pitchblende is not regularly distributed throughout the lode, but occurs in veins and lenses in closest association with the brown or reddish-brown carbonates. It is also frequently found impregnated in slate, such impregnation having proceeded from fractures.

At Joachimsthal, exact determinations are available not only of the uranium content of the ore but also of its radio-activity. This latter varies fairly regularly with the uranium content, being usually, according to the investigations of Professor H. W. Schmidt of Giessen, between 0·233 and

0·373 mgrm. per kilogramme of pitchblende. It may be said that 1 kg. of pitchblende, containing 60 per cent of U_3O_8, develops a radio-activity corresponding to 0·333 mgrm. of radium bromide, popularly known as radium.

These mines are in the position to produce annually 14–20 tons of ore [1] containing on an average 55 per cent U_3O_8. It is noteworthy that the ore is treated first for uranium products which do not contain radium. The radium therefore becomes concentrated in the residues which in consequence possess a radio-activity three or four times that of the original ore.

3. Uranium-Silver-Gold Lodes, Gilpin County, Colorado

LITERATURE

Forbes Rickard. 'Notes on the Vein-Formation and Mining of Gilpin County, Colorado,' Trans. Amer. Inst. Min. Eng., 1898, p. 108.

This highly metamorphosed district consists of gneiss-granite, granitite, protogine-granite, granulite, felsite, and pegmatite, with all intermediate gradations ; and of mica-, talc-, and hornblende-schists with well-defined schistosity. A point of interest in this rock-assembly is, that the gradual passage of granite through gneiss to schistose rocks may be observed. The lodes, which belong to the young gold-silver group,[2] are true fissure-fillings ; along their planes some little faulting has in many cases taken place. The separation from the country-rock is usually well defined and sharp. The lode-filling is characterized by a considerable gold-silver content, these two metals being so arranged that a western zone containing gold more particularly, may be differentiated from an eastern zone containing silver, though all gradations from one to the other are found. According to the strike two intersecting systems may be distinguished, one striking east and the other north-east.

The ore generally consists of argentiferous and auriferous pyrite and some chalcopyrite, with felspar and quartz as gangue. Smelting ore, consisting of compact pyrite with some chalcopyrite, is usually kept separate from milling ore, which is a white or yellow mixture of felspar and quartz impregnated with the same sulphides ; the weight relation between these two classes of ore is roughly as 1 : 20. Native bismuth and arsenic, though they occur, are uncommon ; the former is often associated with tetrahedrite. Arsenopyrite and other arsenical ores likewise occur. The presence of tellurium, apparently in connection with the gold, is significant. The occurrence of pitchblende is just as definite as that of the other primary

---

[1] Ante, p. 682.   [2] Ante, pp. 556, 557.

ores, though its distribution is very irregular. It is found and exploited particularly in the Wood and Kirk mines at Leavenworth Gulch.

Changes in the lode-filling have here been observed to be coincident with change of dip, such changes being followed sometimes by enrichment and sometimes by impoverishment, the general impression being that the deposits when steeper are poorer. Intersections of lodes of the two systems are usually associated with enrichment; not infrequently a sort of pseudo-intersection or deviation is observed where, shortly before the two lodes come together, they diverge again.

With some of these lodes the maintenance of the gold-silver content in depth is remarkable. It would appear as if the fissure-fillings, whether at the surface they were rich or poor, maintain their character in depth. If such be the case, then the cementation and oxidation zones can have caused no material migration of the metal content. The former reaches generally to but 40–80 feet from the surface, though exceptionally, as in the Carr mine, it may reach as much as 200 feet. The tetrahedrite is probably to be regarded as a cementation ore.

It is worthy of remark that several tons of uranium ore from the Wood mine were sold to Swansea, England, in the early days of uranium production, where they fetched a high price; with a larger output following a larger demand, a considerable profit might therefore be obtained. Generally however the pitchblende ore of Gilpin County is not rich, and the United States therefore is under the necessity of importing uranium salts.

Concerning the radium production, at present only the deposits at Joachimsthal may be expected to produce regularly. Experience has shown that the pitchblende treated gives approximately one-third of its weight as residue. In the future these deposits will presumably be in a position to deliver yearly 1·8 grm. of radium salt of maximum activity, of which the present price for 1 grm. is approximately £16,000. Radium, that is to say, its salt, differs from the ordinary metals in that practically speaking it is not consumed by use, while with all ordinary metals a considerable proportion disappears annually. With radium therefore a regular production means an equally regular increase in the amount of radium available. It is of interest to learn that Madame S. Curie recently succeeded in producing the element radium.

THE METASOMATIC LEAD-SILVER-ZINC DEPOSITS

THE formation of the metasomatic lead-silver-zinc deposits has already been discussed in the introductory portion of this work. We know that these occurrences are associated with easily alterable limestones and dolomites, and that they are limited to no particular geological age. The primary ores of this class were formed by a metasomatic replacement of limestone and dolomite by galena and sphalerite, such replacement only very exceptionally being complete. It is generally the case that a more or less far-reaching impregnation has proceeded from channels along which heavy-metal solutions obtained access, such channels being either transgressive fissures, bedding-planes, or joint-planes. Apart from this primary formation, the secondary processes of oxidation have often led to extremely important concentrations of the primarily deposited heavy minerals.

In the chapter upon ores it was explained that at oxidation galena became changed to oxidized lead ores, and sphalerite to oxidized zinc ores, and that among these the carbonate and silicate of zinc played an especially important part in lead-zinc deposits. Galena and sphalerite however do not with equal ease become altered to their respective oxidation products ; when the zinc sulphide has become completely altered to the carbonate and silicate, but a small portion of the galena has been altered to anglesite, cerussite, etc., the larger portion remaining still undecomposed. While with the other lead-zinc deposits the oxidation zone plays no great part, with the metasomatic occurrences it often forms the principal ore-bodies. A cementation zone is usually only present in so far that the galena in that zone is more argentiferous than the ore of the primary zone, which in most cases is unpayable. With these deposits therefore arises the singular case of the oxidation zone forming the actual useful deposit.

This secondary concentration, particularly of the zinc content in metasomatic deposits, results in greater part from ' oxidation-metasomatism.' [1] Krusch understands by this term the subsequent replacement of the country-rock by the action of those heavy-metal solutions which were

[1] Krusch, *Zeit. f. prakt. Geol.*, 1910.

formed in the deposit by the entry of oxygenated meteoric waters. If limestone form the country-rock this limestone by such solutions is metasomatically changed to ore. If this process long continue and large quantities of metal thus be brought to one particular limestone bed, a richly metalliferous deposit may be formed even when the primary occurrence was poor in heavy-metal. It is easy therefore to realize that in the formation of the metasomatic lead-zinc deposits oxidation-metasomatism is of more than ordinary importance. Since neither the primary nor the secondary alteration of the limestone has anything to do with its geological age but these alterations depend entirely upon certain chemical-geological properties of the rock, it is often the case that in a limestone formation only particular layers have been transformed to ore, these layers being limited above and below by others not suited to such transformation.

When studying such deposits the stratigraphist who has busied himself little with the science of ore-deposits is very liable to lay stress upon the conformity, and to regard the deposits as ore-beds, conformity being the essential characteristic of such beds. With metasomatic deposits, on the other hand, from the manner of their formation, an association with limestone and dolomite is the principal characteristic. In all districts therefore where these two rocks are uncommon or confined to one particular geological horizon, such deposits if formed must exhibit a certain conformity, though they are essentially epigenetic. This conformity being generally qualified and limited may be best described as a pseudo-conformity.

The relation between the mineralization and dolomitization is particularly interesting. Although with some metasomatic lead-zinc occurrences the ore occurs exclusively in limestone without dolomite, this latter rock is often present, sometimes even to the exclusion of limestone. The existence of this dolomite as the result of the dolomitization of limestone, may often be established, such dolomitization being effected by the metasomatic action of solutions containing CO_2 and $MgCO_3$ upon the limestone. The extent of such dolomitization in nature is sometimes stupendous and expressive of one of the most far-reaching chemical-geological phenomena known.

In many cases the dolomitization shows itself to be somewhat older than the mineralization, so that two stages in the ore-deposition may be distinguished, namely, the metasomatic replacement of the limestone by dolomite, and the metasomatic replacement of the dolomite, chiefly by galena and sphalerite. Even in such cases however, the time-interval between these two stages was probably so small that they may be considered as part of one and the same chemical-geological phenomenon.

Concerning the shape of the deposit, two types of metasomatic deposit are usually distinguished, namely, that with complete alteration of the original rock, when the shape greatly resembles that of a bed, as often happens with metasomatic iron deposits; and that with incomplete alteration, when the tendency is to form ore-bodies following the channels of access; such deposits may either exhibit transgressive bedding, or they may follow the bedding-planes or joint-planes.

To the second type many of the metasomatic lead-zinc deposits belong, and it is therefore easily understood that rich concentrations of ore are found more particularly where two channel-systems intersect. Though science may formulate types, Nature expresses herself in gradation. Cases of metasomatic lead-zinc deposits do occur where individual layers have in places been completely altered to ore, as for instance in the Beuthen and Tarnowitz synclines in Upper Silesia. With such as these the question has always arisen whether they were sedimentary ore-beds, or whether they were metasomatic occurrences where one bed had been completely mineralized. In any particular case this question can only be settled when, as was the case a few years ago in Upper Silesia, the channels along which the solutions rose, are discovered.

In those cases where limestone alternates with slate the ore generally becomes deposited at the contact of these two rocks, particularly when the lower bed is slate, which being impermeable prevents the further descent of the solutions and holds them in long contact with the limestone. Should the solutions in such cases finally escape along faults, the intersections of such faults with the limestone-slate contact would be particularly good places to look for ore. Generally therefore the shape of the lead-zinc deposits is irregular. When prospecting for these metasomatic occurrences it is accordingly well to remember that on the one hand they are associated with the distribution of limestone, and on the other with zones of fracture and disturbance. This association is illustrated in Fig. 6.

Concerning extension in strike and dip, no generally applicable rule can be formulated. It is realized that fissures of great length are usually also of great depth, nevertheless, the metasomatic deposits though closely connected with fissures need by no means occur along the whole extent of such fissures. Not infrequently these deposits are associated with irregular cavities which, formed before the ascent of the metal solutions by meteoric waters rich in carbonic acid, afterwards became filled with ore. The distribution of these cavities or chambers often appears to be capricious and conformable to no law.

The indications of metasomatic lead-zinc deposits at the surface are of great importance in prospecting and exploration. Where the limestone is not covered by thick detritus, they are relatively easily discovered by means of the distinctive colouring to which they give rise. The sulphides of lead and zinc always contain more or less iron, which by oxidation is converted to limonite. Since, in addition, pyrite and marcasite are generally present and these also become altered to limonite, the limestone in the neighbourhood of such a deposit is often transformed by oxidation-metasomatism into iron ore. In but few cases is the deposit quite free from iron, or without galena, or with the zinc sulphide represented by white schalenblende—fine-grained wurtzite or sphalerite, or a mixture of both minerals, generally exhibiting highly developed crusted structure. In such cases it would be difficult at first to recognize the zinc deposit. When therefore in limestone districts masses more or less earthy and remarkable for their high specific gravity occur, it is advisable to look for zinc. The flame coloration test would then be sufficient.

The filling of the metasomatic lead-zinc deposits so far as ore is concerned, consists chiefly of zinc sulphide, galena, and pyrite or marcasite. The zinc sulphide occurs both in the form of sphalerite as also often in that of schalenblende; indeed, the comparative study of the world's zinc deposits shows that schalenblende is almost limited to deposits of this type. It is noteworthy also that this mineral is not only found as a metasomatic product associated with the alteration of limestone, but also as a chamber-filling. What property of the solution, or what particular circumstance of precipitation caused the zinc sulphide to be precipitated as schalenblende in metasomatic deposits, while in lodes sphalerite is almost exclusively found, has not yet been possible of determination. The colour of schalenblende, like that of sphalerite, varies in proportion to its iron content. It is interesting that with schalenblende particularly, whitish or very light coloured layers are often found practically free from iron. Since with these metasomatic deposits schalenblende is usually more abundant than sphalerite, the oxidized ores associated with them have chiefly resulted from this form of zinc sulphide; schalenblende is also apparently more readily changed to cellular oxidized ores than is sphalerite. Concerning pyrite and marcasite, it is noteworthy that with these deposits marcasite occurs strikingly often and sometimes in large quantity, while pyrite on the other hand recedes. All other ore-minerals are subordinate.

Among the gangue-minerals barite is the most common. Since this mineral is also characteristic of the metasomatic occurrences of other metals, it must doubtless be referable to the barium content of the altered limestone. As was pointed out in the chapter on mineral formation, the smallest amount of barium in the presence of sulphuric acid suffices in the course of time to build up the largest masses. That calcite is frequent and often occurs as the youngest filling, is natural, in view of the calcareous nature of the country-rock. The occurrence of blue anhydrite in some of these deposits is particularly interesting.

The most common rock inclusions are of limestone and dolomite. Experience of these deposits has taught that such inclusions, even with incomplete alteration of the primary bed, can in themselves constitute ore-deposits. On the other hand, when fissure-systems of different strike and dip intersect, at such intersections blocks of limestone more or less large may become involved as rock inclusions and yet be in their original position and remain unaltered.

Everything considered, it may be said that the particular characteristics of the filling of these deposits are, the preponderance of oxidized zinc ores ; the occurrence of the primary sulphides, schalenblende, and marcasite ; and the occasional presence of anhydrite.

In addition to these differences between the filling of the metasomatic lead-zinc deposits and that of lodes, there are others no less definite in the manner the minerals and rocks forming these deposits are intergrown. With the metasomatic deposits the most perfect examples of crustification are met, such as were formed when metal solutions entering cavities deposited ore all round the walls. In addition, concentric crustification of stalactitic and stalagmitic growth, each outside layer being younger than those inside, is found. In the arrangement of the different crusts the frequent alternation of schalenblende with marcasite, such as is illustrated in Fig. 125, is striking. The wide distribution of a cellular structure is also worthy of notice, such structure resulting when schalenblende is altered to cellular oxidized ores. Finally, a pseudo-brecciated structure, resulting from the above-described imperfect alteration of the limestone to ore, is extensive and frequent.

The oxidized iron ores which now and then make their appearance in payable quantity in these deposits, owe their formation to the decomposition of the marcasite or pyrite.

Primary depth-zones play little part with many of these deposits, since often the oxidation zone only is mined, the primary zone being workable in but few cases. It must however be particularly remarked that, as with the lodes, where work is undertaken in the primary zone, an upper zone of lead ore may occasionally be distinguished from a deeper zone of zinc. This, for instance, is often the case in Upper Silesia, where in many mines two beds are worked, whereof in isolated cases the upper carries galena chiefly while the lower carries sphalerite, the intervening limestone having apparently been unsuited to any such alteration. The part played by secondary depth-zones in metasomatic deposits was particularly mentioned at the beginning of this description, when discussing the formation of these deposits.

The silver content deserves one or two remarks. Since mining in most cases proceeds in the oxidation zone only, and the primary deposits are relatively but little worked, the silver content of these ores is subject to great fluctuation. The oxidation zone is usually poor in silver, and consequently in many cases but little of this precious metal is found in oxidized deposits. The silver content is occasionally higher in districts where primary ores are mined, the galena of Upper Silesia for instance having a high silver content. In such deposits it fluctuates in like manner to its occurrence in the galena of the sulphide lead-zinc lodes.

The geological age of the altered limestone, as will be seen from the following statement, varies greatly :

Locality or Name of Deposit.	Formation in which the ores occur.
Laurion in part.	Cretaceous.
Upper Silesia, Wiesloch, and Carinthia.	Triassic.
Aachen, Leadville in part, Missouri.	Carboniferous.
Leadville in part, Iserlohn, Schwelm.	Devonian.
Sardinia, Leadville in part, Mississippi.	Silurian.
Eureka.	Cambrian.
Laurion in part, Sala.	Fundamental schists.

The deposits are mostly considerably younger than the formations in which they occur ; at Laurion, for instance, the occurrences in the fundamental schists are late- or post-Cretaceous, while those at Leadville and Eureka, in Palæozoic limestone or dolomite, are late Mesozoic. On the other hand, the deposits at Sala are very old, namely Archaean, and with this great age the abnormal character of the mineral-association is probably connected. Owing to the close genetic connection between metasomatic lead-zinc deposits and lead-zinc lodes, the same experiences in regard to the age of the primary ore generally apply to both. Thus, with the deposits in Germany there are two periods of ore-deposition, late Carboniferous and Tertiary. The formation of the oxidation ores, so important with the metasomatic lead-zinc deposits, may have begun immediately after the deposition of the primary ore, and have continued right to the present.

Since with these deposits the most important ore, zinc carbonate, is formed by oxidation, it is found almost invariably above the present ground-water level ; the primary sulphide ores existing in greater depth are often unpayable so that many mines cease work at that level, and in consequence most mines working metasomatic deposits reach only to depths of a few hundred metres. At Aachen for instance, mining is proceeding at depths of 300 m., in Upper Silesia at 100 m., at Iserlohn at 200 m., at Bleiberg at about 400 m., and at Sala at about 300 metres. The question of the economic importance of the metasomatic lead-zinc deposits is further discussed at the end of the description of these deposits.

UPPER SILESIA

LITERATURE

KRUG VON NIDDA. ' Über die Erzlagerstätten des oberschlesischen Muschelkalkes,' Zeit. d. d. geol. Ges., 1850, p. 206.—F. RÖMER. Geologic von Oberschlesien, 1870.—E. CAPPELL. ' Über die Erzfuhrung der oberschlesischen Trias nördlich von Tarnowitz, O.-S.' Zeit. f. d. Berg- Hutten- u. Salinenwesen im preuss. Staate, 1887, Vol. XXXV.—FR. BERNHARDI. ' Über die Bildung der Erzlagerstätten im oberschlesischen Muschelkalk,' Zeit. d. oberschlesischen berg- u. hüttenmannischen Vereins, XXXVIII., 1889.—R. ALTHANS. ' Die Erzformation des Muschelkalks in Oberschlesien,' Jahrb. der k. pr. geol. Landesanst., 1891, Vol. XII. p. 37.—H. HÖFER. ' L'Origine des gisements des minérais de plomb, de zink, et de fer de la Haute Silésie,' Revue universelle des mines, Vol. XXX., Liége-Paris, 1895.—F. BEYSCHLAG. ' Über die Erzlagerstätten des oberschlesischen Muschelkalkes,' Zeit. f. prakt. geol., 1902, p. 143.—G. GÜRICH. ' Über die Entstehungsweise schlesischer Erzlagerstätten (Oberschlesien und Kupferberg),' Jahresbericht der Schlesischen Ges. f. vaterländische Kultur, 1902; 'Zur Genese der oberschlesischen Erzlagerstätten,' Zeit. f., prakt. Geol., 1903, p. 202.—A. SACHS. ' Die Bildung der oberschlesischen Erzlager- stätten,' Zentralblatt f. Mineralogie u.s.w., Stuttgart, 1904, p. 40; Die Bodenschätze Schlesiens, 1906.—R. MICHAEL. ' Die oberschlesischen Erzlagerstätten,' Kohle und Erz, 1903, and Zeit. d. d. geol. Ges. LVI., 1904.—FR. BARTONEC. ' Über die erzführenden Triasschichten Westgaliziens,' Österr. Zeit. f. Berg- u. Hüttenwesen, 1906.

The lead-zinc ores of Upper Silesia occur interbedded in the Muschel-kalk, the limestone division of the great German Triassic basin. This limestone forms a belt 10–20 km. wide and more than 80 km. long, which, striking east-west, extends from Krappitz on the river Oder, through Gogolin to Olkusz in Russian Poland, along which extent its surface continuity is in places broken by coverings of younger beds. On the other side of the Russian frontier this belt turns south-east and south, in which direction its extreme outliers are found in the neighbourhood of Czerna in Galicia. Tectonically, it rests with gentle northern dip upon the Bunter of the Triassic and the Rotliegendes of the Permian, these in turn lying unconformably upon Carboniferous beds. It is only the southern outliers of this belt which, on both sides of the frontier as well as in Galicia, are ore-bearing.

At Tarnowitz, the Tarnowitz syncline with a north-south strike and some 20 km. wide, branches from the main belt. To this in turn is attached the Beuthen syncline some 7 km. wide, which, between Mikult-schütz, Miechowitz, and Dombrowka, strikes sometimes east and some-times south-east. This latter syncline, illustrated in Fig. 152, extends over Beuthen, across the frontier to Czeladz, Bendzin, and as far as Klimontow in Russian Poland, when after a short break it continues to Dlugosczyn and Szakowa, situated in the synclinal subsidence of Chrzanow and Trzebinia, in Galicia. To the north, between Ptakowitz and Stollarzowitz, an anticlinal lift of the older beds exposes two secondary synclines, the Trockenberg to the north and the Miechowitz to the south. According to Beyschlag and

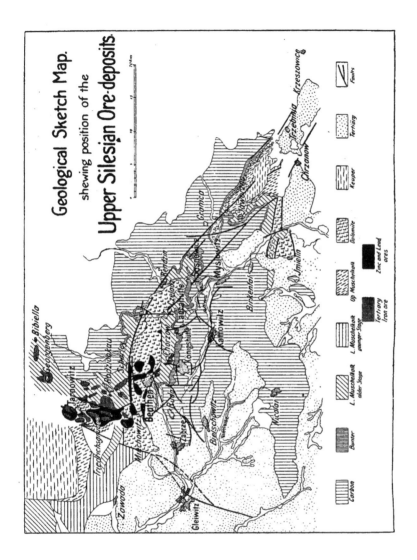

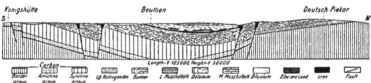

FIG. 344.—Geologic-tectonic map and diagrammatic section of the Upper Silesian lead-zinc district. Michael and Quitzow.

Michael, this synclinal disposition of the beds is caused in greater part by faulting, in which case the expression synclinal subsidence better expresses the tectonics than would the term syncline.

The country between and around Tarnowitz, Miechowitz, Beuthen, Scharley, and Gr. Dombrowka, embraces the principal extension of the deposits. Smaller and less important occurrences are found in the eastern districts wherever dolomite occurs ; in Russia at Boleslaw and Olkusz, and in Galicia at Szakowa and Trzebinia ; the only one of these worthy of note is that of galena and zinc oxidized ore in the Matilde mine west of Chrzanow in Galicia.

The Triassic beds south of the large anticlinal uplift of the Coal formation from Zabrze to Myslowitz, are without apparent connection with the occurrences now being described. In general with them no extensive mineralization is known ; only those beds to the south of Myslowitz the continuity of which is broken by repeated appearance of the Carboniferous basement, show in their dolomitic members any traces of ore.

In the whole Upper Silesian district the surface of the Muschelkalk as the result of erosion is most accidented. In pre-Tertiary or early Tertiary time running waters furrowed this limestone in the most extraordinary manner, so that to-day, freed from the softening contours of any Diluvial or Tertiary covering, it constitutes the feature of an extensively incised landscape. It is probable that at that time also, waters circulating along planes of bedding and fracture effected the alteration of particular beds of the Muschelkalk to ore. Before, however, going more fully into this question it is well to describe the development of that formation itself.

So far as the ore is concerned only the Lower Muschelkalk, some 120 m. thick and known more particularly as the Wellenkalk, comes into question. This is divided into two stages, the Lower or Wellenkalk proper, which practically coincides with the Chorzow beds in the early classification of Eck ; and the Upper Wellenkalk, formerly known as the Schaumkalk. The cavernous limestone formerly regarded by Eck as the base of the Muschelkalk, belongs to the calcareous and dolomitic members of the Upper Bunter, these members occupying a thickness of about 55 metres. The uppermost beds of the Wellenkalk proper, some 7 m. thick and easily recognizable during development by their colour and contained fossils, are often known as the blue floor-limestone. The Upper Wellenkalk in Upper Silesia is variously developed ; in the Oder river district it is calcareous, in that of the Weichsel on the other hand it is dolomitic. In the Beuthen, Tarnowitz, Laurahütte, and Zabrze sections of the Geological Survey the whole sequence of the Muschelkalk is given as follows :

Upper Muschelkalk, about 30 m.

Middle Muschelkalk, about 15 m.

| Upper Wellenkalk | Diplopora dolomite. Karchowitz beds. Terebratella beds. Gorasdz beds. | } Ore-bearing dolomite about 75 m. |

| Lower or Wellenkalk proper | Wellenkalk. Marl limestone. Second Wellenkalk. Conglomerate bands. Cellular limestone. First Wellenkalk. Pecta and Dadocrinus limestone. | } About 45 m. |

The presence of ore is peculiar to the dolomite of the Upper Wellenkalk. This dolomite exhibits so many characteristics in common with the contemporaneous calcareous beds, the Gorasdz, Terebratella, and Karchowitz beds exposed in the Drama valley and at Mikultschütz, that its secondary formation from limestone along fissures carrying ground-water must be assumed.

Contrary to experience in the Carboniferous beds below, faults have been established only in small number in the Upper Silesian Muschelkalk. In addition to the frequent pre-Triassic disturbances in the Carboniferous, however, younger disturbances have also undoubtedly been observed in the Triassic and in the ore-beds, which must be regarded as the upper continuations of similar faults in the Carboniferous beneath.

The ore-bodies, consisting essentially of galena, sphalerite, zinc carbonate, zinc hydrosilicate, and marcasite, are found in the deeper portions of the synclines, and apparently quite irregularly distributed in the altered dolomite bed. Rich sections alternate indiscriminately with others carrying but little or no ore. These rich sections are connected with fissures and are found not only in the neighbourhood of the strike-faults, but particularly along the north-south transverse faults which reach to the Coal-measures below. Often ore-bodies are found at two horizons, one being then either immediately above the blue floor and separated from it only by a bed of pyritic clay, or separated further by a dolomite bed 1–2 m. thick known as the floor-dolomite. The second ore-bed then occurs in the mass of the dolomite, sometimes 20 m. above the first. It is however regular neither in character nor position, while, as pointed out by Althans, in the Trockenberg syncline it is entirely absent; in fact it only appears under especially favourable conditions, and is then always connected with the lower bed by ore-bearing fissures.

When two beds occur no material difference in composition between the two may be observed. Nowhere can any regular recurrence in deposition, any fixed sequence between galena, sphalerite, and marcasite, be recognized.

Both deposits may be purely of lead ore, as in the Tarnowitz syncline, and then seldom more than 1 m. thick and very discontinuous ; or predominatingly of zinc ore, as in the Beuthen syncline, and then up to 12 m. in thickness and more continuous. The whole thickness is in no case however made up exclusively of ore, but generally of a mixture of ore and dolomite. At the outcrop both beds appear to unite to form one body, which in places may reach a width of 20 metres. In such cases the ore consists chiefly of a red ferruginous zinc ore with some cerussite and earthy lead ore. When it descends into the funnels and crevices in the limestone floor, as illustrated in Fig. 166, it becomes more and more clayey, the amount of iron at the same time decreasing. On account of its lighter colour such altered material is spoken of as white zinc ore.

Clean galena occurs more particularly in the Tarnowitz syncline but also in the Trockenberg syncline, partly in the form of narrow compact layers and partly as irregular masses and nests. The thickness of the ore-bearing layer is usually about 0·25 m. to 0·5 m., though exceptionally it may reach 2 metres. Around Tarnowitz, in the Friedrich mine for instance, a soft galena layer is distinguished from a solid galena layer. In the former the galena occurs as plates and masses in clay, filling the bedding- and joint-planes of the dolomite ; in the latter it occurs solidly intergrown with the dolomite, either as a thin bed or as stringers and aggregates. The solid layers represent the original condition, from which along the outcrop the soft layers have resulted by weathering. The silver content of the galena varies between 0·025 and 0·048 per cent. Traces of copper, antimony, and gold have been disclosed by analyses of furnace products. The usual associates of the galena are cerussite and marcasite, and in addition in the Friedrich mine the rare tarnowitzite, a variety of aragonite containing 10 per cent of lead.[1]

The ore-bed consists of friable, earthy, finely-crystalline, and fibrous sphalerite, or of schalenblende, the latter often appearing in stalactitic form in layers alternating with others of galena and marcasite. Not infrequently a mixture of sphalerite with the carbonates of lime, magnesia, iron, and zinc occurs, such being known as dolomitic sphalerite. This mixture is found for instance in the Neue Helene, Cäcilie, and Bleischarley mines. In addition, the red and white zinc ores already mentioned as contaminated by dolomite, clay, and iron oxide, are of great importance. Their structure and intergrowth with sphalerite indicates their derivation from this primary ore. Pyrite and marcasite are constant associates of the zinc, to the detriment of the value of the deposit ; sometimes they predominate.

Beds of cleaner sphalerite are found on the northern limb of the

[1] Websky, ' Über die Kristallform des Tarnowitzits,' *Zeit. d. d. geol. Ges.* Vol. IX. p. 737.

Beuthen syncline, in the Cäcilie and Neue Helene mines. Such beds rarely attain a thickness above 2 metres. They are distinguished by their purity and the considerable amount of contained galena. At the outcrop they are altered to red oxidized ore, some pieces of which though outwardly identical with the others, consist of schalenblende or galena inside. In places on this same north limb an upper ore-bed of varying thickness has also been worked. This lies some 30 m. above the sphalerite bed and reaches about 1 m. in thickness.

Around the margin of the syncline a thick bed of red oxidized ore occurs, which in the Cäcilie, Scharley, and Wilhelmine mines is in places. 20 m. thick, and in the foot-wall passes over to white ore. On the south side of the syncline also, in the Therese, Apfel, Maria, and Elisabeth mines, this bed, though more irregular, is almost as important. In these mines ore is often found in funnel-shaped bodies extending downwards from the main deposit into the limestone.

Beds of earthy limonite, contaminated by clay, lime, and dolomite, and mixed with zinc oxidized ore, galena, and cerussite, occur either as independent layers of varying thickness, or as part of the above-mentioned lead-zinc occurrences. Such limonite is seen more particularly in the funnel-shaped bodies which penetrate the limestone floor, and also at the outcrop. In this connection it must be mentioned that the dolomite in the roof of the ore-bed has in many places been denuded, and that loose Miocene marine deposits or Diluvial formations now cover the bed. At such denudation the ore-bed itself was naturally more or less eroded and disturbed.

Leaving out of consideration the theories formerly advanced that these deposits were contemporaneous with the dolomite, or that they were formed by the concentration of metalliferous material formerly finely distributed throughout the Muschelkalk, it remains only to more closely discuss the present generally accepted theory of the subsequent introduction of the ore. It may be regarded as certain that here, as with the analogous deposits at Aachen, Wiesloch, Monteponi, Raibl, and Laurion, the ores were originally deposited as sulphides, from which, subsequently and by the action of circulating meteoric waters, the oxidized ores were formed ; numerous pieces of ore with sphalerite or galena inside and zinc carbonate or cerussite outside, most clearly demonstrate this derivation. Further, it may be regarded as established that the ores were deposited from solutions which, circulating within the Muschelkalk beds, effected their dolomitization. At the same time a gradual replacement of the limestone by ore took place, as well as a crusted deposition in chambers previously formed by the dissolution of the dolomite and limestone. The not-infrequent occurrence of stalactites and the abundant occurrence of schalenblende indicate this latter manner of formation. The pocket-like occurrence of the ore in holes like pot-holes in the limestone floor, suggests rapidly flowing water. The apparent limitation of the ore-beds to one or two often well-maintained horizons is adequately explained both by the physical behaviour of these particular beds towards the inflowing water, as well as by the presence in them of material which acted reducingly towards the metals in solution. Against solid compact layers such water would be impounded. Where several of such layers occurred several ore-beds would be formed one above the other, in number corresponding to the number of such water reservoirs. At the same time the bitumen, which is still discernible in some beds, may have acted as the agent whereby the sulphates present were reduced to sulphides.

Concerning the question of the source of the solutions, the following considerations are pertinent. The metalliferous occurrences in the Upper Muschelkalk, Keuper, and Jurassic—which Althans regarded as constituting this source—are probably insufficient to account for such an extensive occurrence. It is more probable indeed that these themselves were formed by the same processes, whatever they may have been, as those to which the principal beds are referable. According to Bernhardi and Gürich, the ores were deposited simultaneously with the country-rock. Bernhardi recognizes the precipitant in the gases escaping from the Coal-measures beneath, while Gürich sees it in the organic substances contained in the Triassic sea. According to Carnall, Websky, and Althans, the ore, formerly finely distributed, became subsequently concentrated by meteoric waters, a view which A. Sachs also endorses. The most probable of all, however, is the assumption that the solutions rose from depth along fissures, from which following fractures and cracks they spread laterally through the permeable beds of the Muschelkalk. This view is endorsed by Krug von Nidda, Eck, Kossmann, and lately, as the result of special investigation, by Beyschlag and Michael. It is occasionally urged against it that the channels of access necessary to such an assumption must, if existing, have been disclosed by the extensive mining operations which have been undertaken in the Coal-measures beneath ; or, alternatively, that much lead- and zinc ore should have been found within the fissures of those measures. Both these objections appear no longer to have force ; firstly, because in the Coal-measures the limestone necessary to such deposition of ore does not exist, and the ascending waters not being able to dissolve the insoluble rocks along the fissures were accordingly not in the position to form funnel-shaped cavities ; secondly, because occurrences of lead- and zinc ore, particularly the former, are actually found in the fissures of those measures ; and finally, because in the Upper Bunter and in the transition beds between the Bunter and the Rotliegendes, occurrences of ore such as have recently been discovered may only be explained by deposition from ascending solutions.

The endeavour from above to follow the fissures known in the Triassic into depth, or from below to follow the faults in the Coal-measures into the Triassic above, has for want of available exposures but seldom succeeded. It is also almost always impossible to follow the Muschelkalk fissures into depth because of water, which in this Triassic region is so abundant as to be sufficient for the entire industrial district of Upper Silesia. It may therefore be assumed that the connection of the fissures in the Muschelkalk with those of the Coal-measures is more frequent than might be supposed from the observations so far possible.

According to Gürich, zinc mining in this district began in the sixteenth century. The oxidized ores alone were mined at first, sphalerite having only been mined for about four decades.

PRODUCTION OF UPPER SILESIA

	Zinc Ore.	Sphalerite included.	Lead Ore.	Lead.	Silver.
	Tons.	Tons.	Tons.	Tons.	Kg.
1791	16,688	...	910	...	...
1816	3,230	...	...	266 Lead 628 Litharge	625
1868	290,362	...	11,047	5,580 Lead	6000
1878	?	4,300	...	...	...
1887	552,614	...	28,580	...	...
1897	510,686	270,426	35,847	19,338 Lead 1,719 Litharge	8349

The present production may be gauged from the following figures : In the year 1908 eleven mines employing 9442 hands produced 1,212,366 tons of argentiferous galena and sphalerite, together containing 210,456 tons of zinc and 61,733 tons of lead, equivalent to 17·4 per cent and 5·1 per cent respectively ; and in addition 208,025 tons of zinc oxidized ore and schalenblende, containing 30,850 tons of zinc, equivalent to 14·8 per cent. If these figures be compared with those of Germany as a whole—which for that year were 2,913,150 tons, containing 320,216 tons of zinc and 114,583 tons of lead, or 11 per cent and 3·9 per cent respectively—it is seen that in respect to metallic zinc, Upper Silesia is responsible for three-fourths of the total produced by mines in Germany, and more than one-half of the total lead.

AACHEN

LITERATURE

W. SCHIFFMANN. ' Die geogn. Verhältnisse und die Erzlagerstätten der Grube Diepenlinchen bei Stolberg (Rheinland),' Zeit. f. d. Berg- Hütten- und Salinenwesen im pr. Staate, 1888, Vol. XXXVI. p. 1.—C. DANTZ. ' Der Kohlenkalk in der Umgebung von Aachen,' Zeit. d. d. geol. Ges., 1893, Vol. XLV. ; Beschreibung des Bergreviers Düren, 1902.—CH.

TIMMERHANS. Les Gîtes métallifères de la région de Moresnet, Liége, 1905, p. 1.—F. KLOCKMANN. ' Die Erzlagerstätten der Gegend von Aachen und der Bergbau auf der linken Seite des Niederrheins,' Festschrift zum XI. Allgemeinen deutschen Bergmannstage in Aachen. Berlin, 1910.—F. HERBST. ' Der technische Betrieb des Erzbergbaues,' ibid. —WUNSTORF. Geologische Exkursionskarte der Umgegend von Aachen, published by the Geologische Landesanstalt, Berlin, 1911.

Lying upon the north-west flank of the Cambrian anticline at Hohe Venn, situated on the left bank of the Rhine between Eschweiler and Lüttich, contorted Devonian and Carboniferous beds are found. These Palæozoic beds, which disappear under the Cretaceous of Aachen and Maastricht, occur folded, and to some extent overthrust, into north-east striking anticlines and synclines, the whole effect being that the individual formations, the Upper Devonian, the Carboniferous limestone, and the Coalmeasures, form a succession of narrow north-east striking belts, the older members of each forming the anticlines and the younger members the synclines. This disposition of the beds is illustrated in Fig. 6.

The deposits are connected with the dolomitic limestones of the different formations on the one hand, and with transverse disturbances running north-west or approximately at right angles to the country, on the other. In German territory they are arranged in two districts, one situated to the south-west of Aachen in the neighbourhood of Moresnet, the mining area there being the property of the Vieille Montagne company ; and the other at Stolberg to the east of Aachen, where the important Diepenlinchen mine is worked by the Stolberg company.[1] Altogether, according to the statistics of the mining district of Düren, forty-five lead- and zinc concessions have been granted, upon approximately one-half of which operations have so far been undertaken.

The ore is associated with the Eifel limestone—Middle to Lower Devonian—as well as with the Carboniferous limestone, these two being separated from one another by a thickness of arenaceous slates, the Famennian beds of the Upper Devonian. However much the outward shape and the mineralogical content of these lead-zinc deposits may vary in the individual occurrences, and some of these variations are illustrated ·in Fig. 345, without exception they are connected with faults striking transversely across the formation, and are only found where solutions circulating along fissures and boundary-planes encountered calcareous and dolomitic rocks, such rocks being chiefly Carboniferous, but also Devonian. It is seldom that any ore is found in the arenaceous slates above or below these limestones. Still more unfavourable to the deposition of ore do the Carboniferous slates appear to have been, since with them ore is only known to occur in connection with the Bleiberg fissuresystem. Somewhat more favourable were the foot-wall slates of the Upper

---

[1] *Gesellschaft für Bergbau, Blei- und Zinkfabrikation zu Stolberg und in Westfalen.*

Devonian in which, as for instance in the Schmalgraf, Loutzen, and Prester mines, and at Hammerberg near Stolberg, etc., ore-bodies are occasionally found.

Such faults may be followed right through all the Palæozoic beds, from the Cambrian to the Carboniferous, even though at times they may be represented only by flucans or narrow fractures. The width of the

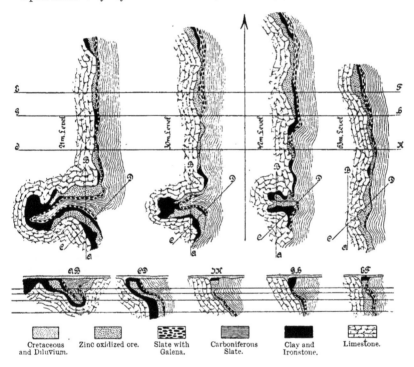

Fig. 345.—Horizontal and cross sections of the St. Pauli lead-zinc mine at Welkenraedt.

fissure in the limestone is different from that in the clastic rocks, while at the same time the mineralization has almost always changed. The strike varies between east by south and south, a bent course being seldom observed. Generally several fissures occur together, in part with their courses parallel and in part diagonal, so that they form a fissure- or linked series. Often again, a fissure, simple and regular in one part of its course, is observed to split up along its continuation.

The length of these transverse faults is quite considerable, the Münstergewand and the Sandgewand faults, for instance, cross not only the

Carboniferous syncline at the river Worm, but also that at the Inde. In cases where fissures pinch out, not far to the right or left others usually begin.   The dip is sometimes to the north-west and sometimes to the south-east, these two directions being occasionally present in different parts of the same fissure.   Generally the dip is steep, but occasionally it may be as flat as 40° or less.   In depth the fissures become smaller, till eventually they die out.   They represent true faults, the throw in particular cases reaching more than 400 metres.

In relation to geological age it may only be said that the first fracturing had begun before the deposition of the Senonian, while subsequent movements in the same fissures continued till after Diluvial time.   The majority of occurrences now being worked are found in the Carboniferous limestone, and particularly in its basal dolomite.   Among these are the famous zinc deposit at Altenberg in the Moresnet syncline, and the deposits in the Schmalgraf, Eschbruch, and Mützhagen mines.   Numerous other occurrences, including that at Diepenlinchen, are found in the Carboniferous limestone of the Werth syncline, to the south-east of Stolberg.

Concerning the arrangement of the fissures into series, four of these may be distinguished, namely : (1) the Welkenraedt series including the deposit of that name ; (2) a series east of this between Ruyff and Herbesthal; (3) the prolongation of the Schmalgraf fissure-system, this prolongation reaching to the neighbourhood of Eupen ; and (4) the Bleiberg Vieille-Montagne series.   Farther to the east comes the poorer district south of Aachen, and then the eastern district with the numerous lode-like occurrences of the Vichbach valley, these being centred around the Münstergewand fault and its parallel associates.   Still farther to the east on the right bank of the Vicht comes the Sandgewand system, which in the Werth syncline cuts the Carboniferous limestone twice and continues into the Eifel limestone.

The deposits form lodes as well as chamber-deposits and metasomatic ore-bodies, these different forms of deposit usually occurring in combination.

The ore consists of zinc sulphide — chiefly schalenblende but also sphalerite — of galena, and exceptionally pyrite and marcasite.   It is illustrated in Fig. 94.   With these deposits the products of oxidation are especially important.   At Schmalgraf the deposits have been proved to be payable to a depth of 175 m. ; at Diepenlinchen to 250 metres.   The oxidized zinc- and lead ores are the oxidation products of sulphide ores, which oxidation, as in the cerussite deposit of Diepenlinchen, may have proceeded so far that the sulphides have been completely replaced.   The most characteristic and frequent minerals are the zinc carbonate, and the hydrosilicate, though the anhydrous silicate also occurs.

The form of these metasomatic deposits is irregular, nor is the separation from the country-rock definite, as all gradations from pure zinc ore to zinciferous and ferruginous limestone may be observed. In texture the ore-bodies are usually porous or cellular. Crystals of secondary zinc carbonate and hydrosilicate are found in druses and fissures. The connection of such purely metasomatic deposits with fissures is often not discernible at first sight, owing to the fact that the processes of metasomatism tend to modify and obliterate the original arrangement.

Concerning the relative age of the different minerals, in the cavity-fillings the recurring sequence, first galena, then schalenblende, and finally marcasite, is remarkable. By subsequent shattering and disturbance a brecciated structure has often been produced, in which the number of slickensides and veins of recent calcite is striking.

The oxidized ores, including those of zinc, represent oxidation-metasomatic deposits. They are associated with the limestone in the neighbourhood of the surface, the alteration of this limestone having been effected by metal solutions formed from the sulphides. The formation of these sulphides probably took place before the deposition of the Senonian.

According to F. Herbst the composition of the ore won at different mines is as follows :

Mine.	Sphalerite.	Galena.	Pyrite.
	Per cent.	Per cent.	Per cent.
Schmalgraf . . .	31·6	5·0	19·5
Eschbruch . . .	32·1	3·8	24·8
Mützhagen . . .	23·0	3·5	21·8
Fossey	31·8	1·5	6·2

The importance of this district may be gathered from the following figures : during 1909 the mines of the *Altenberger Bergwerksgesellschaft*, including the Eschbruch, Schmalgraf, Fossey, and Lontzen mines in Prussia, as well as the Mützhagen mine in Belgium, produced 30,948 tons of ore, from which 938 tons of zinc oxidized ore, 12,289 tons of sphalerite, and 800 tons of galena were obtained. The Diepenlinchen mine belonging to the Stolberg Company [1] in 1910 produced 9611 tons of sphalerite and 1626 tons of galena.

ISERLOHN, WESTPHALIA

LITERATURE

TRAINER. ' Das Vorkommen des Galmeis im devonischen Kalkstein bei Iserlohn,' Verhandl. d. naturhist. Ver. d. pr. Rheinlande, u.s.w. XVII., 1860, p. 261.—v. DECHEN, Die nutzbaren Mineralien u.s.w. im deutschen Reich, Berlin, 1873, p. 627.—EICHHORN.

---

[1] *Aktien-Gesellschaft für Bergbau, Blei- und Zinkfabrikation zu Stolberg und in Westfalen.*

'Die Zinkerzlager bei Iserlohn,' Zeit. f. Berg- Hütten- u. Salinenwesen im pr. Staate, XXXVI. p. 142.—v. DECHEN. Erläuterungen zur geologischen Karte der Rheinprovinz und West-falens. Bonn, 1884.—L. HOFFMANN. 'Das Zinkerzvorkommen von Iserlohn,' Zeit. f. prakt. Geol., 1896, p. 45.—STOCKFLETH. 'Die geogr. geogn. u. miner. Verb. des südlichen Teils des Oberbergamtsbezirks Dortmund,' Verhandl. d. naturhist. Ver. f. Rheinland u. Westfalen, Vol. LII., Bonn, 1895, p. 45.

Between Hagen and Balve the Stringocephalus limestone of the upper Middle Devonian occurs as a belt 32 km. long and more than 1000 m. in width, dipping to the north. Scattered in this belt over an area about 12 km. long, embracing Letmathe, Iserlohn, and Deilingenhofen, some fifteen zinc deposits of varying size occur. These are of irregular shape with semicircular or triangular section, and are found either near or actually at the contact with the Lenne slate. While against this slate their out-line is definite and regular, their extension laterally into the limestone is irregular. The ore-bodies thus formed generally extend in a north-south direction and not infrequently assume a lode-like form. Without doubt they owe their existence to the individually unimportant but numerous transverse disturbances of this neighbourhood. In depth they generally rapidly pinch out, a depth of 205 m. being reached in but one case; with the surface they are usually connected by irregular chimneys filled with Diluvial material.

The ore originally consisted of sphalerite, pyrite, and to a less extent of galena. These minerals are now only seen as kernels within solid masses, the peripheral portions of which consist of zinc carbonate, hydro-silicate, and silicate, of limonite, and more rarely cerussite and pyromor-phite, these being enveloped in a considerable amount of clay. Calcite, often in beautiful crystals, and occasionally quartz, accompany the ore.

Some of the most important of these ore-bodies lie immediately under the town of Iserlohn, and the subsidences of the surface consequent upon the mining of these bodies have led to many complicated legal processes. While the Krug-von-Nidda mine to the east of Iserlohn was in 1893 stopped owing to exhaustion of its deposits, the Tiefbau-von-Hövel situated at the eastern end of the town continued a modest existence to within a few years ago.

The crusted structure of the sulphide ore—the sphalerite and pyrite especially occurring in separate layers—together with the botryoidal and stalactitic forms, point plainly to deposition in cavities formed by water, and to metasomatism. These sulphides subsequently became oxidized by meteoric waters. The clay which almost everywhere accompanies the ore is probably the insoluble residue from marl and limestone, though doubt-less some was subsequently introduced from above.

Upon the observed fact that the Stringocephalus limestone and the Lenne slate possess almost everywhere a low zinc content, the assumption has been based that these ore-bodies were formed by lateral secretion; we, however, are firmly convinced that the pregnant solutions ascended from depth along fissures. The sources of these solutions are probably the same as those from which the lodes of the Velbert anticline [1] between the Rhine and the Ruhr, were filled, this probability arising from the genetic connection established by the Geological Survey between those lode fissures and the fissures at Iserlohn.

In 1894 the zinc mines at Iserlohn worked by the *Märkisch-West-fälischen Bergwerksverein,* produced 7245 tons of zinc oxidized ore, 4182 tons of sphalerite, and 64 tons of pyrite, 350 men being employed in mines and works. In the same year the total production of the district was 8669 tons of zinc oxidized ore, 4185 tons of sphalerite, and 77 tons of pyrite. Since then however and not many years ago, all work was stopped.

Schwelm and Langerfeld, Westphalia

LITERATURE

v. Dechen. ' Über das Eisenstein- und Eisenkiesvorkommen auf der Zeche Schwelm,' Sitzungsber. d. naturhist. Ver. f. Rheinland u. Westfalen, XXXI., 1874.—Stockfleth. ' Die geographischen, geognostischen und mineralogischen Verhältnisse des südlichen Teils des Oberbergamtsbezirkes Dortmund,' Verhandl. d. naturhist. Ver. f. Rheinland u. West-falen, Vol. LII. Bonn, 1895; Der südlichste Teil des Oberbergamtsbezirkes Dortmund, 1896, pp. 57, 58.—H. Mentzel, in the joint work, Entwicklung des niederrheinisch-west-fälischen Steinkohlenbergbaues, Berlin, 1903, Vol. I.—P. Krusch. ' Neue Galmeiauf-schlüsse bei Schwelm,' Zeit. d. d. geol. Ges. Vol. LV., 1903.

Occurrences similar to those at Iserlohn are known in the Rote Berge at Schwelm and at Langerfeld near Barmen. Though these have at present no economic importance they are interesting in that the Stringocephalus limestone, while the form of its corals has been maintained, has itself been altered partly to marcasite and partly to sphalerite, deposits of these two ores occurring close to one another. From these, by the action of meteoric water oxidized iron- and zinc ores subsequently became formed, the former in sufficient quantity to have been the object of mining operations not many years ago.

At Langerfeld the Stringocephalus limestone became likewise replaced by sulphide zinc- and iron ores, which in their turn similarly suffered oxidation by meteoric water. This deposit is also interesting because, presumably in Tertiary time, a mechanical re-arrangement and concentra-tion of the oxidized ore and residual clay took place, so that to-day the upper portion of the deposit represents a sedimentary ore-bed, while the lower portion is an oxidized metasomatic deposit. The occurrence of considerable disturbances in the immediate neighbourhood of the deposit suggests that here also the pregnant solutions ascended from depth.

[1] *Ante,* p. 703.

Wiesloch, Baden

LITERATURE

A. Schmidt. Die Zinkerzlagerstätten von Wiesloch (Baden). Heidelberg, 1883.—
A. Sauer. Blatt Neckargemünd der geologischen Spezialkarte des Grossherzogtums
Baden nebst Erläuterungen. Heidelberg, 1898.

Some 12 km. south of Heidelberg, between Wiesloch and Nussloch, in the Upper Muschelkalk and close to the large Rhine-valley Fault, many irregularly bedded zinc deposits are found associated with the smaller step-like secondary faults.

The section of the Upper Muschelkalk at Kobelsberg consists of the Trochite beds proper and the lower Trochite limestone. The former may again be divided into an uppermost layer about 0·5 m. thick ; then for a thickness of 0·15 m. three limestone layers which in places are replaced by zinc ore or zinciferous clay ; then 3–6 m. of limestone ; and finally 1·5–4·8 m. of yellowish-grey or reddish encrinite limestone alternating with clayey marl. The lower Trochite limestone consists of a bluish-grey limestone alternating with clay and marl.

Mining operations, which go back as far as Roman time and which in the middle of the last century enjoyed a short period of prosperity, have again been given up, doubtless in consequence of the irregularity of the deposits.

The ore consists chiefly of zinc oxidized ore, which, as the exposures to the east on the Kobelsberg demonstrate, has resulted from the decomposition of sphalerite, portions of which are still extant. Schmidt gives the essential minerals occurring at Wiesloch, as sphalerite, galena, marcasite, zinc carbonate, hydrozincite, limonite, iron-ochre, and to a less extent pyrolusite, cerussite, pyromorphite, anglesite, antimony-ochre, realgar, barite, selenite, calcite, dolomite, and clay. The sphalerite occurs in two generations, the older of which, a cryptocrystalline schalenblende arranged in layers with galena and marcasite, forms the principal ore. Not infrequently, in cavities within the deposits, these minerals are deposited in stalactitic form, with a younger generation of sphalerite as the outside envelope. Galena, in addition to occurring in compact form with schalenblende, occurs also crystallized by itself, and, again, as irregular masses in the zinc oxidized ore. This oxidized ore exists as fine-grained, compact, whitish-grey, occasionally also striped, reniform, and botryoidal aggregates, or porous and cellular. In the cleaner portions the zinc content amounts to 40–50 per cent. Limonite and red iron-ochre, the products of decomposition of the marcasite, occur mixed with clay, more particularly in the upper portions of the deposit. In the ochre the silica has become concentrated into well-developed quartz crystals.

Schmidt specifies five ore-beds at Wiesloch, all lying in an unmistakable north and south direction. Three of these are situated on the west slope of the Hessel, forming the Hessel district, the two remaining beds on the south-west slope of the Kobelsberg forming the Baierthal district. All these beds, each of which is made up of a large number of smaller and larger pockets connected by stringers, appear everywhere to belong to the same horizon of the Trochite limestone. The mineralization is associated on the one hand with the bedding-planes, and on the other with the fissures which traverse the country. The thickness of the deposit may be divided into a lower portion consisting predominatingly of zinc ore, and an upper portion consisting substantially of iron ore. The fissures themselves are filled sometimes with ordinary clay, but generally with zinciferous, ferruginous clay, and zinc oxidized ore. That contemporaneously with the process of replacement of limestone by ore, deposition of ore took place in cavities already formed by water, is plainly indicated by the occurrence of sphalerite stalactites on the Kobelsberg. The ore-bed, though usually conformable, occasionally follows the fissures to moderate depths below, reaching in one case even as deep as the Wellenkalk.

In the oxidized zinc deposits the extensive distribution of crystal cavities regarded as the negative crystals of selenite, is striking. These cavities indicate that a sulphatizing action immediately preceded the formation of the oxidized zinc. Further, the occurrence at Wiesloch of many fossils preserved in oxidized ore has long been known, these fossil casts being regarded as simple replacement pseudomorphs. The dolomitization of the Trochite limestone, evidence of which may often be observed in the neighbourhood of Wiesloch, contrary to expectation does not appear to be directly connected with the mineralization, as Schmidt was, generally speaking, not able to establish any intermediate dolomitic zone between the ore and the limestone.

Carinthia

Metasomatic lead- and zinc deposits, such as occur in fissure enlargements or other cavities in close genetic connection with fault fissures, and such as in addition are always associated with soluble rocks, limestone particularly, occur in Carinthia not only in large number but typically developed. It was indeed from these occurrences that the formation of this important and widely distributed type of deposit was first understood, this recognition being more particularly the work of F. Pošepný. It was natural therefore that from them v. Groddeck took his 'Raibl' type. Deposits of this type traverse Carinthia in a broad east-west zone, usually along an impermeable slate bed. The Hauptschiefer, the Bleiberg or

Raibl slate of the Upper Triassic, and the Werfen slate of the Lower Triassic, are all beds of such slate. Pošepný rightly considered the greatest factor in the mineralization to lie not in the geological horizon of the limestone or slate, but in the difference in the permeability of these two rocks. For convenience in description he divided the occurrence into three districts, those of Raibl, Bleiberg, and Lower Carinthia respectively.

RAIBL

LITERATURE

F. Pošepný. ' Die Blei- und Galmeierzlagerstätten von Raibl in Kärnten,' Jahrb. der k. k. geol., Reichsanst. Vol. XXIII., Vienna, 1873 ; ' Über die Entstehung von Blei- und Zinkerzlagerstätten in auflöslichen Gesteinen,' Bericht über den allgem. Bergmannstag zu Klagenfurt, 1893, p. 77.—G. Gürich. Das Mineralreich, 1899, p. 573.—Geological mining maps with sections of Raibl together with drawings of the lead-zinc deposits ; surveyed by the officials of the State Mining Department ; published by the Minister for Agriculture, Vienna, 1903.

The country around this old hill town, which lies in a small valley on the south side of the Gail, forms part of the Alpine Triassic which, striking east-west and dipping south, consists of an alternation of limestone, dolomite, and marl. The ore-bearing white Raibl dolomite and limestone lie upon the Cassian beds of the Middle Alpine Triassic, and are in turn overlaid by the bituminous Fish slate. Above this again follow black marly slate and the Raibl beds proper, these belonging to the Upper Alpine Triassic.

The deposits occur in two different forms and at two different horizons. In the upper portions of the dolomite adjacent to the Fish slate, the ore is connected with transverse faults which in places are enlarged to cavities. In the narrow portions of these the ore forms lode-like masses, while the cavities are filled with concentrically banded galena and sphalerite, accompanied by dolomite as gangue. At a deeper horizon—though still in the limestone which is here but little dolomitized—along the same transverse faults, masses of oxidized ore are found as replacement pseudomorphs of the limestone.

The ore-bodies generally have the form of pipes. In the municipal mines they extend in three main directions, these corresponding respectively to the Johann, Abend, and Morgen faults ; in dip they incline to the south, parallel to the bedding of the limestone. The ore-body associated with the Johann fault lies some 300 m. below the contact between the limestone and the slate ; that following the Abend fault some 150 m. below that contact ; while the Morgen fault approaches ever nearer to that contact, making it probable that in greater depth the ore-body associated with that fault occurs actually at the contact with the slate.

The individual ore-channels are variable in section, being sometimes constricted, sometimes enlarged, while occasionally they split up into branches. With many of them a crusted structure, with layers of different minerals and of varying thickness disposed around a central cavity generally filled with dolomite, is very pronouncedly developed, indicating that the deposition of the ore undoubtedly took place in pre-existing pipe-like channels filled with water. More rarely a brecciated structure has arisen by the collapse of the cavity, such breccia consisting of dolomite- and limestone fragments cemented by ore. A pseudo-brecciated structure is not infrequently found where the limestone has only been partly replaced, kernels of it still remaining.

The primary ore is invariably of sulphides. It consists of schalenblende, galena almost free from silver, and pyrite. Dolomite and barite form the gangue. From these primary minerals, zinc carbonate and cerussite have been formed by oxidation. The ore-pipes are particularly interesting. These occur principally in connection with the Struggl fault-system and consist of pyrite and sphalerite in concentric layers around hollow galena stalactites. Octahedra of galena are occasionally found in these pipe-like aggregates. An alternation of dolomite with sphalerite and galena found in the hanging-wall is known as slate ore.

The oxidized zinc deposits were in greater part formed directly from limestone by oxidation-metasomatism, the original structure of the limestone being maintained. In these deposits zinc carbonate occurs chiefly, hydrozincite more rarely, while the hydrosilicate is uncommon. As with all zinc deposits, limonite, more or less clayey, occurs in the oxidized ore, the material of this limonite having been derived partly from the pyrite and partly from the iron contained in the schalenblende and the limestone. Towards the outcrop particularly, the proportion of iron increases at the expense of the zinc.

Mining at Raibl is of great though unknown age. The credit for the development of this industry is due to the municipality, which in 1762 purchased some of the mines. The present production amounts to about 3000 tons of lead ore and 17,000 tons of zinc ore annually.

BLEIBERG

LITERATURE

Mohs. ' Die Gebirgsgesteine, Lagerungsverhältnisse und Erzlagerstätten zu Bleiberg in Karnten nach den Beobachtungen des k. k. Bergrates Fr. Mohs 1810.' Copy dated Nov. 5, 1830, in the Mine Archives.—K. Peters. ' Die Umgebung von Deutsch-Bleiberg in Kärnten,' Jahrb. d. k. k. geol. Reichsanst., 1856; ' Über die Blei- und Zinklagerstätten Kärntens,' Oesterr. Zeit., 1863, p. 173.—v. Cotta. ' Über die Blei- und Zinkerzlagerstätten Karntens,' Berg- und Hüttenm. Ztg., 1863, Vol. XXII.—P. Potiorek. ' Über die Erzlagerstätten des Bleiberger Erzberges,' Oesterr. Zeit., 1863.—E. Suess. 'Geogn. bergmänn. Skizze von Bleiberg,' Oesterr. Zeit., 1869.—F. Pošepný. 'Über alpine Erzlagerstätten,' Verhandl. d. k. k. geol. Reichsanst., 1870.—v. Mojsisovics. 'Über die tektonischen Verbältnisse des erzführenden Triasgebirges zwischen Drau und Gail (Bleiberg in Kärnten),' Verhandl. der k. k. geol. Reichsanst., 1872.—Brunlechner. 'Die Entstehung und Bildungsfolge der Bleiberger Erze und ihrer Begleiter,' Jahrb. des naturhist. Museums von Kärnten, Vol. XXV., 1895.—Hupfeld. 'Der Bleiberger Erzberg,' Zeit. f. prakt. Geol., 1897. —G. Geyer. 'Zur Tektonik des Bleiberger Tales in Kärnten,' Verhandl. der k. k. Reichsanst., 1901.

Bleiberg and Raibl form together the centre of the long-lived Carinthian lead- and zinc mining. Bleiberg lies 12 km. west of Villach in a deep-cut tectonic valley between the Bleiberg hill 1261 to 1823 m. in height to the north, and the Dobratsch mountain 2167 m. high to the south, as illustrated in Figs. 58 and 346. The tectonics of the district have the greatest bearing upon the extent and distribution of the deposits, this aspect of the subject having been closely studied by Pošepný, Hupfeld, and especially by Geyer.

The oldest formation, exposed at the western end of the Bleiberg valley beyond Kreuth, belongs to the Lower Carboniferous. Upon this to the east lie the Gröden sandstone, the Werfen beds, and the Guttenstein limestone, all of which however have but small extent on surface. Then follows in great development the Wetterstein limestone-dolomite, of which in greater part the Bleiberg and the Dobratsch consist. The Cardita beds and the Main Dolomite. which come next, are only represented at tectonic breaks,— generally in the valley—by such remnants of a former larger extent as have been withdrawn from erosion by faulting. The district is highly disturbed. As indicated in Fig. 346, the Wetterstein limestone-dolomite dips on the Bleiberg to the south and on the Dobratsch to the north, both these directions inclining inwards towards the great Bleiberg Break to which the valley owes its existence. North of this break particularly, many faults along which huge segments of the younger rocks have subsided, have been delineated. The combination of folding and faulting, as illustrated in Fig. 347, produces an extremely complicated tectonic figure.

The ore-bearing horizon occurs in the Wetterstein limestone. It is a light-coloured dolomitic limestone within which, though seldom, darker and presumably bituminous layers are intercalated. The proportion of $MgCO_3$ varies between 0·1 and 40 per cent. According to the section given in Fig. 347 the following beds in addition are represented in the Bleiberg; the Cardita beds which are younger than the limestone; the overlying Main Dolomite; and finally, in the foot-wall of the limestone, the Wetterstein dolomite. The Dobratsch mountain on the opposite side of the valley consists, in the vicinity of the break, of Wetterstein dolomite.

The deposits at Bleiberg are chiefly irregular cavity-fillings and metasomatic deposits, the distribution of these being closely connected

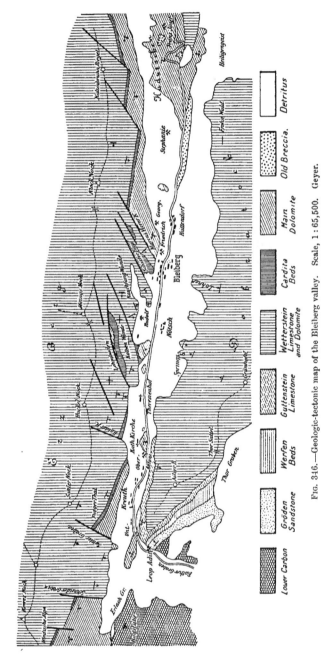

FIG. 346.—Geologic-tectonic map of the Bleiberg valley. Scale, 1 : 65,500. Geyer.

Lower Carbon   Gröden Sandstone   Werfen Beds   Guttenstein Limestone   Wetterstein Limestone and Dolomite   Cardita Beds   Main Dolomite   Old Breccia.   Detritus

with the Bleiberg Break.   They occur almost exclusively in the ore-bearing limestone.   Their form has been aptly described by Pošepný as pipe-shaped. Reference to Fig. 58 shows the striking agreement which exists between the bodies of each individual district, in relation to the direction in which they lie ; the axes of these bodies, as already recognized by Mohs, coincide with the lines of intersection of two planes, namely, the bedding-plane of the lime-stone and that of certain fissures.   At Bleiberg itself these fissures are known as lodes and the bedding-planes as planes, while at Kreuth they are termed cross fissures and beds respectively.   It is particularly noteworthy that these axes are not dependent upon certain limestone bands but upon certain bedding-planes separating the limestone, such planes being

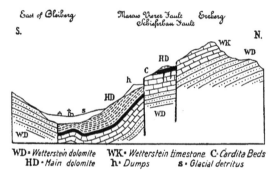

FIG. 347.—Section across the Bleiberg valley. Geyer. `

described as 'kindly.'   These kindly planes are more pronounced than those which are not associated with ore, while in addition they also exhibit traces of a previous water circulation and are accompanied by clay.

In the Fugger valley the fissures strike south-east and dip 50°–70° to the north-east ; in the main district of Kreuth the strike is north-south and the dip 60°–65° to the east ; while around Bleiberg itself the strike is east-west and the dip approximately vertical.   Corresponding to the different strike of the fissures and bedding-planes, the ore-bodies in the Fugger valley and at Kreuth, as indicated in Fig. 58, have a general north-west strike, while in the Bleiberg area proper a north-east strike prevails.   The depth to which these bodies continue at Bleiberg has not yet been determined ; some have already disappeared in depth, while in others the deepest developments—which in the Kreuth district are more than 400 m. below the level of the valley—have proved to be the best and most promising of the whole district.   With regard to the breadth of the area throughout which these deposits occur, Hupfeld

considers there is evidence to show that they are associated with the
Bleiberg slate, from which generally they extend only some 500 m. into the
ore-bearing limestone.

The minerals of the deposits are both primary and secondary. Among
the former are galena, sphalerite, marcasite, barite, fluorite, calcite, and
dolomite. The galena generally occurs in the form of pipes similar to
those mentioned when describing the deposits at Raibl, these pipes here
consisting of long cylinders of galena filled with calcite. This galena is
remarkable for its great purity and the complete absence of antimony,
copper, and silver; it was the raw material from which a celebrated
brand of lead was made.[1] The sphalerite occurs chiefly near Kreuth to
the west and at another point right to the east, while in the central
portion of the district it is subordinate. It generally occurs in the form of
schalenblende with a core of calcite or galena. It is mostly light yellow
in colour, becoming darker as the proportion of contained iron increases.
The intimate intergrowth of barite with these sulphides is especially
characteristic at Bleiberg, aggregates of tabular crystals arranged parallel,
or solid masses with laminated structure, being often found. The mode
of occurrence of the fluorite is interesting, this mineral being found in
rose-coloured, violet, or crystal-clear cubes of at most 7·5 mm., upon galena
and sphalerite. Calcite and dolomite either form veins in the country-
rock, or are of recent formation in druses, etc.

The secondary minerals are in greater part those which have resulted
from the oxidation of the primary minerals. The galena upon oxidation
gives rise to good crystals of cerussite, twinned and retwinned, to plumbo-
calcite, anglesite, and wulfenite, this last mineral, which occurs remarkably
often in druses at Bleiberg, being almost always crystallized in thin plates.
The proximate source of the molybdenum in this yellow and occasionally
grey mineral has so far not been determined; no molybdenite has yet
been found in the deposits. The sphalerite at oxidation passes chiefly to
zinc carbonate which is often contaminated by limonite and clay, while
crystals are uncommon. By absorption of water the carbonate in turn
becomes hydrozincite. The change from sphalerite to the hydrosilicate is
less common, though when it does occur crystals are more frequent. Marcasite becomes limonite, sulphuric acid being liberated, wherefrom the
alteration of the limestone to anhydrite results, the amount of this mineral
at Bleiberg being remarkable. It is sky-blue in colour and often includes
unaltered pieces of limestone. Upon exposure to air it absorbs water,
selenite becoming formed. In the Kreuth district the asbetos-like mineral,
mountain-leather, occurs now and then.

Some idea of the form of these deposits may be gathered from Fig. 60.

<hr>

[1] *Das Kärntener Jungfernblei.*

As illustrated in that figure, the ore, in detail also, follows the bedding-planes on the one hand and the fissures on the other, as solution of the rock proceeded.

The relative ages of the different minerals as revealed by crystal intergrowth and incrustation, has been determined by A. Brunlechner, who came to the conclusion that two generations of sphalerite, galena, calcite, and wulfenite, must be assumed, in the first of which galena predominates, and in the second, sphalerite. The sequence of the first generation he gives as calcite, sphalerite, galena, barite, marcasite, calcite, fluorite ; and that. of the second, calcite, sphalerite, galena, schalenblende, barite, marcasite, calcite, fluorite, dolomite, and anhydrite. To these two generations the different oxidation minerals described above are of course additional. To these, according to Brunlechner, may be added ilsemannite, $Mo_3O_8 + xH_2O$, sulphur, and goslarite, though the occurrence of these minerals at Bleiberg has not yet been fully established.

The genesis of the Bleiberg occurrence has been studied and discussed by many authorities, some of whom have regarded it as syngenetic, others again as epigenetic. Disciples of a syngenetic genesis have been represented at all times, among these being Mohs, Fuchs, Lipold, and Peters. To-day, however, it is generally agreed that these deposits are of epigenetic origin, an origin early suggested, though subsequently abandoned by various authors. The miner, as far back as the end of the eighteenth century, recognized that the ore-pipes were enriched intersections. E. Phillipps, a French mining engineer, in a paper published about the middle of last century, pronounced for their formation after the manner of lodes. In 1863 von Cotta published a long paper upon Bleiberg in which he classified the deposits as secondary and due to metal solutions which, circulating along fissures in the country, found their way into the fractured limestone. Potiorek, for many years the mine manager at Bleiberg, when in the same year describing the occurrence at Bleiberg, disputed the idea of a bedded character, this description being the first made in any great detail. Pošepný, in an annex to his investigation at Raibl, then put forward the view that these deposits were pipe-like cavity-fillings, generally following the lines of intersection between fissures and bedding-planes, a view which Hupfeld endorsed. Brunlechner [1] suggested the possibility of formation by lateral secretion, because, according to his view, the nature and manner of the distribution of the ore in the extensive and massive complex of the Wetterstein limestone, strongly supported this hypothesis. According to him alteration so proceeded from the fissures that some became blocked by deposited minerals, while others became enlarged.

The authors endorse the theory elaborated more particularly by

[1] *Loc. cit.*

Pošepný that the deposits are the fillings of irregular cavities, such fillings having been accompanied by the metasomatic alteration of a portion of the Wetterstein limestone, both processes of deposition proceeding from fissures. At the subsequent oxidation of the ore in the manner explained above in detail, heavy-metal solutions once more came into contact with limestone, whereby a second replacement of the limestone by lead and zinc became possible. This process, which doubtless was accompanied by enlargement of the original width of the deposit, comes within the range of what Krusch terms oxidation-metasomatism.

LAURION, GREECE

LITERATURE

A. Böckh. 'Über die antiken laurischen Silberbergwerke,' Abhandl. d. Akad. d. Wiss., 1814–1815, pp. 85-140. Berlin, 1818.—A. Cordella. Le Laurium. Marseille, 1869. —R. Nasse. 'Mitteilungen über die Geologie von Laurion und den dortigen Bergbau,' Zeit. f. d. Berg- Hütten- u. Salinenwesen im pr. Staate. Vol. XXI. Berlin, 1873.—B. Simonnet. Le Laurium. Bull. de la Soc. de l'ind. min. St-Étienne, 1883, II. Ser. Vol. XII. p. 641.—A. Cordella. La Grèce sous le rapport géologique et minéralogique. Paris, 1878 ; Mineralogisch-geologische Reiseskizzen aus Griechenland. Leipzig, 1883.—R. Lepsius. Geologie von Attika. Berlin, 1893.—Review : 'Die geologischen Verhältnisse der laurischen Erzlagerstätten,' Zeit. f. prakt. Geol., 1896, p. 152.—J. J. Binder. 'Laurion : Die Attischen Bergwerke im Altertum,' Jahresber. d. k. k. Oberrealschule Laibach, 1895.—C. v. Ernst. 'Über den Bergbau im Laurion,' Jahrb. d. k. k. Bergakad., Vienna, 1902, Vol. L. p. 447.

The valley of Legrana, extending north-north-east along a break in the country, separates the south-eastern and metalliferous portion of southern Attica from the north-western portion, which is differently constructed and non-metalliferous. Most of the ore-deposits are found in a disturbed zone represented by an anticlinal break which, traversing the district of Legrana, passes through Kamaresa and the Plaka Pass and eventually reaches the north-east coast at Daskalio-Niki.

The general strike of the beds in this south-eastern portion of Attica, known as the district of Laurion or Ergastiria, is north-north-east, and the dip is to the east. Of these beds Lepsius gives the following sequence :

Quaternary.

Tertiary : Upper, represented by the Pikermi beds ; and Lower.

Cretaceous : Upper limestone, grey in colour ; green slates and marl of Athens ; lower limestone, ferruginous and yellowish-white.

Crystalline schists : Upper Marble of Attica, bluish-grey and fissile ; Kaesariani mica-schist, in places with an interbedded marble layer ; Lower Marble of Attica ; dolomitic and calcareous schists ; calcareous mica-schist with quartz lenses.

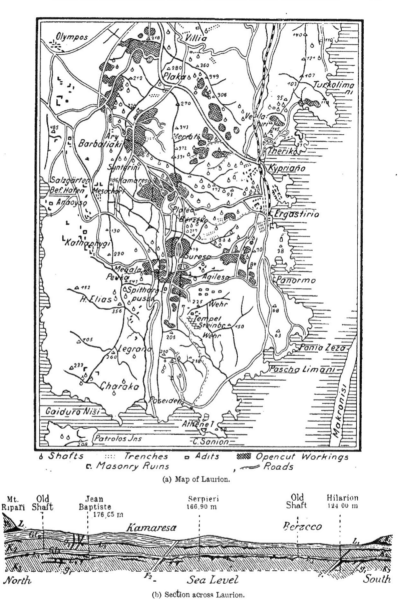

(a) Map of Laurion.

(b) Section across Laurion.

FIG. 348.—Map (a) and section (b) of Laurion.   Lepsius and Cordella.

Gl, mica-schist with lead lodes (g); K, limestone (Marble); L, metasomatic ore-bodies.

The deepest beds exposed in the Laurion hills are those of the Lower Marble, which are several hundred metres thick. It is in these that the zinciferous and argentiferous lead deposits are found. Numerous occurrences of gabbro often serpentinized—as well as the granite of Plaka which with numerous apophyses lies in the course of the anticlinal fault—break through and metamorphose the crystalline schists and the Cretaceous beds. While the western flank of the Legrana anticline is highly disturbed, the eastern flank dips very regularly at an angle of 10°–20° all the way to the coast.

At Laurion three so-called ore-contacts are distinguished, these being disposed as follows : the first or uppermost at the contact between the Cretaceous slate and the Lower Cretaceous limestone ; the second between the Upper Marble and the Kaesariani mica-schist ; and the third or lowest contact between this mica-schist and the Lower Marble. The irregular bed-like sphalerite-, galena-, and oxidized zinc deposits occur almost always at the contacts between the Kaesariani mica-schist and the Upper and Lower Marble, particularly the latter. At Kamaresa, on the steeply dipping western flank, they are found also along the lower surface of the marble bed which is regularly intercalated in the mica-schist at that place. At the contact of the granite and its apophyses with the marble, similar deposits also are often seen, from which rich veins shoot out into the limestone. Against the granite at Plaka the Cretaceous slate is altered to an augite-epidote-garnet rock, termed plakite by Cordella. Upon the Rimbari hills also, the Lower Cretaceous limestone, often dolomitized and sideritized, is in many places altered to manganiferous limonite with some galena. The ores at the first contact are much exploited and exported.

The lower lead-zinc bedded deposits, which have been formed by the alteration of the limestone associated with the crystalline schists, are according to their depth known respectively as the second and third contact ores. These contain no iron but consist of argentiferous galena, sphalerite, and zinc oxidized ore.

In ancient times attention was only paid to the galena, from which silver was obtained by smelting. The dumps left from those times, consisting of rock, ore, and slag, are now in many cases being re-worked, yielding 3–4 per cent of lead and a little zinc. The slags contain 13–14 per cent of lead with 0·5–3 kg. of silver per ton of lead. According to von Ernst the galena, which is usually very compact, has a high percentage of lead and is rich in silver, 2 kg. per ton often being obtained. The percentage of zinc in the oxidized ore is variable ; with the lower deposits, from which at times as much as 3000–4000 tons of such ore has been obtained monthly, it was formerly, after roasting, 65 per cent. Modern mining will however have to deal with ore of much lower value.

According to Cordella, the production of the mines worked by the French company at Laurion in 1901 was 10,730 tons of roasted ore of the qualities known as Nos. 1 to 4 ; 803 tons of plumbiferous zinc ore ; 3942 tons of ferruginous zinc ore ; and 494 tons of sphalerite. In 1906, altogether 22,000 tons of zinc oxidized ore and 12,298 tons of argentiferous lead were obtained from this district.

SARDINIA

LITERATURE

A. DE LA MARMORA. Description de la Sardaigne Picos de l'Europe, 1839.—E. FERRARIS. Memoria geognostica sulla formazione metallifera delle miniere di Monteponi e adiacenti, 1882.—G. ZOPPI. Descrizione geologico-mineraria dell' Iglesiente (Sardegna), 1888 ; Mem. descr. della Carta geol. d. Ital. IV., 1888.—Carta Geologico-Mineraria del l' Iglesiente 1 : 50,000, R. Comit. Geol. d' Italia. Rome, 1888.—MARX. 'Geogn. und bergm. Mitteilungen über den Bergbaubezirk von Iglesias auf der Insel Sardinien,' Zeit. f. das Berg-Hütten- und Salinenwesen im pr. Staate, 1892, Vol. XL.—FUCHS et DE LAUNAY. Traité des Gîtes minéraux et métallifères, II., 1893, p. 387.—A. DANNENBERG. 'Reisenotizen aus Sardinien,' Zeit. f. prakt. Geol., 1896, p. 252.—B. LOTTI. I depositi dei minerali metalliferi, p. 63.

. The geological picture presented by Sardinia is rendered particularly interesting by the variety of both the sedimentary formations and the eruptive rocks. The basement rock consists of granite, upon which in succession lie Cambrian sandstone and slate, Silurian limestone, slate, and grauwacke, and Devonian beds. During the deposition of the Carboniferous limestone and the Permian beds Sardinia was dry land, a condition which altered to allow the Jurassic and the Cretaceous to be deposited, these two formations being separated from one another by an unconformity. Finally, Tertiary beds likewise occur, covering large areas. At the conclusion of the Nummulite epoch the Mesozoic beds in Sardinia were tilted; while at the birth of the Alps a last extensive folding took·place.

While in most places in Europe the Devonian, Carboniferous, Permian, and Tertiary systems are rich in ore-deposits, the famous zinc- and lead deposits of Sardinia, and particularly those of the province Iglesias, are found in by far the greater number in the Silurian, and it is probable that they are of great geological age.

In this island three mining districts may be recognized, namely, a very important one to the south-west in the province Iglesias ; a less important one near Alghero ; and an eastern one in the eruptive massive around Sarrabus, Ogliastre, and Lulla. Of these, only the first will here be described.

The district of Iglesias includes to the north the granite massive of Arbus, upon which lies a mantle of Silurian slate and grauwacke. In this mantle the deposits at Montevecchio, Gennemari, etc., are found. To the south large areas are occupied by Cambrian quartzite and slate, these being surrounded by ore-bearing limestone presumably of Silurian age. In this limestone almost all the famous oxidized zinc deposits of Sardinia occur. To the south-west at Fontanamare, Triassic beds appear, while to the east and north-west large areas are covered by late Eocene and Quaternary formations. Deposits of argentiferous galena and sphalerite are numerous. They are found more particularly at the contact between limestone and slate or between two limestones differing petrographically, or they cut across the formation as veins or lodes. In such lodes irregular masses of zinc oxidized ore with a little galena are often found. As with all metasomatic lead-zinc deposits, the connection between the lodes and the metasomatic deposits is so close that the two classes of deposit must be described together.

The distribution of the ore is in general connected with that of the limestone. In greater detail, the galena in the lodes shows a tendency to be concentrated in ore-shoots, as for instance at San Benedetto, Monteponi, etc., while, in addition, the following general rules may be considered to have become established : 1. The more irregular the deposit, the greater the silver content. 2. The silver content decreases in depth. 3. The sphalerite increases in depth.

The important zinc carbonate and hydrosilicate deposits are found in the upper levels though they do not always reach the surface. In depth, often enough, they end in a clay-filled fissure.

The extension along the strike is often considerable. To the north of the granite massive, for instance, a galena lode is known which, cutting across a wide sequence of old slates for more than 3 km., has been worked under different names in different properties, these including the Montevecchio, Perdixeddosu, Ingustosa, Gennamari, etc. A point worthy of notice is that this lode contains but very little sphalerite. To the south of this granite another lode-system occurs, the members of which strike south-east to east-south-east. As will be seen from Fig. 349, these two systems form together a surround to the granite. In the neighbourhood of Iglesias are found the lead lodes of San Giovanni to the south, and San Benedetto and Malacalzetta to the north ; while to the west the important metasomatic deposits worked in the Nebida and Monteponi mines occur. Finally, there is the no less important zone of oxidized zinc deposits at Malfidano to the north-west of Iglesias.

The lode-series at Montevecchio includes three lodes striking east-west and dipping 65° to the north. Of these the most important, having a width of 60 m., stands out on the surface from the slaty country-rock like a massive quartz wall. The galena, which may contain as much as 80

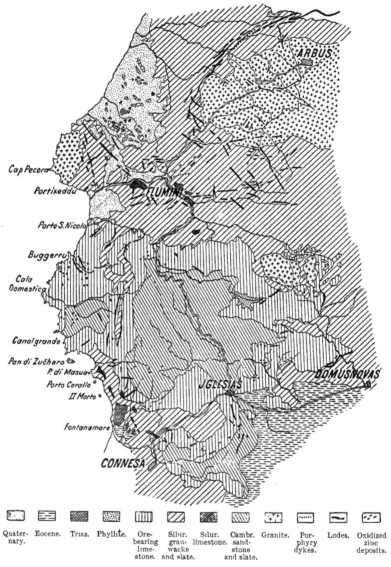

| Quater-nary. | Eocene. | Trias. | Phyllite. | Ore-bearing lime-stone. | Silur. grau-wacke and slate. | Silur. limestone. | Cambr. sand-stone and slate. | Granite. | Por-phyry dykes. | Lodes. | Oxidized zinc deposits. |

Fig. 349.—Geological map of the Iglesias mining district in Sardinia.   Scale, 1 : 240,000.
Testore. Joppi, Lambert, and Deferrari.

per cent of lead and 800 grm. of silver per ton, occurs in lenses sometimes as much as 8 m. in width.   In depth sphalerite becomes more common. The production amounts to over 10,000 tons per year.

The lode worked in the San Giovanni mine lies almost opposite to Monteponi in a valley descending from Iglesias to the sea.   The Silurian slate and the Devonian limestone have here an almost vertical east-west contact, along which runs a lode.   The main lode however occurs between a red dolomitic limestone and an overlying bed of blue limestone, in the immediate vicinity of a quartz lode which separates the limestone from clay-slate.   In addition to these two lodes a third has been exposed in the blue limestone.   The silver content of the first-mentioned lode is on an average 150 grm. per ton.   The second lode, lying between the two limestones, contains occurrences of argentiferous galena with quartzose, calcareous, or argillaceous gangue.   This galena has a high silver content, generally 1200–1900 grm. per ton.   The more irregular the deposit, the higher the silver content appears to be.   Near the surface this lode, which is often characterized by a brecciated structure, contains barite.   All the fractures within the lode are filled with calcite, which in places has subsequently been removed giving rise to cavities containing concretions of calcite.   In one place a galena-sphalerite lode traverses such a cavity, the ore in this traverse being oxidized.   The third lode carries galena with quartzose gangue and a silver content sometimes as high as 1500 grm. per ton.

The argentiferous galena lode at San Benedetto, 7 km. north of Iglesias, is celebrated for its zones of secondary enrichment.   At this place the ore-bearing and presumably Silurian limestone abuts against Cambrian grauwacke and sandstone.   The gangue of the lode consists chiefly of quartz and calcite.   The portions formerly worked contained 1300 grm. of silver per ton; the ore upon the third level however now contains but 400 grammes.   The disposition of the ore-shoots within the plane of the lode is interesting, these forming lenticular ore-bodies approximately parallel to the slope of the hill.   The lode-filling almost everywhere contains some zinc carbonate and cerussite; near the outcrop some considerable masses of the former were met.   These deposits yield some 2000 tons of galena and 2000 tons of zinc carbonate per year. ·

At Malacalzetta north of Iglesias and east of San Benedetto three galena lodes occur in limestone.   The first, known as Monte Novo, consists of several veins filled with galena, calcite, and quartz.   In the second, that of Monte-Cucchedu, a lens 30 m. long and 2·5 m. wide was found consisting of an intimate mixture of the same three minerals.   The third lode has its ore disposed in ore-shoots approximately 100 m. in length, 1·5 m. in width, and but little extent in depth.   In accordance with the nature of the country-rock the intergrowth of gangue and ore is intimate.

The filling consists of galena, cerussite, and traces of malachite as the valuable minerals, with quartz, calcite, siderite, clay, and exceptionally barite, as gangue.   It is worthy of note that the cerussite is generally richer in silver than the galena.   The yearly production is some 1000 tons of galena containing 67 per cent of lead and 1000 grm. of silver per ton.

The Nebida mine lies on the south-west coast in the ore-bearing lime-stone.   The deposits at this place appear very diversified, true lodes, impregnations, and irregular metasomatic masses ·being all represented. The country-rock strikes north-south and consists of an alternation of slate and limestone.   The metasomatic oxidized zinc masses form large, almost vertical columns which were formerly thought to be the fillings of pre-existing cavities, a view supported by the discovery during prospecting of a stalactitic cave.   The principal column, 20 m. diameter and consisting of 45 per cent zinc ore, reaches a depth of 180 metres.   Another, containing much galena and having a width of 8 m., runs parallel to the bedding and has been worked to a vertical depth of 100 metres.   The true lodes are found in the northern part of this particular district.   Some of these display a very irregular filling of quartz, siderite, and abundant galena. In the limestone they are occasionally well developed, while in the Silurian slate they disappear.   The galena on an average contains 7·5 kg.,'though occasionally as much as 11 kg. of silver per ton : it is accompanied by cerussite.   The impregnations are found along the contact of slate and limestone.   With these deposits, from the narrowest of fractures the lime-stone has been so invaded by ore that for lengths of as much as 150 m. the more easily decomposed portions have been replaced for a width of 8–10 m., sometimes by sphalerite and sometimes by zinc oxidized ore. The annual production from this district is some 3000 to 4000 tons of zinc oxidized ore, and 1000 tons of galena containing on an average some 7 kg. of silver per ton.

The famous occurrence at Monteponi, the oldest mine in Sardinia, situated some 2 km. south-west of Iglesias, includes the galena lodes and irregular oxidized zinc deposits which were worked for lead and silver by the Carthaginians and the Romans in their time, and afterwards in the Middle Ages by the Spaniards.   Till 1851 they belonged to the State, passing then into the possession of a private company, by which, since 1867, they have been worked.   These deposits are now in greater part exhausted.   The mine lies upon a hill 360 m. high consisting of ore-bearing limestone overlaid by a calcareous slate containing fossil *Trilobites*.

Two different occurrences may be noted, that of galena ore-bodies to the south, and that of oxidized zinc deposits to the north.   The bodies of galena follow the bedding-planes of the limestone which dip at a gentle angle to the east.   Within the particular beds the ore-bodies pitch at

angles varying between 35° and 55° towards the contact with the slate.
They are distinguished by a notable continuity in length and great diversity
in width and content.   The filling consists in general of galena, with which
some zinc oxidized ore is associated.   The former, containing up to as
much as 82 per cent of lead and on an average 250 grm. of silver per
ton, generally occurs in lenses in which the only impurities are scattered
nodules of pyrite.   Almost every lens is enveloped in a shell of brown or
clayey iron oxide ; in but few cases does the ore come actually into
contact with the limestone, and then the well-known beautiful crystals
of cerussite, anglesite, and phosgenite, are found.

The oxidized zinc deposits at Monteponi occur in the limestone in form
more or less resembling lodes or pipes, though they have no great extension.
The most important occurrence lies to the north.   There the limestone
is decomposed and the ore follows the bedding-planes and crevices in
such a manner as to at once suggest the metasomatic character of the
deposit.   With incomplete replacement of the limestone a pseudo-breccia
results.   In this manner masses of zinc ore are formed, 40 m. or more
in length and width, but which in depth become poor and disappear.
Although most of the deposits are now exhausted, it may still be seen
that some attained dimensions of 200 m. by 120 metres.   The ore is
not particularly pure ; in the raw condition it contains 35 per cent
of zinc, which after roasting becomes increased to 45–47 per cent.   It
is accompanied by iron- and manganese minerals, while cerussite also is
often found intimately mixed with it.

As with most oxidized zinc deposits the question of dressing is an
important matter.   These deposits are worked in opencuts which produce
annually approximately 4000 tons of ore containing 55 per cent of zinc,
and a large amount of poorer ore which upon concentration gives 10,000
to 12,000 tons of marketable material.   The galena production amounts
to 4000 to 5000 tons of ore containing 60–80 per cent of lead and 200–350
grm. of silver per ton.

The oxidized zinc deposits at Malfidano occur near the village of
Bugerru.   At Planu-Sartu the ore-bearing limestone, which strikes south-
south-west and dips 50°–55° to the east, contains five well-developed
deposits which can be followed for 340–350 metres.   These in places have
a lode-like character, the ore being sharply separated from the country-
rock ; more generally however they form beds alternating with limestone.
The ore may be schistose or compact, while the zinc content fluctuates
between 45 and 50 per cent.   Small veins of red ferruginous clay and
zinc ore, ramifying in all directions, are often found in the deposit.
In addition other small veins of quartz and galena occur.   At Malfidano
and Caïtas in the eastern portion of the Malfidano property, the ore-bearing

limestone strikes south-south-west and dips 80° to the east. The ore-bodies here contain but little sphalerite and still less galena ; they gradually merge into normal limestone. This district along its whole length is traversed by a break 30 m. wide, filled with a breccia of limestone and clay, and known for a length of 900 metres. On both sides of the break the oxidized zinc masses form large columns 80 to 100 m. wide and 15 to 20 m. in thickness. That at Malfidano, striking north-south, is crossed by east-west quartzose galena lodes. The deposits at Malfidano and Caïtas produced in 1889 some 42,000 tons of milling ore and 319 tons of sphalerite.

In addition to those mentioned, Fuchs mentions other deposits in the province of Iglesias, at Baueddu, Planu-Dentis, Sedda-Cherci, and Cucuru-Taris. Most of these are typical oxidized zinc deposits. That at Baueddu occurs between slate and Silurian limestone. In width it varies from a few centimetres to 40 m., and its known length is 400 metres. It strikes generally north-south and dips 30°–80°. The filling is very irregular ; quartzose masses with fragments of zinc carbonate, calcite impregnated with zinc ore, and finally, almost pure brown or red carbonate containing ferruginous material, alternate with one another in the northern portion ; the central portion is clayey and contains lenses of zinc hydrosilicate 2–3 m. thick ; while to the south the silicate is less common, and poor clayey and quartzose masses become prevalent. The occurrence at Planu-Dentis and Pira-Roma is in a Silurian limestone, not far from its contact with slate. It consists of a system of crevices parallel with the bedding-planes of the limestone, from which crevices the alteration of that rock into zinc oxidized ore proceeded. The richest ore lies directly at the contact with slate. From a total mineralized length of 250 m., only 60 m. are payable. The deposit at Sedda-Cherchi is quite analogous, though it quickly becomes poor and passes into zinciferous limestone. That at Cucuru-Taris likewise occurs at the contact between slate and limestone ; in the neighbourhood of that contact it includes masses which are almost vertical and which have been formed from fractures.

The participation of the different districts in the lead-zinc production of Sardinia for the year 1889, was as follows :

[TABLE

	Galena.	Zinc Oxidized Ore.
	Tons.	Tons.
Montevecchio . . .	12,100	...
San Giovanni . . .	3,660	...
San Benedetto . . .	1,350	1,068
Malacalzetta . . .	2,900	...
Nebida	1,500	3,800
Monteponi	4,400	15,300
Malfidano	...	60,000
Baueddu	...	3,000
Total . . .	25,910	83,168

These figures illustrate the variable character of the lead-zinc deposits of Sardinia, where all the intermediate stages between purely lead lodes, metasomatic lead- and zinc deposits, and finally metasomatic oxidized zinc deposits are represented. Concerning genesis, it is doubtless the case that oxidation-metasomatism played an important part in the formation of these deposits.

THASOS

The island of Thasos, belonging to the Khedive of Egypt, rises to a height of 1205 m. out of the Aegean Sea south of the Macedonian mainland and facing the bay of Cavalla. According to Herodotus, whose facts however have not always proved to be reliable, this island even in olden times had a considerable metal production. In any case the remains of ancient mining operations and of an extensive exploitation of excellent statue marble are numerous. For many years the *Speidel Minengesellschaft* of Pforzheim, Germany, has mined extensively for zinc oxidized ore, planning its development work on one hand from the disposition of the ancient workings, and on the other from assistance gained by a regular geological exploration of the island. According to this latter the following main features obtain.

The island consists of an alternation of highly crystalline and in places gneissic schists with marble of unknown geological age. At one or two places in the south-west of the island, as for instance at Cape Maries and near Hamidie, granitic intrusions of small extent have become exposed by the sea, these intrusions occurring partly in association with flat thrust-planes, so that no contact phenomena are observable. The Powder Mountain at Hamidie, in the immediate vicinity of the Vouves zinc mine, consists of a young eruptive so highly decomposed that a satisfactory determination of its identity is no longer possible. Many pebbles of andesite have been picked up on the shore, while mineral aggregates such as suggest a granitic contact zone, tourmaline for instance, have also been found, though no contact zone *in situ* is known. The highly complicated tectonics receive best expression in the variety of the mountain shapes, a complexity confirmed by a study of the ore-deposits.

FIG. 350.—Map of the island of Thasos, showing the most important tectonic lines and the principal mining centres.   Scale, 1 : 300,000.

The highest ground occurs in the east of the island in a fairly straight line from the Ipsarion Mountain 1205 m. high, over the Kamena Pétra 1074 m., to the Soussoula 857 m., in a south-south-east direction, as indicated in Fig. 350.   While from these heights the descent to the east is steep and continues to the sea which is little more than 2 km. distant, that to the

west is gentle and to a large flat syncline occupying the central portion of
the island where the villages of Theologos and Castro are situated. Of
this syncline the other flank rises again near the west coast to a height, in
the case of the Aghios Mattis, of 808 metres. This western flank, extending
from the most southerly point, Cape Salonikios, to the most northerly, Cape
Pachys, is traversed by many fault zones, along which an intense dolomitization of the marble and a mineralization recognizable even from a distance
by the strongly ferruginous colour of the ground, are associated. With
these fault zones all the more important ore-deposits upon the island are
connected.

   While in the district extending from Cape Salonikios across the Astris
and Vouves mines and the Maries valley to the Aghios Mattis, a north-west

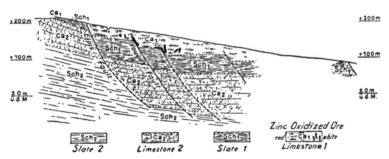

FIG. 351.—Section across the formation at the Vouves mine, Thasos. Beyschlag.

striking fault-system predominates, from that hill into the district where
the Marlou and Corlou mines are situated an east-west system is more
important, this finally giving way to a north-east system in the district
which extends from Metamorphosis through the Sotiros and Casavitti mines
to the Pergaros, Spilio, and Pachys occurrences, these latter being situated
not far from Cape Pachys. It must be remarked however that in any one
place, beside the main fault-system the other two systems are always
represented ; in fact the principal occurrences, those at Vouves, Marlou-
Corlou, and Sotiros, are manifestly at places where numerous intersecting
and converging fissures meet. The course of these faults on surface is
often very distinctly indicated by the uneven character of the contours, and
often also by differences in the vegetation consequent upon the repeated
change from marble to slate. In areas of greater dolomitization and
mineralization however, the tracing of the faults is more difficult.

   With all these occurrences zinc oxidized ore plays the principal part. It
occurs in all varieties, from a completely white, botryoidal, pure mineral-
aggregate through all gradations of admixture with iron and dolomite, to

zinciferous limestone. Schalenblende is very uncommon, while friable galena, with varying silver content and the appearance of having suffered corrosion, is quite common. Calcite occurs to a small extent and generally in fissures, while barite in places is found intergrown with the ore.

As indicated in the longitudinal section of the Vouves opencut given in Fig. 351, the ore is associated with fissures and often concentrated along the impermeable planes where dolomitized limestone lies bedded upon slate. The considerable richness of this mine is due to the white, pure oxidized ore which, having been formed by secondary migration of the zinc content, contains no galena. In this it differs from the original ore.

The occurrence at the Marlou-Corlou mines is associated with disturbed country wherein a wedge-like section of slate is found enclosed in limestone. The solutions ascending along one side of this wedge were so rich in silica and silver that the dolomite along the fissure was completely altered to quartz with a variable and in part high silver content. Farther from the fissure the dolomite was less completely replaced by quartz, though for some distance it continued to be argentiferous. Along the fissure the irregular nest-like metasomatic oxidized zinc deposits occur. In these, near the fissure, a good deal of lead occurs, then purer zinc ore, and finally, farther from the actual fissure, zinciferous and ferruginous dolomite, and limonite. At the Sotiros mine the zinc deposits proceed from the fissures in the form of irregular bed-like masses dipping gently into the hill.

Although to-day these zinc deposits often occur as a surface formation below the gossan, there can be little doubt that they owe their origin to solutions ascending from depth. In addition, it is probable that in greater part they are not the result of the alteration of sphalerite, but were formed primarily as carbonate and silicate. The solutions themselves may probably be referable to the granite magma. Such a genesis is suggested by the occurrence at Marlou, where the ferruginous and zinciferous solutions penetrated far into the rock, while those containing silver and silica remained in the neighbourhood of the fissures. A sample of granite taken by Beyschlag from the neighbourhood of Cape Maries contained 44 grm. of silver per ton, this fact suggesting the probable proximate source of the silver. Silver, galena, zinc carbonate and silicate, and barite constitute therefore the primary sequence ; while from the surface, limonite, white and brown zinc oxidized ores, accompanied first by cerussite and lower down by galena, mark the course of the secondary alteration.

The production of zinc ore from Thasos in 1910 was about 30,000 tons.

THE SILVER-LEAD DEPOSITS AT LEADVILLE, COLORADO

LITERATURE

S. F. EMMONS. 'Geology and Mining Industry of Leadville, Colorado,' U.S. Geol. Survey. Mon. XII., 1886.—CH. M. ROLKER. 'Notes on the Leadville Ore Deposits,' Trans. Amer. Inst. Min. Eng. XIV., 1885.—C. HENRICH. 'The Character of the Leadville Ore Deposits,' Eng. and Min. Jour. XXVIII., Dec. 1879 ; 'Origin of the Leadville Deposit,' Eng. and Min. Jour., May 1888.—A. A. BLOW. 'The Geology and Ore Deposits of Iron Hill,' Trans. Amer. Inst. Min. Eng., 1889, XVIII.—J. F. KEMP. Ore-Deposits, 1900, 3rd Ed.—S. F. EMMONS and J. D. IRVING. 'The Downtown District of Leadville, Colorado,' Bull. 320, U.S. Geol. Survey, 1907.—FR. M. AMELUNG. 'The Geology of the Leadville Ore District,' Eng. and Min. Jour. XXIX., 1880.—FR. T. FREELAND. 'The Sulphide-Deposit of South Iron Hill, Leadville,' Trans. Amer. Inst. Min. Eng. XIV., 1885.

Leadville lies in the Arkansas valley upon a terrace at the foot of one of the western spurs of the Mosquito Range. The mines, which have made this district during the last thirty years one of the most important producers of silver, gold, lead, and zinc, in the western United States, are found two or three miles east of the town. From this situation however, mining operations have of late extended to the west under the terrace upon which Leadville is situated, the last most excellent monograph by Emmons concerning itself exclusively with the occurrence immediately at the town.

The Arkansas valley, which extends in a north-south direction from Tennessee Pass to Salida, owes its origin to a geologically recent depression occurring between the Sawatch Range to the west and the Mosquito Range to the east. The Sawatch Range is an oval massive consisting of gneiss, granite, and schist, considered to be of Archaean age, upon which Cambrian and younger sediments have been so laid down that their combined outcrop mantles the oval completely, though the beds are not on all sides the same. The Mosquito Range is a chain of mountains striking north-south and having individual points which reach to heights of 13,000–14,000 feet. It consists in greater part of Palæozoic beds which to the east are overlaid by others of Mesozoic age. With these old sediments considerable masses of eruptive rocks in the form of sheets and laccoliths are interbedded, these eruptives being older than the tilting of the beds. This tilting appears to have been caused by pressure from the east, which, affecting sediment and eruptive alike, formed a number of asymmetrical anticlines and synclines, having their steeper limbs to the west. These are traversed by a number of north-south faults. Following this orogenic period came a time of erosion to which the present contours are mainly due. The large depression of the Arkansas valley was excavated on the east side of the Sawatch Range, approximately along the old coast line. The geological position of the deposits is represented in Figs. 352 and 353.

Along that portion of the Mosquito Range where the Leadville district

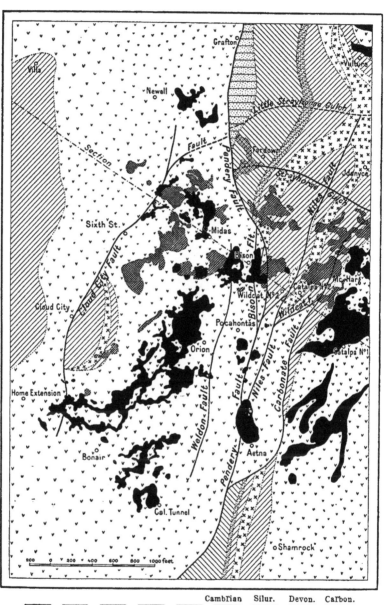

Grafton
Villa
Newell
Fault
Little Stronhouse Gulch
Pendery Fault
Section
Fardown
Vulture
Idanyce
Sixth St.
Pinthorse Fault
Niles Gulch
Midas
Cloud City Fault
Bison
Bison Flt.
Wildcat Nº2
Catalpa Nº2
Mc.Hare
Cloud City
Pocahontas
Wildcat F.
Catalpa Nº1
Orion
Wildcat Fault
Home Extension
Carbonate Fault
Weldon Fault
Pendery Fault
Niles Fault
Bonair
Aetna
Cal. Tunnel
200 0 200 400 600 800 1000 feet
oShamrock

Cambrian   Silur.   Devon.   Carbon.

III. Con-   II. Con-   I Con-   White   Grey   Lower   White   Parting-   Blue
tact   tact.   tact   Porphyry   Porphyry   Quartzite   Lime-   Quartzite   Leadville-
                                                            stone   Limestone
Ore-bodies

Fig. 352.—Geological map of a portion of the Leadville district, showing the extension of the different formations and ore-bodies. Emmons and Irving, Washington, 1907.

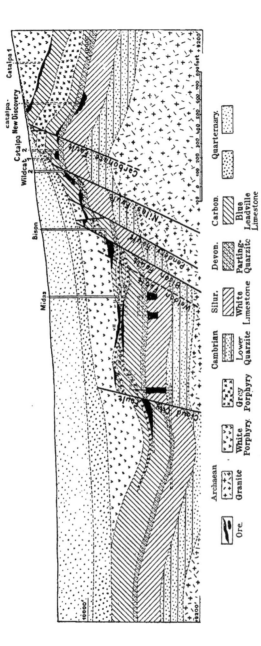

Fig. 353.—Section through the Leadville district along the line indicated in Fig. 352. Emmons and Irving, Washington, 1907.

is situated the eruptive rocks preponderate, and the contour of the district is conditioned rather by faulting than by folding, while the landscape presents itself as an accidented rather than an undulating country.   Thus, a number of steps bounded by faults are formed, the most prominent of these on the surface being known as Breece Hill, Iron Hill, Carbonate Hill, and Fryer Hill, respectively.   Among the sediments participating in this geological complex the more important are, the Lower quartzite, which is probably Cambrian ; the White limestone of the Silurian ; the Parting quartzite of the Devonian ; and the extremely important blue Leadville limestone, which is of Carboniferous age.   Among the eruptive rocks, those known locally as the Grey and White porphyries are the most important. The grey variety is a monzonite- and quartz-monzonite porphyry, while the white is regarded as a rhyolite-porphyry.   From Fig. 352 it is seen that the latter occupies large areas of the surface and usually lies upon the blue Leadville limestone.   The grey porphyry, on the other hand, is generally interbedded in that limestone, though it may also occur in actual contact with the white porphyry.   The basement rock, which here is covered by Cambrian quartzite, consists of granite.   The youngest formation of the district is represented by Quaternary terrace deposits.

The ore-deposits are invariably associated with limestone, appearing more particularly in the blue Leadville limestone between the white porphyry above and the grey porphyry below, this horizon being known as the first contact.   A second and somewhat deeper horizon occurs in the same limestone, between the grey porphyry and the Parting quartzite at the bottom of that limestone.   Finally, ore is also known to occur at a third contact still deeper, at some point between the Parting quartzite and the granite, usually in the White limestone but sometimes also in the Lower quartzite.   The walls of the ore-bodies are not always recognizable underground as ore and country-rock pass gradually into each other.

The primary ore was doubtless deposited as sulphides, chiefly as galena, sphalerite, and pyrite.   These minerals by the action of meteoric waters became altered to oxidized ore down to considerable depths.   From the pyrite, ferric sulphate was first formed, and later iron-ochre and limonite, this latter containing considerable amounts of silver, anglesite, and other minerals.   The ferruginous solutions transformed the surrounding limestone into iron ore with varying proportions of silica and manganese.   The sphalerite upon oxidation appears to have become entirely removed or to have become concentrated below the oxidation zone. Dechenite, the vanadium salt of lead and zinc, though seldom, does occur. Concerning the galena, this mineral being less easily decomposed by oxidation than pyrite or sphalerite is often found in the oxidation zone.   Part of it, however, is altered to anglesite and cerussite, the sulphate and carbonate respectively. Large masses of pure cerussite free from anglesite, on the one hand, and smaller masses of anglesite without cerussite, on the other, are found. Pyromorphite likewise is common. Cerussite occurs sometimes as loose sandy carbonate, and sometimes as hard carbonate. In the first condition it consists of a collection of imperfectly crystallized grains of cerussite of remarkable purity, and as such is found in particularly large masses immediately below, or in the neighbourhood of the porphyry contact. The hard carbonate, on the other hand, is a mixture of quartz and cerussite, somewhat resembling jasper in appearance and doubtless the result of silicification. This hard carbonate is irregularly distributed in masses of iron ore, and particularly in the immediate vicinity of large patches of the sandy carbonate.

The sandy carbonate has usually the lowest silver content in relation to the lead present, this content being 20–40 oz. per ton with 50–70 per cent of lead, while the hard carbonate usually contains one ounce of silver for every one per cent of lead. The galena of the secondary zones is extremely rich in silver, containing occasionally more than 100 oz. per ton. It represents therefore a typical cementation galena. The silver content of the primary sulphides is probably chemically combined with these sulphides. According to the few analyses available it appears to be more abundant in the sphalerite and galena than in the pyrite, though its presence in the latter may always be demonstrated. While fresh galena and sphalerite may contain 50 oz. or more per ton, with pyrite scarcely more than 10 oz. may be expected. In the oxidation zone the silver is found more often as light-green chloride containing a little bromine and iodine. This mineral generally occurs along the cleavage-planes of other minerals, though in places it also appears in small masses by itself. Native silver is occasionally found in the richer portions of the deposits, and particularly along the upper contact. Generally the silver content of the oxidation ore diminishes in depth, this relation of content to depth being also expressed in the fact that the upper contact is the richest. The small amount of gold found at Leadville seldom reaches more than one-hundredth part of an ounce per ton ; it is most intimately associated with the silver. At some places in the district traces of gold telluride have been found.

The difference between the amount of manganese contained in the sulphide ore and that in the oxidized ore is worthy of remark. Rhodonite and rhodochrosite, which with most deposits are primary manganese ores, are not found in the sulphide ore of Leadville. As the result of a large number of analyses the manganese content of this ore was shown to be seldom more than 2 per cent, and on an average not to exceed 1 per cent. Yet the large masses of iron ore occurring in the oxidation zone contain a considerable though variable percentage of manganese oxide, this being reflected in the dark colour of the ore. It would appear as if the upper portion of the oxidation zone near the contact with the overlying porphyry were the richest in manganese, the ore there often containing 15–25 per cent of manganese oxide with 20–30 per cent of iron oxide.

Opinions concerning the genesis of these deposits have greatly altered with the lapse of time. According to Emmons and Irving, the Leadville sulphide ore represents a metasomatic replacement of the country-rock, principally of limestone, this replacement having taken place after the intrusion of the porphyry but before the beds were tilted. The evidence of this posterior limit lies in the fact that the ore-bodies are cut off by the faults brought about at that tilting, which probably took place at the end of the Jurassic and before the beginning of the Cretaceous. The deposits therefore are pre-Cretaceous.

The importance of the Leadville district may be gathered from the following figures : in the year 1908 the Leadville mines produced 33,127 tons of carbonate ore, 117,423 tons of oxidized ore, 162,188 tons of sulphide ore, 70,197 tons of zinc ore, 92,187 tons of quartzose ore, and 1500 tons of manganese ore, making a total of 476,622 tons, which contained 68,135 oz. of gold, 3,509,378 oz. of silver, 9005 tons of lead, 3205 tons of copper, and 16,846 tons of zinc. As in the same year the total lead production of Colorado was some 28,000 tons, of this the Leadville district contributed approximately one-third. For comparison, the total lead production in the United States during that year was 314,067 tons.

The Occurrence of Silver at Eureka, Nevada

LITERATURE

R. W. RAYMOND. 'Eureka-Richmond Case,' Trans. Amer. Inst. Min. Eng., 1877, VI.—J. S. CURTIS. 'Silver-Lead Deposits of Eureka,' Nevada U.S. Geol. Survey, Mon., 1884, VII.—A. HAGUE. 'Geology of the Eureka District,' Nevada, Mon. XX., 1892.

The Eureka mountains consist of limestone, quartzite, sandstone, and slate, of Cambrian, Silurian, Devonian, and Carboniferous age. These formations, having a total thickness of many thousands of feet, have been subdivided in a most detailed manner by Hague. The Cambrian system, which here is of particular interest, is represented by the following sequence ; first a brownish-white quartzite 1500 feet in thickness with intercalated argillaceous beds, this quartzite being known as the Prospect Mountain quartzite ; then an overlying grey compact limestone more than 3000 feet thick, the Prospect Mountain limestone ; then yellow and grey shales, known as the Secret Cañon shales, among the upper members of which thin beds of limestone are found ; next the Hamburg limestone 1200 feet thick ;

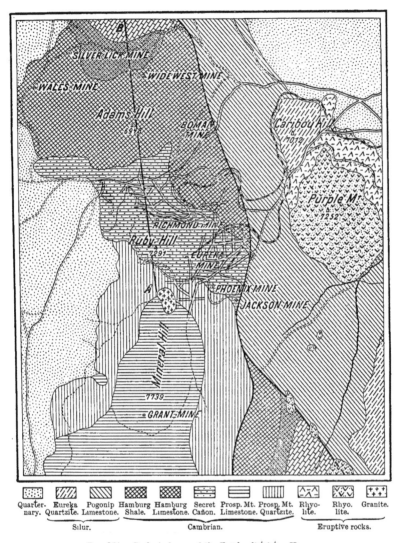

| Quarter-nary. | Eureka Quartzite. | Pogonip Limestone. | Hamburg Shale. | Hamburg Limestone. | Secret Cañon. | Prosp. Mt. Limestone. | Prosp. Mt. Quartzite. | Rhyo-lite. | Rhyo-lite. | Granite. |

Silur.   Cambrian.   Eruptive rocks.

FIG. 354. Geological map of the Eureka district. Hague.

and finally a yellow shale. All these are cut by a large number of faults.

The ore-deposits in their occurrence are limited to the Cambrian lime-stones, none being known below the Prospect Mountain limestone. In that limestone however, from its lowest sections right to the Secret Cañon shales above, deposits are numerous. On the slope of Prospect Mountain, from Mineral Hill southwards to Surprise Peak, the limestone is traversed by fissures and irregular cavities of varying width and extent. Many of these run parallel to the bedding, while others cross it apparently at any angle. In the cavities oxidized ore-bodies are met, many of these being connected with one another by narrow channels or seams more or less filled with ore. The Williamsburg mine to the west exhibits a good example of a deposit filling an irregular cavity, while to the east the Geddes and Bertrand mine lying to the extreme south, works a well-defined fissure-filling. In the latter mine a large east-west rhyolite dyke traverses the limestone and the overlying shale.

The second ore horizon is the Hamburg limestone, which on Adam's

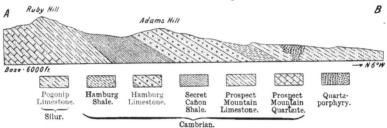

FIG. 355.—Section through Ruby Hill and Adam's Hill in the Eureka district, along the line AB indicated in Fig. 354.

Hill and in the immediate vicinity of the Secret Cañon shales contains many ore-bodies. Those for instance worked in the Price and Davies mine belong to this horizon. Similarly, the Wide-West mine exploits ore occurring in the upper portion of this limestone close under the Hamburg shale.

In isolated cases ore is also found in still younger formations, as in the case of the Ruby Hill deposit. Ruby Hill consists of the three lowest Cambrian beds, the Prospect Mountain quartzite, the Prospect Mountain limestone, and the Secret Cañon shales, which dip north fairly regularly at an angle of 40°. The country is nevertheless greatly affected by different large faults known as the Ruby Hill fault, the Jackson Hill fault, etc., these being in close connection with large rhyolite intrusions which around Ruby Hill are fairly numerous. Since in many cases, in the Jackson and Dunderburg mines for instance, the ore is found along these faults, it is generally speaking younger than the intrusions.

The ore, chiefly galena and pyrite, was primarily deposited in the limestone from a complicated network of fractures. The primary ore thus formed was subsequently subjected to an oxidation so long continued that to-day unaltered sulphides are practically never found above the ground-water level. The numerous carbonates, sulphates, arsenates, molybdenates, and chlorides of the oxidation zone are without exception rich in gold. Wulfenite, in brilliant lemon- or orange-coloured crystals, occurs in comparatively speaking large amount, giving to these deposits a resemblance to those at Bleiberg. An analysis of an average sample of the ore won during the year 1878 gave 35·65 per cent of lead oxide, 34·39 per cent of iron oxide, 2·37 per cent of zinc oxide, and 6·34 per cent of arsenious acid, with 27·55 oz. of silver per ton and 1·59 oz. of gold. The occurrence of tellurium, probably in association with bismuth, is interesting.

The geological age of these deposits can only be Pliocene or post-Pliocene. From their occasional association with rhyolite dykes no reliable conclusion may be drawn that there exists a genetic relation between the two.[1] The primary sulphides however doubtless came from depth, subsequent to which they suffered oxidation from meteoric waters.

The Missouri-Mississippi District in Kansas, Indian Territory, Arkansas, and Illinois

LITERATURE

J. D. Whitney. Report of a Geol. Survey of the Upper Mississippi Lead Region, Albany, 1862.—A. Schmidt. ' Forms and Origin of the Lead and Zinc Deposits of South West Missouri,' Trans. St. Louis Acad. of Sc. III. p. 246.—W. P. Jenney. ' The Lead- and Zinc-Deposits of the Mississippi Valley,' Trans. Amer. Inst. Min. Eng. XXII., 1893.—F. Pošepný. Ueber die Genesis der Erzlagerstätten. Freiberg, 1893.—A. Winslow. ' Lead and Zinc Deposits of Missouri,' Trans. Amer. Inst. Min. Eng. XXIV. ; ' Lead and Zinc Deposits,' Miss. Geol. Survey, 1894.—W. P. Blake. ' Lead and Zinc Deposits of the Mississippi Valley,' Trans. Amer. Inst. Min. Eng. XXII., 1894 ; ' Wisconsin Lead and Zinc Deposits,' Bull. Geol. Soc. Am. V., 1894.—J. D. Robertson. ' Missouri Lead and Zinc Deposits,' Am. Geol., 1895.—A. G. Leonard. ' Lead and Zinc Deposits of Iowa,' Report of a Geol. Survey. VI., 1897.—J. F. Kemp. Ore Deposits of the United States and Canada. —J. C. Branner. ' The Zinc and Lead Region of North Arkansas,' Ann. Rep. Geol. Survey of Arkansas, V., 1900.—W. E. Burk. ' The Fluorspar-Mines of Western Kentucky and Southern Illinois,' The Mineral Industry, 1901, IX. p. 293.—C. R. van Hise and H. F. Bain. ' Lead and Zinc Deposits of the Mississippi Valley,' Trans. Inst. Min. Eng., London, 1902.— H. Foster Bain. ' Some Relations of Paleogeography to Ore Deposition in the Mississippi Valley,' Compte Rendu, Congr. Geo. Intern., Mexico, 1906, I. p. 483.—Ch. R. Keyes, ' Diverse Origins and Diverse Times of Formation of the Lead- and Zinc-Deposits of the Mississippi Valley,' Trans. Amer. Inst. Min. Eng. XXXI., 1902.—Horten. ' Der Zinkerz-bergbau bei Joplin, Missouri, und seine wirtschaftliche Bedeutung,' Zeit. f. d. Berg- Hütten-und Salinenwesen im pr. Staate, 1902, Vol. L.

The low ground to the north of the Gulf Plains and south of the Lake Superior area, is a wide expanse of flat-lying unaltered Palæozoic sediments, which to the west disappear under the Red beds and Cretaceous sandstones

---

[1] *Ante*, p. 557.

of the Great Plain, and to the east are bounded by the Appalachian Mountains which, though themselves consisting of Palæozoic beds, break the farther continuity by their complicated structure, the result of Permian orogenics. The low ground itself has suffered no great disturbance so that the beds lie almost horizontal ; the predominating rocks are dolomite, limestone, slate, and sandstone, while of the coarser sediments there is an almost complete absence. These rocks lie upon a pre-Cambrian crystalline basement which comes to surface all around this area. Eruptive rocks, but for a few dykes, are absent. The sediments range from Middle Cambrian to Permian ; they represent the product of uninterrupted and regular deposition upon an even floor.

In this vast district where mining began in the year 1719, three centres may be distinguished : (1) South-eastern Missouri, with a yearly production of lead, iron, and copper to the value of approximately 120 million dollars ; (2) South-western Missouri and the adjoining portions of Kansas and Indian Territory, with a yearly zinc and lead production of 118 million dollars ; and (3) South-western Wisconsin and the adjoining portions of Illinois and Iowa, the annual lead, zinc, and copper production of which reaches 60 million dollars.

In the Upper Mississippi or Wisconsin lead-silver district the ore occurs exclusively in Silurian beds consisting of Cincinnati slate ; dolomitic galena-limestone, 135 m. in thickness, equivalent to the Upper Magnesian limestone of the Lower Silurian Trenton period ; oil shale ; limestone, 12–30 m. thick ; greenish-brown shale ; St. Peter's sandstone ; and lower dolomitic limestone, 30–75 m. thick. These Silurian beds lie upon the Potsdam sandstone of the Cambrian system. The occurrence of ore is limited to the limestones. The mines have a depth of but 30–60 m. and no ore is found very much below ground-water level. Above that level it occurs either as ' sheets ' filling vertical fissures in a practically undecomposed country-rock, the width of such sheets being seldom more than 3 inches, the length in the most favourable cases 100 feet, and the extension in depth generally 20 to 40 feet ; or as ' openings,' these being the cavity enlargements in which the conditions for the deposition of lead ore were especially favourable ; in these the galena is generally embedded in ferruginous clay. One such ore-body, that known as Levins lode at Dubuque, was 130 feet long, 45 feet high, and 30 feet wide. Where the dimensions are very irregular so that a series of such openings connected with one another by narrow channels exists, the occurrence is termed a ' crevice with pocket openings.' These are chiefly limited to the upper portion of the galena-limestone, while in the lower portions, flat sheets or flat openings, which generally extend horizontally and parallel to the bedding-planes, are the characteristic form of deposit. The difference between the vertical and flat sheets lies entirely in the lay of the long axis. Above the ground-water level galena and zinc carbonate are the usual ores in the gash veins, while in the flats, sphalerite, galena, and marcasite occur, either in intimate intergrowth or in alternate layers.

The principal ore of these deposits is a very pure galena containing but little silver. This galena occurs crystallized chiefly as cubes and often accompanied by sphalerite and zinc carbonate. Pyrite and chalcopyrite are comparatively speaking uncommon ; limonite on the other hand appears invariably to accompany the lead- and zinc ores. Calcite and barite are of small importance, while quartz and the compounds of lead with arsenic-and phosphoric acids, are almost entirely absent. The occurrence of mammoth bones and other bones with galena in cavities, is evidence of the time of formation and of the aqueous origin of the ore.

The most important mine of the district is La Motte, which began work as far back as 1720. The mines generally are irregularly distributed throughout the district in groups separated from one another by large barren stretches. According to Bain it would almost appear as if certain smaller basins were favoured by the ore-deposition. Some five-sixths of the lead-zinc production of the entire region is derived from the Wisconsin district.

In addition, the Missouri-Kansas occurrence with Joplin as centre, and that of Illinois-Kentucky, deserve mention. The geological position of the Missouri deposits is similar to that of the Wisconsin deposits but with this difference, that the ore occurs in the Lower Carboniferous Cherokee limestone which is overlaid by Upper Carboniferous beds and underlaid by slates. In the Illinois district the deposits are galena- and sphalerite lodes remarkable for the amount of fluorite they carry, this mineral being sometimes so abundant as to be mined at a profit.

Concerning the genesis of these deposits there is an extensive literature. Whitney, Chamberlin, and others, advocate a primary metal content in the limestone, which content subsequently became concentrated, the occurrence according to this view representing a special form of lateral secretion. Blake, Van Hise, and others, regard the primary ores as formed from ascending solutions, the former assuming these to have been hot springs and the latter artesian waters. Bain in his last work returns again to the idea of the original deposition of the metalliferous material from sea-water.

We, however, are of the opinion that the occurrences in the Mississippi-Missouri region owe their primary deposition to ascending solutions, the primary ores so deposited appearing to-day unchanged below ground-water level—the occurrence of fluorite in certain lodes, as Beck has rightly pointed out, supports this view—and that subsequently, by the action of meteoric waters, a transformation of these primary ores to oxidized ores took place, such transformation being intimately connected with oxidation-metasomatism. The occurrence of the galena deposits in basin-shaped areas separated from one another by larger stretches of poor country, points in our opinion to conditions similar to those obtaining in Upper Silesia. In the Mississippi-Missouri district, also, fissures play a large part, so that it is not so much the basin-shape which has compelled the deposition of the ore in particular places, but the number of fissures there collected.

The importance of these deposits may be gathered from the following figures of production. In the year 1908 the Joplin district produced 259,609 tons of zinc ore and 38,514 tons of lead ore, distributed over the different States as follows :

	Zinc Ore.	Lead Ore.
Missouri . . .	220,638 tons.	33,335 tons.
Kansas	28,598 ,,	3,455 ,,
Oklahoma . . .	10,373 ,,	1,724 ,,
Totals . . .	259,609 tons.	38,514 tons.

The total production of zinc ore in the United States in the same year was 838,377 tons, to which Missouri-Kansas contributed 273,420 tons and Wisconsin 58,135 tons.

SALA, SWEDEN

LITERATURE

HJ. SJÖGREN. ' The Sala Mine,' Guide to the International Geological Congress, Stockholm, 1910 ; ' Über das Auftreten des Silbers in dem Sala-Erz und über Amalgam von Sala'; ' Über Gediegen Silber, Quecksilber, Amalgam und Zinnober von Sala'; ' Über Boulan-gerite von Sala,' in Geol. Fören. Förh. XIX., XX., XXII., 1897, 1898, 1900 respectively.— J. H. L. VOGT. Manuscript upon Sala, 1905.

As already stated [1] and as indicated in Fig. 241, Sala lies some 100 km. west-north-west of Stockholm in an area of crystalline schists, and more particularly in the youngest Archaean hälleflinta. At Sala itself, a considerable area of limestone-dolomite occurs surrounded by hälleflinta and granite, this area being about 7·5 km. long and in the neighbourhood of the mines, to the south, about 1·5 km. wide. In greater part, and particularly in the neighbourhood of the deposits, this rock has suffered dolomitization, normal dolomite consisting of one part of $MgCO_3$ to one part of $CaCO_3$ having extensively resulted.

Two different ores occur, one a silver-lead ore and the other a zinc

[1] *Ante*, p. 379.

ore. The first, containing but a few per cent of lead, consists of argentiferous galena with a fine-grained, generally microscopic admixture of
argentite, pyrargyrite, etc., some sphalerite, pyrite, arsenopyrite, and rare
antimony- and copper minerals. The second contains chiefly sphalerite
with a little galena and pyrite, but practically no copper. In this zinc
ore particularly, and perhaps exclusively, some cinnabar with native
quicksilver and amalgam occurs, the ore on an average containing
at least 0·01 per cent of quicksilver. Both these ores are found in the
dolomitic limestone over an area 800 m. long and a hundred or two hundred
metres wide. Roughly in the middle of this area a large fault-zone often
several metres wide occurs, which, as indicated in Figs. 356 and 357, is
accompanied on both sides by a number of similar but smaller fissures.

Fig. 356.—Horizontal section of the Sala mine at a depth of 190 metres. Sjogren, 1910.

The filling of this zone consists of crushed fragments of dolomite, with
calcite, talc-, chlorite-, and serpentine minerals, etc. ; metalliferous
minerals, apart from a little pyrite and associated sulphides, are absent.

The silver-lead ore occurs in steep chimneys and net-like impregnations
within the dolomite and in the neighbourhood of the fault-zone. In the
upper levels particularly, which were exhausted centuries ago, the ore was
very argentiferous. In depth the deposit has become poorer, though the mine,
which is now four hundred years old and whereof many of the workings have
collapsed, has only reached a vertical depth of 275–300 metres. The ore-
body pitches at an angle away from the line of dip.

The zinc ore forms in the dolomite flatly-dipping pipes 6–12m. wide,
enclosing many fragments of dolomite and throwing off many branches.

The strike of the limestone-dolomite bed in the neighbourhood of the
mine is north-north-east, that of the main ore-body, north-north-west,
and that of those zinc deposits which have so far been investigated,
N. 5° E. In view of this disposition and the brecciated character of
the ore, the deposits cannot be of sedimentary origin. The differences

between the silver-lead ore and the zinc ore, both in respect to mineral-character and to strike, may, according to Vogt, be because the two classes of ore were probably not formed at exactly the same time.

The ore is intermixed more particularly with the mineral salite—a variety of diopside named after this place where it was discovered—the amphiboles tremolite and actinolite, and some biotite, talc, chlorite, serpentine, epidote, garnet, and tourmaline, the last few minerals being however very uncommon. In one of the zinc ore-bodies a considerable amount of barite was also found.

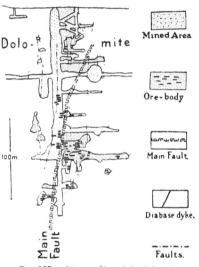

FIG. 357.—Cross section of the Sala mine: Sjögren, 1910.

The mineralization is younger than the dolomitization. The fault-zone and the mineraliza-tion are, according to Sjögren, associated occurrences and ap-proximately contemporaneous. The action of silicated solutions upon the dolomite brought about the formation of the lime-mag-nesian silicates, salite, tremolite, etc., simultaneously with the deposition of the ore. In many respects the occurrence at Sala agrees o e with the ordinary silver-lead-zinc deposits. The occurrence of salite, tremolite, etc., constitutes however a marked difference which, according to Vogt, may perhaps be explained in that the mineralization took place under physical-chemical conditions similar to those obtaining at the formation of contact-metamorphic deposits.

Mining upon this particular deposit began in the year 1500 and reached its zenith in the first half of the sixteenth century. The total production amounts to some 400 tons of silver, made up as to 200 tons obtained from 1510-1600 ; 63 tons from 1601-1700 ; 37 tons from 1701-1800 ; and 87 tons from 1801-1908. Latterly the production of silver has almost ceased, while that of lead has always been small. Zinc on the other hand is now being mined ; at Sala therefore, as with many other lead-zinc deposits, zinc mining has with time taken the place of silver-lead mining.

The World's Production of Lead- and Zinc Ores and their
Distribution among the different Classes of Ore-Deposit

This we begin by a statement of the production of metallic lead and
zinc taken from yearly compilations of the *Metallurgishe Gesellschaft* of
Frankfort :[1]

	Lead.			Zinc.		
	1900.	1905.	1910.	1900.	1905.	1910.
	Tons.	Tons.	Tons.	Tons.	Tons.	Tons.
Germany . . .	121,500	152,600	157,900	154,572	197,184	227,747
Belgium . . .	16,400	22,900	39,600	119,231	145,592	172,578
Holland . . .	...	...	...	6,953	13,767	20,975
Great Britain . .	35,500	23,300	30,500	30,307	50,927	63,078
France . . .	17,000	24,100	21,000	}42,117	50,369	59,141
Spain . . .	154,500	180,700	191,600			
Austria-Hungary .	11,900	13,500	17,500	} 7,086	9,357	13,305
Italy	23,800	19,100	16,000			
Greece . . .	16,700	13,700	16,800	...	...	...
Sweden . . .	1,400	600	300	...	...	...
Russia . . .	200	300	1,200	5,968	7,642	8,631
Asiatic-Turkey . .	2,800	10,400	12,700	...	...	...
United States . .	269,000	312,500	371,600	112,234	183,245	250,627
Mexico . . .	80,000	75,000	126,000	...	...	...
Canada . . .	28,600	25,700	15,000	...	...	...
Japan	1,900	2,300	3,500	...	...	...
Australia . . .	87,100	107,000	98,800	...	650	508
Other states . .	3,000	200	12,900	...	...	...
Totals . .	871,300	983,900	1,132,900	478,500	658,700	816,000

Since 1880 the lowest average yearly price for lead was £9 : 18 : 0 and
£9 : 12 : 0 in 1893 and 1894 respectively ; and the highest £18 : 4 : 0 and
£19 : 12 : 0 in 1906 and 1907 respectively. That for zinc was similarly
£13 : 19 : 6 and £14 : 5 : 0 in 1885 and 1886 ; and £25 : 4 : 0 and £27 : 1 : 0
in 1905 and 1906 respectively.

Unlike the figures of production for metallic lead and zinc, those
relating to the lead- and zinc ores produced by the different countries are
generally very inexact, since in many cases these ores are sent to foreign
works for treatment. This is particularly the case with zinc ores. Belgium
for instance produces much metallic zinc though but little ore, whereas with
Australia the opposite is the case. A good review of this subject has
been given by W. Hotz in a paper entitled ' *Die wirtschaftliche Bedeutung
der Blei - Zinkerzlagerstätten der Welt im Jahre 1907 mit besonderer
Berücktsichtigung der genetischen Lagerstättengruppen* ' in Part 2, 1910,

[1] *Ante*, p. 207.

of the *Bergwirtschaftlichen Zeitfragen* and in the April issue 1910 of the *Bergwirtschaftlichen Mitteilungen*, both published by M. Krahmann, Berlin. According to this authority the twenty most important lead deposits of the world in 1907, produced in that year lead ore to the following values :

Broken Hill	£3,660,000
Missouri-Kansas	2,280,000
Shoshone	1,960,000
Sierra Morena	1,550,000
Utah	1,470,000
Mexico	1,130,000
Leadville-Aspen	1,080,000
Cartagena	400,000
Iglesias	324,000
Upper Silesia	297,000
Rhenish Schiefergebirge	294,000
Mazarron	294,000
Tasmania	280,000
Canada	270,000
Crete	245,000
Laurion	196,000
Harz	186,000
Nordengland	181,000
Commern	162,000
Carinthia	157,000
Other occurrences	1,245,000
Total	**£17,661,000**

Of these occurrences those of Missouri-Kansas, Leadville-Aspen, Upper Silesia, Laurion, Carinthia ; several of those in the Sierra Morena, in Utah, Mexico, Iglesias, Nordengland ; and many others,[1] with a total lead ore production in 1907 to the value of at least 7 million sterling and perhaps even 7·5 million or 8·5 million, belong to the metasomatic group. The occurrences at Shoshone in Idaho, at Cartagena, Mazarron, Iglesias, and Crete, are lodes, though to some extent at any rate they are associated with metasomatic deposits. Typical lodes occur in the Rhenish Schiefergebirge, in the Oberharz, and at many places in the Sierra Morena and in Mexico,[2] etc. The value of the lead ore produced during 1907 from lodes may be estimated at 5·5 to 6·0 million sterling. The Broken Hill deposit we consider [3] as belonging to the contact-deposits, and that at Commern to the impregnations. It may therefore be said that of the total lead ore production in 1907, somewhat less than one-half was derived from metasomatic deposits, roughly one-third from the old and young lodes, while the remainder was divided among the other classes of deposit and more particularly contact-deposits, ore-beds, etc.

The ten most important zinc deposits in 1907, produced in that year ore to the following values :

---

[1] *Ante*, p. 717.   [2] *Ante*, p. 650.

[3] *Ante*, pp. 399-402.

Missouri-Kansas	.	.	.	£2,510,000
Upper Silesia	.	.	.	1,390,000
Broken Hill	.	.	.	735,000
Iglesias .	.	.	.	656,000
Rhenish Schiefergebirge	.	.	.	525,000
Algeria	.	.	.	345,000
Poland	.	.	.	245,000
Mexico	.	.	.	186,000
Cartagena	.	.	.	176,000
New Jersey .	.	.	.	162,000
Other occurrences .	.	.	.	1,728,000
Total	.	.	.	£8,658,000

Beside the two principal occurrences of Missouri-Kansas and Upper Silesia, a number of other important zinc deposits belong to the metasomatic group,[1] so that the value of the total production of this group in 1907 was at least 5·5 million sterling and perhaps even 6·2–6·5 millions. From the lodes of the Rhenish Schiefergebirge, the Oberharz, Cartagena, and many other smaller districts, zinc ore to the total value of 1–1·25 million sterling was produced. In addition are the productions of Broken Hill and New Jersey from deposits regarded as contact-deposits, and that from deposits of other genesis. Altogether, therefore, at least two-thirds of the total zinc-ore production and perhaps even three-quarters, is derived from metasomatic occurrences, one-ninth to one-seventh from lodes, and the remainder from other classes of deposit.

For lead therefore as well as for zinc the metasomatic deposits are the most important class, this being particularly so in the case of zinc.

[1] *Ante,* p. 717.

THE ANTIMONY LODES

As may be surmised from the small yearly production of metallic antimony, lodes carrying this metal are not very numerous. Since antimony sulphide is one of those compounds which form sulpho-salts, the formation of antimony deposits is probably analogous to that of the cinnabar deposits.[1] Such an analogy is further suggested by the fact that stibnite, though only to a subordinate extent, accompanies cinnabar in its deposits. Morphologically however, there is this difference between the two, that stibnite occurs chiefly as a fissure-filling, while cinnabar occurs preferably as an impregnation in sandstone, etc.

Most antimony lodes, and perhaps even all, occur associated with eruptive rocks, being indeed often found within such rocks. Many, as for instance those of the Central Plateau in France, occur in connection with granite; while others are found associated with young eruptives of various composition.

Antimony lodes are usually simple lodes and possess but limited extension in strike. The extension in depth is also generally inconsiderable, a rapid pinching out in depth having often been established. The width of the lodes worked to-day is generally but a few decimetres and seldom as much as one metre. The distribution of the ore along the lode plane is not regular but subject to great irregularity. On account of the relatively small amount of antimony available in the earth's crust for natural concentration, none of the antimony deposits can be described as large.

Among the minerals found in antimony deposits, stibnite preponderates, following which come arsenopyrite, galena, sphalerite, pyrite, chalcopyrite, realgar, and orpiment. The stibnite occurs in fine-grained, occasionally almost compact or fibrous crystalline masses, which, with a little quartz admixed, often make up the entire width. The lodes occur either independently as a special type, or in connection with lead-silver lodes. Only those of great purity are worked as the market

[1] *Ante*, p. 457.

heavily penalizes impurities in the ore for sale.   Such ore for instance should not contain more than 0·25 per cent of arsenic, nor more than 0·75 per cent of lead or copper.   Quartz is the most prominent gangue; limonite, calcite, and barite are less frequent.

Antimony mining has nowhere attained to any great depth so that no definite primary depth-zones have been observed ; the above-mentioned change from rich ore in the upper sections to poor ore lower down, must however be regarded as a primary variation.   Of the secondary depth-zones only an oxidation zone has yet been observed and no secondary enrichment of antimony appears therefore to take place.   The secondary ores resulting from the oxidation of stibnite are stiblite, valentinite, and senarmontite.[1]

The accessory precious metal contained in stibnite is of particular interest.   Gold often occurs, and stibnite should accordingly always be assayed for that metal.   This association is so pronounced that all gradations are found between antimony deposits pure and simple and ordinary gold deposits.[2]   If the gold content increase so that gold becomes the main object of exploitation, the demands which must be satisfied to ensure the payability of the deposit are quite other than those with pure antimony deposits.

The most important antimony deposits as far as the market is concerned are those of China, from which country a considerable amount of ore assaying 50 per cent of antimony is exported.   France produces approximately an equal amount which however is consumed in that country itself.   In addition, there are large occurrences in the United States from which country however there is likewise no export.   Considerable deposits are also known in Australia, though there the costs of production are so high that these may only be worked with advantage during times of high metal price.   Smaller occurrences with unimportant production are found in Spain, Portugal, Algeria, Hungary, and Bohemia. The parts taken by the different countries in the antimony market are illustrated by the following table pertaining to the year 1908.

England	.	.	.	.	about	8,100	tons.
France	.	.	.	.	,,	5,000	,,
Belgium	.	.	.	.	,,	800	,,
Austria	.	.	.	.	,,	6,000	,,
United States	.	.	.	.	,,	3,000	,,
Japan	.	.	.	.	,,	300	,,
Total	.	.	.	about	23,200	tons.

The price of metallic antimony fluctuates greatly; in the year 1907, for instance, it varied from something less than £37 : 10 : 0 per ton to £110 per ton.

[1] *Ante*, p. 101.   [2] *Ante*, p. 601.

Antimony sulpho-salts such as tetrahedrite, pyrargyrite, stephanite, bournonite, etc., are comparatively plentiful in many lead-silver lodes of the older as well as of the younger group; stibnite also is found at times, though only exceptionally in any considerable quantity. In the treatment of these ores the antimony oxide collects, together with lead oxide, in the skimmings and by-products, these being afterwards worked for antimonial or hard lead. In the production of this technically so important antimony alloy, metallic antimony reduced directly from its ore is not used; indeed the application of this metal is in general fairly limited.

INDIVIDUAL OCCURRENCES

LITERATURE

DUFRÉNOY et ÉLIE DE BEAUMONT. Explication de la carte géologique de la France, 1841, p. 173.—HELMHACKER. ' Der Antimonbergbau Mileschau bei Krasnahora in Böhmen,' Jahrb. der Vereinigten Bergakademien zu Leoben, 1874, Vol. XXII.—C. BLOEMECKE. Die Erzlagerstätten des Harzes, 1885.—LÜDECKE. Die Minerale des Harzes, 1892.—F. POŠEPNÝ.· Arch. f. pr. Geol. II., 1895.—A. LACROIX. Minéralogie de la France et de ses colonies, Vol. II., 1897, p. 449.—A. IRMLER. ' Über das Goldvorkommen von Bražna im mittleren Böhmen,' Verhandl. der k. k. Geol. Reichsanst., 1899.—A. HOFMANN. ' Antimonitgänge von Přičov in Böhmen,' Zeit. f. prakt. Geol., 1901, p. 94.—F. KATZER. ' Zur geol. Kenntnis des Antimonitvorkommens von Křitz bei Rakonitz,' Verhandl. der k. k. Geol. Reichsanst., 1904, No. 12.—F. FUCHS et DE LAUNAY. Gîtes minéraux, Paris, 1893, II. 193. —L. DE LAUNAY. ' Excursion à quelques gîtes minéraux et métallifères du plateau central. VIII.,' Internat. Geologenkongress zu Paris, 1900, Report II. p. 953.—Mining in Japan, Past and Present, published by the Bureau of Mines, 1909.

An interesting deposit is that at the Jost-Christian mine on the Wolfsberg in the Harz, where a lode, something more than one metre wide and occurring in the Lower Wiederschiefer, consists of prismatic and compact stibnite, together with federerz and lead-stibnite; while zundererz, boulangerite, and wolfsbergite are less common. The gangue consists of strontianite, calcite, barite, selenite, and fluorite.

Better-known occurrences are found at Přičov in the mid-Bohemian granite, some 4 km. north-west of Selčan, at the foot of the Deschnaberg. In the amphibole-biotite granite there, narrow kersantite dykes occur, with which the antimony lodes appear to be associated, these lodes usually containing decomposed kersantite. The lodes at their outcrop show abundant stiblite as the oxidation product of the stibnite, as do also those in the neighbouring districts of Schönberg and Mileschau. At Přičov, in addition, porous and cellular hornstone is found in casts after stibnite, in which casts the ochreous remains of the original crystals have settled down. Such decomposition reaches to a depth of about 18 metres. The Emil lode, which has been examined to an inclined depth of 62 m., strikes north-south and dips 40°–50° to the west, while in width it varies between 10 cm. and 50 cm. The filling consists of milk-white or bluish hornstone with crystals of stibnite regularly scattered in radial aggregates throughout. Less frequently crusted structure is found, while in other places the stibnite preponderates, the separate crystals matting themselves together to form a compact aggregate. The blue-black colour of the lode material is due to microscopically small stibnite crystals. An interesting feature here is that at recrystallization the hornstone has become transformed to ordinary quartz, a phenomenon which Beyschlag and Krusch also observed in the gold-quartz lodes at Donnybrook.[1] The dark hornstone, according to C. Mann, contains 3·5 per cent of antimony. The other veins in the district form a network, the character of the vein material being very similar to that of the country-rock. In regard to genesis, Hofmann considers that mineral solutions carrying silica and stibnite were responsible for the hornstone, which is the cryptocrystalline variety of quartz, while the ordinary quartz occurring as stringers and nests in the hornstone is without exception secondary. Unlike the neighbouring lodes of Schönberg-Mileschau those at Přičov contain no gold.

The Schönberg-Mileschau antimony lodes have equal right to be considered gold deposits. They occur 55 km. south of Prague in a granite area traversed by kersantite dykes with which in several cases the lodes are associated. In contra-distinction to the occurrences at Přičov, the lodes here are quartz-stibnite lodes remarkable for the gold they contain, this not infrequently being discernible to the naked eye. This gold content however is not everywhere sufficient to warrant mining on its account alone ; according to Hofmann[2] it varies between 4 and 17 grm. per ton. Genetically, Hofmann regards these lodes, as also those at Přičov, as a consequence of the intrusion of the granite, or perhaps of the kersantite. Another antimony lode occurs at Křitz in the neighbourhood of Rakonitz, at the contact of phyllite and diabase.

The antimony occurrence upon the Central Plateau in France is economically important, a large proportion of the lodes there occurring being payable. To the north, though still south of the Colettes granite massive, the Nades mine in the department of Bourbonnais works a deposit which occurs in the mica-schists surrounding that massive. Two lodes are known, which strike south 30° E. and of which the larger has a width of 1·20 metres. The gangue consists of quartz. This deposit was worked uninterruptedly from 1829 to 1837, after which date it remained untouched for fifteen years, when work was again begun.

Farther to the east near Bresnay in the department of Souvigny, two other lodes with similar strike to the above are found in a muscovite granite, this granite being similar to that at Magurka in Hungary. The lode-filling

---

[1] *Zeit. f. prakt. Geol.*, 1900, p. 172.   [2] *Op. cit.*

consists of quartz with stibnite, which latter near the surface is oxidized. Mining, which began here in the year 1763, ceased before the end of that century.

At Montignat in the west of the department Allier, in the rural district of Petite-Marche, another lode is known, which strikes about N. 30° E. and dips 45° to the east, along the contact of granite and gneiss. At Villerange, still farther to the west, an interesting lode occurs in Culm grauwacke, the geological age of this occurrence being consequently more exactly determinable. It is a quartz-stibnite lode striking east-west and dipping north; the quartz and the stibnite appear to be contemporaneous. To the south of Saint Yrieix, in the department of Haute Vienne, mica-schists or amphibole-schists are traversed by numerous granitic dykes which strike north-east and are 0·5–1 m. wide. These often carry quartz with stibnite; one such dyke, one metre wide, exhibits for instance two zones of a grey milky quartz, each 8–10 cm. thick and slightly impregnated with stibnite, enclosing between them a vein of solid stibnite 1–2 cm. thick. Other lodes are known at Chanac and Valfleury. At the former place they are 0·40–0·70 m. wide and occur in clay-slate, while at the latter they are quartz-stibnite lodes in gneiss, this gneiss being regarded as derived from granite.

The more important deposits of the Central Plateau are however those at Freycenet, La Licoulne, etc., in Puy-de-Dôme, Le Cantal et la Haute Loire, respectively. These occur as vertical lodes containing lenses of stibnite separated by barren stretches, in Archaean gneissic mica-schist or in granite; a lens of compact stibnite 20–30 cm. thick and 12 m. long, for instance, will be followed by a barren stretch 10–15 m. long. The average thickness of the payable material is 15–30 cm. The stibnite is sometimes intimately intergrown with quartz. It almost invariably contains iron, a fact which is often insufficiently realized, with the result that too high an idea of the antimony content is obtained. At the outcrop, antimony oxide occurs in crystalline or amorphous masses of variable colour. These were formerly overlooked. Since 1889 however they have been carefully collected for export to England or Germany. At Freycenet a quartz lode contains lenses 30–40 cm. thick and up to 15 m. in length which consist of almost pure stibnite with but 8–10 per cent of quartz. These are separated from one another by quartz so impregnated with stibnite as sometimes to contain as much as 25 per cent of metallic antimony.

The deposits at La Licoulne in Haute Loire occur in an extensive gneiss plateau lying at an average height of 980 m. above the sea and furrowed by valleys eroded to depths of 200 metres. As country-rock are found all gradations from varieties which may be described as granite to such as greatly resemble mica-schist. The numerous stibnite lodes belong to two systems, one striking N. 30° E. and the other N. 60° E. Geographically, in this important district four groups may be recognized, namely, those at Mercoeur, Montel, Valadou, and La Licoulne. At Mercoeur the Bissade lode is the most important. This may be followed for a length of 2500 m. along a strike N. 30° E., with a width of 30–60 cm., of solid stibnite accompanied by a little quartz. In places this lode breaks up into several veins. The separation from the gneiss is fairly sharp though veins and nests are occasionally found beyond the walls. At Montel the deposits have nowhere been exploited. At Valadou, on the other hand, a lode has been opened to a depth of 110 m., along a strike varying from N. 45° E. to N. 60° E. This lode occurs in very tough old slates which its presence leaves quite unaltered. In it the stibnite forms a number of shoots up to 30 cm. in thickness, and separated from one another by barren stretches. The stibnite here appears to contain more silver than that at Mercoeur. This lode is cut by a quartz vein striking N. 30° E. and containing nests of antimony ore. At La Licoulne there are several irregularly disposed lodes, of which the two most important have been investigated to a depth of more than 300 m. on the incline.

At Malbosc in the department of Ardèche, quartz-stibnite lodes with north-east strike traverse mica-schists, not far beneath which lies the La Lozère granite. An interesting feature of these lodes is that they carry some calcite and barite. The stibnite appears either in the form of well-defined nests, or as irregular, veins up to 10–20 cm. thick. The occurrence of these veins is very irregular and uncertain; sometimes they are out in the hanging-wall, sometimes actually on the wall of the quartz lode, while often they split up or die out completely. This irregularity has caused work upon these deposits repeatedly to be given up.

The occurrences in Japan are of great interest. In general these are lodes which traverse Mesozoic and Palæozoic formations, and but seldom are found either in the crystalline schists or in Tertiary rocks. They often occur in sediments near the contact with intrusions of quartz-porphyry, or in that eruptive itself. While in Japan the areas occupied by Mesozoic beds are in general remarkable for their poverty in other useful metals, they are comparatively rich in antimony. The most important deposits occur at Kano in the province of Suwo, where a lode in Mesozoic country is worked ; at Hanta in the province of Yamato, lodes in similar country ; at Taguchihara in the province of Hyuga, in similar country ; at Ichinokawa in the province of Jyo, in crystalline schists and Mesozoic beds ; at Nakase and Nakagawa in the province of Tajima, in Palæozoic beds ; at Arahira in the province of Hyuga, in Palæozoic beds ; and finally, at Amatsutsumi in the same province, but in quartz-porphyry. Geographically these deposits occur more particularly in the bend of southern Japan, and especially along the outer curve. The antimony production of Japan in the year 1907 was 562 long tons, and in 1908, 537 long tons.

The Occurrences in Asia Minor, etc.

LITERATURE

Br. Simmersbach. ' Die nutzbaren mineralischen Bodenschätze in der kleinasia-tischen Türkei,' Zeit. f. d. Berg- Hütten- und Salinenwesen, 1904, Vol. LII. p. 515 ; ' Die wirtschaftliche Entwicklung einiger Bergbaubetriebe in der Türkei,' Verhandl. des Vereins zur Beförderung des Gewerbefleisses, 1905, p. 487.—K. Schmeisser. ' Bodenschätze und Bergbau Kleinasiens,' Zeit. f. prakt. Geol., 1906, p. 186.

The occurrences of antimony in Asia Minor are found in the vilayets Brussa, Smyrna, and Siwas. That in Brussa is represented by lodes 0·1–2·0 m. wide, worked in a mine known as the Gômetschiftlik-Antimon-Madén belonging to the Sultan, situated 24 km. east of Gedis on the south-western slope of the Kysyl-Dagh. The yearly production is about 500 tons of antimony ore. Half a kilometre south of Demirkapu there are other antimony mines at Irvindi and Sülukkoi. In the vilayet of Smyrna, a double lode, the outcrop of which may be followed for 2 km., is worked in the Tschinlikaja mine 20 km. south-east of Oedemisch and 100 km. east-south-east of Smyrna, on the north-west slope of the Baliam-boli-Dagh. The width of this deposit varies from a few centimetres to some metres. In 1898, 500 tons of ore valued at about £6000 were won. In the same vilayet, the mines near Rozsdan and Aidin, and finally the Geramos and Kordelio mines, also occur. In the vilayet of Siwas, antimony ore has been opened up at Karahissar.

The deposit at Allkhar north-west of Salonika in Macedonia is well - known. As however no quartzose gangue is present and the country-rock in the foot-wall consists of limestone, this deposit may be a metasomatic occurrence.

Other antimony lodes are known at Bastia on Cape Corse in the north of Corsica, in Archæan sericite-schists. At Su Suergiu in Sardinia, lenticular masses of stibnite and pyrite occur in a zone several hundred metres long and 40 m. wide, in graphitic schists and calc-phyllites, presumably of Silurian age. In Tuscany, a lode one kilometre long is known to occur in Eocene and Miocene beds ; while finally, in Portugal a number of occurrences are known at Casa Branca, Oporto, and Alcoutim, some of which are auriferous.

Metasomatic Antimony Deposits

LITERATURE

Buff. 'Geogn. Bemerkungen über das Vorkommen von Spiessglanzerzen auf der Grube Caspari bei Wintrop und auf der Grube Unverhofft Glück bei Nuttlar im ehemaligen Herzogtum Westfalen,' Karstens Arch. für Bergbau- und Hüttenwesen, 1827, XVI.—F. M. Simmersbach. 'Das Antimonerzvorkommen auf der Casparizeche bei Arnsberg in Westfalen,' Jahrb. der Bergakademie zu Leoben, 1870, XIX.—H. B. von Foullon. 'Über Antimonit und Schwefel von Allchar bei Rozsdan in Mazedonien,' Verhandl. der k. k. Geol. Reichsanst., 1890, p. 318; Beschreibung der Bergreviere Arnsberg, Brilon und Olpe, sowie der Fürstentümer Waldeck und Pyrmont, 1890, p. 158.—R. Hofmann. 'Antimon. und Arsenerzbergbau "Allchar" in Mazedonien,' Osterr. Zeitschr. f. Berg- und Hüttenwesen, 1891, Vol. XXXIX.—B. Lotti. 'Die zinnober- und antimonführenden Lagerstätten Toskanas und ihre Beziehungen zu den quartären Eruptivgesteinen,' Zeit. f. prakt. Geol., 1901, p. 41.

The geological position of the deposits of this class is uncertain and indefinite. Presumably they have arisen by replacement of limestone and dolomite of different ages. Usually they are distinguished by the absence of gangue. In the neighbourhood of the outcrop they are decomposed to stiblite, while the limestone is often altered to selenite and dolomite.

It is probable that the occurrence at Allkhar in Macedonia, briefly described above, belongs to this class. The hanging-wall of that deposit consists of mica-schist, the foot-wall of dolomite and limestone. The ore occurs in stringers and lenses without gangue, together with arsenic ores. The width of solid ore may at times be as much as 1·50 m., while the occurrence has been proved for a length of 4 kilometres. Near the deposit the dolomite is highly altered under formation of sulphur and selenite. A portion of the ore consists of realgar and orpiment.

The occurrence at Cetine di Cotorniano in the province of Siena, Italy, occurs between Eocene limestone and Permian slate, and consists of hornstone or lydian-like quartz, traversed by long crystals of stibnite, and by pyrostibnite, sulphur, quartz, calcite, etc.

The genesis of the deposit in the Caspari mine near Arnsberg in Westphalia still remains somewhat doubtful. Generally it is described as a bed ; by Bergeat in his *Lagerstättenlehre* it is regarded as a lode-like occurrence ; while according to Krusch it represents a metasomatic occurrence. The ore occurs in Culm limestone which forms the easternmost point of the Arnsberg anticline, this anticline inclining to the east till the limestone is covered by younger beds, more especially the Millstone Grit. An assumed air-anticline in the neighbourhood of the Caspari mine would join the north-west and south-east flanks of this anticline. The latter flank has been opened up by mining operations for a length of 1100 metres. This anticline is accompanied by a large number of secondary anticlines and folds which are traversed by numerous faults. The limestone in the neighbourhood of the occurrence appears dark and decomposed. Five metalliferous beds carrying stibnite in irregular segregations 5–15 cm. thick, are known. From these beds veins proceed into the limestone which itself contains a sprinkling of ore. On the north-west flank the ore, consisting chiefly of stiblite, is so impure as to be unpayable.

THE IRON LODES

IRON deposits of magmatic, contact-metamorphic, metasomatic, and sedimentary origin being usually so large, it is exceptional to find an iron lode satisfying the demands of payability, and the number of districts where such lodes have been exploited is consequently not great.

The formation of iron lodes varies according to the particular ore contained, and in this respect, in general, only hæmatite, specularite, and siderite come into question. In a number of cases, as for instance in Siegerland and in the Harz, the lode-filling can either be proved to be, or put down in all probability as being of great age. The widely distributed Devonian system in Germany, for instance, is remarkable for the number of iron ore-beds found in it, while many of the German iron lodes are of Devonian age. These, in part at least, are genetically related to old eruptive rocks.

Since in many districts hæmatite- and siderite lodes occur together, both these ores were probably formed from the same solutions. In such cases, exactly what were the precipitants which brought these two minerals separately to deposition, has not yet been determined, further investigation is necessary. In the absence of oxygen, siderite becomes precipitated from a solution of ferric bicarbonate when the excess of carbonic acid which would keep it in solution as ferrous carbonate, escapes. Such escape may take place when in the course of circulation the carbonated solution from depth reaches the surface. In depth, the carbonic acid under pressure keeps the ferrous carbonate in solution ; at a higher horizon where the pressure is lower, a portion of this acid escapes, and precipitation of siderite results.

In form, the iron lodes are generally simple lodes, the separation between ore and country-rock being sharp and definite on both walls. In length, they are generally limited to some hundred metres, though they are often collected together in series or groups which, as for instance in Siegerland and in the Zips country, may extend for kilometres. With regard to extension in depth, this, when considering the fissure alone, may

786

be considerable. When however the lode-filling or the payability of the deposit is considered, the case is different. In consequence of the keen competition from the more easily worked deposits of other genesis, and of the increase of cost with depth, work on many iron lodes has had to be stopped, even though the iron content was well maintained. The deepest mines in Siegerland are 425 m. deep, while in the Zips the adit levels penetrate to points 400–600 m. below the summit.

The iron lodes being often of great geological age have been greatly affected by more recent earth movements. It is consequently found that in addition to ordinary faults and overthrusts, which are hardly ever absent, the lodes, which is more rare with them, have also suffered plication and disturbance from vertical and lateral displacements, so much so that in many cases the original ore-body has been completely dismembered.

The distribution of the ore in the lode is not regular. The impression is often given that the original filling was fairly simple, or at all events more simple than that found to-day. Into this first filling, as a consequence of subsequent earth movement, other solutions penetrated again and again, gradually replacing the original filling. Did, for instance, that consist of much siderite and little quartz, at some later period a great part of the siderite would often be replaced by quartz. It is consequently frequently the case with siderite lodes that both in strike and dip bodies of pure siderite alternate with others containing much quartz. By weathering and erosion the siderite bodies of such lodes become easily disintegrated, leaving the quartz to project upon the surface.

Although it is probable that the quartz-carrying solutions which effected this alteration came from depth, it does not necessarily follow that a siderite lode which is silicified at the outcrop is never good in depth, though that statement is often heard. Replacement does not proceed regularly from depth upwards, but begins at points favourable to its inception, the remainder of the lode between the silicified patches thus formed, retaining its original character.

Of greater interest than silicification however are the complex primary deposition of, and the secondary replacement by sulphides. In many cases small amounts of different sulphides, particularly pyrite and perhaps also chalcopyrite, are deposited with the siderite. These two sulphides when present beyond a certain amount spoil the quality of the ore, chalcopyrite in this respect being worse than pyrite. Where the siderite is roasted before being smelted the presence of sulphur matters little, indeed at times it is even welcomed since the ore after roasting is more porous. But apart from the sulphur the copper content of chalcopyrite is prejudicial, so that while ore containing up to 0·4 per cent of copper is accepted, ore containing more than that amount is difficult to market. In some districts there are in consequence mines which, formerly worked for iron, were stopped in depth because the permissible percentage of copper had been exceeded. On the other hand, when describing the copper lodes cases will be instanced where siderite lodes by increase in the copper content have gradually merged into copper lodes with as much as 3·5 per cent of copper, and have thus become very important.

In addition to these primary sulphides, secondary sulphides such as galena, sphalerite, and chalcopyrite, also occur. These, as indicated by the fact that they often occur along the walls or in separate veins within the lode mass, are the products of later solutions. Sometimes the original filling, and especially the siderite, has in greater part been replaced by such veins, it is then more difficult to recognize any such relative age. This replacement may be more or less complete; Krusch, in the lodes at Mitterberg near Bischofshofen, for instance, found pieces of ore which though now consisting of quartz and chalcopyrite, formerly, as indicated by unaltered kernels, belonged to a carbonate lode.

Such sulphides in siderite lodes are often more particularly seen in the upper levels, though the solutions from which they were derived undoubtedly came from depth. In appraising the possibilities of lodes of such great age as these, the varied age of the filling must be most carefully considered. Apart from the sulphides which have been mentioned, others have also been found, though not in amount sufficient to make them important.

The gangue-minerals may also be divided into those which are primary and those which are secondary. Quartz is always the most common. The carbonates are less frequent; where however they occur, isomorphous mixtures of calcium carbonate, magnesium carbonate, and ferrous carbonate, in the most varied proportions, are often found.

Since these lodes are generally simple lodes, rock inclusions do not play any important part; such as do occur represent pieces which have fallen from the hanging-wall into the fissure. The structure of the ore, in consequence of replacement, is complex. Not infrequently a crusted structure consisting of bands of siderite throughout which sulphides are disseminated, appears as the oldest structure. From this, by metasomatism arose either a secondary irregular-coarse structure, or a pseudo-brecciated structure still containing angular pieces of the original filling in considerable number; or finally, and expressive of complete replacement, a simple filling, as when quartz had completely replaced the siderite. A drusy structure is characteristic of the oxidation zone, this zone being further remarkable for its stalactites and reniform structure.

Primary and Secondary Depth-Zones.—Primary depth-zones have been observed in so far that metalliferous bodies are in depth succeeded by others which are siliceous. Since silicification is mostly secondary and in many cases only local, siderite lodes containing much silica may in depth again become workable. The metasomatic replacement which results in silicification is in itself dependent upon the greater or less resistance presented; pure siderite, though chemically fairly uniform, may nevertheless, in consequence of varying internal structure, etc., behave itself variously towards such replacement, this fact being sufficient in itself to explain the possibility that in depth altered zones may alternate with others of clean ore.

Another primary depth-zone may be expressed in the distribution of the small copper- and sulphur contents, since both these metals may in depth increase or decrease; while a similar significance may also be read into the occasional occurrences of cobalt minerals. At Dobschau in Upper Hungary, for instance, iron ore was first found, then copper ore, and deeper still, cobalt and nickel ores. On the other hand, there are lodes in which the cobalt belongs to a higher zone than that containing siderite, and others in which there is a repeated sequence of cobalt-nickel- and iron ores.

When small amounts of lead and zinc occur with the siderite, it must first be determined whether these sulphides represent a later sulphide deposition consequent upon the re-opening of the fissure, or whether they are contemporaneous with the siderite. In the first case pseudo-depth-zones would arise wherein no regularity whatever would obtain, though perhaps even then the sulphides would preferably be deposited in the upper levels. In the second case, the observations recorded hitherto are not sufficient to allow any definite regularity to be stated, except perhaps that in lead-zinc lodes the siderite zone is seldom found above, but often below the lead-zinc zone.

That siderite by atmospheric agencies becomes hydrated and oxidized is of course a well-known fact. The vertical dimension of the oxidation zone so formed is variable, depending upon the climate and upon the relation between the activity of erosion and the advance of oxidation. In general, the alteration of siderite to limonite from the outcrop to depths of 10 m. or even 15 m. is not uncommon. Mineralogically these chemical-geological reactions are particularly interesting in that iron and manganese, which are both common to the siderite deposits, often become separated by oxidation. The ore resulting from these oxidation processes is generally gelatinous, that is to say, amorphous limonite and amorphous manganese minerals, such as psilomelane, wad, etc., are chiefly formed. In these amorphous masses, according to Cornu, the various forms of crystalline limonite, and especially kidney ore, are formed by recrystallization, such crystalline material filling the cavities and crevices. The porous and friable nature of the limonite gossan of such iron lodes is of particular importance to the miner and metallurgist. Similarly, from the amorphous manganese ore, crystallized minerals such as pyrolusite are formed, though crystalline manganite and pyrolusite are sometimes formed directly and without passing through an intermediate gelatinous stage.

Unlike siderite, which generally does not contain more than 2 per cent of moisture, limonite contains at least 8 per cent and may contain as much as 15 per cent or even more. Where therefore the distance between the mine and furnace is great the transport cost of limonite will be considerable. On the other hand, though experience has shown that when the width of the siderite deposit is great the cost of mining in this material is not so high as might be expected, mining in the limonite gossan is easier than in siderite. It must however be remembered that though limonite has a higher iron content than siderite, the above-mentioned difference in the moisture offsets this advantage. When therefore the siderite is pure the difference between the mining cost in the primary and oxidation zones is on the whole not very considerable.

In relation to the iron content no cementation zone is known in these deposits, the primary zone follows immediately after the oxidation zone. When however a small copper content is present in the primary ore or silver occurs in any galena or sphalerite present, then between these two zones a silver- or copper cementation zone may occur, though this would in any case be unimportant, indeed in the presence of undecomposed siderite the concentration of these other metals is usually overlooked. Such a combination of fresh siderite with cementation minerals is however not uncommon, since the formation of the cementation zone demands the absence of oxygen [1] and the activity of reducing agencies only, conditions which are without effect upon the siderite.

On account of the low value of siderite, referred to in introducing this subject, only such lodes are suitable for exploitation as have a considerable width. The fact that but few siderite lodes are exploited does not therefore imply that but few exist; it betokens rather the many conditions to be fulfilled before such lodes become payable. There are, for instance, many large lodes unfavourably situated in regard to communication and far from coal or wood, and therefore unpayable. Where fuel is cheap or the distance between mine and furnace great, the siderite is roasted to an oxide which, chemically considered, approaches magnetite, and of which the iron content as compared with that in the original ore is as 7 : 5 or 13 : 10. This convenience may in itself occasionally be the chief factor in rendering a deposit workable. Even however under the best conditions the successful mining of small siderite lodes cannot be undertaken, since in relation to the value of the ore the cost of mining is too high. Accordingly, in Europe there are only two large iron lode-mining districts

---

[1] *Ante*, pp. 140, 145.

in operation, namely, the Siegerland and Zips-Görmörer districts, the latter being in the Erzgebirge.

The amounts of phosphorus, manganese, and sulphur, contained in siderite are important factors. In Siegerland the phosphorus content for 1,862,244 tons was 0·05 per cent; for another lot of 110,708 tons, 0·05–0·75 per cent; and for a third lot of 9,850 tons, 0·75–1·00 per cent. The manganese content is usually high and generally higher than with most other iron ores. This is because at the deposition of siderite the $MnCO_3$ in solution was simultaneously precipitated, while when oxides are precipitated the oxide of manganese is not precipitated with that of iron. In Siegerland the manganese content reaches 12 per cent, and the line between manganiferous iron ore and iron-manganese ore may therefore conveniently be put at this percentage. Normal Siegerland ore on which prices are based contains 9 per cent of manganese, whereas in that of the Zips there is but 2 per cent. The sulphur content in siderite varies considerably. In this connection those ores which contain so little sulphur that they may be smelted direct, are to be distinguished from those which must be roasted. These latter may naturally contain more sulphur since the greater part would be driven off in the roasting, while the iron associated with the sulphur would remain to benefit the ore.

The price of siderite at the mine varies with its distance from a furnace. By the Siegerland ironstone syndicate, for example, the price is based upon the content of the average ore, and then adjusted according to supply and demand. Under these circumstances the price within the last few years has varied between £9 : 15 : 0 and £6 : 10 : 0 per ten tons.

The lode-like hæmatite- and magnetite deposits must now be briefly described. These belong generally to the class of simple lodes, those of magnetite being much less frequent than those of hæmatite and specularite. The width of these lodes is usually small and generally under one metre. It is seldom that the iron oxide occurs crystalline as specularite, it is generally a compact earthy or fibrous hæmatite. Quartz, hornstone, and jasper form the gangue, carbonates and barite being uncommon. When magnetite and hæmatite occur together in the same lode it is a question—and particularly when eruptive rocks exist in the neighbourhood—whether the magnetite is primary, or whether it is secondary and the result of contact action upon hæmatite, etc.

Primary depth-zones may often be observed when hæmatite and manganese ores occur together. In such cases the manganese occupies a higher zone than the iron. It was formerly supposed that atmospheric agencies had little effect upon hæmatite and specularite. Experience has however shown that both these ores at the surface often become altered to limonite, and that lodes containing these oxides may have the same gossan as siderite lodes ; in fact even magnetite which is so uncommon in lodes, may by meteoric waters become changed to limonite.

SIEGERLAND

LITERATURE

General : K. SCHMEISSER. 'Über das Unterdevon des Siegerlandes und die darin aufsetzenden Gänge, unter Berücksichtigung der Gebirgsbildung und der genetischen Verhältnisse der Gänge,' with an appendix, 'Die Mineralien des Siegerlandes,' Jahrb. der k. pr. geol. Landesanst., 1882, pp. 48-148.—C. LEYBOLD. 'Geognostische Beschreibung des Ganggebietes der Eisenerzgruben Wingertshardt, Friederich, Eisengarten, Eupel und Rasselskaute bei Wissen a. d. Sieg,' ibid. pp. 1-47.—A. RIBBENTROP. Beschreibung des Bergreviers Daaden-Kirchen. Mining Department, Bonn, 1882.—E. BUFF. Beschreibung des Bergreviers Deutz. Mining Department, Bonn, 1882.—FR. L. KINNE. Beschreibung des Bergreviers Ründeroth. Mining Department, Bonn, 1884.—G. Wolf. Beschreibung des Bergreviers Hamm. Bonn, 1885.—HUNDT, GERLACH, RÓTH, and W. SCHMIDT. Beschreibung der Bergreviere Siegen I., Siegen II., Burbach und Müsen. Mining Department, Bonn, 1887.—HAEGE. Die Mineralien des Siegerlandes und der angrenzenden Bezirke Siegen, 1887.—K. DIESTERWEG. Beschreibung des Bergreviers Wied. Mining Department, Bonn, 1888.—E. SCHULZE and others. Beschreibung der Bergreviere Arnsberg, Brilon und Olpe. Mining Department, Bonn, 1890.—ULRICH, HOLZAPFEL, KÖRFER, and others. Beschreibung der Bergreviere Wiesbaden und Diez. Mining Department, Bonn, 1893.—W. BORNHARDT. 'Über die Gangverhältnisse des Siegerlandes und seiner Umgebung,' Archiv für Lagerstättenforschung, Part 1. Berlin, 1910. Part 2 with an appendix by P. KRUSCH. 'Die mikroskopische Untersuchung der Gangausfüllungen des Siegerlandes und seiner Umgebung,' with illustrations by Baumgärtel. Berlin, 1912.

Special : A. DENCKMANN. 'Zur Geologie des Müsener Horstes,' monthly report d. d. g. Ges., 1906, p. 93 ; 'Mitteilungen über eine Gliederung in den Siegener Schichten,' Jahrb. der k. pr. geol. Landesanst., 1906, Vol. XXVII. Book I. ; Über die geologischen Verhältnisse der Grube Kuhlenberg bei Welschenennest, 1906 ; 'Über das Nebengestein der Siegerländer Gänge,' Lecture before Verein Berggeist in Siegen, 1906 ; Die Überschiebung des alten Unterdevon zwischen Siegburg a. d. Sieg und Bilstein im Kreise Olpe. Celebration 70th anniversary of Adolf v. Koenen's birthday, with a survey map ; scale 1 : 500,000, Stuttgart, 1907 ; 'Über das Nebengestein der Ramsbecker Erzlagerstätten,' Jahrb. der k. pr. geol. Landesanst., 1908, Vol. XXIX. p. 243.

Maps of the mining districts : Siegen I., Siegen II., Burbach, Müsen, Daaden-Kirchen, Hamm, Wied, Olpe, und Diez ; scale 1 : 80,000.—Map of the lodes of the Bensberg district in six sections, and one large map to scale 1 : 20,000, by A. Schneider, published by the Mining Department, Bonn, 1882.—Map of the useful deposits of Germany ; Rheinland and Westphalia, 1st Ed. in eight sections, scale 1 : 200,000, 1904 ; 2nd Ed., 1912, Königl. geol. Landesanst.—A map of the Siegerland lodes, scale 1 : 10,000, is being prepared by the Mining Department, Bonn, and will be published by the Geol. Landesanst.

In the Lower Devonian of Siegerland, the southernmost portion of the province Westphalia, iron lodes occur chiefly in three districts, namely, a northern district between Müsen and Olpe ; a central district between Siegen and Altenkirchen ; and a southern district between Altenkirchen and the Rhine.

This Lower Devonian, of which the stratigraphy has not yet been completely determined, consists chiefly of clay-slates more or less arenaceous, sandstone, and grauwacke, these beds having been folded into north-east anticlines and synclines and subjected to a number of disturbances.

According to Denckmann, lodes occur in all the different stages, from the Gedinnian, through the Siegen, and up to the Coblenz, though in these last they are considerably smaller. With the siderite lodes, lead- and zinc lodes carrying siderite also occur, these generally appearing around the margins of the siderite district and in all formations from the lowermost Devonian to the Upper Carboniferous. They have been particularly observed in the east, south-west, and north of this region.

Copper lodes, in which the copper content decreases in depth while quartz increases, are found exceptionally. These occur not only in the Lower but also in the Middle Devonian, cutting the hæmatite ore-beds and the accompanying diabase and schalstein which occur between the Middle and Upper Devonian. Where typical siderite lodes carry considerable copper in the upper levels, such copper generally does not continue below the ground-water level.

The well-known cobalt lodes, the occurrence of which is limited to the country between Siegen and Kirchen on both sides of the river Sieg, are of great interest. They are found only in the Lower Devonian.

The disposition of the siderite lodes is not uniform; they appear rather to occur in zones, swarms, or groups, following a direction more or less parallel to the main strike of the Rhenish Schiefergebirge, though between such zones or groups isolated occurrences are also found. Within the zones themselves the lodes strike most irregularly, this being even the case with the large main lodes, though these may maintain their strike for great distances. The term 'lode-swarm' or 'lode-group' suggested by Leybold and Bornhardt is therefore more descriptive than lode-series, which suggests a more or less parallel strike. The following swarms or groups may be differentiated :

Siderite lodes—
    The Schmiedeberg Group.
    The Gosenbach Group.
    The Kulenwald Group.
    The Eiserfeld Group.
    The Biersdorf Group.
    The Eisener Group.
    The Müsen-Silberberg Group.

Lead-silver-zinc lodes—
    The Johannessegen Group.
    The Niederfischbach Group.
    The Oberfischbach Group.
    The Obersdorf Group.
    The Altenseelbach-Wilden Group.
    The Buchhell Group.

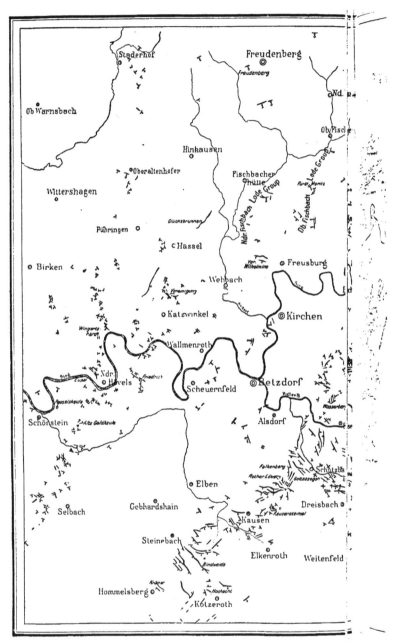

FIG. 358.—General map of the Siegerland siderite l... s,

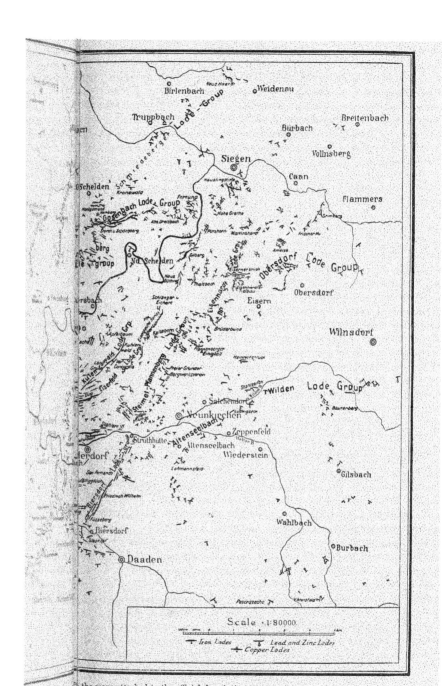

Birlenbach   Neue Haardt   Weidenau

Truppbach   Lode Group   Burbach   Breitenbach

Siegen   Vollnsberg

Scheiden   Kronewald   Caan   Flammers

Gosenbach Lode Group   Forrung

Seren u Schönberg   Hohe Grethe

berg   Pershorn   Martinshard   Frischer Mu

Lie group   Nd Scheiden   Gilberg   Oberdorf Lode Group

Haus   Thalbach

rsbach   Schlanger Eichert   Eisern   Oberdorf

Lode Grp   Brüderbund   Wilnsdorf

Landwehr Louis Grp   Hanenberger

Eiserfeld   Concordia   Peter Gruner

Stromer Flammensberg   Stahlseifen   Wilden Lode Group

Neunkirchen   Salchendorf   Baurenberg

Altenseelbach   Zeppenfeld

rdorf   Struthhütte   Altenseelbach   Wiederstein   Gilsbach

San Fernando   Lohmannsfeld

Friedrich Wilhelm

Fusseberg   Wahlbach

Biersdorf   Burbach

Daaden

Peterszeche

Scale   1:80000

✛ Iron Lodes   ⊤ Lead and Zinc Lodes

✛ Copper Lodes

ff the maps attached to the official descriptions.

According to Denckmann, the limits to these groups are determined by tectonic and stratigraphical factors.   The Müsen-Silberberg group, for instance, occurs within an uplift of geologically recent date ; while a stratigraphical dependence shows itself in the agreement in extension between these groups and certain geological horizons.   In this connection also, the fact that the fissures of these iron lodes coincide with the boundary fissures, branch veins, and cross-courses of subsidences presumably formed in upper Middle Devonian time, is of great significance, the determination of this fact having resulted from laborious research by Denckmann.   A knowledge of the course of these subsidences becomes therefore of prime importance when prospecting for, or following such siderite lodes.

Most of these lodes are steep, dipping from 60° to 90°, and they usually cross the bedding of the country at an acute angle.   Concerning their persistence in depth, it appears that here also the statement holds good that lodes of considerable length along the strike have also considerable extent in depth, though throughout that extent payable and unpayable portions alternate.   Along the strike the lodes generally follow an inclined direction in depth, that is to say, they have a distinct pitch.   This pitch as a rule, and as illustrated in Fig. 359, D, follows the line of intersection between the lode plane and the plane of bedding.

For the greater part these lodes are the fillings of simple fissures, though, as illustrated in Fig. 359, B, parallel lodes and branch veins also occur, and the immediate country-rock may be ramified by numerous veins and stringers of siderite and quartz.   In general the lode-filling appears to be intergrown with the country-rock, and only exceptionally are ore and rock separated by clayey material.   Rock inclusions, in contradistinction to what is the case with the lead-zinc lodes, are seldom seen in iron lodes ; when occurring, they are either scattered throughout the mass or, as may often be observed, collected particularly in the neighbourhood of the walls, in which case not infrequently the lode material gradually merges into country.

The lode width in Siegerland is remarkably great, being in many cases 5–10 m. and sometimes even 20 metres.   In places, as in the Petersbach mine near Eichelhardt, the St. Andreas near Bitzen, and the Neue Haardt near Weidenau, it varies remarkably.

The occurrence of irregular ore-bodies, such as the Stahlbergstock at Müsen, illustrated in Figs. 9 and 359, A, necessitates great caution when speculating as to the width in depth.   Decrease in width may occur either by the approach of the walls to each other, or by the splitting of the lode into branches which quickly die out.   Sometimes, on the other hand, the width may increase, especially where the lode makes a sharp bend,

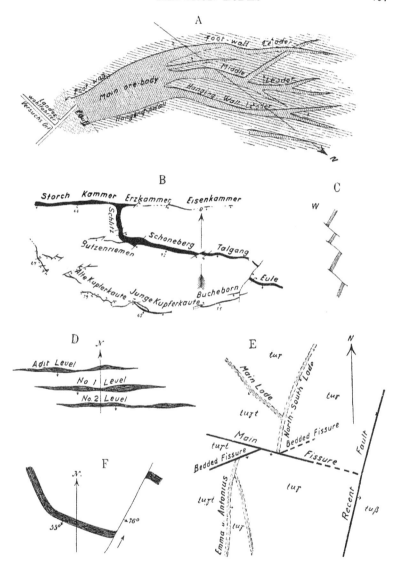

FIG. 359.—Features of the Siegerland iron lodes.

A, the Stahlberg deposit near Müsen at the adit level (Nöggerath); *B*, lodes in the Storch and Schöneberg mines at the adit level (Bornhardt); *C*, transverse section representing the shallow faults, Glucksbrunnen mine near Niederfischbach (Bornhardt); *D*, displacement of a lode in plan (Bornhardt); *E*, diagram illustrating the Kuhlenberg series near Welschennest; *tuγ* = Gedinnien; *tuγt* = Lower Siegen beds; *tuβ* = Birkelbach beds (Denckmann); *F*, slides (Geschiebe) at the south end of the Thomas lode in the Kuhlenberg mine (Bornhardt).

cases of this having occurred at the Alte Lurzenbach, Neue Haardt, and St. Andreas mines. A dependence of the width upon the nature of the country-rock has been noticed in so far that lodes in medium hard or compact rock are the widest, while in the more compact and in the softer slaty rocks they are smaller. The lead-zinc lodes on the whole are narrower than the siderite lodes, solid ore-bodies of large size being seldom seen; payable masses of large width are more often made up of country-rock traversed by numerous ramifying veins.

The country-rock along the siderite lodes has either suffered no alteration at all or but very little, though the occurrence in it of siderite replacing quartzose material may often be observed. With the specularite- and hæmatite lodes, on the other hand, pronounced decomposition of the country-rock accompanied by bleaching, is common so long as the ore continues, and accordingly ore-deposition and the decomposition of the rock must be genetically connected. With the lead-zinc lodes there is likewise considerable decomposition, though bleaching of the country-rock is less frequent.

Along the lodes the occurrence of lustrous black slate pressed into lenses with numerous polished surfaces, is interesting. The black colour, which was formerly ascribed to anthracite or graphite, is in reality due to amorphous carbon, a sample of such slate investigated by Pufahl having shown the presence of 1·3 per cent of that element.

As illustrated in Fig. 359, disturbances are common. These may be classed as follows :

1. Bends, folds, and kinks, in relation to which it is not easy to determine whether they are original, or were subsequently formed.

2. Elongations and flattenings connected with a turning or twisting of the lode, these occurring more frequently with the smaller lodes. Such were caused by movements of the country-rock along planes running obliquely to the lode plane, or cutting directly across it.

3. Tectonic fissures, which are numerous and of the following different natures :

(a) Normal faults, where, as illustrated in Fig. 359, E, the country-rock in the hanging-wall has subsided.

(b) Very flat shallow faults generally inclined upwards towards the north, and along which the hanging-wall has been normally or inclinedly thrust upwards, as illustrated in Fig. 359, C.

(c) The so-called Geschiebe or slides, these being horizontal displacements along planes at the most 20°–30° from the vertical, as illustrated in Figs. 45 and 359, F.

Lode deflections, such as resulted from obstacles in the path of deposition already existing at the time of the lode's formation, are in

Siegerland of little importance. Most of the disturbances formerly considered as belonging to this class have since been proved to be younger than the lodes.

In regard to age, according to Denckmann the various disturbances may be divided into two groups, an older and a younger. To the former belong the lode fissures, which are Middle Devonian. Those produced by the post-Culm folding are likewise of great age; they include the lateral displacements and certainly also the shallow faults; they are moreover contemporaneous with the overthrust of the Lower Devonian over the Middle Devonian in the Lenne slate area.

To the younger post-Palæozoic disturbances belong the east-south-east to south-south-east fissures, the east-west, and the north-north-east to north-east fissures, of which three groups the last has had greatest influence upon the present structure of Siegerland.

The Lode-Filling.—With the iron lodes siderite forms the preponderating mass of the filling, the whole mass sometimes consisting of this mineral. The structure is irregularly coarse- or fine-grained, and only occasionally in any sense banded or crusted. Fragments of country-rock are common, though lodes with few or no such inclusions are characteristic of the district. The gangue consists of quartz, any carbonates present representing a more recent deposition in crevices; barite and fluorite do not occur.

Among the primary minerals, pyrite and chalcopyrite are the most important; sphalerite and galena are more uncommon, being in fact unknown in many of the deposits; while tetrahedrite, chalcocite, bornite, cobaltite, linnæite, nickel-stibnite, gersdorffite, millerite, boulangerite, bournonite, stibnite, and marcasite, are present in places.

In the oxidation zone occur limonite, lepidocrocite, göthite, pyrolusite, manganite, psilomelane, wad, and, very rarely, rhodochrosite and malachite. In the amorphous gelatinous iron resulting from weathering, veins of more recent crystalline content are found.

With the lead- and zinc lodes the filling likewise consists chiefly of siderite or quartz and rock inclusions, galena and sphalerite being in smaller amount. In many lodes the sphalerite greatly exceeds the galena; occasionally however it only appears in depth, when it represents a deeper primary zone. Barite, calcite, and other carbonates, as well as pyrite and chalcopyrite, are more uncommon.

These lodes may be described under two types, namely, those with preponderating siderite and quartz, and those consisting chiefly of fragments of country-rock, these two types being connected by gradations. With most lodes galena and sphalerite represent younger ores formed by the metasomatic replacement of siderite and quartz. In depth these lodes often become poor, passing over to iron- or quartz lodes, cases of this being known at the Lohmannsfeld mine at Altenseelbach, Peterszeche at Burbach, Wildberg at Wildberger Hütte, Bliesenbach at Engelskirchen, etc. In other cases the lead and zinc have continued to greater depth, sometimes to as much as 400 m. or more, the Viktoria mine at Littfeld being 380 m., and the Wildermann-Stahlberg at Müsen 424 m. deep.

In the oxidation zone, in addition to the oxidized iron- and manganese minerals, occur the oxides, carbonates, sulphates, and other combinations of those other metals which in the primary zone are associated with sulphur. Below the oxidation zone a cementation or enriched zone, distinguished by containing good copper ore, often occurs.

The copper lodes in their upper portions contain chalcopyrite chiefly, and chalcocite subordinately. As these in depth give place to siderite containing disseminated chalcopyrite, this cupriferous upper zone may be regarded as a cementation zone.

In addition to these copper lodes associated with iron lodes, there are around Nieder-Dreisbach, on both sides of the Lower Daade valley, others which are independent. These carry chalcopyrite, vuggy quartz, and dolomite, and are younger than the iron- and lead-zinc lodes across which with well-defined walls they often cut. With these lodes also, the cementation zone is the more cupriferous. The only deposit of this kind now being worked is that at the Danielszug mine at Wipperfürth.

The cobalt lodes contain chiefly very fine-grained cobaltite. This mineral occurs chiefly in the upper levels, where its occurrence may be regarded as a primary depth-zone of the siderite lodes. It rarely occurs in solid masses but rather as a cloudy impregnation throughout younger quartz.

Finally, to complete the description, the quartz lodes must be mentioned. These may be older, in which case they approach the siderite lodes in age, having to a great extent resulted from these by metasomatism; or they may be considerably younger, in which case they are often characterized by carrying copper.

Bornhardt divided the fissure-fillings of Siegerland according to their genesis and age and beginning from the oldest, into :

1. Siderite filling.
2. Main quartz filling.
3. Lead and zinc filling.
4. Older copper filling.
5. Younger copper filling.
6. Cobalt filling.

Since the minerals occurring in the Siegerland lodes may, in consequence of subsequent metasomatic replacement, be of varied age, several of these fillings may be represented in one and the same lode. In this respect this grouping by Bornhardt marks a distinct step in advance of the lode subdivisions of earlier authorities.

The siderite filling took place in Middle Devonian time. To it belong, in addition, the primary pyrite scattered throughout the mass; a small portion of the quartz, as far as this is primary; and some uncommon minerals. The bulk of the quartz, chalcopyrite, and galena, the sphalerite and the cobalt minerals, on the other hand, were subsequently introduced and belong therefore to younger fillings. The main quartz filling, to which the silicification of many of the siderite lodes must be ascribed, is however also probably of Devonian age. Such silicated solutions used not only the older siderite lodes as channels but also any fissures unoccupied by ore, pure quartz lodes thereby arising.

The geological age of the lead-zinc filling cannot be determined with certainty. Without exception however, it may be said that the galena is younger than the sphalerite, so that the lead-zinc filling might be subdivided into two generations. Galena and sphalerite have not only replaced siderite but in many cases quartz also. Lodes therefore formed at the siderite deposition may have been changed to quartz lodes, and these again to lead-zinc lodes. The older copper filling took place within the lodes of the three earlier fillings. Its principal minerals, chalcopyrite and tetrahedrite, are replacements, particularly of siderite and quartz. The younger copper filling is remarkable in that it forms independent lodes which cross those of the other fillings, occasionally with well-defined walls. We consider that the independence of the older copper deposition is doubtful since it may well be explained by the action of cementation processes upon the low copper content in the siderite or other lodes.

The cobalt filling is certainly younger than the siderite- and main quartz filling, though its relation to the others has not yet been determined. The cobaltite occurs as an impregnation, seldom in siderite but more often in the main quartz and in slaty country-rock. Being found chiefly in the upper levels, Bornhardt was of opinion that its presence was the result of secondary concentration processes.

We consider it would be of great interest to determine the relation between the cobalt- and copper fillings. At Dobschau, the siderite district in the Carpathians which so greatly resembles Siegerland, the copper- and cobalt minerals form well-defined depth-zones, the copper zone being higher than that of the cobalt.

The nickel minerals in their occurrence at Siegerland differ, according to Bornhardt, from those of cobalt, in that they are not concentrated in the upper levels but occur as mineralogical curiosities impregnated in siderite at various horizons and in various districts.| Nickel-stibnite and gersdorffite have in this connection shown themselves to be contemporaneous with the siderite. The formulation of an independent nickel filling is not therefore possible. In the relation of its cobalt- and nickel minerals Siegerland differs not immaterially from the other sulph-arsenide districts containing nickel and cobalt, where these two metals are so disposed that no sharp line between their occurrence may be drawn. Limonite, specularite, and hæmatite, represent secondary alteration products of the siderite, at the surface.

Concerning structure, the siderite exhibits simple granular structure, net structure, and porphyritic or irregular structure, while ordinary and concentric crusted structures are often seen.

The determination of the age-relation of the Siegerland siderite lodes to the eruptives which cross them, is of great interest. Not only is it the case that in many places the siderite under the influence, or by the action of the Tertiary basalt has been changed to magnetite, but, as in the Glaskopf mine at Biersdorf, the late Devonian diabase has had the same effect. On the other hand, no contact effects of the Lower ·Devonian porphyry are known.

The alteration of the siderite to limonite, that is, the formation of gossan, has in different lodes taken place to very different depths. Cases occur on the one hand where the siderite is found close under the surface, and on the other where the limonite extends considerably below the valley level. In explanation of these differences, not only must the local tectonics be considered but also the variable action of erosion from place to place.

The alteration of siderite to specularite and hæmatite is of particular interest. According to Bornhardt this alteration was direct and took place in water-filled cavities by reagents which arrived there from surface. In this connection he endorses the view of Hornung, according to which, concentrated saline solutions formed on surface during periods of great dryness, in consequence of their high specific gravity, sank along fractures or crevices where, by virtue of contained atmospheric oxygen, they exerted an oxidizing effect. Such saline solutions would for instance form upon the upper Rotliegendes. Krusch, on the contrary, considers the assumption of gelatinous solutions first put forward by Wölbling as being more probable.[1]

Pyrite occurs in the Siegerland lodes both within the ordinary filling as well as in separate veins. Like the other sulphides it appears to prefer the

[1] Bornhardt, *Rotspat und Eisenoxyd*, p. 476.

crushed or pinched parts of the lode, or those where inclusions of country-rock are numerous. It occurs generally in granules bounded by crystal faces, such granules being either aggregated loosely or in compact masses. The pyrite has generally resisted the alteration which has changed the siderite to quartz or chalcopyrite, so that it is found enveloped in these minerals. On the other hand, however, it is seldom seen in sphalerite or galena. Especially fine crystals of pyrite are found in the Heinrichssegen mine.

The varying composition of siderite as it occurs in the ore-deposits of this region may be gathered from the analyses given in the tabulated statement on page 804, while on an average this ore contains the following percentages of the more important constituents :

Iron	.	.	.	37·78	per cent.
Manganese	.	.	.	7·16	„
Lime	.	.	.	0·19	„
Magnesia	.	.	.	2·18	„

The phosphorus content is low, generally amounting to 0·001–0·3 per cent. The copper content fluctuates between traces and 0·6 per cent, and on an average may be taken to be 0·15 per cent. Roasted siderite contains 45·4–51·4 per cent of iron, 6·4–11·2 per cent of manganese, 0·10–0·64 per cent of copper, and 6·38–18·39 per cent of insoluble residue, or on an average :

Iron	.	.	.	48·22	per cent.
Manganese	.	.	.	9·30	„
Copper	.	.	.	0·22	„
Insoluble Residue	.	.	12·35	„	

The importance of iron mining in Siegerland may be gathered from the following brief statement. The industry has been important since the middle of last century. Its development has greatly depended upon the advances in iron metallurgy. The importation of high-grade foreign ores into Germany, which began in the 'sixties, affected it but little, owing to the favourable composition of the siderite. Ten years later, however, the discovery of the Thomas process and the wonderful development of the minette deposits consequent thereupon, had an effect all the more distressing because of the absence of distress hitherto. Prices fell and the mines fell into sore straits. The first distress-tariff was granted in 1886, after which in 1902 followed further favours, till finally in 1911, as the measure of greatest relief, the right to transport the ore to the Upper Silesian furnaces was granted.

Production has been as follows : in 1875 about 720,000 tons ; 1880 about 1,100,000 tons ; 1885 about 1,160,000 tons ; 1890 about 1,500,000 tons ; 1895 about 1,666,000 tons ; 1900 about 1,800,000 tons ; 1905 about

PERCENTAGE ANALYSES OF RAW SIDERITE.

No.	Name of Mine.	FeO.	MnO.	CaO.	MgO.	CO₂.	Insoluble Residue.	Fe.	Mn.	Fe + Mn.	Authority.
	Mining District Müsen										
8	Glanzenberg	43·23	7·69	1·38	5·66	38·26	2·42	33·64	5·96	39·60	Dr. Schwarz, Kgl. Berg-akademie, Berlin
*9	Kuhlenberg Series	45·10	8·74	1·68	3·78	...	(9·52)	35·08	6·77	41·85	Management of Mine; average of 5 analyses
10a	Stahlberg, old ore-body	47·03	10·61	0·51	3·24	39·27	...	36·68	8·22	44·90	Schnabel
10b	Stahlberg, old ore-body	44·79	10·53	0·75	2·74	...	1·08	35·95	8·16	44·11	Fresenius
*10c	Stahlberg, new ore-body	44·72	9·43	1·66	2·94	...	(6·76)	37·11	7·25	44·36	Management of Mine
	Mining District Siegen										
11	Häuslingstiefe	50·37	8·30	0·25	2·15	38·48	0·45	39·20	6·42	45·62	Schnabel
12	Jungo Kessel mine	50·72	7·64	0·40	1·48	38·90	0·48	39·40	5·90	45·30	} Karsten
13	Kirschenbaum	47·20	8·34	0·63	3·75	38·85	0·95	36·70	6·40	43·10	
14	Kammer und Storch	48·69	9·38	...	0·93	36·56	4·44	39·76	7·63	47·39	} Schnabel
15	Alte Thalsbach	48·79	9·66	0·36	1·25	37·43	2·51	37·95	7·48	45·43	
	Mining District Burbach										
*16	Pfannenberger Einigkeit	50·56	9·04	0·74	0·88	...	(8·74)	39·34	7·01	46·35	Heufelder
18	Peterszeche	49·62	9·47	1·10	1·29	38·37	...	38·60	7·28	45·88	Management of Mine
	Mining District Daaden-Kirchen										
20	Stahlert	48·86	8·19	0·32	2·34	37·74	2·54	38·00	6·35	44·35	Schnabel
	Mining District Wied										
26	Eisenhardt	49·37	8·00	0·72	3·20	37·10	0·11	38·40	6·20	44·60	Description by Hamm
27	St. Andreas	46·68	9·87	0·35	3·91	39·19	...	36·31	7·65	43·96	Schnabel
29	Petersbach	49·63	7·36	1·10	2·50	38·20	1·70	38·60	5·70	44·30	Description by Hamm
31	Girmscheid	46·57	9·55	0·60	2·55	...	0·50	36·22	7·40	43·60	
32	Louise	49·54	9·98	0·60	...	38·32	0·32	38·51	7·73	46·24	} Management of Mines
33	Org	49·86	8·60	0·71	1·72	...	0·42	38·75	6·67	45·42	
34	Lammerichskaule	48·91	8·66	0·32	1·94	37·62	1·14	37·84	6·71	44·55	Schnabel

* In these analyses where the residue amounted to more than 5 per cent the percentages were calculated as though no residue were contained, in order to obtain figures more comparable with the others. In these cases the residue percentage is given in brackets.

1,940,000 tons ; and 1907 about 2,360,000 tons. In 1909, according to official figures, the production amounted to 2,075,321 tons containing 34·9 per cent of iron ; and in 1910 to 2,281,000 tons containing 35·1 per cent.

The Siderite Lodes of Zips-Gömör in the Hungarian Erzgebirge

LITERATURE

L. Zeuschner. 'Gangverhältnisse bei Kotterbach und Poracs,' K. Akad. der Wiss., Vienna, 1853, 1855.—F. v. Andrian. Bericht über die Übersichtsaufnahmen im Zipser und Gömörer-Comitat während des Sommers 1858 ; Jahrb. der k. k. geol. Reichsanst., 1859 ; k. k. geol. Reichsanst., 1859, pp. 20, 39, and 79 ; 1867, p. 257 ; 1868, p. 55 ; k. k. geol. Reichsanst., Jan. 22, 1859.—Fr. R. v. Hauer. 'Geological Survey Map of the Austrian Monarchy,' text on section 3, Jahrb. der k. k. geol. Reichsanst., 1869, Vol. XIX.— D. Stur. 'Bericht über die geologische Aufnahme der Umgebungen von Schmöllnitz und Göllnitz,' Jahrb. der k. k. geol. Reichsanst., 1869, Vol. XIX.—G. v. Rath. Bericht über eine nach Ungarn unternommene Reise, Sitzungsber. der niederrheinischen Gesellschaft für Natur- und Heilkunde. Bonn, 1876.—L. Maderspach. 'Der Bergbau von Zsakarócz in der Zips,' Öster. Zeit. f. Berg- und Hüttenwesen, 1876, p. 175.—S. Roth. Der Jeckelsdorfer und Dobschauer Diallag-Serpentin. Földtani Közlöny, 1881, p. 144.—A. v. Groddeck. 'Über die Gesteine der Bindt in Oberungarn,' Jahrb. der k. k. geol. Reichsanst., 1885 ; 'Über Lagergänge,' Berg- u. Hüttenm. Ztg., 1885.—A. Schmidt. 'Mitteilungen über ungarische Mineralvorkommen,' Zeit. f. Kristall. 1886.—R. Helmhacker. 'Die Bergbaue von Slovinka und Göllnitz in Ungarn,' Berg- u. Hüttenm. Ztg., 1895.—F. W. Voit. 'Geognostische Schilderung der Lagerstättenverhältnisse von Dobschau in Ungarn,' Jahrb. der geol. Reichsanst., 1900.—V. Uhlig. Bau und Bild der Karpathen im Bau und Bild Österreichs. Vienna and Leipzig, 1903.—W. Viebig. 'Der Spateisensteinbergbau des Zipser Erzgebirges in Oberungarn,' Glückauf 42, No. 1, 1906.—W. Bartels. 'Die Spateisensteinlagerstätten des Zipser Comitates in Oberungarn,' Archiv für Lagerstättenforschung, Book 5. Berlin, 1910.

This district geologically and stratigraphically constitutes in itself a separate section of the West Carpathians, in which the following sequence of beds is represented :

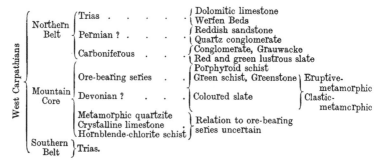

The country-rock of the lodes consists of Devonian schists, quartzites,

and foliated eruptives, distributed in a series of zones among which the
following from hanging-wall to foot-wall are the most noticeable :

1. Sericitic, graphitic, and phyllitic schists.
2. Chloritic and quartzitic schists, and clay-slates.
3. Micaceous schists.
4. Green schists or foliated diorite.

The schist zones, striking east-west, enclose a large number of the
siderite deposits, these being chiefly in the form of bedded lodes which
likewise strike east-west. In the neighbourhood of the lodes the so-called
green schists are altered to a compact rock known as greenstone, which no
longer retains any schistose structure. The Carboniferous beds occurring
in the hanging-wall unconformably to these schists, are likewise petrographically of very varied composition, conglomerates, clay-slates, and
arenaceous slates being the most frequent. At the contact between
the Devonian and the Carboniferous the Kotterbach and Bindt deposits
occur. The later formations, the Permian and the Trias, are not concerned
in the question of these ore-deposits.

Among the eruptive rocks, diorite and serpentine, in irregular bosses,
are particularly important. The porphyroid schist intercalated between
the other schists likewise consists of eruptive material. Most of the
lodes are found within an area 70 km. long and 30–40 km. wide, which,
consisting of green schists, greenstone, and metamorphic rocks, extends
from Dobschau parallel to the northern border of the Erzgebirge, to the
neighbourhood of Kaschau. To the west these rocks are bounded by the
Kohut granite massive, while to the east they become impoverished in the
neighbourhood of Kaschau. Around the boundaries of this metalliferous
region the later sediments occur.

The lodes may be divided into several groups, of which the first includes
the occurrences at Zsakarocz, Krompach, Kotterbach, Bindt, and Rostoken.
The second group, embracing the Zips lodes, occurs around Göllnitz,
Prakendorf, Helczmanocz, Einsiedel, and Slowinka, in a district 40 km.
long and 15 km. wide.

In the whole circumstances of their occurrence, except that they are
bedded lodes, these lodes resemble those of Siegerland. Their lode character
however is indicated by numerous junctions, definite walls, small cross
veins, and large enclosed fragments of country-rock. In greater part
they strike east-west and dip 60°–80° to the south. They have been
affected by the numerous disturbances to which the beds in which they
occur have been subjected, so that faults are common, these often producing
such a dismemberment of the deposit that the underground workings
constitute a regular maze.

The filling consists chiefly of siderite, quartz, calcite, and barite; while sulphides, such as tetrahedrite, chalcopyrite, galena, sphalerite, and arsenopyrite, are more uncommon.   Pyrite occurs both in the lode and in the country-rock.   The lode width varies considerably; it is however seldom more than a few metres, widths of 30 m. being only attained when thicknesses of country-rock are included.   The structure is often granular. The siderite is sometimes coarse- and sometimes fine-grained, while its colour varies between typical pea-yellow and almost white; it usually contains 36–38 per cent of iron, and near the surface is altered to limonite

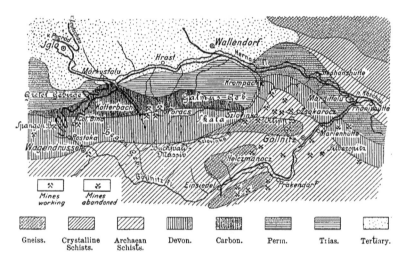

FIG. 360.—General map of the siderite deposits of the Upper Hungarian Erzgebirge.
Scale, 1 : 400,000.

or iron-ochre.   Since in this oxidation zone cinnabar is often met, it is probable that quicksilver-tetrahedrite occurs in greater depth.   Primary depth-zones are observed in so far that while in depth the carbonates increase, the sulphides, on the other hand, decrease.

In regard to copper, it is noteworthy that a secondary enrichment has often taken place, this, as at Dobschau, taking the form of tetrahedrite. Cobalt and nickel, which in Siegerland were fairly regularly distributed throughout, are, in spite of the resemblance between the lodes of the two districts, only found in this district at Dobschau, where they constitute a primary depth-zone below the copper.

The composition of the siderite of this district may be gathered from the following analysis from Rostoken :

Iron	.	.	.	.	34·00–38·00 per cent.
Manganese	.			.	1·50 ,,
Copper		.	.	.	0·13–0·15 · ,,
Calcite	.	.	.	.	1·00 ,,
Alumina	.		.	.	0·50 ,,
Insoluble residue		.	.	.	4–9·00 ,,
Loss on ignition		.	.	.	30·00 ,,

The composition of the limonite may be seen from the following analyses from Göllnitz :

	Hilfe-Gottes Mine.	Ottokar Mine.
	Per cent.	Per cent.
Moisture	1·96	5·36
Iron, moist ore . . .	51·07	45·67
,, dry ore . . .	52·08	48·26
,, roasted ore . . .	57·01	53·11
Loss on ignition . . .	9·60	9·14
Gangue	11·69	16·44
Manganese	2·06	2·18
Phosphorus	0·02	0·05
Sulphur	0·04	0·08
Copper	0·33	0·44
Antimony	0·16	0·01
Arsenic	0·026	...
Lead	...	...
Barium	...	0·46

In regard to genesis also, there exists a striking agreement between these lodes and those of Siegerland. The lodes were formed in early Devonian time by solutions which without doubt were associated with eruptive rocks. By many authors the presence of tourmaline—which increases in amount from Kotterbach through Bindt to Rostoken, while those lodes lying to the east of Göllnitz contain practically none—is regarded as important in regard to the question of genesis. According to present-day knowledge however, tourmaline has a wide distribution in lodes and is no longer considered one of the less common minerals. In addition, the tourmaline in this district occurs indifferently in the ordinary and in the bedded lodes. Bartels has observed that the quartz and tourmaline are older minerals, while all the others belong to a younger generation. The tourmaline is intergrown with the quartz in the most intimate manner, its crystals sometimes protruding from the quartz into the siderite. In addition, drusy cavities lined with tourmaline occur in quartz only and not in siderite.

According to observations made by Krusch at Dobschau, there is, in addition to this older quartz, a younger generation of that mineral, which replaces siderite. The presence of this is doubtless due to a re-opening of the fissure, in which also the conditions here would further resemble those in Siegerland. According to Krusch, in this district also, the main portion of the sulphides is younger than the siderite. Although it is assumed by earlier authorities that the lodes are genetically associated with the known occurrence of greenstone in the district, there is no evidence as to what minerals are contemporaneous with this greenstone, though in the face of an apparent re-opening and re-mineralization all cannot be contemporaneous.

The Upper Hungarian iron ores when roasted are used in great quantities in Upper Silesia. They are poor in manganese, containing but 1·5–2 per cent ; when raw they contain 0·01–0·2 per cent of phosphorus and 0·4–1 per cent of copper, this amount of copper rendering them a little difficult to treat.

Of the total iron ore production of Hungary, which amounts to about 1,700,000 tons, Upper Hungary is responsible for 60–70 per cent, or about 1,000,000 tons. The importance of the individual mines may be gathered from the table on page 810.

In addition to these economically important siderite districts there are many similar occurrences which to-day are unimportant. Among these are those at Lobenstein and Leubetha near Oelsnitz and Röttis respectively. These deposits, which occur in Palæozoic beds, carry siderite and, in part, some copper and nickel.

LITERATURE

A. BREITHAUPT. Paragenesis. Freiberg, 1849.—R. BECK. Explanatory Text, Section Adorf, Geol. Spezialkarte of Saxony, 1884.—E. WEISE. Explanatory Text, Section Plauen-Pausa, 1904.—E. ZIMMERMANN. Section Lobenstein, Geol. Spezialkarte of Prussia, etc.

In the Salzburg hills not far from the well-known copper deposit of Mitterberg near Bischofshofen, a large number of siderite lodes were formerly worked, especially in the Middle Ages. Here also the lodes traverse Palæozoic beds and carry some copper. The passage from siderite deposits to copper deposits carrying siderite is so gradual that it cannot always be said with certainty whether any particular old working was worked for siderite and stopped because of the increase in copper, or whether copper was the metal sought and the work stopped because in depth the proportion of siderite increased.[1]

HAEMATITE LODES

LITERATURE

K. ERMISCH. ' Die Knollengrube bei Lauterberg am Harz,' Zeit. f. prakt. Geol., 1904, p. 160.—H. CREDNER. ' Geogn. Beschr. des Bergwerkdistriktes von St. Andreasberg,' Zeit. d. d. geol. Ges., 1865, XVII.—K. DALMER. Explanatory Text, Section Plmitz-Ebersbrunn, Geol. Spezialkarte of Saxony, 1885.—H. MÜLLER. Die Eisenerzlagerstätten

---

[1] Postea, p. 905.

THE MOST IMPORTANT IRON PRODUCERS OF THE UPPER HUNGARIAN ERZGEBIRGE WITH PRODUCTION FROM 1905.

No.	Owner.	Position of Mine.	Size of Mining Area in Sq. m.	Production in Tons.	Value.
1.	Upper Silesian Eisenbahnbedarfs Railway Plant Company in Friedenshütto-Morgenroth near Beuthen, Upper Silesia	Rostoken * and Zavadka	7,740,027	12,549 raw ore 48,759 roasted ore 1,046 manganese ore	£2,477 19 2 £21,511 3 4 £374 19 2
2.	Witkowitz Mining and Ironworks Company in Witkowitz, Mähren	Kotterbach *	2,890,291	Raw production 129,197 6,258 raw ore 88,178 roasted ore 3,526 roasted tetrahedrite 71 barite 45 quicksilver ore	... ...
3.	Austrian Mining and Smelting Company, Teschen	Zeacarocs *	7,180,421	144,611 raw ore 114,598 roasted ore	£31,034 12 6 £53,192 10 10
4.	Upper Silesian Mining and Smelting Company in Gleiwitz, Upper Silesia	Bindt * Graetl.* Meseny *	3,925,126 ...	38,665 ...	£8,860 16 8 ...
5.	Hernadtal Hungarian Iron Company under technical and administrative direction of the Rima-Murany-Salgo-Tarjan Iron Company	1. Luczeyabánya † 2. Slovinka * 3. Vashegy † 4. Rákos † 5. Rozsnyó † 6. Krompach *	... ...	76,255 raw ore 55,000 ,, 89,480 ,, 55,436 ,, 51,031 ,, 37,400 ,,	... ...

* = Komitat Zips.   † = Komitat Gömör.

des obern Erzgebirges des Voigtlandes, 1856.—H. V. OPPE. 'Die Zinn- und Eisenerzgänge der Eibenstocker Granitpartie und deren Umgebung,' Cottas Gangstudien, 1854, II.

Although hæmatite lodes occur in many districts it is in but few places that they constitute payable deposits.

The Knollen mine at Lauterberg in the Harz, where for a time a hæmatite lode associated with copper- and barite lodes was worked, has a certain reputation. According to Ermisch the country-rock at that mine consists of the supposedly Culm Tanne grauwacke. The lode strikes east-south-east to south-east and dips 80° south-west. The width, which though on an average 1 m. may reach 4 metres, consists of country-rock and hæmatite, while barite and lithomarge are less common. The absence of sulphides and the rare occurrence of quartz are characteristic features of this lode. The hæmatite occurs chiefly in the form of kidney ore containing 96–99 per cent of ferric oxide, while the compact hæmatite occurring more particularly in the neighbourhood of the outcrop, contains 91 per cent of that oxide.

At St. Andreasberg in the Harz the lode-filling of the different lodes varies according to whether the lode occurs between, or to the north or south of the two boundary faults.[1] While the lodes within the wedge of ground pointing westwards contain silver chiefly, those outside are distinguished by containing hæmatite. According to H. Credner the most important iron deposits at St. Andreasberg occur with variable strike and width, either in the granite or at the contact of that rock with hornfels.

Other hæmatite lodes are known in the Saxon Vogtland, between Stenn to the south-west of Zwickau, and Christgrün. These, containing hæmatite and limonite, occur at the contact of diabase and clay-slate, or between limestone and decomposed Lower Silurian diabase. There are hæmatite lodes also in the Saxon Erzgebirge, in the neighbourhood of Schwarzenberg. The lodes there occur either within the granite or at the contact of granite and schist; those occurring at the contact have already been mentioned.[2] The occurrence of copper ore with some of these lodes is noteworthy, the lodes otherwise carrying compact fibrous hæmatite or kidney ore.

Finally, the following occurrences deserve mention: the hæmatite lodes at Johanngeorgenstadt, Platten, Schellerhau; the numerous small iron lodes at Suhl in the Thuringian Forest upon which the armament industry in that neighbourhood was started; and similar occurrences at Zorge in the Harz and at Gleissingerfels in the Fichtelgebirge.

It is of interest to note that there exists a relation between the hæmatite lodes and those of manganese afterwards to be described, in that these latter in depth often gradually merge into hæmatite lodes.[3]

[1] Ante, p. 688.   [2] Ante, p. 352.   [3] Postea, p. 851.

THE METASOMATIC IRON DEPOSITS

As already stated more than once, lodes are closely related to meta-somatic deposits. Deposits of this latter class are in the case of iron of special importance.

In addition, however, there are other deposits occupying an inter-mediate stage between the two, in which though the lode character may be apparent, metasomatism has already attained a certain importance. Such deposits are found for example at Toroczko in Transylvania where the lode fissure may still plainly be recognized and the deposit on the whole gives the impression of a lode-like occurrence, though along the fissure a portion of the limestone has been altered to ore. Apart from these intermediate occurrences there are other numerous well-defined metasomatic iron deposits which are distinctly bedded in character. Since with most metasomatic iron deposits it is a question of the relatively low-priced siderite—though this may be more or less completely altered to the more valuable limonite or hæmatite—there are not many districts where such deposits are worked. Payable deposits naturally occur more frequently in those countries where the means of communication are best developed. The formation of these deposits was fully discussed in the first volume. In form they differ from the analogous lead-zinc deposits in that the alteration of the limestone and dolomite—which in the great majority of cases is what takes place—is generally complete, and accordingly in many cases the shape of the deposit coincides with that of the original bed.

Not infrequently slate becomes thus altered to limonite, though experience shows that such occurrences have no economic importance; at shallow depths the deposit gives way to a ferruginous rock and this in turn to ordinary slate. This type of deposit is known as the Hunsrück type.

In the case of such deposits as have resulted from limestone and dolomite, when the alteration has been complete or secondary migration of the metals has taken place, the channels of access as well as the replace-ment fissures are difficult to recognize. When however the alteration is not complete the intensity of the replacement is seen to diminish with distance from the fissure. Near the fissure iron ore is found, then ferruginous limestone or dolomite, and finally unaltered rock. The deposit has then as a rule the form of a lens, the greatest section of which is along the fissure, and the tapering ends farthest from the fissure.

The extent of the metasomatic iron deposits upon the surface as well as in depth is dependent upon the extension of the original rock and upon the solutions brought to that rock. The greater the volume of these latter in relation to the mass of the original rock, the more completely does the form of the deposit coincide with that of that rock. The ore-bodies accordingly are horizontal where the bedding is undisturbed, and folded to anticlines and synclines or disturbed by faults or overthrusts when the beds in which they lie have been so folded or disturbed. The depth to which a metasomatic iron deposit reaches depends therefore entirely upon the disturbed or undisturbed bedding of the original rock. The distribution of the ore in the deposit may be regular, as for instance at the Hüggel; [1] or it may be extremely irregular, as when the entire bedded complex is shattered and not all the limestone or dolomite has been altered. An irregular distribution may also arise when in the alteration of the limestone not only one mineral, such as siderite, was deposited, but at the same time a carbonate such as ankerite, or the equivalent calcium-iron carbonate. At those points where such occurred the deposit would be unpayable and the regularity of the deposit broken.

The principal minerals found in metasomatic iron deposits are siderite and limonite, at the formation of which calcium carbonate or calcium-magnesium carbonate became removed. Less frequently sulphides such as pyrite and chalcopyrite, formed at the same time, are present. The gangue consists chiefly of carbonates, most of which have a composition intermediate between that of the ore and the original rock. Should the limestone not be completely replaced by ore it remains as the gangue in which the ore is embedded.

The structure of the ore in these deposits is either crusted or irregular. At the boundaries of the deposit a pseudo-brecciated structure is often seen, where kernels of unaltered limestone occur between the fractures from which the alteration proceeded. The frequent occurrence of crusted structure has been the cause of divergence of opinion concerning the genesis of several occurrences, which by some are regarded as true sedimentary beds and by others as having been formed by metasomatism. With such crusted structure the general occurrence of cavities parallel to the crusts is remarkable, this being particularly noticeable for instance at the Hüggel.[2]

---

[1] *Postea,* p. 841.   [2] *Postea,* p. 844.

The composition of the ore varies according as hæmatite, limonite, or siderite, are present. Careful determinations of the average content of the most important elements are available, among others, in respect to the Nassau hæmatites, where however, it must be conceded, true sedimentary occurrences exceed the metasomatic. Here during 1910 about 1,004,000 tons were produced containing on an average 40·9 per cent of iron, almost the whole of this tonnage being ready for smelting without further preparation. This average figure was rendered somewhat low by the inclusion of some partly altered material, which was valuable as ferruginous flux. The phosphorus content of these hæmatite ores is an important factor. Most of the above total, almost 800,000 tons in fact, contained 0·05–0·75 per cent of phosphorus; about 118,000 tons contained still less; while only a small proportion, some 900 tons, containing more than 1 per cent, belonged to the high phosphorus ores. Accordingly, in general the hæmatite of the metasomatic deposits belongs to the low phosphorus ores.

With regard to siderite, in Germany during 1910, including the Osnabrück district, the Schafberg, and the Hüggel, 261,461 tons containing 28·1 per cent of iron, were won, the whole of which could be smelted at once. In this ore the phosphorus content varied between 0 and 1 per cent, the larger portion containing less than 0·06 per cent.

The metasomatic limonite of Germany is derived from the Saxon-Thuringia, the Nassau-Oberhesse, the Taunus, and the Vogelsberg districts, wherein, although deposits of other genesis are worked, the bulk of the production comes from the deposits in question.

With these deposits the manganese content is an important factor, some of them actually yielding ore containing approximately 20 per cent of iron and 20 per cent of manganese. Such ore formerly, for want of fixity in classification, was sometimes classed with the iron ores and sometimes with the manganese ores. Now, however, the term iron-manganese ores has become generally accepted for them.

According to the manganese content the ore from these deposits may be divided into :

(a) Limonite, with less than 12 per cent of manganese.
(b) Iron-manganese ore, with 12–30 per cent of manganese.
(c) Manganese ore, with more than 30 per cent of manganese.

These divisions naturally merge into one another. Only the first two are considered here, the last being described when dealing with manganese ores.[1] Of the total amount of these ores produced in Germany during 1910—this amount being approximately 2,900,000 tons—the greater proportion by far, namely 2,600,000 tons, belongs to the first

[1] *Postea*, pp. 851, 863.

division containing less than 12 per cent of manganese ; somewhat more than one quarter to the second ; while less than 200 tons may be classed as manganese ore.

In determining the market value of these ores the manganese content plays an important part. Generally it may be assumed that 1·5 per cent of manganese becomes lost in smelting, while the remainder is allowed for at the rate of 2 per cent of iron for each per cent of manganese. The ore containing less than 12 per cent of manganese contains on an average 31·6 per cent of iron, while the iron-manganese ore contains 24 per cent of iron and 20 per cent of manganese, on an average.

The second important factor is the phosphorus content, which, like that of manganese, fluctuates considerably. Approximately 746,000 tons of the production in 1910 contained less than 0·05 per cent ; about 863,000 tons contained 0·05–0·75 per cent ; about 128,000 tons, 0·75–1 per cent ; and about 896,000 tons, more than 1 per cent.

The metasomatic limonite deposits are particularly interesting because being of great geological age they have been subject to much subsequent disintegration and alteration. There are in consequence many deposits which, at different stages of exposure by mining operations, might be regarded as either metasomatic or fragmentary deposits. Such is for example the case with the Lindener Mark near Giessen and with many of the occurrences in the Bingerbrück limestone, in which cases Middle Devonian massive limestone was first changed to ore which in turn became disintegrated either by running water or by water circulating in fissures. With some deposits of this nature it cannot be determined whether the limestone was changed directly to limonite, or was first changed to siderite and then by subsequent oxidation to limonite.

Since these deposits are usually of little thickness the opportunity to display primary depth-zones is limited. Not all the limestone beds however are equally suited to this alteration, and in consequence, in a section normal to the bedding, ore often alternates with limestone. Even when the whole of the limestone has been altered it need not necessarily follow that all the component layers have been altered to siderite ; some according to their character may have been altered to siderite, others to ankerite, etc.

The secondary depth-zones are particularly important. In those cases where the limestone was altered directly to limonite no opportunity for subsequent migration of the metal content existed ; but where siderite was first formed such migration would be brought about near the surface by atmospheric agencies in the manner indicated when describing the siderite lodes,[1] and limonite would result.

[1] *Ante*, p. 789.

In this weathering, as Krusch has pointed out, oxidation-metasomatism takes part.[1]

The meteoric waters, which dissolve the iron carbonate doubtless as bicarbonate, sink into the limestone which in the presence of oxygen they change to limonite. The migration of the iron therefore proceeds from above, downwards, so that the thickness of the deposit in the oxidation zone may be considerably greater than in the primary zone. Within a limited vertical measurement in the oxidation zone, by the processes of oxidation and the allied processes of oxidation-metasomatism, an amount of ore may be accumulated which formerly belonged to a much greater vertical extent—in greater part now eroded—of the primary deposit. When therefore one of these deposits appears at surface as a gossan, great care must be exercised in deducing from the chemical and dimensional properties of this gossan any estimates relative to possible content, width, nature, and quantity of ore, in depth.

Where the surface decomposition is incomplete, as for instance at Kamsdorf, its actual occurrence may be demonstrated. Siderite and limonite are then found together, the former when of metasomatic origin often having a different character from that which it has when occurring in lodes. It is finely crystalline and displays a certain scintillation, in consequence of which by the Kamsdorf miners it is aptly termed mica. At Bilbao, however, where also the existence of secondary oxidation is apparent, the siderite is as a rule coarsely crystalline. The passage of siderite to limonite may often be seen in hand specimens, consisting of a kernel of siderite representing the original ore, and an envelope of limonite.

Oxidation is not only apparent in its effects upon pure siderite, but also upon limestone which has only been partly replaced by siderite. When this occurs a limestone saturated with limonite is formed, which in many districts is appropriately known as iron-limestone.

Where small amounts of primary copper- and silver ores were deposited with the iron, a narrow cementation zone becomes formed below the gossan, wherein the copper and silver are concentrated, but not the iron.

For the reasons given when describing the siderite lodes,[2] payable metasomatic iron deposits are not numerous. In determining the question of payability fuel is often an important factor, since only where the roasting of the siderite to the oxide is possible can any considerable transport of the ore be considered.

In countries where wages are high these deposits cannot as a rule be worked. In Europe, on the other hand, large deposits of this genesis are worked at Bilbao on the north-east coast of Spain, and at Erzberg

---

[1] 'Primäre und metasomatische Prozesse auf Erzlagerstätten,' Zeit. f. prakt. Geol., 1910.   [2] Ante, p. 786.

near Eisenerz in Styria ; while smaller occurrences are worked in the Thuringian Forest, at the Hüggel in Westphalia, and at many other places. The connection of these deposits with large fissure-systems through which the alterative solutions circulated, is invariable. Thus, the metasomatic iron deposits of the Stahlberg and the Mommel near Schmalkaden, are connected with the disturbed zone in the south of the Thuringian Forest ; while the occurrences at Kamsdorf are in connection with the northern boundary fissure of that forest.

The size of these deposits varies. The deposits at Bilbao and that at Erzberg are of large dimension, while the occurrences in the Thuringian Forest and at the Hüggel are smaller. The metasomatic Devonian hæmatite in Nassau, which is partly of metasomatic origin, must be reckoned among those of medium size.

The output from some of these mines is variable because around the boundaries of the ore-body there are masses of poorer material which, though they contain too little iron to be regarded as ore, serve well as ferruginous flux. Some of this material however, in consequence of the irregularity of the ore-body, is often reckoned with the ore.·

ERZBERG NEAR EISENERZ, STYRIA

LITERATURE

V. UHLICH. The Iron Ore Resources of the World, XI. Intern. Geol. Congress. Stockholm, 1910.—A. MILLER v. HAUENFELS. ' Die steiermärkischen Bergbaue,' special extract from Ein treues Bild des Herzogtums Steiermark, p. 14. Vienna, 1859.—D. STUR. ' Vorkommen obersilurischer Petrefakte am Erzberg u.s.w.' Jahrb. der k. k. geol. Reichsanst., 1865, p. 267.—F. v. Hauer. ' Geologie der österr.-ungar. Monarchie, 1875, p. 223.—M. VAČEK. ' Über den geologischen Bau der Zentralalpen zwischen Enns und Mur,' Verhandl. d. k. k. geol. Reichsanst., 1886, p. 71 ; ' Skizze eines geologischen Profils durch den steierischen Erzberg,' Jahrb. der k. k. geol. Reichsanst., 1900, p. 23.—M. VAČEK and E. SEDLACEK. ' Der steierische Erzberg,' Guide to IX. Intern. Geol. Congress. Vienna, 1903, V. p. 1.— F. CORNU and K. A. REDLICH. ' Notizen über einige Mineralvorkommen der Ostalpen,' Zentralbl. f. Min. Geol. und Petrogr., 1908, p. 277. Stuttgart, 1908.—K. A. REDLICH. ' Die Genesis der Pinolitmagnesite, Siderite und Ankerite der Ostalpen,' Tscherm. min. und petr. Mitt., 1907, Vol. XXVI. Parts 5 and 6.

In the hanging-wall portion of the large northern grauwacke belt which traverses the Austrian Alps in an east-west direction, numerous siderite deposits occur, the strike of which within northern Styria follows the line Liezen-Eisenerz-Neuberg. Of these deposits the most important is that known as the Styrian Erzberg, the greatest part of which belongs to the *Österreichische Montangesellschaft*. In the report made by Uhlich upon this company it is stated that the occurrences at Aigen, Admont, Krummau, Johnsbach, Radmer, Donnersalpe, Tullech, Glanzberg, Polster, Gollrad, Niederalpe, Neuberg, Bohnkogel, and Altenberg, belong to the same belt. The description here will be limited to that of the Erzberg which at present is the only one working.

This occurrence lies isolated in a wide valley, the sides of which are

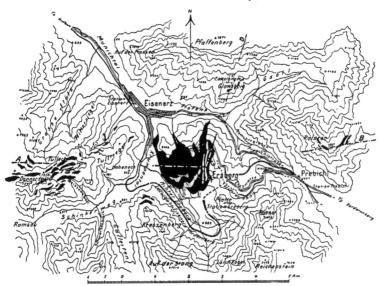

FIG. 361.—Situation plan of Erzberg near Eisenerz. The deposit is coloured black.

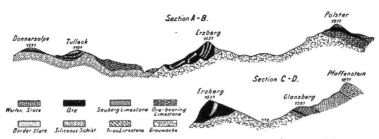

FIG. 362.—Sections of Erzberg near Eisenerz, along lines indicated in Fig. 361.

formed by towering walls of limestone, while to the south it is connected with the Reichenstein massive by an anticline known as the Breite Platte.[1]

In the upper portion of the Erzberg, within a yellow and reddish limestone—the Sauberg or ore-bearing limestone—*Crinoids* are found, indicating that this limestone belongs to the Lower Devonian.

[1] Wide plateau.

The ore-bed, as indicated in Fig. 362, forms a syncline in the foot-wall grauwacke. The youngest bed in the hanging-wall is a red, blue, and greenish-grey slate, correlated with the Werfen slates. This bed is however only found on the eastern slope of the Erzberg, having elsewhere been eroded. The ore-bed comes right to surface, its dislocated, folded, and crushed material indicating the tremendous tectonic forces to which it has been subjected. The deposit, as already mentioned, occurs trough-like

Fig. 363.—Erzberg opencut at Eisenerz. *Iron Ore Resources of the World.*

in the foot-wall grauwacke, yet in consequence of plication it reaches a height of about 730 m. with a thickness of 160–200 metres. The ore-body however does not consist exclusively of ore ; the huge ore-bearing complex consists of an alternation of ore with ankerite, limestone, and slate. The extension along the strike in the Eisenerz portion is 680 m., and in the Vordernberg portion 370 metres.

Mining at Erzberg is an old industry. The ore-body has been attacked both from the surface as well as from underground workings. The large opencut worked to-day has fifty benches, some of which are illustrated in Fig. 363.

The siderite of this deposit is almost free from sulphides ; pyrite, chalcopyrite, galena, tetrahedrite, and cinnabar, are rare occurrences. The ore as mined contains 38–40 per cent of iron, and the roasted ore up to 52 per cent. An average sample of this latter would give 44·6 per cent of iron, 2·12 per cent of manganese, 0·03 per cent of phosphorus, and 0·04 per cent of sulphur. The ankerite though at present not payable may in the future be worth consideration by the smelter, since on an average it contains 15–25 per cent of iron.

The genesis of this deposit has not yet been satisfactorily settled, in fact by some authorities quite different views are held. Bergeat, following the view first suggested by Schouppe [1] that it was a sedimentary deposit, describes it among the ore-beds, though he also reckoned with the possibility of a metasomatic origin. Beck [2] describes it among the epigenetic masses, inclining therefore, as did Redlich and many others before, to the assumption of a metasomatic replacement of limestone by siderite. Redlich was the first to remark the close relationship between the ankerite-iron occurrences of the East Alps and the pinolite-magnesite of Veitsch, this latter also representing replacement of limestone. He advocated the possibility of a close relationship between the iron deposits and the sulphide lodes of the East Alps. He was of opinion that in determining the genesis of the Erzberg, due consideration should be given to the small amount of sulphides in the siderite of this deposit. According to the authors, most of the observed facts betoken a metasomatic origin.

The total production hitherto has been as follows : from 1701 to 1800 about 3·7 million tons ; from 1801 to 1900 about 22·2 million ; and from 1901 to 1911 about 15·1 million tons ; so that including the amount produced before 1700, altogether about 42–43 million tons have been produced.[3] The present ore-reserves are estimated to be 170 million tons at Eisenerz, and 36 million tons at Vordernberg, making a total of 206 million tons. In 1910 the production was 1·7 million tons, and in 1911 about 1·8 million tons.

THE HÜTTENBERG ERZBERG

LITERATURE

M. V. LIPOLD. ' Bemerkungen über F. Münnichsdorfers Beschreibungen des Hüttenberger Erzberges,' Jahrb. der k. k. geol. Reichsanst., 1855, p. 643.—SCHAUENSTEIN. Denkbuch des österreichischen Berg- und Hüttenw., 1873, p. 204, Vienna, Minister for Agriculture. —F. SEELAND. ' Der Hüttenberger Erzberg und seine nächste Umgebung,' Jahrb. der k. k. geol. Reichsanst., 1876, p. 49.—Die Eisenerze Osterreichs und ihre Verhüttung, Minister for Agriculture, Paris Exhibition, 1878.—A. BRUNLECHNER. ' Die Form der Eisenerz-

---

[1] *Jahrb. d. k. k. geol. Reichsanst.* V., 1854.
[2] *Erzlagerstättenlehre*, p. 226.　　　　[3] *Stahl und Eisen*, 1912, I. No. 8.

lagerstätten in Hüttenberg (Kärnten),' Zeit. f. prakt. Geol., 1893, p. 301 ; Die Abstammung der Eisenerze und der Charakter ihrer Lagerstätten im nordöstlichen Kärnten, Carinthia, II., 1894, p. 47.—BRUNO BAUMGÄRTEL. ' Der Erzberg bei Hüttenberg in Kärnten,' Jahrb. der k. k. geol. Reichsanst., 1902, Vol. LII. p. 219.—The Iron Ore Resources of the World, XI., Intern. Geol. Congress, Stockholm, 1910.

In eastern Carinthia, belts of limestone striking south-east occur in young gneiss and old mica-schist. They are known at St. Lambrecht

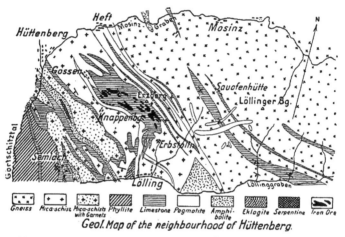

Geol. Map of the neighbourhood of Hüttenberg.

FIG. 364.—Geological map of the neighbourhood of Hüttenberg. Scale about 1 : 75,000.
Baumgartel.

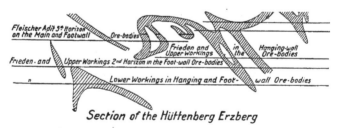

Section of the Hüttenberg Erzberg

FIG. 365.—Section of the Hüttenberg Erzberg. Scale about 1 : 75,000. Baumgartel.

in Styria and at Friesach in Carinthia, at Waitschach, Hüttenberg, Lölling, Wölsch, Loben, Waldenstein, Teissene, etc. In this limestone, beds of siderite and limonite occur which have been worked for centuries at the mines, Geisberg, Zeltschach, Olsa, Waitschach, Zossen, and Hüttenberg. The most important deposit is that at the Hüttenberg Erzberg, illustrated in Figs. 364 and 365.

This metalliferous hill forms the end of a ridge which the north-south chain of the Sau Alps throws off in a westerly direction from Hohewart through Walzofen, Löllinger Berg, and Sauofen to Erzberg. Upon this ridge occur the numerous workings of the districts, Knappenberg, Heft, and Lölling.

The country here consists chiefly of schistose rocks—so-called gneisses —forming a flat south-east striking anticline. Upon these gneisses and with no sharp separation from them, lie mica-schist, phyllite, green schist, and clay-slate. Intercalated in the gneiss as well as in the mica-schist and phyllite, occurs the limestone, which is usually light-coloured and granular, and carries mica, pyrite, and sometimes realgar and arsenopyrite, while in places it merges into a garnet-diopside rock. A second group of intercalations is constituted by the tourmaline-bearing pegmatitic gneisses, some of which exhibit true dyke character. These are particularly numerous in the ore-bearing limestone found in the mica-schist.

The deposit when in unweathered schist consists of siderite, ankerite, pyrite, barite, and, more seldom, löllingite and metallic bismuth, while in the upper levels limonite, manganese ore, and the oxidized products of the sulphides, are found.

According to the extent to which this weathering has proceeded, blue ore, representing the greatest decomposition, may be separated from brown ore, kidney ore, and white ore, this last representing undecomposed siderite. Quartz, mica, pyrite, and barite, occur as impurities, the last-named being often associated with very pure siderite.

It was formerly thought that the deposit consisted of a number of disconnected and irregular lenses; later developments however have shown that the lenses are connected together, forming a continuous many-membered mass which at times bulges out into hanging-wall and foot-wall, or sends out veins into the country-rock. The passage from ore to rock is sometimes gradual, passing through the intermediate stage of ankerite; or it is sudden and with only a clay-parting between. The dip is generally to the south-west, though cases of opposite tendency occur. The ore-bed is not always parallel to the limestone beds, but at times crosses them. Layers of limestone interbedded in the ore are common.

Irregularity of form being characteristic of metasomatic deposits, the question of the genesis of this deposit is more easily settled. It is generally agreed that it is a metasomatic deposit.

The iron content is 43–49 per cent ; the amount of silica varies, though there is always sufficient to make the ore an acid ore. Ore-reserves to 860,000 tons developed, and 800,000 tons estimated, have been declared. The production in 1910 was 14,110 tons of siderite and 33,356 tons of limonite.

Iron Deposits in the Carboniferous Limestone of England

LITERATURE

W. W. Smyth. 'The Iron Ores of Great Britain,' Memoirs, Geol. Survey, 1856, Part 1, p. 15.—J. A. Phillips and Henry Louis. A Treatise on Ore Deposits, 2nd Ed. London, 1896.—J. P. Kendall. 'The Hæmatite Deposits of Whitehaven and Furness,' Trans. Manch. Geol. Soc., 1876, Vol. XIII. p. 231.

Most of these deposits are found in the Carboniferous Limestone region of the north of England, in Durham, Northumberland, Cumberland, etc. At Alston they are associated with lead lodes which cut through the entire thickness of this limestone formation. These lodes occasionally are filled with limonite in the place of lead, as for instance in the productive lode at Rodderup Fell which is 5–6 m. wide, and in the Manor House lode from which large quantities of good limonite have been obtained. Such occurrences are of economic importance. On the northern shoulder of Cross Fell and in Weardale similar iron deposits outcrop, while in the eastern portion of the region the occurrence of siderite in the lead lodes is noteworthy, even though it is of no importance. In the mines at Allenheads the siderite occurs in regular lodes, while at Stanhope Burn, on the other hand, the country-rock is traversed by such a tangle of veins containing iron and lead that the whole mass is quarried. Some of these mines worked the gossan resulting from atmospheric oxidation of the siderite.

The siderite is sometimes white or yellowish-grey, the so-called white ore ; and sometimes dark grey, microcrystalline and then feebly magnetic, the so-called grey ore. As the result of oxidation an envelope of limonite is now sometimes found enclosing kernels of white siderite. These deposits are found in the Great Limestone, particularly where two fissure-systems intersect, as for instance at the Carrick mines.

Unlike the above-mentioned ores, the red hæmatite at Whitehaven in Cumberland and of Furness in Lancashire is a very valuable ore which, containing 50 per cent of iron and but little phosphorus, is admirably suited to the Bessemer process. This ore is found in rocks of Silurian age as well as in the Carboniferous limestone, though only the occurrences in the latter are of any practical importance. Both formations consist of an alternation of limestones with shales and sandstones. While however the ore-bearing limestones display thicknesses of 100 m. or more, the interbedded sandstones and shales are generally only 1 m. in thickness and seldom reach as much as 4 metres.

The ore-bodies are usually fissure-fillings and pockets such as that illustrated in Fig. 367, though some are quite irregular and penetrate deep

into the limestone. They are not always at the same horizon but may
occur in any layer, from the lowest, lying immediately upon the Silurian, to
the highest forming the base of the Grit and Yoredale Rocks at Whitehaven
and Furness respectively. The shape of the deposit varies greatly according
to the degree to which the limestone has surrendered to alteration.

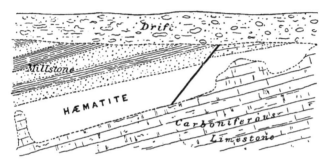

Fig. 366.—Diagrammatic section of the Parkside iron deposit.   J. D. Kendall.

At Bigrigg, Crowgarth, and Parkside, the ore-bodies are irregular
masses immediately under the Millstone Grit, which, as illustrated in Fig.
366, may form the actual hanging-wall. In other cases bed-like bodies
are formed which may be 65 feet or more in thickness.

The superficial extent of these occurrences may sometimes be quite
considerable ; that for instance at Parkside covers 18 acres, or 72,000

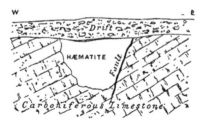

Fig. 367.—Diagrammatic section of the occurrence worked by the
Crossfield Iron Company.   Louis.

sq. m., while numerous others cover from 8000 to 40,000 square metres.
A good example of a pocket-like deposit is that illustrated in Fig. 367, which
is worked in opencut by the Crossfield Iron Company.

The Parkside and Lindal Moor deposits in the Furness district occur in
the lower portion of the Carboniferous limestone. The first of these has a
superficial extent of 60,000 sq. m., and at one point has been proved to a

depth of more than 100 metres. The occurrence at Lindal Moor has a length of 800 m. and a thickness of 21 metres.

The geological position of occurrences bounded by Silurian slate may be gathered from Fig. 368.

The Whitehaven hæmatite, which is generally of a dull red colour, often forms compact masses in which numerous irregular cavities occur. In the Furness district, apart from the occurrences at Lindal Moor, Stank, and Askam, the ore differs materially from that at Whitehaven, in that it is usually soft and friable and consists in greater part of delicate filmy micaceous hæmatite which envelops compact ore having often a concretionary structure.

The harder hæmatite, known locally as 'blast ore,' is smelted direct,

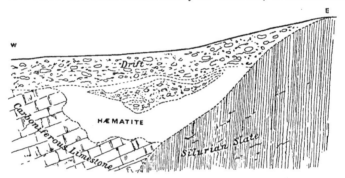

FIG. 368.—Diagrammatic section of an iron deposit in Carboniferous limestone bounded by Silurian.

while the softer variety, the so-called 'smitty ore,' is used for lining the puddle furnaces.

While the soft ore at Furness contains no fossils, numerous fossils, all belonging to the Mountain Limestone, have been found in the compact ore at Lindal Moor.

Concerning genesis, it is very probable that in the metasomatic replacement of the limestone the iron was first deposited as carbonate, which carbonate subsequently became altered to hæmatite by oxidizing meteoric agencies. It is more difficult to settle the question of the source of the iron.

Kendall is inclined to regard the Coal-measures as this source. He points out that the sandstones and shales of these measures contain a considerable amount of iron, and that since probably these rocks formerly covered the Carboniferous limestone of this metalliferous district, it is possible that carbonic acid waters carrying iron percolated to the limestone beneath, which they then replaced.

The production of Cumberland hæmatite in 1881 amounted to 1,615,635 tons, and that of Lancashire for the same year to 1,189,836 tons. In 1882 Cumberland reached its highest production with 1,725,478 tons, since when there has been a gradual but irregular decrease. In 1894, for instance, the production was 1,286,590 tons containing 54 per cent of iron, worth £698,457. In Lancashire the output for the same year was but 870,500 tons with 51 per cent of iron, worth £372,576.

Like the Cumberland district, that of Furness also appears to have reached its zenith in 1882, when 1,408,693 tons were produced. Since that date there has been a gradual decrease, the production in 1890, for instance, being under one million tons.

THE IRON DEPOSITS AT BILBAO

LITERATURE

BOURSON. 'Les Mines de Somorrostro,' Rev. univers. Vol. IV., and Bol. mapa geologico, Vol. VI.—REVAUX. 'Die Eisenerzgruben bei Bilbao,' Génie civil, 1883, Nos. 12 and 13.— A. HABETS. 'Note sur l'état actuel des mines de fer de Bilbao,' Rev. univ. des min. (3) III. 4, 1888.—D. RAMON ADÁN DE YARZA. Descripción física y geológica de la provincia de Vizcaya, Madrid, 1892 ; Mem. de la Com. del Mapa Geol. de España, 1892.—H. WEDDING. 'Die Eisenerze an der Nordküste von Spanien in den provinzen Vizcaya und Santander,' Verh. Ver. f. Gewerbfl., p. 293. Berlin, 1896.—Review in Zeit. f. prakt. Geol., 1897, p. 254. —W. GILL. 'The Present Position of the Iron Ore Industries of Biscay and Santander,' Journ. Iron and Steel Inst., 1896, Vol. II. p. 36.—JOHN. 'Die Eisenerzlagerstätten von Bilbao und ihre Bedeutung für die zukünftige Eisenerzversorgung Grossbritanniens und Deutschlands,' Glückauf, 1910, pp. 2002 and 2045.

This ironfield in northern Spain, so important in supplying ore to England and Germany, lies, as indicated in Fig. 369, in greater part between the Somorrostro and Nervion rivers, and but a short distance from the shores of the Bay of Biscay.

In this situation it extends in a north-westerly direction along the left bank of the Nervion for a length of 24 km. with a maximum width of 10 kilometres. The various occurrences, lying 250–500 m. above the sea, form part of a mountain chain which, beginning approximately 5 km. from the coast, rises to heights of 890 m., 909 m., and 1006 m. at Peña Obieta, Monte Ereza, and Monte Ganerogorta, respectively.

The disposition of the different districts is as follows : That of Galdames extends along the south-west slope of the Peña Pastores at a height of 450–500 m. ; the large occurrences at Triano and Matamoros, forming together the district of Somorrostro, extend upon a hilly plateau between the Pico de Moruecos and the Pico de Mendivil ; the district of Regato occurs south of this plateau, along the left bank of the Regato river which flows into the Rio Galindo ; that of Guenes occurs farther south-east,

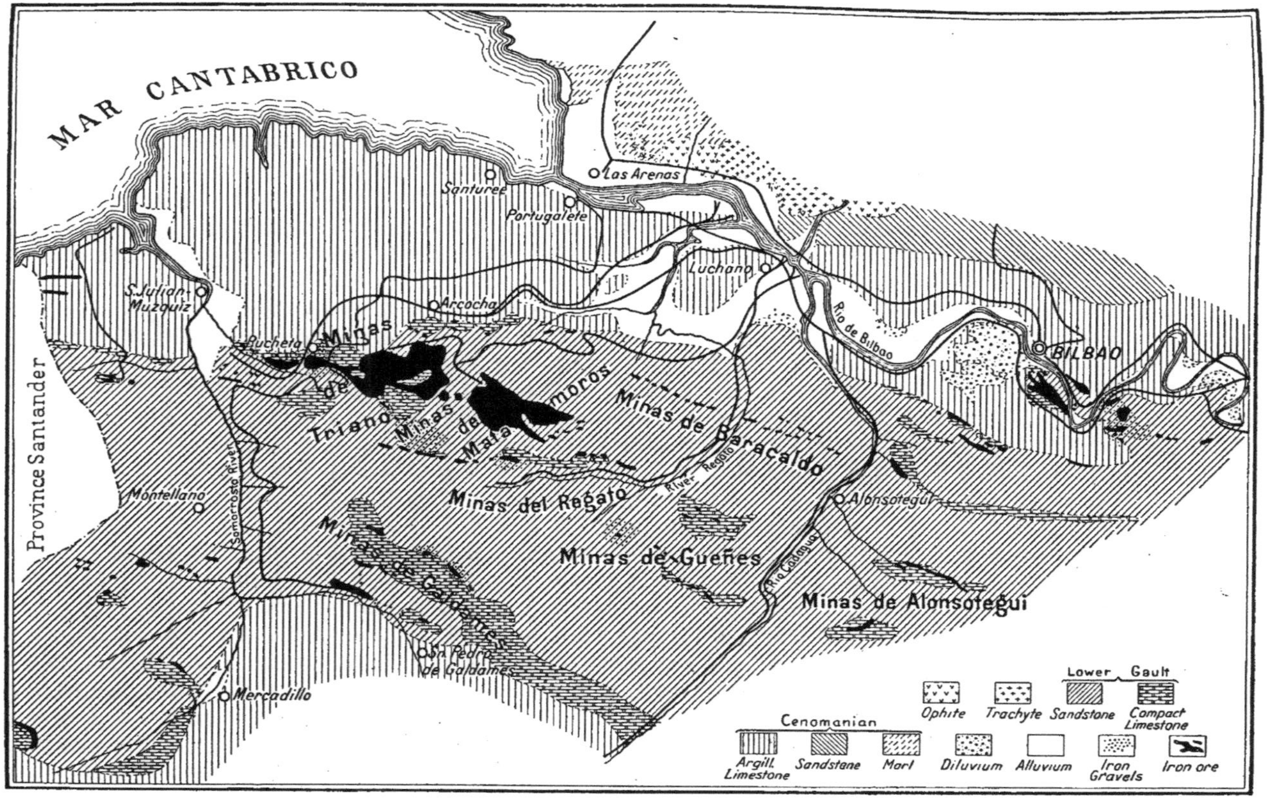

Fig. 369.—General geological map of the Bilbao ironfield.   Scale 1 : 130,000.

between Monte Ereza and the Peña de Espelarli; while that of Baracaldo occurs cast of the last-named hill.   The district of Alonsotegui lies on the right bank of the Rio Cadagua ; the Primitiva mine is situated west of Monte Arraiz ;   while the Iturrigorri district extends east of that hill. Above Bilbao, along the river Nervion, are found the Ollargan, El Morro, and Miravilla mines.   In the western portion of the field, along the west bank of the Rio Somorrostro not far from the boundary with the province of Santander, the mines Arcentales and Sopuerta are found to the south, and Amalia Vizcaya, Asuncion, and Francisco to the north.

In consequence of the short distance from the river Bilbao, which is navigable as far as the town of that name, the situation of these deposits with regard to transport is very favourable, shipment being made both from Portugalete and Luchana.   According to investigation undertaken by Yarza, Collette, Verneuil, Colomb, Triger, and others, the rocks of this district, striking in a south-east direction parallel to the Pyrenees, belong chiefly to the Cretaceous, that is, either to the Gault or to the Cenomanian.   The Gault from below upwards consists when undecomposed of bluish-grey, and when weathered of yellowish-brown, micaceous, fine-grained, and non-fossiliferous sandstone beds followed by fossiliferous limestone in massive beds of variable thickness, traversed by calcite veins. In this limestone the ore-deposits occur, this rock being known by the miners as the mother of the ore.

Above the Gault comes the Cenomanian, this series consisting first of a clayey limestone, which contains fragments of *Acanthoceras Mantelli* and at Triano, for instance, forms the hanging-wall of the deposits ; and then of sandstone and marl in which *Ammonites peramplus* are found.

Outside of the metalliferous district, on the right bank of the Rio de Bilbao, the hill Monte Axpe, consisting of trachyte and ophite, occurs.

Concerning the superficial extent of the individual geological horizons in this district, the sandstones and limestones of the Lower Gault have the widest distribution.   These extend in a south-east striking strip 7 km. wide, alongside of which to the north-east and south-west the clayey Cenomanian limestone ranges itself.   This disposition of the beds is indicated in Fig. 369.   The upper sandstones and marls are no longer present in this district.   The sandstones and limestones in which the deposits occur have been compressed to form a more or less steeply folded mountain chain, the altitudes of the highest points of which have already been given.   Further illustration of this geological position is provided by Figs. 370–373, from which it will be seen that with many of the occurrences the Gault sandstone forms an anticlinal core upon which in most cases there remain only patches of the limestone.   Within these patches

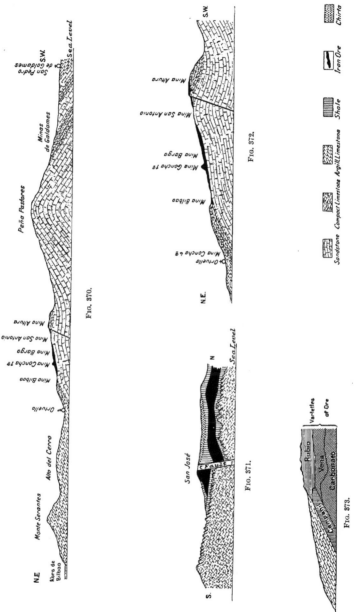

FIGS. 370-373.—Diagrammatic sections through the Bilbao ironfield.

the metasomatic iron deposits find their seat and protection ; they occur chiefly at the contact of limestone and sandstone. Other deposits are found in limestone synclines, such deposits occurring more particularly near the surface and at the contact with the upper sandstone. The deposits consequently are found chiefly along the boundary planes of both the upper and the lower limestone.

A dependence of the deposits upon the presence of faults may often be observed, this being always possible when the deposit occurs in an outcrop of limestone, as for instance at Triano and in the San Francisco mine.

These deposits are generally in the form of elongated and often irregular or indefinite lenses some 300–500 m. long with an average width of 100 metres. They strike roughly south-east, that is, conformably to the country.

The largest bodies are found in the Somorrostro district, these being the two occurrences at Matamoros and Triano which in all probability are parts of what was formerly one continuous deposit. Matamoros is known for a length of 2250 m. and a width of 900 m., and Triano for 3100 m. and 100–1300 m. respectively. Next, in point of size, come the deposits worked in the mines San Louis, Silfide, and Abondonoda, near Miravilla above Bilbao, the lengths of these deposits being approximately 1200 metres. In addition to these larger deposits there are many other smaller and irregular occurrences.

In all cases the thickness varies exceedingly, and in depth particularly the deposits become irregular. At Triano, for example, the thickness in the Barga mine is roughly 40 m., while in the Altura mine it is but 10 metres. Between the deposit and the sandstone or limestone forming the foot-wall, there is often a bed of clay 2–6 m. thick ; at other places, however, irregular limestone protuberances penetrate the ore.

Almost all the known deposits come to the surface, though in the Somorrostro district a thin clayey Cenomanian limestone covers the deposit. Similarly, all the deposits so far developed occur at a considerable height above sea-level, Triano at 250–500 m., Galdames at 300 m., and Sorpresso in the Arcentales-Sopuerta district, at 470–580 m., Primitiva near Castrejana at 300 m., and those at Gueñes at 600 metres.

In the formation of these deposits the limestone was first altered to siderite, which subsequently, in the neighbourhood of the surface and by the action of surface agencies, became altered to limonite or hæmatite, while siderite still existed in depth. It is an interesting fact that hæmatite is only found where the deposit does not directly come to surface.

The siderite is termed *Carbonato*; it is sometimes yellowish-white and typically coarsely-crystalline ; sometimes grey and then granular.

The first condition, representing the better quality, has a higher value than the latter.

As with all metasomatic ores, kernels of undecomposed limestone are not uncommon. The gradation of ore to limestone around the margins of the deposit takes place similarly to that described in the case of Kamsdorf.[1] Important masses of siderite are more particularly found at Triano, in the mines Concha, Inocencia, Trinidad, Buena Fortuna, and Esperanza.

When hæmatite occurs pseudomorphic after siderite in a compact and crystalline aggregate it is known as *Campanil*, while when earthy it is termed *Vena*. In this latter condition it is often found beneath a thickness of limonite or of *Campanil*, but also in veins crossing other ores, hence its name. The largest masses or quantities of both *Campanil* and *Vena* are found in the Triano deposit.

The Bilbao limonite is yellowish or reddish, in consequence of which it is known as *Rubio*. Its structure, in harmony with its character as an alteration product, is generally cavernous, most of the cavities being lined with stalactitic and reniform limonite and with quartz crystals, while the cavities themselves are often to a great extent occupied by clayey material.

When the limonite occurs earthy it often contains pyrite crystals as well as sulphur arising from the decomposition of pyrite. In it also, many kernels of unaltered limestone occur. In addition, a fragmentary ore consisting of clay and limonite fragments and known as *Chirta*, is now and then met.

Limonite occurs exclusively in the neighbourhood of the surface. Of all the ores in the Bilbao district it has the widest distribution. Between siderite, limonite, and hæmatite, there are all sorts of gradations, in most of which however limonite preponderates ; such ores are known as *Rubio Avenado*.

The Bilbao ores are in general of medium iron content, that is, on an average they contain 50–52 per cent of iron. They are almost free from deleterious constituents. The silica content is moderate ; phosphorus and sulphur are almost completely absent. They are consequently ideal Bessemer ores.

The chemical composition of the different ores may be gathered from the following table :

---

[1] *Postea*, p. 835.

.

[TABLE

Carbonato (Siderite)

	Superior. Per cent.	Inferior. Per cent.			Superior. Per cent.	Inferior. Per cent.
Metallic iron . .	41·474	38·780	Lime	1·700	1·560
Manganese .	0·935	0·695	Alumina . .	0·170	0·300
Phosphorus .	0·017	0·019	Carbonic acid .	33·633	32·957
Sulphur . .	0·140	0·270	Silicic acid . .	6·590	8·990
Magnesia . .	0·450	0·870	Combined water .	0·480	1·480

Campanil

	Per cent.		Per cent.
Metallic iron	52·749	Lime	5·530
Manganese	1·333	Alumina	1·840
Phosphorus	0·010	Carbonic acid . . .	0·093
Sulphur	0·014	Silicic acid . .	5·300
Magnesia	1·540	Combined water . .	7·470

Vena

	Per cent.		Per cent.
Metallic iron	56·809	Lime	1·310
Manganese	0·846	Alumina	1·200
Phosphorus	0·015	Carbonic acid . . .	0·100
Sulphur . . .	0·016	Silicic acid . . .	6·210
Magnesia	0·450	Combined water . . .	0·120

Rubio

	Per cent.		Per cent.
Metallic iron	51·065	Lime	0·500
Manganese .	0·492	Alumina	1·700
Phosphorus	0·024	Carbonic acid . . .	0·850
Sulphur . .	0·040	Silicic acid . . ·	9·750
Magnesia . . .	0·250	Combined water . . .	6·950

Rubio Avenda

	Per cent.		Per cent.
Metallic iron	54·959	Lime	0·850
Manganese	0·568	Alumina	1·250
Phosphorus	0·013	Carbonic acid . . .	0·650
Sulphur	0·025	Silicic acid . . .	7·120
Magnesia	0·550	Combined water . . .	4·100

In spacial connection with, as well as in genetic dependence upon these metasomatic and oxidized ores are the fragmentary deposits which yield the ore known as *Chirta*. These are mostly loose, only partly cemented agglomerates of more or less rounded pieces of limonite enveloped in a red clay. The fragments are often but a few millimetres in size, and seldom consist of hæmatite.

These deposits were formed by the mechanical destruction and subsequent natural concentration of the original metasomatic deposits. They are generally found upon ore *in situ*, in which situation they constitute eluvial deposits. Often also they occupy depressions on the surface, in which case they are separated from the main deposit by a layer of clay of variable thickness. These widely distributed detrital or eluvial deposits fluctuate in thickness between a few centimetres and 5 metres. Their position in reference to the main deposit is indicated in Fig. 371.

The ore is recovered from the *Chirta* by washing, when 40–50 per cent of the material washed is recovered as iron ore of a composition corresponding exactly to that of the limonite of the original deposit.

The importance of these *Chirta* deposits is evident from the fact that some mines by washing such material produce 500 tons of ore daily. The most important of such deposits are : in the Triano district, the mines Rubia, Ventura and Josefita, Cerrillo, Marta and Capela ; in the Regato district, the mine Lejana ; in the Galdames district, Elvira and La Buena ; in the Arcentales-Sopuerta district, Catalina and Safo ; and in the Ollargan, El Morro, and Miravilla districts, the mines Segunda and San Pedro.

Fluviatile gravel-deposits in the Bilbao field occur on both sides of the river Cadegal, where from the Vicenta and Maria mines about 60 tons are won daily. The length of such deposits is stated to amount to several kilometres. It is interesting to note that in a vertical section through these deposits porous uncemented ore often alternates with compact ore.

Concerning the genesis of the original deposits at Bilbao, the following factors must be considered : The plication of the Cretaceous beds took place between the Eocene and Miocene periods. At this plication not only did synclines and anticlines result, but a number of other disturbances in addition. One probable consequence of the folding was the ascent of mineral solutions containing carbonic acid and iron. These solutions, more particularly where sandstone was the foot-wall, metasomatically altered the limestone, this alteration having probably taken place in Miocene time. First then, as with almost all metasomatic iron deposits, siderite was formed, which mineral afterwards became changed to limonite and hæmatite.

In Tertiary time also, the disintegration of the primary occurrence began, this continuing into the Alluvium. In this manner the *Chirta* deposits, that is, the eluvial and fluviatile gravel-deposits, were formed, some of which are being worked to-day. Since in the chemical alteration of siderite to limonite cavities result, which by the action of meteoric waters become partly filled with fragmentary ore, the above-mentioned vertical alternation of porous with compact ore, arises. The large amount of these fragmentary ores indicates the tremendous volume of water which must have been active in their formation.

According to the view held by Wedding, which Krusch however controverts,[1] limonite and hæmatite are primary and to be regarded as precipitates from a lake, such lake having been formed after part replacement of the limestone by siderite had taken place. This however is not in agreement with the generally accepted view concerning the origin of these deposits.

---

[1] *Zeit. f. prakt. Geol.*, 1897, p. 254.

The importance of this Bilbao field may be gathered from the following table of tons produced in different years :

	Vena.	Campanil.	Carbonato.	Rubio.	Total.
	Tons.	Tons.	Tons.	Tons.	Tons.
1901	400,000	200,000	750,000	3,273,312	4,623,312
1902	...	57,081	442,237	4,482,500	4,981,818
1903	...	81,634	509,801	4,417,078	5,308,513
1904	...	54,537	801,582	4,983,885	5,620,458
1905	...	33,363	478,122	5,186,163	5,597,648
1906	...	140,000	546,577	4,396,421	5,082,998

From these figures it is seen that in 1902 the production of *Vena* had completely ceased, while that of *Campanil* had diminished considerably. *Rubio* is responsible for the largest percentage of the ore produced, this varying between 70·9 and 91 per cent. The increase in production shown by these figures is due to the working of the fragmentary deposits, this having been first seriously undertaken in 1902.

According to John, from the figures of the annual reports of the British Consul, the relation between the *Rubio* obtained by washing and that won from the oxidation zone has been as follows :

	Total *Rubio*.	*Rubio* obtained by Washing.	Equivalent Percentage of the Total.
	Tons.	Tons.	
1902	4,482,500	330,000	6·7
1903	4,717,078	450,000	9·5
1904	4,983,885	550,000	11·0
1905	5,186,163	700,000	13·0
1906	4,396,421	900,000	20·4

These figures would indicate that a further rise in the proportion of the washed *Rubio* is likely.

The quantity of ore available in this district was in the year 1883 estimated by Goënaga[1] at 48,000,000 tons, while J. Forrest in the same year estimated 55,000,000 tons.[2] Ramon Adán de Yarza[3] in the year 1892 came to the much higher figure of 163,350,000 tons. The latest estimate is that published by Luis M. Vidal in *The Iron Ore Resources of the World*, Stockholm 1910, according to which, the amount of ore produced in the province of Biscay during the 32 years immediately preceding, was 150,000,000 tons, while the amount at that time still to be won was 61,000,000 tons. The greatest proportion of these tonnages is accounted for by the Bilbao district.

[1] *Revista Minera bei Triano und Matamoros.*
[2] *North of England Inst. Min. Mech. Eng. Ante*, p. 197.   [3] *Op. cit.* p. 826.

THURINGIAN FOREST

(a) Kamsdorf near Saalfeld

LITERATURE

K. TH. LIEBE. 'Übersicht über den Schichtenaufbau Ostthüringens,' Treatise with the geol. Spezialkarte of Prussia, etc., Vol. V. Part 4, pp. 69 and 116-119; 'Aus dem Zechsteingebiet Ostthüringens,' Jahrb. d. k. pr. geol. Landesanst., 1884, p. 386.—E. ZIMMERMANN. 'Der geologische Bau und die geologische Geschichte Ostthüringens,' Mitt. a. d. Osterl. N.F. III. p. 79.—F. BEYSCHLAG. 'Die Erzlagerstätten von Kamsdorf,' Jahrb. d. k. pr. geol. Landesanst., 1888, p. 329. — W. KELLNER. 'Nachrichten über Bergbau und Hüttenwesen in Südthüringen,' Berg- u. Hüttenm. Ztg., 1889, p. 157.—F. BEYSCHLAG. 'Geologischer Bau des Thüringer Waldes,' Zeit. f. prakt. Geol., 1895, p. 498 ; 'Bergbau in den Thüringischen Staaten,' Zeit. f. prakt. Geol., 1898, p. 269.—J. LOWAG. 'Mangan- und Eisenerzvorkommen im Thüringer Wald,' Öster. Zeit. f. Berg- u. Hüttenw., 1902, pp. 608, 623, and 635.—G. EINECKE and W. KÖHLER. 'Die Eisenerzvorräte des Deutschen Reiches,' Archiv f. Lagerstättenforschung, Part 1, d. k. pr. geol. Landesanst. Berlin, 1910.

The Palæozoic mountain core of the Thuringian Forest is separated from the Zechstein and Triassic fore-ground to the north, by a well-defined boundary fault. This bayonet-like break, which in places is developed as a flexure, is of great importance in the metasomatic mineralization, in that it served as the circulation channel for the solutions. While it generally conforms to an Hercynian strike, east of the Saale it strikes east-west, losing at the same time its simple character and breaking up into a number of parallel fissures accompanied by step-faulting. From these fissures the Zechstein limestone and dolomite near Kamsdorf, dipping flatly to the north-north-west, were altered to iron ore.

Two different Zechstein horizons were thus attacked. The Zechstein, lying unconformably upon the Culm, consists in its lower portion of a conglomerate, above which comes the Kupferschiefer, and then limestone with intercalated bituminous marly slates ; in its middle portion, of the main dolomite which in part is porous; and in its upper section, of a lower variegated clay, a blocky dolomite, and an upper variegated clay. The beds to suffer alteration are, in the first place the lower limestone, and then subordinately the middle dolomite. The Upper Zechstein contains no iron ore.

The iron solutions to which the replacement is due circulated in the main fissures, producing siderite, or 'mica' as the miners at Kamsdorf call it. This alteration is the most intense along the fissures, from whence the intensity diminishes with distance. With complete replacement the original structure and bedding of the limestone may no longer be recognized. As indicated in Fig. 374, around the borders the ore merges into ferruginous limestone, which in turn gradually gives way to normal limestone.

It is interesting to note that at Kamsdorf not all the limestone layers were equally susceptible to this alteration, and that in consequence the iron content and the extension of the alteration varies with the layers. Generally a lower and an upper deposit are distinguished, these two being separated by a bed of slate. In places however there are other subordinate deposits.

The width of the deposit at right angles to the plane of the fissure is generally 20–50 m. with 80 m. as a maximum. The maximum thickness

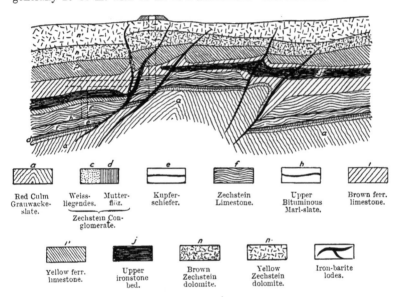

| Red Culm Grauwacke-slate. | Weiss-liegendes. | Mutter-flöz. | Kupfer-schiefer. | Zechstein Limestone. | Upper Bituminous Marl-slate. | Brown ferr. limestone. |

Zechstein Con-glomerate.

| Yellow ferr. limestone. | Upper ironstone bed. | Brown Zechstein dolomite. | Yellow Zechstein dolomite. | Iron-barite lodes. |

FIG. 374.—Section of the iron deposits at Kamsdorf in Thuringia. Scale 1 : 2000. Beyschlag, *Jahrb. d. geol. Landesanst.* 1888.

is usually 4–8 m., from which maximum it gradually diminishes till the deposit disappears.

By the secondary process of oxidation, in which doubtless oxidation-metasomatism played a part, the siderite, and especially that of the upper bed, has become changed to limonite.

Kamsdorf is distinguished from all other similar metasomatic deposits in that copper occurs both in the fissures themselves as well as in the altered limestone in their immediate neighbourhood.

The average composition of the ore may be gathered from the following figures of percentage content :

	Fe.	Mn.	Cao.	MgO.	Al₂O₃.	SiO₂.	P.	S.	Cu.	BaSO₄.
Siderite . .	36·0	3·5	4·7	1·80	0·12	4·2	0·20	0·24	...	0·5
Limonite . .	46·0	5·0	3·6	0·86	0·52	6·7	0·02	0·10	...	1·2
Ferruginous lime-stone . .	15·9	3·6	33·6	0·94	0·05	1·6	...	0·04	0·1	0·3

The ferruginous limestone is used as ferruginous flux.

The annual production of Kamsdorf is about 120,000 tons. The ore is smelted at the Maxhütte furnaces at Unterwellenborn. It is noticeable however that high-grade ore is no longer produced to any great extent, but that ferruginous flux with a low iron content preponderates, this flux being smelted with the neighbouring chamosite and thuringite from Schmiedefeld.

(b) *The Neighbourhood of Schmalkalden, including the Occurrences at Stahlberg, Mommel, and Klinge*

LITERATURE

H. MENTZEL. 'Die Lagerstätten der Stahlberger und Klinger Störung im Thüringer Wald,' Zeit. f. prakt. Geol., 1898, p. 273.—G. EINECKE and W. KÖHLER. 'Die Eisenerz-vorräte des Deutschen Reichs,' Archiv f. Lagerstättenforschung, 1910, Part 1, K. pr. geol. Landesanst.

The deposits in this neighbourhood occur between Liebenstein and Seligenthal, along the south border of the Thuringian Forest, where they occupy a geological position similar to that of the occurrences at Kamsdorf. The southern boundary fault of the Thuringian Forest, which on this side also separates the fore-ground from the western core, is likewise not a simple fissure but one of many components accompanied by step-faulting, giving to the Zechstein in places a large horizontal extent.

One of the two main components, known as the Stahlberg break, runs through Seligenthal north of Schmalkalden, in a north-west direction through the Stahlberg, the Kammerkuppe, etc., to a position north-west of Beierode. Along this fault and its tributary fissures occur : to the east at Seligenthal, the Stahlberg iron mine ; to the west at Herges, the Mommel iron and barite mine ; and in addition some independent barite mines. Not many kilometres to the north occurs the second and parallel main fault upon which the Klinge iron mine is situated.

While the Stahlberg break generally brings Bunter sandstone against the blocky dolomite, along the Klinge fault fundamental gneiss and mica-schist come up against that dolomite.

At all three of the deposits mentioned the Zechstein lies unconformably upon the mica-schist, though at none of them is it completely developed, the Lower Zechstein being fully present only at Asbach and Liebenstein. The Middle and Upper Zechstein are however always complete.

In regard to mineralization the blocky dolomite is most important since the deposits are chiefly associated with it. This thick-layered dolomite is of a grey or yellowish-brown colour and of friable character, while between its different layers rauchwacke-like, porous, and sandy rocks occasionally occur. As at Kamsdorf so also here, siderite was first formed which afterwards by oxidation, partly from the surface and partly from fissures,

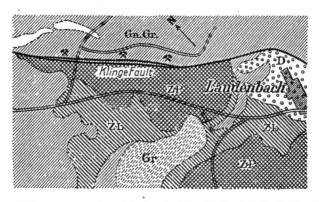

Fig. 375.—Position of the limonite pockets along the Klinge Fault.   Scale 1 : 25,000.   Scheibe.

Gr. = granite; Gn. Gr. = gneissic granite; ZP = Zechstein Blocky Dolomite; ZL = Zechstein Clay; D = diluvial detritus; ⤬ = iron deposits.

became altered to limonite. This limonite often displays the structure of siderite, though it also occasionally occurs in the form of kidney ore.

In consequence of the contraction in volume which takes place at the alteration of siderite to limonite, the oxidized ore is porous and friable but yet stiff enough to smelt in the blast furnace, earthy and loose ore occurring only in the neighbourhood of the outcrop. The siderite is coarsely crystalline and generally of a leather colour; when whitish-grey and finely-crystalline it has a lower iron content. The ferruginous limestone resembles the limestone and dolomite in structure and bedding, though, differing from the occurrence at Kamsdorf, it contains but 5–12 per cent of iron.

More than was the case at Kamsdorf, these Schmalkalden deposits are distinguished by a large number of associated minerals. Of these barite is the most common, this mineral partly occurring massive in veins and nests, rather in the limonite than the siderite. Calcite is not uncommon.

The high manganese content of the ore, which at the oxidation of siderite becomes concentrated in nests of pyrolusite, is particularly notable.

The composition of the Schmalkalden ore may be gathered from the following analyses :

	Fe.	Mn.	CaO.	MgO.	BaSO₄.	SiO₂.	Al₂O₃.	H₂O.	P₂O₅.	CO₂.
Siderite—										
Coarsely-crystalline	39·3	5·2	1·1	1·5	...	4·4	...	...	...	48·2	...
Compact . .	34·9	5·7	0·7	1·4	...	12·3	...	...	...	44·7	...
Raw ore . . .	37·3	7·5	...	...	...	...	...	...	...	...	...
Roasted . . .	52·0	6·7	...	...	...	...	...	...	...	...	...
Limonite—										
Mommel, old average sample .	44·3	5·8	1·9	1·1	BaO 6·2	12 7	2·6	7·3	...	...	...
Mommel, decomposed . . .	40·6	5·1	1·7	3·6	1·4	0·5	0·9	...	0·04	CO₂ + H₂O 27·3	...
Hand specimen .	49·2	6·1	CaO + MgO 9·3		...	17·06	1·9	...	...	11·2	...

From these figures it is seen that the ore must be termed a good ore. The average content of the raw ore is about 40–50 per cent of iron, with 5–6 per cent of manganese. Deleterious substances are present in but small amount, so that only a moderate amount of flux is necessary.

The yearly production of Stahlberg and Mommel—the Klinge is at present not working—is about 5000 tons, part of which is smelted in the Schmalkalden furnace, while the remainder is sent to Westphalia. By smelting, an almost phosphorus- and sulphur-free manganiferous charcoal iron is obtained which is famous for its fine quality.

IBERG NEAR GRUND

LITERATURE

A. V. GRODDECK. 'Lagerungsverhältnisse am Iberg und Winterberg,' Zeit. d. d. geol. Ges., 1878, Vol. XXX. p. 540 ; 'Zur Kenntnisse des Oberharzer Kulm,' Jahrb. d. k. pr. geol. Landesanst., 1882, p. 55.—C. BLÖMECKE. Die Erzlagerstätten des Harzes und die Geschichte des auf denselben geführten Bergbaus. Vienna, 1885.—W. RITTERHAUS. 'Der Iberger Kalkstock bei Grund am Harz,' Zeit. f. Berg- Hütten- u. Salinenwesen in Preussen, 1886.—BANNIZA, KLOCKMANN, LENGEMANN, UND SYMPHER. Das Berg- und Hüttenwesen des Oberharzes. Stuttgart, 1895.—K. HUHN. 'Der Iberg,' Manuscript in possession of the Preuss. Geol. Landesanst., 1897.—E. HARBORT. Über Mitteldevonische Trilobitenarten im Iberger Kalk bei Grund im Harz,' Zeit. d. d. geol. Ges., 1903.—G. EINECKE and W. KÖHLER. 'Die Eisenerzvorräte des Deutschen Reiches,' Archiv f. Lagerstättenforschung, Part 1, d. k. pr. geol. Landesanst. Berlin, 1910.

Iberg, 565 m. above sea-level, and Winterberg, form together a Devonian ellipse, 2·3 km. long in a north-west direction and 1 km. wide, near the western border of the Harz, this ellipse rising as an uplift from the Culm slates and grauwackes of the Oberharz plateau. The sharply defined topographical boundary of the Devonian coral limestone results from a partial sinking of the Culm beds against this Devonian uplift, along fissures belonging to the Oberharz lode-system. The southern boundary for instance is formed by the Silbernaal lode-series with its subsidiary and associated fissures.

Not everywhere, of course, is the boundary of the Devonian against the Culm referable exclusively to tectonic disturbance ; several underground developments have indeed shown that Culm beds were actually deposited upon the limestone; while, occasionally, faults which on surface appear as boundary faults, in depth penetrate the Devonian limestone. From the south-east striking and mostly south-west dipping boundary faults and the numerous subsidiary fissures accompanying them, the limestone became changed by mineral solutions to siderite, while barite and quartz were deposited at the same time. This replacement of the limestone is neither regular nor uniform. It is much more usual to find a succession of small funnel-shaped bodies, in which the limestone appears sometimes silicified or dolomitized, and sometimes altered to siderite, and then again by oxidation to limonite.

At such oxidation the isomorphous substances associated with the siderite—and especially manganese oxides, calcite, and dolomite—became separated in part as well-defined crystals or stalactites. In addition to barite—which often so contaminates the deposit as to make it unpayable —quartz, pyrite, chalcopyrite, bornite, malachite, and asphalt, also occur.

The connection between the metasomatic alteration of the Iberg limestone and the latest fillings of the Oberharz lodes west of the Innerste, is notorious. On the one hand, the boundary faults on the south-west, embracing the Prinz Regent and Ober lodes, carry the same filling as the Oberharz lodes ; while, on the other, the barite of the Iberg ironstone may have been introduced in aqueous solution from the neighbouring Zechstein along the westerly continuations of the boundary faults, just as the Lautenthal saline spring to-day deposits barium sulphate in the pipes underground in the mines.

The disposition and intensity of the water circulation through these fissures is still recognizable in the numerous caves and pot-holes formed, such as may be observed in great number at Iberg. The separate ore-pockets are very irregular in form. Bodies of 1 m. in thickness may suddenly swell out to 40 m. and just as quickly completely disappear. With these it cannot always be said whether the cavities in which they occur were formed from fissures, or whether such bodies resulted from typical metasomatism. Since however at Iberg both types of deposit occur closely associated, the occurrences without doubt belong to the class of cavity-fillings with associated metasomatic deposits.

The composition of the siderite and limonite may be gathered from the following analyses, each of which represents the average of many determinations :

	Raw Siderite.		Limonite.	
	Prinz Regent Lode.	Pfannenberg und Stieg Mine.	I.	II.
Fe	33·04	31·68	43·15	50·03
Mn	6·02	6·03	8·45	8·66
CaO . . .	4·61	3·28	2·11	3·52
MgO . . .	2·66	2·62	0·16	Trace
SiO_2 . . .	11·25	10·29	10·50	9·98
Al_2O_3 . . .	1·17	2·27	3·16	...
S	0·16	0·15	0·05	0·10
P	Trace	0·028	0·05	Trace

The iron content of the siderite fluctuates generally between 25 and 35 per cent.

Formerly, a brisk industry flourished upon these deposits. This however came to an end in the 'eighties, less for want of ore than because of general economic conditions, and particularly because after the starting of the Westphalian furnaces that at Gittelde was obliged to stop. Einecke and Köhler however are of the opinion that in these deposits there is a sufficient ore-reserve to last a single furnace for a long time.

SCHAFBERG AND HÜGGEL

LITERATURE

W. TRENKNER. Die geognostischen Verhältnisse der Umgegend von Osnabrück. Osnabrück, 1881.—v. RENESSE. 'Bergbau und Hüttenindustrie bei Osnabrück,' Nat. Verhandl. Osnabrück, 1885, Vol. VI. p. 46.—BÖLSCHE. 'Die geologischen Verhältnisse der nächsten Umgebung von Osnabrück,' Nat. Verhandl., 1885, Korr. 46.—STOCKFLETH. 'Die Eisenerzvorkommen am Hüggel bei Osnabrück,' Glückauf, 1894.—H. MÜLLER. Der Georgs-Marien-Bergwerks- und Hüttenverein, 1896.—W. HAACK. 'Der Teutoburger Wald südlich von Osnabrück,' Jahrb. d. k. pr. geol. Landesanst., 1908.—E. HAARMANN. 'Die Eisenerze des Hüggels bei Osnabrück,' Zeit. f. prakt. Geol., 1909.—G. EINECKE and W. KÖHLER. 'Die Eisenerzvorräte des Deutschen Reiches,' Archiv für Lagerstätten-forschung, Part 1, K. pr. geol. Landesanst., 1910.

Between the western outliers of the Weser Hills and the Teutoburg Forest occur the Hüggel and Schafberg ridges. These consist of Coal-measures surrounded by a mantle of Zechstein, Bunter, and other younger formations. In this district also, the iron deposits are associated with the Zechstein, of which, along the northern outline of the Hüggel and the southern outline of the Schafberg, the limestone has been altered just as at Kamsdorf and Schmalkalden.

The Schafberg forms a subsidiary fold of the large Hercynian main fold. The Zechstein lying upon the Coal-measures consists of the basal conglomerate, the 0·75 m. thick Kupferschiefer which here only contains traces of copper, and thin-layered dolomitic limestone, this last constituting the principal mass of the formation. The district is traversed by strike- and dip faults, these being particularly noticeable along the southern outline of the Schafberg.

The iron deposits are associated with the thin-layered dolomitic lime-

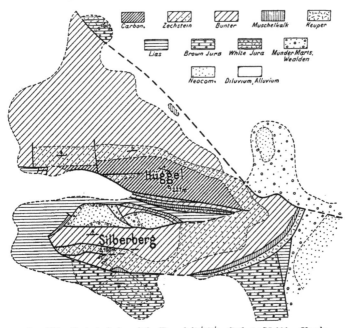

FIG. 376.—Geological plan of the Huggel district. Scale 1 : 75,000. Haack.

stone, from the material of which however they are sharply separated. They consist of irregular pockets of limonite which, though sometimes but a few metres in size, may at times be more than 100 metres. Such are found right around the Schafberg. Concerning their extent in depth very little is known as the deepest workings are but 20 m. below the surface. The limestone in which the masses occur is well bedded, the separate layers being 5–15 cm. in thickness and dipping 20°–35° to the south. Only in the immediate neighbourhood of the ore is it at all ferruginous, displaying then chimneys or pipes from finger to arm's thickness, these being occasionally filled with loose sand. Intercalated with the ore are beds of silica, so that, unlike that at the Hüggel, this ore contains some insoluble residue.

While some of the deposits are connected with one another, most occur isolated in the limestone. There is no regularity in their occurrence ; they do not all lie in the direction of the faults, and it must therefore be assumed that the solutions themselves made their own way out from the fissures. Almost always however, their extension is greater along than across the bedding, probably because the bedding-planes facilitated the circulation.

The deposits at the Schafberg are undoubtedly due to the metasomatic replacement of limestone, though the actual source of the solutions, which some say came from depth and others from surface, is not clear.

The composition of the ore may be gathered from the following analyses :

	I.	II.	III.	IV.	V.
Fe_2O_3 . . .	74·90	69·16	50·51	...	...
Fe . . .	52·48	48·41	35·36	42·94	42·30
Mn_2O_3 . . .	0·37	0·46	7·87	...	...
Mn .	0·26	0·32	5·47	2·27	1·92
CaO . . .	Trace	Trace	1·40	2·10	2·70
MgO . . .	Trace	Trace	Trace	...	3·20
SiO_2 . .	6·00	12·10	23·80	17·80	17·60
Al_2O_3 . . .	3·18	3·31	3·80	5·59	2·53
P_2O_5 . . .	0·22	0·13	0·09	0·90	0·08
P .	0·09	0·05	0·03	...	...
SO_3 . . .	0·34	9·32	0·20	...	...
S	0·13	0·13	0·08	0·16	0·08
ZnO . . .	2·20	2·35	3·00	1·10	1·15
Zn . . .	1·77	1·89	2·41	...	...
Loss on ignition	12·43	12·26	9·51	...	...
Insoluble residue	...	...	...	10·66	12·25

The limonite ore as mined, on account of its porosity, contains 35–50 per cent of water ; the iron content, as with most metasomatic deposits, is comparatively high ; the manganese content fluctuates greatly, as does also the amount of silica present, both however being usually considerable. The ore is smelted at the Georgs-Marien works where it is used to offset the calcareous preponderance of the Hüggel ore.

The Hüggel deposit when the poorer portions are included is exceedingly thick, since practically the whole of the Zechstein limestone, 30–40 m. in thickness, is ferruginous.

The Zechstein formation above the Kupferschiefer consists of a 5–10 m. thick bituminous limestone poor in iron, the so-called Stinkstein, followed by the ferruginous beds, the iron content of which decreases towards the top. The only bed of clean ore is the 8–10 m. on the foot-wall, the 20–30 m. above this bed being ferruginous flux, as which indeed it is mined.

In this case also, the limestone was first altered to siderite which subsequently by meteoric waters became oxidized to limonite, or, in the uppermost portions and in consequence of advanced weathering, to a dark-yellow or brown ochre. The undecomposed light-grey sideritic dolomite in depth is sometimes so ferruginous as to deserve the term siderite.

The developments in the zone of unaltered siderite are particularly interesting. Almost everywhere the yellowish grey siderite is hard and finely-crystalline, occasionally dark bituminous streaks alternate with those of lighter colour, while alternations of siderite and clay-slate have also been observed. In addition, the occurrence of *Styolites* in the ore-bed is noteworthy.

The deposits in general dip and strike uniformly. At places in depth narrow cavities occur between the separate thicknesses, an occurrence which must be regarded as evidence of re-crystallization.

According to Beyschlag these siderite beds were formed metasomatically at the folding and tilting of the Hüggel, when iron solutions penetrated the limestone along fractures and crevices, dissolving that rock and depositing iron carbonate, with the result that where the change was complete clean siderite was formed, and where incomplete, ferruginous limestone.

The parent fissures of these occurrences have not yet been located, and it must therefore be assumed that these in the process of change have been obliterated. The disturbances which have brought the ore-bed into contact with unaltered dolomite are probably of younger age.

The composition of the Hüggel ore may be gathered from the following analyses :

	Limonite.			Siderite.		Flux.	Average of a Calcareous Minette from Lorraine, for comparison.
Fe . . .	37·0	12·3	36·9	35·9	31·8	14·9	33·1
Mn . . .	1·7	2·0	1·9	1·8	1·8	1·3	...
CaO . .	7·7	3·2	8·6	10·8	12·4	29·5	15·7
MgO . . .	1·3	...	...	3·7	3·6	7·8	...
SiO . .	16·2	17·6	15·2	4·8	5·0	2·4	8·1
Al_2O_3 . .	4·9	2·5	3·2	1·1	1·2	2·8	5·8
S . . .	0·1	0·08	0·08	1·5	0·4	...	...
P_2O_5 . .	0·06	0·08	0·08	0·02	0·02	...	1·62
$CO_2 + H_2O$.	24·0	12·2	15·8	25·3	27·3	36·4	8·0
		Zn 0·5	Zn 0·5				

On an average the limonite contains 35 per cent of iron, 12 per cent of insoluble residue, and 24 per cent of water ; the siderite has less water and only 6–10 per cent of insoluble residue. The ore generally contains but little phosphorus or manganese and is consequently suited for the production of Bessemer steel.

Bieber

LITERATURE

W. Bücking. 'Der nordwestliche Spessart,' Abhandl. d. pr. geol. Landesanst., 1892, Part 12, p. 148.—W. Bruhns. Die nutzbaren Mineralien und Gebirgsarten im Deutschen Reiche, 2nd edition of Dechen's work. Berlin, 1906.—Explanatory text with the geol. Spezialkarte of Prussia, etc., Section Bieber and Lorhaupten.

Bieber lies on the north-west border of the Spessart. In this situation crystalline schists constitute the core of a syncline formed after the deposition of the Bunter, this syncline, still in part overlaid by the Rotliegendes, appearing through a Zechstein and Bunter covering ; striking north-west its limbs dip respectively to the north-east and south-west. To the north-east the Bunter continues undisturbed, while to the south-west it is cut by a north-west striking fault, the position of which is indicated on the surface by a ridge. In the foot-wall of this fault, which dips to the south-west, the iron deposits of Bieber occur. These have resulted from the metasomatic alteration of the limestone and dolomite of the lower Middle Zechstein, by iron and silica.

Four payable deposits, known respectively as the Büchelbach, the Streitfeld, the Lager, and the Lochborn beds, have been opened up. Of these only the last, the most important of the whole Spessart, is still being worked. This deposit has been developed for an unbroken length of 2 km., that is, from Galgenberg to the Lochborn valley, while a further continuation of 2 km. has been proved by boring. It runs obliquely to the above-mentioned fault by which to the north it is cut off. Genetically therefore the deposit has nothing to do with this fault, by which indeed it is downthrown 100 m. in the hanging-wall. The width of this 4 km. long Lochborn bed fluctuates, as does also its thickness. The ore moreover does not keep to any fixed horizon, but from west to east it rises higher and higher above the Kupferschiefer, though occasionally the whole thickness between this copper-shale and the Upper Zechstein consists of ore. While the width reaches as much as 450 m., the thickness may therefore be said to vary from nothing to 20 metres. A sharp separation between ore and rock, whether in hanging-wall or foot-wall, is only found when that rock is other than limestone and at the same time not suited to alteration. When limestone forms the country-rock there is a gradual passage from ore to rock.

The ore-bed consists in its lower portion of a bedded, clayey siderite sprinkled with tetrahedrite, galena, and copper minerals, while the upper portion consists of porous ore in compact limonite. As is often the case with such deposits, the iron ore is associated with the manganese minerals pyrolusite, manganite, psilomelane, and wad, which occur either in independent bunches or finely distributed throughout the mass. In smaller amount occur the carbonate, phosphate, and arsenate of lead, copper, and iron, these probably having resulted from the oxidation of the sulphides. It is possible that these heavy metals have come from the cobalt lodes in the vicinity. The three other beds, at Galgenberg and Burgberg, occur along the continuation of the Lochborn deposit ; their thickness however appears to be less, and the ore is more loose, ochreous, and manganiferous.

It will probably not be wrong to ascribe these Bieber deposits to the alteration of limestone by ascending solutions, the channels for which existed in the Zechstein fissures. The first replacement was that by iron ore ; then came the lead-, cobalt-, and copper ores, which are younger ; while finally, the barite solutions saturated alike both fissure and meta-somatic deposit.

The iron content fluctuates between 19 and 34·5 per cent, with a man-ganese content which may rise as high as 17 per cent, while phosphorus may reach 0·246 per cent, copper 0·56 per cent, and sulphur 0·17 per cent. The arsenic content, which may be as much as 0·47 per cent, is an unfavour-able factor.

The ore in spite of its high iron- and manganese content is a difficult one on account of contamination by other heavy metals and by arsenic. In consequence, large blocks are not mined, and only those works which are in the position to dilute the amount of arsenic by admixture with purer ores, can deal with it. Less than 0·1 per cent of arsenic is sufficient to produce a cold-short iron.

The production of Bieber has latterly been 40,000–60,000 tons per year. As to ore available, Einecke and Köhler estimate this at several million tons.

THE UNITED STATES

LITERATURE

E. C. HARDER. ' The Iron Ores of the Appalachian Region in Virginia,' Bull. 380, U.S. Geol. Survey, p. 215.—' Mineral Resources of the U.S. 1891,' Twenty Years' Progress in Iron and Steel Manufacture in the United States.—J. F. KEMP. The Ore Deposits of the United States and Canada, pp. 83-188, eighth impression, 1906.—H. RIES. Economic Geology with Special Reference to the United States. New York, 1910.—CHARLES L. HENNING. Die Erzlagerstätten der Vereinigten Staaten von Nordamerika mit Einschluss von Alaska, Cuba, Portorico, und den Philippinen. Stuttgart, 1911.

Metasomatic iron deposits, so far as known, are not very numerous in the United States. There are, it is true, a large number of deposits in the formation of which metamorphic processes have taken part, yet cases of typical replacement of limestone by iron ore are very seldom met. It is fairly certain however that hitherto but a small proportion of the existing deposits of this class in America have become known, and also that these deposits have received little attention because purely metasomatic iron ores under American conditions are often too poor to render exploitation profitable.

The Appalachian Limonites

Limonite deposits are found in the highly contorted sedimentary beds of the Appalachian mountains, which extend from northern Vermont to central Alabama. The foot-wall of this metalliferous series is formed by the old crystalline schists, and the hanging-wall by Coal-measures.

The deposits occur in limestone and dolomite, sandstone and quartzite, of all ages from Cambrian to Carboniferous.[1] The sandstones and quartzites generally form well-defined ridges, while the calcareous rocks lie in the valley.

The limestone when undecomposed contains small amounts of ferruginous minerals, namely, the sulphide, carbonate, and silicate of iron. From these, by subsequent leaching and concentration, the deposits known under the following names became formed :

(a) Mountain ores.

(b) Valley or Limestone ores.

(c) Oriskany ores.

The Mountain ores are always found along the flanks or at the foot of a sandstone-, hornstone-, and quartzite belt. In Virginia they occur in two narrow zones, the first of which extends along the western slope of the Blue Ridge, from Front Royal in Warren County on the north, to a point 16 km. south of Roanoke County on the south. The second zone, which appears to be a continuation of the first, lies in the New River district in south-western Virginia.

These deposits occur in Lower Cambrian quartzite and in the fragmentary sediments upon it. They form small irregular unconnected beds, and are often associated with Tertiary clays, sands, and gravels. In relation to form, Harder divides the deposits into the following classes : (1) pocket deposits, which may be eluvial as well as fluviatile, and which may be replacement masses or angular fragments and pieces ; (2) beds in marl, which along veins and crevices have become replaced by ore ;

---

[1] Pennsylvanian series.

(3) beds in quartzite, which are either brecciated beds with contemporaneous replacement of the country-rock, or lodes.

The most important deposits consist of irregular bodies of 10–75 m. diameter ; more characteristic however are the pocket-like masses illustrated in Fig. 377, which contain much manganiferous iron, the manganese being generally concentrated into bunches of psilomelane and pyrolusite. The ore may be light or dark in colour.

Genetically it is proved that such Mountain ore has resulted from the action of meteoric waters ; the iron content in the limestone or marl above, and perhaps also partly that of the quartzite, after having been leached,

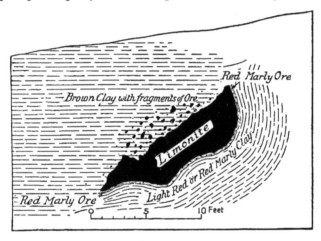

FIG. 377.—Mountain limonite occurring as an irregular mass in clay.
Mary Creek mine near Vesuvius, Pa.   Harder.

was deposited lower down, where more favourable conditions prevailed. The faults and fissures present, would favour such a migration of the iron content.

The Valley or Limestone ores are found, in larger or smaller porous masses closely associated with limestone, in the belt which bounds the Mountain ores to the west and north-west.   Into these masses the limestone sometimes penetrates its craggy points.

These ores are often accompanied by clay ; their quality is however in general better than that of the Mountain ores.   They usually contain 40–55 per cent of iron, 5–20 per cent of silica, and 0·02–0·1 per cent of phosphorus, this last being somewhat less than with the Mountain ores. They likewise are unsuitable for the Bessemer process.

These deposits were formed by solutions descending through the limestone, though, unlike the Mountain ores, these Limestone ores do not reach the quartzite under the limestone.

The Oriskany ores take their name from the Oriskany sandstone; they are replacements in the upper horizon of the Lewistown limestone. As may be gathered from Fig. 379, with comparatively large size they may extend to considerable depth. The iron content of the 5–10 m. deposit amounts to 35–50 per cent, with 3–4 per cent of manganese, 0·06–0·5 per cent of phosphorus, and 10–25 per cent of silica. The resemblance to the Mountain ores is considerable. The iron of the Oriskany ores comes from the overlying Devonian marls, the considerable iron content of which

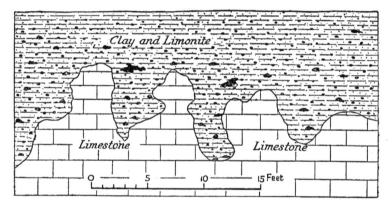

FIG. 378.—Structure of the Valley limonite deposits of the Rich Hill mine near Reed Island, Va. Harder.

filtered down through the Monterey sandstone to the Lewistown limestone beneath, which it replaced.

The genesis of the Appalachian limonites is consequently very similar to that of metasomatic deposits, such as those for instance which occur at Kamsdorf in Germany.

In the case of the Oriskany deposits metasomatism proper took place, while with the other two types subsequent fluviatile re-arrangement played a part.

The West Tennessee limonite deposits appear to be related to the Appalachian. They are found within a wide zone extending from the northern boundary of Alabama and Mississippi through the western boundary of Tennessee and Kentucky. The foot-wall consists of a horn-stone-like Cretaceous limestone. In regard to their composition also, there exists a great similarity between the Tennessee and Appalachian

ores. The principal places of production are Russelville, Mannie, and
Goodrich.

It is likely that a portion of the ores found in the Ozark uplift belong
to this class, this uplift being the dome-like group of hills, almost 2000 m.
in height, occupying the southern half of Missouri and a narrow strip

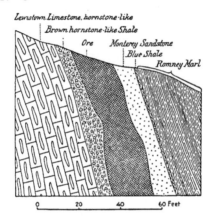

Fig. 379.—Oriskany limonite deposit at the Wilton mine near Glen Wilton, Pa. E. C. Harder.

along the northern boundary of Arkansas. The basement is formed of
Archaean and Algonkian porphyry and granite, overlaid by Cambrian,
Silurian and, in the south-west of the district, also by Lower Carboniferous.

The deposits occurring here in Carboniferous limestone are regarded
as of metasomatic formation, their iron content having percolated from
the Cambrian and Silurian beds into the Carboniferous limestone. These
deposits do not appear to be of any great economic importance.

THE MANGANESE LODES

As already stated,[1] and re-stated when discussing the sedimentation of the iron- and manganese deposits,[2] iron and manganese have chemically and geologically many properties in common, such for instance as the fact that the small amounts of both metals contained in the rocks go fairly readily into solution. Since however in the earth's crust iron occurs to about sixty times the extent of manganese,[3] deposits of manganese, and especially manganese lodes, are much more seldom than those of iron. This relation between the two metals finds expression in the figures of production;[4] while more than 140,000,000 tons of iron ore are produced yearly, approximately but 1,500,000 tons of manganese ore, together with a fairly considerable amount of iron-manganese ore, are marketed.

The most important manganese deposits are the bedded deposits which, in the Caucasus, South Russia, British India, and Brazil, together yield the preponderating portion, some 80–90 per cent, of the total production. Chiefly owing to the competition coming from such deposits, but also because of the stringent conditions as to purity of ore, manganese lodes are but seldom payable.

With regard to the genesis of manganese lodes, it is significant that in many cases it is not only a question of simple fissure-filling but at the same time of subordinate though often striking metasomatic alteration of the country-rock. In isolated cases the fissure-filling is quite insignificant and the deposit is only payable by reason of the concomitant metasomatism. It is evident therefore that the solutions which deposited the manganese were capable of powerfully attacking the country-rock.

At Elgersburg, in the north of the Thuringian Forest, a metasomatic alteration of quartz-porphyry, one of the most resistant rocks, may be observed, this alteration being so far advanced that only the porphyritic quartz individuals remain untouched. To the same complete extent are porphyrites also sometimes replaced.

[1] *Ante*, p. 161. [3] *Ante*, p. 153.
[2] *Postea*, p. 979. [4] *Ante*, p. 161.

In regard to the source of the manganiferous solutions, each case must be specially studied and considered. Small amounts of manganese are admittedly contained in most limestones and dolomites, in the form of the carbonate. At many places, therefore, there exists the possibility for the descending fissure- and ground waters to extract from such dolomite and limestone, not only the carbonates of iron, lime, etc., but also that of manganese. This of course could only take place provided that oxygen were not present or that it had already been elsewhere consumed, since otherwise the manganese would be very quickly precipitated from its solution as oxide.

From solutions charged with carbonic acid and under oxidizing conditions iron is precipitated before manganese; under neutral or reducing conditions, on the other hand, the two carbonates are precipitated simultaneously. Under these latter conditions the siderite lodes, which in general are remarkable for a fairly high manganese content, were formed. Since however the solutions only exceptionally contained much manganese and but little iron, large deposits of rhodochrosite poor in iron are exceedingly uncommon; they are found for instance, as is mentioned in the section dealing with bedded manganese deposits, in the Huelva district of southern Spain. Typical lodes with payable rhodochrosite have however not yet been observed.

In lodes formed under oxidizing conditions, iron oxide—and particularly specularite and hæmatite—is deposited separately from manganese oxide, the deposition of this latter taking place later than that of the iron oxide. In hand-specimens it may often be observed that iron ore or an impregnation with iron appears immediately next to the kernel of unaltered rock, while the manganese occurs farther away as an outside rind. A similar sequence may also be observed in lodes, where in many cases below the manganese a zone of iron ore is found.

Owing to the wide distribution of manganese in calcareous rocks, the opportunity to take up manganese is often afforded to such solutions as penetrate limestone. Silicate rocks also without exception carry some manganese, which by weathering or by the action of heated waters passes fairly readily into solution. The occurrence of manganese as dendrites upon the joints of even slightly decomposed rocks is extremely common.

A considerable number of manganese lodes, among which however not many are payable, occur in eruptive rocks, and preferably in such as are acid, namely, granite, quartz - porphyry, etc.; also often in gneiss.[1] Yet the basic rocks upon the whole are distinguished by higher contents of iron and manganese than the acid. The relation between the two metals is however to this extent disturbed, that the acid

[1] Vogt, *Zeit. f. prakt. Geol.*, 1906, pp. 231-233.

rocks contain a relatively higher proportion of manganese to iron than do the basic. Solutions therefore arising from the decomposition of acid rocks will contain relatively more manganese than those coming from basic rocks. These considerations explain why manganese lodes occur more often in acid than in basic rocks.

The shape of a deposit is closely connected with the conditions of its genesis. With regard to manganese lodes it may be said that though normal lodes may often be seen, an irregular ramification of the rock is more frequent. Observations relative to the persistence in depth of the manganese lodes point to no great extension in that direction.

The distribution of the manganese in the lode is sometimes more, and sometimes less regular. In a payable lode, in any case, the mineralization must to some extent be continuous.

The minerals present in those manganese lodes which have been worked are not very varied, being in general limited to those of manganese and iron. This fact however is largely due to the stringent conditions made by the market as to the quality of manganese ore. There are doubtless occasionally deposits where small amounts of sulphides are contained in the manganese; such as these, however, cannot for economic reasons be exploited.

The primary manganese minerals in lodes occur crystalline, the most common being manganite, braunite, hausmannite, and pyrolusite. Of these the last arises by alteration of the other three, so that pseudomorphs of pyrolusite after manganite, braunite, and hausmannite, are often seen.

Near the surface and as a result of the weathering of these primary ores, compact amorphous manganese minerals such as psilomelane, representing the gelatinous combinations, are often formed; of the associated iron minerals, hæmatite is the most common.

Among the gangue-minerals barite is particularly characteristic, while calcite also is fairly common. The association of manganese with barite is not only noticeable in lodes but, as will be mentioned later, also in metasomatic manganese deposits. In addition to clean compact manganese ore which on account of the market conditions is the class of ore most sought, vuggy and brecciated ores also occur, while crusted ores are fairly uncommon.

The distribution of payable manganese lodes is limited to a few districts in Japan, Central France, and Germany, these last including Ilfeld along the southern border of the Harz, and Ilmenau-Elgersburg in the Thuringian Forest. The German manganese lodes are practically exhausted, while in all cases the production from lodes is small, the total output from this class of deposit representing but a fraction of the world's production.

The German production, including that from metasomatic deposits, is only about 300–400 tons yearly. The phosphorus content being below 0·05 per cent, the average value of this ore, owing to its good quality, is about £4 per ton.

The Manganese Lodes in Thuringia

LITERATURE

Dr. Carl Zerrenner. Die Braunstein- oder Manganerzbergbaue in Deutschland, Frankreich und Spanien. Freiberg, 1861.—K. v. Fritsch. 'Geognostische Skizze der Umgegend von Ilmenau am Thüringer Walde,' Zeit. d. d. geol. Ges., 1860, Vol. XII. pp. 97-155.—H. v. Dechen. Die nutzbaren Mineralien Deutschlands, 2nd Ed., 1906.—E. Zimmermann. Explanatory Text with the sections Ilmenau and Suhl of the geological map of Prussia, etc.

The Rotliegendes of the central Thuringian Forest is crossed in the neighbourhood of Ilmenau and Elgersburg by a large number of Hercynian faults which, particularly on the hill range, are collected together in separate groups. The direction in which they strike, coinciding with that of the faults which bound the range, suggests a Tertiary age for these subsidiary faults, which in greater part are filled with fragments of country-rock, and to a less extent with manganese ore. The three most important groups are those at Mittelberg near Arlesberg, Rumpelsberg near Elgersburg, and Oehrenstock near Ilmenau. Smaller groups are found at the Lütsche, as well as at Oberhof, Klein-Schmalkalden, and many other places. With few exceptions the lodes at Mittelberg and Rumpelsberg occur in quartz-porphyry, and those at Oehrenstock in porphyrite tuff.

A steep dip of 70°–90°, the absence of a defined parting on the hanging-wall, and a sharp separation between ore and country-rock on the foot-wall, characterize these manganese lodes as fissures along which large mountain masses falling through a small vertical distance, appear to have subsided. The hanging-wall limit to the lode is often hard to determine, there being a gradual passage from lode material to compact unbroken rock.

The Mittelberg group includes the occurrences on the Mittelberg, the Wüsttrummey, and the Himmelreichskopf, between the Jüchnitz and Zahmer Gera valleys. Quartz-porphyry sheets belonging to different outflows constitute the country-rock; no influence upon mineralization of the change from one sheet to another has anywhere been observed. The principal lode, worked in the Volle Rose and Wilhelms Glück mines, may eventually be proved to be valuable for a considerable length; it reaches in places a width of 4 m. and even 9 metres. This lode, as well as its parallel and branch associates, consists of fragments of country-rock involved in a network of small metalliferous veins. Psilomelane, locally known as hard manganese, is the principal ore, in comparison with which fibrous pyrolusite is subordinate.   Barite is almost the only gangue-mineral.   Veins of pure manganese ore occasionally reach as much as 0·40 m. in width.   Sometimes even when the vein itself has only the thickness of a knife blade, the porphyry enclosing it is completely coloured black and in part altered to pay-

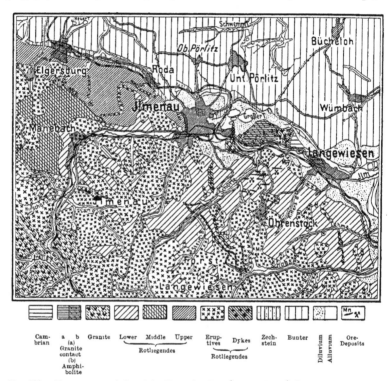

FIG. 380.—The manganese lodes of the Thuringian Forest near Oehrenstock.   Scale 1 : 100,000.
Geological Map of Prussia.

able hard manganese ore, the quartz individuals alone remaining unaltered.

The Rumpelsberg group extends chiefly along the Hohe Warte and the Schnittstein.   In regard to the texture and condition of the country-rock, as well as to the small amount of gangue with the ore, this series resembles the Mittelberg group.   The principal lode, worked in the For-tunatus and Hoffnungs mines, strikes east by south and dips 70°–80° to the north.   Its width is about 0·8–1·5 m., the separation from the country-rock being regular and definite on both walls.   The lode-filling consists of fibrous pyrolusite which sometimes extends in separate seams along the hanging-wall and foot-wall, and sometimes forms the cement of a breccia of angular fragments of porphyry.

To the Rumpelsberg group belongs also the important series which, beginning in the neighbourhood of the lower Stein valley, continues from Jüchnitzgrunde to the Hohe Warte. Upon this series the old mines Gottesgabe, Friedensfürst, and Altes Röderfeld, worked. Only in the lower Stein valley does the conglomerate of the Upper Rotliegendes become the country-rock, this rock consisting elsewhere of different, often spherulitic outflows of quartz-porphyry. Here also a brecciated filling is more common than the occurrence of normal veins of pyrolusite and psilomelane. In the Altes Röderfeld the lode width reached in places as much as 10 metres. Foot-wall veins often occur accompanied on either side by a deep red colouration of the porphyry, the width of this colouration increasing and decreasing with that of the vein. This can only be taken to indicate a saturation of the quartz-porphyry by ferruginous solutions proceeding from the vein fissures.

In the Oehrenstock group, as before mentioned, the country-rock consists of porphyrite tuff. In general the lodes here can only be followed for short lengths, the one exception being that worked in the mines Hüttenholz, Pingen, Luthersteufe, and Beschert Glück, this lode continuing metalliferous for a length of one kilometre. Calcite in the gangue is as characteristic as the occurrence of crystallized hausmannite, braunite, and radial and fibrous pyrolusite, though these minerals, it must be remarked, become less frequent in depth. The lodes in general strike south by east, though numerous branch and re-entering veins often conceal the real strike. They dip some 70° to the south. In the foot-wall the parting is well defined; upon this the 10-50 cm. of brecciated filling rests. With the manganese ores, limonite and hæmatite are occasionally associated, this being also the case in the more isolated occurrences at Oberhof and at the Kehltal fault; indeed, at the last-named place these two iron ores in depth completely replace the manganese.

The coincidence between the strike of these lodes and that of the boundary faults of the Thuringian Forest suggests that these lodes owe their existence to the same tectonic events which brought about the subsidence of the Thuringian fore-ground. They are accordingly probably of Tertiary age, while their filling probably began in Miocene time.

Concerning the source of the solutions from which the ore and gangue were deposited, the occasional decomposition or bleaching of the country-rock along the fissures, as well as the occurrence of manganese dendrites along the joint planes, have by some been quoted as due to lateral secretion or the leaching of the country-rock.[1] It must, however, be pointed out

---

[1] Zerrenner, *op. cit.* pp. 136, 157.

that in most cases there can be no question of extensive decomposition, while the continual change in country, which consists alternatively of porphyry, porphyrite tuff, and conglomerate, completely excludes this view.

The experience that these manganese lodes become poorer in depth—although in the principal lodes the limit to the mineralization has not yet been reached—and the analogy in geological position between these lodes and the manganese lodes at Ilfeld which are not nearly so deep, render it probable that they have received their filling from descending solutions. The circumstance also that in depth along the same fissure a passage from manganese ore to iron ore has in places been observed, favours such a view.

Most certainly the lodes are more numerous in the neighbourhood of the boundary faults and flexures, while impregnation of the beds with manganese in the neighbourhood of subsidences or following the peripheral distribution of the Zechstein is undeniable. At Louisental, the steeply-tilted Zechstein dolomite is altered at the flexure to loose copper-manganese ore ; at the Kehltal fault, sunken wedges of Zechstein dolomite are found altered to manganese ore ; while on the Raubschloss and the Walsberg on both sides of the Wilde Gera valley, manganese occurrences associated with the Zechstein subsidence are known. The discovery of boulders of silicified Zechstein in an elevated position at Oberhof demonstrates, as do also the above-mentioned subsidences and the occurrence of Zechstein remnants at Scheibe near Rennsteig, that formerly the middle portion of the Thuringian Forest was completely covered by Zechstein.

The distribution of the Thuringian and Harz manganese lodes consequently appears to be genetically dependent upon a previous covering of Zechstein which became removed during post-Tertiary denudation.

THE MANGANESE LODES AT ILFELD IN THE HARZ

LITERATURE

W. HOLTZBERGER. ' Neues Vorkommen von Manganerzen bei Elbingerode am Harz,' Berg- u. Hüttenm. Ztg., 1859, No. 42, p. 383.—F. NAUMANN. ' Über die geotektonischen Verhältnisse des Melaphyrgebietes von Ilfeld,' Neues Jahrb. f. Min., Geogn. u. Geol., 1860, p. 1—C. ZERRENNER. Die Braunstein- oder Manganerzbergbaue in Deutschland, Frankreich und Spanien. Freiberg, 1861.—H. v. DECHEN. Die nutzbaren Mineralien Deutschlands, 2nd Ed., 1906.—O. SCHILLING. Explanatory Text with the geological map of Prussia, etc., Section Nordhausen by E. Beyrich and H. Eck, 2nd Ed. p. 10, 1893.

The manganese lodes at Ilfeld occur in Rotliegendes beds unconformably overlying the Elbingerode grauwacke, and in turn unconformably overlaid by Zechstein. These Rotliegendes beds form an extensive complex wherein conglomerates alternate with sandy and clayey sediments, and which includes sheets of melaphyre, porphyrite, and felsite-porphyry, some of which are of considerable thickness.

This complex may be divided into three sections, of which the lowest, formerly considered by E. Weiss as belonging to the Upper Carboniferous, is characterized by conglomerate of exclusively Hercynian origin. The middle section, formerly considered as the Lower Rotliegendes, contains the intercalated sheets of melaphyre and porphyrite, and otherwise consists of conglomerate and tuff-like rocks which, in addition to Hercynian material,

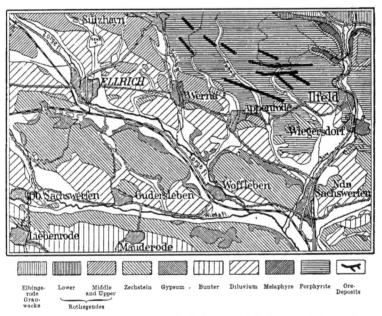

Elbinge-rode Grau-wacke	Lower	Middle and Upper	Zechstein	Gypsum	Bunter	Diluvium	Melaphyre	Porphyrite	Ore-Deposits
		Rotliegendes							

Fig. 381.—The manganese lodes at Ilfeld in the Harz.   Scale 1 : 100,000.   Survey by the
Pr. Geol. Landesanst.

also include fragments of porphyrite and melaphyre.   For the upper Rotliegendes only a narrow horizon is set apart, which, as in the other German districts, contains no eruptives.

In connection with the manganese lodes the porphyrites particularly are concerned, these being separated from a melaphyre below by sediments, while, in part at least, they are covered by slates, etc.   Both these eruptives, as Naumann pointed out, are bedded occurrences.   The porphyrites are distinguished by large segregations of felspar, by decomposed hornblende, and more seldom by specularite and garnet, in a compact, brown, reddish-grey, or green coloured ground-mass, consisting presumably of the same constituents. This eruptive forms in the Ilfeld district an irregular confusion of steep and rocky hills.

Most of the iron- and manganese lodes are found within this occurrence of porphyrite, to the west of the Behre. The manganese lodes vary from a few inches up to 2 feet in width ; they strike from east by north to south-south-east, and dip generally 60°–80°. At the Mönchenberg near Ilfeld, the entire porphyrite mass is traversed by manganese ore in such a manner that the whole is worked in a quarry-like opencut. The depth to which these lodes reach is generally but 10–12 m. and exceptionally as much as 60 metres.

All the lodes are accompanied by subsidiary veins, the ore of which, generally without gangue, is fast attached to the country-rock. The ore is sometimes compact and sometimes crystalline. The crystals of manganite presented from this district to so many mineral collections occurred very extensively in the upper levels. The principal minerals are, manganite, pyrolusite, varvizite, braunite, hausmannite, psilomelane, and wad; while barite, subordinate felspar in part coloured by manganese, and, in the Silberbach mine, rhodochrosite, are the principal gangue-minerals.

An important factor in regard to the genesis of these manganese lodes is the companionship of iron veins, which here attain a width of 1–2 m., so that approximately one-third of the filling consists of iron ore.

The separation from the country-rock is sharp. The red gouge on the walls is of use as a colouring material. When barite predominates the lode-filling is generally coarsely-crystalline. The iron ore, which in depth increases in amount, consists sometimes of pure, and sometimes of impure clayey hæmatite or of kidney ore, in either case intimately intergrown with barite. These mines are now regarded as exhausted.

The iron lodes may also occur in amygdaloidal melaphyre, in which case compact hæmatite, yellow and red jasper, are found.

Genetically there is much to suggest that the solutions to which the manganese- and iron lodes owe their existence were descending solutions, which probably derived their metalliferous content from the ferruginous rocks in the Zechstein and Rotliegendes.

THE JAPANESE MANGANESE DEPOSITS

LITERATURE

Mining in Japan, Past and Present. Published by the Bureau of Mines, 1909.

For a long time rich manganese deposits, the ore from which fetched good prices in the German market, have been known to exist in

Japan. Unfortunately however, as the technical descriptions were all in Japanese it has hitherto been almost impossible to get any exact information concerning these deposits. The appearance in English of the above-cited work wherein the principal Japanese deposits are described, is therefore all the more welcome. The position of these deposits is given in Fig. 310.

Manganese ores are among the most widely distributed of the useful minerals in Japan, being found in not less than forty-nine of the seventy-nine provinces.

These ores occur in deposits which according to their age may be divided into two main groups, namely, the Palæozoic or still older group ; and the Tertiary or still younger group.

From the descriptions in the above-cited work it is not always clear whether the different deposits are epigenetic or syngenetic ; in but few cases only is lode character distinctly stated, though, as will be shown below, in other cases it may be assumed that the ore-bodies occurring conformably with the strike and dip of the country, are epigenetic occurrences in connection with fissures. We therefore for the time consider it proper to consider the Japanese manganese deposits here, leaving future investigation to decide whether or not they should in reality be classed with the ore-beds.

The older deposits have the greatest distribution. They occur in gneiss, sericite-schist, quartzite, Radiolarian schist, schalstein, and clay-slate, generally of Palæozoic age. Whether or not some are of Mesozoic age cannot at present be determined.

Most of these deposits are irregular lenticular bodies with their greatest length approximately parallel to the bedding. In size they vary between wide limits ; sometimes they are small pockets, sometimes large masses out of which 100 tons or more of clean ore may be obtained, while sometimes again they may even be large enough to sustain prolonged operations. The deposits are often accompanied by disturbed zones.

Usually the ore occurs exclusively in the oxidation zone, that is, above the ground-water level. In such case they are worked by opencut or from small shafts. Mineralogically the ore consists of various mixtures of the oxides, psilomelane, pyrolusite, wad, etc. ; more seldom manganite occurs in small prismatic crystals lining cavities in the ore, as on the islands Mutsu and Echigo ; while not infrequently the psilomelane forms compact, radial or fibrous aggregates, as upon Ugo and Noto. In depth, rhodonite is often found with the oxides, from which fact Japanese investigators are inclined to assume that these latter represent the decomposition pro-. ducts of rhodonite and other manganese silicates.

At Toba in Shima the ore is occasionally associated with a light yellow phosphoric mineral found enveloping pieces of manganese ore, than which consequently it must be younger.

The younger manganese deposits occur in Tertiary and Quarternary rocks. Large deposits belonging to this group are found upon the islands Noto, Ugo, Mutsu, and Hokkaido. At Searashi in Noto irregular pieces from nut- to barrel size are found in a bleached green breccia-tuff. The ore is usually accompanied by jasper, this having probably resulted from concomitant silicification.

Along the coast at Koiji in Sudzumizaki, Noto, isolated occurrences are found in a district which is probably underlaid by basalt.

At Fuku-ura, Mutsu, there is among others a 2–4 foot thick bed-like manganese deposit intercalated with Tertiary shales. In addition, between the ore-bodies concretionary masses are often found arranged almost parallel to the bedding. The metalliferous sections of the complex are bleached and in consequence known as *Shabontsuchi* or soap-clay, while the concretions are known as *Shabonkui* or soap-eaters. Here also the ore is associated with jasper, which is often found coated with manganese oxide.

At Owani in the southern portion of Mutsu, an approximately 3 foot wide manganese lode occurs in altered liparite, this rock on both sides of the lode being altered to soap-clay.

In the western portion of the island Hokkaido, mines are at work in the provinces Shiribeshi and Oshima. The Pirika mine in Shiribeshi lies approximately 8 miles north-west of Kunnui, a station on the Hakodate-Otaru railway. There the ore occurs in irregular masses, 1–3 feet in thickness, in a tuffaceous bed made up of sandstone, slate, and breccia, which strikes N. 70° E. and dips 75° north-west. In this the metalliferous horizon is underlaid by a coarse-grained hornblende-granite, and overlaid by a presumably diluvial gravel. In the last-named gravel, manganese ore also occurs in irregular layers. Since the ore encrusts boulders of granite and andesite as well as tree roots, it is evident that the deposit is a secondary and recent formation.

Japanese manganese ores contain :

	Per cent.	Per cent.
Manganese . . .	45–57, and generally	50–55
Iron	1–5 „ „	3
Silica	1–21 „ „	2–7
Sulphur	Nil or traces	...
Phosphorus . . .	0·02–0·7, and generally	0·08–0·16
Copper	Nil, traces, or very little	...
Water	3–21, and generally	4–6

In addition to the occurrences which have been described there is another group concerning which, though the deposits present a bed-like arrangement of large and small ore-bodies, it is expressly stated

that the immediate country-rock is bleached and that boulders of a
supposedly diluvial gravel are coated with manganese oxide.

In spite therefore of the bed-like form it is probable that the Japanese
manganese deposits are epigenetic occurrences formed presumably by
mineral solutions which saturated the rocks.   The possible source of such
solutions is not clear; in part they may have been meteoric waters
percolating down through porous beds, and in part waters circulating in
fissures.   There are however some which are possibly true ore-beds.

The manganese ore production of Japan in 1907 was 20,327 long tons,
made up of 874 tons from Shiribeshi, 3237 tons from Tamba, 2259 tons
from Mutsu, 1671 tons from Bungo, and the remainder from seventeen smaller mining or prospecting districts.   In 1908 the production fell
to 10,890 tons, and in 1909 to 8708 tons, while in 1910 it rose again to 11,120
tons.

THE METASOMATIC IRON-MANGANESE AND MANGANESE
DEPOSITS

As already remarked when describing the metasomatic iron deposits, the
passage between the ores of these two groups is very gradual, and in one
and the same deposit sometimes limonite and sometimes manganese cre
may be won.   In consequence, the distribution of the metasomatic manganese
deposits arising from the alteration of limestone and dolomite coincides in
general with that of the similarly formed limonite deposits.   As previously
stated, iron-manganese deposits are such as contain 12–30 per cent of manganese in an iron ore.   The occurrences in the Bingerbrück limestone belt
and in the Lindener Mark near Giessen are therefore classed as such.

The genesis of these deposits is the same as that of the metasomatic
limonite deposits.   In addition, they too have often been disintegrated by
meteoric or fissure waters, and their substance re-arranged and occasionally
re-concentrated in fragmentary deposits by running water.

The form of the deposits when the replacement has been complete is
bed-like, and when incomplete, irregular and pockety.   The extension in
depth is dependent upon the distribution of the limestone.   The mineralization is seldom so regular as with the siderite- and limonite deposits.

Wherever the limestone contained clayey matter, this remained as
residual clay.   Such clay is often now so mixed with the ore that this
latter as it comes from the mine must be washed.

The minerals present are the same as those with the analogous limonite
deposits.   The manganese is generally distributed as concretions or nodules

embedded in loose limonite. Earthy amorphous ores, resulting from the alteration and concentration produced by weathering, are common. Compared with these the crystalline manganese enrichments appear to be younger.

It has already been stated that with the manganese lodes barite was the characteristic gangue-mineral; the same mineral is almost always present in the metasomatic iron-manganese- and manganese deposits. This constant occurrence is attributed to the original low barium content of the limestone, this content being in the form of carbonate. At the alteration of the limestone the barium was probably taken into solution as bicarbonate, and afterwards, though apparently somewhat before the manganese and iron, deposited as barite. The crystals of this mineral, which occur intergrown in the ore, may sometimes be of considerable size.

The structure of the ore is generally irregularly coarse, though when only the greatest portion but not all of the limestone has been altered, a brecciated structure occurs in which unaltered limestone-kernels and fragments are contained between the fractures from which the alteration proceeded.

In view of the small thickness of these deposits, primary depth-zones have only in so far been observed that it is not every limestone bed which is suited to such alteration, and that consequently an alternation of limestone with iron-manganese- and manganese ore, may occur. Secondary depth-zones do not occur.

As an accessory constituent the presence of copper is occasionally noticed, this metal also being found in the fissures and veins from which the metasomatism proceeded. In such cases, though no enrichment of manganese may occur, there may nevertheless be a secondary cementation zone in relation to copper. Such secondary copper enrichment would be found both in, and immediately under the manganese oxidation zone.

Owing to economic conditions already mentioned, metasomatic iron-manganese- and manganese deposits are worked in but few places. Such workable deposits are represented principally by the Lindener Mark near Giessen, the deposits in the Bingerbrück limestone, and those in the Zechstein mantle in the eastern Odenwald.

Concerning the composition of these ores the analyses given for the lodes may be taken to be applicable, though with the qualification that the metasomatic ores are not so pure. They have consequently little application in the chemical industry and fetch but a low price.

The Iron-Manganese Deposits along the Southern Border of the Taunus and the Soonwald

(a) Oberrossbach, Biebrich, and Bingerbrück

LITERATURE

E. Dieffenbach. Geological Map of Hesse, Section Giessen, with explanatory text. Darmstadt, 1856, p. 21.—F. Beyschlag. ' Die Manganeisenerzvorkommen der " Lindner Mark " bei Giessen in Oberhessen,' Zeit. f. prakt. Geol., 1898, p. 94.—C. Chelius. ' Eisen und Mangan im Grossherzogtum Hessen und deren wirtschaftliche Bedeutung,' Zeit. f. prakt. Geol., 1904, p. 356 ; ' Der Eisenerzbergbau in Oberhessen an Lahn, Dill und Sieg,' Zeit. f. prakt. Geol., 1904, p. 53.—R. Delkeskamp. ' Die technisch nutzbaren Mineralien und Gesteine des Taunus und seiner nächsten Umgebung,' Zeit. f. prakt. Geol., 1903, p. 265. —W. Venator. ' Die Deckung des Bedarfs an Manganerzen,' Stahl und Eisen, 1906, p. 65.— P. Krusch. Die Untersuchung und Bewertung der Erzlagerstätten. Stuttgart, 1907.— Jüngst. ' Die Manganeisenerzvorkommen der Grube Elisenhöhe, bei Bingerbrück,' Glückauf, 1907, p. 993.—Bodifée. ' Über die Genesis der Eisen- und Manganerzvorkommen bei Oberrossbach im Taunus,' Zeit. f. prakt. Geol., 1907, p. 309.—W. Venator. ' Zur Deckung des Bedarfs an Manganerzen,' Stahl und Eisen, 1908, p. 876.—Einecke and Köhler. ' Die Eisenerzvorräte des Deutschen Reiches.' Archiv für Lagerstättenforschung, Part 1, K. pr. geol. Landesanst. Berlin, 1910.

Along this hill range which consists principally of quartzitic rocks, iron-manganese deposits occur chiefly at three places, namely, between Oberrossbach and Köppern, between Bingerbrück and Biebrich, and between Bingerbrück and Stromberg.

According to v. Reinach the deposits at Oberrossbach-Köppern are found in a subsidence running parallel to the Taunus in a south-west direction, in which subsidence a strip of contorted Middle Devonian limestone has dropped down between pre-Devonian and Devonian beds. Since however the surface is covered by Tertiary beds, the limestone never outcrops, and the tectonic connection between the sunken beds, the adjoining plateaus, and the two Tertiary coverings, is not apparent. Nor has this Stringocephalus limestone yet been exposed in all the mines.

Two types of ore - bed may be distinguished, namely, one lying immediately upon the limestone, as for example that occurring north of Oberrossbach ; and a second intercalated between clay-slate and Tertiary beds, as for example that occurring at Köppern where no limestone has yet been met.

At Oberrossbach, where mining operations are still only upon a small scale, the dolomitized Stringocephalus limestone, which has so far been exposed for a length of one kilometre, dips on an average about 50° southeast. This limestone is irregularly fractured and forms the immediate foot-wall of the deposit, the hanging-wall of which consists of Tertiary sands and clays. The ore occurs in pockets, bunches, and valley-like

channels, along the crushed and steeply inclined boundary between limestone and slate, the intercalated slaty, clayey, and sandy layers being
evidence of its transported character. The thickness of the deposits
varies greatly ; while along the channels this reaches 10–20 m., on the
crests of the limestone the deposit is often entirely absent. The ore-bed

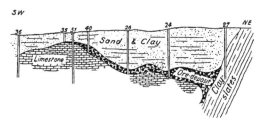

FIG. 382.—Transverse section of the manganese deposits north of Oberrossbach.
Scale 1 : 2200. Einecke and Köhler.

consists of limonite with variable manganese content, and of pocket-
like segregations of pyrolusite and psilomelane. The geological position
is indicated in Fig. 382.

A second bed, having apparently no connection with limestone,
occurs about 1 km. to the west. This is probably the westerly continuation
of the steeply inclined portion of the eastern deposit. It has been opened
up for a length of 300 m. and to a depth of 100 metres. It is distinguished

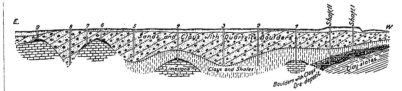

FIG. 383.—Transverse section of the manganese deposits south of Oberrossbach. Vertical scale
1 : 2500 ; horizontal scale 1 : 7500. Einecke and Köhler.

from the first occurrence by its thickness and pronounced bedding.
In the hanging-wall the ore is earthy and usually mixed with
quartzite pieces from the Tertiary and Diluvial country-rock. It is
probable therefore that it is no primary deposit from solutions, but a
fragmentary deposit of which the primary occurrence appears to have
been of metasomatic origin. The fragments of slate and quartzite contained in it have been crushed and disintegrated, with the formation of
clayey and sandy material around them ; the gradation between such
pieces and ore is gradual. The geological position of this deposit is illustrated in Fig. 383.

The deposit at Köppern along the southern border of the Taunus, lies upon quartzite, from which it is separated by disintegrated quartzite and clay, while in the hanging-wall it is covered by an alternation of clay and sandy beds. Here also the deposit is secondary, and high-grade concretions are scarce. The thickness on an average is 4 m. and the dip 15°, while the deposit has been opened up to a depth of 50 metres.

In the Biebrich-Bingerbrück district the deposits occur along the southern slope of the Rhine valley, from Biebrich to Bingen, where the Rhine reaches and breaks through the Taunus, at the subsidence there existing. In this situation, though payable in but few places, deposits are distributed over the entire quartzite ridge. Payable deposits have been known in the mines Hörkopf, Kons. Schlossberg-Dachsbau, Hollgarten, and Neudorf, of which however only Kons. Schlossberg is now working. The foot-wall of the deposit consists of quartzite, or sandstone and conglomerate resulting from the disintegration of the quartzite; between this and the ore there is occasionally a thickness of clay. The hanging-wall consists of a boulder clay 10–30 m. thick.

The size and number of deposits vary considerably. The average thickness is 2–3 metres. At the Kons. Schlossberg limonite occurs with psilomelane and pyrolusite, while at Hörkopf hæmatite is found. The deposits appear to be more manganiferous where ridges of quartzite protrude upwards from the foot-wall into the deposit, or even into the Tertiary hanging-wall.

The Stromberg-Bingerbrück deposits occur between these two places on the left bank of the Rhine. Here are found the mines Concordia at Stromberg, Amalienhöhe at Waldalgesheim, Elisenhöhe at Weilerwest, and the Bingerloch adit.

The geological position is similar to that which has been described in the previous cases. The deposit at Amalienhöhe, which has been explored for a depth of 115 m., pitches to the south-west and is both wide and long. While upon the 18 m. level the ore-body had a section 60 × 50 m., on the 85 m. level the section increased to 200 × 120 m., though the quality had deteriorated. Except for an exposure in a prospecting drive to the south on the 85 m. level, limestone has not been encountered here. The deposit contains clay, shale, quartz-breccia, boulders, and sand, and shows numerous clayey pressure surfaces. Without doubt therefore it has been accumulated by the mechanical agency of water.

In the Elisenhöhe mine the deposit lies on dolomite from whence it extends to the Tertiary beds above. It pitches to the north, the shape and content continually altering. At the adit level it had a section of 1600 sq. m.; on the 20 m. level, 6000 sq. m.; and on the 60 m. level, 4000 square metres. The ore is similar to that described in the other deposits. As mined it contains 20 per cent of manganese, 30 per cent of iron, and 15 per cent of insoluble residue.

Genetically, all these deposits are connected with zones of disturbance, and in all probability the first stage in their formation was the replacement of Stringocephalus limestone by iron- and manganese solutions circulating through fissures. In the second stage the metasomatic bodies thus formed became disintegrated by water and then re-formed as fragmentary deposits chiefly in the neighbourhood of the original deposit, but also some distance therefrom.

The disturbances, which are met fairly often in these deposits, are the result of movement along faults in more recent time. While the age of the primary deposits is uncertain, the fragmentary deposits are probably Tertiary.

Einecke and Köhler estimate the ore available in the present known deposits of the three districts at 1,500,000 tons.

(b) The Iron-Manganese Occurrence at Lindener Mark near Giessen

This occurrence, the most important of this class in Germany, lies about half an hour's journey from Giessen. It is exploited in several large opencuts. The ore consists of an irregular bed of Middle Devonian Stringocephalus limestone covered by thick, light red to white clays supposedly of Tertiary age, and by fluviatile gravels. In but few places does this limestone preserve its original character; generally it has been altered to a ferruginous and manganiferous dolomite, while its surface has been so incised by running water that numerous pot-holes now exist.

In many places the boundary between payable loose ore and the foot-wall dolomite is not definite, while at other places where a light yellow dolomite occurs, there is a sharp separation; a crust of high-grade manganese ore, however, usually covers the irregular dolomite surface. Against the hanging-wall clay also, the outline of the ore-body is not well defined, this clay descending into irregular, pocket-like holes in the ore. The thickness varies greatly, reaching sometimes as much as 30 m., though on an average it is but 8 metres.

The quality of the ore also varies considerably; poor clayey masses of lens- or funnel shape occur in the deposit. The ore consists principally of loose limonite, in which patches of pyrolusite, polianite, wad, and manganite, occur. The average composition of the ore won is approximately 20 per cent of iron and 20 per cent of manganese, so that it is a typical iron-manganese ore. On the market it is known as Fernie ore after the name of the former owner of the mine.

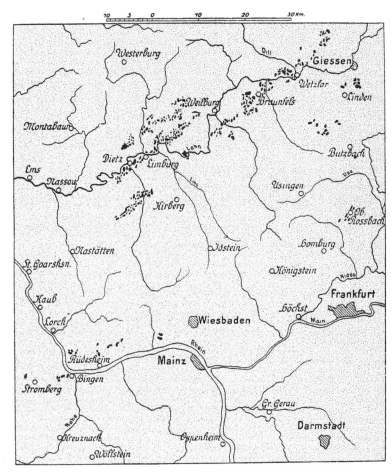

FIG. 384.—Map showing the position of the metasomatic limonite and iron-manganese deposits of the Taunus and the Soonwald.

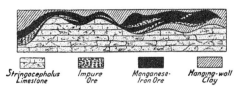

Stringocephalus Impure Manganese- Hanging-wall
Limestone Ore Iron Ore Clay

FIG. 385.—Transverse section of the manganese-iron deposits at the Lindener Mark near Giessen. Beyschlag.

Genetically, the occurrence in the first place resulted from the meta-somatic replacement of Stringocephalus limestone, subsequently to which and probably in Tertiary time, it became in greater part disintegrated and finally re-assembled in the present fragmentary deposit.

The economic importance of the above-described occurrences in the Taunus district, including that at the Lindener Mark, may be gathered from the following figures. In 1910, ten mines at work produced 278,055 tons of ore, of which the average iron content was 23·5 per cent, the manganese content being variable. The average value of this ore was about eight shillings per ton, the whole of it being saleable. It is interest-ing to compare this figure with the total production of iron ore in Germany during the same year, which was 22,964,765 tons.

THE COPPER LODES

COPPER lodes, in which sulphide copper ores occur almost exclusively, represent fissure-fillings similar to the lead-zinc-silver lodes. Of the gangue-minerals, quartz, sometimes accompanied by tourmaline, is the most frequent. In some cases calcite, at times associated with barite and fluorite, predominates, while only very exceptionally is this the case with the two latter minerals. With regard to gangue, a particular type of lode is represented by those in which, apart from quartz, siderite is by far the most characteristic mineral, forming as it then does a large part of the lode-filling.

Copper lodes frequently contain some galena and sphalerite; since however some of the lead-zinc lodes carry some copper in addition to silver, galena, and sphalerite, such lodes may be taken to represent gradations between these two types of lode. Beside those lead-zinc lodes wherein the copper content may be considerable, there are isolated districts where such lodes carry copper in particularly large amount. These constitute the cupriferous facies of the sulphide lead lodes, representative occurrences of which are found at Himmelfahrt and the Junge Hohe Birke, etc., near Freiberg;[1] at Oberschlema and Schneeberg in the Erzgebirge; and at Stolberg-Neudorf and Lauterberg in the Harz. The tetrahedrite lodes at Schwaz and Brixlegg in the Tyrol, and the silver-copper lodes at Cerro de Pasco in Peru,[2] constitute another special type.

Again, the gradations on the one side to the gold-silver lodes, and on the other to the tin lodes, are of great interest. For instance, the gold lodes at Altenberg in Silesia, to the north of the Bergmannstrost lode, carry copper to such an extent that prior to the discovery of their considerable gold content they were regarded as copper lodes. Furthermore, the copper-bearing tin lodes of Cornwall and the Herberton district, Australia, have long been famous. In these the occurrence of the copper is dependent

[1] *Ante*, p. 674.　　　　　　　　　　[2] *Ante*, p. 580.

upon the nature of the country-rock and upon the position relative to the primary zones, though even among the tin lodes proper others occur in which copper predominates.

Since however such gradations are only of small economic importance, the copper lodes proper may be regarded as a distinct, well-defined group of deposits in which a type containing clean copper ore may be distinguished from one of copper-bearing siderite.

Among the primary ores of this group chalcopyrite is the most important. This in most cases is accompanied in more or less large amount by pyrite and sometimes also by pyrrhotite. Only exceptionally, as at Telemarken in southern Norway, is pyrite completely absent ; at that place, beside chalcopyrite the lodes carry only bornite or chalcocite. In the case of the very productive chalcopyrite-quartz lodes at Moonta in South Australia, the pyrite content, however, is low.

Chalcopyrite occurs secondarily in the cementation zone of deposits in which cupriferous pyrite constitutes the primary ore. In almost every case where the genesis of the deposit has been definitely determined, bornite and chalcocite are products of the cementation zone, though this fact does not exclude the possibility of both also occurring as primary ores. In each such case this question must be settled on its own merits. Vogt in this connection refers to the Aamadal mine in Telemarken,[1] where the surface has suffered extensive erosion from land ice. There the decomposition zone in consequence of this erosion is only about 1 m. in depth, below which follows a chalcopyrite-quartz zone which down to a depth of 150 m. carries neither pyrite nor bornite. Only at very great depth was bornite, intimately intergrown with chalcopyrite, found in the main lode, so that a secondary formation of bornite in this case appears to be excluded.

In many copper lodes the sulph-antimonides and sulph-arsenides are completely absent, or occur quite subordinately. Cases do however occur where such minerals are especially characteristic. Economically, the most important among these ores is the mineral enargite, Cu_3AsS_4, which at Butte, Montana,—the most important copper lode-district of the present day—represents not less than some 30 per cent of the minerals present ; without doubt this occurrence is in greater part secondary. Enargite is also found in the copper lodes at Tintic in Utah ; at the Morococha Lagoon in the Peruvian Cordilleras ; in the silver-copper lodes at Cerro de Famatina in Argentina ; in some of the deposits in Chili ; at Mancayan in the Philippines ; and at Bor in the east of Serbia. Bergeat and Beck in their text-books have placed these copper lodes characterized by containing enargite, in a class by themselves.

[1] *Postea*, p. 901.

Tetrahedrite is not uncommon in copper lodes, this mineral in fact being sometimes the most important of those present, as for instance in the occurrences at Schwaz and Brixlegg in the Tyrol, where it contains a little silver and occasionally also a little quicksilver. Similar deposits containing an argentiferous quicksilver-tetrahedrite with 35 per cent of copper, are found at Mascara and Kresevo in Bosnia; and with tennantite, though almost without silver, at Teniente in Chili. The other copper sulpho-salts are only interesting mineralogically, as are also selenium-copper, selenium-silver-copper, etc.

The various primary and secondary copper ores, together with their characteristics, have already been described when dealing with the ores of the different metals.[1]

The world's copper production in 1909, which was approximately 840,000 tons, was contributed to by the different ores as follows:

Native copper, from Michigan, United States, about 101,000 tons; Corocoro, Bolivia, about 2000 tons; or, including a small amount from other mines, in all about 12 per cent of the total production.

Carbonate-oxide ores are estimated to have yielded 150,000 tons, or 15-20 per cent.

Enargite alone is responsible for about three-tenths of the Butte production, or 40,000 tons; or, including some other deposits, in all about 5 per cent.

Tetrahedrite and other sulpho-salts are estimated to have produced at most 1-2 per cent.

Chalcopyrite, bornite, and chalcocite yielded together about 60–65 per cent. It may in fact be assumed with a fair amount of certainty that about one-half the copper produced comes from chalcopyrite and cupriferous pyrite.

Stibnite, arsenopyrite, and silver minerals, are found in most copper lodes in much the same subordinate amount as galena and sphalerite. The matte obtained upon smelting, apart from a few unimportant occurrences, usually therefore contains small amounts of lead, zinc, antimony, arsenic, etc.

Some silver also is invariably present. Even where silver minerals themselves are completely absent, furnace copper seldom contains less than 0·025 per cent of silver as well as a small amount of gold. This low precious-metal content is derived in greater part from the sulphide copper ores, chalcopyrite, bornite, and chalcocite. The gold, together with a small portion of the silver, comes perhaps from pyrite. The silver content frequently reaches 0·1 per cent or more, the Bessemer copper from the Butte district—which is subsequently refined by electrolysis—containing

---

[1] *Ante*, pp. 89-92, 198-201.

for instance an average of 0·23–0·24 per cent, in addition to 0·0008 per cent of gold. This relatively large amount is however an exception. Similarly, copper matte almost invariably contains some nickel and cobalt, though in 100 parts of copper there are as a rule only about 0·2 parts of nickel-cobalt.

Leaving iron out of consideration, the metals in the copper lodes are found in approximately the same relative quantities as in the intrusive pyrite deposits.[1] In this connection it is particularly noteworthy that both types of deposit exhibit almost the same proportions of copper to silver and of copper to nickel-cobalt.

Apart from the association of copper with tin in Cornwall and in the Herberton district, tin is almost completely absent from copper lodes. Exceptionally, it is found in the lodes at Katharinaberg, south of Sayda in the Bohemian-Erzgebirge, not far from the tin lodes of that district ; and at Boccheggiano in Tuscany.[2]

Of the gangue-minerals, quartz generally predominates in most copper lodes. It is frequently accompanied by some calcite, while barite and fluorite are absent or occur subordinately.

The presence of tourmaline—probably connected with a tourmaliniza-tion of the country-rock — remarked by A. von Groddeck, 1887, and A. W. Stelzner, 1897, in connection with some copper lodes in Chili, is important. This mineral has also been found at Svartdal in Telemarken, Norway; at Copper Mountain in British Columbia;[3] in the Blue Mountains;[4] at Sonora, Mexico ;[5] in the Knisib Valley, German South-West Africa ;[6] and in some small lodes at Monte Mulatto near Predazzo in the southern Tyrol.[7] In some of these lodes molybdenite, scheelite, and wolframite, also occur, while cassiterite is absent ; in others the gold content is so considerable that such lodes may be taken to represent gradations to the tourmaline-bearing gold lodes.

It would appear that a fair number of these tourmaline-bearing quartz-copper lodes occur in association with acid or intermediate eruptive rocks, such as granite, quartz-monzonite, quartz-diorite, etc. They therefore in many respects resemble the tin lodes associated with granite, and especially the tin-copper lodes of Cornwall, from which however they are distinguished by the complete absence of cassiterite. Several of

[1] *Ante*, pp. 163, 302.   [2] *Postea*, p. 911.
[3] Catherinet, *Eng. Min. Journ.*, 1905, Vol. LXXIX. p. 125-127.
[4] W. Lindgren, XXII. *Ann. Rep. U.S. Geol. Survey*, 1900–1901, II. p. 629.
[5] W. H. Weed, *Trans. Amer. Ins. Min. Eng.* XXXII., 1902.
[6] R. Scheibe, *Zeit. d. d. geol. Ges.*, 1888, Vol. XL. p. 200.
[7] A. Hofmann, *Sitzungsber. d. böhm. Ges. d. Wiss.*, Prague, 1903 ; O. Stutzer, ' Über turmalinführende Kobalterzgänge von Mina Blanca in Chili,' *Zeit. f. prakt. Geol.*, 1906, p. 294 ; and K. A. Redlich, 'Turmalin auf Erzlagerstätten' in Tschermak's *Min. Petr. Mitt.* XXII., 1903.

the quartz-copper lodes of Telemarken, occurring in granite, carry potash-mica along the walls, this occurrence being similar to that of zinnwaldite in some tin lodes.[1] These lodes contain in addition some fluorite, while the granite immediately along the lode fissure is altered to a greisen-like rock.[2]

For these reasons, therefore, Vogt[3] described such copper lodes as 'tin lodes with copper in the place of tin.' This analogy to tin lodes however is, even with lodes in granite, only observable in exceptional cases, it being actually the case that the most important copper lodes in granite—such as those at Butte where the granite or quartz-monzonite contains 64 per cent SiO_2—display no traces of a greisen formation.

The crusted structure so frequently met in other lodes is seldom found in copper lodes. In many cases the structure has been rendered complex by the repeated opening of the lode fissure and the consequent re-entry of mineral solutions.

The Relation of Copper Lodes to Eruptive Rocks

A differentiation is made between :

(a) The tin-copper lodes of the Cornwall type, genetically associated with granite.[4]

(b) The quartz-copper lodes characterized by tourmaline and the formation of greisen, and occurring partly within and partly in the immediate neighbourhood of acid or intermediate eruptive rocks, with which, as will be indicated in the description of the occurrences in Chili and Telemarken, they are likewise genetically associated.

(c) The quartz-copper lodes not characterized by the presence of tourmaline, yet occurring in principally acid or moderately acid eruptives. To these among others belong the lodes of Butte, Montana, occurring in quartz-monzonite ; many lodes in Chili, principally in acid and intermediate rocks ; those of Ashio, in liparite, and a large number of other Japanese copper lodes, in acid or intermediate rocks ; the deposits of Moonta in South Australia, in quartz-porphyry ; and the occurrences in Shasta County north of Redding in California, in presumably Triassic alaskite-porphyry.[5] In addition to these more important occurrences the following copper lodes occurring in principally acid or moderately acid eruptive areas are worthy of mention : those in the Robinson District,

---

[1] *Ante*, Fig. 263.   [2] *Ante*, Figs. 145, 263.
[3] *Loc. cit.*, 1887.   [4] *Ante*, pp. 431-436.
[5] Ries, *Econ. Geol. of U.S.*, 1910, p. 419 ; Diller, *U.S. Geol. Surv.*, Bull. 213, 1903, and 225, 1904.

Nevada, in a moderately acid porphyry ;[1] lodes and impregnations in granite, Llano County, Texas ;[2] auriferous copper lodes in gneiss-granite and quartz-porphyry, Gilpin County, Colorado;[3] many copper lodes in the neighbourhood of Sherbrooke in Quebec, in the vicinity of schistose porphyritic andesite;[4] and those at Tilt Cove, Newfoundland, in a mica- and quartz-bearing propylitized porphyrite.[5]

Similar occurrences in Mexico have been described by J. G. Aguilera,[6] who emphasizes the fact that these occur mostly in Tertiary rocks, partly acid, such as granite and rhyolite, and partly intermediate, such as quartz-diorite, andesite, etc. ; but not in basic eruptives.

Similar quartz lodes containing chalcopyrite, pyrite, etc., occur at Cobre on the island of Cuba, in volcanic breccias and lavas ; on the island of Haiti, in diorite dykes and in the contact aureole of such dykes and others of andesite, and in melaphyre.[7]

Among the copper deposits in New South Wales, according to J. E. Carne,[8] five occur in granite, five in porphyry, nine in andesite, and seven in and near serpentine. Most of the occurrences however, and among them the important one at Cobar, occur in slate belonging chiefly to the Silurian.

Quartz lodes with bornite, chalcopyrite, tetrahedrite, etc., occurring in granite penetrated by dykes of olivine-diabase, are found at the Albert Silver mine, 50 miles north-east of Pretoria in the Transvaal.[9] Similar lodes are known in quartz-porphyry at Tschudack in the Altai ;[10] and in diabase at the Sünik mines at Katar in Trans-Caucasia.[11] The analogous occurrence at Kedabek is described a little further on.

Copper lodes, containing among other minerals enargite and covellite, occur at several places near Bor in eastern Serbia, and are associated with kaolinized or propylitized andesite.[12] Those at Imsbach in the Rhenish Palatinate occur in quartz-porphyry and melaphyre.[13]

Many other copper lodes so situated might be enumerated.

(d) The contact copper deposits of which, as already mentioned,[14]

[1] A. C. Lawson, 'The Copper Deposits of the Robinson Mining District, Nevada,' University of California, Bull. of Geol., 1906, Vol. IV. No. 14.
[2] J. F. Kemp, Ore Deposits of U.S., 1900, p. 204.
[3] Ibid. p. 203.
[4] J. A. Dresser, 'Copper Deposits of the Eastern Townships of Canada,' Econ. Geol. I., 1906, pp. 445-453.
[5] Bergeat, loc. cit. p. 823.
[6] Trans. Am. Inst. Min. Eng., 1902, XXXII. pp. 510-512.
[7] H. H. Thomas and D. A. MacAlister, Geology of Ore Deposits, 1909, p. 168.
[8] The Copper-Mining Industry of New South Wales, Sydney, 1899.
[9] F. W. Voit, Zeit. f. prakt. Geol., 1908, p. 137.
[10] B. v. Cotta, Berg- u. Hüttenm. Zeit., 1870, XXIX. No. 7, p. 29.
[11] K. Ermisch, Zeit. f. prakt. Geol., 1902, p. 88.
[12] F. Cornu and M. Lazarević, Zeit. f. prakt. Geol., 1908, p. 153.
[13] O. Krauth, see Beck, 1909, p. 334. [14] Ante, pp. 354, 396-398.

there are a large number, among them being many of great economic importance.

Within one and the same district, true contact occurrences containing copper with garnet, augite, scapolite, wollastonite, etc., are very frequently associated with ordinary copper lodes. Examples of such combined contact- and lode occurrences have been recorded at Bingham Canon in Utah ; in the Clifton-Morenci district, Arizona ;[1] at Cananea in Mexico near the boundary with Arizona ;[2] and at Concepcion del Oro in Zacatecas, Mexico. All these occurrences are of considerable importance, and in all of them both types of deposit appear to be intimately associated with each other. Moreover, according to Bergeat, Boutwell, Emmons, Lindgren, Weed, and others, both were formed during the waning phases of eruptive activity, though according to Boutwell the lodes at Bingham may be somewhat younger than the contact occurrences, which, according to Emmons, is also the case at Cananea.

The native-copper deposits associated with basic volcanic flows, occurring at Lake Superior, are treated in a special chapter.[3]

(e) Finally, many admittedly mostly small occurrences of sulphide copper ores, with a frequently high metal content, in serpentine and its associated eruptive rocks.

To these belong the deposits at Monte Catini in Tuscany and Liguria;[4] that at Riparbella in Tuscany, described by R. Delkeskamp ;[5] the deposits at Rebelj and Wis in serpentine, in north-west Serbia ;[6] and those at Kemenica in Bòsnia, also in serpentine.[7] Delkeskamp mentions in addition some apparently similar occurrences, as for instance the well-known deposit at Arghana Maden in Asia Minor, and several in the north of Corsica. To this class also belongs the small pocket-like deposit of bornite and chalcopyrite in serpentine, at Hatfjeldalen in Norway.

A numerical statement of the various types of copper deposit shows that by far the greater number of these occur in association with eruptive rocks, and that bed-like occurrences, such as the copper-shale in the Zechstein of Germany, the pyrite bed at Rammelsberg near Goslar, and the Permian copper-sandstone, are exceptions.

The copper lodes proper and those connected with contact-metamorphic occurrences—these two types together yielding considerably more than one-half of the total copper production—upon examination show themselves in the majority of cases to be associated with eruptive rocks or eruptive periods. The magmatic-intrusive pyrite deposits[8] are con-

---

[1] *Ante*, p. 396.   [2] *Ante*, p. 398.
[3] *Postea*, p. 928.   [4] *Ante*, pp. 300-301.   [5] *Zeit. f. prakt. Geol.*, 1907.
[6] R. Beck and W. v. Fircks, *Zeit. f. prakt. Geol.*, 1901.
[7] Fr. Katzer, *Leobener Berg- u. Hüttenm. Jahrb.*, 1905, Vol. LIII. Part 3.
[8] *Ante*, pp. 301-337.

nected with basic plutonic rocks. The well-known occurrences in the Lake Superior district are found in close association with basic volcanic rocks, though with these the controlling chemical-geological processes were entirely different.

From the nature of the last-mentioned occurrences many investigators came to the conclusion that the copper lodes also were genetically connected principally with basic eruptives. This conclusion is indeed true in the case of the afore-mentioned unimportant occurrence in serpentine at Monte Catini, and a number of unimportant lodes scattered all over the world in gabbro-diorite, diabase, etc. The greater number of copper deposits occurring within or in the immediate neighbourhood of eruptive rocks—and among them the most important economic deposits—are however closely connected with acid and intermediate eruptives, such as granite, quartz-monzonite, grano-diorite or quartz-diorite, quartz-porphyry, liparite, andesite, etc. Such is the case at Butte, Bingham, Clifton-Morenci, Bisbee, Cananea, Chili, Japan, Moonta, etc.

According to Möricke, in Chili it is principally the quartz and quartz-tourmaline-copper lodes which are associated with acid or intermediate eruptive rocks, while the lodes carrying a preponderating amount of calcite and some barite are usually connected with more basic rocks. It cannot however yet be said that this holds good in general.

The geological significance of the bedded copper lodes appears particularly difficult to determine, since these are related to the intrusive and other pyrite deposits, not only morphologically but often also by their high content of pyrite and pyrrhotite. The occurrence at Ducktown in Tennessee, for instance, is regarded by American authorities as a fissure vein connected with replacement. Vogt, however, from descriptions considers that this occurrence is probably an intrusive magmatic deposit. According to this authority also, the large pyrite deposits at Mount Lyell in Tasmania may be similarly regarded.

It would appear likely therefore that many occurrences formerly regarded as formed by heated waters belong to the intrusive deposits. Doubtless however there are bedded copper lodes which have been formed by heated waters, and which must be reckoned with the lodes.

THE AGE OF COPPER LODES

Many copper lodes, being associated with the extensive Tertiary eruptive or at times late Cretaceous eruptive activity, are comparatively young. Among these are the following :

Butte in Montana : Tertiary, probably early Tertiary ;

Bingham in Utah : late Mesozoic or early Tertiary ;

Clifton-Morenci, Bisbee, and others in Arizona, Cananea in Mexico near the boundary with Arizona : late Cretaceous or early Tertiary, these deposits being combined contact- and lode occurrences.

Several other important lodes in Mexico ; most of the occurrences in Chili ; and most of the lodes in Japan, these latter occurring in Tertiary liparite, andesite, etc. The Japanese pyrite deposits, on the other hand, as for instance those occurring at Besshi, etc., are probably of greater age.

Massa Marittima, etc., in Tuscany : Tertiary, namely, Eocene to perhaps Upper Miocene.

Kedabek in the Caucasus : possibly Tertiary.

Bor in Serbia, Cobre on the island of Cuba, Boleo in Mexico, etc., are also young.

Of the world's copper production, which in 1909 amounted to 840,000 tons, at least 400,000 tons, or about one-half, came from Tertiary and late Cretaceous deposits. The young eruptive epochs are therefore exceedingly important not only for silver, gold, and quicksilver, as has already been demonstrated, but also for copper.

A large number of lodes, such as those in Cornwall, the Urals, Mitterberg, Telemarken, etc., are on the other hand of considerably greater age. Concerning the non-lode-like copper deposits, the intrusive pyrite deposits are mostly exclusively Palæozoic ; the Lake Superior deposits, Cambrian-Algonkian ; and the German copper-shale, Permian, etc., that is to say, they are of greater geological age.

THE CLASSIFICATION OF COPPER LODES

Both Beck and Bergeat in their text-books, conforming to the views of the Freiberg school, classify the copper lodes according to the characteristic ore- and gangue-minerals. Both these authorities place the native-copper deposits of Lake Superior among the copper lodes. These deposits however differ so much both mineralogically and geologically from ordinary lodes that in this work they are treated as an independent class. In this work also, the controlling factor in the classification of copper lodes has not been the mineral-association in any particular deposit, but rather the sum of the general characteristics of that deposit.

Following this idea—though in many cases detailed description is lacking—the following classification, admittedly capable of improvement in many respects, may be formulated :

(1) Copper-tin lodes, regarded geologically as a facies of the tin lodes, associated with granite, e.g. lodes in Cornwall ; at Herberton in Queensland ; and others long exhausted in the Erzgebirge, Saxony.

(2) Quartz-copper lodes containing tourmaline, in part also containing other minerals characteristic of tin lodes, associated with granite and other acid and intermediate eruptive rocks, *e.g.* Telemarken, Norway.

(3) Quartz-copper lodes without tourmaline but often containing some calcite, etc. :

(*a*) within, or in contact with principally acid or intermediate eruptive rocks, *e.g.* Butte, Montana ; Moonta, South Australia ; Japan, etc. ; more rarely in basic eruptives ;

(*b*) in slates, without any apparent association with eruptive rocks.

(4) Copper lodes within, or in contact with principally acid or intermediate eruptive rocks and in association with contact-deposits, *e.g.* Telemarken, Norway.

(5) Copper lodes in association with metasomatic deposits, the latter principally in limestones, *e.g.* Massa Marittima in Tuscany, Otavi.

(6) Copper lodes with preponderating carbonates, including calcite, dolomite, siderite, and some quartz, barite, etc. :

(*a*) in basic eruptive rocks, *e.g.* many occurrences in Chili ;

(*b*) in slates, sometimes chiefly with siderite, *e.g.* occurrences in . Siegerland.

(7) Copper lodes, masses, pockets, etc., in serpentine rocks ; so far, economically speaking, not particularly important, *e.g.* Monte Catini, Tuscany.

(8) Bedded lodes in crystalline schists, *e.g.* Mitterberg in Salzburg, Aamdal in Norway.

The present-day economically most important copper lodes belong more particularly to classes 2, 3*a*, and 4.

THE GENESIS OF COPPER LODES

The alteration of the country-rock along the lode fissure, as descriptions of occurrences in Telemarken, Chili, at Butte and Massa Marittima, indicate, may in different districts take very various form. This alteration is due to hydrothermal processes and is dependent chiefly upon the composition of the circulating solution.

The hydrothermal character of the lodes is expressed beyond question in the mineral-association and the alteration of the country-rock. In the case of many copper lodes associated with eruptive rocks—in Cornwall, Telemarken, at Butte, Clifton-Morenci, Bisbee, Cananea, etc., for instance —it may be proved that their formation took place during the last stages of the eruption. From similar considerations to those put forward when discussing the tin lodes [1] and the young gold-silver lodes,[2] the

[1] *Ante*, pp. 418-423. [2] *Ante*, p. 534.

conclusion may be drawn also for copper lodes associated with eruptive epochs, that the ore was derived from the particular magma.

With lodes occurring in slate and having no apparent connection with eruptive epochs the source of the metal is indeterminate. . Most of these lodes however were probably formed similarly to those at Butte, Clifton-Morenci, etc., though any generalization to that effect should be guarded against.

With regard to the formation of the copper deposits associated with serpentine, such as those at Monte Catini, Rebelj in Serbia, etc., geologists still hold very varied opinions. Lotti considers the deposit at Monte Catini and similar deposits in Tuscany and Liguria to be the products of magmatic differentiation,[1] while Beck and v. Fircks, Katzer, Delkeskamp, and others, on the other hand, consider them to be of secondary formation. In this connection Beck and v. Fircks [2] express themselves as follows : ' The copper belongs primarily to the serpentinized eruptive rocks. It however probably no longer exists in its original condition, the ore-bodies now found being the results of concentration which took place during the complete chemical alteration of the original olivine rock.'

Primary and Secondary Depth-Zones in Copper Deposits

As previously mentioned, secondary alterations of sulphide copper deposits of every genesis take various form in different districts.[3] In some districts, as for instance Butte in Montana and Huelva in Spain, the copper content at the outcrop has been practically completely removed.[4] In other districts such as Burra-Burra, Moonta, and Wallaroo, in South Australia ; Bisbee and Clifton-Morenci in Arizona ; at many places in Chili, at Mednorudiansk in the Urals, and at Katanga in the Congo district, on the other hand, particularly large quantities of secondary copper ores are found at the outcrop, close to the surface. At Burra-Burra, for instance, astonishingly rich carbonate and oxide ores were found near the surface, though in depth no payable sulphide ores existed.

The complete removal of the copper from the oxidation zone is connected on the one hand undoubtedly with the occurrence of much pyrite, which upon weathering yields the necessary acid, and on the other with the absence of carbonates to neutralize this acid. Sulphuric acid becomes formed as the final product of the oxidation of FeS_2, the reactions being as follows :

$$FeS_2 + H_2O + O_7 = FeSO_4 + H_2SO_4,$$
$$2FeS_2 + H_2O + O_{15} = Fe_2(SO_4)_3 + H_2SO_4.$$

[1] Ante, p. 301. [2] Zeit. f. prakt. Geol., 1901, p. 322.
[3] Ante, p. 216. [4] Ante, p. 321; postea, p. 886.

Chalcopyrite, bornite, and chalcocite, by oxidation are in part altered directly to sulphates, and in part dissolved by the ferric sulphate formed at the decomposition of the pyrite, thus :

$$Cu_2S + 2Fe_2(SO_4)_3 = 2CuSO_4 + 4FeSO_4 + S,$$
$$Cu_2S + 5Fe_2(SO_4)_3 + 4H_2O = 2CuSO_4 + 10FeSO_4 + 4H_2SO_4.$$

In the process of this oxidation ferric oxide or hydrate are formed either directly or indirectly, to an extent dependent upon the amount of iron sulphide present.

Chalcopyrite [1] by itself does not upon complete oxidation yield sufficient sulphuric acid to form $CuSO_4$ and $Fe_2(SO_4)_3$, this being also the case with bornite and chalcocite. Generally speaking, therefore, complete removal of the copper content is not found where sulphide copper ores occur without, or with only a small admixture of pyrite and pyrrhotite. The occurrence of limestone at the outcrop—as for instance with many contact-deposits and with some lodes—or of calcite or other carbonates in the gangue, would neutralize the acid formed and the carbonate and oxide of copper, etc., would be precipitated, thus :

$$2CuSO_4 + 2CaCO_3 + H_2O = CuCO_3.Cu(OH)_2 + 2CaSO_4 + CO_2.$$

The most widely distributed of the secondary minerals is malachite, with which mineral, azurite, cuprite, native copper, chrysocolla, atacamite, chalcanthite, brochantite, tenorite, some phosphates, arsenates, etc., are found associated.

The conditions under which native copper is formed have already been discussed.[2] The subject is however again taken up a little later when describing the Lake Superior deposits.[3]

Atacamite, $CuCl_2.3Cu(OH)_2$, is found in large quantities in the oxidation zone not only in Atacama in Chili, where it occurs to a depth of some 100 m., but also in South Australia. Chalcanthite and brochantite have also been encountered in large amount in Chili, in such places as have a low rainfall.

In addition to copper deposits associated with limestone or calcite, oxidation ores also occur to a large extent where lime could not have had a neutralizing effect, as for instance in friable sandstones and slates, when originally sulphide copper ores were associated with relatively little iron sulphide. In such cases the friable nature of the country-rock and a dry climate appear to have been the principal factors.

The reason that in some districts free from lime the copper content has been practically completely removed, while in others likewise free the

[1] Cu : Fe : 2S.　　　[2] *Ante*, pp. 139-140.　　　[3] *Postea*, p. 928.

copper content has been collected to form rich oxidation ores, has not yet been satisfactorily determined.

In the case of large occurrences and where rapid erosion has not taken place, oxidation ores reach to a depth of 10–100 m. ; in Arizona even to 200 metres. As already pointed out,[1] their extent is dependent principally upon the position of the ground-water level, but also upon other factors. In many districts, immediately under the oxidation zone a very rich cementation zone is met, the copper content of which was derived from descending solutions, according to reactions expressed in the following equations :

$$CuSO_4 + 2FeS = CuFeS_2 + FeSO_4,$$
$$CuSO_4 + 2FeS_2 + 4O = CuFeS_2 + FeSO_4 + 2SO_2,$$
$$CuSO_4 + CuFeS_2 = 2CuS + FeSO_4,$$
$$CuSO_4 + 2CuFeS_2 + O_2 = Cu_3FeS_3 + FeSO_4 + SO_2,$$
$$2CuSO_4 + CuFeS_2 + SO_2 + 2H_2O = Cu_2S + CuS + FeSO_4 + 2H_2SO_4,$$
$$11CuSO_4 + 5CuFeS_2 + 8H_2O = 8Cu_2S + 5FeSO_4 + 8H_2SO_4.$$

The demarcation between the oxidation and cementation zones is often strikingly sharp, as for instance at Huelva,[2] Butte, and Ducktown.[3] At Butte, the rich cementation zone extends to a depth of some hundred metres below the oxidation zone, while in other deposits, as for instance at Ducktown, its extent is limited to a few metres.

In consequence of the discontinuance of cementation ores in depth, many mines working lodes containing sulphide copper ores have become considerably poorer in depth. Such variations in content, before the secondary depth-zones became appreciated, were regarded as primary depth-zones.

The deepest copper lode-mines yet known are those at Butte and Moonta in South Australia with depths of about 900 m. and 800 m. respectively ; and some in Chili 600–800 metres deep. In the case of the two first-mentioned the copper content in depth has considerably decreased, though to what extent this must be attributed to the cessation of cementation ores cannot be gathered from existing descriptions of the occurrences.

In Cornwall,[4] a primary depth-zone occurs, in so far that copper ores in depth are replaced by cassiterite. In the silver-copper lodes at Cerro de Pasco in Peru, the silver content diminishes in depth while that of the copper increases.[5]

The copper lodes at Dobschau in Hungary are also remarkable, in that beneath a siderite zone, chalcopyrite, and tetrahedrite occur, and beneath this again another primary zone containing cobalt and nickel.[6]

[1] *Ante,* p. 213.   [2] *Ante,* pp. 10, 321.   [3] *Postea,* p. 889.
[4] *Ante,* p. 434.   [5] *Ante,* p. 580.   [6] *Ante,* p. 807 ; *postea,* p. 903.

The distribution of the copper upon the lode plane in the primary zone though still very variable is not nearly so irregular as is the case with gold- and silver lodes. Rich ore-shoots nevertheless occur, the tendency being for them to be found at lode intersections.

Economically, these lodes represent the most important of all the different classes of copper deposit; as will be indicated later, they are responsible for approximately one-half of the world's copper production. From 100 tons of ore, usually 1–1·5 tons, occasionally 2 tons, and exceptionally a somewhat larger number of tons of metallic copper, are obtained.

The most productive lode district of the present day is that at Butte, Montana.

The District of Butte, Montana

LITERATURE

W. H. Weed, S. F. Emmons, and G. W. Tower, jun. 'Butte Special Folio,' Geol. Atlas of the U.S., 1897, Folio 38.—W. H. Weed. 'Ore-Deposits at Butte,' in U.S. Geol. Surv., Bull. No. 213, 1903, pp. 170-180 ; Journ. of Geol. VIII. pp. 773-775.—Emmons and Weed. 'The Secondary Enrichment of Ore-Deposits,' Trans. Amer. Inst. Min. Eng. XXX., 1901, pp. 177-217.—Weed. 'Enrichment of Mineral Veins by later Metallic Sulphides,' Bull. Geol. Soc. Am. XI., 1900, pp. 179-206.—Emmons. 'The Secondary Enrichment of Ore-Deposits, in Genesis of Ore-Deposits,' Trans. Amer. Inst. Min. Eng. XXX., 1902, pp. 433-472.—H. V. Winchell. 'Synthesis of Chalcocite and its Genesis at Butte Montana,' Bull. Geol. Soc. Am. XIV., 1903, pp. 269-276.—J. F. Simpson. 'The Relation of Copper to Pyrite in the Lean Copper Ores of Butte,' Econ. Geol. III., 1908, pp. 628-636.—R. H. Sales. 'The Localisation of Values in Ore-Bodies, etc., at Butte,' Econ. Geol. III., 1908, pp. 326-331 : 'Superficial Alteration of the Butte Veins,' ibid. V., 1910, pp. 15-21.

The town of Butte in south-west Montana is situated in 46° north latitude, in the central portion of the Rocky Mountains, at a height of 1800 m. above sea-level. The comparatively small mining district to which it gives its name occurs within a large Tertiary eruptive area, about 70 miles long and 40 miles wide.[1]

The most important lodes are found in a relatively basic granite, containing only 64 per cent SiO_2 and rich in hornblende and mica, known as Butte Granite, but which is more correctly a quartz-monzonite.

In this granite, intrusions of granite-aplite under the name of Bluebird Granite containing 77 per cent SiO_2, and of quartz-porphyry under the name of Medoc Porphyry with 70 per cent SiO_2, occur as later differentiated products ; while, as the last eruptive in the sequence and, in all probability of Neozoic age, follows rhyolite with 74 per cent SiO_2, partly in the form of dykes and partly as flows. The disposition of these rocks is indicated in Fig. 386. The rhyolite constitutes the Big Butte mountain which

---

[1] Many analyses of eruptive rocks from this area are published in F. W. Clarke, *Analyses of Rocks, from the Laboratory of the U.S. Geol. Surv.*, Bull. No. 228, 1904, pp. 132-134.—W. H. Weed, *Journ. Geol.* VII. p. 737.

rises 250 m. above the valley-level, and gives its name to town and
district.

The granite is in all probability of early Tertiary age.

The lodes, which are invariably steep, belong to three systems :

(a) The oldest and economically most important are the east-west
lodes, in which a repeated re-opening of the lode fissure may frequently be

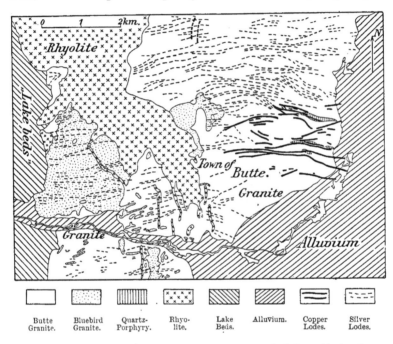

Fig. 386.—Map of the Butte Field.   Emmons, 1897.   The Lake beds consist of sand,
rhyolite, tuff, etc.

observed, and in this sense they may be described as composite lodes.
Their length along the strike frequently reaches several kilometres, as
for instance in the Anaconda, Parrot, Mountain View, West-Colusa, and
Syndicate lodes.

(b) Younger north-west lodes, which cut and dislocate those of the
first system.   Some of these are rich.

(c) Still younger north-east lodes, which dislocate the other two
systems.

These lodes cut the aplite- and quartz-porphyry dykes occurring in
the granite, but on the other hand are older than the rhyolite.

According to the nature of the filling, in addition to copper lodes silver lodes may also be differentiated, these being however of little economic importance. All these quite rich copper lodes, with some minor exceptions, are found in a small area, 3 km. long and 1·5–2 km. wide, or about 5 sq. km. in extent. The silver lodes, now no longer worked, occur to the north, west, and south-west of this area. They contain sulphide silver ores such as argentite, proustite, pyrargyrite, tetrahedrite, stephanite, etc., with some native silver, sphalerite, pyrite, and galena; but in general and except a few in the immediate neighbourhood of the copper district, they contain practically no copper.

The gangue-minerals in the silver lodes are quartz, rhodochrosite, rhodonite, and hübnerite, $MnWoO_4$. In this connection it is interesting to recall the fact that rhodochrosite in many silver lodes, and rhodonite in several of the Hungarian gold-silver deposits, play important parts. In addition, the silver lodes, unlike those of copper, exhibit a pronounced crusted structure and vuggy character.

The principal gangue-mineral with the copper lodes is quartz; calcite and barite occur seldom, while fluorite is practically absent. The separation from the highly decomposed granite is in general not sharp, the solutions circulating in the lode fissure having in part metasomatically replaced the country-rock, such lodes being spoken of by American authors as 'replacement veins.' On either side, the granite is traversed by numerous veins so that the average payable width is 15 m., while in the case of the Anaconda lode it may even reach 30–40 metres. The length along the strike is likewise considerable, frequently reaching one or more kilometres.

The metalliferous material of the copper lodes consists on an average of about 60 per cent of chalcocite, 30 per cent enargite, 8 per cent bornite, and 2 per cent chalcopyrite, covellite, and tetrahedrite; pyrite also occurs in the primary ore.

The occurrence of fresh undecomposed granite, though met in the central portion of the district, is very uncommon, from which fact an intense thermal activity must be presumed.

. The formation of the lodes took place after the consolidation of the granite, aplite, and quartz-porphyry, but before the eruption of the rhyolite. Their formation belongs therefore to one of the last phases of the eruption, and one of intense tectonic movement. Weed in 1903 came therefore to the conclusion that the lode material was derived from the original magma.

Economically as well as scientifically the secondary depth-zones in this district, represented by extensive oxidation and cementation zones, are of great importance.

The gradual development of copper mining in this district may be gathered from the following brief description: In the 'sixties some alluvial gold was won from the neighbourhood of Butte. Then silver mining in the oxidation zone of silver-quartz lodes characterized by large quantities of manganese oxide, was carried on fairly extensively, until the middle of the 'nineties when the price of silver fell. Some of the copper lodes also were originally worked near the surface exclusively for silver, the small amount of silver contained in the primary ore having remained in the oxidation zone, while the copper had been carried almost completely to the deeper cementation zone. In such cases a very small amount of native copper occurring now and then was the only indication of the true character of these deposits, copper carbonates being completely absent.[1]

Generally speaking, in this district oxidized copper ores are rare, though from one place in the oxidation zone of the Bullwhacker mine about 25,000 tons of silicate ore, chrysocolla, were obtained. In general the oxidation zone, reaching to a maximum depth of 100 m., is so completely leached of its copper content that in the early days it was impossible to conclude that rich copper ores existed in depth. Upon further sinking however, at the beginning of the 'eighties, rich cementation ores, constituting what was termed the sulphide enrichment, were found, the separation of these from the oxidation ores being so strikingly sharp that the passage from one to the other occupied at the most not more than a few feet. From that time production increased very rapidly. In the autumn of 1910 the deepest mine had reached 2900 feet, while the deepest drives of the more important mines are now on an average about 2000 feet. The ore-bodies in depth consist principally of quartz and of pyrite with a low copper content. According to micrometallographic studies by Simpson,[2] the pyrite contains mechanically admixed within itself, chalcopyrite, enargite, bornite, and chalcocite, this sequence representing the order of their age.

The composition of the primary mineralization has not yet been definitely settled. It is fairly certain however that the principal minerals, chalcocite, enargite, bornite, and covellite, are secondary. These minerals form crusts upon, and veins in the pyrite, which latter mineral in many places they have completely replaced. Since copper carbonates are entirely absent the conclusion has been drawn that the descending solutions were acid, and not alkaline.

The east-west lodes are characterized by extensive enrichment zones of more than 1 km. in length, while the youngest fault fissures are in part filled with large quantities of crushed clayey material, whereby the mineralization is limited to ore-shoots or bonanzas of very rich ore. Sales,[3] in explanation of this, considers that the descending cupriferous

[1] *Ante*, p. 882   [2] *Loc. cit.*, 1908.   [3] *Loc. cit.*, 1908.

solutions, owing to this impervious clayey material, were only able to use certain channels. According to more recent investigation, chalcocite and enargite, which minerals in general diminish in depth, are to some extent also primary ; of the two, chalcocite being found deposited upon enargite, is the younger.

In this district the ore has become poorer in depth ; the average copper content of the smelting ore from the first thousand feet below the oxidation zone was 8–10 per cent, while in the second thousand feet it was not more than 6 per cent. It is however a question whether in depth, owing to continued improvements in mining and smelting, ore of such low content as formerly to be considered unpayable, has not latterly been worked at a profit.

The boundary between the sulphide enrichment zone and the primary ore is still but little known. According to Sales, the structural geologist of Butte, the cementation zone extends at most 1000 feet below the oxidation zone.

The development of copper mining at Butte is best shown by the following table by Merton, in which the Lake mines and those of Arizona are compared.

COPPER PRODUCTION IN LONG TONS

	1882.	1885.	1890.	1895.	1900.	1905.	1909.	1911.
Montana . .	4,045	30,270	49,560	82,589	114,144	142,490	140,105	122,070
Calumet and Hecla .	14,300	21,075	26,250	34,454	34,745	37,950	40,000	35,000
Other Lake mines .	11,140	11,135	18,200	23,582	24,396	59,820	61,450	61,995
Arizona . . .	8,030	10,135	15,945	21,429	49,447	99,490	130,375	141,490
Others in United States	2,955	1,435	6,370	10,246	40,800	49,370	118,350	132,095
Total of United States .	40,470	74,050	116,325	172,300	263,500	389,120	490,280	492,650
World's total .	181,600	225,600	269,500	334,500	479,500	682,125	839,425	873,460

The copper production of the Butte district increased very rapidly from the beginning of the 'eighties up to 1905, since when it has remained at approximately the same figure. This in part is probably due to the fact that in several of the principal mines operations have latterly had their seat in the intermediate zone between the rich cementation and the poorer primary zone. To-day Butte produces about one-seventh of the world's production. According to Weed, up to the end of 1901 a total of 1,282,000 metric tons, worth £79,375,000, were produced in this district ; and from 1901 to the end of 1911, 2,528,000 tons, equivalent to a value of £160,000,000, the value of the silver and gold being additional.

This enormous amount of copper was practically all obtained from an area only 5 sq. km. in extent. Taking the now existing copper reserves into consideration it is evident that Butte represents the most gigantic copper enrichment hitherto known in any circumscribed district.

The metallic copper produced contains on an average 0·23–0·24 per cent of silver and 0·0008 per cent of gold,[1] while in addition the Bessemer copper contains 0·008 per cent of tellurium. The ore is smelted for the production of matte, this being then refined by the Bessemer and electrolytic processes. The most important works are those of Anaconda, Parrot, Boston, the Butte Reduction Works, and the Montana Ore Producing Company.

In Arizona, which state surpasses Montana in its copper production, the three principal districts, Bisbee, including the famous Copper Queen mine, Clifton-Morenci, and Globe,[2] situated not far from the Mexican border, are all to be regarded as exhibiting a combination of contact-deposits and lodes. At Clifton-Morenci, in the neighbourhood of an occurrence of quartz-monzonite porphyry,[3] there are, in addition to contact-deposits characterized by contact minerals, a number of copper lodes, these being responsible for a considerable portion of the production. Similar circumstances attend the occurrences in the other two districts. Hitherto, in all three districts operations have been carried on chiefly in the oxidation zone.

In the Jerome or Black Range copper district, likewise in Arizona, and at Mineral Creek, the geological position of the deposits appears to be somewhat different.

The most important copper-producing region of the United States, apart from Arizona, Montana, and Michigan, is Bingham Cañon in Utah, 30 km. south-west of Salt Lake City. At that place, as at Clifton-Morenci, contact-metasomatic deposits occur in limestone in the neighbourhood of late Mesozoic or early Tertiary monzonite, on the one hand, and lodes accompanied by impregnation zones in monzonite, on the other. The monzonite first effected the contact-metamorphism of the limestone and its replacement by copper sulphides, and then, when the upper portion of the eruptive rock had become in part consolidated, north-west fissures were formed which became filled with hot aqueous solutions probably derived from the deeper and still molten monzonite magma. These solutions altered not only the limestone but also the monzonite, enriching them both with copper, silver, gold, and some molybdenite.[4]

The copper production of the United States in 1908 was distributed among the various states as follows :

[1] *Ante*, pp. 163, 165, 873.   [2] *Ante*, pp. 396-398.   [3] *Ante*, Fig. 252.
[4] Bouthwell, Keith, and Emmons, *U.S. Geol. Surv.*, Prof. Pap. 38, 1905 ; H. Ries, *Econ. Geol.*, 1910.

	Metric tons.				Metric tons.
Arizona . . .	129,700	Nevada		. .	7,800
Montana . . .	114,200	Idaho		. .	4,600
Michigan . . .	101,600	Colorado . .		.	4,500
Utah	39,400	New Mexico .		.	2,800
California . . .	17,600	Alaska . .		.	2,100
Tennessee . . .	8,800	Wyoming . .		.	1,100

To these are to be added a number of states having a very small production, so that altogether, according to 'Mine Returns,' 434,000 tons, or according to 'Smelters Returns' 427,000 tons, were obtained. The deposits of the Lake Superior district in Michigan are discussed in a subsequent chapter.

DUCKTOWN, TENNESSEE

This district, so often cited in the older American literature, is remarkable among the copper deposits. Here occur lenticular pyrite beds in general conformably intercalated in crystalline schists, presumably pre-Cambrian, and more particularly in mica-quartz schists. These beds are usually 20 m., or in exceptional cases 50 m. thick, and carry principally pyrrhotite with pyrite, chalcopyrite, some sphalerite, and galena. Hornblende, augite, garnet, zoisite, quartz, calcite, etc., occur as gangue. At the outcrop of the principal occurrences, the deposits down to a depth of 30 m. are altered to a gossan, beneath which follows a 1-3 m. thick cementation or sulphide-enrichment zone consisting in greater part of amorphous copper sulphide with some malachite and azurite. This zone at a still greater depth gives place to primary pyrite.

W. H. Weed[1] regards this deposit as a true fissure-filling—containing principally pyrrhotite and pyrite, and practically free from quartz—which derived its material from the alteration of a rock zone consisting chiefly of metamorphic minerals.

J. F. Kemp[2] likewise regards the deposit as a lode-like replacement product, though he believes the original rock to have been limestone.

H. Ries[3] describes the deposit among the true fissure-fillings in mica-schists, this filling being accompanied by metasomatism of a rock consisting of garnet, zoisite, actinolite, epidote, pyroxene, etc., these minerals according to him having resulted from the alteration of calcareous slate.

Vogt points out that according to the descriptions the pyrite occurs as idiomorphous crystals embedded in the pyrrhotite, and that the mineral-association is identical with that of the Norwegian pyrite deposits. He is of opinion that this occurrence, like the intrusive pyrite deposits, owes its existence to magmatic differentiation.

The Ducktown mine is one of the oldest copper mines in the United

[1] 1900.   [2] 1901.   [3] *Econ. Geol.*, Text-book, 1910.

States, operations having been carried on since 1850, when the rich cementa-tion zone was first exploited. Subsequently pyrite was mined, this in 1908 containing on an average 1·55 per cent of copper, as well as some silver and gold. The sulphuric acid obtained from the pyrite is used in the development of the phosphate deposits occurring in the neighbourhood.[1]

MEXICO

The copper production of this country from 1880 to 1911 may be gathered from the following figures :

	Boleo.	Other Mines.
	Tons.	Tons.
1880	...	400
1885	...	375
1890	3,450	875
1895	10,450	1,170
1900	11,050	11,000
1905	10,185	54,255
1909	12,230	44,095
1911	12,165	41,865

The deposit at Boleo in the lower half of the Californian peninsula, which has been worked since the middle of the 'eighties, occurs bed-like in Tertiary tuffs. It is discussed in a subsequent chapter.

Concerning the already-mentioned [2] deposits at Cananea, a paper by S. F. Emmons [3] has recently been published. These deposits occur near the boundary with Arizona in a presumably early Tertiary petrographic province consisting from oldest to youngest of :

1. Diabase.
2. Rhyolite.
3. Mesa-tuffs and andesite.
4. Syenite and syenite-porphyry.
5. Diorite-porphyrite.
6. Granite-porphyry, grano-diorite, and quartz-porphyry.
7. Gabbro-diabase.

In the contact aureole along the diabase-porphyrite, grano-diorite, and quartz-porphyry, and principally within a zone 10 km. long and 1·5-3 km. wide, numerous important contact-deposits containing bornite,

---

[1] C. Heinrich, *Trans. Amer. Inst. Min. Inst.* XXV., 1896 ; J. F. Kemp, *ibid.* XXXI., 1902 ; W. H. Weed, ' Types of Copper-Deposits in Southern United States,' *ibid.* XXX., 1901 ; *Bull. Geol. Soc. Am.* XI., April 1900.

[2] *Ante*, p. 398.

[3] ' Cananea Mining District of Sonora,' *Econ. Geol.* V., June 1910.

chalcopyrite and sphalerite, pyrite, etc., are found.   In addition, a number of likewise very productive quartz lodes containing chalcopyrite, pyrite, sphalerite, etc., occur, principally along fault fissures.   Geologically therefore, Cananea is fairly closely related to the not far distant occurrences of Bisbee and Clifton-Morenci.

According to Emmons, mineralization was effected by strongly heated solutions emanating from the cooling magma.   Two classes of deposit representing two stages in the deposition are to be differentiated, namely, in the first place, contact occurrences, which were formed above the critical temperature of the solutions ; and secondly, lodes, these having been formed by solutions below the critical temperature.

The occurrence of copper contact-deposits and copper lodes in granodiorite at Concepcion del Oro has already been mentioned.[1]   At San José in Tamaudipas, near the boundary with Texas, according to J. F. Kemp,[2] copper deposits also occur which are in part contact-metamorphic and in part lode-like.

CHILI

LITERATURE

J. DOMEYKO (Santiago). Numerous Treatises, Ann. des Mines, Paris, 1840, 1841, 1846 ; Essaye sobre los dépositos metaliferos de Chilé, 1876.—F. MOESTA. Über das Vorkommen der Chlor-, Brom- und Jodverbindungen des Silbers in der Natur. Marburg, 1870.—v. GRODDECK. ' Über Turmalin enthaltende Kupfererze von Tamaya in Chile,' Zeit. d. d. geol. Ges. XXXIX., 1887, p. 237.—W. MÖRICKE. ' Einige Beobachtungen über chilenische Erzlagerstätten und ihre Beziehungen zu Eruptivgesteinen,' Tscherm. Min.-Petrog. Mitt., 1891, p. 121 ; ' Vergleichende Studien über Eruptivgesteine und Erzführung in Chile und Ungarn,' Ber. d. Naturforscherges. Freiburg in Br., Vol. VI., 1892 ; ' Die Gold-, Silber- und Kupfererzlagerstätten in Chile und ihre Abhängigkeit von Eruptivgesteinen,' ibid. X., 1897, p. 152 ; ' Geologisch-petrographische Studien in den chilenischen Anden,' Sitzungsber. d. preuss. Akad. d. Wiss. Vol. XLIV., 1896.—A. W. STELZNER. ' Über die Turmalinführung der Kupfererzgänge von Chile,' Zeit. f. prakt. Geol., 1897, p. 41.—Several works of L. M. CROSNIER. Ann. des Mines, 1851, 1859 ; D. FORBES, 1861, 1863, 1866 ; R. A. PHILIPPI, 1860 ; A. PISSIS, 1875 ; L. SUNDT, 1895, are cited in Stelzner's and in Möricke's treatises.—OTTO NORDENSKJÖLD. ' Über einige Erzlagerstätten der Atacama-wüste,' Bull. Geol. Inst. Upsala, III., 1897, and IV., 1898.—L. DARAPSKY. Das Departement Taltal. Berlin, 1900.—A. ENDTER. ' Das Kupfererzlager von Amolanas im Departement Copiapó,' Zeit. f. prakt. Geol., 1902, pp. 293-296.

The disposition of the most important copper deposits in Chili is indicated on the map constituting Fig. 387, from which it is seen that these deposits are widely distributed in numerous districts between 19° and 35° south latitude, south of which there are but few.

In this region lodes are the chief form of deposit, though according to Möricke some quite subordinate contact-deposits and a few unimportant

[1] *Ante,* p. 398.   [2] *Trans. Amer. Inst. Min. Eng*, May 1905.

occurrences of the Lake Superior type are also found. This authority in 1897 divided the copper-, silver-, and gold-occurrences of Chili, which are so closely connected with eruptive rocks, into the following groups :

Ch = Chuquicamata ; P = Paposo ; G = Guanaco ; R = Remolinos ; C = Carairal ; H = Higuera.

FIG. 387.—Map of the Chilian copper deposits.

1. Gold-copper deposits : Lodes and impregnations containing gold and usually some auriferous copper ores in moderately acid and acid eruptive rocks, such as quartz-gabbro or quartz - augite - diorite, quartz-diorite, syenite, amphibole-granitite, quartz-porphyry or liparite. The principal gangue-mineral is quartz ; tourmaline(!) is frequently present.

A. Gold deposits proper, e.g. at Guanaco in Antofogasta ; Inca de Oro, Cachiyuyo, and Jesus Maria in Atacama ; Talca, Andacollo, and Los Sauces in Coquimbo ; Chivatos in Talca, etc.

B. Deposits containing rich copper ore with, as a rule, a very variable gold content, free gold being found here and there, e.g. at Remolinos and Ojancos in Atacama ; Tamaya and La Higuera in Coquimbo ; Caleu, Las Condes, and Peralillo in Santiago, etc.

These subordinate groups of deposit, being connected by every possible gradation, are not easily differentiated from each other.

2. Silver-copper deposits : Deposits containing silver minerals—with no considerable gold content—associated with argentiferous copper ores, in basic plagioclase-augite rocks, such as diabase, augite-porphyrite, and augite-andesite, or in Mesozoic sediments, especially limestones, penetrated by those eruptive rocks. The principal gangue - minerals are calcite, barite, and quartz. Zeolites are frequently present, while tourmaline is completely absent. These deposits may be divided into the following subdivisions :

A. Deposits with copper ores containing no gold but as a rule some silver. Native silver is sometimes present, *e.g.* at Puquios and Checo in Atacama ; Mercedes de Algodones in Coquimbo ; Catemo in Aconcagua ; Lampa in Santiago, etc.

B. Deposits containing silver. Copper ores occur here more or less subordinately, *e.g.* at Tres Puntas, Cabeza de Vaca, Los Bordos, Chañarcillo, San Antonio in Atacama ; Algodones, Rodaito, Argueros, Quitana in Coquimbo, etc.

These two subordinate groups are likewise most closely connected with each other.

3. Silver lodes with a high gold content. These occur in both basic and acid eruptive rocks. Free gold as well as silver chlorides occur not infrequently, *e.g.* at Lomas Bayas in Atacama, and Condoriaco in Coquimbo.

This type of precious-metal deposit may be taken to represent at the same time a combination of groups 1A and 2B.

4. Deposits containing galena, sphalerite, tetrahedrite, enargite, etc. Though silver minerals are rarely found these deposits are argentiferous throughout, and usually somewhat auriferous. They are associated with Tertiary andesite and liparite and their respective tuffs, *e.g.* at Cerro Blanco and La Coipa in Atacama ; Las Hediondas, Vacas Heladas and Rio Seco in Coquimbo, etc.

The last-mentioned deposits, situated in the Andes 3000–4000 m. above sea-level and worked more particularly for silver and gold but not for copper, belong to the Tertiary gold-silver group.[1] This is also the case with the silver-gold deposits mentioned in group 3, while from descriptions it would appear that very many of the silver deposits of group 2B and of the gold deposits of group 1A are also of Tertiary age.

Those eruptive rocks with which the copper lodes are associated are invariably fairly young, that is, Upper Cretaceous or early Tertiary, some perhaps still younger.

Economically, the copper deposits of Chili are more important than those of silver and gold. This may be gathered from the following figures of production for 1910 :

Copper .	.	. 35,235 tons, worth £2,000,000
Silver .	.	. 44,479 kg.,	„ 150,000
Gold .	.	. 1,268 „	„ 175,000

The Chilian copper lodes carry quartz as principal gangue. Tourmaline, as pointed out by Groddeck,[2] Stelzner,[3] and Möricke,[4] is highly characteristic of many. Such for instance is the case with the auriferous copper lodes in a massive syenitic rock at Taltal in northern Chili ; the auriferous copper deposits in quartz-gabbro and quartz-diorite at Tamaya in Coquimbo, where the country-rock is often completely tourmalinized ;

---

[1] *Ante*, pp. 515-600.   [2] 1887   [3] 1897.   [4] 1897.

the likewise auriferous copper lodes in gabbro-diorite and quartz-diorite in the important mining district of Copaquire in Coquimbo ; and many other deposits cited in the works of Stelzner and Möricke. In addition, actinolite frequently occurs and, according to information received from Professor Schneider, occasionally also garnet. Specularite and molybdenite occur fairly extensively ; scheelite, cuproscheelite, anatase, and zircon, are found as rareties ; while cassiterite, fluorite, and topaz, are absent. According to Möricke, tourmaline occurs only in lodes connected with acid or moderately acid eruptive rocks, such as granite, syenite, diorite, quartz-gabbro, quartz-porphyry, etc., and is absent from those associated with basic eruptives.

The most important ores of the Chilian copper deposits are, chalcopyrite, bornite, and chalcocite ; and in the upper levels, cuprite, malachite and azurite, native copper, covellite, tenorite, brochantite, and in the northern provinces, atacamite. Occasionally much chalcanthite has been found at the outcrop of some deposits, as for instance at Copaquire high up in the Andes.[1]

Tetrahedrite and domeykite are uncommon ; enargite, which occurs fairly extensively in the young lodes mentioned under group 4, appears to be absent from the copper lodes. These copper lodes frequently contain a small amount of silver and gold, the silver being found principally in the bornite and chalcocite, the gold chiefly in the chalcopyrite. Though tetrahedrite is uncommon, it frequently has a high though variable silver content. It is interesting to note on the other hand that this mineral, as in the Teniente mine, may occur practically free from silver.

In these Chilian copper lodes also, the cementation or sulphide-enrichment zone is particularly rich ; at Tamaya, for instance, under a bright-coloured oxidation zone[2] followed a zone containing bornite and chalcocite,[3] this continuing to a depth of 220 m. ; under this again came the primary zone in which chalcopyrite predominated.[4]

Copper mining in Chili began on a very moderate scale at the commencement of the seventeenth century ; it increased considerably about the middle of the nineteenth century so that from 1855 to 1880 Chili ranked first among the copper-producing countries ; but subsequently was surpassed by other large copper-producing countries till now it occupies the sixth place, the sequence being :

1. United States.   4. Japan.
2. Mexico.   5. Australia.
3. Spain.   6. Chili.

[1] H. Oehmichen, *Zeit. f. prakt. Geol.*, 1902, p. 147.
[2] *Metal de color.*   [3] *Metal de color bronceada.*   [4] *Bronce amarillo.*

The importance of the copper production of Chili at different periods, including the copper content of the exported ore and the by-products from smelting, may be gathered from the following table :

	Long Tons.			Long Tons.
1650 . . .	50	1875 . . .		47,670
1700 . . .	100	1880 . . .		39,580
1750 . . .	· 750	1885 . . .		39,800
1800 . . .	1,250	1890 . . .		26,650
1830 . . .	3,000	1895 . . .		22,075
1840 . . .	6,500	1900 . . .		25,700
1850 . . .	12,340	1905 . . .		29,165
1860 . . .	34,120	1910 . . .		35,235
1865 . . .	41,210	1911 . . .		29,595
1870 . . .	44,200			

The total copper production of Chili has been :

before 1840 . . .	estimated at	250,000 tons.
1840–1854 . . .	about	250,000 ,,
1855–1894 . . .	,,	1,501,000 ,,
1895–1911 . . .	,,	484,000 ,,
	Total nearly	2,500,000 tons.

The sources from which the above statistics were compiled were : For the recent periods, H. R. Merton, *Annual Copper Statistics* ; for the earlier periods, A. Herrmann, *La Produccion de Oro, Plata i Cobre en Chile*, Santiago, 1894.[1]

The production for 1903 was distributed among the various provinces as follows :

	Tons of Copper.
Tacna i Arica, in the north	462
Prov. Tarapacá, 19°–21° south lat.	1,496
Dept. Antofagasta, 21°–25° south lat. . . .	3,647
,, Tocopilla ⎰ 25°–26° south lat.	⎰ 1,588
,, Taltal ⎱	⎱ 517
,, Chañaral ⎰ 26°–28° south lat.	⎰ 4,821
,, Copiapó ⎱	⎱ 6,606
,, Vallenar i Freirina ⎰ 28°–30° south lat. . .	⎰ 3,524
,, Elqui i la Serena ⎱	⎱ 2,014
,, Coquimbo i Ovalle, 30° south lat. . . .	4,679
,, Combarbalá e Illapel	426
,, Petorca i Ligua	880
,, Putaendo, etc.	1,152
Prov. Valparaiso ⎰ 33°–34° south lat. . . .	⎰ 773
,, Santiago ⎱	⎱ 2,052
Dept. Talca, 36° south lat.	8
Total .	34,645

According to the *Estadistica Minera de Chile en 1903*, Santiago, 1905, Merton reckons the copper produced in Chili during that year to have been 30,930 tons.

The most productive mines lie between 26° and 30° south latitude, in part in the Atacama district. At its zenith, from 1860 to 1885, Chili

[1] Vogt, ' Die Statistik des Kupfers,' *Zeit. f. prakt. Geol.*, 1896, p. 89.

produced some 40,000 tons annually, a production which subsequently decreased to 20,000–25,000 tons ; only however to rise again considerably of late years. The great fall in production from 1885 to 1900 was due to the particularly low price of copper, and to political disturbance. The statement sometimes heard that the copper lodes rapidly pinch out in depth or become completely impoverished, is incorrect. According to Schneider in 1910, none of the long-established mines were then exhausted, though some had reached a vertical depth of 600–800 metres. The more important mines are : Collahusai in Tarapacá ; Chuquicamata in Antofagasta ; Guanaco with very auriferous copper ores, and Paposo with a vertical depth of 400 m., in Taltal ; Descubridora de Carizalillo, 650 m. deep, and Fortunata de las Amimas, 430 m. deep, in Chañaral. In Copiapó, the important Dulcinea mine, 800 m. deep, and the Cerro Blanco and Ojanco mines, 380 m. deep, near Remolinos, with auriferous copper ore. In Vallenar i Freirina, Carizal-Alto, 414 m. deep. In Elqui i la Serena, the celebrated La Higuera mine, 350 m. deep, and the Brillador, 550 m. deep. In Coquimbo i Qualle, Zannhillo, 200 m. deep, and Tamaya with the Rosario mine, 590 m. deep. And finally, Desengaño in the neighbourhood of Santiago.

The famous Chañaricillo silver mine, first discovered in 1832 though now practically exhausted, reached an approximate depth of 700 metres. The value of the silver produced from this mine was according to Möricke about £60,000,000, though Nordenskjöld estimated it to be only £22,500,000.

JAPAN

LITERATURE

Les Mines du Japon, Paris Exhibition, 1900.—' Mining in Japan,' Past and Present, Bureau of Mines, 1900.—Résumé statistique de l'empire du Japon, Tokio.—L. de LAUNAY. La géologie et les richesses minérales de l'Asie, Paris, 1911.

A map indicating the disposition of the useful deposits of Japan is presented in Fig. 310. Of late years this country has ranked fourth among the copper-producing countries. In addition to deposits of cupriferous pyrite, copper lodes are also known. To the first-mentioned belongs the deposit occurring in crystalline schists under similar circumstances to the deposits of Norway and Spain, at the Besshi mine on Shikou, one of the southern islands, this mine being 544 m. deep and responsible for about one-eighth of the copper production of Japan. Of greater economic importance however are the copper lodes occurring principally on the main island of Nipon. In the north of that island and in the provinces Echizen and Kaga somewhat farther south, such lodes are found at Kosaka, Osarusawa, Ani, Hisanichi, Arakawa, Nagamatsu, Ogoya, and Yusenyi.   Almost all these occur either in, or in the neighbourhood of Tertiary propylitized liparite and andesite.   Exceptionally, the lodes at Omodani occur in Mesozoic beds in association with quartz-porphyry, and others at Mizusawa in granite, though liparite exists in the neighbourhood.   At Kosaka, now on account of its high silver content the most important copper deposit in Japan, the lodes are found in andesite or trachyte and attendant tuffs.

The deposits at the famous Ashio mine, which are responsible for approximately one-sixth of the present Japanese copper production, occur principally in liparite, dacite, and andesite ; a few only, have as country-rock the slates adjacent to these eruptives.   The ore of these lodes consists chiefly of chalcopyrite and pyrite with some bornite, a little sphalerite, and galena ;   while quartz, calcite, and barite, occur as gangue.   L. de Launay compares this district with that of Butte, Montana, manganese minerals being common to both.

In Chugoku, the deposits traverse Palæozoic as well as Tertiary beds penetrated by eruptive rocks.   Although that at Sasagatani is regarded as a contact occurrence, others are undoubtedly typical fissure-fillings in quartz-porphyry and liparite.

From this brief description it is seen that many of the Japanese copper lodes, and among them the important deposits at Kosaka and Ashio, are of Tertiary age.

Copper mining is the most important mining industry of Japan.   The value of the production of the different heavy metals in 1908 was as follows :

Copper	.	.	.	.	£2,350,000	Antimony	.	.	.	.	£5,500
Gold	.	.	.	.	715,000	Tin	.	.	.	.	3,600
Silver	.	.	.	.	462,500	Zinc ore	.	.	.	.	31,500
Iron	.	.	.	.	400,000	Pyrite	.	.	.	.	17,500
Lead	.	.	.	.	41,500	Manganese ore	.	.	.	.	15,500

The copper production of Japan, whose art industry in this metal is universally recognized, is many centuries old.   The deposit at Besshi, discovered in 1690, produced in 1698 some 1500 tons of copper ; while the Ashio deposit discovered in 1610, from 1676 to 1688, produced on an average 1375 tons yearly.

At Joshioka mining operations began as far back as 807, and at Omodani between 1342 and 1344.

From the following table it is seen that the copper output of this country has of late years increased considerably.

			Tons.				Tons.
1875	.	.	2,400	1895	.	.	19,100
1880	.	.	4,700	1900	.	.	25,300
1885	.	.	10,500	1905	.	.	35,500
1890	.	.	18,100	1911	.	.	52,000

In 1908 there were at work no less than forty-three copper mines of importance, among which the following are worthy of particular mention : Kosaka in Rikuchu, Ashio in Shimotsuke, Besshi in Jyo, Osaruzawa in Rikuchu, Ani in Ugo, Ikuno in Tajima, and Kano in Iwashiro.

AUSTRALIA

LITERATURE

J. A. PHILLIPS and H. LOUIS. A Treatise on Ore Deposits. London, 1896.—Reports of the various colonies for the Mining Exhibition, London, 1890, and other exhibitions.— L. GASGUEL. Ann. des Mines, 10 Ser. Vol. VII. pp. 544-562, 1905.

One of the most important copper-fields in this country is that of Moonta-Wallaroo in the Yorke peninsula, near Adelaide, South Australia. At Moonta, first discovered in 1861, copper lodes occur in quartz-porphyry. Seventeen kilometres away the well-known Wallaroo mine, discovered a year earlier, exploits lodes in Cambrian mica-schists with some limestone, etc. The lodes of both districts carry principally chalcopyrite and some bornite, with quartz as gangue. Wallaroo carries in addition some pyrite, arsenopyrite, cobaltite, etc. At Moonta the pyrite content is very low. Barite, fluorite, and tourmaline, apparently do not occur at either place ; some felspar on the other hand is found at Moonta.

In this latter district, within an area 1·5 km. square, a considerable number of lodes are known. In 1903 the deepest shaft had reached a depth of 800 metres. The lodes become poorer in depth. At Wallaroo a steep lode of considerable extent and usually at least 2 m. wide is worked. This lode cuts through the schists. In 1890 the deepest shaft was 400 metres.

At the outcrop of these lodes large quantities of malachite, azurite, cuprite, atacamite, native copper, etc., occurred, the carbonates and atacamite predominating. The sulphides were first met in greater depth. At Wallaroo for instance, down to a depth of 30 m. only carbonates, etc. were known ; below this and to a depth of 50 m., black copper sulphides ; and below this again, chalcopyrite mixed with quartz. The black copper sulphides are in all probability cementation ores. At Moonta on the other hand, where in the neighbourhood some limestone comes to the surface, chalcopyrite began to appear at a depth of 30–40 metres. From 1861 to 1903 this mine produced about 120,000 tons of metallic copper ; to this must be added the production of Wallaroo, and later that of the united Moonta-Wallaroo mine, this being about 6000 tons per year.

Burra-Burra, in South Australia and about 160 km. north-north-east of Adelaide, was formerly famous for its rich oxidation ores. At this place two lodes occur in a complex consisting of slate, limestone, and sandstone—sometimes described as serpentinized limestone—these lodes near the surface carrying malachite, azurite, cuprite, native copper, etc. These oxidation ores reach to a depth of 75–90 m. free from sulphides ; not till a depth of 170–190 m. is reached do sulphides exclusively occur, and then such are represented almost entirely by chalcopyrite.

This deposit was discovered in 1845, and exploited with considerable profit up to 1877 ; altogether 234,648 tons of ore containing 51,522 tons of copper worth £4,749,224, were produced, a considerable portion of this sum being net profit. In this case, in spite of the high content of the oxidation ore, the sulphide ore proved to be unpayable.

Most of the other Australian copper deposits, as for instance the important occurrence at Cobar in New South Wales, likewise occur in the form of lodes. The pyrite deposits at Mount Lyell on the west coast of Tasmania are however exceptions. These deposits were discovered in 1886, though not till the middle of the 'nineties was exploitation begun. They occur in metamorphic Silurian slates, and in many respects resemble the Huelva pyrite deposits. The ore-bodies have an average width of 200 feet or a maximum of 300 feet, and may be 950 feet long. The ore consists on an average of 83 per cent of pyrite, 14 per cent chalcopyrite with 4·5 per cent of copper, 2 per cent barite, and 1 per cent of quartz. Mining operations began in 1896, resulting later in a yearly output of 8000 tons of copper. From 1896 to 1908 the total copper production of Tasmania amounted to 94,923 tons, worth £7,771,830, this production having been derived almost exclusively from Mount Lyell.

In addition to copper, Mount Lyell pyrite contains 0·25 part of silver to 100 parts of copper, and 1 part of gold to 52 parts of silver, so that it is more argentiferous and auriferous than other intrusive pyrite deposits.

The approximate value of the copper and copper ore exported from Australia may be gathered from the following table :

South Australia, 1843 to 1855	.	.	.	£2,077,300
,, ,, 1856 to 1895	.	.	.	18,603,655
New South Wales, 1859 to 1895	.	.	.	6,483,929
Queensland, 1860 to 1895	.	.	.	1,987,074
Victoria, 1895	.	.	.	206,395
Total		.	.	£29,358,353

Including the home consumption and the small amount from Tasmania before 1895, the total output of copper in Australia up to and including that year may be estimated to have been about 400,000 tons.

The annual production of copper in Australia, including the copper contained in the ore exported, has at different periods been as follows :

1860	.	.	about	4,500	long tons.	1895	.	.	about	10,000	long tons.
1870	.	.	,,	9,500	,,	1900	.	.	,,	23,020	,,
1880	.	.	,,	9,700	,,	1905	.	.	,,	33,940	,,
1885	.	.	,,	11,400	,,	1910	.	.	,,	40,315	,,
1890	.	.	,,	7,500	,,	1911	.	.	,,	41,840	,,

From 1896 to 1911 the total production was 440,000 tons. Adding this to the figures given above for the production before this period, Australia up to the end of 1911 had altogether produced 840,000 tons of copper. Of this total, some 120,000 tons came from the Mount Lyell pyrite deposits, about 51,000 tons from Burra-Burra during the period 1846–1877, and probably some 250,000 tons from Moonta-Wallaroo.

RUSSIA-SIBERIA

Copper mining in Russia and Siberia, as will be seen from the following figures of production, is centuries old :

	Tons.		Tons.
1700	3276	1880	3,200
1820	3500	1890	4,800
1830	3870	1895	5,280
1840	4120	1900	8,220
1850	6450	1905	8,700
1860	5020	1910	22,310
1870	5050	1911	25,570

According to de Launay, the production in 1908 was distributed as follows :

	Tons.		Tons.
The Urals .	8560	The Kirghiz Steppes .	1100
The Caucasus	4840	Miscellaneous deposits .	2200

In the year 1905 there were 7 copper mines in the Urals, 8 in the Caucasus, 1 in the Altai Mountains, and 1 in the Kirghiz Steppes.

The Permian ores, which are described later when dealing with impregnations, are no longer exploited.

The famous copper deposits in the Urals, at Bogoslovsk 60° north latitude, at Mednoroudiansk near Nishne Tagilsk,[1] well known for the blocks of malachite employed in the manufacture of ornaments, at Goumeshevsk, Simonovsk, etc., are probably of contact-metamorphic origin.[2]

The Kedabek deposit on the south side of the Caucasus is also regarded by most investigators as contact-metamorphic. According to A. Oehm, whose manuscript is quoted by Beck, in connection with this deposit a large number of eruptive rocks—quartz-porphyry or liparite, keratophyre, quartz-diorite ; dykes of diabase and diabase-porphyrite ; and other, effusive rocks —and tuffs occur. According to P. Nicou,[3] whom de Launay quotes, a micro-granulite or quartz-porphyry probably of Jurassic age, occurs under sheets of andesite and in the neighbourhood of diorite. He mentions in addition andesite dykes which cut through the ore-deposits. The ore-bodies, of which seventeen are known, are lenticular and appear to occur

---

[1] *Ante*, Figs. 150, 231.   [2] *Ante*, pp. 360-366.
[3] *Ann. des mines*, 10th Series, Vol. VI., 1904.

along and within the quartz-porphyry.[1] This deposit since 1865 has been worked by Siemens Brothers.[2]

In the Kirghiz Steppes, numerous lead-copper lodes, capable of division into several groups, are found in slate, limestone, and porphyry or porphyrite. In the Altai district also, lead-silver-copper lodes are known around Smeinogorsk. In these regions mining operations, formerly so extensive, have of late years considerably decreased.[3]

TELEMARKEN IN NORWAY

LITERATURE

J. H. L. VOGT. ' Den Thelemark-Sätersdalske Ertsformation. Norske Ertsforekemster III. and IIIb,' Archif f. Mathem. Naturv. X., 1886, and XII., 1888 ; also Zeit. f. prakt. Geol., April 1895, in which papers by T. Dahll, 1860, Th. Scheerer, 1844, 1845, 1863, B. M. Keilhar, 1850, and P. Herter, 1871, are cited.

Telemarken consists of late Archaean or Algonkian slates, conglomerates, and quartzites, all of which are penetrated by later pre-Silurian granite. In one place at Svartdal quartz-diorite occurs. In these rocks a large number of ore-deposits are found, most of which carry chalcopyrite, bornite, and chalcocite without any accompaniment of pyrite or pyrrhotite. These occurrences may be grouped in the following manner :

1. Lodes in granite, principally along vertical joint-planes, as illustrated in Fig. 145 ; occasionally, as at Svartdal, in quartz-diorite.

2. Lodes along vertical joint-planes in granitic dykes, as illustrated in Fig. 388.

3. Lodes along the walls of granitic dykes in slate, as illustrated in Fig. 389.

4. Lodes in slate.

5. Bedded lodes and impregnations resembling fahlbands, in slate.[4]

The two last-named types occur principally in the neighbourhood of the granite contact ; the Hoffnung bedded lode, for instance, at Aamdal, which is at least 1400 m. long, is only about 40 m. distant from gneissgranite.

In addition to copper ores, molybdenite occurs in some places so abundantly that it is exploited, as for instance at Langvand in Sätersdal, and, farther west, at Knaben in Fjotland ; a few lodes contain galena, sphalerite, arsenopyrite, etc. ; at Dalane in Kvitseid, an impregnation of native copper with native silver occurs ; in the 5 km. long quartz-diorite

---

[1] Or andesite ?

[2] L. de Launay, La Géologie et les richesses minérales de l'Asie, 1911; and in Russian, E. Fedorow, Ann. géol. et minér. de la Russie, 1901, IV. Section I. ; Mém. Ac. Imp. de Science de St-Pétersbourg, Ser. VIII., Vol. XIV., 1903.

[3] L. de Launay, etc.   [4] Fig. 54, Zeit. f. prakt. Geol., 1895.

area at Svartdal some gold lodes. containing chalcopyrite, pyrite, etc., and a fair amount of bismuthinite, occur ; while finally tetradymite has been observed in several places.

Quartz is the most important gangue, this mineral in some lodes being accompanied by tourmaline, and particularly so in the gold-bismuthinite lode at Svartdal. Muscovite frequently occurs, as for instance at Klovereid

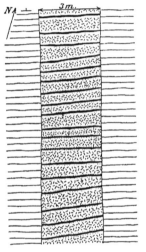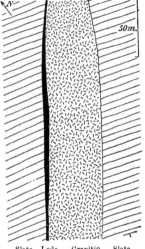

Slate.   Granitic dyke with   Slate.
Veins of ore.

FIG. 388. — Sketch - plan of the occurrence of copper ore in the Nasmark mine. Vogt, *Zeit f. prakt. Geol.*, 1895, p. 149.

Granitic dyke traversing quartz-schists and containing transverse copper veins following one another at regular intervals of 0·3-0·4 m., these veins being accompanied by narrow greisen zones.

Slate.   Lode.   Granitic   Slate.
dyke.

FIG. 389. — Sketch - plan of the copper lode in the Moberg mine in Telemarken. Vogt, *Zeit. ·f. prakt. Geol.*, 1895, p. 149.

Granite dyke traversing quartz-schists and having on one wall a wide quartz lode with muscovite crystals on the walls and nests of copper ore in places.

illustrated in Fig. 145, Näsmark illustrated in Fig. 388, and Svartdal. Similarly to the zinnwaldite of the tin lodes at Zinnwald, illustrated in Fig. 146, this muscovite is in part arranged perpendicular to, and on the walls. Calcite and other crystalline carbonates are subordinate. Fluorite is occasionally found in some lodes ; at Dalen in the neighbourhood of Bandakvand it occurs in such large amount that some thousand tons have been won. Epidote, hornblende, beryl, etc., also occur.

In many mines the granite along the lodes is altered to a rock resembling greisen.[1]

_____
[1] *Ante,* Figs. 145, 261, 262, and pp. 874 and 879.

The lodes generally are closely associated with the granite, or at Svartdal with the quartz-diorite. Their occurrence along joint-planes in granite or in granitic dykes points to a genetic association with the granite eruption. In addition, the presence of tin minerals, tourmaline, fluorite, molybdenite—in one place also wolframite, scheelite, molybdenite, and exceptionally, uranium and beryl—together with the greisen-like alteration of the country-rock, indicates that the genetic conditions at the formation of the lodes at Telemarken were quite similar to those of the tin lodes.[1] Vogt therefore described these deposits as 'tin lodes with copper in the place of tin.'

Some of the deposits were exploited as far back as the sixteenth century. The bedded lodes at Aamdal have so far produced copper ore containing about 7500 tons of copper, while mining operations upon them continue to expand.

THE CUPRIFEROUS SIDERITE LODES

These constitute a special type of copper lode of which the most important representatives are found at Mitterberg in the Salzburg Hills, Dobschau in Upper Hungary, and in Siegerland.[2] They occur frequently in early Palæozoic beds, and agree in their general characteristics with the siderite lodes of Siegerland. The width, as at Dobschau, may be considerable and reach several metres. The lodes belonging to this class carry principally siderite, quartz, and sulphides, of which latter chalcopyrite and pyrite are the most important, the pyrite frequently increasing in depth.

Bornhardt, in respect to Siegerland, was the first to point out that in general a great difference in age exists between the different minerals of the filling, a fact which Krusch by the microscope was not only able to confirm but also to establish in respect to Mitterberg and Dobschau. With these deposits the first filling consists of siderite together with a small amount of sulphides, such as pyrite and chalcopyrite. The siderite overwhelmingly predominating, the lodes in general represent siderite lodes. The entry of quartz solutions to effect the more or less complete replacement of the siderite took place at a later period and after repeated re-opening of the fissure, this process being accordingly described as a silicification of the lodes. Although such silicification took place undoubtedly from depth upwards, nevertheless in a vertical section silicified and unsilicified sections may alternate, replacement of the material in the original fissure having depended not only upon the exact course of the re-opened fissure but also

---

[1] *Ante*, p. 365.　　　　　　　　　　[2] *Ante*, p. 792.

upon the different degrees of resistance which the variable structure of the siderite offered to metasomatic replacement. As a third stage new mineral solutions entered, which in their turn in part replaced both siderite and quartz. The lodes belonging to this group are excellent examples therefore of internal lode metasomatism.

The original siderite filling being usually of great age, these lodes often exhibit evidence of intense tectonic activity and now frequently occur folded and disturbed by all manner of faults.

MITTERBERG IN THE SALZBURG ALPS

LITERATURE

F. M. STAPFF. Berg- u. Hüttenm. Ztg. Vol. XXIV., 1865.—F. POŠEPNÝ. Archiv f. prakt. Geol. Vol. I., 1880, pp. 274-293.—A. v. GRODDECK. 'Zur Kenntnis einiger Serizit-gesteine, welche neben und in Erzlagerstätten auftreten,' N. Jahrb. f. Min. B.-B. II., 1883, pp. 72-183 ; 'Zur Kenntnis des grünen Gesteins von Mitterberg,' Jahrb. d. k. k. geol. Reichsanst. Vol. XXXIII., 1883, pp. 397-404 ; 'Über Lagergänge,' Berg- u. Hüttenm. Ztg. Vol. XLIV., 1885 ; 'Studien über Thonschiefer, Gangthonschiefer und Serizitschiefer,' Jahrb. d. pr. geol. Landesanst., 1885, pp. 1-52.—W. v. GÜMBEL. 'Geologische Bemer-kungen über die Thermen von Gastein und ihre Umgebung,' Sitzungsber. d. bayr. Akad. d. Wiss , 1889, pp. 341-408.—C. A. HERING. Berg- u. Hüttenm. Ztg. Vol. LIV., 1895, p. 215. —A. W. G. BLEECK. 'Die Kupferkiesgänge von Mitterberg in Salzburg,' Zeit. f. prakt. Geol., 1906, pp. 365-370.—MUCH. Das vorgeschichtliche Kupferbergwerk auf dem Mitter-berg bei Bischofshofen, 1879 ; Die Kupferzeit in Europa, 1886.—KRUSCH. Some Investiga-tions.

The Bischofshofen copper district near Salzburg, situated about 1500 m. above sea-level, has long been known not only for its copper deposits but also for those of siderite. The district consists in greater part of Alpine Werfen beds,[1] greatly disturbed by subsequent tectonics. According to Groddeck and Bleeck who have petrographically examined these beds, they consist chiefly of sericite-schist with ottrelite, quartz, and crystalline carbonates ; Bleeck is of opinion that they represent contact-metamorphic clay-slates, sandstones, and quartz-porphyry.

In these beds a large number of steep bedded lodes, usually 1–3 m. thick, occur, these being principally cupriferous siderite lodes. These lodes when possessing a low copper content were worked for iron, and when the content was high, for copper. In the case of old mines therefore, it is not always possible to determine whether such are old iron mines abandoned owing to increase of copper in depth, or copper mines in which the copper in depth gave place to siderite.

The conformity between the lodes and the frequently transversely schistose rocks is often so pronounced that these lodes have been occasion-ally regarded as sedimentary beds. Their epigenetic character is apparent

---

[1] Triassic.

however, partly in small penetrations of the schists by the lodes, and partly in the occurrence of numerous fragments of country-rock in the lode mass.

The disturbances which these lodes have suffered are just as varied as those of Siegerland, folds, faults, lateral displacements, overthrusts, and vertical displacements, all being present. Folds arose when after the formation of the lode the entire complex suffered further plication. When such are present the lode mass is usually traversed by numerous fracture planes, since this mass offered more resistance than the schists of the country-rock. The unravelment of folds is simple because as a rule no break in the continuity of the lode occurs. Faults and overthrusts call for no particular mention. Lateral displacements on the other hand, brought about by horizontal pressure and usually exhibiting no concomitant subsidence, are as interesting as the vertical displacements. These latter present phenomena similar to the shallow faults of Siegerland.

The primary ore is chalcopyrite, which, where the lode-filling has suffered no subsequent alteration, occurs in quartziferous siderite or ankerite, with a variable pyrite content. Chloantite, erythrite, arsenopyrite, and some quicksilver-tetrahedrite which readily decomposes with the formation of cinnabar, occur subordinately.

According to Bleeck two types of lode may be differentiated, namely, quartz-chalcopyrite lodes and quartz-ankerite lodes, the last-named being the younger. In both cases the country-rock is impregnated with chalcopyrite and pyrite.

Krusch considers it may be conclusively proved that a repeated re-opening of the lode fissure took place, and that the first filling consisted principally of ankerite and siderite, and subordinately of quartz, chalcopyrite, and pyrite. During a second period of tectonic disturbance, silicic acid solutions circulating through the re-opened fissures entered the ore mass and metasomatically replaced the carbonates. As a result of this period of silicification, in addition to younger quartz veins formed directly from these solutions, all gradations between the old carbonate lodes and the quartz lodes formed by replacement of the carbonates, are found. From the disposition and individual arrangement of carbonates remaining in the quartz, it may frequently be proved that originally the entire filling was carbonate. A part of the sulphides, and presumably also of the chalcopyrite, being still younger than the quartz must owe its existence to a third opening of the lode fissure. In this connection therefore there exists a great analogy between the districts of Siegerland and Mitterberg.

The geological position is especially complicated when a younger quartz vein not owing its formation to internal metasomatism, intersects

an older secondarily formed quartz lode. In such case the most careful observation is necessary in order to distinguish between the two occurrences.

With regard to the sericite, ottrelite, etc. of the country-rock, it may
be said with some certainty||that these were formed by hydrothermal
processes.

In addition to the lodes worked in the principal mine at Mitterberg,
analogous deposits have been developed to the south, the most important
of which are the Brand and Buchberg lodes.

Mining at Mitterberg, as already indicated, was an industry in very
ancient times. From the second century A.D. however, and for a period
of about 1500 years, the district lay idle. Present operations, which with
an annual production of about 1200–1800 tons make this occurrence the
most important copper deposit in Austria, date back to 1827.

The silver content, which may reach 150 grm. per ton of ore, is also
interesting.

The Bedded Lodes of Kitzbühel in the Tyrol

LITERATURE

P. ·M. Stapff. 'Geognostische Notizen über einige alpinische Kupfererzlagerstätten,'
Berg- u. Hüttenm. Ztg. Vol. XXIV., 1865.—F. Pošepný. 'Die Erzlagerstätten von
Kitzbühel in Tirol und dem angrenzenden Teile Salzburgs,' Archiv f. prakt. Geol., 1880,
Vol. I. p. 257.—G. Dörler. Bilder von den Kupferkieslagerstätten bei Kitzbühel und den
Schwefellagerstätten bei Swoszowice, Minister for Agriculture, Vienna, 1890.—Geological
map by Professor v. Joachimsthal, 1891.

The Silurian formation in the Kitzbühel district consists of grauwacke-
slates, slaty grauwackes, clay-slates, and clay-mica-slates, all of which
generally speaking strike east-west. The carriers of the copper are the
clay-slates. Of the numerous occurrences in this district, till a short
time back only those on the Schattberg, in the Kupferplatte, and on the
Kelchalpe, were worked.

The lodes on the Schattberg lie conformably to the clay-slates, which
dip 25°–80°, strike east-west, and are overlaid with thick detritus and
diluvial beds. These lodes, according to Dörler, represent a complex of
lode-like fillings of fissures following the variable strike and dip, and
adapting themselves to the bends and folds of the slates. Like the
slates the lodes are so crushed, contorted, faulted, and thrust against one
another, that pieces of one and the same lode have at times been regarded
as separate parallel deposits.

The principal lodes, which in width vary from a few centimetres
up to 4 m., are sharply separated on both walls from the country-rock,
into which however they send many veins or leaders. The filling consists
in general of ankerite with large masses of grey and black slate, between
which are found nests and veins of milky-white quartz and chalcopyrite.

The ore occurs, either as an impregnation, or in compact masses on the hanging-wall and foot-wall, or, again, distributed irregularly in stringers and pockets. Tetrahedrite and millerite are uncommon.

The lodes in the Kupferplatte at Lochberg likewise in general coincide in strike and dip with the country-rock.

The geological position of those on the Kelchalpe is similar. The country consists of Silurian clay-slates and clay-mica slates. The so-called *Falkenschiefer* are especially interesting, these being the yellowish-grey, light-coloured, and seldom reddish clay-slates with quartz flakes parallel to the cleavage, associated exclusively with the ore-occurrences. The copper deposit reaches a width of 4 m., strikes north-east, and dips 30° towards the east. The lode-filling consists principally of ankerite, quartz, and *Falkenschiefer*, as gangue, and pyrite, niccolite, chloantite, sphalerite, and galena, as the valuable minerals.

Mining at Kitzbühel, now no longer of any importance, dates back to the eighteenth century. The copper production of the Tyrol has in recent years amounted to about 700 tons annually.

THE COPPER LODES OF THE KAMSDORF DISTRICT

LITERATURE

F. BEYSCHLAG. 'Die Erzlagerstatten der Umgegend von Kamsdorf in Thüringen,' Jahrb. d. k. pr. geol. Landesanst., 1888.

In the Zechstein area east of the river Saal, between Saalfeld and Könitz, continuations of the northern Thuringian Forest boundary-faults are found. Such fissures, striking south-east and dipping 50°–80° north-east, have Triassic, Permian, and Culm, as country-rock. Mineralization is however limited to that portion of their extent between the faulted portions of the Weissliegendes and the Zechstein dolomite, the richest section being between the faulted Kupferschiefer terminals. The ore, occurring in irregular pocket-like accumulations, consists of tetrahedrite, chalcopyrite, cuprite, malachite, azurite, and, principally on the Roter Berg, of cobalt- and nickel ores. The gangue-minerals are, siderite, limonite, barite, and calcite. Asphalt occurs as the result of pressure upon bituminous slates.

From these lodes the bed-like metasomatic alteration of the Zechstein limestone and dolomite, which has been more fully described in the chapter on metasomatic iron ores,[1] proceeded. In these metasomatic deposits, not far from the lodes, nests and pockets of compact tetrahedrite have been found.

The production of this district is small.

[1] *Ante*, p. 835.

THE METASOMATIC COPPER DEPOSITS

In connection with copper lodes metasomatic deposits may be formed, especially when the country-rock consists, in part at least, of limestone or dolomite. Since this condition however is more rarely fulfilled in nature than those necessary to the formation of lodes, there are but few places where copper deposits of this type are exploited. With such deposits the replacement of the limestone is more or less complete, though as yet no case is known where a limestone bed has been entirely replaced.

The form of these deposits is consequently always irregular; pipes and chimneys exist which in regard to their extent along the strike are dependent upon fissures, fractures, and bedding-planes. Since the thickness of the bed undergoing alteration, especially if it be limestone, is usually limited, in the case of undisturbed bedding no material extension in depth is, generally speaking, possible. Where extension in this direction is encountered, it is due to tilting of the beds, such as may have taken place before or after the formation of the deposit.

The distribution of the ore is irregular, the most important copper enrichments being not infrequently found in the vicinity of lode fissures.

The primary minerals are as a rule pyrite and chalcopyrite, though other sulphides such as galena and sphalerite also occur. In keeping with the nature of the origin of these deposits, carbonates predominate among the gangue-minerals. The structure is generally irregular-coarse.

As with most metasomatic deposits, the primary depth-zones are generally limited to an alternation of poorer and richer zones, this alternation being frequently connected with the varying degree of replacement the individual beds comprising the limestone formation have suffered.

Secondary depth-zones, on the other hand, are of great importance, since copper ores are pronouncedly prone to migrate [1] and the country-rock, consisting in part at least of limestone, favours such migration. Oxidation and cementation zones of considerable thickness, and in striking disproportion to the size of the primary deposit, may therefore be formed. In such

---

[1] *Ante*, pp. 89-92, 216.

cases not only is the limestone practically completely replaced, but also any eruptive rock associated with it, as for instance aplite and kersantite at Otavi. Should other sulphides occur together with the copper in the primary zone, these act reducingly upon the secondarily-formed descending heavy-metal solutions. In this manner and over a long period a pronounced copper cementation zone may be formed, even with such metasomatic deposits as originally carried principally galena and sphalerite and but little copper.

The accessory precious-metal content in copper deposits may be very considerable. While such accessory silver content may be considerable and constitute a substantial part of the production, the gold content is less important. It has already been pointed out [1] that gradations between copper- and silver deposits exist.

From the close connection between metasomatic copper deposits and copper lodes it follows that in the principal lode districts subordinate metasomatic deposits also occur, such as a rule having been formed by the alteration of limestone. Unavoidably therefore, a number of metasomatic deposits were mentioned when describing the copper lodes; further description of this subordinate class will accordingly here be confined to the deposits at Massa Marittima, Boccheggiano, Otavi, and Katanga.

Massa Marittima, Boccheggiano, and the Adjacent Deposits in Tuscany

LITERATURE

B. Lotti. Descrizione geol.-miner. di Massa Marittima, Geol. Survey Dept., Italy, 1893 ; Geologia della Toscana, Geol. Survey Dept., Italy, 1910 ; I Depositi dei minerali metalliferi, Turin, 1903.—K. Ermisch. ' Die gangförmigen Erzlagerstätten der Umgebung von Massa Marittima in Toskana auf Grund der Lottischen Untersuchungen,' Zeit. f. prakt. Geol., 1905, pp. 206-239.—L. de Launay. La Métallogénie de l'Italie, Report of the 10th Intern. Geol. Congress, Mexico, 1906. A number of treatises are cited in these works including those by Savi, Pilla, Cailleux, Meneghini, G. v. Rath, Serpieri, Novarese, Corteze, etc.

The deposits of the ' Massetana Metal Province ' lie within an area of 450 square kilometres. North of this district are found the boracic acid springs of Sasso and Monte Rotondo ; to the south-east, the quicksilver district of Monte Amiata ; [2] and to the west, the contact-deposits and tin occurrence of Campiglia Marittima,[3] the copper mines of Monte Catini,[4] and the contact-deposits of Elba, etc.[5]

The Massetana district consists principally of folded and faulted Permian phyllites, Rhaetic, Liassic limestone, and Eocene, with a small extent of Tertiary eruptives, including gabbro—euphotide—serpentine, etc.

---

[1] Ante, pp. 163, 164.   [2] Ante, pp. 471.474.   [3] Ante, pp. 409-411.
[4] Ante, pp. 300, 301.   [5] Ante, pp. 369.372.

The numerous deposits may be grouped as follows :

1. True fissure lodes, which are often faults and frequently display internal metasomatism of the original brecciated filling. With these the country-rock is more or less highly altered.

2. Lode-like deposits connected with metasomatic occurrences, true fissure-fillings being subordinate.

3. Non-lode-like deposits of lenticular or pocket form, these being in greater part purely metasomatic. Such deposits are most closely associated with calcareous rocks — the ' metalliferous limestone ' of the Rhaetic or Lias, for instance — and are principally ferruginous zinc-carbonate deposits.

Groups 1 and 2 especially are closely associated with each other. Most of the deposits, or perhaps all, occur along four large N.N.W.–N.

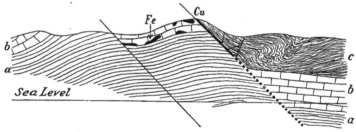

a = lustrous slate, Permian ; b = cavernous limestone, Rhaetic ; c = calcareous clay, Eocene.

FIG. 390.—Section of the copper lode accompanying metasomatic limonite deposits at Boccheggiano. Lotti ; see also Ermisch, *Zeit. f. prakt. Geol.*, 1895, p. 229.

striking faults, namely, (a) the south Serrabottini, (b) the Capanne Vecchi, (c) the Montoccoli, and (d) the Boccheggiano. The section at Boccheggiano given in Fig. 390, illustrates this connection between fault and deposit. Among other deposits of this type, those at Serrabottini, Capanne Vecchie, Carbonaie - Valdaspra, Montieri, and Montevecchio deserve mention.

Many lodes, as for instance those at Boccheggiano, belong to the type of quartzose copper deposits ; they carry cupriferous pyrite, some galena, sphalerite, etc. Others have more the character of the sulphide lead-zinc deposits, though, containing a large amount of chalcopyrite; they represent the cupriferous facies of that group.

The latter, as pointed out long ago by Stelzner [1] and later by Ermisch, frequently exhibit a striking resemblance to the sulphide lead group at Freiberg, even though Archaean gneiss there constitutes the country-rock whereas in the case of the Italian occurrences that rock consists of young

[1] *Berg- und Hüttenm. Ztg.*, 1877, No. 11.

clay-slate. This fact is the plainest proof that the metalliferous filling could not have been derived from the country-rock.

The best known deposits are those at Massa Marittima and Boccheggiano.

At Massa Marittima and at Poggio Guardione normal lodes occur in Eocene marls, such lodes carrying pyrite, chalcopyrite, a little sphalerite, and galena, in a quartzose gangue containing coarse-grained calcite in subordinate amount.

The famous Boccheggiano deposits belonging to the Monte Catini company are situated in the neighbourhood of the important town of Montieri, in the Massetana hinterland. The 1–25 m. wide quartz lode, impregnated with pyrite and chalcopyrite, strikes N.N.W. and dips 40° to the east. It extends along the contact of Eocene with Permian and Rhaetic, and may be followed for a length of 3 km., from the Farmulla river to the Merse Savioli. The ore in general has a banded structure.

According to the mineral-association two belts may be differentiated, namely, a southern stretching from the Farmulla river to Boccheggiano, a distance of 1·3 km. ; and a northern from Boccheggiano to the Merse Savioli river, 1·7 kilometres.

The mineralization of the former is very variable. While formerly sphalerite, argentiferous galena, and zinc carbonate were won, to-day two chimneys containing clean chalcopyrite, galena, and sphalerite, are known. The greater portion of the work is now centred upon three inclined ore-shoots. The difference in depth between the highest and lowest levels is approximately 250 metres. A small amount of tin to the extent of 1 of tin to 80 of copper is interesting. In April 1910 a boracic acid spring, having a temperature over 40° C. and emitting 30 litres per second, was encountered in depth. The hindrance caused by this spring finally led to the abandonment of operations.

An invariable metamorphism of the country-rock is especially characteristic, this rock being frequently silicified, decarbonated, and impregnated with chalcopyrite and pyrite, etc. In other places epidotization is encountered ; and finally, though infrequently, an alteration to pyroxene, garnet, and epidote, accompanied at times even by lievrite. This mineral-association has much in common therefore with contact-metamorphism.

The average yearly production of Boccheggiano, according to Lotti, amounts to 35,000–36,000 tons. The different qualities of ore with their average content may be gathered from the following table :

AVERAGE YEARLY PRODUCTION OF THE BOCCHEGGIANO MINE, 1895–1904

Rich copper ore	with 31·97 % S	and 10·16 % Cu :	4,328 tons.
Rich sulphur-ore (cupriferous pyrite)	„ 40·48 % S	„ 3·44 % Cu :	6,863 „
Poor ore (quartzose)	„ 24·19 % S	„ 2·65 % Cu :	25,570 „

General Manager Marengo gave the production for 1904 as follows :

Rich copper ore with 9·0 % Cu, 32 % S and 28 % SiO₂ : 3,600 tons.
Rich sulphur-ore   „   3·3 % Cu, 40 % S „ 18 % SiO₂ : 12,000 „
Poor ore   „   2·5 % Cu, 24 % S „ 45 % SiO₂ : 21,000 „

The genesis of these deposits is, according to Lotti, as follows : As after-effects of the great mountain-forming movements at the end of the Eocene period, fissures and fissure-systems were formed in Massetana and in the neighbouring districts of Campiglia, Elba, etc., along which eruptive magmas in part, but heavy-metal solutions principally, entered. These circulated either along the fissures, bringing about the formation of the lodes, or penetrated laterally into the country-rock principally along the bedding-planes, thereby metasomatically replacing the alterable calcareous rocks. The whole process, according to Lotti, was completed in the Miocene period.

Mining in Massetana is extremely old, its zenith having lasted from 1200 to the time of the Great Plague in 1348, after which followed a period of decline. Operations were however renewed with energy at the close of the nineteenth century, but the industry is again in decline and the Boccheggiano mine was recently shut down.

The copper production of Italy from 1879 to 1883 was at the rate of 1200–1600 tons per year, and from 1884 to 1895 roughly 2500 tons ; this rate increased later to 3000 tons, till operations in Boccheggiano were discontinued, when it fell considerably, this mine having been responsible for a considerable portion of the annual output. In 1911 the total output was 2600 tons.

THE OTAVI DEPOSIT, GERMAN SOUTH-WEST AFRICA

LITERATURE

FRANCIS GALTON. Travels in Tropical Africa, 1852.—H. SCHINZ. Deutsche Südwest-afrika. Oldenburg, 1891.—P. A. WAGNER. ' The Geology of a Portion of the Grootfontein-District of German South-West Africa,' Trans. Geol. Soc. S. Africa, 1900, Vol. VIII.—J. KUNTZ. ' Kupfererzvorkommen in Südwestafrika,' Zeit. f. prakt. Geol., 1904, p. 402.—W. MAUCHER. ' Die Erzlagerstätte von Tsumeb im Otavi-Bezirk im Norden Deutsch-Südwestafrikas,' Zeit. f. prakt. Geol., 1908.—P. KRUSCH. ' Die genetischen Verhältnisse der Kupfererzvorkommen von Otavi,' Zeit. d. d. geol. Ges., 1911, p. 240, Part 2.

Otavi is situated in the Otavi hills in the north of Hereroland, about 550 km. from the coast. These hills consist chiefly of dolomite in east-west folds, the flanks of which incline sometimes to the north and sometimes to the south. Certain beds of the Otavi dolomite favour the formation of caves, these latter being mostly empty but occasionally filled with water. The Otjikoto lake, south-west of Tsumeb, which supplies the necessary water for the mine, owes its existence to one of the largest of such caves. The Palæozoic Otavi formation, to which the dolomite belongs, consists, according to Wagner, from hanging-wall to foot-wall, of the Fish River sandstone, the Otavi dolomite, and the Nosib series. The Otavi dolomite is presumably Devonian and the equivalent of the Black-Reef dolomite and the Pretoria formation of the Transvaal.

In the Otavi hills copper ore is found in four different places, namely, on the northern slope at Tsumeb, and on the southern slope at Gross Otavi, Klein Otavi, and Guchab, these places being indicated in Fig. 391.

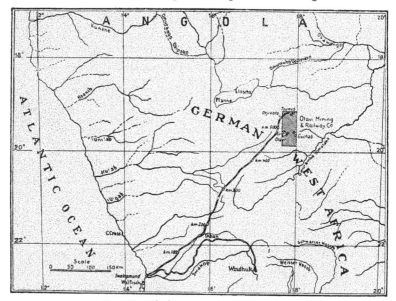

FIG. 391.—Situation of the Otavi copper district.

The least important of these deposits is that at Gross Otavi, where nests and net-like veins of ore occur in dolomite, the beds of which dip steeply to the south. The width of the principal ore-zone is approximately 1 m. at the centre, but diminishes on both sides. As with several of the Otavi deposits, the sandstone-like masses so frequently mentioned in descriptions occur here also. Scheibe assumed these to be eruptive, while Krusch has proved those at Tsumeb to be aplite. The mineral-association consists of chalcocite with much malachite and galena. The pockets of ore vary from those having the smallest dimensions to bodies of more than 1 km. in length.

The deposits at Klein Otavi and at Guchab in the Otavi valley, near

Kilometre 54 of the Otavi-Grootfontein railway, are more compact. According to Kuntz, at these places one particular dolomite bed appears to have been especially suited for alteration to ore.

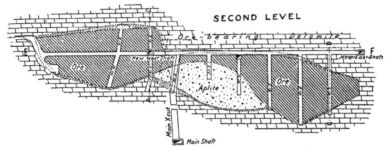

FIG. 392.—Plan of the copper deposit on the second level of the Otavi mine.

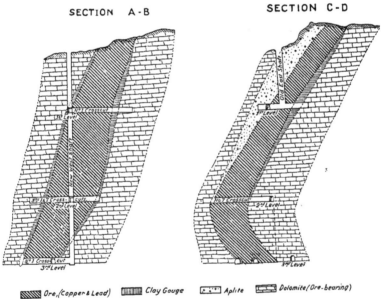

FIG. 393.—Sections of the Otavi deposit.

Without doubt the most important occurrence is that at Tsumeb on the northern slope of the Otavi hills, where a green copper-stained hill rises abruptly out of the Otavi dolomite. Kuntz at the time of his visit estimated the ore-bearing area to be 200 paces along the strike and 40 paces to the dip. Both dolomite and deposit, as illustrated in Fig. 393, dip steeply to the south, though the eastern portion of the latter turns in depth to dip towards the north. Here also the ore favours an apparently less resistant dolomite bed. As indicated in Fig. 392, two ore-bodies may be distinguished, namely, an eastern and a western. These are connected in the centre by a contracted width of ore, the remainder of the width being occupied in greater part by the above-mentioned aplite. The separation of the ore and aplite from the dolomite is formed by a clay-parting or gouge. The contraction of the ore-body increases from the surface downwards. The length along the strike of both eastern and western bodies diminishes somewhat in depth, where also, in 1911, between the third and fourth levels a fault was met, though this was subsequently unravelled. From the latest reports of the Otavi company it would appear that the aplite likewise diminishes in depth. The Otavi dolomite, the ore-body, and the aplite, are all crossed by olivine-kersantite.

In regard to the genesis of these deposits two features are particularly worthy of attention, namely, the gouge which separates the ore and aplite from the dolomite ; and the gradual passage from aplite to ore.

According to R. Scheibe the deposit strikes on the whole east-west, at an oblique angle with the limestone beds, and dips 50°–70° to the south.

The mineralization, so far known to a depth of about 100 m., is not simple. The ore occurs chiefly at the contact of aplite and dolomite, where, on the one hand, compact masses of ore contain veins of aplite and silicified dolomite, and on the other, veins of ore penetrate both dolomite and aplite. On the third level in the eastern portion of the deposit, a rich bunch of ore was found in the centre of the aplite.

The width of the ore-body, though subject to considerable variation, is often 20 metres or more. The eastern portion is richer in lead but poorer in copper, ore containing 6–14 per cent of copper predominating ; the western portion, on the other hand, is richer in copper, the ore containing 12–15 per cent.

The mineral-association, as first pointed out by Maucher, consists of sulphide ores and their oxidation products. The latter he divides into secondary ores, the immediate result of the oxidation of the primary, and tertiary ores. These tertiary ores represent a chemical alteration of the secondary ores, in effecting which distinct evidence of the participation of the country-rock may be observed. The oxidized ores, among which malachite and azurite occur most frequently, contain on an average 12·9 per cent of copper and 4·4 per cent of lead. Among what Maucher considers as primary ores, are massive intergrowths of chalcocite, enargite, galena, sphalerite, and pyrite, though Krusch considers these in greater part to be cementation ores.

The distribution of the ore-minerals is such that the middle portion of the deposit, the ore-body proper, consists of compact sulphide exhibiting no drusy cavities whatever. The ore-minerals in the aplite, as long as that eruptive rock is still recognizable, are found in greater part along fractures ; among them linarite occurs most frequently. While in the foot-wall portion of the deposit malachite and azurite predominate. .

Concerning genesis, Macco considers that at Otavi only two possibilities exist, namely, the deposits are either fissure - fillings accompanied by metasomatic replacement of the dolomite, or the fillings of irregular cavities. Maucher on the other hand endeavours to prove that the deposit at Tsumeb is a magmatic segregation. In determining the seniority of the various minerals he considers exclusively the question of the fusion-point, namely, that this becomes substantially lower when sulphide components are dissolved in each other. Stutzer concludes that meta-somatic replacement took place and that mineralization is to be ascribed to aqueous solutions. W. Voit, after mentioning those characteristics which appear to be opposed both to magmatic segregation and metasomatism, finally sees no reason for doubting Maucher's conclusion. Range favours the idea of a cavity-filling accompanied by metasomatism as that term has hitherto been understood, while P. A. Wagner holds a similar opinion, believing a replacement of the dolomitic limestone to have taken place.

According to Krusch, metasomatism undoubtedly played an important part, though he emphasizes the necessity of distinguishing between primary and secondary metasomatism.

Metasomatism as hitherto understood—that is primary metasomatism, consisting principally of a replacement of the limestone and dolomite—is possible at Tsumeb, though proof is not at present available and can only be obtained by development in greater depth. In any case, according to microscopic slides, the principal portion of the chalcocite does not belong to the primary metasomatic ore-minerals, and, accordingly, secondary metasomatic processes, that is, cementation- and oxidation-metasomatism, must be responsible. The former at Otavi is considerably more subordinate than the latter, and is expressed chiefly in the replacement of olivine-kersantite and to a less extent of aplite and dolomite, by malachite and azurite. The bulk of the chalcocite however was undoubtedly formed by cementation-metasomatism, that at Tsumeb being deposited on galena, sphalerite, and pyrite to such an extent that these primary sulphides are almost completely replaced. The replacement in the eastern body was less complete and a plumbiferous ore resulted ; in the western body the copper ore is purer. In this process the influence of the country-rock is shown by the replacement of dolomite and aplite, by ore. In the case of the alteration of the aplite, adsorption must have played an im-portant part since the resultant chalcocite contains kaolinized felspars.

Reviewing all the facts there appears to be at Tsumeb a zone of fracture along which apparently an aplite body so subsided as to form in depth a wedge within the Otavi dolomite ; the amount of such subsidence need not necessarily have been very great. The absence of all contact phenomena speaks against an intrusion of magma *in situ*.

The occasionally disturbed bedding of the dolomite supports this assumption of a fracture zone, as does also the oblique angle, observed by Scheibe, which the ore-body makes with the strike of the dolomite. Along this zone the heavy-metal solutions to which the primary minerals—the number of which is not yet fully known—owe their existence, probably circulated. These minerals presumably are likewise in part metasomatic. The oxidation and cementation processes resulted from the action of meteoric waters, and as far as oxidation is concerned such processes are still proceeding. It is interesting to note that the chalcocite is younger than the frequently-observed silicification of the dolomite.

The economic importance of the Otavi deposit may be gathered from the following statements. At the end of the book-year 1907–1908, apart from the irregular dolomite- and aplite ore-bodies—the so-called sandstone ores—containing about 7–8 per cent of copper and 5–6 per cent of lead, 313,000 tons of ore containing 16 per cent of copper and 25 per cent of lead were proved. In the same year 25,700 tons of ore were raised, of which 60 per cent was export ore containing about 18 per cent of copper, approximately 30 per cent smelting ore containing 12 per cent, leaving 10 per cent as ore placed on the dump.

In that year about 15,000 tons of Tsumeb ore containing 0·035 per cent of silver,[1] 19 per cent of copper, and 23 per cent of lead, were shipped ; while 3500 tons were smelted on the spot, this total including 2100 tons assaying about 10 per cent of copper and 18 per cent of lead, and 1400 tons of plumbiferous ore assaying 55 per cent of lead and 12 per cent of copper.

During the year 1908–1909, the 13–15 m. wide deposit produced 44,250 tons of ore, of which 27,000 tons were export ore containing 17 per cent of copper, 30 per cent of lead, and 0·033 per cent of silver ;[2] while during the next year, 1909–1910, the production reached 49,500 tons, of which 44,770 tons came from Tsumeb. In that year also, 33,500 tons containing 16 per cent of copper, 26 per cent of lead, and 0·028 per cent of silver,[3] were exported.

Guchab in the year 1907–1908, produced 1800 tons of argentiferous copper ore containing on an average 0·04 per cent of silver [4] and 33 per cent of copper ; while in the following year the production was 500 tons contain-

---

[1] 350 grm. per ton.　　　　　　[2] 330 grm. per ton.
[3] 280 grm. per ton.　　　　　　[4] 400 grm. per ton.

ing 29 per cent of copper and 0·032 per cent of silver.[1] Klein Otavi in the year 1908–1909 yielded 200 tons with 27 per cent of copper and 0·029 per cent of silver ;[2] while Gross-Otavi in the following year yielded some ore containing 40 per cent of copper.

The Copper Deposits at Katanga in the Belgian Congo

LITERATURE

L. Cornet. 'Die geologischen Ergebnisse der Katanga-Expedition,' Petermanns Mitteilungen, 1894, p. 121 ; 'Les Formations post-primaires du Bassin de Congo,' Ann. de la Soc. Géol. de Belg., 1897 ; 'Observations sur les terrains anciens du Katanga,' Ann. de la Soc. Géol. de Belg., 1897 ; 'Les Gisements métallifères du Katanga,' Bull. de la Soc. Belge de Géol., 1903.—O. Stutzer. 'Die Kupfererzlagerstatten Étoile du Congo im Lande Katanga, Belgian Congo,' Zeit. f. prakt. Geol., 1911, p. 240.

The district of Katanga in the Belgian Congo, around and within which numerous exploring expeditions are at present active, has become better known since the construction of the railway from Rhodesia to Elizabethville. It is in the neighbourhood of this latter town that the principal deposits are found.

The geological conditions of the district were first investigated by Cornet, who from 1891 to 1893 accompanied the Bia-Francqui expedition. The fundamental rocks consist of folded and in part metamorphosed sediments and eruptives, the exact age of which, owing to the absence of fossils, has not yet been possible of determination.

The most important copper deposit in this district is the Étoile du Congo near Elizabethville. This occurs, striking north-south, in non-metamorphosed sediments, which in the southern portion of the deposit dip steeply to the west, a dip however which towards the north passes gradually over to one steeply to the east. The country-rock consists chiefly of slate, within which, marking the centre of the deposit, a porous quartzite some 10 m. thick is intercalated. Dolomite occurs farther in the foot-wall and hanging-wall. The slate is sometimes typical clay-slate and sometimes more arenaceous or calcareous. It is almost invariably much decomposed, this decomposition consisting in kaolinization or an impregnation with ore. Its colour accordingly varies between white, black, green, blue, and red.

The most remarkable of these rocks, according to Stutzer, is the extremely hard, grey-coloured, porous quartzite which strikes parallel with the slates. The cavities in this rock are sometimes as large as an egg. They are filled with malachite and other copper minerals. Crystal casts are evidence of pre-existing carbonates. The dolomite is coarsely-crystalline,

---

[1] 320 grm. per ton.   [2] 290 grm. per ton.

merging on the east side of the deposit and in the vicinity of the ore-bearing slate, into compact calcareous schist. On the west side it is covered by irregular pocket-like depressions of so-called ' black ore.'

The most important minerals are chalcocite and malachite ; chryso-colla occurs frequently, while azurite is subordinate. The chalcocite occurs in the form of masses and compact veins in all rocks, but preferably in the kaolinized white slate. It is also the principal constituent of the so-called ' black ore,' this being a dark or black earthy mass coloured by the chalcocite and containing in addition, cobalt, iron, manganese, and some nickel. Stutzer explains the ' black ore ' as being in part a decom-position product of the impregnated clay-slate, and in part old mine-filling. This deposit in former times was worked in the most primitive manner by the natives, who by means of small shafts sought the mala-chite only, while using the chalcocite-ore as filling. This became mixed with the other fine waste material eventually to form ' black ore.'

The second most important mineral is malachite, which occurs along fractures and in crevices, or as impregnations. Occasionally in the larger cavities it forms kidney-shaped or stalactitic, beautifully banded masses. At the southern end of the deposit a bright-coloured ore-breccia, consisting of fragments of red slate with dark green glass-like chrysocolla and light green malachite as cementing materials, occurs in the red arenaceous slates. From the banded structure it is seen that the chrysocolla is of somewhat greater age than the malachite.

Azurite and chalcopyrite are uncommon. The latter in the upper levels occurs only in minute grains, while in greater depth it is more plentiful and associated with pyrite. In addition, numerous secondary copper minerals which have not yet been determined, also occur.

Concerning genesis, this deposit, according to Stutzer, is one where the minerals as now found are not in the place of their original deposition, though any considerable migration can hardly have taken place. The deposit consists of an extensive oxidation zone, under which follows a zone of cementation. The primary ores were undoubtedly sulphides, though the determination of the parent rock is difficult. Stutzer considers the quartzite to have been the parent rock, the occurrence of this quartzite according to him being lode-like.

If this view be correct the Étoile du Congo deposit represents the gossan of a bedded lode containing primary sulphides and carrying as gangue, quartz chiefly and carbonate subordinately. At the decomposition of the primary ores solutions became formed which sank principally into the kaolinized clay-slate, where adsorption of their heavy-metal content took place. Since, according to Krusch, with adsorption, replacement also plays an important part, this deposit is closely related to that at Otavi ;

both are large accumulations of oxidation and cementation ores, and in both the country-rock has in the highest degree been replaced by the heavy-metal constituents of solutions resulting from atmospheric agencies.

The large quantities of copper which this deposit has been estimated to contain need confirmation.

THE PYRITE AND ARSENOPYRITE LODES

As previously stated,[1] the principal pyrite deposits belong not to the lodes but to the intrusive pyrite group. Although in both cases these are cavity-fillings, the intrusive deposits differ from the lodes in two respects ; firstly, in the form of the occurrence, which in the case of lodes is tabular, while with the intrusive deposits the lenticular form is common ; and secondly, in regard to genesis. While the lode-filling owes its existence to aqueous solutions, the intrusive deposits represent the entrance of sulphide magma. Gradations between these two geneses naturally occur, as when for instance the sulphide magma contained large quantities of water. Arsenopyrite is frequently associated with pyrite in lodes, the relative quantities of the two minerals being very variable, the former sometimes predominating.

Owing to the enormous ore-reserves sometimes associated with intrusive deposits—the Huelva district alone produces about 3,200,000 tons yearly, and mines having a yearly production of 50,000 tons are classed among the smaller occurrences—the conditions imposed upon pyrite lodes in respect to their payability press very hard. The width must be great and the material very pure in order that the occurrence may compete with the gigantic intrusive deposits.

In consequence of the small number of pyrite- and arsenopyrite lodes which have yet been exploited, few facts concerning their extent in depth are available. Seeing however that other lodes having but little pyrite in the upper levels contain a larger amount of this mineral in depth,[2] the conclusion may be drawn that clean pyrite lodes also may continue ore-bearing down to a considerable depth. Since the market only accepts fairly clean material, the exploitation of pyrite lodes can only be entertained when the filling consists almost exclusively of this mineral. The usual gangue-mineral is quartz, crystalline carbonates being very subordinate. The structure is mostly irregularly coarse.

With these lodes secondary depth-zones are important. In the

---

[1] *Ante*, p. 301.   [2] See copper lodes, p. 905.

neighbourhood of the surface and down to a depth of 20–30 m. or more, the pyrite is frequently altered to limonite characterized by great purity. The result is, that when developments have not proceeded sufficiently in depth, evidence that this limonite has resulted from the alteration of the sulphide ore existing in depth is only forthcoming in the characteristic decomposition of the country-rock.   In the process of oxidation the sulphur of the pyrite is removed, while no sulphide-enrichment zone between the oxidation and primary zones is known. The gossan, it is true, may still contain some sulphur, though often it is so completely free from this element that without further thought it may be smelted for iron.   Arsenopyrite is similarly oxidized to limonite and, again, no arseno-cementation ores are known.

The subordinate gold, silver, copper, and tin, existing as accessory constituents of these lodes, are of great interest. Minimal amounts of gold are often found. Should the gold content increase, say, to at least 5 grm. per ton, and sufficient reserves be available, such pyrite and arseno-pyrite lodes may become payable gold deposits. The silver content of both these pyritic sulphides frequently reaches 30–50 grm. per ton, and may sometimes be so high that silver becomes the chief object of the mining operations. The pyrite lodes must then be reckoned with the silver lodes. That this may also be the case in respect to the copper content of the lode was indicated when discussing the copper deposits ; a pyrite lode con-taining 2·5 per cent of copper must without doubt be reckoned among the copper deposits.   In Bolivia, tin is largely associated with pyrite, a pyrite lode containing 7 per cent of tin being regarded as a tin lode.   There are therefore numerous gradations between pyrite lodes on the one hand, and certain precious-metal-, copper-, and tin lodes on the other.

In the formation of secondary depth-zones this original accessory heavy-metal content becomes concentrated as a cementation zone between the oxidation zone—which in all cases consists of limonite—and the primary zone, this cementation zone carrying either much gold or silver, or rich copper ores such as bornite and chalcocite. Exceptionally, the tin content in the form of wood-tin appears to be confined to the oxidation zone.

In many cases the pyrite is somewhat arseniferous, and since arsenic greatly depreciates the value of pyrite any analyses should always be extended to include that element. Similarly, with arsenopyrite the occur-rence of pyrite, which often happens, is harmful.

Large pyrite lodes are not, so far as is known, at present being worked, since large concentrations of this mineral are only exceptionally found in lode fissures. The mines working such deposits are therefore mostly small.

The pyrite produced to-day in order to be first-class must contain 45-50 per cent of sulphur ; low-grade pyrite contains 40-45 per cent, while poor pyrite contains 35-40 per cent. Pyrite with less than 35 per cent of sulphur only finds a purchaser in special cases, as for instance when the occurrence is exceptionally favourably situated in respect to the place where it would be applied.

The price of the pyrite depends upon the arsenic content ; in the case of very good pyrite it varies between 3½d. and 4½d. per unit of sulphur. Arseniferous pyrite as a rule does not fetch more than 3½d. per unit.

ROTHENZECHAU IN THE RIESENGEBIRGE

LITERATURE

H. TRAUBE. Die Minerale Schlesiens, 1888.—J. ROTH. Erläuterungen zu der geognostischen Karte vom niederschlesischen Gebirge und den umliegenden Gegenden, 1867.— v. FESTENBERG-PACKISCH. Der metallische Bergbau Niederschlesiens, 1881.—A. SACHS. Die Bodenschätze Schlesiens, Erze, Kohle, nutzbare Gesteine, 1906.—v. DECHEN, revised by BRUHNS. Die nutzbaren Mineralien und Gebirgsarten im Deutschen Reiche, 1906, 2nd Ed.—LANGER. ' Die geologischen Verhältnisse des Kiesvorkommens von Rothenzechau im Riesengebirge,' 1911, Archiv d. k. geol. Landesanst.

Upon the granite core of the Riesengebirge rests a belt of crystalline schists overlaid by so-called green schists and clay-slates. This belt has been so much altered by contact-metamorphism that the geological age of its members can no longer be determined with certainty, though the discovery of *Graptolites* in some places would indicate Silurian. Within this belt the deposits at Kupferberg,[1] Rohnau, 5 km. farther south, and Rothenzechau, 11 km. still farther south, are found. The last of these three, worked in the Evelinens Glück mine, is marked upon the surface by a large number of old workings which may be followed for about 500 m. along the strike.

The Rothenzechau hills consist of mica-schist, hornblende-schist, crystalline limestone, cordierite-gneiss, quartzite, and schistose conglomerate. All these, as may be observed in the new workings, strike north-east, are steeply inclined, folded, and crossed by numerous disturbances.

Five different ore-bodies are known, all of which, generally speaking, occur in the hornblende-schist. Those in the Deep Adit lie between mica-schist and contact-metamorphosed rocks on the foot-wall, and hornblende-schist on the hanging-wall. The strike and dip coincide in general with that of the hornblende-schist. Sometimes the separation between ore-body and country-rock is sharp, while at other times these

[1] *Ante,* p. 402.

two merge gradually into each other.   The width of the payable ore-body varies between 0·5 and 3 metres.

The valuable content is irregularly distributed and consists principally of arsenopyrite and pyrrhotite.   The richest arsenopyrite contains 45 per cent of arsenic.   The ore as it is mined contains on an average 35 per cent of mineral with 27–28 per cent of arsenic, 2–4 grm. of gold, and 40–60 grm. of silver per ton.   Pyrite and marcasite are scarce ; galena and sphalerite very rare.   The pyrrhotite contains no cobalt or nickel, but some chalcopyrite.

The nature of the deposit has not yet been definitely determined.   It is probable that two occurrences occur here side by side, one lode-like and the other contact-metamorphic.   If this be so the pyrite deposit at Rothenzechau would occupy an intermediate position between the lodes and the true contact-deposits at Reichenstein.

THE METASOMATIC PYRITE DEPOSITS

Metasomatic arsenopyrite deposits are unknown.   The economic factors mentioned at the beginning of the discussion of pyrite and arsenopyrite lodes apply in general also to this group.   These deposits were formed by the replacement of limestone and dolomite—and to a small extent also of other rocks—constituting the country-rock of pyrite lodes.

In this metasomatic origin lies their material difference from the lodes.   Where the replacement of the limestone was extensive, large bodies of pyrite were formed, which, as regards size, may, to some extent at least, be compared with the smaller intrusive pyrite deposits.   The form of these occurrences is that of the beds they have completely replaced. From the view of genesis, all gradations between the lode- and the bedded form may occur.   Since however it is economically only possible to exploit the large occurrences, the incomplete non-bedded replacements are of no significance.

From the nature of these deposits no great extent in depth is as a rule possible, though a subsequent tilting of the beds, either before or after their alteration, may take place and bring greater depth into question. Possible primary depth-zones can however only have reference to the conditions of original bedding.

The material of the deposit varies but little.   Frequently pyrite or marcasite is found exclusively ; other sulphides are rare or very subordinate.   The most frequent gangue-mineral is barite, which occurs either intimately intergrown with the pyrite, or, as is the case at

Meggen, replaces it locally, the pyrite in places gradually merging into barite. In such cases the difficult question arises, whether the barite is younger than the pyrite, or whether the limestone was first replaced by barite and this in turn by pyrite; or whether both are contemporaneous.

Of the secondary depth-zones, the formation of gossan is only known in such cases where the deposit comes to surface. Should there be a low precious-metal content in the pyrite, a precious-metal cementation zone may occur immediately under the gossan. Such a content has generally otherwise no significance. The gossan may have been considerably increased by oxidation-metasomatism, so that its thickness must not without further examination be taken as indicative of the strength of the primary deposit below.

An important deposit of this kind is now being worked in two mines at Meggen in Westphalia. The ore there is not particularly rich but owes its payability rather to the favourable economic position of the deposit. Similar occurrences are found at Schwelm in Westphalia.

MEGGEN

LITERATURE

M. BRAUBACH. 'Der Schwefelkiesbergbau bei Meggen an der Lenne,' Zeit. f. B.-, H.- u. S.-Wesen im preuss. Staate, 1888, Vol. XXXVI. p. 215.—R. HUNDT. 'Das Schwefelkies- und Schwerspatvorkommen bei Meggen a. d. Lenne,' Zeit. f. prakt. Geol., 1895. p. 156.— A. DENCKMANN. 'Das Vorkommen von Prolecaniten im Sauerlande,' Zeit. d. d. geol. Ges., 1900, Vol. LII. ; 'Goniatitenfunde im Devon und Carbon des Sauerlandes,' ibid., 1902, Vol. LIV. ; 'Über die untere Grenze des Oberdevon im Lennetale und im Hönnetale,' Zeit. d. d. geol. Ges., 1903, Vol. LV.—Description of the districts of Arnsberg, Brilon, and Olpe, and of the principalities Waldeck and Pyrmont, 1890, Kgl. O.B.A. Bonn.—W. HENKE. 'Zur Stratigraphie des südwestlichen Teils der Attendorn-Elsper Doppelmulde,' Inaug.-Diss. Göttingen, 1907.—R. BÄRTLING. Die Schwerspatlagerstätten Deutschlands. Stuttgart, 1911, Ferd. Enke.

The pyrite deposit at Meggen occurs in the Attendorn-Elspe double syncline of the Lenne slate in Middle Devonian limestone, the geological investigation of these beds having been taken up principally by Denckmann and Henke.

According to Denckmann, the limestone forming the hanging-wall of the deposit consists of two horizons, of which the lower, from fossil evidence, belongs undoubtedly to the uppermost Middle Devonian, while the upper is representative of the lowest Upper Devonian; within this limestone complex therefore lies the boundary between Middle and Upper Devonian. Above the lowest Upper Devonian, which is calcareous in nature, follow the Büdesheim slates of the lower Upper Devonian;

while above these again, unconformably, comes the so-called Fossley formation, which embraces the clay-slates and sandstones of the upper Upper Devonian. The foot-wall of the deposit consists of the Lenne slate, the relative age of which here has not yet been definitely determined.

Tectonically, this ore-bed belongs to a fairly complex system of folds, the double syncline being accompanied by a number of secondary anticlines which have suffered such considerable denudation that the corresponding synclines now frequently appear as independent units.

The folds strike north-east, the limbs to the south-east being often steep in the neighbourhood of the deposit, while those to the north-west are often flat. In Fig. 394 a ground plan and section showing the geological position of this deposit are given ; in strike and dip it follows the country.

The filling consists partly of pyrite and partly of barite, these minerals appearing to replace each other to such an extent that the barite, which occurs principally in the hanging-wall, may sometimes occupy the entire thickness. Generally speaking the pyrite predominates in the middle portion of both north-east striking flanks, west and east of Halberbracht, while the south-west and north-east continuations of this middle portion consist of barite. The pyrite mass, usually about 4 m. thick, has been followed along the strike for more than $2\frac{1}{2}$ kilometres. While the lower portion is irregularly coarse in structure, the upper appears to be banded and to consist of an alternation of pyrite with thin clay-slate seams. One of these seams, 10–30 cm. thick with pyrite finely impregnated throughout, frequently occurs at the contact of the pyrite bed with the hanging-wall limestone. Chalcopyrite and galena occur subordinately ; they are found principally to the east where numerous transverse fractures occur, along which the younger lead-copper solutions probably circulated.

The replacement of the pyrite by barite has taken place in the east, south-east, south-west, and west, to such an extent that the pyrite bed gradually pinches out, and a barite bed, which increases in thickness till it finally reaches 6 m., takes its place. The lowest barite layers in the immediate neighbourhood of the pyrite contain narrow pyrite stringers. The barite here is either dense or spherulitic. Along the steep south flank decomposed pyrite occurs at the surface, the border of barite, owing to erosion, being absent. To the north-west the ore-bed is overlaid by younger beds, so that in this direction the line between the pyrite and barite is not yet known.

The deposit is traversed by numerous small disturbances which, since they affect the general position but little, cannot be expressed in the section given in Fig. 394. On the other hand, a powerful overthrust somewhat north-east of the deposit and indicated in that figure, is of great importance.

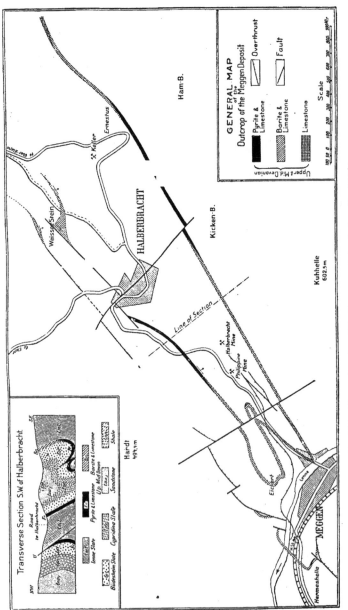

FIG. 394.—Plan and section of the Meggen pyrite-barite deposit. Henke.

At Bonzel, a 4 m. thick limestone mass containing disseminated pyrite crystals occupies the place of the pyrite bed. With an increase in the amount of pyrite this bed in places probably becomes a body of clean pyrite.

With regard to genesis, Henke's conception deserves consideration. This authority sees in the deposit a large pyrite lens bordered by barite, which has been folded with the country. According to this idea the possibility exists that the barite was formed . later than the pyrite, and the deposition of the barite followed upon the decomposition of the pyrite around its borders. Henke regards it as conceivable that this replacement took place simultaneously with the limestone, which on the Eickert lies 4·3 m. above the deposit and contains 1–2 per cent of finely distributed barite. If this were so, the occurrence would be a true ore-bed accompanied by subsequent surface decomposition and replacement.

The possibility of such a genesis is not to be denied. It is still more probable however that the deposit is a metasomatic replacement by pyrite and barite of an original Middle Devonian limestone bed. The gradual increase of the barite while the thickness of the bed remains practically constant, and an occurrence at Bonzel where in the place of the pyrite bed a 4 m. thick limestone with disseminated pyrite crystals occurs, support this view ; in addition, experience has shown that elsewhere in the Rhenish Schiefergebirge, Middle Devonian limestones have been altered to pyrite. Furthermore, it must be considered that in nature barite occurs far more frequently as lode-filling and in metasomatic deposits than in true beds. Finally, the fact that in this occurrence pure barite only occurs associated with pyrite, while the other decomposition products usually found as the result of processes in which sulphuric acid and sulphurous salts were active, are absent, opposes Henke's theory of the barite formation.

The deposit at Meggen has been worked since the year 1845. The two principal mines at the present day are Sizilia and Siegena in Müsen, which in 1910 yielded together 185,328 tons of pyrite. This occurrence in spite of its small size is of considerable importance in Germany, since in that country only about 250,000 tons of pyrite are produced annually. In relation to the world's production, however, it plays but a small part.

The average pyrite content of the ore as it is mined is low, reaching only 34 per cent. This can be raised by hand-sorting to about 42·6 per cent, though by so doing about one-fifth of the weight is lost.

The deposit on the Rote Bergen at Schwelm belongs also in part to the metasomatic pyrite deposits. The massive limestone at that place has been altered not only to zinc oxidized ore but also in part to marcasite, the replacement having been so gradual that the form of the corals is retained. In this alteration the faults occurring at the contact of the massive limestone and Lenne slate undoubtedly played an important part.

THE NATIVE COPPER DEPOSITS

To this group belong the calcite- and zeolite-bearing occurrences of native copper associated with basic eruptive sheets—diabase and melaphyre —at Lake Superior in Michigan, as well as a number of mineralogically and geologically similar but unimportant deposits. The copper sandstone of Corocoro, etc., in Bolivia, which in many respects resembles the copper conglomerates of the Lake Superior district but carries no zeolites, occupies a place by itself.

THE LAKE SUPERIOR DISTRICT

LITERATURE

R. D. IRVING. 'The Copper-bearing Rocks of Lake Superior,' U.S. Geol. Survey Mon. V., 1883 ; in which, among other works, the following are mentioned : J. W. FOSTER and J. D. WHITNEY, 1850 and 1851.—H. CREDNER. Neues Jahrb. f. Min., Geol., Pal., 1869, and Zeit. d. d. geol. Ges., 1869.—R. PUMPELLY. 'The Paragenesis and Derivation of Copper and its Associates on Lake Superior,' Amer. Journ. Sc. III., 1871, Geological Survey of Michigan, Part 2, 1873 ; 'Metasomatic Development of the Copper-bearing Rocks of Lake Superior,' Proc. Amer. Acad. XIII., 1878.—M. E. WADSWORTH. 'Bull. Mus. of Comp.,' Zoology at Harvard College ; Geol. Ser. I., 1880 ; Proc. Boston Soc. Nat. Hist. XXI., 1880.

Of the numerous other works which appeared after 1883 the following deserve mention : 'Keweenaw Point,' Geol. Survey of Michigan, VI. Part 2, 1899.—C. ROMINGER. 'Copper Regions of Michigan,' Geol. Survey of Michigan, V., 1895.—L. L. HUBBARD. Ibid. V.— A. C. LANE and A. E. SEAMAN. 'Notes on the Geol. Section of Michigan, Lansing,' Mich. Ann. Rep. for 1908.—A. C. LANE. 'Notes of the Geol. Section of Michigan, I.' (with A. E. Seaman), Journ of Geol. XV., 1907 ; II. ibid. XVII., 1910 ; 'The Theory of Copper Deposition,' Amer. Geol., Nov. 1904; also Econ. Geol. IV., 1909.—'Mine Waters,' 13th Ann. Meeting of the Lake Superior Min. Industry, June 1908.—G. FERNEKES. 'Precipitation of Copper from Chloride Solutions by means of Ferrous Chloride,' Econ. Geol. II., 1907.—H. L. SMYTH. Science, 14th Feb. 1896.—C. R. VAN HISE. Several articles in Some Principles Controlling the Deposition of Ores ('Genesis of Ore-Deposits,' Trans. Amer. Inst. Min. Eng. XXX. and XXXI., 1902), and 'A Treatise on Metamorphism,' U.S. Geol. Survey Mon. XLVII., 1904.—C. R. VAN HISE and C. K. LEITH. 'The Geology of the Lake Superior Region,' U.S. Geol. Survey Mon. LII., 1911.—J. F. KEMP. Ore-Deposits of the United States and Canada.—M. E. WADSWORTH. Trans. Amer. Inst. Min. Eng. XXVII., 1897.—P. GROTH. 'Über das Vorkommen des Kupfers am Lake Superior,' Lecture on 18th Nov. 1895, Bayr. Industrie- u. Gewerbebl. (Munich).—EBELING. 'Das Berg- und Hüttenwesen in den Kupferbezirken am Oberen See und bei Bingham' (Nordamerika), Zeit. f. d. B.-, H.- u. S.-Wesen im preuss. Staate, LVIII., 1910.

The Keweenawan series, named after the peninsula of that name, belonging to the Algonkian, consists in its lower portion of conglomerates, sandstones, and slates, free from eruptives. In the Middle Keweenawan a period of violent eruptivity began which left its impression upon the whole Lake Superior district. During this period plutonic rocks such as quartz- and orthoclase-bearing gabbros were formed, though not to so great an extent as tremendous eruptive sheets consisting essentially of basic rocks, but subordinately also of such acid rocks as felsite- and quartz-porphyry. These basic rocks contain mostly 48–50 per cent of SiO_2; they carry olivine, augite, and plagioclase, and may in greater part be described as diabase and melaphyre. The thickness of the individual sheets varies from ten metres to several hundred metres; their surface is characterized by the presence of numerous vesicles due to an original high gaseous content. Between these eruptive sheets, conglomerates containing boulders of felsite-porphyry and diabase generally 1–8 m. in diameter though some-times larger, were deposited.

At Portage Lake, where the most important mines are situated, this eruptive and conglomerate formation reaches a thickness of 13,680 feet, of which 2125 feet are occupied by twenty-two conglomerate beds, while the eruptive sheets are responsible for 11,555 feet. At other places these sheets are still thicker.

The Upper Keweenawan consists principally of sandstones and slates, which in the neighbourhood of Portage Lake reach a thickness of about 9000 feet. The whole Keweenawan series on Keweenaw Point is separated from the Cambrian sandstones by a huge fault.

The eruptive sheets with the accompanying conglomerates form together a syncline, the central portion of which is covered by Lake Superior. Including the extent under the bed of this lake and some areas now denuded, the Keweenawan formation has a superficial extent of roughly 75,000 sq. km., or an area equal to that of Bavaria. The total volume of the eruptive masses included therein has been estimated by Van Hise and Leith at some 54,000 cubic miles, and this district accord-ingly is one of the largest eruptive regions in the world.

The large copper deposits of this district are associated with the Middle Keweenawan which, as shown, consists so largely of these lava flows. The copper occurs in various forms, these being connected with one another by gradations. The following forms of occurrence may be differentiated :

1. Lodes containing chiefly calcite, usually 0·5–1 m. wide, which traverse the diabase, etc., vertically. Blocks of copper up to 420 tons in weight have been encountered in these; as a rule however such fissure-fillings are poor and irregular. They are found principally on the extreme point of the Keweenaw Peninsula, where, especially in former years, they were vigorously worked. The chief mines were, Central Cliff, with a vertical

depth of 1600 feet, Phoenix, and Copper Falls. To-day the exploitation
of this type of deposit has practically ceased.

2. Cavity-fillings in amygdaloidal melaphyre or diabase, chiefly where
these are scoraceous and coarsely vesicular. Such sections are locally known
as ' ash-beds ' ; they represent the original surfaces of the lava flows. The
eruptive rock in the neighbourhood of the deposits is always greatly decomposed. The copper is associated with calcite, quartz, prehnite, etc., which
minerals frequently not only fill the vesicles but occur also in veins. The
mineral-association of the amygdaloid beds, which are frequently 2–5 m.

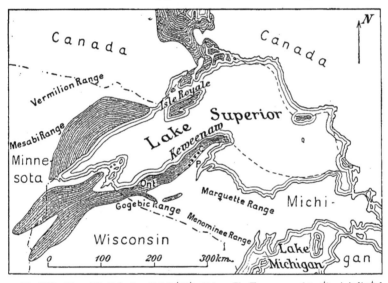

Fig. 395.—Map of the Lake Superior district. Irving. The Keweenawan formation is indicated
by lines roughly parallel to the strike, while the assumed boundaries in the lake ore are indicated
by dotted lines.

C = Calumet ; P = Portage Lake ; Ont. = Ontanogon ; while the different ranges indicate the iron districts.

and exceptionally 12 m., thick, is somewhat variable, but not so variable
as that of the lodes. The country around Houghton and Calumet in the
centre of the Keweenaw Peninsula, is especially characterized by this
type of deposit. In this locality the Middle Keweenawan forms a north-
east striking zone, with a horizontal width of 8–10 km., the beds of which
dip 35°–55° to the north-west. In this zone, from hanging-wall to
foot-wall, the following ' amygdaloid lodes ' are known : the Atlantic,
Pewabic, Osceola, Kearsarge, Arcadian, and Baltic amygdaloids, these
being separated from one another by bands of unpayable material 50–
200 m. in thickness.

3. Impregnations in the porphyry conglomerates lying between the lava flows. The copper in these deposits occurs with calcite, epidote, chlorite, etc., as the cement or matrix between the pebbles, these mostly varying from pea- to nut size. It is interesting to note that these pebbles in the process of impregnation have in part become completely replaced by copper. Flakes of native silver are found here and there, though as the native copper itself is almost entirely free from silver the relative quantities of these two metals is at most 1 of silver to 1000 of copper. Small amounts of whitneyite, Cu_9As, and domeykite, Cu_3As, occur as mineralogical rarities ; a few sulphides have also been found in minute quantities. The furnace copper, among other things, contains some arsenic, which in the case of the Calumet and Hecla increases somewhat in depth.

The native copper is accompanied principally by calcite, quartz, prehnite, and laumontite, but also by analcime, apophyllite, natrolite and

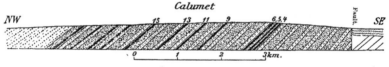

Calumet

NW   *15*   *13*   *11*   *9*   *6,5,4*   Fault.   SE

0   1   2   3km.

Upper Keweenawau ; red sandstones and conglomerates.   Lower Keweenawan ; eruptive sheets, thickly dotted ; conglomerates, lightly dotted ; and copper deposits, dark lines, the most important being numbered.   Eastern Sandstone.

FIG. 396.—Section through the middle section of the Keweenaw peninsula near Calumet. Irving, 1883.

other zeolites, orthoclase, datolite, epidote, and delessite, a chlorite mineral ; minerals other than these occur in quite subordinate amount. It is particularly associated with what is known as ' green earth,' this being a mixture of delessite and ferric oxide.

The copper content of the above-mentioned porphyry conglomerates, though usually not very high, is particularly interesting. The famous Calumet and Hecla conglomerate, in respect to its tenor, is an exception, as is also the Allouez conglomerate—known sometimes as the Boston and Albany—this being payable in places. The first-mentioned had near its outcrop a thickness of about 13 feet, which increased in depth to 20 feet. The average copper content in the upper levels was 2–3 per cent and occasionally somewhat more ; below the 750 m. level however this diminished, till below the 1000 m. level the average content was only 1·3–1·7 per cent. These conglomerates, as illustrated in Fig. 396, dip 35°–55°.

By far the greater number of the Lake mines at present working, exploit the above-mentioned amygdaloid lodes. Of the total production, the Calumet conglomerate has latterly been responsible for about 27 per cent, while the amygdaloid lodes have contributed almost all the remainder. Nearly all the present-day mines are situated near the towns Houghton and Calumet in the neighbourhood of Portage Lake, within an area 25 km. long and 3–4 km. wide ; some are found farther north on the Keweenaw Peninsula and at Ontanogon south-west of Portage Lake.

Similar deposits are found on Isle Royale in the Upper Lake, and at many other places within the Keweenawan formation, as for instance in the Douglas county of the neighbouring state of Wisconsin, and in Minnesota. Geologically, therefore, this type of deposit has a wide distribution. Only the mines on the Keweenaw Peninsula however are of any importance.

Some of the Lake deposits were worked by the Indians before the coming of the white race. Exploration work on a small scale was carried on by Europeans in the seventeenth and eighteenth centuries. The present industry began in the middle of the 'forties. It developed so rapidly that, soon afterwards, the Lake district ranked among the most important copper districts of the world. Its total production, and that of the Calumet and Hecla mine which began in 1867, have at different periods been as follows :

	Total Copper Production of Lake Mines.	Calumet & Hecla.
	Long Tons.	Long Tons.
1845	12	...
1850	572	...
1860	5,388	...
1870	10,992	6,277
1880	22,204	14,140
1890	45,273	26,722
1900	63,461	34,715
1905	102,874	37,950
1910	99,545	35,000

The Lake district, which now produces about one-ninth of the world's copper, has of late years been surpassed by Butte in Montana, and by Arizona. The total production of the Lake district up to the end of the year 1910 amounted to 2,122,000 tons, worth about £120,000,000. Of this amount, 1,045,000 tons, or almost exactly one-half, were contributed by Calumet & Hecla, the largest mine. Apart from this, the most important mines are : Tamarack, now united with Calumet & Hecla, production in 1908, 5800 tons ; Osecola, a few kilometres distant from Calumet, production in 1908, 9700 tons ; Quincy, 9300 tons ; Baltic, 8700 tons ; Champion, 8100 tons ; and Mohawk, 4700 tons. To these must be added fifteen other companies with a total production of 16,800 tons in 1908. The ore won from all the mines in 1906, 1907, and 1908, had an average copper content of 1·26, 1·10, and 1·06 per cent, respectively. The Calumet conglomerate during these years contained 2·00, 1·69, and 1·60 per cent ; and the mines working cupriferous amygdaloids 0·96, 0·88, and 0·88 per cent of copper, respectively. Some mines have exploited ore containing only 0·6 and 0·7 per cent. Gradual diminution of the copper content in depth has been the invariable experience.

Owing to the large production, the relatively low copper content of the ore mined, and the comparatively short distance the ore-body extends along the strike, the principal mines quickly attained considerable depth. The deepest shafts of the Calumet & Hecla and Tamarack mines in July 1909 were 4920, 5253, and 5363 feet, respectively. Several of the shafts are inclined at an angle of about 40°, so that their depth along the dip may reach as much as 8500 feet. The Quincy mine, which exploits one of the amygdaloid lodes, reached in 1910 an inclined depth of 5280 feet, equivalent to a vertical depth of 4008 feet.

In this district the temperature increases in depth remarkably slowly, namely, only about 1° C. for every 200 feet of vertical depth. At a vertical depth of 5000 feet the temperature of new development ends is only about 38°, and that of well-ventilated stopes only 27°–30° C.

The ore is treated in greater part for concentrate containing 65–90 per cent of copper, which is refined direct in the reverberatory furnace. In process of this concentration a considerable amount of slime containing 25 per cent of copper is recovered, which after being briquetted is first smelted for blister copper and then refined. The cost of production in the more important mines has of late years been 8·77–9·5 cents per lb. of copper produced, or £40 : 10s.–£44 per ton. Of this cost, 62 per cent is incurred in mining, 23 per cent in dressing, and 15 per cent in smelting.

The Calumet & Hecla Company had up to 1910 altogether paid £23,200,000 in dividends, or roughly £22 : 10s. per ton of copper. Financially speaking, the good time for the Lake mines is now probably over, the Calumet & Hecla mine having almost completely exhausted its conglomerate down to a vertical depth of 3600 feet, while the Quincy mine is now working at depths of 2750 to 3250 feet. The Calumet & Hecla, which formerly produced at the rate of 10,000 tons of copper per metre of vertical depth, now at a depth of 3600 feet only produces roughly 1000 tons, equivalent to about 700 tons per metre of inclined depth.

The conglomerate bed is at present worked for a horizontal length of 2500 m. and an average horizontal thickness of about 7 m., while on an average it contains about 2 per cent of copper. From these data also, and assuming 2·9 tons of ore per cubic metre, almost exactly 1000 tons of copper per metre of vertical depth would result.

Mineralogically and geologically similar occurrences, though containing but little copper, have been discovered in many other places. Native copper, for instance, has been found on at least six of the Faroe islands, these islands consisting principally of flat late Tertiary basalt sheets. In this occurrence the copper occurs associated with calcite and different zeolites—such as heulandite, desmine, mesolite, more rarely chabasite, apophyllite, and gyrolite—in vesicles within the basalt. According to Cornu, the native copper was formed first and the zeolites later. Occasionally the copper occurs also as a secondary deposition in a basalt breccia, and in tuffs between the flows. It is never accompanied by sulphides. On the other hand, some cuprite, malachite, and chrysocolla, have often been formed secondarily from the copper.[1] This occurrence of copper in the Faroe islands was first mentioned by L. J. Debes, Copenhagen, 1673.

Similar occurrences of native copper, associated with calcite and prehnite in narrow veins and in vesicles, are found in the sheets of essexite-porphyrite and essexite-melaphyre at Guldholmen near Moss, Lövöen near Horten, Skredhelle near Skien, etc., in the Christiania district.

Further, the occurrence of native copper and prehnite in melaphyre at Oberstein on the Nahe, in diabase at Stirling in Scotland, in amygdaloidal trap in Alaska,[2] and in andesite in eastern Serbia,[3] are also worthy of mention.

Concerning genesis, it is accordingly worthy of note that the considerable number of native copper occurrences scattered throughout the world are mineralogically and geologically very similar. It is common to all that they are associated with basic eruptive rocks, and especially with flows of melaphyre, diabase, essexite-porphyrite, dolerite, basalt, and exceptionally of andesite. Such an association justifies the general conclusion that the copper must in some way or other be genetically connected with these basic eruptives. The characteristic mineral-association, including as it does, calcite, epidote, chlorite, prehnite, and other zeolites, points to deposition from an aqueous solution the temperature of which did not reach the critical temperature of water, 365° C.

Van Hise and Leith in 1911 pointed out that many of the associated minerals, such as prehnite, epidote, chlorite, and laumontite, contain alumina, and that alumina is either not soluble in cold water or only in very small amount. From this fact they conclude that the solutions had

---

[1] F. Cornu, ' Über das gediegen Kupfer in den Trappbasalten der Faröeinseln,' *Zeit. f. prakt. Geol.*, 1907, p. 321.

[2] A. Knopf, ' The Copper-bearing Amygdaloids of the White River Region, Alaska,' *Econ. Geol.* V., 1910, p. 251.

[3] M. Lazarevic, ' Ein Beispiel der " Zeolith-Kupfer-Formation " im Andesit-Massiv Ostserbiens,' *Zeit. f. prakt. Geol.*, 1910, pp. 81-82.

a high temperature. These authorities come likewise to this conclusion from a study of the decomposition of the country-rock; with these deposits the usual phenomena of weathering are not found, but decomposition such as can only be explained by the action of hot solutions. Pumpelly believes the copper to have been originally leached from the sandstone forming the hanging-wall, and that the cupriferous solutions represent descension solutions. With the latter portion of this view Lane agrees. Against this, Van Hise and Leith emphasize the fact that deposition from hot solutions postulates ascension, such as must have taken place immediately after the extrusion of the large basic sheets. When it is also remembered that these sheets, not only in the Lake district but also in the small analogous occurrences elsewhere, are characterized by a strikingly large number of vesicles due to the high gaseous content of the magma, and, furthermore, that the large mines of the Lake district exhibit distinct ore-shoots which usually do not coincide exactly with the dip, it may, with Van Hise and Leith, be concluded that the solutions were in greater part juvenile and arose at the consolidation of the basic rocks; meteoric waters which would have become heated during their passage through the heated rocks probably played but a subordinate part. Such juvenile solutions would account also for the boron- and fluorine contents of the datolite and apophyllite, etc., two minerals especially characteristic of the copper occurrence at Lake Superior.

The native copper of the Lake district and of analogous occurrences was in all probability deposited fairly simultaneously with the associated minerals, though in different districts a certain paragenetic seniority may be formulated. For the Lake Superior district Pumpelly gives the following : (1) chlorite and some laumontite, (2) laumontite, (3) laumontite, prehnite, and epidote, (4) quartz, (5) calcite, (6) copper and calcite, (7) calcite, analcime, apophyllite, datolite, and orthoclase; in this sequence the different generations somewhat overlap. In the Faroe islands the zeolites, according to Cornu, are younger than the native copper. For the andesite occurrence in Serbia, Lazarevic gives the following sequence : (1) copper, (2) chabasite, (3) apophyllite, (4) calcite, (5) cuprite, (6) chrysocolla and malachite, and finally several minerals formed secondarily from the copper. Here therefore the copper is older than the zeolites, while in the Lake district some of the zeolites are older than the copper.

In regard to the mineral-association, the vesicular filling of the 'ash-beds' of the diabase, etc. in the Lake district, and of the associated rocks of the other occurrences, is generally speaking identical with the ordinary vesicular filling of basic eruptives, the only difference being the copper content. The formation of these copper deposits is therefore a special case of the zeolitization of basic eruptive sheets.[1] In harmony with this conception Beck later termed this group the zeolitic copper deposits.

The basic eruptive rocks in general invariably contain copper,[2] though naturally in very small amount, and some sulphur. From the mutual affinity of these two elements Vogt considers that in the magma and in the consolidated rock the copper occurs preferably as a sulphide, such as chalcopyrite, chalcocite, etc., and only subordinately associated with the silicates. This small sulphide copper content would easily be extracted by solutions containing ferric salts, such as $FeCl_3$, $Fe_2(SO)_3$, etc., according to the formula : $CuS + 2FeCl_3 = CuCl_2 + 2FeCl_2 + S$,[3] while any silver present would also go into solution. A very low copper content in the basic magma or rock would under these conditions be quite sufficient to account for even large deposits.

The total thickness of the copper-bearing conglomerates and amygdaloid beds at Portage Lake is some 25 m., the average copper content of this thickness being 1·5 per cent. The basic eruptive sheets on the other hand have together a thickness of nearly 4000 metres. Had the copper-bearing beds an uninterrupted extent, an average copper content equivalent to 0·0094 per cent in the basic sheets would be sufficient to supply the whole of the copper in the deposits. But, since these beds in relation to the whole complex extend only a short distance along the strike, a small fraction of 0·0094 per cent would amply suffice for the enormous quantity of copper in these deposits.

With regard to the precipitation of the copper in the native state, it was formerly considered that electrolytic processes had brought this about.[4] Such a view however, would necessitate the assumption that a cathode existed in each individual vesicle, etc., which seems somewhat unreasonable.

The view put forward by Pumpelly in the 'seventies that the copper was precipitated by reduction, is probably more apt. Minerals containing ferrous oxide, such as magnetite, augite, etc., have a reducing effect. The native copper frequently found in the oxidation and cementation zones of various deposits was formed in this manner. In the case of the Lake occurrences this view is supported by the kernels of magnetite found here and there enclosed in native copper, and by the large amount of iron oxide, in the form of friable hæmatite, found in the immediate vicinity of the copper. This oxide is so considerable in amount that it colours the battery water brick-red and forms large deposits in the lake below the battery.

A few years ago H. N. Stokes [5] by means of experiment established the fact that ferrous sulphate, $FeSO_4$, at a temperature of 200° C. pre-

[1] Vogt, *Zeit. f. prakt. Geol.*, 1899, p. 13.   [2] *Ante*, p. 158.   [3] *Ante*, p. 216.
        [4] Foster and Whitney, 1850.   [5] *Econ. Geol.*, 1906, Vol. I. p. 647.

cipitates metallic copper from a solution of copper sulphate, iron oxide being simultaneously formed by hydrolysis of the ferric sulphate.[1]   G. Fernekes [2] later showed that chlorides at a temperature of 200°–250° behave similarly, the following formulae probably representing the reactions :

1. $2FeCl_2 + 2CuCl_2 = 2CuCl + 2FeCl_3$.
2. $2FeCl_2 + 2CuCl = 2Cu + 2FeCl_3$.
3. $FeCl_3 + 2H_2O = Fe(OH)_2Cl + 2HCl$.

The basic iron chloride after a time becomes altered to iron oxide and hydrochloric acid. In the experiments, in order to neutralize this acid a small amount of easily affected silicate, such as prehnite, was added.

From these facts the general conclusion may be drawn that all ferrous compounds, not only minerals but also dissolved salts, at a sufficiently high temperature and given time enough, precipitate metallic copper from solutions, with the simultaneous formation more particularly of iron oxide. Thus Pumpelly's theory becomes expanded.

The flakes of native silver occurring on the copper are probably the result of galvanic precipitation, according to the formula : $Cu + Ag_2SO_4 = Ag_2 + CuSO_4$.[3]

The whole process resulting in the deposition of native copper may accordingly be depicted in the following manner : Heated solutions containing copper salts, carbon dioxide, silica, alkalies, etc., as well as some boron and fluorine, penetrated certain porous beds, sometimes taking advantage of fissures along which the country-rock then became decomposed. At first, chlorite more especially, and some epidote, prehnite, etc., were formed ; then, during a second stage, the copper more particularly was precipitated, principally by ferrous oxide; while finally, the deposition of the bulk of the alkaline zeolites and calcite followed.

The decrease in the copper content in depth is a primary factor depending upon the physical-chemical laws governing the precipitation of copper from solution ; it has nothing to do with secondary migration.

The mine waters associated with the Lake occurrences are characterized by containing a striking amount of the chlorides $CaCl_2$ and $NaCl$. This would indicate that the solutions from which the copper was precipitated contained much chloride, and that the metal was originally present in greater part as chloride. The almost complete absence of sulphides in these occurrences indicates that the solutions contained no sulphates, or only traces. Lane has put forward the hypothesis that the chloride waters of the Lake mines represent fossil lake- or desert waters, which filtered from above and on their way extracted the copper from the basic rocks. On general geological grounds this view cannot be endorsed.

---

[1] $Fe_2(SO_4)_3$.   [2] *Econ. Geol.*, 1907.   [3] *Ante,* p. 139.

Corocoro in Bolivia

LITERATURE

D. Forbes. Quart. Journ., 1861; Philadelphia Mag., 1866.—H. Reck. Berg- u. Hüttenm. Ztg., 1864 —E Mossbach. Berggeist, 1873.—J. Domeyko. Ann. des mines, 1880.—L. Sundt. Bol. Soc. Nacion. Min. Santiago, 1892.—A. W. Stelzner. Zeit. d. d. geol. Ges., 1897.—G. Steinmann. Die Entstehung der Kupferlagerstätte in Corocoro und verwandte Vorkommen in Bolivien,' Rosenbusch Celebration. Stuttgart, 1906, pp. 335-367.

In addition to the occurrence at Lake Superior, native copper is also found in considerable amount in copper-sandstone at Corocoro in Bolivia—17° south latitude and about 4000 m. above sea-level—as well as at numerous other places in that country between 16° and 22° 42′ south latitude. The principal district lies in the inter-Andean plateau of the Bolivian Cordilleras. Here the copper occurs in a red ferruginous sandstone—described by Steinmann as Puca sandstone, ' puca ' meaning red— belonging to the Cretaceous. In addition to native copper, some native silver, domeykite, subordinate chalcocite, and a few other sulphur-, arsenic-copper-, and silver ores, also occur. Some selenite, barite, and a sparing amount of calcite, are found as associated minerals.

The copper at Corocoro—coro meaning copper—occurs principally as an impregnation in some twenty sandstone and conglomerate beds, mostly 1–2 m. thick. This cupriferous complex occurs in a zone 3–4 km. long and 2 km. wide following a large fault-fissure, and has a total thickness of several hundred metres. The deposits however are neither conformable in general nor in detail. Narrow veins of native copper and pseudomorphs of native copper after aragonite, according to Sundt and Steinmann, prove undoubtedly that the copper is epigenetic. It is evident that, just as with the conglomerates and amygdaloids at Lake Superior, the copper was deposited from solutions which saturated certain porous beds.

In the neighbourhood there is an occurrence of diorite, and Steinmann suggests that this eruptive rock also extends under the copper deposit. Like the mine waters of the Lake mines which, according to Lane, in depth contain chlorides, those at Corocoro also are rich in chlorides of the alkalies and alkaline earths.

The copper-sandstone is red and ferruginous, a fact which points to a reduction process similar to that postulated for the Lake mines. Curiously enough, however, the sandstone in the immediate neighbourhood of the copper is bleached or faded, a reduction from Fe_2O_3 to FeO having taken place. Steinmann supposes a partial oxidation of Cu_2S by Fe_2O_3 to have taken place, whereat the sulphur only was oxidized, the copper remaining as metal. Such an explanation however appears very questionable.

Corocoro was originally worked by the Incas, an Indian race. The mines then lay idle for some time, and only after 1832 was continuous work resumed. In 1867 they were 100 m. deep. The yearly production of copper during the last thirty years, as given in the copper statistics of H. R. Merton & Co., has been 2000–2500 tons. According to Steinmann it has at times been somewhat higher than this ; in 1902 for instance it was 4200 tons.

Copper-sandstone is also being worked for copper at Chacarilla, 50 km. south of Corocoro, and at Cobriros, situated in 20° 15′ south latitude.

THE COPPER PRODUCTION OF THE WORLD AND ITS DISTRIBUTION AMONG THE VARIOUS CLASSES OF DEPOSIT

LITERATURE

For the period after 1879 the annual copper statistics of H. R. Merton & Co. of London were principally used ; and for the period before that time, chiefly B. NEUMANN, Die Metalle, Halle, 1904 ; cited on p. 644.—J. H. L. VOGT. ' Die Geschichte des Kupfers,' Christiania, 1895, with résumé under the title of ' Die Statistik des Kupfers,' Zeit. f. prakt. Geol., 1896, pp. 89-93. H. Wencker's work, ' Die wirtschaftliche Bedeutung der Kupfererzlagerstätten der Welt in den Jahren 1906-1910 mit besonderer Berücksichtigung der genetischen Lagerstättengruppen,' Bergwirtschaftlichen Zeitfragen, Part 3, Berlin, 1912, only appeared when this book was in the press.

The copper statistics from 1850 onward are regarded as fairly accurate, and reliable statistical material also exists for the earlier period, at least for the majority of the large works. The total production before 1600 was certainly more than 1–1·5 million tons, and may be put at about 2 million tons.

TOTAL COPPER PRODUCTION AND AVERAGE PRICE OF COPPER

Years.	Total Copper Production.	Average Price of Copper per Ton, Chilian Bar.
	Metric Tons.	£
1901–1910	6,945,000	65½
1891–1900	3,770,000	51½
1881–1890	2,252,000	56¼
1871–1880	1,050,000	74
1861–1870	760,000	79
1851–1860	510,000	91
1801–1850	about 1,400,000	...
1701–1800	„ 1,300,000	
1601–1700	„ 750,000	
Still earlier	2,000,000	...
Total, about	20,750,000	...

This is roughly 20,000,000 tons, worth some £1,400,000,000. The total value of the gold production up to date has been reckoned at about £2,900,000,000,[1] and that of silver at about £2,500,000,000.[2] As much copper was produced in the last twenty-five years, that is since 1885, as in the previous eighty-five years of the nineteenth century, or in all the former centuries together.

The copper production of the individual countries may be gathered from the following table :

ANNUAL COPPER PRODUCTION (output from mining operations)

	In Long Tons.				
	1880.	1890.	1900.	1905.	1910.
Germany { Mansfeld	9,800	15,800	18,390	19,565	19,995
Germany { Elsewhere	1,000	1,825	2,020	2,595	4,715
Austria	470	1,210	865	1,175	2,130
Hungary, including Bosnia, Serbia	820	300	490	150	4,955
Norway { Sulitjelma	...	...	2,220	3,195	4,925
Norway { Elsewhere	2,426	1,390	1,715	3,110	5,500
England	3,662	935	650	715	500
Sweden	1,074	830	450	550	2,000
Russia-Siberia	3,300	4,800	6,740	8,700	22,310
Spain-Portugal	36,313	51,700	52,872	44,810	50,255
Italy	1,380	2,200	2,955	2,950	3,220
Turkey	...	...	520	700	600
Canada	50	3,050	8,500	20,535	25,715
Mexico { Boleo	...	3,450	11,050	10,185	12,795
Mexico { Elsewhere	400	875	11,000	54,255	46,030
Newfoundland	1,500	1,735	1,900	2,280	1,080
United States	25,010	116,325	263,502	389,120	484,890
Lake District	22,200	44,450	54,111	97,770	97,770
Montana	...	49,560	114,144	142,490	128,770
Arizona	...	15,945	49,447	99,490	132,625
Elsewhere	2,810	6,370	40,800	49,370	125,725
Argentina	300	150	75	155	300
Bolivia, Corocoro	2,000	1,900	2,100	2,000	2,500
Chili	42,916	26,120	25,700	29,165	35,235
Peru	600	150	8,220	8,625	18,305
Venezuela	1,800	5,640	...	...	...
Cuba	...	...	...	...	3,475
South { Cape Copper Co.	5,038	5,000	4,420	5,025	4,405
Africa { Namaqualand	...	1,450	2,300	2,300	2,500
Remaining part of Africa	500	120	...	415	8,300
Japan	3,900	15,000	27,840	35,910	46,000
Australia	9,700	7,500	23,020	33,940	40,315
Totals	154,000	269,500	479,500	682,000	852,950
Price per ton, Chilian bars	£63	£54	£73¼	£69	£57¼

The total production up to the end of 1910 is distributed among the different countries as follows :

---

[1] Ante, p. 644.   [2] Ante, pp. 647, 648.

United States, since 1845	.	.	.	6,900,000 tons.
Spain and Portugal	.	.	about	2,600,000 ,,
Chili [1]	.	.	,,	2,500,000 ,,
Japan	.	.	,,	1,250,000 ,,
Cornwall	.	.	,,	1,100,000 ,,
Germany	.	.	,,	1,000,000 ,,
Russia-Siberia	.	.	,,	1,000,000 ,,
Australia, since 1844 [2]	.	.	,,	800,000 ,,
Mexico, only since 1879	.	.	.	660,000 ,,
Sweden, since about 1220	.	.	about	550,000 ,,
Austria-Hungary	.	.	,,	333,333 ,,
Canada, only since 1879	.	.	.	300,000 ,,
Norway, since 1631	.	.	.	225,000 ,,

South Africa ⎫
Peru   ⎬ in general since 1879 ; Italy and ⎰ 200,000 ,,
Bolivia   ⎮   Bolivia also earlier   ⎱ 125,000 ,,
Italy   ⎭   70,000 ,,
                                               85,000 ,,

This table makes a total of about 19,750,000 tons. To it must be added the earlier production of Italy, Bolivia, Cape Colony, etc., as well as the production of Turkey, Serbia, Newfoundland, Cuba, Venezuela, etc.

The UNITED STATES during the period 1845–1850 produced only about 2500 tons ; 1851–1875, 214,000 tons ; 1876–1900, 2,840,000 tons ; 1901–1910, 3,880,000 tons ; or altogether and up to 1910 a total of 6,935,000 metric tons. Of this amount 2,410,000 tons came from Montana ; [3] 2,160,000 tons from the Lake District ; [4] and 1,450,000 tons from Arizona.[5]

In the Huelva district of SPAIN AND PORTUGAL [6] in ancient times 800,000 or 1,200,000 tons, according to various estimates, were produced ; from the eighth century A.D. to 1850, only some 10,000 or 20,000 tons ; 1851–1880, 142,000 tons ; 1881–1900, 925,000 tons ; 1901–1910, 510,000 tons ; making a total of say 2,600,000 tons.

JAPAN, during the period 1901–1910, produced 403,000 metric tons ; and during 1881–1900, 315,000 tons; so that including earlier figures the production may be estimated at some 1,250,000 tons.[7]

CORNWALL, according to some estimates [8] produced from 1501 to 1725 about 20,000 tons, and from 1726 to 1905 some 883,000 tons, making a total of roughly 900,000 tons. Other estimates [9] give somewhat more, some as much as 1,300,000 tons ; in round figures, therefore, 1,000,000 tons may be taken. The production has considerably decreased of late years.[10]

GERMANY from 1901 to 1910 produced 220,000 metric tons ; 1875 to 1900, 525,500 tons ; 1851 to 1875, 87,710 tons ; 1826 to 1850, 26,500 tons ; and before 1825 perhaps 100,000 tons or at most 200,000 tons. A total of 950,000 or 1,050,000 tons, or roughly 1,000,000 tons, is thus reached. Of this figure the Mansfeld copper-shale produced undoubtedly the bulk, namely, from 1779 to 1910, 650,000 tons ; and from 1688 to 1779, 25,500 tons ; [11] or including the earlier period, a total of not quite 700,000 tons. Adding to this the other old German mines working copper-shale it may be reckoned that this deposit has produced hitherto roughly 750,000 tons.

[1] *Ante,* pp. 895, 896.   [2] *Ante,* pp. 899, 900.   [3] *Ante,* pp. 887, 889.
[4] *Ante,* p. 933.   [5] *Ante,* pp. 398, 887, 899.   [6] *Ante,* pp. 199-200, 327.
[7] *Ante,* pp. 200, 897.   [8] *Ante,* p. 436.   [9] *Ante,* p. 200.
[10] *Ante,* p. 436.   [11] Schrader, *Pr. Zeit. f. B-., H- u. S-wesen,* 1869.

RUSSIA AND SIBERIA [1] from 1901 to 1910 produced 134,000 metric tons ; 1876 to 1900, 127,000 tons ; 1851 to 1875, 122,800 tons ; 1826 to 1850, 109,000 tons ; or since 1825 a total of 493,000 tons. In the eighteenth century and in the second half of the seventeenth the production was very high, in 1700 for instance it was 3276 tons. The total production up to 1910 may accordingly be estimated at about 1,000,000 tons.

MEXICO.—Boleo from the start of work in 1887 to 1910 produced 219,000 metric tons. Other Mexican mines from 1901 to 1910 produced 404,000 tons ; 1891 to 1900, 35,900 tons ; 1879 to 1890, 4500 tons ; or since 1879 a total of about 660,000 tons. Previous to 1879 the production was very small.

SWEDEN.—Fahlun since about 1200 has produced roughly 500,000 tons. The zenith of production was reached in the middle of the seventeenth century.[2] Åtvidaberg from 1764 to about 1900 produced 35,000 tons ;[3] to this must be added the output of some other small mines. The total production is therefore some 550,000 tons.

AUSTRIA-HUNGARY, including of late years Bosnia and Serbia, from 1901 to 1910 produced 27,000 metric tons ; 1876 to 1900, 32,000 tons ; 1851 to 1875, 56,300 tons ; 1828 to 1850, 65,200 tons ; or since 1825 a total of 180,000 tons. Including the earlier period this would become some 333,330 tons.

CANADA from 1901 to 1910 produced 230,000 metric tons ; 1891 to 1900, 53,000 tons ; 1879 to 1890, 15,000 tons ; or since 1879 a total of 298,000 tons. Previous to this date the production was small.

NORWAY.[4]—The home production from 1901 to 1910 amounted to 14,100 metric tons ; 1815 to 1900, 45,000 tons ; and 1624 to 1814, about 70,000 tons ; or altogether from 1634 to 1910, 129,000 tons. Since about 1860 roughly 4,000,000 tons of pyrite with a net content of about 96,000 tons of copper have been exported. Including this the total output would be about 225,000 tons.

ITALY from 1879 to 1910 produced 85,000 tons, in addition to a large amount won earlier, especially in the Middle Ages.

SOUTH AFRICA from 1879 to 1910 produced 199,000 tons, in addition to a small amount earlier.

BOLIVIA from 1879 to 1910 produced about 70,000 metric tons, in addition to a large amount much earlier.[5]

PERU from 1898, the beginning of copper mining in that country, to 1910 produced 119,000 metric tons ; 1879 to 1897 about 7500 tons ; or, from 1879 to 1910, a total of about 125,000 tons. The early production is not very important.

The average yearly copper outputs of the most important districts during the period 1906–1910 were according to Wencker [6] as follows :

	Tons.			Tons.
Butte, Montana	124,600	Jerome, Arizona		19,600
Lake Superior	101,500	Globe, Arizona .		17,200
Bisbee, Arizona	59,000	Shashta, California		15,100
Huelva .	51,000	Japan .		14,700
Cananea, Mexico	34,000	Boundary, Canada		14,000
Chili	33,200	Queensland		13,700
Clifton-Morenci, Arizona .	31,800	Ely, Nevada .		12,200
Bingham. Utah	31,500	Cerro de Pasco, Peru		12,100
Mansfeld	19,100	Boleo, Mexico .		12,000

---

[1] *Ante*, p. 900   [2] *Ante*, pp. 198, 314-315.   [3] *Ante*, p. 340.
[4] *Ante*, pp. 198, 313.   [5] *Ante*, p. 938.   [6] *Loc. cit.*

Between them, these eighteen districts produced 615,300 tons per year, or 78·20 per cent of the world's copper production. The remaining 171,000 tons, or 21·8 per cent, came from a large number of smaller districts, each responsible for less than 10,000 tons yearly.

THE DISTRIBUTION OF THE COPPER PRODUCTION AMONG THE VARIOUS CLASSES OF DEPOSIT

Copper Lodes.—The whole production of Butte and several other districts in the United States, almost the whole output of Chili and Peru, and a subordinate part of the production of Mexico, are derived from this class of deposit; further, some three-fourths of the Japanese, and about two-thirds of the Australian production; and finally, the output from a large number of other, mostly small districts all over the world. On the basis of the figures for 1910 the copper produced from lodes represents about 40 per cent of the world's production.

Contact-Deposits and Combined Lode- and Contact-Deposits.—To this class belong at least three of the principal occurrences in Arizona, producing together some 100,000 tons of copper yearly; as well as Bingham in Utah and Cananea in Mexico, each of these two producing about 35,000 tons per year; and in addition several other occurrences in the United States and Mexico, and a number in the Urals, the Caucasus, the Banat, and at Traversella in Piemont, etc. This class is responsible for at least 25 per cent, and perhaps even 30 per cent of the total production.

Native Copper Deposits.—About 12 per cent comes from this class.[1]

Magmatic-Intrusive Pyrite Deposits.—The Huelva occurrence,[2] producing now about 50,000 tons of copper yearly; nearly all Norwegian deposits, producing about 8500 tons of copper in pyrite and copper ore yearly; Fahlun, now producing little; and Sain Bel, Agordo, Schmöllnitz, etc., belong to this class. Including Mount Lyell in Tasmania with about 8000 tons yearly, and Ducktown in Tennessee, this class is responsible for not quite 10 per cent of the total production.

Magmatic Nickel-Pyrrhotite Deposits.—About 7000 tons of copper are produced annually from the Sudbury deposit belonging to this class.[3] To this must be added some hundred tons from other deposits, making a total of not quite 1 per cent.

Magmatic Bornite Deposits.—The deposits of this class typically represented by the occurrence at Klein Namaqualand in South Africa,[4] have hitherto been very little investigated. About 0·5 per cent (?) of the yearly copper production comes from them.

Metasomatic Copper Deposits.[5]—About 3 per cent may be reckoned

---

[1] *Ante,* p. 928.   [2] *Ante,* p. 327.   [3] *Ante,* p. 293.
       [4] *Ante,* p. 300.   [5] *Ante,* pp. 908-920

as coming at present from these deposits ; in the future perhaps this percentage will be higher.

Copper-Shale Deposits.—The German Kupferschiefer yields to-day 2·3 per cent of the world's copper production. Other copper ore-beds— among them Boleo in Mexico and the fahlbands, the genesis of which latter has not yet been satisfactorily determined—produce together some 2 per cent.

Finally, a small amount of copper, namely, some 1 per cent, is won as a by-product from gold-silver- and lead-silver-zinc lodes.

The above percentages may be tabulated as follows :

PERCENTAGES OF THE WORLD'S PRODUCTION OF COPPER PRODUCED BY THE DIFFERENT CLASSES OF DEPOSIT

Magmatic deposits { Nickel-pyrrhotite group, some 1 per cent / Bornite group, 0–5 per cent / Pyrite group, 9–10 per cent	10–11 per cent.
Contact-deposits and combined contact- and lode-deposits	25–30 ,,
Copper lodes	some 40 .,
Metasomatic deposits	,, 3 ,,
Native copper deposits	12 .,
Kupferschiefer	2·3 ,,
Other ore-beds	some 2 ,,
By-products from deposits of other metals . .	,, 1 ,,

This table in the case of several of its figures is naturally only approximate. It is sufficient however to indicate that by far the greater number and the most important copper deposits are connected in some way or other — by magmatic differentiation, by contact-metamorphism, or by subsequent thermal action, etc.—with eruptive phenomena.

THE NICKEL-COBALT ARSENIDE LODES

Nickel- and cobalt ores are completely absent from, or occur only in small amount in most of the lodes already described, which carry gold, silver, lead, zinc, quicksilver, tin, copper, iron, etc. Thus, such ores, which are associated principally with basic eruptive rocks, are represented only by traces in the tin lodes connected with granite, in the quicksilver- and gold lodes, as well as in most of the gold-silver-, silver-, and silver-lead lodes. For instance, nickel and cobalt occur so sparingly in the Schemnitz lodes [1] that their presence can only be inferred from minute quantities found in those furnace products where in smelting they have become concentrated.

In the Freiberg lodes, it is true, both metals are found, but only in very small amount. In the Beschert Glück, according to Stelzner,[2] every square metre on the lode plane yields 386 grm. of silver, 747 grm. lead, and 16 grm. nickel; in the Himmelfahrt sulphide lodes, 230 grm. of silver, 61·45 kg. lead, and 1 grm. nickel-cobalt; and in the Himmelfahrt dolomite lode, 1052 grm. of silver, 2103 kg. lead, and 189 grm. nickel-cobalt. Generally speaking, however, these lodes yield less than 20 grm. of nickel per square metre on the lode plane.

In most copper lodes nickel and cobalt occur only to the extent of about 0·1 of nickel-cobalt to 100 of copper. In other lodes carrying silver, lead, copper, bismuth, etc., these two metals occur in somewhat larger amount, as, for instance, in the silver- or silver-lead lodes of the Frohnbach valley in the Schwarzwald, Baden; at Markirch in southern Alsace; at Sarabus in Sardinia; and in the copper lodes at Kitzbühel in the Tyrol, and Mitterberg in Salzburg, etc.

More important however is the nickel- and more especially the cobalt-content at Schneeberg, in association with silver and bismuth, and at Annaberg, in association with silver,[3] while the large silver-cobalt lode-district of Temiskaming in Ontario,[4] discovered only a few years ago, must be regarded as a particularly important occurrence.

---

[1] *Ante,* p. 541.
[3] *Ante,* pp. 677-683.
[2] *Zeit. f. prakt. Geol.,* 1896, pp. 401-402.
[4] *Ante,* pp. 666-669.

In the Oberharz, the St. Andreasberg lodes [1] carry only a small amount of cobalt and nickel, while those of Clausthal [2] in general carry still less. Exceptionally however, in the Grossfürstin Alexandra mine in the Schleifstein valley, 5 km. south of Goslar and about 10 km. north of Clausthal, a lode discovered about twenty years ago and belonging geologically to the Clausthal group, contains in places quite considerable quantities of nickel in the form of gersdorffite, NiAsS, though the other lodes in the same mine are of the ordinary Clausthal type.[3]

The presumably metasomatic deposit at Mine La Motte [4] in Missouri, containing preponderating lead ore and in places much linnæite (Ni, Co)$_3$S$_4$, is also worthy of mention. This occurrence and the magmatic nickel-pyrrhotite deposit at the Lancaster Gap mine in Pennsylvania [5] are the most important sources of nickel in the United States. From these two occurrences approximately 2000 tons of nickel have been produced up to date, each contributing roughly one-half.

Finally, there are lode districts where nickel-cobalt arsenide- and sulphide ores predominate economically, though in such cases, without exception, other ores—copper, silver, lead, zinc, or iron—are present, to some extent in large amount. Experience shows that with lodes of this type copper is always represented, while silver and lead are mostly absent.

The most important nickel ores in lodes are, chloantite, NiAs$_2$, niccolite, NiAs, gersdorffite, NiAsS, and millerite, NiS. The most important cobalt ores are, smaltite, CoAs$_2$, and in some occurrences linnæite, Co$_3$S$_4$ or (Co, Ni)$_3$S$_4$, together with the secondary minerals formed from these. Cobaltite, CoAsS, is found only very exceptionally in lodes, but is especially characteristic of the cobalt fahlbands ; the same is the case with cobalt-arsenopyrite (Fe, Co)AsS. With the exception of some Norwegian apatite lodes, nickel-pyrrhotite is absent from lodes, this being also the case with pentlandite.

Wherever they are found in lodes these two closely associated metals, nickel and cobalt, accompany one another, the former frequently predominating, but sometimes the latter.

Districts where the lodes contain preponderating nickel-cobalt ore are rather uncommon. In such lodes, so far as they are yet known, such ore occurs only in relatively small quantity. The arsenide lodes, owing to the unfavourable conditions prevailing in the last decades for nickel and cobalt, have either not been worked at all, or only on a very small scale. In former times some of these lodes were actively worked, though only exceptionally were operations extensive.

[1] *Ante*, pp. 687-692.   [2] *Ante*, pp. 684-687.
[3] F. Klockmann, *Zeit. f. prakt. Geol.*, 1893, p. 385.—Dr. Söhle, *Naturwissenschaftliche Wochenschr.*, 1900, Vol. XV. No. 7.
[4] *Ante*, p. 770.   [5] *Ante*, p. 293.

The most important among the nickel-cobalt lodes is that at Dobschau in Hungary, now unfortunately exhausted. This lode during the period 1840–1880 yielded about 26,000 tons of nickel-cobalt ore containing some 1000 tons of nickel and a little cobalt.

According to H. Laspeyres,[1] the total production of the Rhenish Schiefergebirge from 1841 to 1890 was as follows :

District.	Ore hoisted.	Value.
Dillenburg	10,259 tons.	£34,326 12 0
Wetzlar	1,346 ,,	3,014 17 0
Hamm on the Sieg . .	92 ,,	1,901 9 0
Deutz	116 ,,	1,574 14 0
Siegen II. . . .	78 ,,	1,289 3 0
Müsen	41 ,,	1,000 18 0
Six other places . . .	87 ,,	977 19 0
Total	12,019 tons.	£44,085 12 0

This production is equivalent to some 400 tons of nickel.

The Schladming mine in Styria was productive from 1832 to 1880, during which period, however, little more than 800 tons of dressed nickel ore, equivalent to about 80 tons of nickel, were won.[2] The time of greatest activity at this mine was from 1840 to 1847. A few other old nickel mining districts have likewise had a total output of only a hundred or so tons of nickel.

The production of cobalt also was formerly very small, the total yearly production having been only some 50 tons. To this total the silver-cobalt- and silver-bismuth-cobalt lodes of the western Erzgebirge as well as the cobalt fahlbands, contributed not inconsiderably. Modum in Norway, for instance, from 1856 to 1898, according to Vogt, yielded cobalt products containing 257 tons of cobalt, and from 1821 to 1856 probably somewhat more, the total during the whole period being therefore some 500–600 tons of cobalt.

Some nickel-cobalt-arsenic lodes carry calcite or ankerite as the principal gangue, others barite, and others again quartz. The presence of tourmaline has also been recorded in a cobalt- or cobalt-copper lode at Mina Blanca in Atacama, Chili.[3]

INDIVIDUAL DEPOSITS

As already mentioned, one of the most important deposits occurs at Dobschau in the Carpathians, where a number of partly lode-like and

[1] ' Das Vorkommen und die Verbreitung des Nickels im Rheinischen Schiefergebirge,' *Verhandl. d. naturhist. Vereins*, Vol. L., Bonn, 1893.

[2] C. Schmidt and I. H. Verloop, *Zeit. f. prakt. Geol.*, 1909, p. 271.

[3] O. Stutzer, *Zeit. f. prakt. Geol.*, 1906, p. 294.

partly chimney-like deposits are found at the contact of a diorite-horn-blende-granitite with slate. The lodes contain chiefly siderite, and in the Hollopatak district [1] are still worked for iron; in the Zemberg district they carry copper-, nickel-, and cobalt ores. The following primary depth-zones were observed in one of the most important lodes : near the surface, siderite; then, copper ore; and finally, nickel-cobalt ore—chloantite and smaltite—which yielded concentrate containing 17·5 per cent of nickel and 6·5 per cent of cobalt. This nickel-cobalt ore in turn ceased at a depth of 180–200 metres.[2]

The siderite lodes of Siegerland here and there also carry cobalt ore, which is occasionally mined for itself. In Nassau and the neighbouring districts some copper lodes are found in diabase and schalstein—the so-called schalstein schists representing, according to Laspeyres, serpentinized tuffs. In one of these, developed in part as a bedded lode, the Hilfe Gottes mine at Nanzenbach in 1841 opened up a nickeliferous ore-body containing pyrite, chalcopyrite, millerite, niccolite, chloantite, etc., with calcite, some dolomite, siderite, and quartz, as the gangue-minerals. This ore-body, which had a length of about 40–80 m., and reached a depth of roughly 100 m., produced about 10,000 tons of nickel ore containing some 3 per cent of nickel.[3]

The cobalt fissures of the Zechstein formation, which are no longer of any economic importance and to which further reference is made in the chapter on the copper-shale group, are lodes carrying ore principally between the two faulted portions of the Kupferschiefer, though mineralization also continues a few metres above the hanging-wall and sometimes considerably below the foot-wall of that bed. The filling, which may be several decimetres in width, consists of sulph-arsenide cobalt- and nickel ores associated sometimes with predominate barite, and sometimes with crystalline carbonates. The ore-body is discontinuous along the strike as well as in dip. The principal occurrences were at Schweina in Thuringia and at Riechelsdorf in Hesse.

At Schladming in Styria 2000 m. above sea-level, fahlbands occur in gneiss- and mica-schists, these fahlbands owing to the brown weathered zone they form at the surface, being known as *Branden*. They are traversed by vertical quartz lodes. In the neighbourhood of these lodes and within the fahlbands, nickel ores are found in the form of nodules and nests, consisting in greater part of niccolite and nickel-arsenopyrite, but also of

---

[1] *Ante*, p. 805.

[2] *Ante*, p. 810, A. Gesell, ' Montangeol. Aufn. auf dem v. d. Dobsinaer südöstlichen Stadtgrenze südl. gel. Gebiete,' *Jahresber. d. kgl. ung. geol. Anst. f. 1902.*—F. W. Voit, ' Geologische Schilderung der Lagerstättenverhältnisse von Dobschau in Ungarn,' *Jahrb. d. k. k. geol. Reichsanst.* Vol. L., 1900.—K. A. Redlich, *Zeit. f. prakt. Geol.*, 1908, and H. v. Böckh, *ibid.*   [3] Laspeyres, 1893, *loc. cit.*

smaltite, native bismuth and arsenic, etc. Confined generally speaking to the lodes are tetrahedrite, galena, chalcopyrite, stibnite, and pyrite. The favourable influence of these transverse quartz lodes recalls the well-known influence of the afore-mentioned cobalt fissures upon the Kupferschiefer. These deposits at Schladming were worked in the fifteenth and sixteenth centuries and from 1832 to 1880, the latter period however being of only small importance.[1]

Fairly analogous deposits occur also in the Annivier and Turtmann valleys of Valais, Switzerland. At the first-mentioned place two siderite lodes were worked, which contained ores particularly rich in nickel where sulphide disseminations occurred in the gneissic country-rock. The total output was very small.[2]

The following deposits belonging to this group are worthy of simple mention. The Leo lode near Salzburg, containing pockets and nests of niccolite and smaltite and numerous other minerals, occurs in a Silurian dolomite ; the well-known copper deposits at Mitterberg and Kitzbühel occur in the neighbourhood.[3] Quartz- and crystalline carbonate lodes containing nickel-, cobalt-, and copper ores have been remarked in the greenstone on Monte Cruvin in Bruzolo, on the Besighetto near Balme, and at Usseglio in Piemont.[4] Quartz lodes containing smaltite, etc., are also known in the Rebota valley in the Transvaal.[5] Analogous smaltite lodes containing some gold and silver occur also in the north-eastern portion of the Transvaal,[6] while quartz lodes containing millerite have been recorded at Benton in Arkansas.[7] Smaltite lodes containing quartz as gangue, and some gold and silver, are also known at Cerro de Famatina in Argentina.[8]

Important mining operations for cobalt were formerly prosecuted in the cobalt fahlbands at Modum, Tunaberg, etc. These deposits are discussed in a subsequent chapter.

[1] C. Schmidt and J. H. Verloop, ' Notiz über die Lagerstätte von Kobalt- und Nickel-erzen bei Schladming in Steiermark,' Zeit. f. prakt. Geol., 1909, p. 271.

[2] Zeit. f. prakt. Geol., 1909, pp. 36, 274.

[3] Lipold, Jahrb. d. k. k. geol. Reichsanst., 1854, Vol. V.—F. Pošepný, Archiv f. prakt. Geol. I., 1880.—Buchrucker, ' Mineralogisches,' Zeit. f. Krist. Min. XIX., 1891.

[4] D' Achiardi, I Metalli, II. p. 22.—Strüver, ' Die Minerallagerstätten des Alatales in Piemont,' Neues Jahrb. f. Min. Geol. Pal., 1871, p. 345.

[5] Hatch and Corstorphine, The Geology of South Africa, 1905, pp. 178, 193.

[6] H. Oehmichen, ' Goldhaltige Kobaltgänge in Transvaal,' Zeit. f. prakt. Geol., 1899, p. 271.

[7] J. L. Fletscher, Min. Res. U. St., 1887, pp. 128-129.

[8] Eriksson, Eng. Min. Journ. LXXVII.

THE NICKEL-SILICATE, OR GARNIERITE DEPOSITS

LITERATURE

H. B. v. FOULLON. 'Über einige Nickelvorkommen,' Jahrb. d. k. k. geol. Reichsanst. XLIII., 1892.

Lodes and veins of nickel-magnesium hydrosilicates are found all over the world under practically the same conditions, in more or less serpentinized olivine rocks, and exceptionally also in associated augite rocks. Such deposits occur very extensively on the island of New Caledonia, where they are of great economic importance. Similar occurrences are those at Frankenstein in Silesia, Riddles in Oregon, Webster in North Carolina, and Revda in the Urals. The occurrence at Malaga in Spain, already described,[1] which extends below the ground-water level, occupies a place by itself.

Of the mostly light-green nickel silicates characteristic of this group, garnierite—so named after Jules Garnier the discoverer of the New Caledonian ore; or numeite, after Numea the capital of New Caledonia—having a specific gravity about 2·9, is the most frequent. The composition of this mineral is approximately $H_2(Mg, Ni)SiO_4$, with variable amounts of NiO and MgO. Some samples have given as much as 45 per cent of NiO,[2] though the average content of the clean mineral is only 15–25 per cent. In addition, genthite, which is very similar to garnierite, nickel-gymnite, revdanskite, pimelite, schuchardite, etc., are worthy of mention. All these minerals are amorphous or cryptocrystalline, and usually badly defined as mineral species. The so-called nickel-chocolate—a chocolate-coloured ore of somewhat variable composition, and obviously consisting of a mixture of different minerals—also occurs very extensively in New Caledonia. At Frankenstein,[3] the so-called 'grey ore,' a somewhat decomposed serpentine containing a few per cent of nickel, occurs in large amount. In New Caledonia, at Revda, and at other places, the nickel-silicate deposits are accompanied by separate deposits of asbolane, that is, cobalt-manganese ore, the two classes of deposit being, however, always distinctly separated from each other.

The composition of these ores may be gathered from the following analyses taken from works afterwards cited:

---

[1] *Ante*, p. 299.   [2] Analyses, No. 1a–1c.
[3] *Postea*, p. 961.

[TABLE

	New Caledonia.							Riddles.	Revda.
	Garnierite.			Chocolate Ore.		Meer-schaum.	Export Ore.		
	1a.	1b.	1c.	2a.	2b.	3.	4.	5.	6.
SiO$_2$	35·45	37·49	42·61	33·70	37·05	41·80	42	48·82	54·15
Al$_2$O$_3$				1·40	0·63	...	1	...	0·23
Fe$_2$O$_3$	0·50	0·11	0·89	19·09	16·92	1·26[1]	15	0·06	0·27
NiO	45·15	29·72	21·91	31·28	17·60	...	9	19·04	27·61
CoO	...	...	...	...	...	...	0·15	...	...
MnO	...	...	...	0·20	...	...	0·7	...	...
MgO	2·47	14·97	18·27	3·22	16·03	37·38	22	18·49	6·82
CaO	...	...	...	0·63	0·48	...	0·1	...	...
H$_2$O	15·55	17·60	15·40	9·21	10·51	20·39	10	12·29	7·74
Chromite	...	...	...	1·20	1·21	...	...	...	...
Totals	99·12	99·89	99·08	99·93	100·43	100·83	100	98·70	96·82

	Frankenstein.		
	7.	8.	9.
SiO$_2$	47·49	33·28	60–65
Al$_2$O$_3$	1·53	14·62	6–8
Fe$_2$O$_3$	0·48	3·83	
FeO	...	3·56	...
NiO	20·01	5·68	2·9–4·5
MgO	10·18	23·72	8·5–12
CaO	...	1·47	...
H$_2$O	18·82	13·91	8–15
Totals	98·51	100·07	...

	Asbolane, New Caledonia.			
	10a.	10b.	10c.	10d.
SiO$_2$	3·0	2·20	16·40	23·09
Al$_2$O$_3$	...	14·29	14·60	10·30
Fe$_2$O$_3$	10·6	8·91	15·50	16·06
Mn$_3$O$_4$	46·7	33·62	12·07	17·59
CoO	15·0	7·76	3·00	5·56
NiO	...	1·64	1·48	1·48
MgO	4·8	2·38	...	2·23
CaO	...	...	...	...
H$_2$O	17·5	29·20	36·95	23·69
Totals	97·6	100·00	100·00	100·00

Nos. 1–4, from New Caledonia; 1a–1c = Garnierite.—Nos. 2a–2b = Chocolate ore, rendered impure by admixed chromite.—No. 3 = Meerschaum; [1] is FeO.—No. 4 = Average of the export ore, dried at 100°.—Nos. 5, 6 = Genthite from Riddles and Revda.—Nos. 7–9 from Frankenstein; 7 = Pimelite, 8 = Schuchardite, 9 = Smelting ore.—Nos. 10a–d, Asbolane from New Caledonia; 10a = Clean picked, 10b–d = Impure.—Manganese is reckoned as Mn$_3$O$_4$ and cobalt as CoO, though possibly chiefly MnO$_2$ and Co$_2$O$_3$ are present.

The parent rock of these deposits is in general a more or less serpentinized olivine rock, peridotite, this being principally, as is the case in New Caledonia, Oregon, Webster, etc., a dunite—olivine with some chromite and picotite—though this is frequently accompanied by saxonite or harzburgite—olivine with some enstatite or bronzite—as for instance in New Caledonia and in Oregon. Exceptionally, as at Revda, the parent rock consists of an augite rock serpentinized to antigorite.

The dunite-saxonite, which represents an anchi-monomineral eruptive, in a non-serpentinized state usually contains 40–47 per cent SiO_2, 40–48 per cent MgO, 4–8 per cent $FeO + Fe_2O_3$, 0·5–2·5 per cent Al_2O_3, and 0–2 per cent CaO, that is, remarkably little Al_2O_3 and CaO; furthermore, some Cr_2O_3,[1] frequently 0·1–0·5 per cent MnO, and always some NiO as silicate, this latter being chiefly in the olivine and subordinately in the pyroxene minerals.[2] CoO occurs only in traces. This dunite-saxonite is richer in nickel than is any other rock. The NiO content varies usually between 0·1 and 0·5 per cent, though occasionally it reaches 1 per cent or more. According to analyses by Glasser, the New Caledonian peridotite in particular is remarkably rich in NiO, this rock sometimes containing over 1 per cent.

The hydrated nickel silicates of the German deposits are accompanied principally by hydrated magnesium silicates poor in, or free from nickel, such as gymnite, kerolite, and meerschaum ;[3] also by iron-ochre, quartz, hyalite, opal, chalcedony, and chrysoprase containing some nickel ; exceptionally also by some magnesite. The chrysoprase of Frankenstein is used as a precious stone ; it is also found at Riddles and at Revda. This mineral-association indicates with certainty a deposition from aqueous solution.

The occurrences in New Caledonia, at Frankenstein, Riddles, Webster, and Revda, undoubtedly did not result from the decomposition of arsenide or sulphide nickel ores, such ores having nowhere been found at those places. Looking elsewhere for the genesis of these deposits, from their constant connection with rocks relatively rich in nickel, from their mineral-association, and from their morphology, it may be assumed with certainty, as was demonstrated years ago by Sterry Hunt, Garnier, Foullon, Diller, and others, that these garnierite deposits were formed by leaching of the country-rock.

The view has occasionally been put forward that the origin of the garnierite was connected with a particularly intense serpentinization. Against this, however, it must be remarked that the rock in which garnierite is found, not infrequently, as for instance in New Caledonia and at Riddles in Oregon, consists of a comparatively little serpentinized,

[1] *Ante*, pp. 153, 244.   [2] *Ante*, pp. 153, 287.   [3] Analysis No. 3.

fairly fresh peridotite containing but 1–5 per cent of H_2O. Further, the serpentine, according to numerous analyses, contains about the same amount of nickel as the rock from which it resulted, and accordingly the small nickel content of the peridotite was, generally speaking, not disturbed in the process of serpentinization. Some investigators [1] considered the occurrences in New Caledonia to have been formed by ascending heated waters, which having extracted nickel, magnesia, silica, etc., from the peridotite and serpentine in depth, deposited them afterwards near the surface in the form of hydrated nickel-magnesium silicates. Such an explanation however is not apt for these occurrences; on the contrary, Glasser in his detailed description gives conclusive evidence that they were formed by weathering processes limited to the surface, while for the deposits at Frankenstein, Beyschlag and Krusch have been able to produce similar evidence.

In New Caledonia, the silica and magnesia of the more or less serpentin-ized peridotite have near the surface in greater part been removed by the warm tropical rain-water, leaving behind large eluvial deposits, often 5–15 m. thick, of a red earth—*terre rouge*, frequently mistakenly described as *l'argile rouge* or *l'argile ferrugineuse*, though Al_2O_3 is absent—consisting in greater part of iron-ochre. One analysis of this red earth gave, 18·4 per cent of SiO_2, 69·3 per cent Fe_2O_3, 0·5 per cent Al_2O_3, 1·6 per cent NiO, 0·4 per cent $MgO + MnO$, and 9·8 per cent of H_2O.

In process of weathering, the nickel, cobalt, and manganese, together with some magnesia and silica, went into solution, from which solution asbolane on the one hand and garnierite on the other, were subsequently precipitated. The latter in New Caledonia is not accompanied by magnesite or other carbonates; it occurs principally along fractures and crevices in firm rock near the surface, but also as incrustations in the lower part of the eluvial deposits. The asbolane is limited exclusively to the eluvial deposits, though in all probability the solutions which deposited both it and the garnierite were the same. The asbolane, which possesses a certain similarity to manganiferous lake ore, contains 1 part of nickel to some 2–5 parts of cobalt; the garnierite on the other hand contains 1 of cobalt to about 100 of nickel. These two elements have therefore in these deposits become separated, though analytically the separation is not sharp.

Examination of the juxtaposition of the asbolane and garnierite indicates that the cobalt was precipitated earlier than the nickel. The asbolane consists in greater part of highly oxidized manganese oxides, chiefly MnO_2, from which fact, precipitation by oxidation—probably by means of atmospheric oxygen acting upon a neutral solution—may

[1] Heurteau, 1876; de Levat, 1892; and Benoit, 1892.

be assumed. By such means cobalt would be precipitated as oxide with but very little nickel, the greater quantity of which would remain in solution. In this connection we would recall the earlier process of separating cobalt from nickel, which consisted in the addition of chloride of lime to a neutral cobalt-nickel solution, when cobalt oxide with very little nickel oxide was precipitated, while the greater part of the nickel remained in solution.[1]

The peridotite always carries much more nickel than cobalt, namely, about 1 part of cobalt to some 20–50 parts of nickel, and accordingly the asbolane occurs subordinately to the garnierite. At the decomposition of the peridotite or serpentine the small amount of ferrous oxide is in greater part oxidized to ferric oxide—red earth—and but little iron goes into solution; the low iron content of the hydrated nickel silicates is thus explained. The same is approximately the case with the small amount of alumina which occurs in the parent rock. The dunite and saxonite are on the whole exceedingly poor in sulphides, titanic acid, and phosphoric acid, and such compounds are consequently practically absent from the garnierite deposits.

The occurrences at Riddles and Revda appear from the descriptions available also to have been formed by surface weathering. Concerning the deposits at Frankenstein, some investigators assume leaching by heated waters, which, having extracted the nickel from the country-rock in depth, deposited it again near the surface. This explanation does not however agree with the observed phenomena.

The New Caledonian deposits, owing to the large extent of the peridotite, are of great economic importance. In the formation of these deposits the tropical climate may possibly have had something to do with the processes concerned, such a climate greatly promoting surface weathering. In that case a certain analogy to lateritization would exist.

New Caledonia

LITERATURE

J. Garnier. 'La Géologie et les ressources minérales de la Nouvelle-Calédonie,' Ann. d. Mines, 6e sér. t. XII., 1867; various treatises in Compt. rend. 82, 1876; Soc. d. Ing. Civils 5, 1887; Revue scientifique, 1895.—E. Heurteau. 'Sur la constitution géologique et les richesses minérales de la Nouvelle-Calédonie,' Ann. d. Mines, 7e sér. t. IX., 1876.— L. Pelatan. 'Les Mines de la Nouvelle-Calédonie' (with geological map of the island), Génie Civil 19, 1891, II.—D. Levat. 'Progrès de la métallurgie du nickel' (together with description of the New Caledonian mines), Ann. d. Mines, 9e sér. t. I., 1892.—F. Benoit. 'Les Mines de nickel de la Nouvelle-Calédonie,' Bull. de la Soc. de l'Ind. Min. St. Etienne, III. 6, 1892.—A. Bernard. L'Archipel de la Nouvelle-Calédonie, Paris, Hachette & Cie, 1895, p. 458, with 2 maps.—Review in Zeit. f. prakt. Geol., 1897, p. 257.—Fr. D. Power.

---

[1] Vogt, Geol. Fören. Förh. XVII., 1892.

THE NICKEL-SILICATE, OR GARNIERITE DEPOSITS   955

'The Mineral Resources of New Caledonia,' Inst. Min. and Met. Abstract of Proceedings,
Vol. VIII. p. 44.—Review in Zeit. f. prakt. Geol., 1901, p. 24.—M. PIROUTET. 'Sur la
géologie d'une partie de la Nouvelle-Calédonie,' Bull. de la Soc. Géol. de France, 4e sér. t.
III., 1903.—E. GLASSER. 'Rapport au ministre des colonies sur les richesses minérales
de la Nouvelle-Calédonie,' Ann. d. Mines, 12e sér. t. IV., 1903, and t. V., 1904.

The island of New Caledonia extends from 20° 5′ to 22° 24′ south
latitude ; it is 400 km. long, 65 km. wide, and has an area of 16,117 square

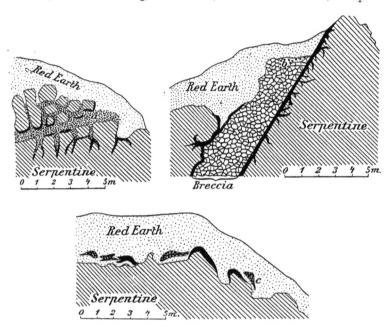

FIG. 397.—Detailed sections of the garnierite deposits of New Caledonia.   Glasser.

In the two upper sections the garnierite veins are indicated black ; (a) is impure garnierite ore, (b) is red
earth with 4-5 per cent of nickel. In the lower section the black crusts on the serpentine indicate asbolane
ore ; (c) are kidneys of asbolane in red earth.

kilometres. Of this no less than about 6000 sq. km. is occupied by the
presumably post- or late Cretaceous, more or less serpentinized peridotite—
dunite and saxonite—one field alone covering 3500 sq. km.   The serpentine
and peridotite areas which, as illustrated in Fig. 156, lie distributed
over practically the whole of the island, here perhaps reach the greatest
known extent in the world.   In these eruptive areas—the more important of which are known under the names of Thio, Canala, Honailou,
Mount Kaale, Neporri, etc.—a large number of garnierite deposits occur.
At the time of the Paris Exhibition in 1889, no less than about 1200

nickel occurrences had been discovered, and in addition approximately 300 asbolane deposits.

According to Glasser, the hydrated nickel-magnesium silicates are found partly in lodes, partly as the cementing material of a breccia, partly in veins and veinlets in the serpentine, and partly as incrustations within detrital matter consisting principally of red earth. There are no sharp lines between these different modes of occurrence, which on the contrary merge gradually into one another. Principally the very irregular and branched garnierite lodes were first exploited. Upon these lodes some mines reached depths of 108 and even 145 m., though mineralization diminished in depth, and generally speaking the garnierite quickly pinched out. Accordingly, mining was subsequently carried on chiefly in open-cut and only in exceptional cases to a depth of as much as 50 m., the ore usually disappearing at 10–20 metres. In this method of mining great quantities of the eluvial red earth have to be removed. The garnierite lodes only exceptionally reach 1 m. in width; more usually they are narrow stringers which traverse the country-rock in all directions. Of such rock 5–10 cubic metres must frequently be worked in order to obtain 1 ton of ore.

That the garnierite is a result of surface weathering follows, according to Glasser, with certainty from the mode of occurrence. This view is supported by the discovery of insect remains coated with garnierite in the eluvial deposits, as well as by the occurrence of recent garnierite stalactites, which prove that the formation of this mineral still continues.

The asbolane occurs occasionally in small amount in the form of lumps and crusts in the red earth, as illustrated in Fig. 397 c.

Mining operations in New Caledonia began in the middle of the 'seventies, the garnierite having been discovered in 1867 and examined in 1873. . The extent of operations may be judged from the following figures of ore won, most of which was exported :

From 1875 to 1879	.	.	.	8,300 tons.
,, 1880 ,, 1884	.	.	.	35,400 ,,
,, 1885 ,, 1889	.	.	.	42,400 ,,
,, 1890 ,, 1894	.	.	.	200,300 ,,
,, 1895 ,, 1899	.	.	.	312,600 ,,
,, 1900 ,, 1904	.	.	.	539,800 ,,
,, 1905 ,, 1909	.	.	.	591,800 ,,
Total to end of 1909	.	.	.	1,730,600 tons.

From this tonnage of ore, after deducting the loss in smelting, some 105,000 tons of metallic nickel resulted,[1] equivalent to a net content of roughly 6 per cent of nickel. The amount of ore, 960,000 tons, won up

[1] Glasser, *Ann. d. Mines*, 1903, p. 512 ; and statistics of the *Metall- und Metallurgischen Gesellschaft*, Frankfort-on-the-Maine.

to the year 1902 contained according to Glasser 60,700 tons of nickel, equivalent to 6·3 per cent when deducting nothing for loss in smelting.

The ore won contains much chemically combined water and a considerable amount of mine moisture, which partly disappears when dried at 100°. Dried ore mostly contains at least 7 per cent of nickel, equivalent to 5·4–5·8 per cent in the wet or raw ore. One kilogramme of nickel in dry ore is paid for at the rate of 60–70 centimes in a New Caledonian port; reckoning 60 centimes and assuming the nickel content of the wet ore to be 6 per cent, the price of the ore in such a port would be 36 francs per ton. To this must be added the freight to Europe, which is some 40 francs per ton, so that the value in an English, French, or German port may be taken to be 76 francs, equivalent to 1·27 francs, or roughly one shilling per kg. of nickel.

In raising the value of the ore by hand-sorting to at least 7 per cent in the dry ore, much low-grade ore is lost. For this reason and on account of the high freight to Europe, it is intended to smelt a part of the production at Thio in New Caledonia.

Mining operations, which up to the end of the year 1909 had produced roughly 1,750,000 tons of ore, have exhausted numerous deposits, in spite of which, owing to the exploitation of new deposits, production continues to increase. At the present time some forty mines or opencuts are working.

A few years ago 2000–6000 tons of asbolane containing mostly 3–4 per cent of cobalt were won annually, and up to the year 1904, or approximately in twenty years, some 60,000 tons with an average of 3·5 per cent of cobalt had been exported.[1]

In the New Caledonian serpentine-peridotite many chromite occurrences also occur, some of which are now being exploited.[2]

FRANKENSTEIN IN SILESIA

LITERATURE

J. ROTH. Erläuterungen zu der geognostischen Karte vom Niederschlesischen Gebirge und den anliegenden Gegenden. Berlin, 1867.—TH. LIEBISCH. 'Mineralogisch-petrographische Mitteilungen aus dem Berliner mineralogischen Museum,' Zeit. d. d. geol. Ges., 1877.—H. TRAUBE. Die Minerale Schlesiens. Breslau, 1888.—H. B. v. FOULLON. 'Das Vorkommen nickelhaltiger Silikate bei Frankenstein in Preussisch-Schlesien,' Jahrb. d. k. k. Geol. Reichsanst. Vienna, 1892.—B. KOSMANN. 'Die Nickelerze von Frankenstein in Schlesien,' Glückauf, 1893, No. 57 and 59.—ASCHERMANN. 'Beiträge zur Kenntnis der Nickelerze von Frankenstein,' Inaug.-Diss. Breslau, 1897.—ILLNER. 'Das Nickelerzvorkommen bei Frankenstein,' Zeit. f. d. B-, H-, u. S-wesen, 1902.—ALBRECHT. 'Das Nickelerzvorkommen bei Frankenstein in Schlesien,' Archiv d. k. Geol. Landesanst. Berlin, 1902.—BATTIG. 'Das Nickelerzvorkommen bei Frankenstein in Schlesien,' Archiv

---

[1] B. Neumann, in Die Metalle, gives 58,730 tons with 3·6 per cent of Co, from 1875-1901.

[2] Ante, p. 249.

d. k. Geol. Landesanst. Berlin, 1906.—FLEGEL. 'Das Nickelerzvorkommen bei Frankenstein in Schlesien,' Archiv d. k. Geol. Landesanst. Berlin, 1909.—P. KRUSCH. 'Über
die Genesis einiger Mineralien u.s.w. von Frankenstein,' Zeit. d. d. geol. Ges., 1913.

Of. the Sudetic foothills which rise through the Diluvial covering
between the Eulengebirge and the Zobten-Strehlen mountains, the
Baumgarten-Grochau group, south-west of Frankenstein, and the range
running through Dittmannsdorf, Prozau, Gläsendorf, Kosemitz, and
Disdorf, north of Frankenstein, are unique by reason of their contained
nickel deposits.

The group south-west of Frankenstein, owing to the smallness of
the deposits, may however hardly be expected ever to become the seat
of mining operations. This group, according to Roth, consists of gneiss
and of hornblende-schist in part augitic and decomposed to serpentine. In the serpentine and more particularly on the southern slope of
Grochberg, magnesite occurring as fissure-filling is exploited. The veins
of this material reach 0·5 m. in width and are characterized by great
purity. Some nickel ore, the discovery of which led to the grant of several
concessions, was found associated with these veins.

The more important nickel deposits at Frankenstein lie north of that
town, being confined to four serpentine hills, indicated in Fig. 398, which
rise about 377 m. above the Diluvial country and probably belong to one
and the same north-south belt. The country-rock of the serpentine
consists of gneiss, blue-grey graphitic quartzite-schists with white and
red quartz bands, and, particularly in the north and west, of a generally-
speaking coarse-grained syenite.

Mining operations for nickel, which were revived in 1891, are being
prosecuted upon the four hills, Kosemitz or Mühlberg, Tomnitz, Gläsendorf,
and Gumberg. These Silesian nickel mines, in addition to opencuts, have
four underground levels at depths of 15–80 m. below the ridge of the
Gläsendorf hill.

The serpentine constituting the immediate country-rock of the ore
consists of a macroscopic, almost compact mass with a smooth conchoidal
fracture, which is usually blackish, olive-green or canary-green in colour,
and exhibits stains due to magnetite. It is always very much fractured
and more or less decomposed. It contains 41–42·5 per cent of SiO_2, 36–42
per cent MgO, 0·25 per cent Cr_2O_3, and a small amount of nickel which
Foullon, from a sample taken from the west slope of Gumberg, gives as
0·34 per cent.

This serpentine, as illustrated in Fig. 399, is traversed by a system
of north-northwest quartz veins which split and re-assemble so that their
width is very variable, and which provide the resistance to erosion to
which the range of hills owes its existence. From these veins as well as

from the surface the serpentine became highly altered, to red earth, etc. In this alteration Beyschlag and Krusch differentiate the white decomposition, consisting of a network of magnesite and kerolite, from the grey and green mineralization and the formation of red earth. This last has the greatest extent. It is a red-brown, friable, highly decomposed material, which usually has no sharp separation from the serpentine but often merges gradually into that rock through the less intense decomposition products represented by green and grey ore. The greater part of the red earth occurs immediately at the surface and along the quartz veins. When it occurs silicified it consists, according to microscopic examination by Krusch, in greater part of a mixture of quartz and chalcedony, these minerals exhibiting a banded structure similar to that formed at the gradually widening replacement of a rock by quartz, along a fracture. The cavities between the quartz layers are filled with ferric hydrate.

Green ore consists of masses of red earth or highly decomposed serpentine, traversed by veins of nickel silicate. When the metasomatic replacement of the rock is still more advanced, green knotted ore arises. The term grey ore is applied to a very nickeliferous serpentine which, though under-

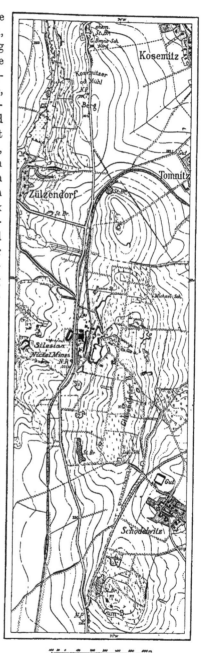

FIG. 398.—The serpentine belt north of Frankenstein in Silesia.

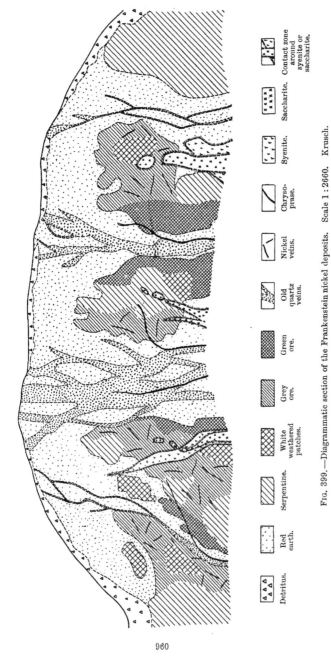

Detritus.   Red earth.   Serpentine.   White weathered patches.   Grey ore.   Green ore.   Old quartz veins.   Nickel veins.   Chryso-prase.   Syenite.   Saccharite.   Contact zone around syenite or saccharite.

FIG. 399.—Diagrammatic section of the Frankenstein nickel deposits.   Scale 1 : 2660.   Krusch.

going decomposition, still retains very clearly the character of serpentine, being distinguishable from that rock only by its higher nickel content, which is 1–2 per cent.

The white decomposition of the serpentine is independent of, and older than the formation of the nickel ore ; kernels of white decomposed serpentine are consequently frequently found in the nickel ore and in the red earth, while a more complete replacement of the serpentine has resulted in the formation of white patches.   Of the characteristic white minerals, the kerolite is apparently younger than the magnesite, which it replaces.

Chrysoprase is the younger chalcedony coloured a bright green by the small amount of nickel it contains ; in addition, opal and pras-opal also occur.   These are in part contemporaneous with, and in part older than the nickel ore.

Saccharite is a white, more seldom grey, mineral aggregate, concerning the origin of which, opinions have hitherto differed.   While Glocker believed it to be a special felspar, Liebisch regarded it as a fine-grained assembly of plagioclase crystals in which small green hornblende individuals and blue-black tourmaline were in small amount included.   Von Lasaulx and Krusch by microscopic examination found in addition, orthoclase, garnet, diopside, epidote, mica, and a little quartz.   One analysis gave 58·9 per cent of SiO_2, 23·5 per cent Al_2O_3, 5·67 per cent CaO, and 7·42 per cent Na_2O. This mineral aggregate occurs vein-like or apophysis-like in the serpentine and its weathered products ; the width of its occurrence in the district of Benno-Süd was proved by boring to be about 50 metres.   According to Krusch the saccharite is a differentiation product of the syenite, like which it exhibits contact phenomena.

The green nickel ores represent hydrated nickel-magnesium silicates of very variable qualitative and quantitative composition.   All contain principally silica, water, ferrous oxide, ferric oxide, magnesia, and many also alumina.   The nickel content varies between very wide limits, the following minerals being distinguished : pimelite, with 2·78–32·66 per cent NiO ; schuchardite, with 5·16–5·78 per cent; garnierite (?), with 38·61 per cent ; and the above-mentioned green knotted ore.   The brown-green variety of this last constitutes the rich ore of the occurrence.

Grey ore, containing 1–2 per cent of nickel and representing a more advanced stage in the decomposition of the serpentine, is a discovery of more recent years.

The nickel ores therefore are associated with red decomposition zones in the serpentine, these in their turn owing their great extent to the quartz veins.   As has already been explained,[1] these garnierite occurrences have been formed by immediate lateral secretion, and probably by similar

[1] *Ante*, p. 191.

processes to those which were active in New Caledonia, that is, by decomposition proceeding from the surface and from quartz veins.

The economic importance of Frankenstein, though, it is true, not great, cannot however be ignored. About 10,000–12,000 tons of low-grade ore are produced yearly, such ore being smelted with high-grade ore from New Caledonia.

Riddles in Douglas County, Oregon

LITERATURE

F. W. Clarke and J. S. Diller. 'Some Nickel Ores from Oregon,' Am. Journ. XXXV., 1888.—B. H. v. Foullon. Loc. cit., 1892.—W. L. Austin. Nickel, Second Paper, 'The Nickel Deposits near Riddles, Oregon,' Colorado Scien. Soc. Denver, Jan. 6, 1896.

Within a fairly small area of in part serpentinized saxonite—an olivine-enstatite rock in which the olivine contains 0·26–0·32, and the enstatite or bronzite 0·05 per cent of ferrous nickel—genthite deposits in the form of numerous mostly narrow fissure-fillings carrying some quartz or chrysoprase, occur; these deposits are limited to the neighbourhood of the surface. In addition, nickel hydrosilicate is found in the loose detritus.

In the exploration work prosecuted about the year 1890, impoverishment became apparent at a depth of 15 metres. The country-rock consists partly of serpentine and partly of a slightly-serpentinized saxonite; in this case therefore the nickel silicate cannot be the end product of ordinary serpentinization. In this district also there are at the surface extensive deposits of an impure iron-ochre. The formation of nickel silicate by surface weathering, already discussed when describing New Caledonia, would consequently appear to apply here also.

Austin assumed the action of heated waters. Foullon considered he was able to recognize successive stages in the deposition of the ore. According to him, thin layers, varying in colour from white to sap-green, in which silica, magnesia, very little iron, and some nickel, were contained, were the first to be deposited. Later, in Tertiary time, these in turn became altered, with the result that genthite became deposited in the larger fissures.

Up to 1896 about 3000 tons of ore containing some 5 per cent of nickel had been produced, this production being spread over many years in which operations were only on a small scale. The deposits are of no great importance. At Webster in North Carolina, narrow unpayable stringers of genthite and gymnite, accompanied by some talc, occur in a partly serpentinized dunite.[1]

[1] Clarke and Diller, loc. cit.; S. H. Emmons, Eng. and Min. Journ., 1892, p. 476; P. H. Wurtz, Amer. Ass. Adv. Sc. XII. p. 24, and Am. Journ. Sc. 2, XXVII. p. 24.

Revda or Revdinsk in the Urals, in the neighbourhood of Ekaterinburg, differs from the above occurrences in that the nickel hydrosilicates occur in an antigorite-serpentine, resulting, according to Foullon, from an olivine-free augite rock.   At the surface the country-rock is completely decomposed and consists largely of ochreous products.   This deposit justified only a small amount of exploration work, which reached to a depth of 50 metres.[1]

Green hydrated nickel-magnesium silicates in serpentine and associated rocks, have also been remarked in many other places, as for instance in Western Australia and in Madagascar.[2]

At Foldal in Norway, a thin layer of nickel silicate on steatite occurs.[3] Nickel silicate is also known along joint-planes in serpentine-asbestos at Meigern in Valais, and as a coating along fractures in chromite at Texas in Lancaster Co., Pennsylvania, where nickel-smaragd, that is, basic nickel carbonate, also occurs.   Finally, an occurrence at Alt Orsova in Hungary must be mentioned.[4]   These small occurrences also suggest lateral secretion from peridotite or serpentine.

Of late years several ship-loads of hydrated nickel-magnesium silicates containing 5–7 per cent of nickel and exactly resembling the New Caledonian ore, have been exported from Greece as well as from Egypt, though details concerning the deposits are not known.

The Nickel Production of the World and its Distribution among the Various Classes of Deposit

The production of nickel for industrial purposes began in the early 'twenties.   The world's production of metallic nickel and, in the early years, that of nickel in copper-nickel, has been as follows:[5]

1906–1910	.	.	.	about 80,400 tons.
1901–1905	.	.	.	„ 52,000 „
1896–1900	.	.	.	„ 32,000 „
1891–1895	.	.	.	„ 18,000 „
1881–1890	.	.	.	„ 11,000 „
1871–1880	.	.	.	„ 6,000 „
1861–1870	.	.	.	„ 3,000 „
1851–1860	.	.	.	„ 2,500 „
about 1825–1850	.	.	.	„ 2,500 „

The total to 1910 was accordingly some 207,400 tons, or roughly

---

[1] Foullon, loc. cit., in which a Russian treatise by A. Karpinsky, Gorni-Journ., 1891, p. 10, is cited ; Helmhacker, Berg- u. Hüttenm. Ztg., 1895, p. 142.

[2] Villiaume, Berg- u. Hüttenm. Ztg., 1899, p. 380.

[3] Chr. A. Münster, Archiv f. Math. u. Naturw., 1890, XIV. p. 240.

[4] Hintze, Lehrbuch der Mineralogie.

[5] Ante, p. 207.

200,000 tons of nickel. In the year 1910 itself, the production had risen to 20,100 tons. The price per kg. of pure nickel has been:

1900–1911	. . .	mostly	2s. 11d.– 3s. 3d.
1895–1899	. . .	„ about	2s. 6d.
1893–1894	. . .	„ „	3s. 6d.
1887–1892	. . .	„	4s. 5d.– 5s. 0d.
1883–1886	. . .	„	5s. 0d.– 5s. 6d.
1879–1882	. . .	„	5s. 6d.– 8s. 0d.
1877–1878	. . .	„	7s. 0d.–10s. 0d.
1876	. . .	.,	12s. 0d.–20s. 0d.
1873–1875	. . .	„	15s. 0d.–20s. 0d.
1872	. . .	„ about	11s. 0d.
1867–1871	. . .	„	7s. 6d.– 8s. 6d.

Sources of figures for the earlier years: principally J. H. L. Vogt, ' Nikkelforekomster og Nikkelproduktion,' *Norweg. Geol. Unters.* No. 6, 1892 ; and B. Neumann, *Die Metalle, etc.*, Halle, 1904. For later years: the annual statistics of lead, copper, etc., of the *Metallgesellschaft*, etc., Frankfort-on-Maine.

Before these dates and approximately till 1840, nickel was obtained exclusively from nickel-arsenic ore[1] and from the nickel-speiss obtained as a by-product in some smelting works. The production of nickel at that time, however, was at most 250 tons and on an average not more than 100 tons yearly. Later, so far as arsenic ores were concerned, the production sank still lower. The treatment of nickel-pyrrhotite[2] began at Klefva in Sweden[3] in the year 1838, and in Norway[4] at the end of the 'forties. From 1850 to the end of the 'seventies this group of deposits was responsible for the greater portion of the nickel output, Norway alone in the middle of the 'seventies being responsible for about one-half. Since the discovery and exploitation of the garnierite deposits of New Caledonia at the end of the 'seventies, and the nickel-pyrrhotite deposits of the Sudbury district in Canada[5] about the middle of the 'eighties, these two important districts have practically divided the world's production of nickel between them. To their production however must be added small amounts from Frankenstein in Silesia, Mine la Motte in the United States, Schneeberg in Saxony, and from Norway, where production has of late years somewhat increased.

The total production up to the present day, as stated before, amounts to something over 200,000 tons, this being distributed among the various classes of deposit as follows :

1. Garnierite deposits:

New Caledonia, since 1875	.	.	.	about 115,000 tons.
Frankenstein .	.	.	.	„ 2,500 „
Revda, Riddles, etc., at most	.	.	.	1,000 „

Total about 120,000 tons.

[1] *Ante,* p. 948. [2] *Ante,* pp. 280-300. [3] *Ante,* p. 297.
[4] *Ante,* pp. 294-297. [5] *Ante,* pp. 289-293.

2. Nickel-pyrrhotite deposits :

Sudbury, since 1885	about	80,000 tons.
Norway, since 1848	„	5,000 „
Klefva in Sweden, 1838–1888 . . .	„	900 „
Other occurrences, together 1000 or at most .	„	2,000 „

Total about 87,000 tons.

3. Nickel-arsenide, including millerite and niccolite, in nickel-arsenide lodes and as a by-product when smelting ore from other lodes : A total estimated at only a few thousand tons and not more than 4000 tons.

4. Metasomatic deposits, Mine la Motte, etc., estimated at about 1000 tons.

Nickel is a pronounced basic element [1] and accordingly the two most important types of nickel deposit, the nickel-pyrrhotite group and the garnierite group, are found genetically associated with basic eruptive rocks. There is no doubt also that several of the nickel arsenide- and sulphide lodes, as for instance those at Dobschau and Dillenburg, are closely associated with basic eruptives, in which case the nickel contained in these may also be traced to the low nickel content of basic magmas.

The manufacture of cobalt-blue or smalt, an art which existed even in ancient times, began in Saxony as far back as the beginning of the sixteenth century. The amount produced however has always been comparatively small ; for instance, that produced in Saxony, Prussia, Sweden, Norway, and the United States from 1891 to 1900 amounted to only 500–600 tons yearly. During the nineteenth century the metallic cobalt in smalt and other cobalt products may accordingly be put down as only some 50, or at most 100–200 tons yearly. Latterly however, owing to the large production at Temiskaming in Canada,[2] there has been a considerable increase.

About the middle of the 'forties, owing to the discovery of artificial ultramarine, the price of cobalt fell. Of late years a further drop was consequent upon the exploitation of the silver-cobalt deposits at Temiskaming. The price per kilogramme of cobalt in ore containing on an average 5 per cent of cobalt, was a few years ago about seven shillings.

The exploitation of cobalt ores takes place principally in the following classes of deposit :

1. Silver-cobalt- and silver-cobalt-bismuth lodes at Temiskaming in Canada and at Schneeberg in Saxony.

2. Cobalt- or cobalt-nickel arsenide- and sulphide lodes, to which the cobalt fissures mentioned on page 948 also belong.

3. Asbolane deposits, principally in New Caledonia.[3]

4. Fahlbands, upon which operations are no longer prosecuted.

---

[1] *Ante*, pp. 153, 158.   [2] *Ante*, p. 668.

[3] *Ante*, p. 950.

THE GENESIS OF LODES, RÉSUMÉ

A. G. Werner, 'the Father of Geology,' in his *Neue Theorie von der Entstehung der Erzgänge*, published at Freiberg in 1791, stated that lodes in general, and more particularly those best known to him at Freiberg, were filled from above. General impoverishment and disappearance of the lodes in depth were the corollaries of this descension theory.[1] In consequence of the high repute in which Werner was held, for decades and even generations this view exercised an unfavourable influence upon the development of mining.

In isolated cases the applicability and correctness of the descension theory have been demonstrated.[2] As pointed out particularly by F. C. Beust in 1840,[3] it is however entirely untenable for the explanation of the origin of by far the majority of lodes. This old descension theory has nevertheless often reappeared, and always in some new guise. Charles Moore, for instance,[4] maintained that the lead lodes in the Carboniferous, Rhætic, and Liassic limestone in North Wales and the north of England, were filled from above. Chr. Münster, again, in 1894[5] discussed the possibility of the silver of the Kongsberg lodes having been derived from a Palæozoic sheet originally overlying these lodes but subsequently removed by erosion ; while many Swedish mining engineers even in the last decade defended the view that the iron deposits at Persberg, Långban, etc.,[6] had resulted from transport of iron by meteoric waters from surface into depth. From this theory it must logically be concluded that deposits rapidly pinch out in depth.

The crusted and banded structure of many lode-fillings ; the nature of the characteristic gangue-minerals, quartz, calcite, barite, zeolites, etc. ; and the secondary formation of minerals such as chlorite, sericite, epidote, calcite, etc., in the country-rock, compelled a generation ago the assumption that the filling of ordinary lodes, as for instance those in the Freiberg and Clausthal districts, was by deposition from aqueous solutions, these being principally heated waters or hot springs.

So far as is known, only the so-called a-quartz of prismatic habit is found in lodes, and not the β-quartz which occurs as dihexahedra in eruptive rocks. Under the pressure of one atmosphere the dividing line between the formation of these two varieties occurs at 570° C., and this figure accordingly gives an approximate idea of the maximum temperatures

---

[1] *Ante*, p. 189.
[2] *Ante*, pp. 189, 190, and the coming description of the Lake Superior Iron Ore Occurrences.   [3] *Kritische Beleuchtung der Wernerschen Gangtheorie*.
[4] *Report of the Brit. Ass. Acad. of Science*, 1869, p. 360.
[5] *Loc. cit.*, p. 660.   [6] *Ante*, pp. 378-394.

which existed at the formation of lodes. Most lodes characterized by zeolites were probably formed at temperatures of 100°–200°, more rarely at 300°, and generally speaking the temperatures were below the critical temperature of water, namely, 365° C.

About the middle of last century most authorities concluded that the solutions with their metal content ascended from depth.[1] A few, including J. G. Forchhammer [2] and G. Bischoff,[3] considered that the metal contents were leached from the country-rock and carried laterally to the fissure. This lateral secretion theory in the 'seventies and 'eighties found in F. v. Sandberger a supporter as skilled as he was enthusiastic. His thesis, however, after a long discussion was most zealously and successfully controverted by A. W. Stelzner. The most important works dealing with this question are :

Fr. v. Sandberger, 'Zur Theorie der Bildung der Erzgänge,' Berg- u. Hüttenm. Ztg., 1877, No. 44-45.—' Untersuchungen über den Gehalt an schweren und edlen Metallen in Augiten, Hornblenden und Glimmern,' Neues Jahrb. f. Min. Geol., Pal., 1878.—' Über die Bildung von Erzgängen mittels Auslaugung des Nebengesteins,' Berg- u. Hüttenm. Ztg., 1880, No. 38 ff., and Zeit. d. d. geol. Ges., 1880.—Untersuchungen über Erzgänge, I., 1882, and II., 1885.—' Neue Beweise für die Abstammung der Erze aus dem Nebengestein,' Verhandl. d. phys.-med. Ges. Würzburg, 1883.—' Über die von der k. k. österreichischen Regierung veranlassten Untersuchungen an den Erzgängen von Přibram,' Sitzungsber. d. Würzburger phys.-med. Ges., 1886.—'Bemerkungen über den Silbergehalt des Glimmers aus dem Gneise von Schapbach u.s.w.,' Neues Jahrb. f. Min., 1887, I.—' Silberbestimmungen in Glimmern aus Freiberger Gneisen. Untersuchungen über das Nebengestein der Přibramer Erzgänge,' Neues Jahrb. f. Min., 1888, I.—' Bemerkungen über die Resultate der Untersuchungen von Nebengestein der Přibramer Erzgänge,' Verhandl. d. k. k. geol. Reichsanst., 1888, No. 3. Also : ' Untersuchungen von Nebengesteinen der Přibramer Gänge mit Rücksicht auf die Lateralsekretionstheorie des Prof. Dr. v. Sandberger,' Berg- u. Hüttenm. Jahrb. d. österr. Bergakad., 1887, XXX.—Publications by : M. Ritter v. Friese, Österr. Zeit. f. Berg- u. Hüttenwesen, 1887, XXXV.; J. Gretzmacher, ibid., 1888, XXXVI.; A. Patera, Verhandl. d. k. k. geol. Reichsanst., 1888, No. 11.—E. Carthaus. ' Die Sandbergersche Erzgangtheorie,' Zeit. f. prakt. Geol., 1896, pp. 107-112.

A. W. Stelzner, ' Die über die Bildung der Erzgänge aufgestellten Theorien,' Berg- u. Hüttenm. Ztg., 1879, No. 3, and Zeit. d. d. geol. Ges., 1879.—' Die Lateralsekretionstheorie und ihre Bedeutung für das Přibramer Ganggebiet,' Jahrb. d. österr. Bergakad., 1889, XXXVII.—'Beiträge zur Entstehung der Freiberger Bleierz- und der erzgebirgischen Zinnerzgänge' (posthumous), Zeit. f. prakt. Geol., 1896. — F. Kolbeck, ' Über die Untersuchung eines Glimmers durch die trockene Probe ' (auf Silber), Jahrb. f. d. Berg- u. Hüttenwesen, Saxony, 1887, II.

The frequent occurrence of calcite veins in limestone, of small quartz veins in quartzite, etc., of fissure-fillings containing gypsum in selenite beds, and of calcite- and zeolite amygdaloids in many eruptive rocks, point to deposition from aqueous solutions, which have leached the constituents in question from the immediate country-rock and deposited them later in open spaces within that rock.

In such manner, as already mentioned, the genesis of the garnierite

---

[1] Ascension theory, ante, p. 190.
[2] Pogg. Ann. XCI., 1854, pp. 568-585, and Vol. XCV., 1885, pp. 60-96
[3] Chemische Geologie, 1st Ed., 1855, II. pp. 2109, 2121-2126.

veins in peridotite and serpentine may be explained.[1] Several manganese-
and iron lodes have a like origin.[2] Many investigators similarly explain
the occurrences of native copper accompanied by calcite, zeolites, etc., in
basic sheets.[3] Sandberger and the other extreme defenders of the lateral
secretion theory went however much farther; they applied the same
theory to the genesis of the lead-silver lodes in the Erzgebirge and at
Přibram, etc., and even to the tin lodes in granite.

Sandberger considered he had proved that the relatively uncommon
heavy metals, such as copper, zinc, lead, etc., and even silver, occur in
fairly appreciable amount in the ferro-magnesian silicates of the eruptive
rocks, gneisses, etc., particularly. This metalliferous content he considered
would at the decomposition of these silicates be released from its combination with silica and carried to the lode fissure, to be deposited there in
combination with sulphur, arsenic, or antimony. He proposed therefore
a chemical prospecting for lodes, which should consist in testing the rocks
or their minerals for their possible heavy-metal content. Such a detailed
examination was actually undertaken by a Government commission
about the middle of the 'eighties in the lead-silver district of Přibram,
but without economic result.

As already mentioned,[4] on different grounds the conclusion is justified
that even the relatively uncommon heavy metals are contained in many
rocks in small amount, though Sandberger greatly overestimated the percentage. Further, his opinion that the heavy metals, copper, lead, silver,
etc., occur as constituents of mica, hornblende, and augite, was also in the
main mistaken. As the result of most careful analyses undertaken at the
instigation of Stelzner, it was demonstrated that the micas in question, so
far as the methods of analysis allowed, were free from copper, silver, etc.,
but that frequently a small amount of mineral sulphide had migrated from
the lode fissure into minute cracks in the country-rock. This possibility
had been overlooked by Sandberger, who consequently in his analyses had
determined not the original heavy-metal content in the rock but the small
quantity which had migrated from the lode fissure. J. R. Don,[5] by
numerous analyses of the gold-quartz lodes in Australia, determined that
in general the presence of gold in fresh country-rock is not demonstrable
at a great distance from the lodes, though gold is frequently found as a
migrated constituent in their neighbourhood. For these reasons and others
mentioned elsewhere, the lateral secretion theory in the sense of Sandberger
is not applicable to the explanation of the origin of the silver-lead-zinc- or
the copper lodes.

[1] *Ante*, p. 953.   [2] *Ante*, p. 852.   [3] *Ante*, p. 934.
[4] *Ante*, pp. 153-158, and elsewhere.
[5] ' The Genesis of Certain Auriferous Lodes,' *Trans. Amer. Inst. Min. Eng.* XXVII.,
1897.

A new interpretation of the genesis of lodes, though really a strong amplification of the lateral secretion theory, has been put forward by C. R. Van Hise, whose principal works are the following :

C. R. VAN HISE, ' Some Principles controlling the Deposition of Ore,' Trans. Amer. Inst. Min. Eng. XXX., 1900 ; and in Genesis of Ore-Deposits, 1902, pp. 282-432.—'A Treatise on Metamorphism,' U.S. Geol. Surv. Monogr. XVII., 1904.
Reference must also be made to the discussion on Genesis of Ore-Deposits, Meeting of Amer. Inst. of Min., Washington, Feb. 1900 (printed 1902), with copy of Fr. Pošepný's The Genesis of Ore-Deposits, and contributions by C. R. Van Hise, S. F. Emmons, W. H. Weed, W. Lindgren, J. H. L. Vogt, J. F. Kemp, W. P. Blake, T. A. Rickard, etc. : as well as to a similar discussion before the Geological Society of Washington at the beginning of 1903, with contributions by S. F. Emmons, W. H. Weed, J. E. Spurr, W. Lindgren, J. F. Kemp, F. L. Ransome, T. A. Rickard, C. R. Van Hise, and C. W. Purington, under the title, ' Ore-Deposits,' printed in the Eng. Min. Journ., New York, May 1903.

According to its behaviour under the pressure existing at different depths, Van Hise divides the lithosphere into three zones :

1. Zone of fracture, in which fissures may result from pressure.

2. Zone of fracture and flowage, an intermediate zone.

3. Zone of flowage, in which the rocks under the exceedingly high pressure are plastically deformed without the formation of fissures.

The first zone, Van Hise estimates, probably reaches a depth of some 10 kilometres, though this depth will vary with the character of the rock.

Leaving out the deposits formed by magmatic differentiation and sedimentation, Van Hise considers that in general the origin of other deposits may be explained by the action of underground waters, in that meteoric waters, rain, etc., first dissolved the small heavy-metal content of the rocks and subsequently deposited it elsewhere. The movement of the underground waters is illustrated in a number of diagrams, two of which are reproduced in Fig. 400.

At descent, the water, which may even reach the boundary between the first and second zones, becomes somewhat warmed, whereby its mobility, electrolytic dissociation, and consequently its chemical action upon the country-rock, increase. Such warm water accordingly is in a particularly favourable condition to take up heavy metals, etc. When such solutions afterwards ascend, principally through the larger fissures, the dissolved components—chiefly owing to decrease in temperature and pressure or to the mingling with other waters—gradually become precipitated, and ore-deposits result. It is apparent therefore that according to Van Hise's interpretation also, lodes were formed in general by heated waters which leached the heavy metals from the country distant from the lode. This theory differs from that of Sandberger in that the latter believed in the leaching of the immediate country-rock, while Van Hise traces the ore from distances of as much as one or more kilometres.

Van Hise arrived at his conclusions from a study of the iron deposits in the Lake Superior district, in the case of which his theory is applicable. In more uncommon cases other lodes and metasomatic occurrences, principally of iron and manganese, may be similarly explained. Although in our opinion the heated waters in most cases are the after-effects of eruptive phenomena, we admit without question that descending water may bring to these heated waters such constituents as silica, carbon-dioxide, lime, etc. ; in this manner, for instance, the barium in the barite of some German lodes is traceable to the low barium content of the Zechstein or Bunter.

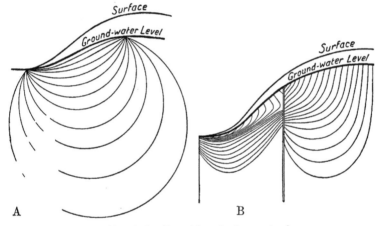

FIG. 400.—Ideal sections of flow of underground water.

A, ideal vertical section of the flow of underground water entering at one point on a slope and issuing at a lower point ; *B*, ideal vertical section of the flow of underground water entering at a number of points upon a slope and passing to a valley below interrrupted by two vertical channels.

Van Hise's new theory is in some cases correct, but we cannot agree to its general application to all lodes, and particularly not to those of gold, silver, lead, copper, quicksilver, etc. Such lodes in numerous cases are obviously closely associated in genesis with eruptive phenomena, and have been formed almost immediately after the eruption. The best examples of this association are afforded by the young gold-silver- and the young quicksilver deposits. This manner of formation cannot be made to harmonize with the theories of Van Hise.

According to his hypothesis also, the character of lodes or of lode-groups must to some extent be dependent upon the chemical composition of the country-rock, or upon the geological structure of the district. As was however mentioned when describing the young gold-silver- and the old lead-zinc lodes, no such general dependence exists, either on a large or on a small scale. Nor does Van Hise's hypothesis explain the frequent occurrence of the most varied lode-groups, sometimes chemically far removed from one another, within one and the same limited area, as for instance at both Freiberg and Schmenitz.

Furthermore, we consider that in the case of tin lodes convincing evidence is forthcoming that the characteristic elements, tin, wolfram, and other heavy metals, as well as lithium and fluorine, were extracted from the magma by a kind of acid extraction,[1] and in our opinion it is highly probable that the heavy-metal content—gold, silver, quicksilver, lead, zinc, copper, etc.—of ordinary lodes was similarly concentrated.[2]

The Deposition of Minerals in Lodes

Very frequently, and especially in lodes containing sulphides, the richest ore is met in the neighbourhood of lode junctions. This phenomenon is probably due to the fact that the waters circulating in one lode fissure contained heavy metals in dilute solution, while those circulating in another contained the precipitant, as for instance sulphuretted hydrogen, likewise in dilute solution. When then two such solutions met and mingled, precipitation took place.

It may be assumed that the heavy-metal sulphides, such as, PbS, Ag_2S, etc., were not contained as such in solution, but that the cathions, Pb, Ag, etc., were present in one solution, and the anions, S, etc., in another.

V. M. Goldschmidt has pointed out that the laws of mineral-association from the view-point of the phase-rule,[3] may be employed in explanation of mineral-association in lodes. He states that : ' The silver-antimony sulpho-salts form a case in point. In addition to the two simple combinations argentite and stibnite there are five different stoichiometric combinations. Of these minerals, two may occur together and be stable, though in exceptional cases three may be present. The seniority of the minerals in all such cases would appear to indicate a system which, while undergoing transition, became fixed by cooling before equilibrium was attained. Generally speaking therefore, when $n = 2$—that is, when only two end members exist, as for instance Ag_2S and Sb_2S_3—only two minerals may be expected, since it is little probable that during crystallization the exact temperature of the transition-point existed. We come then to the conclusion that : If a mineral may be regarded as a summation product of two other minerals, it will only be stable when it occurs together with one or other of those two minerals.'

In point of fact, argentite, Ag_2S, and stephanite, $5Ag_2S \cdot Sb_2S_3$; stephanite and pyrargyrite, $3Ag_2S \cdot Sb_2S_3$; pyrargyrite and miargyrite,

---

[1] *Ante*, p. 421.   [2] *Ante*, pp. 532-534.
[3] *Zeit. f. anorg. Chemie*, Vol. LXXI., 1911.

$Ag_2S . Sb_2S_3$; miargyrite and stibnite; galena and boulangerite, $5PbS . 2Sb_2S_3$; boulangerite and jamesonite, $2PbS . Sb_2S_3$; zinckenite, $PbS . Sb_2S_3$, and stibnite, are frequently met in lodes as contemporaneous deposits, but not the combination argentite and stibnite, or galena and stibnite. Similarly, chalcocite and bornite, or chalcopyrite and pyrite or pyrrhotite, are frequently found occurring together, but not chalcocite and pyrite or pyrrhotite.

The study of paragenesis is rendered extremely difficult by the fact that, as indicated by primary crusted structure, the solutions in lodes frequently change. It is therefore often difficult and even impossible to make a fine distinction and to separate the contemporaneous processes in mineral-formation from the consecutive.

The physical chemistry of lode-minerals has hitherto been but little investigated, though in all probability in the future its aid will be requisitioned in the endeavour to understand complex lode occurrences.

ORE-BEDS

THE sediments deposited in seas and lakes may be divided into :

1. Clastic sediments :
   - Macro-clastic : conglomerates, breccias.
   - Micro-clastic : sandstones.
   - Crypto-clastic : clays.
2. Organic sediments :
   - Coral limestone, Muschelkalk, etc.
   - Nullipore limestone, Diatom earth, coal, etc.
3. Chemical sediments : Rock-salt, potassium salt, gypsum, iron-ochre, etc.

According to their deposition in shallow water or in depth, the present-day, chiefly mechanical marine deposits may be grouped as follows : [1]

A. Coastal deposits :
  1. Upper beach zone, pebbles, coarse sand, etc.
  2. Zone of finer coastal sand, reaching to a depth of some 200 metres.
  3. Zone of coastal mud or clay, reaching to a depth usually of 700 m., but occasionally 900 metres. In this zone the following varieties of sediment may be differentiated :
     - (a) Blue clay, the colour being derived from ferrous sulphide.
     - (b) Green clay.
     - (c) Red clay.
     - (d) Volcanic sand and clay.
     - (e) Coral sand and clay.

B. Deep-sea or pelagic deposits, beginning at some 700–900 m. and reaching, so far as observation goes, down to a maximum depth of roughly 9000 metres :
  1. Deep-sea sands.
  2 Organic clays.
     - (a) Globigerina ooze.

[1] E. Kayser, *Lehrbuch der allgemeinen Geologie*, 1909, Part 1.

973

(b) Pteropod ooze.

(c) Diatom ooze.

(d) Radiolarian ooze.

3. Red deep-sea clay, particularly extensive in several deep oceans and characterized among other things by peculiar concentric encrusted manganese- and iron oxide nodules.

On physical-chemical grounds associated with adsorption phenomena, a fine slime sinks considerably quicker to the bottom of a solution containing a salt or electrolyte, than in clear sweet water. The fine clayey mud carried by rivers to the sea becomes deposited therefore in greater part near the coast, only an exceedingly small portion reaching the great ocean depths. In earlier geological periods also and for the same reasons, coastal deposits predominated quantitatively, while true pelagic deposits formed in great depth, played only a subordinate part.

As emphasized particularly by E. Kohler,[1] adsorption processes are of the greatest importance as factors in the genesis of ore-deposits and in lithogenesis. Adsorption rests upon the fact [2] that over the surface of a solid body, as for instance a clay particle, in a mineral solution, a different concentration of the dissolved material, say for instance of copper sulphate, exists than in the free solution. Such salts therefore as possess the property of becoming particularly concentrated over the surface of any body in suspension, are thereby extracted from the body of the solution in which they are present. Certain extremely finely compacted bodies which in proportion to their weight have a large surface, such as charcoal, animal charcoal, bone charcoal, many other organic bodies, kieselghur, gelatinous silica, kaolin, clay, etc., accordingly produce a strikingly intense adsorbent effect. The two last, for instance, have a considerably stronger effect than quartz grains, which if under a certain size are never precipitated.

The process of adsorption is explained by the following examples : if an aqueous 10 per cent, copper sulphate or lead nitrate solution be filtered through 20–30 grm. of refined kaolin, the filtrate will be practically non-metalliferous, the copper or lead being almost completely retained by the kaolin ; again, if a pulp of kaolin or clay be added to a dilute solution of copper sulphate or zinc chloride, the metalliferous salt is in greater part adsorbed by the kaolin or clay and with it precipitated to the bottom.

Between the adsorbed metalliferous salt and that portion of it which remains in solution there exists an equilibrium. The settlement of slimy material in suspension is not only promoted by salts or electrolytes, but such material itself may by adsorption precipitate metallic salts and

[1] Zeit. f. prakt. Geol., 1903, pp. 49-59.

[2] W. Ostwald, Grundriss der Kolloidchemie, Dresden, 1909.

oxides. From some salts the metals are adsorbed by the slime in part in oxidized form as bases, the acid radicles going back into solution.

The different adsorbents such as kaolin, clay, coal, organic substances, etc., generally exhibit special or specific adsorption, that is, they each adsorb only one material or only one to an unusual extent, while all other materials are either not adsorbed at all or only subordinately. Arable land, for instance, adsorbs or collects potassium salts in a striking manner, while to a great extent it allows sodium salts to filter through. The accumulation of iodine in certain marine algae is similarly explained. It is however a question whether this is adsorption alone or adsorption in connection with chemical action.

The composition of sea-salt and of the ashes of marine algae[1] exploited for iodine on the coast of Norway, may be gathered from the following analyses :

	Na.	K.	Mg.	Ca.	Cl.	Br.	J.	SO₄.
Sea-salt . . .	30·63	1·10	3 76	1·19	55·23	0·19	[0·02]	7·69
Seaweed ashes . .	11·87	22·96	2·09	4·12	30·23	...	1·56	9·68

The figures for sea-salt are taken from F. W. Clarke, *The Data of Geochemistry*, 1908.— Sea-water contains on an average 10–12 times as much bromine as iodine ;[2] on this basis the iodine has been given as 0·02 per cent in the above table.

The analysis of the seaweed is the average of five analyses placed at our disposal by H. Bull of Bergen. Bromine, which occurs to the extent of about 0·5 per cent, was not determined ; the other undetermined constituents consisted of insoluble SiO_2, P_2O_5, Al_2O_3, Fe_2O_3, etc.

In sea-salt there is about 2800 times as much chlorine as iodine, and roughly 28 times as much sodium as potassium ; in seaweed, on the other hand, there is roughly only 20 times as much chlorine as iodine, and only half as much sodium as potassium. Compared with chlorine therefore, the iodine has been concentrated 140 times ; and compared with sodium the potassium has been enriched 55 times. Similarly, manganese becomes largely concentrated by the algae *Padina pavonia*, the ashes of which contain 8·19 per cent of manganese oxide.[3] Copper is taken up by living algae, low organisms, fish, etc., even from extremely dilute solutions. Copper compounds are so poisonous however that death ensues before measurable quantities can thus become collected.

The formation of sulphides by the reducing action of organic matter upon sulphates, has long been known.[4] Pyrite has been met, for instance,

---

[1] *Laminaria digitata*.　　　　　　[2] J. H. L. Vogt, *Zeit. f. prakt. Geol.*, 1898, p. 228.

[3] J. G. Forchhammer, *Ann. des Phys. u. Chem.*, 1885, p. 84.

[4] *Ante*, p. 140.

on old mine timber, and in one lead-zinc mine, sphalerite.[1] The body of
a mouse put into a bottle containing a solution of ferrous sulphate became
covered with a coating of small pyrite crystals.[2] In the decaying beds of
peat bogs and similar marshes pyrite or marcasite, in part as a covering of
vegetable remains, has repeatedly been observed.[3] J. G. Forchhammer[4]
on the coast of Bornholm found pyrite formed by the action of decaying
fucoids, a discovery made use of by Bischof as far back as the year 1862
in explanation of the presence of pyrite in alum slates, etc. It is also
worthy of mention that pyrite and occasionally other sulphides occur as the
fossilizing material of *Ortoceratites, Ammonites*, etc. In the Kupferschiefer
the bodies of ganoid fish,[5] as well as the leaves and fruit of *Ullmannia* and
Voltzia, reduced copper and silver from solution, so that sulphides of these
metals were deposited on fish scales and fruit peel.

In enumerating the recent deposits of coastal clay, mention was made of
the blue clay, which consists of grains of quartz and felspar up to 0·5 mm.
in size and is found at some distance from land and at a depth usually
about 700–900 metres. The colour of this clay is due to contained ferrous
sulphide, precipitated by sulphuretted hydrogen. This clay has been
found more particularly in the bed of the Black Sea at depths between
540 and 1290 metres.[6] By dredging, two kinds of such sulphide clay are
obtained, namely, black and blue. The former under the microscope
shows ferrous sulphide, FeS, partly in the form of small isolated pellets, and
partly as an impregnation between the grains of sand. Among other
places, these pellets are also found contained in diatoms. Occasionally
nail-shaped aggregates of ferric disulphide, FeS_2, are found. The dark
blue clay contains many diatoms, together with some $CaCO_3$ and FeS.

The occurrence of sulphuretted hydrogen, H_2S, in the water of the
Black Sea, which being comparatively stagnant gets no air and is therefore
poor in oxygen, is connected with the sulphide content of the clay. This
gas is first observable at a depth of 140 m., beyond which depth it increases
considerably, 215, 570, and 655 c.c. having been found in 100 litres of
water from depths of 365, 900, and 2200 metres, respectively. This sulphuretted hydrogen is formed by the reduction of the sulphates of certain
anaerobic bacteria which require for their existence organic matter but not

---

[1] G. Bischoff, *Lehrbuch der chemischen Geologie*, 1863, I. p. 559.

[2] C. W. C. Fuchs, *Die künstlich dargestellten Mineralien*, p. 55, Haarlem, 1872.

[3] B. E. Palla, ' Rezente Bildung von Markasit im Moore von Marienbad,' *Neues Jahrb.
f. Min.*, 1887, II. p. 5 ; C. Ochsenius, ' Ganz junge Bildung von Schwefelkies,' *ibid.*, 1898,
II. p. 232.

[4] Bischof, *loc. cit.* I. p. 926.   [5] *Palaeoniscus*, etc.

[6] N. Androussow, ' La Mer Noire,' *Guide des excursions d. VI. Congrès Géol. internat.*,
St. Petersburg, 1897, No. 29 ; M. Jegunow, *Ann. géol. et minér. de la Russie*, 1897 ; Review in
Zeit. f. prakt. Geol., 1902, p. 105 ; Sidorenko, *Mém. de la Soc. des naturalistes de la nouvelle
Russie*, XXI., 1897, Book II. ; Review in *Neues Jahrb. f. Min.*, 1900, I. p. 224.

free oxygen. It can moreover be formed from sulphur bacteria,[1] which as part of their vital functions segregate sulphur in their cells, from which by secondary action sulphuretted hydrogen arises. With regard to the iron bacteria, reference to these is made later.

Among recent deposits which have thrown light upon the origin of sedimentary ore-deposits,[2] first place must be given to : the mechanical deposits of iron sand, especially titaniferous iron sand, in rivers and on sea shores, and other recent gravels ; the chemical deposits of iron- and manganese oxides, and especially the lake- and bog ores ;[3] and finally, the above-mentioned blue sulphide clays. Upon this subject also, a further study of the phenomena of adsorption and of the iron- and sulphur bacteria, etc., would prove most illuminating.

The bituminous slates of different geological formations are frequently and particularly characterized by containing a fairly large amount of sulphide minerals, and especially pyrite. The alum slates, which often contain 3–10 per cent of carbonaceous matter together with about 2–5 per cent of pyrite, were formerly used in many countries, as for instance in Germany, Austria, Sweden, Norway, etc., for the production of alum.[4] A considerable portion of the pyrite of these alum slates is primary, though a small amount may be due to secondary impregnation. In addition to pyrite and sometimes pyrrhotite, a small copper content may exceptionally be found in alum slates.[5] Several investigators in explaining the organic matter and the pyrite in these slates have aptly referred to the above-described recent blue clay, which is similarly constituted.[6] This question is further discussed when describing the Kupferschiefer.

The recent mechanical deposits, such as ferruginous gravels, magnetite- and titaniferous-iron sand, as well as the recent chemical deposits in shallow water, for instance the lake ores, are characterized by great variation in thickness. The lake deposits in particular have this characteristic, though even deposits assembled in greater depth possess as a rule no great conformity either in thickness or bedding.

The normal sediments, such as conglomerate, sandstone, limestone, clay-slate, etc., all of which, apart from the presence of some iron and manganese, are practically free from heavy metals, are formed in sea- or lake water of normal character. The formation of metalliferous sediments containing a fairly high metal content must therefore have resulted from some special transport of heavy metal to the sea- or lake water, though

[1] F. Lafar, *Handbuch der techn. Mykologie*, Vol. III. pp. 214-243, Jena, 1904–1906.
[2] *Ante*, pp. 14, 192-194.   [3] *Postea*, p. 982.   [4] *Ante*, pp. 73, 74.
[5] F. Slavik, ' Über die Alaun- und Pyritschiefer,' *Bull. internat. de l'Académie des Sciences de Bôhmen*, 1904.
[6] F. Pompecky, *Geogn. Jahresheft*, 1901, Vol. XIV. p. 185, and R. Delkeskamp, *Zeit. f. prakt. Geol.*, 1904, p. 296.

the increment in general would be relatively limited. Metalliferous deposition would also frequently be influenced by local phenomena, as for instance by floods, and by the quantity or quality of any adsorbent present. Generally speaking, therefore, the regularity associated with ordinary non-metalliferous sediments may not be expected with ore-beds.

Formerly, conformity with the enclosing beds and a bedded or banded structure, were regarded as sufficient proof of the sedimentary origin of an ore-bed. According to observation in the last decades, however, the same characteristics are shown by many other classes of deposit. As the result of injection, pressure-metamorphism, etc., magmatic deposits may frequently have the same morphological character as ore-sediments ; in fact, the magmatic intrusive pyrite deposits, of which the Norwegian pyrite deposits are typical representatives, were formerly regarded by several authorities as examples of metalliferous sedimentation. Contact-deposits, which in most cases are formed from limestone, may likewise, when only certain layers have been altered, present a bedded character. Such deposits formerly were frequently regarded as sediments. Similar circumstances have also obtained with many metasomatic iron- manganese- and lead-zinc ores. A solution containing heavy metals may occasionally penetrate several rock thicknesses, though precipitation need only take place in certain beds. Such precipitation may depend upon the presence of a reducing carbonaceous substance, or, as pointed out by E. Kohler,[1] it may be brought about by the adsorbent action of finely distributed kaolin or clay substance present only in certain beds.

There are accordingly numerous different processes by which epigenetic deposits with bedded structure may be formed ; indeed it can often be incontestably proved that an apparently conformable, bedded, and banded deposit, is in fact of epigenetic origin. It is much more difficult as a rule to produce conclusive evidence of the sedimentary nature of a deposit. We would recall the prolonged discussion concerning the origin of the Kupferschiefer, the Rammelsberg pyrite bed, the auriferous conglomerate of the Witwatersrand, the different fahlbands, and many iron deposits in the fundamental rocks, etc. With all these deposits there are even now many questions which have not yet been answered to general satisfaction. In the chapter on ' Ore-Beds ' therefore, we are compelled to treat not only undoubted sediments, but also such occurrences as generally speaking exhibit the principal characteristics of sedimentary beds, namely, conformity and bedding, though their genesis may not yet have been finally determined.

The tectonic disturbances—folds, overthrusts, faults, etc.—which ore-beds have suffered, have already been dealt with.[2] It need only be

---

[1] *Zeit. f. prakt. Geol.*, 1903.   [2] Pages 17-34.

repeated that ore-beds in certain formations have been altered by pressure and occasionally also by contact-metamorphism. Reference to the latter has already been made.[1]

Now and again, and independent of the formation of gossan or of oxidation and cementation, subsequent chemical alterations are found to have taken place with ore-beds. Among these belong the concretionary formations described later when dealing with the clay - ironstone and blackband ironstone, but more important is such tremendous secondary transport of metal as, for instance, has taken place with the large iron deposits of Lake Superior.

THE IRON ORE-BEDS

Chemistry of the Sedimentation of Iron- and Manganese Ores

The springs, drainage water, etc., arising in the present zone of weathering carry in general some iron, manganese, lime, magnesia, alumina, and alkali, as well as more or less carbonic acid, silicic acid, phosphoric acid, sulphuric acid, etc., in dilute solution. The iron occurs principally in the ferrous condition and subordinately in the ferric ; the manganese is found only in the manganous condition and not in the manganic. In bog water, with these salts and acids, more or less humic acid, to which reference is made later,[2] is associated.

When a spring containing no humic acid or other organic substance emerges at the surface, it suffers oxidation from atmospheric oxygen, while at the same time the dissolved carbonic acid escapes. In the case of a solution containing relatively much iron, some manganese, etc., but relatively little lime and magnesia, the stages in the resultant precipitation are as follows :

1. Firstly, the bulk of the iron becomes precipitated, mainly as hydrate, together with a fairly large amount of the dissolved silicic, phosphoric, and arsenic acids, but with relatively little manganese, lime, or magnesia. In this first stage the greater portion of the solids in suspension is likewise deposited.

2. Later, the bulk of the manganese is precipitated, more particularly as the dioxide, MnO_2, or its hydrate.

3. Finally, and presuming bicarbonate solutions, comes the deposition of the carbonates, and especially those of lime and magnesia, following upon the escape of the carbonic acid.

These stages are illustrated in the accompanying graph, Fig. 401, where zero on the left marks the beginning of oxidation, where the abscissae record

[1] *Ante*, pp. 350, 351.   [2] *Postea*, p. 986.

the times from the beginning of precipitation, and the ordinates the various amounts precipitated. This graph was drawn largely from analyses by Berzelius, Bischof, Bromeis and Ewald, Fresenius, Justus Roth, and others, of depositions from springs at different distances from their emergence at surface.[1]

Ferrous oxide oxidizes relatively quickly to ferric oxide, which is precipitated from a neutral or weak acid solution, much of the silicic and phosphoric acids being precipitated at the same time. The manganous salts, on the other hand, are not oxidized in an acid solution, though oxidation gradually takes place in a neutral or weakly basic solution.

These reactions pertain to oxidizing and purely inorganic precipitation.

With neutral or reducing precipitation, and especially at escape of carbonic acid from solutions containing ferrous oxide and manganous bicarbonate, the procedure is substantially different, and $FeCO_3$ and $MnCO_3$ are precipitated approximately simultaneously and together. The siderite

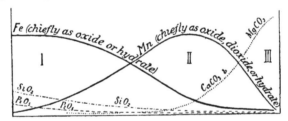

Fig. 401.—Graphic representation of oxidizing precipitation from ferruginous and manganiferous solutions. I, II, III representing different stages in this precipitation. Vogt, *Zeit. f. prakt. Geol.*, 1906, p. 227.

lodes, which are usually characterized by a considerable manganese content, are evidence of this community in precipitation.

In relation to iron and manganese it must therefore be noted that with neutral or reducing precipitation both are precipitated simultaneously, while with oxidizing precipitation the iron is first precipitated with some manganese, and later the manganese with some iron. There are numerous examples of this separate precipitation; thus, the deposits of iron- and manganese ochre; that for instance at Glitrevand in Norway described later,[2] where near the point of emergence principally iron-ochre with clay, gelatinous silica, and relatively much phosphoric acid, are precipitated, while at a distance principally manganese ochre together with relatively little iron, little gelatinous silica, and little phosphoric acid, are found. Again, with many occurrences in the ferruginous mica-schists or itabirite of northern Norway and of Brazil separate deposits of iron poor in man-

[1] Vogt, *Salten og Ranen*, 1890–1891, pp. 139-156 ; *Zeit. f. prakt. Geol.*, 1906, pp. 226-227.
[2] *Postea*, p. 984.

ganese on the one side, and of manganese with more or less iron on the other, may be observed. Other examples will be found in the description of manganiferous lake ores.

In the rocks of the earth's crust iron occurs the most extensively of all the heavy metals, while manganese occupies the second place,[1] the relation between the two amounts being on an average 1 part of manganese to 40–70 parts of iron. Chromium, nickel, cobalt, etc., occur only subordinately. Iron and manganese in the processes of weathering and decomposition go equally readily into solution. For this reason manganese in general plays a more or less prominent part in such iron deposits as have been formed by hydro-chemical processes, and a transition sequence between the sedimentary iron- and manganese deposits can be established.

Of late years roughly 140,000,000 tons of iron ore with an average of about 46 per cent of iron and 1–2 per cent of manganese have been produced, as against 1,600,000 tons of manganese ore with roughly 50 per cent of manganese and a small percentage of iron. The relation between iron and manganese in the total ore production accordingly coincides approximately with that in the rocks of the earth's crust. In individual cases, it is true, somewhat more manganese is found in the ore won, since manganese ore and manganiferous iron ore are more valuable than ordinary iron ore.

Chromium, nickel, cobalt, tin, etc., unlike manganese, are almost completely absent from sedimentary iron deposits; at most they occur here and there in traces. This is accounted for in the first place by their much more scanty distribution in the crust. It must however further be considered that chromium occurs principally in peridotite and its associated rocks, where it is contained in greater part in the resistant spinel, picotite. Concerning nickel, it must also be taken into account that this metal, as experience with the garnierite deposits has shown, is precipitated fairly quickly when contained in a silicated solution. Cobalt is chemically more closely related to iron and manganese than is nickel; it occurs in the asbolane deposits,[2] which in their character are closely allied to the manganese occurrences.

Some sedimentary iron deposits, as for instance the clay- and blackband ironstones, contain considerable admixtures of clayey material, while numerous others contain but little of such material.

Most sedimentary iron ores are mixed with more or less SiO_2, and some CaO, MgO, and Al_2O_3, the lime and magnesia occurring partly with the silicates and partly with the carbonates. Alkalies occur sparingly, and titanic acid is as a rule completely absent. Phosphoric acid on the other hand always occurs, and often in large amount.[3]

[1] Ante, pp. 152, 153. [2] Ante, pp. 951, 957
[3] Ante, p. 980, and the following chapter on Lake Ores.

LAKE- AND BOG-ORE BEDS

LITERATURE

F. M. STAPFF. 'Über die Entstehung der Seeerze,' Zeit. d. d. geol. Ges., 1866, Vol. XVIII., and in Jernkontorets Annaler. Stockholm, 1865.—OSSIAN ASCHAN. 'Die Humusstoffe . . . und ihre Bedeutung für die Bildung der Seeerze,' Helsingfors, 1906; extract in Zeit. f. prakt. Geol., 1907, pp. 56-62.—A. E. ARPPE. Finska Vet.-Soc. Förh. XI., 1868-1869.—C. E. BERGSTRAND. 'Über Vivianit in Eisenocker u.s.w.,' Geol. Fören. Förb. II., 1875.—A. F. THORELD. 'Fragen bezüglich der Regenerationszeit der Seeerze,' ibid. III., 1876.—A. W. CRONQUIST. 'Seeerze in Södermanland,' ibid. V., 1881, and Eisenocker in Helsingland, VIII., 1886.—JOHS. ASCHAN. 'Über manganreiche Seeerze in Finnland,' Teknikern, Helsingfors, 1906.—J. H. L. VOGT. 'Über Manganwiesenerze und über das Verhältnis zwischen Eisen und Mangan in den See- und Wiesenerzen,' Zeit. f. prakt. Geol., 1906.—F. SENFT. Die Torf-und Limonitbildungen u.s.w. Leipzig, 1862.—VAN BEM-MELEN. 'Über Siderit und Vivianit in Wiesenerzen u.s.w.,' Arch. Néerlandaises, 1896, XXX., and Zeit. f. anorg. Chemie, 1900, XXII.—GAERTNER. 'Vivianit und Eisenspat in mecklen-burgischen Mooren,' Arch. des Ver. der Freunde der Naturgesch., Mecklenburg, 1897, LI.— P. H. GRIFFIN. 'The Manufacture of Charcoal-Iron from the Bog- and Lake Ores of the Three Rivers District, Canada,' Trans. Amer. Inst. Min. Eng., 1892, XXI.—For the other American literature see J. F. KEMP, The Ore-Deposits of the United States and Canada. —H. MOLISCH. Die Pflanze in ihren Beziehungen zum Eisen. Jena, 1892.

Lake- and bog ores mostly contain much iron and little manganese, sometimes the iron content is about the same as that of manganese, while only exceptionally does the manganese preponderate. Between the two extremes the passage is gradual, and the iron lake- and bog ores and the manganese lake- and bog ores may therefore be discussed together.

Lake ores occur in shallow lakes and, according to experience in Finland and Scandinavia, more particularly at a depth between 1 and 5 m., and seldom at depths of 10 m. or more. They have not been observed at depths of less than 1 m., that is, they do not occur immediately at the water line. Bog ores are found, as the name indicates, in meadows, morasses, or other marshy depressions containing the stagnant water in which they were formed. Now and then lake ores from dried-up lakes are also found in marshes. The most extensively occurring variety of lake ore is usually fairly poor in manganese, exhibits a characteristic oolitic or pisolitic concretionary structure, and consists in greater part of fairly compact grains 2-7 mm. in diameter. From this structure the terms bean-, pea-, or powder ore, often given to this ore, are derived.

Pea ore, consisting almost entirely of grains of pea size, is particularly extensive. When these peas are cemented together by iron-ochre to form small discs the ore is known in Sweden as 'penny ore.' Lake ore frequently occurs as an incrustation on vegetable remains. Often too it occurs in somewhat larger masses spoken of as 'fragmentary ore.' The grains are partly embedded in clay, and partly free, in which latter case they form more or less continuous masses. The thickness of the bed is usually one or two decimetres, and rarely as much as 0·5 metre. The clay adhering to the grains, etc. is easily removed by washing.

At times, with the yellowish-brown manganese-poor ores, darker, loose, and friable patches rich in manganese dioxide occur, made up of grains varying in size from a pin's head to an egg. Here and there, as for instance

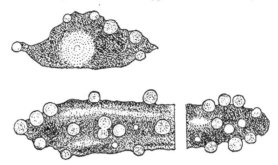

FIG. 402.—Nodules of lake ore from Storsjö ; natural size. Vogt.

The lightly dotted and mostly round areas are of iron-ochre poor in manganese ; the dark dotted base is ferruginous manganese dioxide.

at the Storsjö lake in south-east Norway, the yellowish-brown manganese-poor grains are cemented together, as illustrated in Fig. 402, by a hard, dark material very rich in ferruginous manganese dioxide. In such cases a definite sequence in precipitation can be established, namely, first,

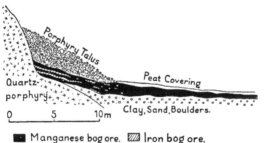

FIG. 403.—Manganese bog ore in the Borvik valley near Glitrevand. Vogt, *Zeit. f. prakt. Geol.*, 1906.

iron and a little manganese ; and afterwards, manganese with more or less iron. This sequence indeed may fairly frequently be observed.

It is often asserted, and particularly in the earlier text-books, that lake ores after they have been removed become formed again in the course of some twenty to thirty years. According to more recent investigation however, this re-formation occupies as a rule a considerably longer period.

The ordinary, usually very cavernous, frequently fairly hard, but occasionally earthy bog ore, containing preponderating iron together with a little manganese, is usually much contaminated by clay, vegetable remains, etc. It is found more particularly immediately under the peat covering. Those deposits which in Germany are known as *Eisenortstein* or *Klump*, approach bog ore in a chemical-geological sense, but are not to be completely identified with it. Bog ore also frequently contains a considerable amount of manganese, when it passes gradually over into manganese bog ore. Of such manganese ore the occurrence at Glitrevand in southern Norway may be taken as an example. There, as illustrated in Fig. 403, a usually 0·7–1 m., but in places 2·75 m. thick layer of porous wad, which near the bed-rock is contaminated by a fairly large amount of iron-ochre,[1] occurs in marshy valleys in an extensive area of quartz-porphyry. A large number of similar occurrences of manganese bog ore in different countries are described by Vogt.[2]

In Fennoscandinavia lake- and bog ores occur frequently in districts where granite and gneiss preponderate ; they are accordingly not associated preferably with districts containing basic ferruginous rocks. Lake ore is found in particularly large amount in Finland, 'the Land of the Thousand Lakes,' and in Norway and Sweden ; also in Germany, more particularly in the lakes of the North German diluvial plain.[3] Other deposits are found in Holland, Belgium, Russia,[4] the United States and Canada,[5] Africa, Australia, etc. These deposits therefore occur in the arctic and temperate zones as well as in the tropics. Bog ore also occurs in all parts of the world.

Although the lake ores in northern Europe were deposited in the geologically speaking very short period since the Ice age, these mostly thin but very extensive beds represent quite considerable ore-bodies. In Finland, for instance, about 2,250,000 tons of lake ore were obtained in the fifty years from 1858 to 1908, to which must be added the production of previous centuries. It is interesting to note that according to Ossian Aschan about 1,400,000 tons of humic 'sols' are carried by the rivers of Finland to the Baltic Sea every year.

The following selection of analyses will serve to indicate the general composition of these ores :

---

[1] See analyses Nos. 3*a*, 3*b*.   [2] *Op. cit.*
[3] v. Dechen, 2nd Ed. of W. Bruhn's *Die nutzbaren Mineralien und Gebirgsarten im Deutschen Reiche*, 1906, pp. 467-471.
[4] *The Iron Resources of the World*, Stockholm, 1910.
[5] J. F. Kemp, *loc. cit.*

[TABLE

	Swedish Lake Ore		Finnish Lake Ore					Manganese Bog Ore			Impure Vivianite, Wemdalen	
								Glitrevand		Nevada		
	Min.–Max. (1a, 1b)	Average (1c)	2a	2b	2c	2d	2e	3a	3b	4	5a	5b
Fe_2O_3	43·2 – 75·7	62·6	69·5	57·0	67·7	26·3	20·5	15·2	2·7	3·3	43·2	43·2
Mn_2O_3	0·5 – 34·7	5·6	3·3	7·9	9·6	32·9	40·3	58·6	80·6		1·7	2·1
MnO_2										80·4[2]		
SiO_2	5·5 – 41·3	12·6	8·7	14·4	7·1	19·3	12·9	7·4	1·1	1·7	10·6	8·5
Al_2O_3	1·2 – 7·9	3·6	1·9	3·2	2·0	0·9	1·1	2·2	Little	0·3	1·5	1·8
CaO	0·3 – 3·1	1·4	0·5	Trace	Trace		0·8	0·9	Little	3·4	0·03	0·6
MgO	0·02 – 0·7	0·19	0·4				0·3	0·3	Trace	1·3		
P_2O_5	0·05 – 1·21	0·48	1·25	1·4	0·77	0·45	0·53	0·03	0·10		7·01	9·64
SO_4	Trace – 0·4	0·07			Trace			0·15[1]	0·07[1]			
Loss on ignition	7·6 – 17·8	13·5	12·9	15·2	12·6	17·8	21·9			4·2	36·6	33·7
Total		100·0								98·97	100·6	99·69
Fe		43·8	48·6	39·9	47·4	18·4	14·3	10·6	1·9	2·3	30·2	30·2
Mn		4·3	2·1	5·0	6·1	20·8	25·5	37·2	51·0	50·9	1·3	1·6
P		0·21	0·55	0·62	0·33	0·19	0·23	0·013	0·04		3·1	4·2

Nos. 1a, b, c, minimum, maximum, and average of thirty analyses of lake ore and two analyses of bog ore from Sweden, after Svanberg, see Stapff, loc. cit.—Nos. 2a-e, a selection from eighteen analyses of lake ore, after Johs. Aschan, loc. cit.; see also Zeit. f. prakt. Geol., 1906, p. 223.—Nos. 3a, b, from Glitrevand, after Vogt, see Fig. 403; 3a, manganese ochre with some iron-ochre about 10 m. from the rock wall; 3b, at a still greater distance, pure manganese ochre; [1] is S, in addition 1·94 and 2·40 per cent of Zn.—No. 4, from Galconda, Nevada, after Penrose, Journ. of Geol. I., 1893, in addition 5·65 per cent of BaO, 2·78 per cent of WO_3, 0·35 per cent of K_2O; [2] is 65·66 per cent MnO+10·31 per cent disposable O.—Nos. 5a, b, blue earth, impure vivianite, from Wemdalen in Sweden, after Cronquist, loc. cit. The manganese in the above analyses is sometimes reckoned as Mn_2O_3 and sometimes as MnO_2. Actually several analyses of the oxidation zone, for instance No. 4, show that the manganese exists chiefly as MnO_2, frequently, it is true, with a small amount of a lower oxide.

In addition to iron as ferric oxide, this occurring principally as gelatinous limonite, some ferrous oxide has occasionally been established, and in
exceptional cases even crystallized siderite. The ferrous oxide, according
to v. Bemmelen, may have been formed secondarily from ferric oxide by
the reducing action of organisms. The greater part of the silica is readily
soluble in acids. From microscopic examination it would appear that
with lake ores, mechanically associated quartz, felspar, etc., play in general
only a very subordinate part. The old idea that a grain of sand or such-
like always exists in the centre of lake-ore concretions is, according to Vogt,
very questionable. Lake- and bog ores have a characteristic content of
phosphoric acid, though in this respect the manganiferous varieties apparently form an exception.[1] Further, V, Ti, Cr, Mo, Ni, Co, Zn, As, Cl, etc.,
have also in places been established. The iron content of air-dried lake
ore poor in manganese is mostly 38–42 per cent, though occasionally it
reaches 47–49 per cent. For data concerning the sequence of gradations
between the two extreme members, with a preponderating amount of iron—
99 Fe : 1 Mn—on the one side, and with a preponderating amount of manganese—3·6 Fe : 96·4 Mn—on the other, see Vogt, *Zeit. f. prakt. Geol.*, 1906,
p. 223.

The genesis of lake ore has long been the subject of discussion. C. G.
Ehrenberg considered he was able to show that these ores were deposits
by algae or diatoms, and more particularly by *Gallionella ferruginea* which
stores ferric hydroxide in its cells.[2] This view has however proved to be
incorrect. According to more recent investigation, and especially that by
the bacteriologist S. Winogradsky in Petrograd,[3] there are certain iron
bacteria which in the process of living precipitate ferric oxide containing
some manganese oxide in their envelopes, from water containing ferrous
oxide. In this manner he believed it was possible to explain the formation
of lake- and bog ores. These iron bacteria were subsequently investigated
by H. Molisch[4] in Vienna, with the result that it was established
that they do precipitate ferric oxide, though the part they play in the
formation of iron ore appears to be of little importance ; Molisch, for
instance, in the examination of many lake- and bog ores could only establish
the presence of iron bacteria in three samples.

For many years the formation of the ordinary lake- and bog ores has
been associated with humic acid. This formation has recently been
studied very carefully by Ossian Aschan in Helsingfors. The various
humus materials—the so-called humic 'sols,' consisting of about 50 per cent
C, 43 per cent O, 4·5 per cent H, 2 per cent N, some P and S—formed in

[1] See analyses 3-4.   [2] *Pogg. Ann.* XXXVIII., 1836.
[3] *Botan. Ztg.*, 1888, Vol. XLVI. p. 261.
[4] *Die Pflanze in ihren Beziehungen zum Eisen*, Jena, 1892 ; *Die Eisenbakterien*, Jena,
1910.

large amount in marshy water, promote the decomposition of the under-lying rocks. The humic 'sols' are not precipitated by ferrous, but by ferric salts. When a solution containing ferrous salts meets one containing humic 'sols' a soluble ferro-humate is first formed, which upon oxidation passes gradually over to ferri-humate. This is in part immediately pre-cipitated, forming then the thin, often iridescent coating which when present in spring courses indicates the ferruginous character of the water. A part of the ferri-humate remains for a time in colloidal condition, giving the water a more or less dark colour, and is carried to the next water course, and thus gradually deposited. The amount of oxygen dissolved in the water of inland seas is amply sufficient to oxidize the soluble ferro-humate to insoluble ferri-humate. The precipitated ferri-humate, which can be decomposed by lower organisms, is in turn gradually destroyed, ferric oxide remaining.

That the ordinary lake- and bog ores have been formed in greater part by this process follows, according to Aschan, from the discovery of remains of humus material in all the lake ores of Finland yet examined. The phosphoric acid always present in lake ore, and mostly in very considerable amount, is probably referable to phosphorus in the humic 'sols.' The concretionary structure of the ore is presumably due to the coagulation of the ferri-humate in conjunction with its slow decomposition by lower organisms. The often exceedingly troublesome deposition of iron-ochre in water pipes may be compared with the formation of lake ore; in this deposition thread-like bacteria are frequently found.

Purely inorganic chemical processes also may frequently have played a part in the deposition of these ores, in fact, sometimes the whole process has been entirely inorganic and completed without the assistance of humus material, etc. In this connection the important deposits of iron-ochre in the marshy districts around large pyrite deposits, such as Fahlun and Rio Tinto,[1] are worthy of mention. With these deposits the iron first went into solution as sulphate, to be afterwards precipitated by oxidation in the manner already described.[2] The formation of blue-earth or vivianite[3] may also have been purely inorganic. The same applies to manganese bog ore, as for instance that at Glitrevand, as well as to the manganiferous cement binding together lake ore poor in manganese.

In ancient times and in the early Middle Ages iron was obtained almost exclusively from lake- and bog ores. In most countries however, these ores as far back as the seventeenth and eighteenth centuries lost all their importance in the iron industry. In isolated districts the winning of lake ore particularly and of bog ore exceptionally, was continued into later times, some iron still being produced from these ores in Finland, in Olonez in

---

[1] *Ante*, pp. 314, 321.   [2] *Ante*, p. 975.   [3] See analysis No. 5.

northern Russia,[1] and possibly in other places. The average yearly
production of lake ore in Finland and Sweden may be gathered from the
following table :

	Finland.	Sweden.
	Tons.	Tons.
1860–1869	38,000	12,400
1870–1879	51,500	8,900
1880–1889	36,000	3,200
1890–1899	60,100	1,300
1900–1904	43,250	900
1905–1908	28,600	1,000

In Finland, during the thirty-two years from 1858 to 1889, 492,369 tons
of pig iron were produced from 1,330,727 tons of lake- and bog ore—practically speaking only lake ore—and only 45,550 tons of mine ore. On an
average therefore about 36 per cent of pig iron is obtained from lake ore,
equivalent to about 38 per cent of iron in air-dried ore. Of late years the
long-established exploitation of lake ore in Finland has rapidly declined.
In Sweden, the use of lake ore in isolated blast-furnaces ceased completely
about the year 1880. Subsequently some of this ore was produced in
Sweden and other countries for use in purifying sulphur- and cyanide
products obtained in gas works. An attempt was also made by washing
to prepare lake- and bog ore for use as pigment.

NODULAR IRON AND MANGANESE BEDS

LITERATURE

J. MURRAY. Proc. Roy. Soc. Edinburgh, 1876, IX. p. 255.—CHALLENGER ' Reports,'
Deep-Sea Deposits, 1891, pp. 341-378.—J. MURRAY and R. IRVINE. Trans. Roy. Soc. Edin-
burgh, 1895, XXXVII. p. 721.—J. Y. BUCHANAN, Chemist to the Challenger Expedition.
Proc. Roy. Soc. Edinburgh, 1890, XVIII. p. 19.—R. IRVINE and J. GIBSON. Ibid., 1876,
IX. p. 255.—A. E. NORDENSKIÖLD, 'Vega Expedition I.,' and G. LINDSTRÖM, 'Analysen der
Gesteine und Bodenproben von dem Eismeere.' Stockholm, 1884.—C. W. GÜMBEL, 'Über
die im Stillen Ozean auf dem Meeresgrunde vorkommenden Manganknollen von der Challenger Expedition,' Sitzungsber. math.-phys. Kl. k. bayr. Akad. d. Wiss. VII., Munich,
1878 ; see also Neues Jahrb. f. Min., Geol., Pal., 1878, p. 869.—J. B. BOUSSINGAULT. Ann.
chim. phys. sér. 5, XXVII., 1882, p. 289.—L. DIEULAFAIT. Compt. rend., 1883, LXLVI.
p. 718.—CHUN. Aus den Tiefen des Weltmeeres ; Schilderungen von der deutschen
Tiefsee Expedition, 1900, pp. 162-163.—ANDRUSSOW. ' Von dem Schwarzen Meere,' in
Guide des excursions du VII. Congrès géol. intern. XXIX. 13. Petrograd, 1897.—J.
WALTHER. Einleitung in die Geologie, p. 700. Jena, 1893-1894.—E. KAYSER. Lehrb.
d. Allgem. Geol. 3rd Ed., 1909, I. p. 511.—F. W. CLARKE. ' The Data of Geochemistry,'
U.S. Geol. Surv., Bull. 491, 1911, p. 121.—J. MURRAY, in J. Murray and J. Hjort. Depths
of the Ocean, London, 1912, especially p. 189.

In sequence with lake ore deposited in shallow water it is pertinent

---

[1] *The Iron Ore Resources of the World*, Stockholm, p. 517.

to discuss the manganese- or manganese-iron oxide nodules which occur frequently in large amount in many places on ocean beds and particularly in the pelagic red clay.[1]

Manganese oxide or hydrate is found in all deep-sea deposits, occasionally in the form of grains in the clay or in the Radiolarian, Diatom, Globigerina, and Pteropod oozes, but also as incrustations upon pumice stone, corals, calcareous algae, shells, bone-fragments, etc.; or finally as separate manganese nodules. These last have been occasionally met in relatively shallow waters, as for instance in Loch Fyne off the coast of Scotland; they are however especially characteristic of great ocean depths—between some 1000 m. and 6000 m.—and have been encountered in numerous places in the Pacific Ocean, the Indian Ocean, the Atlantic Ocean, and the Caribbean Sea. Sometimes they occur in strikingly large amount. Thus, R. v. Willemoes-Suhm, the zoologist of the Challenger Expedition, writes : [2]

" The nature of the bed in these in part great ocean depths [3] is remarkable, since in addition to the non-calcareous reddish slime and the many fragments of pumice which we encountered here, the bed in places must have been completely covered with large nodular manganese concretions."

Similar concretions up to a foot in diameter, several inches thick, and weighing as much as 10 kg., have been dredged in the region of the Gulf Stream. These nodules are frequently concentrically thin-crusted and usually enclose in the centre a foreign body, such as a fragment of whale- or fish bone, ear bones, shark's teeth, etc. Their general composition may be gathered from the following analyses :

	Fe_2O_3.	MnO_2.	SiO_2.	Al_2O_3.	CaO.	P_2O_5.	H_2O.
I. From the Pacific Ocean	27·46	23·60	16·03	10·21	0·92	0·023	17·82
II. From the Caribbean Sea	16·63	24·17 [1]	27·84 [2]	1·32	2·04	2·22 [3]	20·95

I. From the Challenger Expedition, after Gümbel; contains in addition 2·36 Na_2O, 0·40 K_2O, 0·18 MgO, 0·66 TiO_2, 0·48 SO_3, 0·05 CO_2, 0·02 CuO, 0·01 CoO, NiO, 0·009 BaO, as well as traces of different elements.—II. From the Vega Expedition, after Lindström; [1] Mn_2O_3; [2] insoluble; [3] possibly bone material.

The manganese content varies considerably, namely, between 4·16 and 63·23 per cent of manganese oxide.

Several hypotheses have been put forward in explanation of this peculiar deposit. Buchanan [4] suggested a reduction by organic substance from manganese sulphate to manganese sulphide, which subsequently, and

[1] Ante, p. 974.   [2] Zeit. f. wissensch. Zoolog. XXVII. CIV.
[3] The sea between Japan and the Sandwich Islands.
[4] Loc. cit., 1890.

together with iron sulphide, became oxidized. This explanation however is little probable. Gümbel[1] explained the occurrence as having been formed by submarine manganiferous springs from which the manganese became precipitated upon contact with sea-water. This interpretation however does not explain the strikingly large extent of the deposit. According to Murray[2] the manganese, like the iron, was formed in a similar manner to the red clay, and like it therefore must be attributed to volcanic ash, etc., in part decomposed by the sea. It must be reflected however that though in the course of recent geological periods immense quantities of iron and manganese have been carried by the rivers to the ocean, these two heavy metals can now only be demonstrated to exist in traces in sea-water. From this fact it follows that the iron and manganese so brought to the ocean must in greater part have been precipitated, probably by slow oxidation due to the oxygen dissolved in the water. It may therefore be considered that the iron- and manganese oxides of the deposits upon ocean beds are in general derived in greater part from the sea-water itself, though an adsorption by descending volcanic ash, bone-fragments, etc., may also have been operative.

BEAN-ORE BEDS

LITERATURE

GRAV F. V. MANDELSLOH. Geognostische Profile der Schwäbischen Alb, 1834.—A. v. MORLOT. 'Über die geologischen Verhältnisse von Oberkrain,' Jahrb. d. k. k. geol. Reichsanst., 1850, p. 389.—DEFFNER. 'Zur Erklärung der Bohnerzgebilde,' Jahreshefte d. Ver. f. vaterl. Naturkde. in Württemb., 1859, XV. p. 257.—FRAAS. 'Die Bohnerze,' ibid. p. 38.—ACHENBACH. 'Über Bohnerze auf dem südwestlichen Plateau der Alpen, ibid. p. 103.—QUENSTEDT. Geologische Ausflüge in Schwaben, 1864, p. 136.—C. W. v. GÜMBEL. Geognostische Beschreibung der bayerischen Alpengebirges, Munich, 1861, p. 647.—TECKLENBURG. 'Über die Bohnerze in Rheinhessen,' B- H- u. S- Wesen, Vol. XXIX. pp. 210-217. Berlin, 1881.—DE GROSSOUVRE. 'Étude sur les gisements de minéral de fer du centre de la France,' Ann. des Mines (8), X., 1886.—H. B. v. FOULLON. Verhandl. d. k. k. geol. Reichsanst., 1887, No. 10, p. 219.—L. BUCHRUCKER. 'Die Montanindustrie im Grossherzogtum Baden,' Zeit. f. prakt. Geol., 1894, p. 169 ; 1895, p. 393 ; 1896, p. 6.—W. BRANKO. Die menschenähnlichen Zähne aus dem Bohnerz der Schwäbischen Alb. Stuttgart, 1897.—L. DE LAUNAY. Compte rendu du VIII. Congrès Géol. Intern., 1900, pp. 939-947.—L. ROLLIER. 'Über das Bohnerz und seine Entstehungsweise,' Vierteljahrsschr. d. naturf. Ges. Zürich, 1905, pp. 1-2.—v. DECHEN, revised by W. BRUHNS. Die nutzbaren Mineralien und Gebirgsarten im Deutschen Reiche. Berlin, 1906.—C. GEIGER. 'Die Eisenerzlagerstätten und die Eisenindustrie Württemburgs,' Stahl u. Eisen, 1907, p. 592.—C. SCHMIDT (Basel). 'Asphalt,' 'Steinsalz,' 'Erze' im Handwörterbuch der Schweizer Volkswirtschaft u.s.w. Vol. III. p. 113. Bern, 1907.—R. FLUHR. 'Die Eisenerzlagerstätten Württembergs und ihre volkswirtschaftliche Bedeutung,' Zeit. f. prakt. Geol., 1908, p. 1.—F. KLOCKMANN. 'Die eluvialen Brauneisenerze der nördlichen Fränkischen Alb bei Hollfeld in Bayern,' Stahl u. Eisen, 1908.—G. EINECKE and W. KÖHLER. 'Die Eisenerzvorräte des Deutschen Reiches,' geol. Landesanst. Berlin, 1910.—E. HOLZAPFEL.' Die Eisenerzvorkommen in der Fränkischen Alb,' Glückauf, 1910, Part 10. —L. DE LAUNAY. The Iron Ore Resources of the World, Stockholm, 1910, Vol. II.—

---

[1] Loc. cit., 1878.　　　　　[2] Loc. cit., 1876.

C. Schmidt. 'Bericht über die Eisenerzvorräte der Schweiz,' The Iron Ore Resources of the World. Stockholm, 1910. — Tecklenburg. Die Mineralschätze Württembergs. Stuttgart, 1912.

The name ' bean ore ' is applied to-day to occurrences of very varied genesis. Many bean ores are closely related to the present lake ore. On the other hand, most Tertiary bean ores may be compared with the recent Carlsbad ' pea rock,' not only in regard to structure but also in regard to the size of the grains ; the term ' pea ' has been applied to Alluvium ore, and ' bean ' to Tertiary ore. Occasionally also a certain similarity between bean ore and laterite is indisputable.

Most bean-ore deposits are limited to limestone districts, where spring waters probably ascended along fissures in the limestone and deposited oolitic iron where they emerged. The process was presumably similar to that which may still be observed in the case of the pea rock at Carlsbad, with this difference however, that at Carlsbad principally calcium carbonate in the form of aragonite is deposited, while with bean ore, owing to the large amount of iron in solution, chiefly iron compounds and especially ferric hydrate, were precipitated. At the same time the limestone country-rock was attacked by the processes of weathering, and apparently, since in the Eocene period the climate was considerably warmer than it is to-day, concentrations of iron similar to the laterite of the present-day tropics, became formed. This analogy is the more complete in that, just as with laterite, bauxite is frequently found in association with bean ore. With these processes of weathering metasomatic replacement of the limestone by iron oxides proceeded concomitantly. Residual clay became formed in which concretions of limonite found their seat. The metasomatic iron incrustations were again broken up by the action of meteoric water, and the pieces washed into the clay. Thus, concretions and fragments of limonite are found side by side in clay, which, when red, is known as bolus.[1]

Since limestone districts are usually traversed by a complex fissure- and channel system, detritus from these bean-ore beds, which are surface deposits, has sometimes been washed into channels, in which case bean ore occurs partly in lode-like form.

The above-described chemical-geological processes being independent of the geological age of the limestone, bean ore is found in the Jurassic, the Triassic, and the Cretaceous. In some cases it is concentrated so as to form thick beds, while in others it is limited exclusively to the filling of fissures and channels. In conformity with the genesis of these deposits, the ore in almost every case is impure and characterized by much insoluble residue. With many deposits therefore the question of pre-

[1] A coarse red pigment.

paration or dressing is of substantial importance, because, before smelting, the amount of this residue should be reduced as much as possible.

Seeing that the economic importance of bean ore is at present small, the following few descriptions of occurrences will probably suffice :

MARDORF.—North of Homberg in Niederhessen, geological position shown in Fig. 404. Diluvium and Tertiary formations here occupy shallow

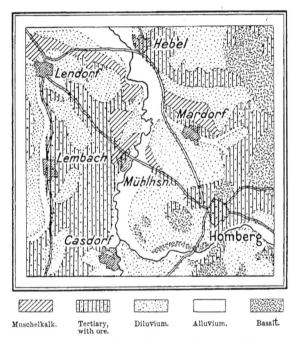

| Muschelkalk. | Tertiary, with ore. | Diluvium. | Alluvium. | Basalt. |

FIG. 404.—Iron bean-ore deposits associated with Tertiary sediments near Mardorf.
Scale 1 : 40,000. Dechen.

synclines in a Muschelkalk depression. The foot-wall of the ore-bed is formed by a milky-white, plastic, Pliocene clay containing numerous pea-sized limonite concretions often in contact with one another. In both hanging-wall and foot-wall a gradual merging into non-metalliferous clay may be observed. The bed itself strikes north-south, dips towards the west, and is 1–2·5 m. thick. The hanging-wall consists of a 3–10 m. thick, greasy, white clay ; quicksand ; and Diluvium. Lepsius regards the clay as the residue from dissolved limestone, to which limestone also he would refer the iron. In any case these are syngenetic deposits occurring in the form of beds, and presumably they are of considerable extent. According to an early analysis, the iron content of the unwashed ore is about 48·4 per cent, with 13 per cent of insoluble residue ; the phosphorus is low.

RHEINHESSEN.—In spite of their great extent these deposits of this

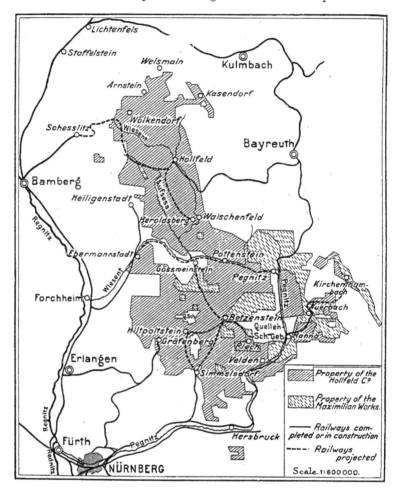

FIG. 405.—The Albian bean ore iron-field in the Franconian Alb.

ore are probably only payable in few places. They are found as primary deposits upon the Tertiary *Cerithium-*, *Corbicula-*, and *Hydrobia* lime-stones, and the *Cyrena* marls of the Mainz basin. Gravel-deposits from them are met in the valley alluvions and in the boulder accumulations, etc., which occur at the base of the loess, a calcareous surface clay typically developed on the Rhine. The deposits appear to be poor throughout and consequently have been but little investigated. The ore upon the limestone has been worked in places for a thickness of 0·3 to 2·5 m. ; the Alluvium deposits are much ·thinner. The ore exhibits no concretionary structure, but occurs rather as aggregates of large and small nodules cemented together by a sandy iron ore. The beds on the Wiesberg plateau near Spredlingen, on the Westerberg near Appenheim, and on the Kloppberg, are relatively the most important, these being up to 2·4 m. in thickness. The ore always occurs in association with variegated bolus-like clays frequently traversed by nodular layers of limestone and marl. Since these limestones contain 0·3–2·5 per cent of iron, they may have been the source of the iron in these beds.

HOLLFELD IN THE FRANCONIAN ALB.—This district to some extent

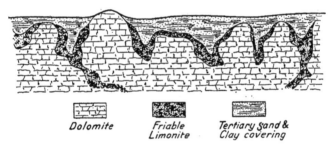

Dolomite   Friable   Tertiary Sand &
                Limonite   Clay covering

FIG. 406.—Diagrammatic section of limonite deposits in cavities of various size in the Franconian Alb ; length 50 metres. Klockmann.

was known in former centuries, after which it lay forgotten. The deposits of the so-called Albian covering extend over the northern Franconian Alb to the western bend of that range near Regensburg. This district, delineated in Fig. 405, consists of Jurassic beds which slope gently towards the north-east and are cut by deep valleys. The Albian formation, which is characteristic only of the high plateau and consists of Diluvial loam, clay, and sand, covers the deposits. Residual clay such as remained after the decomposition of the limestone and was washed into local pockets, fissures, and channels, is conspicuous ; it is mixed with sand and limestone fragments. The Albian covering is the result of deformation of Jurassic and Cretaceous beds during the Tertiary and Quaternary periods. Its thickness is usually only 1–5 m., but may be as much as 20 metres. In its immediate foot-wall in the neighbourhood of Hollfeld near Regensburg the iron deposits occur. They are absent from the dolomite ridges and crests, having been removed from thence by erosion.

Between the foot-wall dolomite and the ore, finely-crystalline dolomite, sand, and red clayey masses, are frequently found. The geological position of the deposits is well illustrated in Figs. 406, 407, and 408.

The shape of the under surface of the ore-body depends upon the irregular surface of the dolomite. The ore follows the cavities and undulations of this surface in a more or less thick layer, the filling of these irregularities being completed by residual clay and Albian material. Numerous planes in the clay of these funicular ore-bodies point not so much to tectonic movement as to local action due to varying plasticity. The deposit consists chiefly of loose limonite and pieces of compact ore as large as a man's head, this ore being flaky and frequently exhibiting a reniform

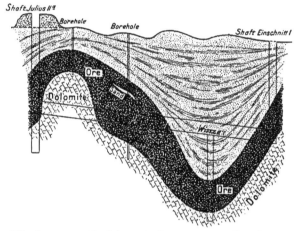

Fig. 407.—Large ore-syncline between the Julius and Einschnitt shafts at Hollfeld.
Scale 1 : 400.   Einecke and Köhler.

structure. A large number of angular ore-fragments varying from pea- to nut size occur more or less regularly distributed throughout the deposit, though such are also frequently concentrated at its base.

The source of the iron is attributed by Köhler and Einecke to waters circulating through fissures. These mineral solutions, however, in all probability metasomatically replaced the limestone in part only, the greater part of the iron having been precipitated at the surface, by the action of meteoric waters. In this action, weathering during the Tertiary period and, in places, undoubted iron enrichment during the Eocene period, played parts. The subsequent sinking of the ore into cavities in the limestone and the concomitant formation of numerous movement planes must be referred to the later cavity formation.

The average composition of the ore based upon hundreds of analyses
gives 38 per cent of iron, though the content varies from 28 to 54·46 per
cent.   The manganese content is from 0·26 to 8·07 per cent, that of alumina
1·05 to 9·31 per cent, and that of phosphorus 0·23 to 0·4 per cent.   Across
a section of the ore-body, the loose clayey material occurring chiefly in the
upper portion contains 30–32 per cent of iron, the granular clayey material
38–40 per cent, and the compact ore up to 50 per cent or more.   The
insoluble residue in the ore amounts to about 20 per cent.

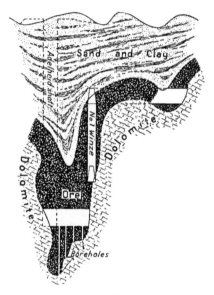

Fig. 408.—Funnel-shaped iron deposit at Hollfeld due to gradual deepening of the cavity.
Scale 1 : 400.   Einecke and Kohler.

The proportions of these differently aggregated varieties is expressed
in the fact that the average ore consists of 30–35 per cent of compact ore,
17–20 per cent of fine ore, 40–50 per cent of powdered ore, and contains on
an average 40 per cent of iron.   Generally speaking, however, these proportions are very irregularly distributed, patches of compact ore occurring
alongside others of powdered ore, while a mixture of both in most varied
proportions may occur.

The use of this ore for industrial purposes is limited by its silica content
of 20 per cent or more, though much of this may be removed by dressing.
The reserves are considerable, having been estimated by Köhler and
Einecke at 165,000,000 tons.

SOUTHERN SLOPE OF THE SWABIAN ALB.—These Würtemberg iron deposits occur in Tertiary beds consisting of conglomerates of Jurassic material, limestone, clay, and sandstone. The conglomerate is frequently represented by bean ore, the so-called clay ore, such occurring either in seams or filling depressions. At subsequent destruction of these seams fragments of ore resulted, which together with clay were washed into cavities in the Jurassic limestone, there occasionally forming payable deposits—the so-called rock ore. Such deposits occur up to 8 m. in thickness. Fluhr gives their iron content as 31·82 per cent, with 29 per cent of insoluble residue. At present they are not worked, though according to Köhler and Einecke the reserves are considerable.

In the case of the bean ore occurring in the Eocene of southern Baden the same two kinds are differentiated, namely, clay ore which occurs very extensively in clays, and rock ore which fills cavities in the limestone. Strictly speaking, only the first of these, occurring principally in the Klettgau and in the lake district of Baden, may be classed as bean ore. The deposits of this ore have Jurassic for their foot-wall, and are always associated with red clays which frequently contain patches of sand. The ore consists of greenish-grey to light-brown nodules of flaky structure and varying from pea- to nut size. It is washed before smelting, containing then some 50 per cent of iron, 17 per cent of insoluble residue, 0·6 per cent of lime, and 9 per cent of water. The phosphorus content is mostly low. Up to the year 1860 this ore was mined and smelted in small charcoal furnaces, as for instance at Mülheim and Oberweiler. The economic importance of these deposits appears to be small.

SWITZERLAND.—In this country also bean ore is an early Tertiary formation, lying upon calcareous Mesozoic beds and associated with bolus, *Huppererde*, phosphorite, etc. Since in this case also funnel-shaped cavities in the limestone are frequent, the thickness of the ore-bed varies considerably. The extent of this ore in Switzerland has been greatly disturbed by younger orogenics and erosion. The metalliferous formation consists principally of red ferruginous clay or bolus, which may reach 100 m. in thickness, all gradations existing from rich plastic material to sterile sandy clay. This clay is traversed by patches or bands of fine quartz-sand. Where the quartz increases simultaneously with diminution of the clay, a sandy facies known as *Huppererde* is formed. In the bolus the alumina presumably is partly contained in the form of bauxite.

Where the complete metalliferous formation is present the upper part appears to be poor in limonite. The beds are characterized by boulders of limestone frequently corroded and in part silicified. Fibrous gypsum also occurs. In depth the clay gradually becomes brownish-red to scarlet, and grains of limonite begin to occur, these finally forming a true limonite bed up to 1 m. in thickness. This bed often lies immediately upon the Jurassic limestone, occupying its uneven surface. Since ore and gangue replace each other along the strike, the ore-bodies have a columnar or lenticular form in plan, while in section ore is frequently absent.

The bean ore contains concretions varying in size from a pea to a hazel-nut ; they are partly compact and sometimes have a concretionary structure, though according to Schmidt typical pisolitic structure is exceptional. Concretions with a hard ferruginous crust and a soft argillaceous dark-yellow kernel, are frequent. In order to separate the limonite from the bolus the deposit is washed. About 50 per cent of ore containing approximately 42 per cent of iron is obtained in this manner, while the remaining 50 per cent of bolus contains only 2–8 per cent of iron. The insoluble residue in the ore before it is washed is always considerable, generally exceeding 20 per cent. The P_2O_5 content is 0·19–0·22 per cent.

Mining operations upon these deposits are being undertaken in many places. The industry began in the seventeenth century in the Schaff-hausen Canton, where in the year 1850 about 1000 tons of raw ore were produced. The best deposits are found in the Jurassic of the Berne and Solothurn Cantons, Birstal being the principal district. There the bean ore is found upon the upper beds of the White Jurassic. From 1854 to 1904 about 765,000 tons of ore were won, equivalent to 320,000 tons of pig iron. Owing to the great variations in thickness it is difficult to make any reliable estimate of the ore-reserves.

These Swiss deposits also are presumably a result of tropical weathering early in the Tertiary period, such deposits having subsequently suffered both mechanical and chemical deformation during more recent orogenics.

L. de Launay [1] describes a very extensive area of Tertiary bean ore at Berry in France.

The Limonite Deposits at Vogelsberg

LITERATURE

H. Tasche. ' Toniger Brauneisenstein, dessen vormalige und jetzige Gewinnung im Vogelsberge,' Neues Jahrb. f. Min. Geol. Pal., 1852, p. 897.—G. Würtenberger. ' Die diluvialen Eisensteine im Regierungsbezirk Kassel, verglichen mit den Basalteisensteinen des Vogelsberges,' ibid., 1867, p. 685 ; ' Eisenerzbergbau am Vogelsberg,' Glückauf, 1891, p. 45.—F. Beyschlag. ' Die Eisenerze des Vogelsberges,' Zeit. f. prakt. Geol., 1897, pp. 337-338.—H. Münster. ' Entstehung der Vogelsberger Eisenerze,' Zeit. f. prakt. Geol., 1905, XIII. p. 413.

The limonite deposits at Vogelsberg are of somewhat different origin from bean ore. These deposits, which are genetically and spacially most intimately associated with basalt, have in the Vogelsberg district a very

[1] Loc. cit., 1900.

considerable extent. All of them lie near the surface, in which position they are usually overlaid by shallow Diluvial and Alluvial beds. Only exceptionally do several beds occur one above the other, and when that is the case such are separated from one another by layers of tuff poor in iron. Two types of deposit are known, the relatively primary deposit, and the secondary deposit which has resulted from the disintegration and reconcentration of the primary deposit. Since the Vogelsberg deposit, a section of which is illustrated in Fig. 409, has entirely resulted from the weathering and decomposition of the basalt and its tuffs, in the case of the primary

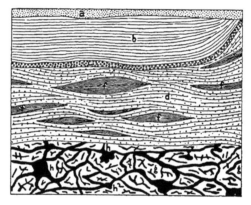

FIG. 409.—Section of the limonite deposit at the Ernestine mine near Niederohmen not far from Giessen in the Vogelsberg. *Zeit. f. prakt. Geol.*, 1887.

Alluvium   (a) earth, soil.
Diluvium   (b) different coloured loams.
            (c) river rubble, gravel and bauxite pebbles.
            (d) clayey impure limonite.
            (e) poor clayey patches in d } Secondary deposit.
Tertiary   (f) basalt tuff.
            (h) limonite veins, and crusts } Primary deposit. coloured dark by manganese.

deposits the much deformed basalt exhibits a covering of ferric hydroxide, while the centre is clayey and decomposed. The individual ore-bodies are separated from each other by clay.

The river-concentrated deposits are found in terraces along the present river valleys. They consist of ferruginous, argillaceous masses with a few boulders and numerous limonite concretions in the form of lumps, nodules, and geodes, varying in size from a pea to a man's head. The deposits display distinct fluviatile structure. The upper and lower surfaces are not as a rule parallel planes, the ore-body often sinking far into the basalt along channels and shallow basin-like depressions. The upper surface also is irregularly undulating, and characterized by the frequent occurrence of bauxite boulders. The secondary Diluvial beds hardly ever come to the surface, but remain covered with Diluvial gravel and marshy loam.

The best deposits are found in the neighbourhood of Mücke, Lich, Grünberg, Laubach, and Niederohmen, where they have a thickness of 5-20 metres. These deposits chiefly produce washing ore. In the foot-wall basalt of many mines, fragmentary ore occurs in flat cakes of limited horizontal extent and of a thickness which may at times be as much as 1 metre. The occurrences at Mücke and Niederohmen in the Vogelsberg probably lie in a north-south subsidence, along which the streams Seen-bach and Ohm have their course.

In relation to genesis two opposite theories have been advanced, namely, deposition as the result of weathering, and deposition from springs. The former attributes the ore to a process of weathering having much in common with the lateritization now proceeding in the tropics and sub-tropics. The second theory, which of late years has gained ground, assumes on the other hand that the decomposition of the basalt was brought about by springs. This theory is supported by the occurrence of recent chalybite springs. Whether however these springs carry iron of them-selves, or whether they have extracted their iron entirely from the basalt, cannot be said. If the Vogelsberg ore were formed by weathering in the Tertiary period—which is very probable—it is then very closely associated with bean ore.

The raw ore is freed from clay in washing trommels, and then sorted on tables. By this means 30 per cent of the material hoisted is obtained as ore containing on an average 45 per cent of iron. The phosphorus content is 0·2 per cent; the manganese 0·8-1·2 per cent. In the case of particularly good deposits the iron content of the sorted ore reaches 50 per cent.

OOLITIC IRON BEDS

The oolitic iron ores consist in greater part of concretionary nodules bound together by a cementing material. Under the microscope these nodules are frequently seen to contain in the centre a mineral particle, such as a grain of quartz or a small crystal of calcite or siderite. The formation of these nodules may be explained, in that the mineral particles now found in the centre were formerly loose in the mineral solution from which by deposition they became incrusted with iron. This process may be still observed in operation in the formation of the Carlsbad *Sprudelstein*.

The layers which make up these nodules consist in the case of iron

deposits, generally of brown or red limonite or of hæmatite, but sometimes of iron silicate. With limonite it is not always easy to determine
whether the ore is primary or an oxidation product of siderite or of the
silicate.

The cementing material of these oolitic ores likewise varies greatly.
It may consist of calcite, as in the case of calcareous minette ; of quartz
material, as with siliceous minette ; or of ore. Only very exceptionally
is it of clayey material. Phosphate of lime frequently shares in the
composition of the nodules as well as in that of the cementing material.
On this account therefore, before the discovery of the Thomas basic
process it was not possible to make use of the phosphoric minette.

The geological age of the oolitic iron deposits is very variable. Since
they are distinguished from other iron beds solely by their structure, like
these, they may belong to all formations. They are particularly extensive in the Silurian of the United States, where the Clinton hæmatite
deposit extends over whole government districts. Of late years also the
occurrence of these ores in the Coal-measures of Germany and England has
attracted the attention of some geologists, and though these occurrences are
at present of no economic importance they are nevertheless of significance
in relation to the genesis of oolitic iron ore-beds. The dull, grey, fine-
grained deposits are intercalated in shales which also include marine beds,
and, accordingly, fluviatile and marine beds exist here side by side. It
would appear therefore that with these Carboniferous oolites an origin in
great sea depth were impossible. The Jurassic formation—the middle
horizon of which, the Brown Jurassic, owes its name to its high iron content
—is particularly rich in such oolitic deposits. The minette occurrences in
German Lorraine, Luxemburg, and Meurthe-et-Moselle in France, are
situated, for instance, at the base of the Middle Jurassic. The Cleveland
ore in England is Middle Lias, the same horizon as that to which the minette
of the lower Rhine around Xanten and Bislich, belongs. Among the
Jurassic deposits, the minette of German Lorraine, etc. and the oolite of
Cleveland are characterized by particularly large extent.

In the Cretaceous no oolitic iron ore has yet been recorded, though
such ore frequently occurs in the Tertiary. In a shallow bore-hole at
Winterswyk on Dutch territory near the German frontier, a red oolitic ore
consisting of hæmatite nodules cemented together by a clayey material,
was found at the base of the Tertiary, presumably in the Eocene.
Further efforts to find the continuation of this ore-bed were unfortunately
unsuccessful. The term minette was also applied to this ore.

The Tertiary oolitic bean-ore deposits, the nodules of which, unlike
those of minette which are fine-grained throughout, may reach relatively
large dimensions—pea- and bean size—have already been described. They

occur in part in funnel-shaped cavities upon limestone plateaus of the most varied geological age, and are mostly associated with clay and fragments of older ore-beds. With them also a fast cemented matrix is quite exceptional.

In the Quartenary period oolitic iron ore is replaced by lake ore, which likewise consists principally of nodules.[1] The nomenclature of the oolitic ores is therefore not simple. The term minette is used for one large group of ore-beds, while the ore of another group is known as bean ore. The following nomenclature is suggested :

1. Minette, a more or less fine-grained oolitic iron ore, the matrix of which may consist of different materials. It is further advisable also to limit this term to Jurassic ores.

2. Bean ore, a loose oolitic iron ore, the grains or nodules of which have mostly a larger diameter. Such ore is frequently associated with clay and ore-fragments and is generally of Tertiary age.

3. There remains then a number of ore-beds which in regard to structure are very similar to minette, but which as a rule possess a greater geological age.

ECONOMIC IMPORTANCE.—The oolitic iron ores have not in general a high metal content. In the majority of cases the average content varies between 28 and 35 per cent. The Clinton ore with 48 per cent is an exception. In spite of this generally-speaking low metal content the oolitic ores, in consequence of their great extent, are exceedingly important. The total reserves of the Lorraine minette have been reckoned at about 2,000,000,000 tons of ore, equivalent to 700,000,000 tons of iron. In the year 1910 out of 22,900,000 tons, the total German production, this minette contributed no less than 16,600,000 tons.[2]

With regard to genesis, in the first place it must be pointed out that in relation to structure the oolitic iron ores exhibit a great similarity, frequently even identity, to recent lake ores. Both varieties of ore consist in greater part of concretionary nodules, the difference in many cases lying only in the cementing material, which with the oolitic ores is as a rule hard, and with the lake ores, soft. Speaking generally, the majority of metalliferous oolites of earlier geological periods have a genesis very similar to that of the lake ores.

Additional evidence of a sedimentary origin exists in the fact that the iron oolites often form fairly regular beds of strikingly large extent, and that frequently within one and the same district several ore-beds follow one above the other, each bed over a large area displaying a constant or almost constant composition. We would also point out that metalliferous oolites in some districts are found with almost undisturbed bedding.

[1] *Ante*, pp. 982-988.   [2] *Ante*, p. 1001.

Against the theory of sedimentation the objection has occasionally been raised that so large a quantity of iron, as for instance in the seam-like Jurassic deposits, cannot be explained by increment from surface solutions. Since however the recent lake ores demonstrate the transport of iron in large amount by surface solutions, this objection appears to lack force.

According to the view defended by some authorities, the iron oolites, or at all events some of them, do not represent primary sediments but metasomatic replacements of calcareous oolites, by percolating iron solutions. In the case of individual occurrences the possibility of such an origin cannot be denied. For many districts, however, the primary origin is as good as proved. This question is dealt with more fully when describing the minette of Lorraine-Luxemburg. The formation of bean ore by lateritization, etc., is likewise discussed later.

THE MINETTE DISTRICT OF LORRAINE-LUXEMBURG AND THE NEIGHBOURING FRENCH DEPARTMENT OF MEURTHE-ET-MOSELLE

LITERATURE

L. VAN WERVEKE. ' Bemerkungen über die Zusammensetzung und Entstehung der lothringisch-luxemburgischen oolithischen Eisenerze ' (Minetten), Ber. über die 34. Vers. d. oberrhein. geol. Ver. Diedenhofen, 1887, p. 19.—WANDESLEBEN. ' Das Vorkommen der oolithischen Eisenerze (Minette) in Lothringen, Luxemburg und dem östlichen Frankreich u.s.w.,' Stahl u. Eisen, 1890, p. 677.—L. VAN WERVEKE. ' Magneteisen in Minetten,' Zeit. f. prakt. Geol., 1895, p. 497.—L. HOFFMANN. ' Magneteisen in Minetten,' Zeit. f. prakt. Geol., 1896, p. 68.—E. SCHRÖDTER. ' Über die Eisensteinvorkommen in Lothringen,' Zeit. f. prakt. Geol., 1897, p. 295.—E. W. BENECKE. ' Beitrag zur Kenntnis des Jura in Deutsch-Lothringen,' Abh. Geol. Landesanst. Alsace-Lorraine, N.F. Part 1, 1898.—FR. GREVEN. ' Das Vorkommen des oolithischen Eisenerzes im südlichen Teile Deutsch-Lothringens,' Stahl u. Eisen, 1898, p. 1.—W. KOHLMANN. ' Die Minette-formation Deutsch-Lothringens nördlich der Fentsch,' Stahl u. Eisen, 1898, p. 593.—W. ALBRECHT. ' Die Minetteablagerung Deutsch-Lothringens nordwestlich der Verschiebung von Deutsch-Oth,' Stahl u. Eisen, 1899.—O. LANG. ' Die Bildung der oolithischen Eisenerze Lothringens,' Stahl u. Eisen, 1899, p. 714.—General map of the Iron Ore Fields of West-German Lorraine, 1 : 80,000, 3rd Ed., Strassburg, 1905; Geol. Landesanst., Alsace-Lorraine.—H. ANSEL. 'Die oolithische Eisenerzformation Deutsch-Lothringens,' Zeit. f. prakt. Geol., 1901, p. 81.—L. BLUM. ' Zur Genesis der lothringisch-luxemburgischen Minette,' Stahl u. Eisen, 1901, XXI. p. 1285.—L. VAN WERVEKE. ' Bemerkungen über die Zusammensetzung und Entstehung der lothringisch-luxemburgischen oolithischen Eisenerze,' Zeit. f. prakt. Geol., 1901, p. 396.— E. W. BENECKE. ' Gliederung der Eisenerzformation in Deutsch-Lothringen,' Mitt. d. geol. Landesanst., Alsace-Lorraine, 1902, Vol. V. p. 139.—L. HOFFMANN. ' Das Vorkommen - der oolithischen Eisenerze (Minette) in Luxemburg und Lothringen,' Neues Jahrb. f. Min., 1902, II. p. 88; Stahl u. Eisen, 1896, No. 23, 24, Glückauf, 1899, p. 640.—W. KOHLMANN. ' Die Minettenablagerungen des lothringischen Jura,' Stahl u. Eisen, 1902, pp. 493-503, 554-570, 1273-1287, 1340-1351.—KOHLMANN. ' Über das deutsch-französisch-luxemburgische Minettevorkommen nach den neueren Aufschlüssen,' Zeit. Ver. Deutsch. Ingen., 1902, 46, p. 358; Chem. Ztg., 1902, 26, p. 218.—L. VAN WERVEKE. ' Über das Vorkommen, die mineralogische Zusammensetzung und die Entstehung der deutsch-lothringischen und Luxemburger Eisenerzlager,' Bull. mens. Luxemburg, 1902.—O. LANG. ' Das lothringische Eisenerzlager,' Glückauf, 1903, p. 649, 687.—KRELL. Übersicht über die Eisenindustrie in Lothringen und Luxemburg, sowie im angrenzenden Longwyer und Nancyer Erzbecken,

zum 9. Bergmannstage überreicht, 1904.—BENECKE. 'Die Versteinerungen der Eisenerzformation in Deutsch-Lothringen und Luxemburg,' Abh. zur geol. Spezialkarte von Elsass-
Lothringen, N.F., 1905, Part 6.—G. EINECKE and W. KÖHLER. 'Die Eisenerzvorräte
des Deutschen Reiches,' Pr. geol. Landesanst. Berlin, 1910.—G. ROLLAND. 'Sur les
gisements de minerais de fer ool. du nouveau bassin de Briey,' Comptes Rendus Ac. Sc.,
17 jan. 1898.—A. PIRARD. 'Sur la partie nord du bassin minier lorrain-luxembourgeois,'
Rev. univers. des mines, Vol. LV., 1901.—F. SCHMIDT. 'Le Gis. des mines de fer du bassin
de Briey et de la Lorraine allem.,' Rev. univers. d. Mines, Vol. LV.—LAUR. Et. du bass.
fer de Briey.—VILLAIN. 'Le Gis. de min. de fer ool. de la Lorraine,' Ann. d. Mines févr. et
mars, 1900, and Comptes Rendus, 1901.—P. NICOU. 'Le Bassin ferrifère de Briey,' La
Nature, 28, XII., 1907.—'Les Ressources de la France en minerais de fer,' The Iron Ore
Resources of the World, XI., Internat. Geol. Congress. Stockholm, 1910.

The tableland of Lorraine is bounded by the Ardennes, the Eifel, the
Hunsrück, and the Vosges mountains, which together form a large semi-
circle open towards the west. To the western portion of this tableland
the Briey plateau belongs, this extending from the southern border of the
Ardennes to the south of Metz, as illustrated in Figs. 17, 158, and 410.
In the structure of this tableland the minette beds take part, these beds
occupying a belt of country in the east, 18-20 km. wide and 55-60 km.
long from north to south. How far this occurrence extends to the west
beyond the width already known, remains for subsequent developments
to determine. A small minette district, beginning south and extending
20 km. north of Nancy, occurs along the southern continuation of this
belt, being separated from the principal district by an unpayable portion
25 km. in extent.

The principal district covers a narrow strip of German territory to the
east, a wide strip of French territory to the west, and finally a border zone
in Luxemburg to the north. Important industrial towns within it are,
Gross Moyeuvre, Fentsch, Algringen, Aumetz-Friede, Düdelingen, Rüme-
lingen, German Oth, Differdingen, Redingen, Hussigny, Longwy, Briey,
and Nancy. The geological structure of the district is exhibited most
clearly in the deep erosion valleys which cut the Briey plateau, among
these being the valleys of the Mosel, Orne, Fentsch, Algringen, Meurthe, etc.

The deposits are limited to the Upper Lias and the Dogger. In its
lower portion the Upper Lias consists of thin bituminous beds, while in the
upper portion black marls predominate. Above these comes the metalliferous formation, which palæontologically cannot be sharply separated
from the Lias, at least, German and French geologists have hitherto not
been able to come to an agreement concerning this very important boundary.
Van Werveke correlates the ore-bearing beds with the Lower Dogger, the
horizon of the *Trigonia navis* and *Ammonites Murchisoni*. French authorities, on the other hand, include the minette beds with the Lias, so that
according to their correlation the Dogger first begins above the minette.

In the hanging-wall of the ore a 20-30 m. thick impervious marl-slate
complex of dark blue colour and known by the miner as the ' hanging-wall

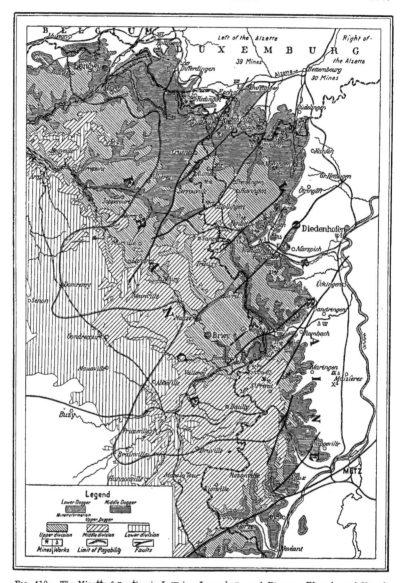

FIG. 410.—The Minette formation in Lorraine, Luxemburg, and France. Werveke and Krusch.

marl,' occurs. Van Werveke correlates this with the Charennes beds, the higher portion of which consists of alternating beds of limestone and marl.

Although all the beds of the Lower Dogger are ferruginous, the individual minette beds have sharp boundaries against the hanging-wall and foot-wall, so that they may rightly be regarded as independent seams or beds, distinct from the country-rock not only by their greater iron content but also by their definite walls. The minette beds and the country-rock together are known in Germany as the ' Minetteformation,' and in France as the ' Formation ferrugineuse.'

To the south of the Briey plateau and along its eastern border this formation has a thickness of 15–20 m., a thickness which increases considerably going north and west, till at Esch and Bollingen it reaches a maximum of 60 metres.

Smaller, more or less sharply defined areas within the whole district have special names. Thus, in Luxemburg the Esch - Dudelange basin around Esch, Rüdelingen, and Düdelingen, and the Belvaux-Lamadelaine basin around Belvaux - Lamadelaine, are distinguished. In German Luxemburg the Fentsch and the Orne divide the district into three parts, namely, the Aumetz-Arsweiler tableland north of the Fentsch, the country between Fentsch and Orne, and

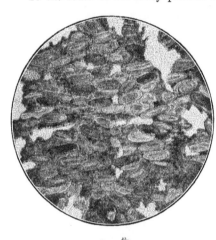

Quartz. Lime-
stone.

Fig. 411.—Microscopic picture of a red sandy minette from the Moyeuvre mine in Lorraine. Scale 1 : 50.

finally, the district south of the Orne. In France the boundaries of payability are made use of to distinguish three areas, the Longwy basin to the north, the Central basin in the centre, and the Orne basin to the south.

The different minette beds, separated from each other by blue and grey marls, are distinguished and known by their respective red, brown, black, yellow, and grey colours. All of them are oolitic. The individual grains or oolites are on an average 0·25 mm. diam. and are cemented sometimes in a calcareous, sometimes in a clayey or siliceous matrix. They are round, flat, or irregular in shape; when one axis is longer than the other the longer axis lies in the plane of the bedding. The iron content is contained essentially in the oolites and consequently is roughly proportional to them. The mineralogical composition of these oolites has not yet been completely determined. According to investigation by Blum, they consist not only of limonite but also of ferrous carbonate, ferroso-ferric oxide, and ferrous silicates. Kohlmann and van Werveke consider that, in addition, they contain iron-aluminium- and iron-magnesium-aluminium silicates, that is to say, compounds similar to thuringite and chamosite. The ferric hydrates are found more particularly in the neighbourhood of the outcrop, and in the vicinity of faults. Other iron compounds also occur in large amount, ferric oxide being found more particularly in the upper levels and ferrous oxide in the lower. Whether, as Kohlmann assumes, this constitutes a case of primary depth-zones, or whether secondary processes are responsible, has not yet been determined.

Within the ore-bed, limestone bands made up of nodules, patches or irregular elongated masses, frequently occur, occupying sometimes as much as two-thirds of the thickness of the bed. The passage between these limestone bands and ore is gradual. Not infrequently marls occur in thin layers alternating with ore. While the subsequent separation of the ore from such limestone is easy, the marl, being inclined to crumble, frequently contaminates the ore. Around the margins the minette beds gradually pass into limestone poor in iron, marl, and sandy marl. Although the distribution of the different minette beds is not regular, the boundaries of the richer beds coincide approximately with those of the payable areas. Within these areas the petrographical characters of the ore-bed differ substantially from those of the intercalated waste material. This material, however, may merge into ore, while, on the other hand, ore may merge into waste. The calcareous and siliceous bands characteristic of individual beds are subject to considerable variation within one and the same bed.

Kohlmann formulated the following sequence of minette beds :

Hanging-wall bed	Hanging-wall marl. Red sandy bed.
Middle calcareous group	Red calcareous bed. Yellow bed. Grey bed.
Lower siliceous group	Brown bed. Black bed. Green bed. Foot-wall marl.

In the yellow and red calcareous bed van Werveke differentiates several ore-bearing horizons. The three lowest beds, the green, the black, and the brown, are characterized by the amount of silica they contain, and accordingly are classed by him as the lower siliceous group, this being separated from the hanging-wall red sandy bed by the middle calcareous group.

The thickness of the separate beds may reach 7 metres. The most important is the grey bed, which in regard to composition, thickness, and extent, is the most constant. This bed appears to be well developed throughout the Luxemburg and German areas. The ore is mostly grey, but contains also reddish, brownish, and greenish patches. The maximum thickness of 7 m. is not always payable throughout. It contains on an average 28–40 per cent Fe, 10–15 per cent CaO, and 5–10 per cent SiO_2. Since the ore-beds may become impoverished and the intercalated material may become enriched, there are in the minette formation considerable quantities of ferruginous flux in addition to ore.

The tectonics of the district are simple. The beds dip mostly 3°, rarely as much as 7°, towards the west, as illustrated in Fig. 17. In many districts a gentle folding may be observed, the anticlinal and synclinal axes of which strike north-east. While at the eastern limit a dip towards the west occurs, to the north the beds are inclined south-east to south. While again in the north-west they occur at the surface with almost horizontal bedding, in the south-west they lie relatively deep. The regular bedding is disturbed by numerous faults which in one and the same bed may suddenly produce considerable variations in depth. These faults strike mostly north-east, that is, parallel to the shallow anticlines and synclines ; usually they dip 60°–70°, sometimes north-west and sometimes south-east.

Going from north-west to south-east the principal faults are those of Gorcy, Saulnes, Differdingen-Godbrange, German-Oth, the Middle Fault, and those of Oettingen, Fentsch, Avril, Hayingen, Gross Moyeuvre, Rombach, Verneville-Flavigny, Gravelotte, and Metz. The throw of these faults varies from a few metres to 150 metres. They frequently form uplifts and subsidences, while step-faulting is particularly noticeable. These faults in consequence of their usually small throw and the great distance between them, have no great effect upon mining operations. They produce their worst effect in disturbing the impervious marl- and clay beds in the hanging-wall.

The genesis of the minette deposits has not yet been fully agreed. In general two views are advanced. According to the one they were formed syngenetically by precipitation from sea-water, while according to the other, ferruginous solutions gradually effected the metasomatic replacement of pre-existing lime oolites. Against this latter theory however, no channels along which these solutions could have come and which should be found in the foot-wall of the deposit, are known, nor in the course of mining operations has any support for this view been found. The numerous faults cannot be taken to represent such channels, since the ore-beds in the

COMPOSITION OF THE GERMAN LORRAINE MINETTE[1]

		Thickness.	Fe.	CaO.	SiO₂.	Al₂O₃.
North-west portion between German-Oth and Bollingen.	Grey bed at the St. Michel and Red Rock mines .	3–5	29–33	15–17	5–10	...
	Brown bed at the St. Michel and Red Rock mines . .	3–3·5	38–39	4–9	3–13	...
	Brown bed proved by Boring in the August claims	3·5	39	5	16	7
South-east of Bollingen.	Lower red calcareous bed	...	35	12	7	...
	Yellow bed . . .	3–4	24	13	24	6
	Grey bed	3–4	33	18	10	4
North-east portion of the plateau of Aumetz-Arsweiler.	Red sandy bed at the Oettingen mine . .	...	36	2–3	26–27	...
	Lower red calcareous bed at the Oettingen mine	...	34	15	8	...
	Yellow bed at the Oettingen mine . . .	...	33–36	10–12	9–10	...
	Grey bed	3	32	15	9	...
Algringen Valley.	Grey bed	2–3·5	31	18	8	2·3
	Black bed at the Friede mine . . .	...	36–37	6–7	14–15	...
	Grey bed average . .	2–4	28–40	10–15	5–10	...
Neuchef Mine between Feutsch and Orne.	Red calcareous bed .	1·20	40	9·5	9·5	5
	Grey bed	4·20	40	7	9	...
	Black bed . . .	2·20	30	6	24·5	6
Hayingen Mine.	Grey bed . . .	2–4	30–32	12–15	6–7	4
St. Paul Mine south of the Orne.	Yellow bed . . .	2·20–2·50	36	12	8·4	...
	Grey bed . . .	2·60–3·80	37·4	9·2	6·8	...
	Brown bed . .	2·00	34·3	8·6	16·6	7
	Black bed . . .	1·50–1·80	30	6	24·5	8
Maringen Mine.	Black bed . . .	1·50–9·5	34	9–10	12–14	...

[1] After Klockmann, loc. cit.

neighbourhood have been found by experience to be poorer in iron than at a greater distance. Further, adjacent beds of non-oolitic and less resistant limestone exhibit no alteration to ore ; yet did such channels exist, these beds, as much as the minette beds, should be traversed by them. These deposits might of course have been formed by descending ferruginous solutions, but against this, among other things, is the fact that the strata in the hanging-wall are not characterized by any pronounced iron content. We regard the beds as true sediments.

In consequence of their favourable situation, their regular composition, and their texture, the economic importance of these minette beds is immense, though minette in itself represents the lowest grade of all payable iron ores. From an average iron content of 33–35 per cent, approximately 26–28 per cent is necessary to cover the costs of production.

The composition of the minette deposits may be gathered from the table on p. 1009, which the following average figures, taken from the text to the geological map of western German Lorraine, serve to complete.

	Fe_2O_3.	Fe.	CaO.	MgO.	P_2O_5.	SiO_2.	Al_2O_3.	SO_3.	CO_2.	H_2O.
Red sandy bed	44·5	31·2	5·3	0·5	1·6	33·6	4·2	0·1	4·1	6·6
Red bed . .	60·6	42·4	6·2	0·5	1·8	9·9	5·5	0·1	4·9	10·1
Grey bed . . .	45·5	31·9	19·0	0·5	1·7	7·9	2·3	0·1	14·3	8·0
Black bed .	57·5	39·9	5·9	0·5	1·7	15·1	5·2	...	4·5	9·3

The red calcareous bed, containing some 40 per cent ore, is the richest ; it is however in greater part exhausted. In the case of several beds, with an increase in the iron there is at the same time an increase in the silica, and the cost for flux rises. The grey bed however has a favourable composition ; it produces ore which can either be smelted by itself or with the assistance of a small amount of flux from adjacent beds. The phosphorus throughout is sufficient to allow the application of the Thomas process ; the manganese content on the other hand must be artificially increased, 2s. 6d.–3s. per ton being reckoned as the cost of the pig iron necessary as manganiferous flux. The porosity of the ore allows it to be easily crushed and smelted, while the lime content in many of the beds still further facilitates smelting. In spite of these generally speaking favourable conditions, the limit of payability is capable of further reduction by saving in freight and cost of coke, etc., and it is hoped in future to be able to work at a profit calcareous minette containing but 25 per cent of iron.

The following table gives the production of the Lorraine minette district for the years 1901–1911 :

PRODUCTION AND EXPORT OF THE LORRAINE MINETTE

| Year. | Iron Ore Output. | Export. | | Proportion of output accounted for by area within German Customs Union. |
		To Germany : Saar, Westphalia, etc.	To Belgium and France.	
	Tons.	Tons.	Tons.	Per Cent.
1901	7,595,000	2,272,000	528,000	46
1902	8,793,000	2,944,000	556,000	49
1903	10,683,000	3,347,000	644,000	50
1904	11,135,000	3,327,000	662,000	50
1905	11,968,000	3,486,000	869,000	51
1906	13,834,000	4,309,000	839,000	52
1907	14,208,000	4,425,000	808,000	54
1908	13,282,000	4,581,000	716,000	54
1909	14,443,000	4,558,000	733,000	54
1910	16,652,000	5,480,137	894,278	65
1911	17,754,571	5,335,288	858,212	61

With regard to the ore-reserves, Kohlmann, after most careful examination, considers that under the present economic conditions about 2,130,000,000 tons may be reckoned in German Lorraine and Luxemburg.

The exploitation of the minette deposits has only been possible since the introduction of the Thomas process. The German Lorraine minette district came into German possession at the conclusion of peace in 1871. This district in the year 1873, that is, at the beginning of mining operations under German government, produced 809,541 tons, worth £129,699 from 33 mines with 1905 employees. Since that time the importance of the district has rapidly increased. In 1903 a total of 10,683,042 tons, worth £1,406,500, was produced from 50 mines with 11,010 employees. For the year 1909 the official statistics were 14,441,208 tons containing on an average 28·5 per cent of iron and worth £1,929,900 ; while the corresponding figures for 1910 were 16,652,143 tons containing 28·8 per cent of iron and worth £2,257,750. In these two years operations were carried on in 45 and 46 mines respectively. This district from 1873 to 1912, that is, during a period of forty years, has probably produced about 280,000,000 tons of ore.

The Luxemburg minette district was declared open in the year 1870, though the beginning of mining operations dates back to 1860. From 1868 to 1909 altogether about 150,000,000 tons were won. The production in 1909 amounted to about 11,100,000 tons, and in 1910 to about 8,900,000 tons.

In the French minette district mining operations developed very similarly. In spite of the low content of the ore this district also is of great importance. The following are the figures of production :

1905	.	.	.	6,399,000 tons.
1906	.	.	.	7,399,000 „

1907	.	.	.	8,822,000 tons
1908	.	.	.	8,452,000 ,,
1909	.	.	.	10,673,000 ,,
1910	.	.	about	13,000,000 ,,

The French total ore-reserves are reckoned at 3 milliard tons.

THE OOLITIC LIAS IRONSTONE OF NORTH GERMANY

LITERATURE

V. STROMBECK. ' Der obere Lias und braune Jura bei Braunschweig,' Zeit. d. d. geol. Ges., 1853, Vol. V. p. 81.—VÜLLERS. ' Die Eisenerzlagerstätten des Juras des südlichen Teutoburger Waldes und die dortigen bergbaulichen Verhältnisse,' Berggeist, 1858, IV. p. 558. —U. SCHLÖNBACH. ' Über den Eisenstein des mittleren Lias im nordwestlichen Deutschland,' Zeit. d. d. geol. Ges., 1863, p. 465.—SCHLÜTER. ' Die Schichten des Teutoburger Waldes bei Altenbeken,' Zeit. d. d. geol. Ges., 1866, p. 35.—K. EMERSON. ' Die Liasmulde von Markoldendorf bei Einbeck,' Zeit. d. d. geol. Ges., 1870, p. 271.—G. KÜLPFEL. ' Der Lias-Eisenstein von Harzburg,' Berg- u. Hüttenm. Ztg., 1871, p. 21.—HANIEL. ' Über das Auftreten und die Verbreitung des Eisensteins in den Juraablagerungen Deutschlands,' Zeit. d. d. geol. Ges., 1874, Vol. XXVI. p. 59.—A. v. GRODDECK. Die Lehre von den Lagerstätten der Erze. Göttingen, 1879.—HEUSLER. ' Ein neu aufgeschlossenes oolithisches Eisenerzvorkommen in der Juraformation des Teutoburger Waldes,' Naturh. Verh., 1882, p. 114.—FR. KUCHENBUCH. ' Das Liasvorkommen bei Volkmarsen,' Jahrb. d. pr. geol. Landesanst., 1890, p. 74.—A. ROTPLETZ. ' Über die Bildung der Oolithe,' Bot. Zentralbl., 1892, No. 35.—SIMMERSBACH. ' Der Harzburger Eisenerzbergbau,' Stahl und Eisen, 1894, p. 968.—L. V. WERVEKE. ' Oolithische Eisenerze,' Zeit. f. prakt. Geol., 1894, p. 400.— F. M. STAPFF. ' Oolith und ähnliche Eisenerzkörner,' Zeit. f. prakt. Geol., 1894, p. 326. —v. KOENEN. Erläuterungen zur geologischen Spezialkarte von Preussen, Section Westerhoff, 1895.—KNACKSTEDT. ' Geologisches und Bergmännisches vom Harzburger Eisenstein,' Berg. u. Hüttenm. Ztg., 1902, pp. 169, 181.—BORNEMANN. ' Über die Liasformation in der Umgegend von Göttingen,' Zeit. d. d. geol. Ges., 1904.—GRUPE. ' Präoligozäne und jungmiozäne Dislokationen und tertiäre Transgressionen im Solling und seinem nördlichen Vorlande,' Jahrb. d. pr. geol. Landesanst., 1908.—Geological maps of the districts, Alfeld, Altenbeken, Dassel, Driburg, Einbeck, Gandersheim, Gr. Freden, Gronau, Göttingen, Harzburg, Lichtenau, Lindau, Möringen, Nörten, Westerhof, Willebadessen, Pr. geol. Landesanst. Berlin.—G. EINECKE and W. KÖHLER. ' Die Eisenerzvorräte des Deutschen Reiches,' Pr. geol. Landesanst. Berlin, 1910.

These deposits occur principally in the Middle Lias, and more particularly in the zone of *Ammonites Jamesoni*; sometimes in the Lower Lias and more particularly in the beds of *Ammonites Bucklandi* or *Arietites*; and now and again in still higher horizons of the Lias, though these last are of no great importance. With several of these deposits, in addition to the oolitic, the fine clastic structure is found, the ore having then the appearance of a fine-grained angular breccia. The distribution of the different districts has been greatly influenced by Mesozoic folding, the importance of this relation having been first demonstrated by Stille. As a consequence of this folding, denudation was considerable and the ironstone beds of to-day represent but the remnants of larger sheets which in greater part have been demolished. It must be remarked, however, that these deposits in their original condition were in all probability hardly so extensive as the occurrences of minette in Lorraine. The most important of them are herewith briefly described.

In the Altenbeken-Langeland Lias syncline several oolitic ironstone seams occur along the east border of the Eggegebirge—the southernmost. portion of the Teutoburger Wald—dipping 35°–37° to the west, into the hill. Among these are a grey-brown oolitic seam of 1·1 m. thickness,. a red oolitic seam of 2·2 m., a grey-brown oolitic seam of 1·2 m., and another grey-brown oolitic seam of 1·3 metres. Then in the hanging-wall, after 80 m. of barren material, come the Fund seam of black-brown oolitic ore, 1·2 m., and the Antonius seam of red oolitic ore, 4·5 m., so that within a thickness of 280 m. the total thickness of ore amounts to 11–12 metres. The length along the strike in the case of the Antonius seam has been put by Heusler at 6 kilometres. The beds form a flat syncline. The ore contains 19–40 per cent, and on an average perhaps 28 per cent of iron, with traces of manganese, 7·3–26 per cent CaO, 2–2·95 per cent MgO, 0·04–0·9 per cent P, 9–25 per cent SiO_2, 6·1–10·7 per cent Al_2O_3, and 0·07–0·1 per cent of S. It is accordingly low-grade and contains considerable insoluble residue. These deposits were formerly worked by the old Altenbeken works ; at present the mines lie idle.

The deposit at Bonenburg belongs to the second Lias syncline of the Eggegebirge, which extends from Neuenheerse to Bonenburg, is much broken up, and in its southern portion reaches a width of 0·5 kilometres. To the east it is bounded by Keuper and Muschelkalk, and to the west by Cretaceous. In the Lias, in addition to a large number of spharosiderite seams of little importance, there occurs a red oolitic seam, 5–6 m. in thickness, belonging to the Middle Lias. This seam dips 34°–37° south-west and extends but a short distance along the strike. According to con-tained fossils and to its petrographical nature, etc., this seam is identical with the Antonius seam at Langeland. Two other oolitic seams 1·10 and 1·94 m. thick respectively, which formerly also were worked, occur in the hanging-wall. The composition of the ore resembles that at Langeland. It was formerly smelted at the Teutonia works near Borlinghausen. For economic reasons operations were suspended as the Westphalian industry came to the fore.

The occurrence at Welda-Volkmarsen belongs to a third Lias syncline extending south-west of Warburg, from Wethen through Welda to Volk-marsen, for a length of about 8 km. with a width of a few hundred metres. This oolitic iron ore-bed belongs to the Middle Lias, is 2–5 m. thick, and dips about 35° west-south-west. It resembles the Antonius seam at Lange-land and the main bed at Bonenburg. Although this is but a narrow Lias subsidence bounded on either side by Keuper and Muschelkalk, and large reserves cannot therefore be expected, prospecting has latterly been carried

on to a large extent in the southern part of the syncline, where the deposit
has proved to become richer in depth. The character of the ore appears
to be more favourable than that at Langeland and Bonenburg, the silica
being below 14 per cent. The ore formerly smelted at the Veckernhagen
works contained on an average 30 per cent of iron.

The Markoldendorf Lias syncline, illustrated in Fig. 412, contains
rich iron ore west of Einbeck. In this case also the deposits, which may
reach a thickness of 3–4 m., are found in the Middle Lias. The beds, which
greatly resemble minette, are sap-green, bluish, or brown in colour. The

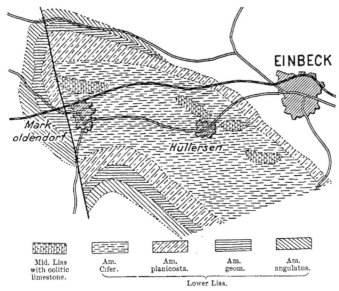

Mid. Lias with oolitic limestone.	Am. Cifer.	Am. planicosta.	Am. geom.	Am. angulatus.

Lower Lias.

FIG. 412.—Oolitic iron deposits in the Markoldendorf Lias syncline.
Scale 1 : 100,000. Emerson.

iron content is given as 30–36 per cent. Since the Lias here rises like
islands through the surface Diluvium, the further extension of these deposits
will have to be determined by boring.

At the Kahlberg east of Kreiensen and in the neighbourhood of Kahle-
feld, Echte, Oldershausen, and Wiershausen, the Jurassic, as illustrated in
Fig. 413, occurs completely developed. The middle division comes to
surface in the south-east, the beds dipping as much as 30°, but as a rule
much flatter, into the hill. The mining area covers several square kilometres. According to Schloenbach, the dark, red-brown, and very fine-
grained ironstone is oolitic, contains many fossils, and reaches a thickness

of 1·5 to 2 metres. It belongs to the horizon of *Ammonites Jamesoni*. The old mines at Oldershausen likewise worked an ironstone bed, 1·5–2 m. in thickness, the ore from which was smelted at Karlshütte near Alfeld. The early figures of the iron content of the ore, 34–38 per cent, are probably unreliable ; it is more likely that the composition was similar to that of

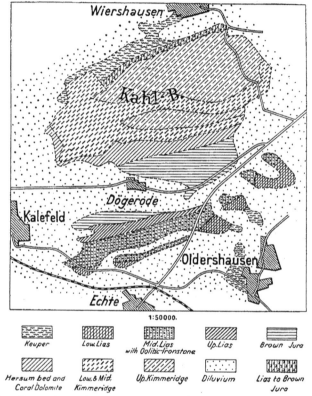

FIG. 413.—Oolitic ironstone in the Kahlefeld-Echte Lias syncline. Scale 1 : 50,000.
Einecke and Kohler.

the foregoing associated deposits. In any case the bedding conditions appear favourable for exploitation.

The ironstone bed on the Heinberg near Salzgitter likewise belongs to the Middle Lias. It comes to surface and dips at a low angle to the east. The bedding conditions are regular for some 5 kilometres. The thickness of this oolitic ironstone, which has a calcareous matrix and belongs to the Margaritatus beds, is 1–2 metres. The iron content is very low.

The most important of these Middle German occurrences are those at Harzburg on the north border of the Harz, illustrated in Fig. 414. These consist of : (1) the Lias oolitic ironstone at the Friederike mine ; (2) the oolitic ferruginous limestone at the Hansa mine, belonging to the White Jurassic ; (3) the limonite- and phosphorite conglomerates of the Lower Cretaceous or more particularly the Neocomian ; (4) the ironstone conglomerates of the Upper Cretaceous or more particularly the Emscherian.[1] Of these however, only the first-mentioned are economically important. The general geological position is indicated in Fig. 414. A section along a north-west line exposes from west to east : Bunter, Muschelkalk, Keuper,

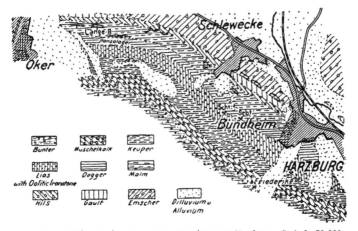

FIG. 414.—Oolitic ironstone in the Lias formation near Harzburg. Scale 1 : 50,000. Einecke and Kohler.

Lias, Dogger, Malm, and Lower- and Upper Cretaceous. The beds are overturned and dip about 50° to the south, that is, into the Harz. With regard to mineralization, only the Lias comes into question, and not only the afore-mentioned *Ammonites Jamesoni* horizon of the Middle Lias, but also the Arietites beds of the Lower Lias. The ironstone seam in the Middle Lias has a thickness of 2 m. ; at a distance of 80 m. in the geological foot-wall, but owing to the overthrust in the actual hanging-wall, follows the rich ironstone horizon of the Lower Lias with four beds having thicknesses of 2 m., 2·5, 6·0, and 1 m. respectively, these being separated from one another by several metres of clay and black slate. The Friederike mine works three beds of 3, 2·50, and 6 m. thickness respectively. The total thickness in the Middle Lias being 14·50 m., this occurrence is the largest in the north-west German Lias.

[1] *Postea*, p. 1046.

The shortness of the length along the strike is noteworthy, this to the east and west being 1100–1200 m., after which the beds become calcareous. The Friederike mine yields annually 50,000–60,000 tons of ore, which is smelted at the Mathilde works.

The occurrence at Rottorf on the Kley, illustrated in Fig. 415, is situated in the north-west portion of the Helmstedt Lias syncline between Helmstedt and Fallersleben, which syncline has a length of more than 40 km. in a north-westerly direction between Oschersleben and Fallersleben. The

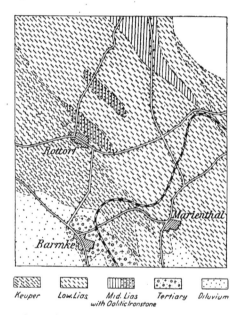

Keuper   Low.Lias   Mid. Lias   Tertiary   Diluvium
with Oolitic Ironstone

FIG. 415.—Oolitic ironstone in the Lias syncline near Rottorf. Scale 1 : 100,000.
Einecke and Köhler.

section there, beginning from the bottom, consists of Lias sandstone, above which, in addition to clays, comes the ironstone horizon of the Arietites beds, these in turn being covered in several places with still later sediments of the Middle and Upper Lias. In the ironstone horizon and principally east of Rottorf, the minette-like ore-beds may be followed for a considerable distance and width. They display distinct oolitic structure, and as a rule are violet or red, and only in the lowermost beds, green. The thickness, hitherto determined in but few places, varies between 2 and 3 m.; the iron content is on an average 30 per cent. In other respects the character of the ore is favourable, more particularly since the phosphorus content, owing to the abundance of phosphoric fossils, is high. The ores were formerly smelted at the Hedwig works at Helmstedt.

South-east of this Lias occurrence the horizon of the iron oolites occurs again between Helmstedt and Oschersleben, around Marienborn, Sommerschenburg, and Badeleben. This deposit is usually named after Sommerschenburg. The beds belong here to the *Ammonites Bucklandi* and *Ammonites bisulcatus* zones of the Lias ; they are gently undulating and dip, generally speaking, 9°–10° west-south-west. The separation between the ore-bed and country-rock is not sharp. The ore in the foot-wall and hanging-wall gradually merges into a sandy ferruginous marl or into clayey-calcareous weathered material. The ore is calcareous, argillaceous, or arenaceous. Only the calcareous portions, which consist almost entirely of coarse, lustrous, oolite grains with very little cementing material, are payable. These oolites are flat, lenticular, elliptical, or angular

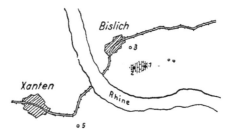

FIG. 416.—Oolitic ironstone at Bislich on the Lower Rhine.   Scale 1 : 100,000.
Einecke and Kohler.

in shape. The thickness in the neighbourhood of Sommerschenburg is 2–6 m. and the length 10 km., the original length having been reduced to some extent by erosion. The ore contains 30–32 per cent of iron, as well as $CaCO_3$ and $MgCO_3$, Al_2O_3, and SiO_2. Its character on the whole is very similar to that of the Harzburg occurrence. It was formerly smelted at the Hedwig works at Helmstedt.

In the year 1903 an oolitic ore-bed, 8·5 m. thick, was accidentally found at a depth of about 470 m. in deep bore-holes for coal on the lower Rhine west of Wesel, at Bislich. This occurrence is illustrated in Fig. 416. The discovery was all the more sensational since the ore in structure and composition was very similar to Lorraine minette. According to the late state geologist Dr. Müller, this ore also belongs to the Middle Lias. Further prospecting work by means of bore-holes has demonstrated the presence of a limited extent of iron-bearing Lias measures, bounded by faults. The length of this extent has been reckoned by Einecke and Köhler at 2 km., and the width at 300 metres.

The economic importance of all these minette-like Lias iron ore-beds of north-west Germany, as will have been gathered, is at present not great. It must be remembered however that the present period is one of prospecting, and that mining operations formerly prosecuted on most of the occurrences mentioned, were stopped not because the deposits were exhausted, but for other good reasons. A favourable factor is that the ore in many cases is self-fluxing. The phosphorus content is usually sufficient for the Thomas process. In the case of ore which is not self-fluxing, the relation between the silica and lime is fairly good so that a large amount of lime flux is not necessary. Comparing the ore of these minette-like occurrences with that of Alsace-Lorraine, after taking everything into consideration it may be said that the minette proper is somewhat superior to the north-west German iron oolite.

The total reserve of all these deposits, not including those of Heinberg and Bislich, has been estimated by Einecke and Köhler at 45,000,000 tons. The individual deposits according to their importance would be placed in the following sequence : Harzburg, Kahlberg, Langeland, Welda-Volkmarsen, Sommerschenburg, Rottorf-am-Kley, Markoldendorf, and Bonenburg.

THE WESERGEBIRGE

LITERATURE

F. Römer. Die jurassische Weserkette. Bonn, 1858.—D. Brauns. Der obere Jura im nordwestlichen Deutschlands. Brunswick, 1871 and 1874.—F. v. Dücker. ' Über oolithische Eisenerze aus der Gegend von Minden,' Nat. Verh., 1875, Corr. p. 57.—H. v. Dechen. Erläuterungen zur geologischen Karte der Rheinprovinz und Westfalens, 1884, Vol. II. p. 368.—F. v. Dücker. ' Eisensteinbergwerke am Wesergebirge,' Nat. Verb., 1884. —E. Harbort. ' Die Schaumburg-Lippesche Kreidemulde,' Neues Jahrb. f. Min., 1903. —Th. Wiese. ' Die nutzbaren Eisensteinlagerstätten, insbesondere das Vorkommen von oolithischem Roteisenstein, im Wesergebirge bei Minden,' Zeit. f. prakt. Geol., 1903, p. 217.— J. Schlunck. ' Die Jurabildungen der Weserkette bei Lübbecke und Pr. Oldendorf,' Jahrb. d. pr. geol. Landesanst., 1904, p. 75. — G. Einecke and W. Köhler. ' Die Eisenerzvorräte des Deutschen Reiches,' Archiv f. Lagerstättenforschung, Kgl. geol. Landesanst. Berlin, 1910.

The Wesergebirge, a range of hills about 100 km. in length, and running approximately west-north-west, consists of Jurassic beds dipping 18°–30° to the north. Upon dark Liassic shales—Amaltheus clays and Posidonia slates—lies the extensively developed Brown Jurassic, consisting of sandy clays, limestones, and marls. Above this again comes the White Jurassic with the coralline oolite—the Coral Rag—the parent rock of the Wesergebirge oolitic hæmatite. This sequence is illustrated in Fig. 417. The following deposits are known : (1) the spharosideritic clay-ironstone of the Upper Lias and Lower Dogger, on the south slope of the range ; (2) the

clayey iron oolite seam in the hanging-wall of the Macrocephalus sandstone
—Upper Dogger—west of the Porta; and (3) the oolitic hæmatite in the
Coral Rag, east of the Porta.

1. The clay-ironstone nodules of the Upper Lias form several beds
of variable thickness lying one above the other; one sample contained
31·5 per cent of iron. These deposits are at present of no economic
importance.

2. Immediately above the Macrocephalus sandstone comes a clayey
iron oolite which, beginning in the neighbourhood of the Porta, reaches
its greatest thickness of 2 m. at Häverstedt, and pinches out west of
Lübbecke. This Wittekind ore-bed dips about 30° north-north-east. It
consists of a calcareous sandy rock which when fresh is blue-black, and
when weathered is loose and reddish- or yellowish-brown. It is mostly
a fine-grained oolite with a clayey and sandy matrix; frequently it

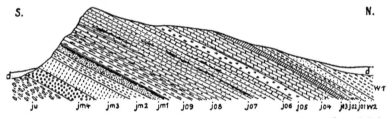

S. N.

ju   jm4   jm3   jm2   jmf   jo9   jo8   jo7   jo6 jo5   jo4   jo3 jo2 jo1 w2

ju = Lias; jm4 = Lower Dogger; jm3 and jm2 = Cornbrash; jm1 = Kelloway; jo9 and jo8 = Lower Oxford;
jo7 = Upper Oxford, blue oolitic Jurassic limestone; jo6—jo3 = Kimmeridge; jo2 = Portland; jo1 = Purbeck;
W2 and W1 = Wealden; d = Diluvium.

FIG. 417.—Diagrammatic section through the Wesergebirge east of the Porta Westfalica.

is spotted with lime oolites varying in size from a pea to a bean. The
upper portion of the seam exhibits spharosideritic segregations in the form
of large, flat, irregular nodules. Average analyses of the raw ore give
29 per cent of iron and considerable insoluble residue. It is exploited by
the Georgs-Marien works at the Porta I. mine in the Wallücke valley. The
reserves are estimated at about 10,000,000 tons of payable ore.

3. The hæmatite of the Coral Rag belongs to the Upper Oxfordian.
The ferruginous oolitic limestone of this Rag merges gradually into oolitic
hæmatite beds which, though generally small, occasionally reach large
dimension. There is little conformity of bedding, so that prospecting for
these deposits is difficult. The thickness in the valley near Lerbeck is
0·6–5 m., at Levernsiek in the former Viktoria mine 0·6–8 m., and in the
valley at Nammern 0·5–1·4 metres. The iron content across the thickness is very variable. Two analyses by the Dortmund Union gave 41 to 43
per cent of iron and about 11 per cent of insoluble residue. The ore-bed of
the Wohlverwahrt mine in the gorge near Kleinenbremen is particularly

well opened up. It is 2–2·5 m. in thickness, and along the dip of 16° has been proved for 1000 metres. The iron content is 37–38 per cent. Thirteen metres below this seam lies the Nammern-Klippen ore-bed, the most extensive deposit in the Coral Rag. This bed continues for a length of 10 km. along the strike, and is 4–12 m. in thickness. The iron content is 25 per cent.

Mining operations upon these hæmatite beds, which began in the years 1875–1877 in the Levernsiek valley 3 km. east of the Porta, have in the case of some of these occurrences continued ever since, though with variable results. In addition to large ore-reserves about 50,000,000 tons of ferruginous limestone flux are estimated to be present.

KRESSENBERG AND SONTHOFEN

LITERATURE

W. V. GÜMBEL. 'Geologie von Bayern,' Vol. II. Geologische Beschreibungen von Bayern. Kassel, 1894.—O. M. REIS. Erläuterungen zu der geologischen Karte der Vorderalpenzone zwischen Bergen und Teisendorf I. Stratigraphical Part, Geognost., Jahresheft 8, 1895, Kassel, 1896 ; 'Zur Geologie der Eisenoolithe führenden Eozänschichten am Kressenberg in Bayern,' Geognost. Jahresheft 10, 1897, p. 24, Munich, 1898.—W. v. GÜMBEL. Geognostische Beschreibung des Bayrischen Alpengebirges, 1861, p. 647.— H. v. DECHEN. Die nutzbaren Mineralien und Gebirgsarten im Deutschen Reiche, revised by W. Bruhns in collaboration with H. Bücking, Berlin. 1906.

In the foothills of the Bavarian Alps, Nummulitic beds belonging to the middle Eocene carry oolitic iron deposits which continue uninterruptedly from the Mattsee and the Haunsberg north of Salzburg, through Kressenberg on the Chiemsee, to the Grünten near Sonthofen in the Algau, and still farther. The most important occurrences are those at Kressenberg and on the Grünten, where mining operations, at present still active, have at times been animated. Reis describes that at Kressenberg as two steep ore-beds, the Upper black bed and the Lower red bed, separated by measures containing some Middle beds. The bedding is complicated by numerous transverse and longitudinal disturbances as well as over-thrusts. These disturbances are responsible for duplications, from which, formerly, the mistaken idea of a large number of beds arose. The occurrence of kinks or pleats along the strike is characteristic of these beds. The ore-beds are 1–2 m. thick, and consist of oolitic limonite which contains quartz, glauconite, etc., and gradually merges into ferruginous sandstone. The cementing material in the black and red beds differs ; that of the former is rich in ferrous oxide, that of the latter in ferric oxide. The black seam contains on an average 35 per cent of iron with 0·55 per cent of phosphorus ; the red seam 18–22 per cent of iron. The operations formerly carried on by the State, were abandoned in the year 1881. The present yearly production is probably some 1000 tons.

On the south slope of the Grünten a large number of oolitic iron beds up to 1 m. thick were formerly opened up. It is probable that in reality the number of independent beds is small, but that these by faults and overthrusts have become duplicated, giving the present appearance of a large number. The iron content of the ore is low.

ENGLAND : CLEVELAND, AND NORTHAMPTON

LITERATURE

W. FAIRLEY. Journ. Iron and Steel Inst., 1871, p. 154.—G. BARROW. Proc. Cleveland Inst. Eng., 1877 to 1880, p. 108.—H. BAUERMANN. Metallurgy of Iron, 1882, p. 89.—

FIG. 418.—Map of the Cleveland iron district. Scale 1 : 600,000.

Sir LOWTHIAN BELL. 'On the American Iron Trade,' Journ. Iron and Steel Inst., 1890, p. 119.—A. L. STEAVENSON. 'The Last Twenty Years in the Cleveland Mining District,' Journ. Iron and Steel Inst., 1893, XLIV.—J. D. KENDALL. The Iron Ores of Great Britain, 1893.—PHILLIPS and LOUIS. A Treatise on Ore Deposits. London, 1896.—H. LOUIS. 'The

Iron Ore Resources of the United Kingdom of Great Britain and Ireland,' in the Iron Ore Resources of the World, Stockholm, 1910, Vol. II. p. 630.—H. B. WOODWARD. 'Lias of England and Wales,' Mem. Geol. Survey, 1893.

Cleveland

This important iron deposit, delineated in Fig. 418, occurs in the Cleveland Hills in north-east Yorkshire, and is at present the most productive in Great Britain. It belongs to the Middle Lias, and generally consists of several beds separated by barren layers of slate and sandstone. Where best developed it exhibits a total thickness of more than 20 feet. The two principal beds, after the abundance of the particular fossils in them, are known respectively as the Pecten and Avicula seams. The ore is usually dark blue-green in consequence of contained iron silicate; it exhibits oolitic structure and contains numerous fossils. Though subject to considerable variation at different places, the following average section will be of interest:

	Thickness of Barren Material.	Thickness of Ore.
Lower Oolite—		
Eller Beck bed	...	1' to 2' 6"
Shales and sandstones [1] about .	100'	...
Dogger ironstone or Top seam of ironstone . .	...	1' to 4'
Upper Lias—		
Alum shale .	115'	...
Jet Rock .	50'	...
Grey shales .	30'	...
Middle Lias—		
Shale . .	0 to 10'	...
Main seam .	...	6' to 12'
Shale .	3' to 4'	...
Pecten seam	...	2' to 4' 6"
Shale .	3'	...
Two-Foot seam	...	1' 6" to 2'
Shale .	25'	...
Avicula seam	...	2'
Shale .	35'	...
Sandy series	40'	...
Lower Lias—		
Shales .	over 400'	...
Rhætic beds .	...	...

The 6–12 ft. Main seam is richest and widest at Eston Nab, from which point in all directions it gradually becomes poorer. At present it is the only bed being worked. Its best ore is found in the neighbourhood of the outcrop, where the thickness reaches 11–12 feet. At Eston Nab this bed lies immediately on the Pecten seam so that both can be

[1] Dogger.

worked together, making a combined thickness of 15 ft. 6 inches. On an average the thickness of the Main seam by itself may be put at 8–9 feet. In the south and east of the district a bed of hard slate, 2 ft. thick, occurs in this seam. The best ore, consisting of impure iron carbonate, is bluish in colour. The iron content varies from 30 to 35 per cent, the phosphoric acid from 1 to 3, and the silica from 6 to 10 per cent. The average composition of the ore may be gathered from the following analysis :

Fe, reckoned as Fe_2O_3	. . .	41·14 per cent.
Al_2O_3 .	. . . . .	9·01 „
MnO .	. . . . .	1·02 „
CaO .	. . . . .	5·90 „
MgO .	. . . . .	3·67 „
SiO_2 .	. . . . .	12·20 „
S .	. . . . .	0·10 „
P_2O_5 .	. . . . .	1·38 „
Loss on ignition	. . .	23·76 „
Total .	. .	100·09 per cent.
Fe .	. . . . .	30·20 „

The Top seam varies greatly in thickness and composition, and is seldom payable. At Rosedale formerly, it was worked in conjunction with some magnetite above it. The Pecten seam is worked only occasionally, as for instance at Grosmont and Eston, where it contains 27–28 per cent of iron, and has a thickness of 3–4½ feet. The Two-Foot seam lies 2–7 ft. under the Pecten seam ; its thickness in the north-west of the district is seldom more than 2 ft., while at Grosmont it is only 10 inches. The iron content is on an average 29 per cent. The Avicula seam occurs about 25 ft. under the Two-Foot seam and varies in thickness from 1 ft. 6 ins. to 3 ft. 9 inches. At Grosmont, where it formerly was worked, it possesses a relatively favourable composition. The iron content is 26–27 per cent.

Louis regards the Cleveland ore as a metasomatic replacement of lime oolite. Convincing evidence in support of this theory is however not yet forthcoming. We regard it as very probably a primary metalliferous sediment.

The Cleveland deposit has been worked for over five hundred years. Ore was first won in the thirteenth century from the outcrop and smelted with charcoal. Operations on a large scale, however, only began in the years 1846 and 1850 at Grosmont and Eston respectively, in mines belonging to Bolckow and Vaughan. The rapid development of this district may be gathered from the following figures of production :

	Tons.			Tons.
1855 . . .	865,000	1885 . . .	5,932,000
1860 . . .	1,471,000	1890 . . .	5,617,000
1865 . . .	2,762,000	1895 . . .	5,286,000
1870 . . .	4,073,000	1900 . . .	5,494,000
1875 . . .	6,122,000	1905 . . .	5,944,000
1880 . . .	6,487,000	1908 . . .	6,073,000

The largest production was reached in the year 1883 with 6,756,000 tons, since which date it has varied between 5,000,000 and 6,000,000 tons. The total output of the Cleveland district up to the present day is reckoned at 250,000,000 tons, to which the Main seam has contributed by far the largest quantity. The ore-reserves still available have been estimated by Louis to amount to 3,000,000,000 tons, equivalent to about 1,000,000,000 tons of iron.

It is interesting to note that a similar ironstone bed occurs in practically the same geological horizon on the island of Raasay off the west coast of Scotland. The thickness of this seam is 5 ft., while the outcrop may be followed for more than a mile. Though it contains 29 per cent of iron this deposit has never been worked.

Northampton

The Northampton iron ore was formerly worked over a wide area which, with breaks, extended from Steeple Aston in Oxfordshire on the south, to Greetwell near Lincoln on the north, a distance of nearly 100 miles. This area is illustrated in Fig. 419. In its southern portion the metalliferous belt strikes north-north-east, and in its northern portion almost due north. The maximum width reaches 20 miles. The ore-bed belongs to the Inferior Oolite, to the zone of *Ammonites Murchisonae*, which extends from near Banbury throughout the whole of Northampton. It lies at the base of the Oolite in a bed known as the Northampton sands. The petrographical character of this bed varies greatly ; sometimes it is ferruginous, sometimes calcareous, and sometimes slaty. It is tilted and possesses in general an easterly dip. It appears fairly certain that the ore-bearing beds pinch out south of Steeple Aston, so that the probability of finding ore in this horizon still farther south, is small. Similar conditions prevail north of Lincoln.

The iron content of the beds, in harmony with their frequently changing character, varies very considerably ; the beds often pinch out or merge into ferruginous sands or thin iron seams. In some places the ore-bed is as much as 30 ft. thick, though only exceptionally is more than 12 ft. payable ; an average of 9 ft. throughout the whole district may be taken. The following thicknesses are mined at the places mentioned :

Waltham-on-the-Wolds	4 feet.
Kettering	6 ,,
Finedon and Eston	12 ,,
Irthlingborough Glebe, near Wellingborough . .	17 ,,

At the outcrop the ore occurs in the form of a sandy brown hæmatite, often consisting of yellowish-brown sand grains cemented together by limonite. When such is the case an oolitic, impure, grey, greenish, or

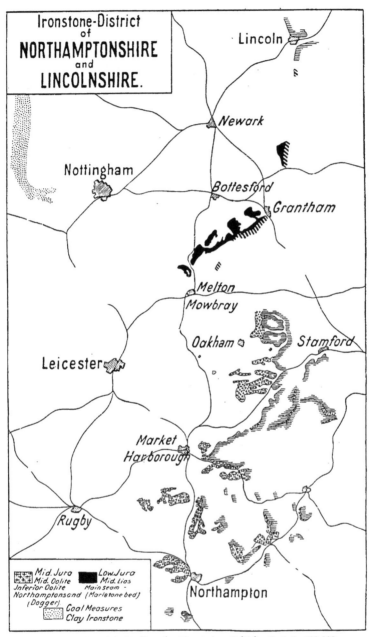

bluish iron carbonate with considerable silica, frequently occurs in depth. The oxidized ore at the surface contains 25–40 per cent of iron, though rarely more than 35 per cent, with 15–25 per cent of silica. The iron content of the carbonate from greater depth varies between 30 and 35 per cent with 12–14 per cent of silica. The phosphorus content is generally 0·5 per cent and exceptionally 1 per cent. Phosphate concretions occur in places. The typical composition of the ore may be gathered from the following analyses :

	Northampton.	Rutland.
Fe	32·00 per cent	32·00 per cent
Mn . . .	0·15 ,,	0·15 ,,
SiO₂ . . .	10·00 ,,	14·00 ,,
CaO . . .	2·6 ,,	1·00 ,,
P	0·65 ,,	0·52 ,,

Mining in Northampton began in the time of the Romans and continued throughout the Middle Ages so long as the adjacent forests provided charcoal for smelting. Subsequently this deposit was almost forgotten, till in the year 1851, or approximately the same time as Cleveland was resuscitated, its existence became remembered. Unfortunately, however, Northampton is unfavourably situated in relation to the coalfields. The production in 1870 amounted to 500,000 tons of ore ; in 1890 to 1,000,000 tons ; and in 1907 to 2,500,000 tons. According to W. Fairley, up to 1870 some 3,500,000 tons had been won, and an area of roughly ⅓ sq. mile thereby exhausted. Louis estimates the ore-reserves at 1,000,000,000 tons.

Lincoln

In addition to the Northampton ironstone of the Lower Oolite which, as indicated in Fig. 419, is also worked at places in Lincoln, other iron ore-beds occur in this latter county at various geological horizons. The most important of these is a deposit situated approximately in the middle of the Lower Lias, this deposit, attaining its best development in the neighbourhood of Frodingham, being known as the Frodingham ironstone. The thickness of this bed is 10–25 feet or on an average about 12 feet. It lies at a gentle inclination to the east. Over wide areas it is only covered by a small depth of surface material so that opencut mining may well be undertaken. The length along the north-south strike, from the southern shore of the Humber to the village of Solter, is about 14 miles. Owing to the low dip, a strip about 1 mile wide can be mined by opencut, a further strip of 2 miles has been proved by boreholes, while it is probable that the deposit extends at a regular dip to still greater depth.

The ore is essentially a calcareous limonite containing up to 20 per cent of CaO. It is often green in colour, consisting then of calcareous iron carbonate. It may be assumed with fair certainty that this limonite is an oxidation product. The ore contains on an average 22–25 per cent Fe, 10 per cent CaO, and about 10 per cent of SiO_2; the phosphorus content varies between 0·2 and 0·5 per cent. Although the iron content is not high the considerable lime content makes this ore very suitable to mix with the siliceous ore from Northampton.

The surface area of this ore-bed can be put down as 40 square miles. Each square mile contains on an average 3,000,000 tons of ore, and the total reserve is accordingly estimated at 100,000,000 tons. The deposit was discovered in the year 1859, and the first smelting works were erected in 1864.

THE CLINTON ORE-BED IN THE UNITED STATES

LITERATURE

R. D. IRVING. 'Mineral Resources of Wisconsin,' Trans. Amer. Inst. Min. Eng. VIII. 1879–1880, p. 478.—Geol. of Wisconsin Survey of 1837–1879, Vol. I. p. 625.—A. H. CHESTER. 'The Iron Region of Central New York,' Address before the Utica Merchants and Manufactures Association. Utica, 1881.—C. H. SMYTH, jun. 'On the Clinton Iron Ore,' Amer. Journ. Sc., June 1892, p. 487; 'Die Hämatite von Clinton in den östlichen Vereinigten Staaten,' Zeit. f. prakt. Geol., 1894, p. 304.—H. H. STOEK. 'Ores at Danville Montour County,' Trans. Amer. Inst. Min. Eng. XX. 1891, p. 369.—J. B. PORTER. 'Iron Ores and Coal of Alabama, Georgia, and Tennessee,' Trans. Amer. Inst. Min. Eng. XV., 1886–1887, p. 170.—J. F. KEMP. The Ore-Deposits of the United States and Canada. New York, 1906.— J. J. RUTLEDGE. 'The Clinton Iron Ore Deposits of Stone Valley, Huntingdon, C.Pa.,' Trans. Amer. Inst. Min. Eng., 1908, No. XXIV. p. 1056.—J. F. KEMP. 'Iron Ore Reserves in the United States,' The Iron Ore Resources of the World, Vol. II. Stockholm, 1910.—ERNEST F. BURCHARD, CHARLES BUTTS, and EDW. C. ECKEL. 'Iron Ores, Fuels, and Fluxes of the Birmingham District, Alabama, with Chapters on the Origin of the Ores,' Bull. 400 U.S. Geol. Surv., 1910.—H. Ries, Economic Geology, with special reference to the United States. New York, 1910.—Charles L. Henning, Die Erzlagerstätten der Vereinigten Staaten von Nordamerika u.s.w. Stuttgart, 1911.

The Clinton stage of the Upper Silurian, delineated in Fig. 420, contains, practically throughout, one or more beds of red hæmatite intercalated between slates and limestones. These deposits possess considerable extent in Wisconsin, Ohio, and Kentucky, in New York, Pennsylvania, West Virginia, in eastern Tennessee, in north-west Georgia, and finally in Alabama, that is to say, in places situated 1500 or even 2000 km. distant from one another. The name of this ore is taken from the town of Clinton in New York, where the deposits are typically developed. The principal mining operations are carried on at present in the Birmingham district, Alabama.

The term 'Clinton group' has since the beginning of last century been applied to an alternation of Upper Silurian clay-slates, sandstones, and impure limestones, which lie on the Medina sandstone and are in turn overlaid by the Niagara slate. All three horizons belong to the lower Upper Silurian. The Clinton beds are obviously a shallow-water deposition, and accordingly subject to great petrographical variation. In New York

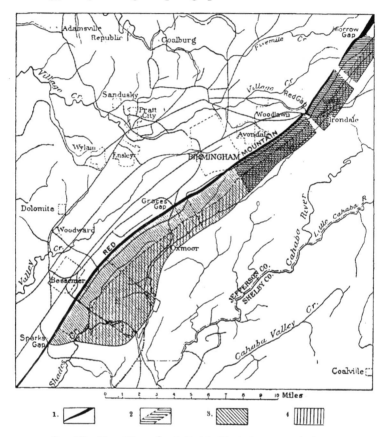

FIG. 420.—Map of the western half of the Birmingham iron district, Ala.
Burchard and Butts, *U.S. Geol. Survey.*

1, outcrop of the Clinton bed on the Red Mountain; 2, area where the Irondale seam is worked; 3, area where the Big seam is worked; 4, area of Irondale and Big seams not payable under present conditions.

state and to the west, they lie with the other Palæozoic formations almost horizontal; farther south, conforming to the folds of the Appallachian Region, they are so disturbed that their outcrops are often sinuous and irregular.

The occurrence of the hæmatite deposits in these beds is, generally speaking and taking into consideration the nature and character of the occurrence, uniform, though the number, thickness, and exact position of the beds, vary. In the eastern districts they are well developed, while in the west, with the exception of Wisconsin, they are almost completely absent.

At Clinton in New York three ore-beds occur, these together making a thickness of 6–10 ft., of which however only about 2 ft. can be worked. The beds have in general a lenticular form and are divided into fossil- and oolitic ore. The former contains abundant remains of Bryozoa, Crinoids, Corals, and Brachiopoda. The oolitic ore consists of an aggregate of flat grains lying with their larger surfaces parallel to the bedding, and cemented either by iron oxide or calcite. The separate oolites frequently contain a grain of sand around which thin layers of iron oxide, silica, and alumina have been deposited.

In Pennsylvania, where the thickness of the Clinton beds increases considerably, C. H. Smyth, jun., distinguishes six different ore-beds, of which only two or three are oolitic. The ore consists of amorphous hæmatite with a variable lime-alumina content. Where it outcrops the lime is leached and a material rich in iron and silica has resulted. The calcareous unaltered ore is known as ' hard ore,' while the leached ore is termed ' soft ore.' In the important Birmingham district four horizons in the Clinton series may be differentiated, these being known as the Hickory Nut, Ida, Big, and Irondale seams, respectively, the positions of which are illustrated in Fig. 420. In Alabama the ore-beds have been followed by bore-holes to a depth of 240 metres. The iron content varies between 30 and 48 per cent ; the phosphorus content, with the exception of the Birmingham district, is high, varying between 0·5 and 1 per cent. The sulphur content is low.

The average thickness of the Clinton ore-beds varies between 0·66 and 4 feet.

Of the United States iron ore production, the Lake Superior ores in Michigan and Minnesota yield at present by far the largest proportion. The Clinton ores come second with a production of some 4,000,000 tons in Alabama alone, and with large productions, not, it is true, reaching a million tons, in some of the other states. The Clinton ore-reserves are estimated by J. F. Kemp at about 505,000,000 tons of positive ore and 1,368,000,000 tons of probable ore.

The genesis of the Clinton deposits has long been the subject of discussion in the United States. Some authorities regard them as ferruginous limestone beds, the calcium carbonate of which has, in the neighbourhood of the surface, been leached. This is known as the residual enrichment theory. Since however nowhere in the deep mines has any change from ore to limestone been noted, this theory must be abandoned. Other authorities have advocated a metasomatic origin, and others again a primary sedimentation. Against metasomatism and in favour of original sedimentation, is the fact that no fissures along which mineral solutions could have penetrated have been recognized, and the further fact that the ore-beds are conformable over large areas. Fragments of ore have also been met in the hanging-wall beds, from which it follows that the ore had already been formed before the sedimentation of these hanging-wall beds. According to C. H. Smyth,[1] the Clinton group of sediments was deposited in a shallow basin into which rivers draining large districts of crystalline schists, flowed. The oxide ore he considered was precipitated in the manner accepted for oolitic ores. Other authorities consider that greenalite [2] or glauconite was first formed, which subsequently became altered to iron oxide.

THE BLACKBAND AND CLAY-IRONSTONE BEDS

There are many districts in which blackband- and clay-ironstone deposits occur, though, since such ironstone rarely fulfils the conditions imposed by the market, but few of these districts are of any importance to-day.

In general two forms of occurrence may be differentiated, namely :

1. Concretionary deposits of spharosiderite.—The individual concretions are more or less loaf-shaped ; they either lie so close together as to form fairly compact beds, or occur more or less numerously embedded in one particular bed. This latter condition hardly ever results in an economic deposit.

2. The second form is that of the true ore-bed, in which the entire thickness consists of ironstone. These so-called seams may extend without interruption over large areas and even through whole coal basins.

The thickness is very variable, beds from one to several feet in thickness being worked. Only exceptionally do the beds lie horizontally ; mostly, as the result of subsequent tectonic disturbance, they are more or less steeply inclined.

The distribution of the ore in the concretionary deposits is very variable, while with the seams, on the other hand, the composition is often uniform over great areas. In the case of beds containing ironstone-concretions it must be carefully determined what fraction of the total thickness these

---

[1] *Loc. cit.*, 1892.
[2] See chapter on Lake Superior ores.

concretions constitute. Ore-beds, also, may alter their petrographical character along the strike to such an extent as to merge quite gradually into ferruginous slate or ferruginous limestone, etc.

These deposits consist of siderite, clay - ironstone, and blackband ironstone. The siderite only exceptionally is clean; usually it is mixed with much clay, such a mixture being known as clay-ironstone. Some beds consist of limonite and clay, this mixture being likewise termed clay-ironstone. The composition varies according to the relation of the siderite or limonite to the barren material; while clean siderite may contain as much as 48 per cent of iron and the average content of the ore as produced sometimes reaches 40 per cent, the clay-ironstone beds contain mostly only an average of about 30 per cent. The term blackband ironstone is applied to an admixture of clayey siderite and carbonaceous material, the siderite varying between 35 and 78 per cent.

Throughout these ores the primary iron ore is siderite, $FeCO_3$, which generally contains a small isomorphous admixture of manganese-, calcium-, and magnesium carbonates. These ores accordingly differ mineralogically from the oolitic ores and the recent lake- and bog ores, in that with these latter the iron originally was deposited exclusively or preponderatingly as ferric oxide or ferric hydrate. Clay-ironstone, again, differs mineralogically from the chamosite and thuringite mentioned later, in that these latter represent silicates.

The clay- and blackband ironstone deposits occur in the Coal-measures in closest association with coal seams. In former years especially, not only coal but ironstone also was produced in many coal mines. Such ironstone is found in the Coal-measures of nearly all coal basins. Though the most important deposits of this ironstone belong to the Coal-measures, some are also found in the Permian, Triassic, Jurassic, and Cretaceous.

In relation to genesis, these ores are partly secondary concretions and partly primary depositions. The former are to be compared, at all events in part, with those of the Scandinavian glacial clay deposits, the so-called *Marleker*, consisting of clay and calcium carbonate, this latter having been precipitated by decaying organisms. Similarly, in the case of the concretionary spharosiderites ferrous carbonate was precipitated from solution in concretionary form before the consolidation of the country-rock, small spharosiderites or pennystones and somewhat larger lenses being formed. The more continuous beds of blackband- or clay-ironstone, on the other hand, are probably in the main a primary deposition from ferrous carbonate, though subsequent replacement may also have played a part.

The close association with coal seams proves that the ironstone horizons are in general shallow-water deposits. A repeated alternation of coal seams, ironstone beds, and beds containing Goniatites, occurs in several places, indicating changes of level during the filling of a large subsiding area.   The Stigmaria horizons, these being the plant-root beds in the immediate foot-wall of the seams, are ferruginous, indicating that deposition of iron continued in most cases after elevation to above sea-level.

The deposition of the iron exclusively or in greater part as ferrous carbonate proves that precipitation took place from solutions poor in oxygen.   This was probably because the oxygen had already been consumed by the large number of organisms present in the metalliferous solution.

In the Coal-measures the ironstone bed occasionally lies immediately on the coal seam, the ironstone and the coal along the strike or dip sometimes even merging into each other.

Primary depth-zones may be observed in the ironstone beds in so far that the hanging-wall portions are usually richer in carbon and clay than the foot-wall portions, while the phosphorus content diminishes towards the hanging-wall.   Secondary depth-zones come into question when, as the result of subsequent tilting, the beds come to the surface, the siderite by atmospheric agencies then becoming altered to limonite.

Of the accessory constituents of the primary ironstone, principally two are to be mentioned, namely, manganese and phosphorus.   The manganese content is seldom considerable, 1 per cent being exceeded in relatively few cases.   Phosphorus, on the other hand, occurs in relatively large amount in many seams, layers of phosphorite $\frac{1}{2}$ inch to several inches thick sometimes occurring associated with the ironstone.   In such cases the phosphate of lime does not form continuous beds but lenticular and nodular masses.   The appearance of the phosphorite may then be so similar to that of the ironstone that it is not always easy to distinguish one from the other.   Clay- and blackband ironstone free from phosphorite does not as a rule exhibit a particularly high phosphorus content, a usual figure being 0·1–0·3 per cent.   As far as the carbonate is concerned, these ores may be enriched by roasting, after which they frequently contain 40–45 per cent of iron.

WESTPHALIA

LITERATURE

v. CARNALL. 'Kohleneisensteine in Westfalen,' Zeit. d. d. geol. Ges., 1851, p. 3, ff. 383 ; 'Sphärosiderit in der Steinkohlengrube Concordia im Bergamtsbezirk Essen,' Zeit. d. d. geol. Ges., 1855, p. 304.—R. PETERS. 'Der Spateisenstein der westfälischen Steinkohlen-formation,' Zeit. d. Ver. Deutscher Ing., 1857.—NOEGGERATH. 'Sphärosiderite aus dem westfälischen Steinkohlengebirge,' Naturw. Verh., 1860, p. 64.—BÄUMLER. 'Über das Vorkommen der Eisensteine im westfälischen Steinkohlengebirge,' Zeit. f. B.-, H.- u. S-. Wesen, 1869, p. 426.—W. RUNGE. Das Ruhrsteinkohlenbecken, Berlin, 1892, p. 70 ; Die Entwicklung des niederrheinisch - westfälischen Steinkohlenbergbaues, Vol. I. Berlin,

1903.—P. Krusch. 'Der Südrand des Beckens von Münster u.s.w.,' Jahrb. d. pr. geol.
Landesanst., 1908.—Köhler and Einecke. ' Die Eisenerzvorräte des Deutschen Reiches,'
Archiv für Lagerstättenforschung, Part 1, Pr. geol. Landesanst., 1910.

The ironstone beds in Westphalia represent but a small percentage of
the Carboniferous formation there. They consist of clay-ironstone and
siderite, and occur, relatively speaking, most frequently in the lower portion
of the Coal-measures, that is, in the lean coal. The rich coal is poorer,
while the gas coal and bituminous coal have proved to be the poorest of
all. This ironstone is particularly interesting, in that one and the same
bed may be developed partly as coal and partly as ironstone, while parts
of a bed and even whole ironstone beds may merge into coal seams, and
vice versa.

Leaving out of consideration the oxidation zone above ground-water
level where limonite has been formed by meteoric waters, there are, according to Bäumler, three kinds of primary ore, namely :

1. Yellowish- to blackish-grey, crystalline, mostly unstratified,
granular siderite.—This is almost pure and occurs comparatively seldom ;
it is known only in the lowermost lean coal.

2. Blackband ironstone. — All horizons of the Westphalian Coal-
measures in one place or another carry this ore. It is found particularly
in the lean coal, while in the rich coal there are but few deposits of any
great thickness.

3. Clayey spharosiderite.—This occurs as more or less large nodules
in shale, the number of such nodules being at times so great that a fairly
compact bed results. Such beds however have nowhere been regularly
mined as they quickly pinch out.

The richer varieties of the blackband ironstone have a specific gravity
of 2·8–3, and a hardness of 3–4. The fracture is slaty to flat-conchoidal.
Across the bedding a banding due to alternating lighter and darker
layers is frequently seen. All gradations occur between pure coal, blackband ironstone, and clean siderite, while the hardness varies according
to the proportion between siderite and coal. The most metalliferous
portion of the ironstone bed is usually the lowest ; in the hanging-wall the
ore occasionally merges into ferruginous slate. These beds are frequently
associated with phosphorite beds which however occupy no special horizon
in relation to the ironstone. This phosphorite apparently diminishes
towards the hanging-wall of the Coal-measures.

The ironstone beds are rich in organic remains. They are often
identical with the marine horizons, more rarely with those containing
fresh-water fossils. Fossils are mostly found in the upper layers of the
ore-beds, at the contact of the coal with the oil-shale or immediately in the
hanging-wall of that contact. Vegetable remains are also frequently present.

Most of the blackband ironstone deposits lie in greater part upon the coal; no preference for any other Carboniferous rock, as for instance slate or sandstone, has been observed. One great drawback connected with these ore-beds is the sudden change in strike and dip to which they are subject, which may often be accompanied by complete disappearance of the bed. This renders their working very difficult. In addition, the individual beds are liable without change in thickness to pass from iron-stone into coal, oil-shale, or ferruginous slate. According to Bäumler, it has repeatedly been observed that ironstone cut off on one side of a fault continues on the other side as a coal seam. Probably in such cases also it is rather a question of sudden diminution in thickness, upon which the fault had no influence whatever. It would not be far wrong to attribute the formation of typical ironstone beds to springs, and to assume that the greatest thicknesses occurred where such springs emerged.

From the foot-wall to the hanging-wall the best known beds are the Herzkämper, the Kirchörder, and the Hattinger siderite seams. The un-roasted ore contains on an average 41–45·6 per cent of iron with 1 per cent of manganese and 0·68 per cent of phosphoric acid; roasted ore has been obtained containing 58 – 65 per cent of iron. The production in 1857 amounted to 675,000 tons; in 1865 it reached its maximum of 1,154,000 tons; in 1906 about 42,000 tons were won; while to-day mining operations are completely suspended.

According to analyses, these Coal-measure deposits should be worthy of beneficiation; unfortunately however, the cost of mining is so high that any idea of working them, even in the future, can hardly be entertained.

ENGLAND AND SCOTLAND

LITERATURE

' Iron Ores of Great Britain,' Memoirs of the Geological Survey, 1856.—J. A. PHILLIPS and HENRY LOUIS. A Treatise on Ore Deposits. London, 1896.—HENRY LOUIS. ' The Iron Ore Resources of the United Kingdom of Great Britain and Ireland,' in The Iron Ore Resources of the World, Vol. II. Stockholm, 1910.

Half a century ago in England and Scotland the clay-ironstone of the Coal-measures was almost the only source of iron ore.

As in Westphalia, this ore occurs in continuous beds or in numerous scattered concretions embedded in slate or clay. The concretions consist in greater part of ferrous carbonate mixed with carbonate of lime, carbonate of magnesium, and, now and then, some carbonate of manganese. The compact beds carry either blackband, that is, a mixture of coal and ore, or siderite. They are found in varying number and thickness in almost every coalfield. At present they are worked but little in England.

In Scotland the annual production amounts to about 800,000 tons of blackband. The ore is intimately associated with coal so that both may be mined together. It occurs both in the Upper and in the Lower Coal-measures, the latter corresponding to the Bernician series of northern England. These two formations are separated by the so-called Limestone series. The Scottish ore-beds extend over large areas, though they are very variable in character; blackband, for instance, merges into siderite on the one hand and into coal on the other. The thickness also varies considerably. In most cases another bed begins where one becomes unworkable or pinches out, so that practically speaking the ironstone extends throughout the whole coal district. The Upper Coal-measures contain on an average 4 ft. of ironstone, and the Lower Coal-measures, 2 feet. One square mile of solid blackband 1 ft. in thickness is equivalent to about 2,000,000 tons of ore. Louis reckons the ore-reserves of Scotland at 8,000,000,000 tons, of which 110,000,000 tons, representing the most easily worked portion, have already been mined.

The coalfield of Northumberland and Durham likewise contains ironstone beds, the best-known of which occurs not far above the Millstone Grit. This ore contains on an average 33–35 per cent of iron. A blackband bed at Chesterwood produces 20,000–25,000 tons of ironstone annually. The reserves of Northumberland and Durham are estimated by Louis at 1,500,000,000 tons.

In the Derbyshire and Yorkshire coalfield the thickness of the ironstone beds varies between 6 inches and 1 foot. In spite of this small thickness Louis estimates 6,000,000,000 tons of ore. In the North Staffordshire coalfield four beds are regarded as workable and important, namely, the Half Yard, the Redshagg, the Red Mine, and the Bassey Mine seams. These vary in thickness from a few inches to several feet and may, according to Louis, contain about 5,000,000,000 tons of ore. These figures of ore-reserves by Louis are based in part upon somewhat uncertain or insufficient data, and consequently they are not comparable off-hand with the figures given for the iron deposits, say, of the German Empire.

Generally speaking, the iron content of the English blackband beds varies between 23 and 40 per cent, mostly it is 26–35 per cent, while on an average 31 per cent may be taken. The insoluble residue is 10–20 per cent, and the phosphorus generally 0·5–1 per cent. The ore is roasted whereby a product containing on an average 45 per cent of iron is obtained.

Although the amount of ore in England and Scotland is very considerable, the economic importance of the individual occurrences must in the present position of the iron industry be regarded as small.

THE CRETACEOUS CLAY-IRONSTONE OF BENTHEIM-OCHTRUP, OTTENSTEIN, AND AHAUS

LITERATURE

A. HILBCK. 'Geognostische Darstellung des Eisensteinvorkommens in der älteren Kreideformation von Ahaus,' Zeit. f. B.-, H.- u. S.-Wesen, 1867, p. 108.—A. HOSIUS. 'Über marine Schichten im Wälderton von Gronau,' Zeit. d. d. geol. Ges., 1893, p. 34.— G. MÜLLER. 'Die untere Kreide im Emsbett nördlich Rheine,' Jahrb. d. pr. geol. Landesanst., 1895, p. 60.—B. KOSMANN. 'Über die Toneisensteinlager in der Bentheim-Ochtruper Tonmulde,' Stahl u. Eisen, 1898, pp. 357 and 623, and Zeit. d. d. geol. Ges., 1898, p. 127.—H. KETTE. 'Das Eisenerzvorkommen von Ochtrup-Bentheim,' Glückauf, 1898, p. 436.—B. KOSMANN. 'Die Toneisensteinlager des Münsterlandes in Westfalen,' Lecture : Zs. Luft, 8, Berlin, 1902, pp. 260, 271.—G. MÜLLER. 'Die untere Kreide westlich der Ems und die Transgression des Wealden,' Zeit. f. prakt. Geol., 1903, p. 72 ; 'Die Lagerungs-verhältnisse der unteren Kreide westlich der Ems,' Jahrb. d. pr. geol. Landesanst., 1903, p. 184.—E. HARBORT. 'Ein geologisches Querprofil durch die Kreide-, Jura- und Triasformation des Bentheim-Isterberger Sattels,' Abstract from Festschrift für A. v. Könen. Stuttgart, 1907.—R. BÄRTLING. 'Die Ausbildung und Verbreitung der unteren Kreide am Westrande des Münsterischen Beckens,' Zeit. d. d. geol. Ges., 1908, Monthly Report, pp. 36, 41.—WILLERT. 'Das Toneisensteinvorkommen von Ahaus und Coesfeld und seine wirtschaftliche Bedeutung,' Glückauf, 1908, p. 304 ; Der Erzbergbau, 1907.— R. GOEBEL. 'Die Rentabilität der Toneisensteingewinnung des Münsterlandes,' Der Erzbergbau, 1909, Parts 16 and 17.—KÖHLER and EINECKE. 'Die Eisenerzvorräte des Deutschen Reiches,' Archiv für Lagerstättenforschung, Part 1, published by the Kgl. pr. geol. Landesanst., 1910.

These ironstone deposits occur along the western border of the northern portion of the Münster basin. The metalliferous district is bounded to the east by the Upper Cretaceous which, in a gentle curve convex towards the west, extends from Westphalia through Stadtlohn and Oeding to Holland. Within this curve are found isolated anticlinal uplifts of older Mesozoic beds from which the Cretaceous covering has been removed by erosion.

The Lower Cretaceous, with which exclusively the ore-deposits are concerned, forms shallow synclines, between Bentheim and Ottenstein, within this area. Of these synclines that between Bentheim and Ochtrup, so hidden by a thick Diluvial covering that only around the border is the Neocomian seen at the surface, is particularly extensive.

The Wealden formation constituting the foot-wall, consists principally of an alternation of dark shales and limestones ; only in the upper portion at the contact with the Valanginian is clay-ironstone found, and this has hitherto been of no importance. The ironstone of the overlying Lower Neocomian is likewise at present of no importance. The Hauterivian also carries ironstone. This stage consists of sandstone or loose sand, and the iron ore as a rule occurs in the form of grains of sand cemented together by limonite. Such ore owing to its high quartz content is economically unimportant. Of greater importance is the clay-ironstone found in the grey-blue to blue-black Aptian clays belonging to the Upper Neocomian. Farther still in the hanging-wall, the Gault beds belonging to the Albian

contain no deposits worth mentioning.   The formation of the ironstone
accordingly terminated in the Gault.   The designation of this ore as Gault
ironstone, though often applied, should therefore be dropped, since the
principal horizon belongs to the Aptian, that is, to the Upper Neocomian.

Fig. 421.—Clay-ironstone deposits between Bentheim and Stadtlohn.   Einecke and Köhler.

In the mineralization, a strip lying between the Hauterivian sandstone
ridge and the Gault greensand, as illustrated in Fig. 421, is concerned.
This strip varies in width according to the dip of the beds, and is in greater
part covered by Diluvium, with which on the Dutch frontier, Miocene and
Oligocene are associated.

The Aptian clay-ironstone occurs in the form of beds, some of which consist of separate nodules, and some of compact material. These latter however may at the margin merge into beds of separate nodules, to which condition they may also at times be reduced by weathering. Such nodules vary in size from a man's head to a fist, and lie with the long axis parallel to the bedding. The beds vary in thickness up to 40 centimetres. They are separated from one another by clay or shale, which though firm in depth is plastic in the neighbourhood of the surface. The thickness of this intercalated material may reach several metres. The quantitative relation between it and the ore is a very important factor. According to the earlier workings this relation was as 1 of ore to 10 of barren material; more recent work has however given more unfavourable figures.

The ore when fresh is blue-grey, but upon weathering soon acquires a reddish-brown colour, the primary carbonate always tending to alter to red limonite. The iron occurs principally in the form of ferrous carbonate, contaminated by silica, lime, alumina, and magnesia, and in part containing considerable phosphorus. An average analysis of eighteen beds gave: 38·42 per cent of Fe, 0·19 per cent Mn, 6·05 per cent CaO, 1·15 per cent MgO, 11·75 per cent residue, 0·71 per cent P, and 0·32 per cent of S. The loss on ignition amounted to 24·63 per cent, so that such ore by roasting could be enriched approximately twenty-five per cent. The roasted ore has so far contained 44–49 per cent of iron. It is very porous though nevertheless firm and transportable, belonging therefore to the easily reducible ores.

The payability of the ore-beds depends greatly upon advances in the methods and means of stripping. The reserves have been estimated by Köhler and Einecke at 15,000,000 tons.

THE CHAMOSITE- AND THÜRINGITE BEDS

These two varieties of ore, which differ petrographically from the ordinary iron ores, occur in large amount in but a few places. Both belong to the lepto-chlorites and accordingly exhibit a crystalline structure, though crystals are never freely developed.

Thuringite, named after Thuringia, is a hydrous iron-aluminium silicate, the formula for which, as given in mineralogical text-books, is $H_{18}Fe_8(Al,Fe)_8Si_6O_{41}$. Since however alumina and silica are frequently present, this formula gives only an approximate representation of the composition of the mineral. It consists of a compact light- to dark green mass with a scaly or fine-grained structure. Its hardness is 2–2·5, and its specific gravity 3·2. It frequently contains 31–35 per cent of FeO and 12–18 per cent of Fe_2O_3, with 23 per cent of SiO_2. Chamosite is likewise a hydrous iron-aluminium silicate, the composition of which is even more variable than that of thuringite. This greenish-grey to green-black mineral occurs compact or finely oolitic in structure and contains 36–42 per cent of FeO. The name is derived from the bed-like masses in the Jurassic at Chamosen, Canton Valais, Switzerland.

These two ores, so far as they can be exploited, form beds of very variable thickness, though it is seldom that any large figure is reached. Since the ore-bed represents only a special facies of normal sedimentation, it may gradually merge into ordinary slaty rock.

The secondary formation of gossan in the thuringite- and chamosite beds is interesting. Both ores by the action of meteoric waters become changed to limonite, which mineral is substantially richer than the primary ore. Such a change may reach various depths.

The extent of these beds may be considerable, as is the case with the Bohemian occurrences. Owing, however, to the low quality of the ore the extent of operations is always dependent upon the prices fixed by the Convention. The two best-known chamosite- and thuringite occurrences in Thuringia and Bohemia are of Silurian age. Other deposits are younger, that of Chamosen in Valais, for instance, being Jurassic. The ore-beds are everywhere conformable and often alternate with other beds of undoubtedly sedimentary origin. The structure of the ore is in part oolitic. Chamosite has also been found in the Lorraine minette. There are accordingly gradations between the oolitic occurrences on the one hand and the chamosite- and thuringite deposits on the other. Typical representatives of both deposits can nevertheless be readily distinguished from each other.

NUČITZ IN BOHEMIA

LITERATURE

DR. F. KATZER. Die Geologie von Böhmen, 1892.—M. V. LIPOLD. ' Die Eisenstein-lager der silurischen Grauwackenformation in Böhmen,' Jahrb. d. k. k. geol. Reichanst., 1863.—JOS. VALA and R. HELMHACKER. Das Eisensteinvorkommen in der Gegend von Prag und Beraun. Prague, 1873.—C. FEISTMANTEL. ' Die Eisensteine in der Etage D des böhmischen Silurgebirges,' Abhandl. d. kgl. böhm. Ges. d. Wiss. VI., 1875 and 1876 ; ' Über die Lagerungsverhältnisse der Eisensteine u.s.w.,' Sitzungsber. d. kgl. böhm. Ges. d. Wiss., 1878.—V. UHLIG. ' Die Eisenerzvorräte Österreichs,' Report of the Geol. Ges. in Vienna, in The Iron Ore Resources of the World. Stockholm, 1910.

The Silurian stage D of Barrande is divided into the zones d_1–d_5, the extension of which coincides with that of the two flanks of the Brdagebirge. Zones d_1, d_4, and d_5, are metalliferous.

Zone d_2, that is, the lowermost zone of the D horizon, is formed of grauwacke and slate, with diabase and ironstone. This last is an oolitic hæmatite up to 17 m. in thickness, though for a great portion of its extent represented by indefinite stringers in which swellings and contractions continually alternate. Often several smaller beds occur, separated from each other by diabase or a clayey schist. This ironstone horizon forms the uppermost division of zone d_1, and has a thickness of $40-100$ metres. Uhlig mentions a large number of iron deposits in this horizon, though to-day these are only of subordinate importance.

The ironstone of zone d_4, that is, of the Zahořan beds, is considerably more important, including as it does the important ore-bed at Nučitz. This zone consists of quartzose - clayey, micaceous, grauwacke schists with intercalated quartzitic or fine-grained grauwacke. The schists when fresh are dark grey, while at the outcrop they are brownish.

The ironstone bed at Nucitz, worked in opencut, occurs in schistose country - rock on the northern flank of this Silurian area. It reaches 22 m. in thickness, from which maximum it pinches out gradually both east and west. The tectonics of this bed, which dips 50°-60°, are complicated by reason of six large faults. The throw of these faults individually reaches 900 m., the amount increasing from east to west. By these faults the disposition of the six shafts, from which the deposit has so far been developed for a length of 8000 m., has been determined. This deposit

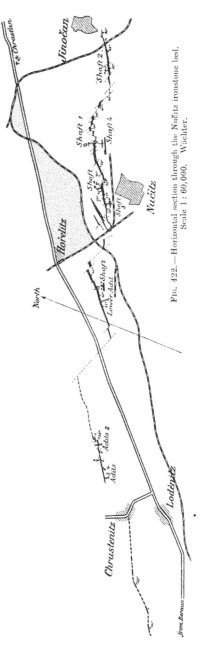

Fig. 422.—Horizontal section through the Nučitz ironstone bed. Scale 1 : 60,000. Wächter.

is known to the north-east as far as Jinonic near Prague, and to the south-east through Lodenitz, Vráž, Beraun, to Knizkowic. Here at Zditz it occurs at the surface with a width of 11 m., so that it can again be mined by opencut. Altogether this ore-bed is known for a length of 40 kilometres.

The dark greenish-grey, more seldom bluish-grey ore is principally a ferrous silicate of oolitic structure with a more or less sideritic or schistose matrix. It is generally described as chamosite, though mineralogically

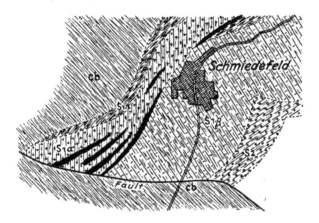

FIG. 423.—The chamosite deposit at Schmiedefeld in Saxe-Meiningen.
Scale 1 : 100,000. Einecke and Köhler.

cb = Cambrian ; $S1\pi$ = Lower Silurian quartz-schists ; $S1a$ = Lower Silurian Griffelschiefer ; $S1\beta$ = overlying Silurian dark slate. The Lower Silurian, together with the deposit, is by the Lichtentann Main Fault thrown against the Cambrian.

it differs from this mineral in its high siderite content. In the neighbourhood of the outcrop it is altered to limonite.

On the southern flank of the Silurian basin this deposit is not known, though an oolitic ore-bed 0·75–1·1 m. thick and having the same characteristics as that at Nučitz is found in the horizon d_5 between Vráž and Řevnitz.

The average iron content of the Nucitz raw ore is 35·5 per cent with about 20 per cent insoluble residue and 20 per cent loss on ignition. The roasted ore contains 44·3 per cent of iron.

The Bohemian iron production amounts to about 700,000 tons, of which 650,000 tons are contributed by the Prague *Eisenindustriegesellschaft* and the *Montangesellschaft* at Nučitz. The developed ore-reserves at Nučitz, from Jinočan to Chrustenitz, a distance of 8 km., are estimated at 11,000,000 tons.

SCHMIEDEFELD IN THE THURINGIAN FOREST

LITERATURE

H. LORETZ. ' Beitrag zur geologischen Kenntnis der kambrisch-phyllitischen Schie-
ferreihe in Thüringen,' Jahrb. d. pr. geol. Landesanst., 1881, p. 175; ' Bemerkungen
über die Untersilurschichten des Thüringer Waldes und ihre Abgrenzung vom Kambrium,'
Jahrb. d. pr. geol. Landesanst., 1884, p. 24; ' Zur Kenntnis der untersilurischen Eisen-
steine im Thüringer Walde,' Jahrb. d. pr. geol. Landesanst., 1884, p. 120.—' Die Eisenerz-
lager Oberfrankens,' Berggeist, 1886, p. 245.—W. KELLNER. ' Nachrichten über den
Bergbau und Hüttenbetrieb in Südthüringen,' Berg.- u. Hüttenm. Ztg., 1889, p. 157.—
E. ZIMMERMANN. Geologie von Sachsen-Meiningen. Hildburghausen, 1902.—ZALINSKI.
' Untersuchungen über Thuringit und Chamosit aus Thüringen und Umgebung,' Neues
Jahrb. f. Min., 1904, p. 40.—Geological map of Prussia : Sections Grafenthal, Probstzella,
Saalfeld, Schmalkalden, Lehesten, Schleusingen, Masserberg, together with text.—G.
EINECKE and W. KÖHLER. ' Die Eisenerzvorräte des Deutschen Reiches,' Archiv für
Lagerstättenforschung, Part 1, published by the Pr. geol. Landesanst., Berlin, 1910.

In the Lower Silurian at Schmiedefeld in Saxe-Meiningen two
conformable ironstone horizons occur, the extent of which over many
square miles has been delineated by the Prussian Geological Survey. A
quartzite zone with intercalated grey-green micaceous clay-slate is regarded
as uppermost Cambrian ; above this, in the Lower Silurian, the clay-slate
rapidly increases, forming dark, soft, blue-black beds, often developed as
Griffelschiefer. In these beds the ironstone horizons are situated, the lower
occurring between the *Griffelschiefer* and the Cambrian quartzites, while
the upper and economically more important occurs in the hanging-wall of
the *Griffelschiefer*, from which it is often separated by a blue-grey or blue-
green quartzite. Above this again, follows an alternation of quartzite with
clay-slate, which in the direction of the Tannenwald merges into the Lower
Silurian main quartzite.

At Schmiedefeld in the eastern Thuringian Forest, three ore-beds are
known in the Lower Silurian ironstone horizon. The lowest of these, which
is now exhausted, consists of an alternation of hæmatite, quartzite, and
Griffelschiefer, making a total thickness of 3 metres. The middle bed, the
position of which is indicated in Fig. 424, displays oolitic structure and has
a thickness of 2 metres. The main bed, occurring in the hanging-wall as
indicated also in Fig. 424, is 15–20 m. thick, strikes north-east, dips about
60° south-east, and is known for 1 km. along the strike. It carries both
thuringite and chamosite. The former contains some 30 per cent of iron
with 10 per cent of water, is olive-green to dark green in colour, and
scaly, compact, or oolitic in structure. The chamosite, which occurs in
larger quantities than the thuringite, is dark silver-grey to black in colour,
and in its oolitic structure greatly resembles minette. It is considerably
harder than the thuringite. Upon weathering chamosite alters less readily
to limonite than does thuringite.

Chemical analysis of the primary ore shows a considerable amount of ferrous carbonate, while chloritic silicates recede. This carbonate content is not in agreement with the composition of chamosite as given in text-books on mineralogy. The chamosite at Schmiedefeld accordingly is rather a siderite with which a small amount of silicate of alumina is associated ; according to Loretz it contains 22 per cent of CO_2. Roasted chamosite, after a loss of about 20 per cent in weight, contains 43–46 per cent of Fe, 20–31 per cent of insoluble residue, and 0·7–1·2 per cent of phosphorus. The occurrence of TiO_2 in chamosite, reaching in the unroasted ore 1·63 per cent, is interesting.

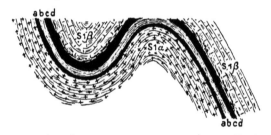

FIG. 424.—Diagrammatic section of the Schmiedefeld chamosite bed.   Einecke and Kohler.

$S1a$ = Griffelschiefer ; a = leader bed ; b = micaceous quartzitic schist ; c = main bed ; d = quartzite alternating with clay-slate ; $S1\beta$ = overlying dark slate.

Concerning genesis, these are true sediments with distinct stratification. The Schmiedefeld mines belong to the Maximilian works at Unterwellenborn, where the ore is smelted with the Kamsdorf calcareous ores.

Iron ore-beds are widely distributed in the Silurian of eastern Thuringia, though but very few are workable. Although the present production is small, a reserve of about 100,000,000 tons exists.

THE DETRITAL IRON BEDS

These may be of very varied formation. They are formed firstly as eluvial gravels.[1] When for instance a ferruginous rock becomes decomposed by surface agencies, the ore contained settles down *in situ*, where it gradually becomes concentrated by the removal of barren material. Only the resistant iron ores such as magnetite and specularite are liable to concentration in this manner. These occur principally as magmatic segregations in eruptive rocks, and as contact-deposits and sediments in crystalline rocks. To such detrital deposits belong, among others, the scattered blocks of Uifak iron on Disko island ;[2] the large blocks of

[1] *Ante*, p. 16.   [2] *Ante*, p. 341.

magnetite occurring at places in Angola, Portuguese West Africa ; the magnetite found with the iron deposits in the Urals, and which first gave rise to iron mining in that district ; and finally, the so-called *Canga* of Minas Geraes [1] in Brazil.

In fluviatile deposits fragments of older iron deposits may also experience enrichment by natural concentration, though no occurrence is known where this has resulted in a workable deposit.

Marine gravels are formed principally by encroachment of the sea. They are accordingly found as the basal conglomerates of coastal formations, and frequently contain fragments of even the less resistant iron ores, such for instance as limonite. In these gravels concentration in many cases reaches such a degree that useful deposits are formed. Basal ferruginous conglomerates frequently extend over wide areas but have the disadvantage of a very irregular aggregation, the action of the surf being to fill the depressions in the bed-rock with conglomerate, leaving large stretches of the more level portions without any such covering. Since the fragments broken off by sea action, as well as those carried by the rivers to the sea, may have the most varied composition, the iron content of such deposits is often very variable. Moreover, if such deposits are of great geological age they may contain the most varied cementing material, calcium carbonate or calcium phosphate being often found. Extensive ferruginous basal conglomerates occur at the base of the Cenomian in the Münster basin, and at the base of the Senonian in the neighbourhood of Peine. Admittedly, prospecting operations have proved the former too poor in iron to allow exploitation ; the latter on the other hand are worked at considerable profit.

The association of such ironstone beds with phosphate nodules and concretions is noteworthy. The phosphate nodule beds are likewise found preferably in basal conglomerates, where, though the principal material may be iron ore, the calcium phosphate occurs not only in the matrix but also as nodules. Such detrital deposits yield therefore an ideal material for the Thomas process. The formation of recent ferruginous detrital deposits, and especially of iron sand, is more fully described in a subsequent chapter.[2]

The shape of these deposits is in general that of a bed. While the eluvial gravels in accordance with their origin depend for their shape upon the contours of the bed-rock, the recent marine magnetite sands form gently-inclined even beds. The important older shore deposits, on the other hand, containing the limonite fragments of earlier formations, are associated with anticlines, synclines, faults, etc., their extension in depth depending upon the inclination given them in process of folding.

[1] *Ante,* p. 620. [2] *Postea,* pp. 1052-1054.

The structure of these deposits when cemented is generally that of a breccia or that of a conglomerate. Primary depth-zones are discernible in so far that in a vertical section the deepest portion, owing to the high gravity of the iron ore, is frequently the more ferruginous, while the upper layers as a rule contain more barren material. Since these deposits always consist of oxidized or hydrated ores, where they come to the surface secondary alteration is rare, though magnetite may be altered to limonite or hæmatite, and limonite to hæmatite, and *vice versa*.

The economic importance of these deposits is on the whole not great. In Germany at the present day only the occurrences at Bülten and Adenstedt near Peine play any great part.

Peine

LITERATURE

BEYRICH. ' Über die Zusammensetzung und Lagerung der Kreideformation zwischen Halberstadt, Blankenburg, und Quedlinburg,' Zeit. d. d. geol. Ges., 1849, p. 288.—A. v. STROMBECK. ' Die Gliederung des Pläners im nordwestlichen Deutschland zunächst dem Harze,' Neues Jahrb. f. Min., 1857 ; ' Über die Eisensteinablagerung bei Peine,' Zeit. d. d. geol. Ges., 1857, p. 313.—D. A. BRAUNS. ' Die obere Kreide von Ilsede und Peine und ihr Verhältnis zu den übrigen subherzynen Kreideablagerungen,' Naturhist. Verhandl., Bonn, 1874 ; ' Über die Eisensteinlager bei Peine,' Zeit. für d. gesamte Naturwissensch., Halle, 1874, p. 280.—FR. KOLLMANN. ' Die Erzlagerstätten für Thomasroheisen in Hannover und Braunschweig,' Stahl u. Eisen, 1886, p. 787.—W. DAMES. ' Senone Phosphoritlager bei Halberstadt,' Zeit. d. d. geol. Ges., 1886, p. 915.—G. MÜLLER. ' Beitrag zur Kenntnis der oberen Kreide am nördlichen Harzrand,' Jahrb. d. pr. geol. Landesanst., 1887, p. 372.—A. DENCKMANN. ' Über zwei Tiefseefazies in der oberen Kreide von Hannover und Peine und eine zwischen ihnen bestehende Transgression,' Jahrb. d. pr. geol. Landesanst., 1888.—G. EINECKE and W. KÖHLER. ' Die Eisenerzvorräte des Deutschen Reichs,' Archiv für Lagerstättenforschung, Part 1, published by the Prussian Geological Survey. Berlin, 1910.

The ironstone horizon from which the supply of ore for the Ilsede smelting works at Peine is drawn, though of considerable extent along the northern foreground to the Harz, appears, so far, to be well developed only at Gross-Bülten and Adenstedt near Peine, and at Lengede about 9 km. to the east of Peine. The position of these places is indicated in Fig. 20. The ore-bed belongs to the extensive Emscherian or lowest Senonian conglomerate horizon, which lies unconformably upon early Cretaceous. The petrographical character of the horizon on the whole varies considerably. The limonite conglomerate has a calcareous or marly, seldom argillaceous, cementing material, in which much phosphorite occurs. In other districts conglomeratic marls or sandstones with a little limonite- or phosphate conglomerate, hardly to be regarded as payable, predominate. Such however have often been worked for phosphorite. Conglomerates quite free from limonite or phosphorite are rare in the so-called Ilsede horizon.

This horizon in the Halberstadt syncline and at Zilly, as well as at Goslar and Oker in the Harz, is developed in sandy facies, as distinguished from the more marly and calcareous development in the neighbourhood of Peine and Ilsede.

With regard to tectonics, it is noticeable that the Peine conglomerates have been subject to the great Tertiary or late Cretaceous folding, this having produced south-east striking anticlines and synclines. By erosion the Cretaceous beds on the anticlines have been removed, permitting Jurassic and Triassic to appear at surface; the ironstone conglomerate is accordingly confined to the synclines.

Concerning genesis, the source of the ore-fragments is interesting. Since Gault fossils, *Ammonites Milletianus*, etc., have been found in large amount in the iron ore, the assumption is justified that the primary ore occurred in the Gault. Occurrences such as clay-ironstone beds [1] separated from one another by soft clay, were certainly such as would in their destruction give rise to the productive Ilsede ferruginous conglomerates. That in process of destruction the original carbonates became oxidized to limonite

FIG. 425.—Ideal section through the iron ore-bed of the Bülten mine and that at Lengede.

h = Upper Bunter ; g = Muschelkalk ; f = Keuper ; e = Jurassic ; w = Necomian ; c = Gault ; cl = Planer ; b = ironstone bed ; a = Senonian.

is easily conceivable. The matrix consists of calcium carbonate which, having originally been contained abundantly in the foot-wall Pläner beds, was subsequently taken up by the sea, to be precipitated again between the ironstone fragments. In consequence of the encroachment of the Senonian sea, the ironstone lies sometimes on Gault, as at Bülten, and sometimes on Pläner, as at Lengede.

The principal bed at Ilsede occupies an extensive closed syncline, around the entire outline of which it comes to surface. This syncline dips steepest in the south-west, and is disturbed by many faults. Twelve kilometres south-east of the steeper flank lies the occurrence of Bodenstedt-Lengede which, as illustrated in Fig. 425, is separated from the above-mentioned syncline by an air-anticline. At Lengede also, the detrital ore-bed belongs to the lowest Senonian.

The Bülten-Adenstedt syncline has been explored by a large number of bore-holes, in all of which Senonian formed the hanging-wall of the ore-bed, and Gault the foot-wall. The syncline is 11 km. long. In the east the beds dip 15°–18°, and in the west 3°–4°. The thickness varies,

[1] *Ante*, p. 1039.

being generally greatest in the east, where it reaches 20 metres. In the middle of the syncline the thickness is not known, while in the west the outcrop shows it to be small. The greatest depth hitherto reached by boring is 205 m.; at that particular place the ore was 7 m. thick. The second occurrence has been proved at Bodenstedt-Lengede for 1600 m. along the strike, the thickness being 5–6 metres. This deposit dips south-east and is separated from the Pläner limestone in the foot-wall by a thin layer of calcareous clay. The hanging-wall, so far as it was not removed during the Diluvial period, consists of beds with quadrangular jointing.

The average iron content of the bed is 28–35 per cent, this being contained in greater part in the nodules and to a less extent in the matrix. According to their colour and the nature of the matrix, grey, white, and yellow limestone- and clay ores are distinguished. The phosphorite nodules contain as much as 12 per cent of phosphorus, this, before the introduction of the Thomas process, having prevented the use of the ore for industrial purposes. The manganese content occurs in the form of rhodochrosite, polianite, and pyrolusite. From place to place the composition of the ore varies considerably. Although that in the Bülten syncline on the whole resembles that at Lengede, the latter, as the following table shows, contains somewhat more silica and alumina.

	Fe.	Mn.	P.	SiO_2.	Al_2O_3.	CaO.	MgO.	Loss on Ignition.
Bülten ore, per cent .	32·9	4·5	1·1	4·4	0·8	17	0·6	20
Lengede ore, per cent .	34·4	0·8	1·6	7·8	3·7	14·6	0·08	17

This same limonite conglomerate is also known at other places, as for instance at Gehrden Berg south of Hanover, and at Isernhagen to the north of that town. In the case of the former occurrence, immediately above the Gault comes a compact fine-grained conglomerate 40 m. thick, containing 15 per cent of Fe, which in turn is followed by a bed, 2·4 m. thick, containing 32 per cent of Fe, 14 per cent $CaCO_3$, and 20 per cent of Al_2O_3, while an argillaceous conglomerate 90 cm. thick forms the hanging-wall. Between Degersen and Bönningen the ore-bed is 1·85 m. thick and workable. At Isernhagen the ferruginous conglomerate is 2–3 m. thick and occurs under 2 m. of Diluvial material. This deposit in regard to iron, manganese, and phosphorus, resembles the Ilsede ore-bed, though silica and alumina are somewhat more abundant.

At Harzburg a phosphorite and limonite conglomerate, fully described by H. Schröder in the explanatory text to the geological map of Harzburg, is found along the Ilsede horizon, the outcrop of which for a considerable distance runs parallel to the Harz. Along it at Scharenberg, from the foot-wall to the hanging-wall the following section obtains : 0·40 m. phosphorite and limonite conglomerate ; 8·0 m. conglomeratic sandstone with 0·2 m. phosphorite and limonite ; 3·0 m. conglomeratic marl with several limonite and phosphorite layers ; 1·5 m. conglomeratic marl almost without limonite and phosphorite ; 8·0 m. conglomeratic sandstone with little limonite and phosphorite ; and finally, 5·8 m. solid and in part conglomeratic sandstone. At this place also, the ore, as indicated by numerous fragments of Gault fossils, was derived from the Gault. Northeast and east of Zilly the horizon was worked for phosphorite. Similar deposits are known south of Halberstadt and Quedlinburg on the Hercynian Upper Cretaceous plateau.

The economic importance of the ironstone deposits of the Ilsede horizon, and particularly those at Peine, is considerable. In spite of the low iron content other factors are so favourable that the Ilsede smelting works makes a handsome profit. The ore contains so much lime in itself that additional lime flux is not necessary. Although self-fluxing ore with lime and silica in right proportion are present only in small amount, the necessary silica flux is obtained from neighbouring mines working the Salzgitter horizon where the ore is rich in silica. The combination of these two occurrences therefore produces a self-fluxing mixture. The high phosphorus content is responsible for an additional profit in the sale of Thomas slag. The manganese content is likewise a favourable factor, as is also the chemically combined water, this latter rendering the ore porous and easy of reduction. About 800,000 tons of ore are produced annually by a staff of 970 employees. The ore-reserves of Bülten-Adenstedt, Lengede, etc., are estimated by Einecke and Köhler at 218,000,000 tons.

SALZGITTER

LITERATURE

WÜRTTEMBERGER. 'Eindrücke an den Protinerzen von Salzgitter und den Petrefakten,' Zeit. d. d. geol. Ges., 1865, p. 232.—H. v. DECHEN. Die nutzbaren Mineralien und Gebirgsarten im Deutschen Reiche, 1873, p. 587.—G. BOEHM. 'Beiträge zur geognostischen Kenntnis der Hilsmulde,' Zeit. d. d. geol. Ges., 1877, p. 215.—M. NEUMAYR. 'Über das Alter der Salzgitterer Eisensteine,' Zeit. d. d. geol. Ges., 1880, p. 637.—A. DENCKMANN. 'Über die geologischen Verhältnisse der Umgegend von Dörnten nördlich Goslar,' Abhandl. d. pr. geol. Landesanst., 1887, Vol. VIII. Part 2.—G. MÜLLER. 'Beitrag zur Kenntnis der Unteren Kreide im Herzogtum Braunschweig,' Jahrb. d. pr. geol. Landesanst., 1885, p. 95.—A. v. KOENEN. 'Über die Untere Kreide Norddeutschlands,' Zeit. d. d. geol. Ges., 1896, p. 713.—KLOOS. 'Über die geologischen Verhältnisse des Herzogtums Braunschweig mit besonderer Berücksichtigung der sogenannten Hilsmulde,' Verhandl. d. Ver. d. Naturforscher u. Ärtzte, 1897, p. 214.—G. MAAS. 'Die Untere Kreide des subherzynen Quadersandsteingebirges,' Zeit. d. d. geol. Ges., 1899, p.243.—A. v. KOENEN. 'Über die Gliederung der norddeutschen Unteren Kreide,' Nachricht d. kgl. Ges. d.

Wissensch. Göttingen, 1901.—H. Schröder. Explanatory text to section ' Harzburg,'
1908.—G. Einecke and W. Köhler. ' Die Eisenerzvorräte des Deutschen Reiches,'
Archiv für Lagerstättenforschung, Part 1, published by the Geological Survey. Berlin, 1910.

This ore-bed, belonging to the Neocomian, extends north of Goslar in
a north-westerly direction to constitute what is described as the Salzgitter
ironstone belt. The Salzgitter Range may be followed for more than
20 kilometres. It consists of three low mountain chains separated from
each other by two valleys. Upon these hills occur a large number of
opencuts and bore-holes, the evidences of previous prospecting operations.
The geological section, illustrated in Fig. 426, is not everywhere the same.
In general the beds form an anticline which, having resulted by horizontal
pressure from the north-east, is in part overturned. The tectonics have
been further complicated by the occurrence of overthrusts and faults, while
the bedding has been obscured by far-reaching erosion.

The core of the range consists of Triassic and Lias ; upon the latter

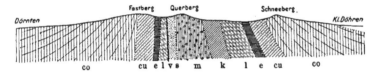

FIG. 426.—Diagrammatic section through the Salzgitter Range, showing the geological
position of the ironstone beds. Schröder.

s = Bunter ; m = Muschelkalk ; k = Keuper ; l = Lias ; cu = Lower Cretaceous ; e = ironstone ;
co = Upper Cretaceous ; v = fault.

comes Lower Cretaceous containing the ironstone ; this in turn is overlaid
by Upper Cretaceous. While Muschelkalk forms the central chain, soft
beds of Keuper, Lias, and Lower Cretaceous form two longitudinal
valleys, one on either hand, leaving the hard limestone of the Upper
Cretaceous to form the two outside chains.

The ore horizon does not lie conformably to the older beds as would
appear from Fig. 426, but transgressively, filling the depressions in their
uneven surface, great variations in thickness thereby resulting. Occasionally the bed consists of a series of elongated and thick lenses. Owing
to the considerable variations in thickness, figures obtained at the outcrop
and in the neighbourhood of the surface are not necessarily applicable in
depth. While the west flank of the Salzgitter Range dips 40°–60° and at
Othfresen contains an overthrust, the east flank appears to be flatter.
The bedding conditions of the ore-bed have been determined by numerous
exposures, which have shown that in many cases a 1–2 m. clay layer, or
Diluvium of considerable thickness covers the older beds. The thickness
of the ore at Altenhagen north of Gustedt, in the extreme west, is 12 m.,

a figure which a little farther south increases to 60 m., only to decrease again to 5 m., and then finally east of Steinlah to reach a thickness of 12 metres. This thickness does not consist entirely of clean ore, but includes a proportion of ferruginous and arenaceous clay. Nevertheless, north-west and west of Salzgitter, as illustrated in Fig. 427, ore-bodies of considerable thickness occur.

The ore is a conglomerate, the nodules of which exhibit the most varied size and shape. While in some layers they are scarcely discernible to the naked eye, in the Georg-Friedrich mine at Eisenkuhlenberg they reach the size of a man's head. The occurrence of this deposit as a basal conglomerate explains the approximately uniform size of the nodules of individual layers. These nodules are mostly incompletely rounded. The matrix is always very ferruginous and mixed with clay and limestone or marl. Now and then oolitic structure is exhibited, a feature which distinguishes the ore of Salzgitter from that of Ilsede. Phosphorite nodules are found in both. Such occur in almost all parts of the bed,

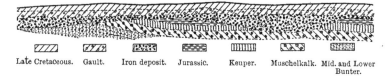

| Late Cretaceous. | Gault. | Iron deposit. | Jurassic. | Keuper. | Muschelkalk. | Mid. and Lower Bunter. |

FIG. 427.—Ideal section through the Salzgitter iron ore-bed and its country-rock.

though in but few places, as for instance in the Fortuna mine east of Gross Döhren and in the Segen Gottes mine near Salzgitter, do they occur in striking amount. At times phosphorite has been mined from this deposit for the manufacture of superphosphate. It is not unusual to find this ironstone horizon divided up by ferruginous clay layers, though intercalated layers of ferruginous limestone are exceptional.

In relation to genesis, the Salzgitter ore-bed is a shore formation by a Neocomian sea. The ironstone presumably was derived entirely, or in greater part from Jurassic beds ; in any case Jurassic beds with their large iron content were the main source of the material. The occurrence of oolite at Salzgitter is particularly interesting. While at Ilsede only the fragments of older geodes were concentrated and cemented, at Salzgitter the marine concentration was more intense ; in addition to mechanical degradation a complete dissolution of the ironstone geodes to an ironstone slime from which oolites became formed, must in places be assumed.

The chemical composition of the ore is best indicated by that won at the Georg-Friedrich mine, which is delivered to the Ilsede works as siliceous flux for the calcareous Peine ore. The iron content varies between 35·2 and 39·5 per cent, though this is lowered by the inclusion of poorer material mined for flux. From exposures west of Salzgitter analyses gave : 35–40·8 per cent of iron with 17·2–32·5 per cent SiO_2, 6·8–8·8 per cent Al_2O_3, 4·0–9·0 per cent $CaCO_3$, and 0·3–0·5 per cent of phosphorus. The high percentage of insoluble residue is a great disadvantage. Use of these ores depends therefore a good deal upon the possibilities of dressing, in which direction considerable difficulties have still to be overcome.

Similar limonite conglomerates are found also in the outlying districts of Salzgitter ; to the west at Alt-Wallmoden and Brodenstein, to the south at Harzburg, to the north at Flachstöckheim, and to the east at Osterwiek. Still more distant are the occurrences at Schandelah, east of Brunswick ; Gross-Vahlberg, Berklingen, and Oesel, east of Wolfenbüttel ; Achim and Rocklum, east of Börssum ; Delligsen, west of Alfeld ; and Börnecke, south of Halberstadt.

Operations are at present proceeding only at the Georg-Friedrich mine east of Dörnten, where the Ilsede works with 130 employees produces some 100,000 tons per year. It has already been stated that this ore is used to form a smelting mixture with the calcareous ore of Ilsede. The reserves are estimated at 60,000,000 tons.

RECENT DEPOSITS OF IRON OR TITANIFEROUS-IRON SAND

The eruptive rocks and the crystalline schists of the fundamental rocks, as is well known, contain on an average about 4·5 per cent of iron,[1] and the basic eruptives by themselves some 6–8 per cent, or more. This iron occurs sometimes in silicate- and sometimes in oxide form, the latter being more particularly represented by magnetite and ilmenite and to a less extent by specularite ; in addition, a small amount occurs also in sulphide form. Upon disintegration—especially of basic and ferruginous eruptive rocks such as gabbro, norite, labradorite, etc.—and subsequent concentration by rivers and shore-water, large deposits of iron sand are frequently formed which, owing to their titanium content, are as a rule known as titaniferous-iron sand. In this sand the iron grains are associated chiefly with quartz and felspar, but also with garnet, augite, hypersthene, hornblende, etc.

At different places, especially in countries far removed from large industrial centres, it has occasionally been attempted on a small scale to use these iron- or titaniferous-iron sands in blast-furnaces. Latterly also, concentration of this material by magnetic separators has been tried. One large bulk sample of such sand from the northern shore of the Gulf of St. Lawrence, near the mouth of the Moisie river,[2] gave, by

---

[1] *Ante*, pp. 149, 152.   [2] J. H. L. Vogt, *Teknisk Ugeblad*, 1908.

magnetic separation without previous crushing, a concentrate containing 66·3 per cent Fe and 3·9 per cent TiO_2, from material containing 29·4 per cent Fe and 7·9 per cent TiO_2. By crushing this first concentrate and submitting the crushed material to a second magnetic separation, a final concentrate containing 69·6 per cent Fe and 2 per cent TiO_2 was obtained. Similar results have also been obtained from several titaniferous-iron sands in Norway.[1]

From these results it follows that the titaniferous-iron sands derived from gabbro and similar rocks consist partly of separate grains of magnetite and ilmenite, and partly of a mixture of these two. Since every sand contains some iron, the term iron sand is only applied when the iron content amounts to at least some 15 per cent. Higher percentages, such as 30–40 per cent, occur only exceptionally, though occasionally within the beds narrow strips containing 50–60 per cent are found. The thickness of beds moderately rich in iron is only in rare cases more than one or two feet.

Some native gold has occasionally been found in iron sands, in fact there are all gradations between the ordinary gold gravels, which always carry some iron, and auriferous iron sands.

Iron sands are only exceptionally found in rivers, but are more frequent on sandy coasts, where the waves effect a natural concentration. The following occurrences are worthy of mention : auriferous iron sands on the coast of Tierra del Fuego in the extreme south of South America, which in the 'eighties were worked for gold ;[2] titaniferous-iron sand containing, according to one analysis, 60 per cent of iron and 8·14 per cent TiO_2, in a coastal zone, roughly 100 km. long, in the south-west corner of North Island, New Zealand ;[3] extensive beds of titaniferous-iron sand near the mouth of the Moisie river, and at other places on the north shore of the Gulf of St. Lawrence, where the country-rock consists of norite and labradorite ;[4] and titaniferous-iron sands on the coast of Sugaya, Department Shimané in Japan, which were formerly smelted.[5]

In volcanic districts also, similar deposits are occasionally met, as for instance on the coast of Naples, on Réunion, and on Celebes.[6] Finally,

[1] J. H. L. Vogt, 'Norges Jernmalmforekomster,' *Norw. Geol. Survey*, 1910, No. 51, pp. 33-34.
[2] Ant. Sjögren and C. Jul. Carlsson, 'On Recent Beds of Iron Ore, etc., in Tierra del Fuego (in Swedish),' *Geol. Fören. Förh.*, 1892, XIV.
[3] E. Metcalf Smith, 'On the Treatment of New Zealand Magnetic Iron Sands,' *Journ. of the British Iron and Steel Inst.*, 1896, Vol. I. p. 65.
[4] J. F. Kemp, 'A Brief Review of the Titaniferous Magnetite,' *School of Mines Quarterly*, July 1899, pp. 331-333.
[5] 'Les Mines du Japon,' *Official Report for the World Exhibition in Paris*, 1900, pp. 327-332.
[6] V. Drasche, *Jahrb. d. k. k. geol. Reichsanst.*, 1876, XXVI. p. 42 ; F. Rinne, *Zeit. d. d. geol. Ges.*, 1900, LII. p. 343.

W. Deecke described ferruginous sands formed from glacial detritus, on the Baltic coast and on the shores of lakes in the north-German lowlands.[1]

THE IRON ORE-BEDS CONSISTING PRINCIPALLY OF SPECULARITE AND MAGNETITE IN THE FUNDAMENTAL ROCKS AND IN EARLY PALÆOZOIC CRYSTALLINE SCHISTS

The determination of the genesis of the iron deposits occurring conformably in the crystalline schists is particularly difficult. Chiefly because of this conformity, but also because of frequent banded structure, these occurrences were formerly regarded by many, though not by all authorities, as altered sediments. This view in its inclusiveness however was wrong, since conformity and banded structure are not in themselves sufficient evidence of sedimentation. Both characteristics, notoriously, are often associated with many deposits of contact-metamorphic or metasomatic origin, as well as with foliated magmatic segregations.

The bed-like iron occurrences in the fundamental rocks at Arendal are, according to the authors' view, contact-metamorphic deposits, with which class also, though with restrictions, the Persberg-Dannemora deposits of Middle Sweden are reckoned. Other occurrences, such as those at Kiirunavaara, Gellivare, etc., are undoubtedly of eruptive character and have been formed by magmatic differentiation. The banded or striped titaniferous-iron ores in gabbro, hornblende-schist, etc., in the fundamental rocks, are likewise the undoubted products of magmatic differentiation. Finally, conformable iron deposits may also have been formed in the fundamental rocks by ordinary metasomatic processes, without association with eruptive rocks.

Accordingly, a large number of deposits formerly often regarded as sediments must now be omitted from that class. A considerable number nevertheless remain, and among these, many particularly important deposits must be regarded as altered sediments.

The ferruginous mica-schist, itabirite, occurs fairly extensively in the fundamental rocks and perhaps still more in the early Palæozoic schists. Deposits of this material have been explored principally in northern Norway [2] and in Minas Geraes, Brazil, but they are also known in many other places, of which the following are worthy of mention : Soonwald between Gebroth and Winterburg ; Borsa in the Marmaros, Bukovina ; Villefranche in the Department Aveyron, France ; South Carolina ; and

[1] ' Über den Magneteisensand der Insel Ruden,' *Mitteil. d. Naturw. Verf. f. Neuvorp und Rügen*, 1888, XX. 4, VII.
[2] *Postea*, p. 1056.

Okandeland, West Africa.   The itabirite on the African Gold Coast, similarly to that in Brazil, is auriferous.[1]   Kriwoj Rog in Russia is to some extent comparable with these occurrences.   The primary ore in the Lake Superior district, where siderite- and greenalite silica-schists predominate, is of somewhat different character.[2]

With all these ore-beds the iron ore is accompanied chiefly by quartz, which in the poorer varieties occurs in large amount.   In several districts, but not in northern Norway, the ferruginous mica-schist with an increase of quartz merges into ferruginous quartz-schist.   Analyses of all these ores show that, even with high silica, as a rule only fairly small amounts of Al_2O_3, CaO, and MgO are present, while the alkalies are practically absent.   In addition to quartz, the associated minerals are chiefly hornblende and chlorite or talc, epidote and garnet, occasionally also augite, while felspar occurs either not at all or only exceptionally.   Titanic acid is completely absent or occurs only in traces.   Phosphoric acid is sometimes very low, as at Kriwoj Rog, Minas Geraes, etc., and sometimes somewhat higher, as at Dunderland.   Pyrite is observed only subordinately.   The manganese content is as a rule very low ; occasionally however some layers or seams of the deposit carry somewhat more manganese, as for instance in northern Norway ; while here and there, as at Minas Geraes, distinct manganese ore-beds occur in close connection with the iron deposits.

The ore-beds here to be described, including the typical representatives at Dunderland, Minas Geraes, Kriwoj Rog, and the primary ores of Lake Superior, occur within formations of undoubted clastic sediments.   In many districts chiefly quartz-schists—altered sandstones—but also mica-schists, phyllites, etc., predominate ; while at other places, as in northern Norway and in Minas Geraes, the ore-beds occur in the immediate vicinity of limestone.

These deposits, as is more fully set forth when describing the occurrences of northern Norway, are in general to be regarded as chemical sediments.   Occasionally, as with the siderite- and greenalite silica-schists of the Lake occurrences, the iron was precipitated primarily as ferrous carbonate, $FeCO_3$, or ferrous silicate, $FeSiO_3$; in most places, on the other hand, an original precipitation as oxide or hydrate, similar to that which takes place with ordinary lake ores, may be assumed, while their present character was that impressed during subsequent regional-metamorphism.

In the case of the Lake occurrences and perhaps also of Kriwoj Rog, a subsequent metalliferous accretion played an important part.   No such accretion appears however to have taken place with other deposits, such for instance as those of northern Norway.

[1] *Postea*, p. 1060.   [2] *Postea*, p. 1062.

Some ore-beds, as for instance those of Marquette and Vermilion in the Lake district, occur in the more ancient fundamental rocks ; the majority however are probably connected with the Algonkian, Cambrian, or other early Palæozoic crystalline schists.

THE OCCURRENCES OF FERRUGINOUS MICA-SCHIST AND MAGNETITE-QUARTZ SCHIST AT DUNDERLAND, SALANGEN, ETC., IN NORTHERN NORWAY

LITERATURE

J. H. L. VOGT. Salten und Ranen, 1891 ; The Iron Ore Field of Dunderland, 1894 ; Norwegian Marble, 1897 ; The Iron Ore Occurrences of Norway, 1910 ; all in publications, Nos. 3, 15, 22, and 51 of the Norwegian Geological Department, with German résumé ; Die Erzvorkommen und der Bergbau des nördlichen Norwegens, Christiania, 1902 ; ' Die regional-metamorphosierten Eisenerzlager im nördlichen Norwegen,' Zeit. f. prakt. Geol., 1903, pp. 24-28, 59-65. See also The Iron Ore Resources of the World, Stockholm, 1910.

The northern Norwegian strongly regional-metamorphosed Palæozoic area already briefly described [1] is, according to Vogt, divided into the following three main divisions : on top, the Sulitjelma schists ; in the middle, a young gneiss ; and at the bottom, a thickness of mica-schist and marble, with ore-beds in its middle and upper portions. This last is not Archaean but probably Cambrian.

The bottom division is characterized more particularly by mica-schists — especially brown garnetiferous schists and occasionally staurolite- and disthene-schists—and by thick carbonate beds consisting partly of calcareous and partly of dolomitic marble. Conglomerates, quartzites, phyllites, etc., occur subordinately. At numerous places between Vefsen, $65\frac{1}{2}°$ north latitude, and Tromso, $69\frac{1}{2}°$ north latitude, this division carries intercalations of specularite- and magnetite-quartz schists, some of which are very thick. The greatest distance in a straight line between any two such occurrences is 520 kilometres. The best known are those in the Dunderland valley, the position of which is indicated in Fig. 188, though the occurrences at Näverhaugen in Salten, at Bogen, and many others in Ofoten, Salangen, Sörreisen, Tromsösundet, etc., are also worthy of mention.

The ore-beds occur in the schists, though in the immediate neighbourhood of limestone beds often several hundred metres in thickness. As illustrated in Figs. 428 and 429, between the ore and the limestone, or exceptionally the dolomite, comes mica-schist generally 1–20 m. thick. The deposits, as at places in the Dunderland valley, occasionally reach a length of one or more kilometres and a thickness of 50 m. or more ; mostly however the thickness is smaller, being 30, 20, 10, 5 m., and even less.

[1] *Ante*, p. 304.

Often an alternation of ore and non-metalliferous mica-schist occurs, and within a schist formation a hundred or more metres in thickness, more than one-half of the section may consist of ore.   Along the strike also, the ore-beds may often be followed for several kilometres before, finally, they disappear.   The beds are frequently contorted and crumpled.   In several districts they are traversed by granitic dykes.

NW

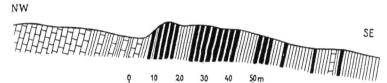

SE

FIG. 428.—Section of the occurrence of iron ore at Dunderland, showing the alternation of ferruginous mica-schist with mica-schist and limestone.   Vogt.

The ore-beds consist of ore—partly specularite and partly magnetite— and quartz, in fine layers or closely intergrown, some hornblende and epidote, while occasionally garnet, pyroxene, etc., occur subordinately. Within each ore-bed a number of smaller beds containing from 20 to 50 per cent of iron, and alternating with each other, are frequently met.   On an average these ore-beds contain 30–36 per cent of iron, and only exceptionally 40 per cent or more.   The association with quartz is so

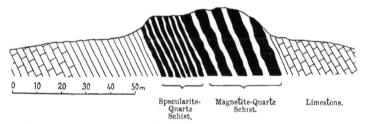

Specularite-   Magnetite-Quartz   Limestone.
Quartz   Schist.
Schist.

FIG. 429.—Section of the occurrence of iron ore at Urtvand in the Dunderland valley.   Vogt.

intimate that the iron content cannot be increased by sorting.   The beds are often developed as typical ferruginous mica-schists without, or with but little magnetite ; in places, however, pure magnetite without specularite is met.

The usual composition of the ore won is 40–52 per cent of Fe_2O_3 and Fe_3O_4, 36–45 per cent SiO_2, 0·5–1·5 per cent Al_2O_3, 1·5–5 per cent CaO, and 0·5–1 per cent of MgO.   The phosphorus content is on an average 0·20–0·25 per cent, this occurring as apatite, principally in the quartz bands but also in the metalliferous layers.   Titanic acid is absent ; sulphur is very low in amount, being mostly only 0·01–0·025 per cent. Manganese is low, being usually only 0·2–0·5 per cent, but higher when beds poor in manganese alternate with others which are manganiferous.[1]

These ferruginous mica - schists and magnetite - quartz schists must be regarded as members of the schist complex in which they occur. From their conformity, their association with the same geological horizon over extensive areas, their independence of eruptive rocks, the alternation with non-metalliferous schists, the bedded structure of different layers, and from chemical analogy with recent ferruginous sediments, it must be considered that these occurrences are of sedimentary origin.

In the Dunderland valley the sectional area of the deposits, perpendicular to the dip, according to Vogt, amounts to about 1,000,000 sq. m., while the other occurrences make together an equivalent area, or altogether a total of about 2,000,000 square metres. Were all these deposits put together they would form one bed 100 km. long and 20 m. wide. From these occurrences it is estimated that about 250,000,000 tons can be exploited by opencut, though such ore would only contain on an average at most 30–36 per cent of iron. In order to make use of the quartzose and low-grade ore, magnetic separators have been erected in three places, namely, at Dunderland, Bogen, and Salangen. At the two last-named places, where the ore is chiefly magnetite, in 1912 about 50,000 tons of iron concentrate and briquets containing about 65 per cent of iron and about 0·03 per cent of phosphorus, were produced. In the Dunderland valley, where the ore consists chiefly of specularite and only subordinately of magnetite, great technical difficulties in making such a separation have still to be overcome.

KRIWOJ ROG IN SOUTH RUSSIA

LITERATURE

K. BOGDANOWITSCH. The Iron Ore Deposits of the World, Stockholm, 1910, Vol. I. pp. 501-511.—P. PIATNITZKY. Trav. de la Soc. des Natural. à l'Univ. de Kharkow.— T. TRASENTER. Revue universelle des mines, etc., 1896, p. 34 ; reviewed by P. KRUSCH, Zeit. f. prakt. Geol., 1897, pp. 182-186.—TSCH. MONKOWSKY. Zeit. f. prakt. Geol., 1897, pp. 374-378.—A. MACCO. Ibid., 1898, pp. 139-149.

Kriwoj Rog is situated at the boundary of the two governments Ekaterinoslav and Cherson, south-west of the town Ekaterinoslav, in the Inguletz valley which at Cherson debouches into the Dnieper.

By the Kriwoj Rog basin is understood an approximately north-south zone of metamorphic rocks about 60 km. long and at its maximum only 6–7 m. wide, in Archaean granite or granite-gneiss. These rocks are

[1] *Ante*, p. 1055.

distinctly stratified; they are correlated by different authorities with the Huronian, Algonkian, or Cambrian, and represent three petrographical groups characterized respectively by quartzites, clayey-schistose rocks, and iron-quartzites. The tectonics of the entire zone are very complex and have not yet been completely determined. Some investigators assume three synclines overturned towards the east and pressed together, while others assume one isoclinal fold with many secondary folds and other disturbances. The beds dip mostly 35°–55° towards the west. In the metalliferous complex, with its iron-quartzite schists and ore-beds, the quartzites by increment of kaolin, talc, and chlorite, merge not infrequently into talc- and chlorite-schists, and in places even to actinolite- and grünerite-schists. In the hanging-wall of the metalliferous beds a series of principally clayey-schistose character occasionally occurs, at times associated with carbonaceous schists.

The ferruginous quartzites form beds often of great thickness and many kilometres long. These include the lenticular ore-bodies which generally contain but little quartz and often assume considerable dimensions, workable thicknesses of 25–50 m. or more being not uncommon. In depth these ferruginous lenses diminish in thickness and often rapidly pinch out. Only those portions containing at least some 50 per cent of iron are regarded as ore. The poorer parts containing 45, 40, 35 per cent or still less, are locally known as quartzite.

The ore consists chiefly of hæmatite or specularite, most of which is pseudomorphic after magnetite and is therefore martite. Macco in his treatise states that some 8 per cent of the iron content is still present as magnetite. As gangue, the iron-quartzites as well as the richer lenses carry practically only quartz. As the iron increases, the silica diminishes. Ore containing 62·5 per cent of iron carries roughly 7·5 per cent of SiO_2, while ore containing 69 per cent of iron contains only 0·7–0·9 per cent of SiO_2, together with very little Al_2O_3, Cao, and MgO. The manganese content is very low, that of sulphur is practically nil, while that of phosphorus is mostly only 0·013–0·020 per cent.

Kontkiewicz, Monkowski, and other authorities regarded these ores as metamorphosed sediments which consisted originally of iron-ochre with more or less sand, etc. On the other hand, the association of the ore-beds with hornblende rocks and grünerite has long been recognized. Piendel [1] considered the iron oxides to have arisen at the chloritization and epidotization of the ferruginous hornblende. Grünerite appears to accompany the beds in all parts of the basin. As is known, the formation of this mineral must be attributed to metamorphism of carbonates in depth; the decomposition of the grünerite again to iron oxide and silica is a process associated

---

[1] *Mém. de la Soc. des Natural. de la Nouvelle-Russie,* 8, 1, 1882.

with surface phenomena. In the deposits at Kriwoj Rog, patches containing bands of highly contorted quartzite and hæmatite completely analogous
to the Jaspelite stage of the Marquette district, are often observed. It
is very probable therefore that the formation of these deposits is just as
complicated a sequence of primary sedimentary formation and subsequent
alteration as may be demonstrated with the Lake Superior deposits.[1]

The ore at Kriwoj Rog was already known to the ancient Greeks,
and it is probable that the famous Scythian iron came from this district.
Mining on a large scale however began only in 1881, after which it developed
rapidly. In 1900 the production amounted to 2,800,000 tons, and in
1906, including that of other deposits in South Russia, to 3,650,000 tons.
These deposits, which with their high iron- and low phosphorus content
are worked principally in opencut, have in conjunction with the coal of the
Donetz basin given rise to the important iron industry of South Russia.
A small quantity of ore is exported to Russian Poland and Upper Silesia.
The total ore-reserves were very carefully estimated by Bogdanowitsch in
1910 at 86,000,000 tons with 62 per cent of iron, that is, an iron content
of 53,200,000 tons.

The Ferruginous Mica-Schists or Itabirite of Brazil

LITERATURE

W. v. Eschwege. Geognostisches Gemalde von Brasilien, 1822, nebst anderen Arbeiten
über Brasilien vom Anfange des 19. Jahrhunderts.—H. K. Scott. The Iron Ores of Brazil;
Iron and Steel Inst., 1902.—E. Hussak. Zentralbl. f. Min. Geol., 1905, and Zeit. f. prakt.
Geol., 1906, pp. 237-239.—O. A. Derby. The Iron Ore Resources of the World, Stockholm,
1910, II.

The most important, or at all events the best known of the itabirite
districts in Brazil is situated in the central portion of Minas Geraes, north
of Rio Janeiro. A railway, 493 km. long, connects the capital with this
district. The occurrences, striking north and north-east, lie scattered over
an area, illustrated in Fig. 320, some 150 km. long and approximately 100
km. wide.

The ore-beds belong to the very thick, highly contorted and regional-
metamorphosed, probably Cambrian, itacolumite-itabirite formation, consisting principally of quartzites and crystalline sandstone—itacolumite
—with altered clay-slates and limestones. This formation lies upon
Archaean gneiss, mica-schist, etc., with granite. The ore-beds occur
conformably to the quartzites and slates in which they lie, and generally, as in Norway, in the immediate neighbourhood of limestone.
Eschwege in 1822 applied the name itabirite to the irregularly coarse,

---

[1] Bogdanowitsch, *ante,* p. 504.

clean ore from the summit of Itabiro do Campo. This name subsequently became used not for the ore more or less free from quartz, but for the ordinary quartz-containing ferruginous mica-schists.

In Minas Geraes all possible gradations occur from quartzite containing only a few flakes of specularite, through ferruginous mica-schists containing specularite and quartz in fairly equal amount, to beds of almost pure specularite. Beds containing up to 99·5 per cent of specularite often reach very considerable thickness. For the same reasons as those given when discussing the similar occurrences in Norway,[1] a sedimentary origin may also be presumed for these Brazilian deposits.

Owing to lack of solidity in the ore and to the hot and rainy climate—annual rainfall 1500–2000 mm.—these Brazilian ore-beds have suffered great disintegration, wherefrom recent mechanically-formed deposits have resulted. Derby accordingly differentiates :

1. Ore-beds *in situ*, the ore in places being thick, rich, and hard enough to form prominences.

2· Loose rubble ore at the denuded outcrop.

3. Gravel-deposits in the valleys below the ore-beds.

4. *Canga*, which occurs over extensive areas and consists of ore-detritus cemented by limonite to form an ironstone conglomerate.

As already mentioned,[2] the itabirite is occasionally auriferous ; certain auriferous jacutinga lines occur which, more particularly in former years, were actively worked.

Derby in his map of the Minas Geraes district indicates no less than fifty-two large deposits *in situ*. In the case of nine of the most important he reckons an ore-reserve—only at the outcrop without taking into account the assumed continuation in depth—of nearly 1000 million tons. The other deposits probably contain about the same amount. At the outcrop alone therefore, there are altogether some 2000 million tons of ore. This estimate includes only rich material with 67–99·5 per cent of iron oxide, equivalent to 47–69·5 per cent of iron, the remainder being quartz. The ore contains very little phosphorus or sulphur, and is free from titanium. Ordinary itabirite is very poor in manganese, sometimes however in association with the itabirite, distinct beds consisting partly of manganiferous and iron ore partly of pure manganese ore, occur ; the two ores of manganese and iron accordingly constitute geologically one complete occurrence. Further reference to this subject is made when discussing manganese ore-beds. The ore-reserves of the *Canga* are estimated by Derby at some 1700 million tons with about 50 per cent of iron.

The iron deposits of Minas Geraes are, so far as is yet known, the most extensive in the world. Similar ore-beds are also found in other

---

[1] *Ante*, p. 1058. [2] *Ante*, p. 619.

districts of Brazil, some of which, according to information received, are of colossal dimension. In spite however of the large reserves it has hitherto been impossible to work the deposits on a large scale owing to difficulties of transport ; for the local smelting works but little ore is required. These deposits therefore, which in all probability will in future attain immense economic importance, are as yet comparatively but little known.

THE LAKE SUPERIOR IRON ORE DISTRICT IN MICHIGAN, MINNESOTA, AND WISCONSIN

LITERATURE

These occurrences have given rise to so many works that we must content ourselves here with a short summary. Formerly the most important work was : The Marquette Iron-Bearing District of Michigan, with atlas by C. R. VAN HISE and W. S. BAYLEY, together with a chapter by H. L. SMYTH, U.S. Geol. Surv. Monogr. XXVIII., 1897. Latterly appeared, C. R. VAN HISE and C. K. LEITH. The Geology of the Lake Superior Region, U.S. Geol. Surv. Monogr. LII., 1911.—In both these works the many earlier studies, particularly on Marquette, the oldest mining field, are cited and discussed, among which the following are worthy of mention : J. W. FORSTER and J. D. WHITNEY, 1850, 1851, and later ; J. P. KIMDALL, 1865 ; H. CREDNER, 1869 ; A. WINCHELL, 1871 and later ; T. B. BROOKS, 1873 and later ; C. ROMINGER, 1873 and later ; J. S. NEWBERRY, 1874 ; R. PUMPELLY, 1875 ; T. STERRY HUNT, 1878 and later ; R. D. IRVING, 1879 and later ; M. E. WADSWORTH, 1880 and later ; N. V. WINCHELL 1888 and later ; G. H. WILLIAMS, 1888 and later ; C. R. VAN HISE, 1891 and later ; N. H. WINCHELL, 1893 and later ; H. L. SMYTH, 1894 ; and many others. Especially detailed monographs were produced by J. M. CLEMENTS, SMYTH, BAYLEY, and VAN HISE, on Crystal Falls, 1899 ; by BAYLEY on Menominee, 1904 ; by C. K. LEITH on Mesabi, 1903 ; and by CLEMENTS on Vermilion, 1903 ; in U.S. Geol. Surv. Monogr. Nos. XXXVI., XLII., XLIII., and XLV.

More recent works are by VAN HISE. U.S. Geol. Surv. 21 Ann. Rep., 1901, III. pp. 305-434, in his chief work, ' A Treatise on Metamorphism,' U.S. Geol. Surv. Monogr. XLVII., 1904, especially pp. 824-853 and 1193-1198.—VAN HISE and LEITH. ' Pre-Cambrian Geology of North America,' U.S. Geol. Surv. Bull. 360.—LEITH. Trans. Am. Inst. Min. Eng. XXXV., 1904 ; Econ. Geol. II., 1907 ; Can. Min. Inst. XI., 1908.—J. F. KEMP. Ore-Deposits, 1906.—H. RIES. Economic Geology of U.S., 1910.—J. BIRKENBINE. The Mineral Resources of U.S.—CH. L. HENNING. Die Erzlagerstätten der Vereinigten Staaten von Nordamerika. Stuttgart, 1911, Ferd. Enke.

The Lake Superior iron deposits, which to-day are responsible for no less than one-third of the iron ore production of the world, occur both on the southern and on the northern shore of Lake Superior, as illustrated in Fig. 396. Of the five most important districts or ranges, Mesabi and Vermilion are situated in Minnesota ; Marquette, Menominee, and Gogebic, in Michigan. In addition, there are several smaller districts, such as Penokee in the neighbourhood of Gogebic, and Baraboo, both in Wisconsin ; Crystal Falls in Michigan ; Cuyuna in Minnesota ; and finally, Michipicoten in Ontario, Canada. The distance in a straight line between the occurrences lying farthest apart is roughly 450 kilometres. The correlation of the formations concerned, formulated by a committee of Canadian and North American geologists in the year 1905, as well as the distribution of the deposits in the different horizons, may be gathered from the following summary :

Cambrian—Potsdam sandstone ; without iron ore.

Algonkian $\begin{cases} \text{Keeweenawan}^{1} \\ \text{Upper Huronian} \\ \text{Middle Huronian} \\ \text{Lower Huronian} \end{cases}$ Mesabi, Penokee, Vermilion, Marquette, Crystal Falls, Menominee, Cuyuna.

Archaean $\begin{cases} \text{Laurentian : Marquette, Vermilion.} \\ \text{Keewatin : Vermilion.} \end{cases}$

The deposits accordingly belong to different geological horizons. These 'iron formations,' consisting principally of ordinary clastic sediments —more particularly quartzite and slate—and reaching a considerable thickness, have different names in different districts. The Archaean iron bed of the Vermilion district is named the Soudan formation, and the iron complex in the Middle Huronian of the Marquette district, the Neganee formation. The Upper Huronian iron series of the Penokee-Gogebic district is described either as the Iron-bearing member, or the Ironwood formation. In the Menominee district the iron horizon is called the Vulcan formation, and in the Crystal Falls district, the Groveland formation, while the ore-bearing Upper Huronian of the Mesabi district is known as the Biwabik formation.

According to later American authorities these are poor primary sediments, which have been contorted and crumpled, and have locally suffered such far-reaching chemical alteration that gigantic accumulations of ore have resulted. Among the primary ores, a siderite-silica schist frequently still existing and in places somewhat gritty, is especially notable. This material, of which a thin section is illustrated in Fig. 431, exhibits siderite in fine alternation with schistose quartz, etc. It generally contains only 40–60 per cent of siderite, equivalent to an iron content of 20–30 per cent, the remainder being chiefly silica. At the outcrop the siderite is in greater part oxidized to limonite. Greenalite-schist, illustrated in Fig. 432, also occurs very extensively. This rock, which is fairly poor in iron, is characterized by so-called greenalite, an amorphous or colloidal iron silicate of the composition $(Fe, Mg) SiO_3 . nH_2O$ with a little Mg, in very small round grains. These ferruginous schists, which occasionally reach a thickness of 1000 feet, have in several places in the neighbourhood of later gabbro intrusions been altered, principally by contact-metamorphism, to magnetite-amphibolite schists.

Much more important however is the hydro-chemical alteration of the ferruginous schists containing on an average only about 25 per cent

[1] *Ante*, p. 928.

of iron, to rich ore-bodies of frequently gigantic dimension, on the
one hand, and to ferruginous slates, hornstones, and jaspelites without
economic value, on the other.   The rich ore-bodies lie upon more or less
impermeable beds consisting partly of schists and partly of highly altered
eruptives—chiefly original diabase often altered to so-called soapstone.

FIG. 430.—Map showing the position of the principal iron deposits in the United States.

As a rule these impermeable beds constitute the foot-wall of the deposit,
being frequently arranged in basin shape.   The deposits have then the
form of large troughs, the axes of which are often much inclined constituting
the so-called 'pitching troughs' of which the sections by Van Hise and
Leith given in Fig. 433 and 434 afford typical examples.   From the comprehensive monograph by Van Hise and colleagues upon Marquette, 1897,

Fig. 435, representing diagrammatically the following positions of rich ore, has likewise been taken:

1. In V-shaped troughs between the schist and a dyke of soapstone; (2) along decomposed eruptive dykes in jasper; (3) in jasper, immediately below overlying quartzite; (4) at the contact of jasper and the underlying soapstone, where the contours of the latter form a trough.

Van Hise and Leith emphasize the fact that the siderite- and greenalite-schists, which frequently reach a thickness of 100–300 m., occur with sharply defined contacts either upon quartzite, conglomerate, and altered clay-slate,

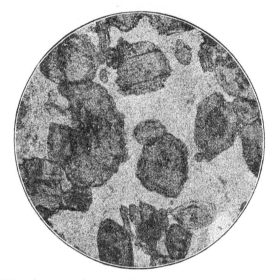

FIG. 431.—Thin section of siderite from the Penokee district; magnified 40 times. Van Hise.

or upon basic eruptive sheets. In the ferruginous schists there are no clayey sediments. On this and other grounds it follows that these are not mechanical, but chemical sediments. Deposition took place in fairly shallow water, possibly in large lagoons, though owing to the abundance of $FeCO_3$ and $FeSiO_3$, that is, of iron originally chiefly in the ferrous and not in the ferric condition, these ferruginous schists may not be compared with lake ore.

The siderite- and greenalite-schists are invariably in the hanging-wall of particularly extensive and thick sheets of principally basic eruptives—basalts—which must be regarded as submarine outpourings. From these submarine sheets the iron content of the schists has probably in greater part been derived. In addition to iron, a little manganese, and silica, some alkali was also conveyed to the water from the basic eruptive. The formation of the greenalite is then explained as follows :

$$FeCl_2 + Na_2SiO_3.nH_2O = FeSiO_3.nH_2O \text{ (greenalite)} + 2NaCl.$$
$$FeSO_4 + Na_2O.3SiO_2.nH_2O = FeO.3SiO_2.nH_2O + Na_2SO_4.$$

The precipitate resulting from the latter reaction consists actually of greenalite, $FeSiO_3.nH_2O$, and free silica.

The siderite-schist, which consists principally of $FeCO_3$ and SiO_2, may

FIG. 432.—Thin section showing greenalite, the dark spheres, in fine-grained quartz, from the Mesabi district ; magnified 40 times. Leith.

in part have been directly precipitated, and in part indirectly formed from the greenalite, thus :

$$FeSiO_3 + CO_2 = FeCO_3 + SiO_2.$$

Though in exceptional cases these deposits may represent primary, relatively rich deposits in the formation of which secondary enrichment played but a subordinate part, as a rule they have resulted from far-reaching secondary processes by which (1) the silica of the original ferruginous schists was removed, or (2) this removal of silica took place simultaneously with an entry of iron. These processes depended upon the action of descending meteoric waters containing oxygen, carbon-dioxide, etc., and in addition alkali salts, probably derived from the basic eruptive sheets.   Such aqueous solutions collected over the impermeable beds, there producing the above-mentioned occurrences of rich ore-bodies.   In all probability they contained alkali carbonates capable, fairly

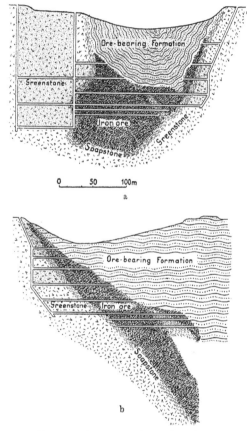

FIG. 433.—Transverse (*a*) and longitudinal (*b*) section north-south and east-west respectively of the iron deposit in the Chandler mine.   Van Hise.

readily, of converting the silica to soluble alkali silicate. In this manner, according to Van Hise and Leith, the enormous migration or removal of silica in the original deposit is explained. By these same solutions iron was also frequently taken up in the ferrous condition, $FeCO_3$ or $FeSiO_3$, to be subsequently precipitated elsewhere as ferric oxide, at

oxidation by solutions containing oxygen.   These secondary processes took
place fairly near the surface, and accordingly the deposits in general do
not continue to great depth.

This extensive migration of material took place in greater part in

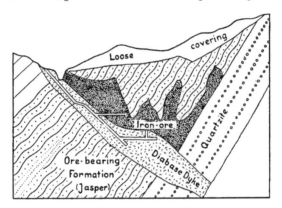

FIG. 434.—Transverse section through the iron deposit at the Colby mine in the
Penokee-Gogebic district.   Van Hise and Leith.

pre-Cambrian time.   Subsequently the district became covered in part by
Palæozoic formations, which later became disintegrated ; then followed
a Cretaceous deposition which likewise surrendered to denudation ; till
finally came the Pleistocene glacial erosion.   Migration of material resulting
in the formation of ferruginous deposits took place also, though only on

FIG. 435.—Diagrammatic representation of the different positions assumed by iron deposits
in the Lake district.   Van Hise.

a limited scale, in those later geological periods during which the ore-
beds after denudation of the hanging-wall sediments lay in the immediate
neighbourhood of the surface.   These briefly are the genetic conclusions
reached after very extensive research by Van Hise, Leith, Bayley, and other
officials of the Survey, and now generally accepted.   Formerly, other
authorities, as for instance Foster and Whitney in 1851, and Wadsworth in
1880, assumed an eruptive formation, while others, including Credner in

1869 and Brooks in 1873, assumed an exclusively sedimentary origin. The far-reaching secondary displacement of material within the original deposition cannot however be denied.

The ore of the Lake Superior deposits is sometimes hard and sometimes fairly soft. Hæmatite with a mostly fairly small admixture of limonite occurs particularly extensively; occasionally, and especially in the Mesabi Range, friable limonite is strongly represented; while in some places, as for instance at Marquette, magnetite occurs. The ore is accompanied more particularly by quartz. Formerly, fairly rich ore containing roughly about 60 per cent of iron was chiefly worked, but of late years somewhat poorer ore has proved to be payable. The average iron content of the total output has therefore gradually decreased, till now it is probably some 50 per cent. Phosphorus as a rule is very low, the minimum being 0·008 per cent, though here and there it is somewhat higher, in places reaching even 1·28 per cent. The sulphur content is also very low.

The composition of the ore may be gathered from the following table of analyses, after H. Ries :

	Marquette.	Menominee.	Gogebic.	Vermilion.	Mesabi.
Fe . . .	56·5	55·2	56·3	61·4	56·1
SiO_2 . . .	4·6	6·8	3·4	4·3	3·5
P . . .	0·035	0·059	0·034	0·037	0·037
S	0·009	...	...	...	...
H_2O . . .	11·9	6·5	10·8	4·6	12·3

The ore is as a rule poor in manganese, though manganiferous ore occurs in places. The average composition of the total ore production for the year 1909 [1] amounted to :

Moisture, loss in drying at 100° . . 11·28 per cent.

Analysis after drying at 100°
Fe	.	.	.	.	58·45	,,
Mn	.	.	.	.	0·71	,,
SiO_2	.	.	.	.	7·67	,,
Al_2O_3	.	.	.	.	2·23	,,
CaO	.	.	.	.	0·54	,,
MgO	.	.	.	.	0·55	,,
P	.	.	.	.	0·091	,,
S	.	.	.	.	0·060	,,
Loss on Ignition	.	.	4·12	,,

The ore mined contained therefore not quite 53 per cent of iron.

The average mineralogical composition of the total production in the same year was as follows :

[1] Van Hise and Leith, *loc. cit.*, 1911, p. 477.

Hæmatite and limonite, with some magnetite	86·45	per cent.
Quartz	4·87	,,
Kaolin	5·25	,,
Chlorite, etc.	1·01	,,
Dolomite	0·81	,,
Apatite	0·48	,,
Miscellaneous	1·11	,,
Total	100·00	per cent.

Originally the ore was mined only by opencut, this method, especially at Mesabi, being still much used. In that district in the year 1908 no less than 63·7 per cent of the total output, equivalent to 42 per cent of the total production of the Lake Superior district, was won from opencuts. The Mesabi ore, the beds of which are so little inclined that the horizontal thickness transversely to the strike is particularly great, is so loose at the surface that it can frequently be mined by steam-shovels. Of late years underground mining has been resorted to, and at the present day there are mines of 300–400 m. and more in depth. An idea of the rapid development of the industry and the gigantic size of the occurrences may be gathered from the following statistical summary, in which the dates of the commencement of work are given in brackets.

Mill. Tons.	1870.	1885.	1895.	1900.	1905.	1910.	Total to end of 1910.
Marquette (1849)	...	...	2·1	3·9	3·8	·4·4	96·3
Menominee (1877)	...	...	1·9	3·7	4·5	4·2	75·4
Gogebic (1885)	...	...	2·5	3·1	3·3	4·3	65·2
Vermilion (1889)	...	...	1·1	1·7	1·6	1·2	30·3
Mesabi (1892)	...	...	2·8	8·1	20·2	29·2	224·9
Baraboo	...	...	...	...	...	0·1	0·7
Total	0·9	2·5	10·4	20·5	33·4	43·4	492·8

The ore is transported by railway to the ports, Duluth, Two Harbours, Ashlanno, and Marquette, on Lake Superior, and Escanaba and Gladstone on Lake Michigan ; and from there by steamer to the large smelting centres in Illinois, Ohio, Pennsylvania, and New York. The cost per ton of ore from the mine to the smelting works in the year 1907 amounted to on an average 2·14 dollars = 9 shillings, out of which the transport companies made a good net profit. Of late years almost exactly 80 per cent of the total iron ore production of the United States has been derived from the Lake district.

With regard to ore-reserves, the following estimates respectively by Van Hise and Leith in 1911, and by J. F. Kemp in 1910,[1] have been

[1] *The Iron Resources of the World*, Stockholm, 1910.

published. They relate partly to 'available ore' and partly to 'not
available ore.'   In the case of Van Hise and Leith's estimate the ore
contains at least 35 per cent of iron.

	Available.		Not available.	
	Van Hise.	Kemp.	Van Hise.	Kemp.
	Million tons.			
Crystal Falls	...	...	1,500	...
Marquette	100	110	16,000	15,900
Menominee	75 [1]	80	3,500	7,360
Gogebic	60	95	1,250	3,900
Menominee and Gogebic . .	...	40	...	4,525
Vermilion	30	60	1,025	1,005
Mesabi	1600	3100	30,000	39,000
Cuyana, etc.	40 [2]	15	...	310
Other occurrences . . .	...	...	14,360	...
Totals	1905	3500	67,635	72,000

Economically, the most important of these districts is the Mesabi
Range, roughly 130 km. in length, with Duluth as the port.   In recent
years this district alone has been responsible for more than one-half of the
North American pig-iron production.

Hitherto it has been usual at these mines—the number of which up
to the year 1910 amounted to 335—to dump the poor ore as valueless, a
procedure which has been strongly criticized from a political economical
standpoint.   In this connection Van Hise rightly draws attention to the
fact that many of the Lake occurrences have become smaller and poorer
at a depth of only 1000 feet, while at 1500 feet few are still payable.
Fifteen years ago all the ore smelted contained at least 60 per cent of iron,
while of late years large quantities containing 40–50 per cent have been
used.   In view of these facts Van Hise advises that the poorer ore be
mined and raised now, so that later as the rich reserves decrease it may
be available.

The total production from 1891 to 1900 amounted to 114,000,000 tons,
while that from 1901 to 1910 was 322,000,000 tons, indicating that in the
last decade production had almost trebled.   Assuming a further increase in
the future, the rich ore will in greater part be exhausted in a few generations.   Of the poorer ore however there is sufficient to last for centuries.

[1] Includes Crystal Falls.   [2] Only Cuyana.

IRON ORE DISTRICTS CONTAINING ORE-BEDS PRINCIPALLY, AND METASOMATIC DEPOSITS SUBORDINATELY

As already stated when dealing with the metasomatic iron deposits, some districts contain principally true ore-beds but also subordinate metasomatic deposits. This combination arose at the formation of ferruginous sediments in districts consisting in greater part of limestone, under which circumstances the ferruginous solutions circulating along the fissures and on the surface, were able to form metasomatic deposits in addition to sediments. The important part played by sedimentation in these districts has only been recognized in more recent years. Formerly, too much importance was given to metasomatism, and these occurrences accordingly were mostly treated with the metasomatic deposits.

In Germany the most important deposits belonging to this class are those in the Lahn-Dill district, and those of the Oberharz Devonian belt at Elbingerode and Hüttenrode.

THE HÆMATITE DEPOSITS IN THE LAHN-DILL DISTRICT

LITERATURE

W. RIEMANN. Beschreibung des Bergreviers Wetzlar. Bonn, 1878.—FR. WENCKEN-BACH. Beschreibung des Bergreviers Weilburg. Bonn, 1879.—W. RIEMANN. Der Bergbau- und der Hüttenbetrieb der Lahn-, Dill-, und benachbarten Reviere (Nassau). Wetzlar, 1894 ; ' Das Vorkommen der devonischen Eisen- und Manganerze in Nassau,' Zeit. f. prakt. Geol., 1894, p. 50.—R. DELKESKAMP. ' Die hessischen und nassauischen Manganerzlagerstätten und ihre Entstehung durch Zerstörung des dolomitisierten Stringocephalenkalkes bezw. Zechsteindolomites,' Zeit. f. prakt. Geol., 1901 ; ' Die mutmassliche Dauer des Fortbestehens des Eisenerzbergbaues der Lahn- und Dillreviere,' Stahl und Eisen, 1902, p. 278.—H. LOTZ. ' Die Dillenburger Rot- und Magneteisenerze,' Zeit. d. d. geol. Ges., 1902.—F. KRECKE. ' Sind die Roteisensteinlager des nassauischen Devons primärer oder sekundärer Bildung ? ' Zeit. f. prakt. Geol., 1904, Vol. XII., p. 348.—CHELIUS. ' Der Eisenerzbergbau in Oberhessen, an der Lahn, Dill und Sieg,' Zeit. f. prakt. Geol., 1904, Vol. XII., p. 53.—BÖHM. ' Die Erzlagerstätten des konsolidierten Bergwerkes Stangenwage bei Haiger' (Bergrevier Dillenburg), Zeit. f. Berg-, Hütten-, und Salinenwesen, 1905, p. 259.—C. HATZFELD. ' Die Roteisenlager bei Fachingen a. d. Lahn,' Zeit. f. prakt. Geol., 1906, Vol. XIV. p. 351.—R. BRAUNS. ' Der oberdevonische Deckdiabas, Diabasbomben, Schalsteine, und Eisenerz,' Neues Jahrb. f. Min. 1905, p. 302.—G. EINECKE. Der Eisenerzbergbau und der Eisenhüttenbetrieb an der Lahn, Dill und in den benachbarten Revieren, eine Darstellung ihrer wirtschaftlichen Entwicklung und gegenwärtigen Lage. Jena, 1907.—ROSE. ' Zur Frage der Entstehung der nassauischen Roteisensteinlager,' Zeit. f. prakt. Geol., 1908, p. 497.—J. AHLBURG. ' Die Tektonik der östlichen Lahnmulde,' Zeit. d. d. geol. Ges., 1908, p. 300 ; ' Die Buderusschen Eisenwerke zu Wetzlar,' Stahl u. Eisen, 1909, p. 1633.—H. BEHLEN. Die naussischen Roteisensteine. Wiesbaden, 1909.—J. AHLBURG. ' Die Grube Schöner Anfang bei Breitenbach (Kreis Wetzlar), ein Beitrag zur Tektonik der östlichen Lahnmulde,' Jahrb. d. pr.geol. Landesanst., 1909.—Geological maps of sections, Dillenburg, Oberscheld, Herborn, Ballersbach, Braunfels, Hadamar, Girod, Eisenbach, Limburg, Schaumburg, Ems. Published by the Königliche pr. geol. Landesanstalt.

The Dill and Lahn synclines are separated from each other by a north-east striking Silurian anticline, the northern Dill syncline coinciding approximately with the Scheldc valley and its southern continuation,

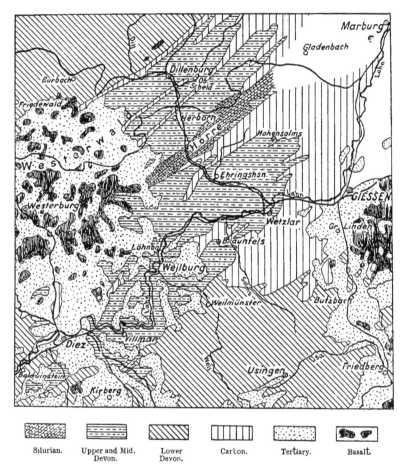

| Silurian. | Upper and Mid. Devon. | Lower Devon. | Carbon. | Tertiary. | Basalt. |

Fig. 436.—Geological map of the Lahn-Dill district. Scale 1 : 530,000.
Einecke and Kohler.

while the Lahn syncline reaches in the south-west almost to Laurenburg, and in the north-east to Giessen. The synclinal flanks of both lie upon outliers of the Taunus and the Westerwald, which here abutt. This geological position is illustrated in Fig. 436.

Both these synclines are known for their valuable hæmatite beds, which occur chiefly at the contact of Middle and Upper Devonian, where the

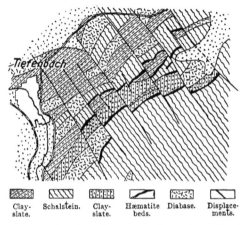

Clay-slate. Schalstein. Clay-slate. Hæmatite beds. Diabase. Displacements.

FIG. 437.—Folded and faulted iron ore-beds in the Lahn syncline.   Einecke and Kohler.

opportunity for the formation of such ore-beds was particularly favourable, and a ferruginous horizon extending for a great distance and carrying principally hæmatite became formed.   Many of these occurrences are

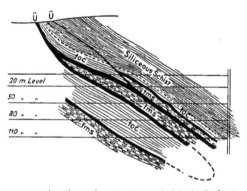

FIG. 438.—Folds and overthrusts affecting the iron ore-bed in the Raab mine near Wetzlar.
Einecke and Kohler.

tms = Schalstein of the upper Middle Devonian ; *toc* = Cypridina slate of the Upper Devonian ;
Ü = overthrusts.

undoubtedly true sediments.   In addition to these sediments, however, hæmatite deposits occur subordinately within the Devonian diabase sheet, and also at the contact of Upper Devonian diabase and Culm grauwacke, these latter having resulted from the replacement of limestone lenses and diabase. The limonite deposits associated with those of hæmatite are considerably younger, and were formed principally by metasomatic alteration of limestone in Tertiary time. Finally, alluvial deposits consisting of fragments of disintegrated older deposits exist, these filling depressions in the slate, schalstein, and limestone.

In consequence of the textural and tectonic changes which different parts of this region have separately experienced, in these two Devonian

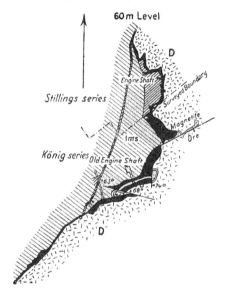

FIG. 439.—Plan of the iron deposit on the 60 m. level of the Königzug and Stillingseinen mines near Oberscheld.   Scale 1 : 4000.   Einecke and Köhler.

synclines the following areas may be differentiated, namely, Dillenburg, Wetzlar, Weilburg, and Dietz. Of the geo-tectonics of these areas, the following brief description of the Dillenburg syncline may be taken as representative.

Upon the Silurian lie the lowermost beds of the Lower Devonian, including the Coblenz stage with its three divisions ; upon these in turn follow those of the Middle Devonian, with the fine clayey Orthoceras- and Tentaculite slates and overlying calcareous sediments. In the upper Middle Devonian began the mighty eruptions to which the diabase and schalstein owe their existence, this eruptivity continuing into the Upper Devonian. In this period likewise the extensive hæmatite beds were formed, the iron of which, as appears probable from the Geological Survey, was derived from exhalations and springs representing the after-effects of the eruptive phenomena. The ironstone horizon extends from the Lahn to the Dill and belongs to the Middle Devonian; its foot-wall always consists of schalstein. The conformity of this horizon is so marked that it can everywhere be taken as the parting between Middle and Upper Devonian. In the Dillenburg syncline the iron ore-beds, apart from some thin and unimportant occurrences in the foot-wall schalstein, are limited to this horizon.

The bed material consists of hæmatite arranged in bands. This ore when containing much iron contains also much silica, and when little iron much calcite. The calcareous development prevails throughout the entire west of the syncline, while in the middle around Oberscheld siliceous areas occur; in addition, sudden changes from calcareous to siliceous ore are often observed. In the neighbourhood of the surface the ore is decarbonated and altered to limonite, frequently also silicified.

The hanging-wall of the ironstone horizon is variously constituted. In the centre of the Dill syncline it is formed by the Upper Devonian diabase, while in the north and north-east, calcareous, siliceous, and Cypridina slates occur in the place of this diabase, or between it and the deposit. Where the bedding is undisturbed the Culm slates occur upon the Devonian.

All these formations in late Carboniferous time were highly disturbed by folding and dislocation.

Many authorities are inclined to regard the hæmatite within the diabase as fragments from the main bed, either primarily enveloped by the diabase or subsequently forced within it. Such deposits are lenticular in form, and the silica content is generally considerable.

The occurrence of ore at the contact of diabase and grauwacke, to which reference has already been made, owes its existence to the metasomatic replacement of limestone lenses or the replacement of diabase or schalstein. The size of these occurrences is mostly small, nor is there anything pronounced about the nature of the ore. They have not hitherto been the object of mining operations.

In the eastern Lahn syncline the tectonics are rendered complex by extensive overthrust coverings, beneath which the country is exceedingly disturbed. In this syncline Middle Devonian predominates, this formation being calcareous, as at the Dill. Here also the eruptive period responsible for the iron began with the schalstein beds of the upper Middle Devonian, these beds being particularly calcareous. At the same time was formed the massive limestone of the Lahn district, which, confined to no particular horizon, may be followed from Giessen to Balduinstein. Above this comes schalstein again. With this powerful limestone range, which is absent from the Dillenburg district, the Tertiary limonite- and manganese ores are associated. The principal hæmatite horizon in this case also, occurs at the contact of Middle and Upper Devonian, though it is broken by numerous disturbances both along the strike and dip. In addition, extensive iron ore-beds, which probably belong to one and the same horizon, occur in the schalstein.

The iron ore of the Lahn syncline is calcareous, and, by the action of meteoric waters, still more decomposed than that of the Dill syncline ; in this decomposition the lime has been removed and the hæmatite altered to limonite. Frequently the ore-beds contain intercalated beds of slate and schalstein as well as intrusive layers of diabase.

In the middle Lahn syncline, that is, in the Weilburg district, no main ferruginous horizon exists, but a large number of mostly limited horizons, one above the other. Among these, the contact horizon between Upper and Middle Devonian is not prominent. Beds of hæmatite reach right into the Upper Devonian, where they are nearly always accompanied by Upper Devonian schalstein. In this portion of the Lahn syncline also, mineralization of the massive limestone is an important feature, in that to the north two other parallel series occur, between Hadamar and Merenberg on the one side, and Weilburg and Dehrn on the other, and, accordingly, metasomatic Tertiary manganiferous ores are frequent. In this district occur also the fragmentary deposits which have resulted from the destruction of older beds and, in synclines on the plateau, have remained protected from erosion. These exhibit bedded structure and occasionally resemble the Tertiary limonite deposits.

To the west, the Lahn syncline dies away in secondary synclines separated from each other by Lower Devonian anticlines which strike in the direction of Holland. Thus, from north to south occur the Hadamar-Niedererbach, the Balduinstein-Ruppachtal, and the Hahnstätten-Katzenelnbogen synclines. Of these, the first carries hæmatite only in the Middle Devonian schalstein ; the second is characterized by a close succession of overturned Middle Devonian limestone, schalstein, and diabase with attendant iron deposits ; while the third is occupied by Middle Devonian limestone and schalstein containing very promising ore-beds. Finally, the entire Lahn district is bounded by the large Ruppach Fault which occurs in the Ruppach valley at Balduinstein.

With regard to the character of the ore in the Lahn-Dill district, calcareous ore, hæmatite, and limonite, must be distinguished. The iron content of the first varies between 22·8 and 44 per cent ; the silica reaches more than 20 per cent, though as a rule it does not exceed 10 per cent ; the $CaCO_3$ may reach 33 per cent, and the phosphorus 0·36 per cent, though the latter as a rule is much less. The iron content of the hæmatite gener ally varies between 45 and 55 per cent, the silica is 12–25 per cent, the lime 9–10 per cent, and the phosphorus 0·06–0·34 per cent. The limonite, so far as this is an alteration from hæmatite, contains as a rule 33–35 per cent of iron with 0·09–0·38 per cent of phosphorus, and 11–25 per cent insoluble residue. All these ores have quite a low manganese content which seldom exceeds 0·5 per cent and does not reach 1 per cent. The sulphur content is likewise low.

The manganiferous limonites of the massive limestone have mostly 30–40 per cent of iron with 13 per cent of manganese and up to 2·1 per cent of phosphorus. The silica content varies between 5 and 19 per cent, the lime between 0·3 and 1·8, and the insoluble residue between 7 and 30 per cent. The concretions of manganese ore found in the limonite contain up to 5 per cent of iron and 41 per cent of manganese.

The economic importance of this district may be gathered from the following figures : In the year 1910 Germany produced 22,964,765 tons of iron ore, of which the Lahn-Dill district with 116 mines contributed 1,004,263 tons containing on an average 40·9 per cent of iron. Einecke and Köhler estimated the reserves which come into present consideration at 166,000,000 tons. Of this figure approximately 100,000,000 tons are hæmatite, and 60,000,000 tons limonite.

The Magnetite, Specularite, and Hæmatite Occurrences in the Harz

LITERATURE

W. Hauchecorne. ' Die Eisenerze der Gegend von Elbingerode am Harz,' Berg-Hutten-, und Salinenwesen, 1867.—H. Wedding. ' Beiträge zur Geschichte des Eisen-hüttenwesens im Harz,' Zeit. d. Harzvereins, 1884.—C. Blomeke. Die Erzlagerstatten des Harzes und die Geschichte des auf demselben geführten Bergbaues. Vienna, 1885.—F. Klockmann. ' Übersicht über die Geologie des nordwestlichen Oberharzes,' Zeit. d. d. geol. Ges., 1893, p. 253.—M. Koch. ' Zusammensetzung und Lagerungsverhältnisse der Schichten zwischen Bruchberg-Acker und dem Oberharzer Diabaszug,' Jahrb. d. pr. geol. Landesanst., 1894, p. 185 ; Cypridinenschiefer im Devongebiet von Elbingerode und Hüttenrode.—Banniza, Klockmann, Lengemann and Sympher. Das Berg- und Hüttenwesen des Oberharzes. Stuttgart, 1895.—M. Koch. ' Gliederung und Bau der Kulm- und Devonablagerungen des Hartenberg-Büchenberger Sattels nördlich von Elbin-gerode im Harz,' Jahrb. d. pr. geol. Landesanst., 1895, p. 151.—L. Beushausen. Das Devon des nordwestlichen Oberharzes, 1900 ; ' Das Devon des nördlichen Oberharzes u.s.w.' Abhandl. d. pr. geol. Landesanst., 1900, N.F. Part 30.—E. Harbot. ' Zur Frage nach der Entstehung gewissen devonischer Roteisenerzlagerstätten,' Neues Jahrb. f. Min., 1903, p. 179 ; ' Über einige Trilobitenfunde bei Grund im Harz und das Alter des Iberger Kalkes,' Zeit. d. d. geol. Ges., 1903.—Einecke and Köhler. Die Eisenerzvorrate des Deutschen Reiches. Published by the Prussian Geological Survey. Berlin, 1910.

1. *The Hæmatite Beds of the Oberharz Devonian Belt.*—The Ober-harz Culm plateau is interrupted to the south-east by a 300–1000 m. wide Devonian strip, the Oberharz Devonian or diabase belt, which, with

the exception of a short gap in the neighbourhood of Altenau, extends from the neighbourhood of Osterode on the south border of the Harz, to near Harzburg on the north border. This situation is illustrated in Fig. 35.

The beds from oldest to youngest consist of Middle Devonian Wissenbach slate, diabase with tuffs and amygdaloids, Stringocephalus limestone, and Upper Devonian diabase and Cypridina slate. The schalstein— diabase tuff—which in greater part lies immediately upon the Wissenbach slate and in turn is overlaid by the Stringocephalus limestone, comes most into the question of these deposits, though the Upper Devonian diabase, occurring immediately above the Stringocephalus limestone, likewise contains some hæmatite beds. As a result of intense plication and overthrusting the beds form south-east striking anticlines and synclines, many times repeated. The tectonics are further complicated by a large number of transverse fissures which strike north-west and dip mostly south-west.

The iron ore-beds consist of numerous more or less large lenses which pinch out both along the strike and the dip, and in consequence of step-faulting are frequently duplicated in transverse section. Often, together with the Stringocephalus limestone, they occur enfolded in the schalstein. By old workings on the surface several such beds may be followed, seven separated by amygdaloid being sometimes found parallel and close together, while hardly ever are less than three present.

The thickness of these ore-bodies may reach 6 metres. The ore is mostly a calcareous hæmatite, though it often becomes siliceous. The iron content according to former assays amounts to 20–50 per cent; sulphur occurs only in traces. Unfortunately the data preserved are no longer sufficient to fix the distribution of the calcareous and siliceous ores. The composition of the ore may be gathered from the following analyses:

	Julius Mine.		Hohebleek Mine.	Other Analyses.			
	Per cent.		Per cent.	p.c.	p.c.	p.c.	p.c.
Fe	28–40		25–35	47·6	51·6	39·6	40·3
Mn		0–2		0·2	0·3	1·3	0·6
P	...		...	0·18	0·13	0·17	0·12
MgO . . .		0–4		4·8	5·7	...	3·2
CaO . . .	5–13		9–17	2·1	1·8	...	21·4
SiO₂ . . .	8–13		5–9	23·6	13·5	18·2	9·2
Al₂O₃ . . .		1–5		...	...	...	...

It would appear as though the richer ores were found south of Buntenbock, in the Hut valley, on the Polsterberg, and east of Lerbach.

Einecke and Köhler give the following table of the thickness and iron content of the calcareous ore-beds :

Name and Position of Bed and of Old Mines.	Thickness.	Iron Content.
	Metres.	Per cent.
Segen-Gottes bed on the Polsterberg; Mines: Segen-Gottes, Grüne Linde, Georg-Andreas .	2–4	24–30
Schwan bed in the Hut Valley; Mine: Weisser Schwan	0·5–1	37
Weinschenke bed at Buntenbock; Mines: Wein- schenke and Goldener Adler	0·3–2	22
Blaue-Busch bed at Lerbach; Mines: Blauer- Busch, Kranich, Julius (exhausted above valley level)	1·5–10	44
Mühlenberg bed at Lerbach; Mine: Mühlenberg	0·5–1	24
Rote Löwe bed at Lerbach; Mine: Rote Löwe .	0·5–1	25
Hohebleek bed at Osterode; Mine: Hohebleek.	1–2	24

In addition, there are in this district considerable quantities of ferruginous flux.

The genesis of these deposits has not yet been satisfactorily determined. Formerly, all ironstone beds associated with schalstein, diabase, and limestone, were regarded exclusively as metasomatic replacements of the Stringocephalus limestone, such metasomatic alteration having been brought about by ferruginous solutions subsequent to the mountain-folding. The geological position of the ore-beds of the Oberharz diabase belt agrees so exactly however with that of the Lahn-Dill district that in both cases the modern view of a combination of ore-beds and metasomatic occurrences must hold good. The time of formation must be regarded as that of the intrusion of the Middle Devonian eruptives.

It is an interesting fact that the extreme north-east outliers of the Oberharz diabase belt, in spite of the similarity in the geological position of their contained ore-beds with those just described, carry not hæmatite but magnetite. Here also, presumably, sedimentary hæmatite was first deposited, which subsequently by contact with the Brocken granite became altered to magnetite.[1] The geological position may be best observed on the Spitzenberg near Altenau; much exploration work however has not yet been done. The ore there, according to one analysis, contains 58 per cent of iron with 13 per cent insoluble residue, and according to another 43 per cent of iron with 30 per cent of residue. Several parallel beds occur.

Mining in the Oberharz diabase belt was active principally from 1650 to 1875 during which period more than 100 mines were working. Owing to the closing down of the smelting works at Altenau, Osterode, and Lauterberg, operations were subsequently abandoned. Einecke and Köhler estimate the ore-reserves, including those of magnetite, at 3–4 million tons.

------

[1] *Ante*, p. 350.

2. *The Hæmatite Beds at Elbingerode and Hüttenrode.*—These deposits are very similar to those of the Lahn-Dill district as well as to those of the Oberharz diabase belt. Here also a similar sequence of formations participates in the geo-tectonics of the district, namely : first, the Middle Devonian Wissenbach slate ; then the older schalstein and amygdaloidal diabase, the keratophyre with diabase tuffs, and the Stringocephalus limestone ; and finally, representing the Upper Devonian, the younger schalstein with amygdaloidal diabase, the Iberg- or Clymenian limestone, and the Cypridina slate. The Devonian is then overlaid by Culm. M. Koch[1] distinguishes three Devonian anticlines, namely, the Büchenberg-Hartenberg, the Hornberg-Elbingerode, and the Hüttenrode-Neuwerk, these being illustrated in Figs. 440 and 441. The anticlinal core in each case consists of schistose schalstein or amygdaloidal diabase. The keratophyre occurs as numerous more or less thick sheets separated by tuffs. The Stringocephalus limestone lies as a mantle around the schalstein. At Elbingerode the anticlines and synclines strike north-east, and consist only exceptionally of simple folds, but are frequently overturned, so that the geological position is not always easy to determine. In this connection M. Koch points out that the Stringocephalus limestone and the ironstone beds on the Büchenberg and the Hartenberg do not belong, as formerly assumed, to an overturned synclinal limb, but to an anticlinal limb. The presence of many overthrusts has caused the duplication of some beds and the disappearance of others. Well-defined transverse faults accompanied by lateral displacement render the bedding conditions still more complex.

The ore-beds, though found in association with the keratophyre- and diabase tuffs of the Middle Devonian, occur principally in the Stringocephalus limestone, which in greater part is so ferruginous that its entire thickness may be regarded as ore. The hæmatite beds of the diabase tuffs are separated from each other by beds of calcareous and chloritic tuff. The ore-beds of the Stringocephalus limestone are the more important, both in relation to number as well as to the quality of the ore. On the Büchenberg and Hartenberg one such bed, 5–40 m. thick and dipping 70°, may be followed for 4 kilometres. This bed is accompanied by many parallel beds which may be 1–10 m. thick. At the outcrop an alteration to limonite has taken place. In the eastern portion of the Büchenberg anticline magnetite has been formed as the result of contact action.

The hæmatite beds of the other anticlines are less important. South-west of Elbingerode, on the Grosser Graben, a bed with a limonite gossan occurs as a mantle around an occurrence of keratophyre. This bed is 6–25 m. thick and dips 12° to the north-east.

[1] *Jahrb. d. pr. geol. Landesanst.*, 1895.

On the Susenberg, limonite chiefly has been mined. Along both sides

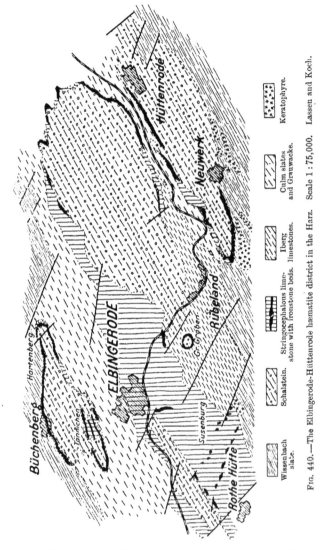

Fig. 440.—The Elbingerode-Hüttenrode hæmatite district in the Harz. Scale 1 : 75,000. Lassen and Koch.

of the Grosser Hornberg, which is an elongated occurrence of diabase and schalstein, workable hæmatite- and limonite beds extend for more than 1700 m. in length and up to 12 m. in thickness. West of these lies the Ahrenfeld bed-series, this having a length of 800 m. and a thickness of 4 m., while in the extreme west the Bastkopf and Wormke deposits occur. Generally speaking, towards the west the silica content increases and the thickness decreases ; the deposits at Lodenblek, Holzberg, Mühlenweg, Gallberg, and Drahl, all these deposits being near Hüttenrode, are, for instance, the most favourable. Of these, the two first occur respectively on the north and south limbs of an anticline, the third and fourth form respectively the south and north flanks of an overturned syncline, while the fifth lies in a secondary anticline of these synclines. The extent of these beds along the strike is 400–600 m. while the thickness is very variable. The compact calcareous hæmatite contains 50–60 per cent of iron, and the ferruginous flux 20–30 per cent.

With regard to genesis, the ironstone beds of Elbingerode-Hüttenrode, like the other Middle Devonian ore-beds of the Harz, were regarded as

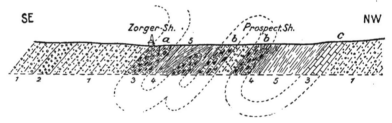

Fig. 441.—Section through the Mühlenweg (a), the Drahl (b), and the Gallberg (c) deposits. Koch.

1, schalstein and amygdaloidal diabase ; 2, keratophyre ; 3, Stringocephalus limestone and ironstone deposits ; 4, Cypridina slate ; 5, Adinole and Wetz slate.

metasomatic occurrences, the iron being considered as derived from the schalstein and diabase. As the result of more recent examination, however, sedimentation has become acknowledged as the most likely genesis for most of these occurrences, which accordingly are of the same age as the Stringocephalus limestone.

The economic importance of the Elbingerode-Hüttenrode deposits depends, in addition to the quantity, upon the character of the ore. According to available analyses the iron content of the hæmatite varies between 30 and 49 per cent ; manganese is very low ; the ore being calcareous lime may reach 30 per cent ; magnesia is low ; phosphorus is given as 0·12–0·8 per cent ; while finally the silica, though it somewhat increases towards the west, in most cases remains below 16 per cent. The limonite contains 35–59 per cent of iron, very little manganese and lime, 0·2–0·7 per cent of phosphorus, and 3·1–18 per cent of silica. The magnetite contains 40–55 per cent of iron, 0·6–6·3 per cent of lime, and

0·05–0·15 per cent of phosphorus, while the insoluble residue varies between 10 and 30 per cent.

The ore-reserves of these deposits down to a depth of 200 m. below the adit level, are estimated by Einecke and Köhler at 40,000,000 tons.

THE IRON ORE PRODUCTION OF THE WORLD AND ITS DISTRIBUTION AMONG THE VARIOUS CLASSES OF DEPOSITS

LITERATURE

The Mineral Industry, New York, and The Mineral Resources of the United States.—
B. NEUMANN. Die Metalle, u.s.w. Halle, 1904. Other works cited in their respective places.

The total production of pig-iron has been as follows :

	Million Tons.			Million Tons.
1911	. . . 63·25	1831–1840	. .	some 24
1901–1910	. . . 528	1821–1830	. .	„ 15
1891–1900	. . . 317	1811–1820	. .	„ 12
1881–1890	. . . 223	1801–1810	. .	„ 10
1871–1880	. . . 148	1701–1800	.	perhaps 50
1861–1870	. . . 99	1601–1700	. .	„ 40
1851–1860	. . . 65	1501–1600	. .	„ 30
1841–1850	. . . 36			

During the period 1801–1911 therefore, some 1540 million tons of pig-iron were produced. Including the production of earlier centuries as far as this relates to ore *in situ*, that is, without taking into account the lake- and bog ores, the total production of pig-iron up to the end of 1911 may be estimated at about 1700 million tons. Of this, somewhat more than one-half has been produced in the last twenty-five years. Reckoning the average iron content of the ore since 1890 at 45 per cent, from 1860 to 1890 at 40 per cent, and still earlier at 35 per cent, the total pig-iron production to the end of 1911 indicates something more than 4000 million tons of ore.

[TABLE

PRODUCTION OF PIG-IRON IN 1900 AND 1910, TOGETHER WITH PRODUCTION, EXPORT, AND IMPORT OF IRON ORE IN 1910

In Millions of Metric Tons

	Pig-iron.		Iron Ore, 1910.		
	1900.	1910.	Production.	Export.	Import.
United States . . .	14·0	27·64	53·27	0·66	2·63
Germany	8·5	14·79	28·71	2·95	9·82
Great Britain and Ireland .	9·1	10·38	15·47	0·01	7·13
France	2·7	4·03	14·61	4·89	1·32
Russia	2·9	2·74	5·70 [1]	...	...
Belgium	1·0	1·80	0·12	0·60	5·18
Austria	} 1·3	{ 1·50	2·63	} 0·14	0·31
Hungary		{ 0·50	1·91		
Canada	...	0·75	...	...	...
Sweden	0·5	0·60	5·55	4·43	...
Spain	0·3	0·37	8·67	about 8	...
Italy	0·02	0·22	0·55	0·01	0·02
Algeria	...	...	1·06	...	...
Greece	...	...	0·54	...	...
Norway	...	...	0·10	0·09	...
India	...	...	0·06	...	...
China	...	...	...	0·13	...
Japan	...	0·06	...	...	...
South Australia . . .	...	...	0·05	...	...
Cuba	...	...	0·66	...	...
Other countries . . .	0·6	0·47	...	...	...
Totals . . .	41·0	65·86	about 142	...	...

[1] = for 1907.

Taking into account that some 1,000,000 tons of pig-iron are produced from roasted pyrite, old slags, etc., the average iron content of the ore produced in 1910 may be reckoned at 46 per cent. In relation to iron content, that ore may be divided approximately as follows :

Some	1	million tons with	65–70	per cent Fe.
,,	7	,, ,,	60–65	,, ,,
,,	19	,, ,,	55–60	,, ,,
,,	30	,, ,,	50–55	,, ,,
,,	20	,, ,,	45–50	,, ,,
,,	15	,, ,,	40–45	,, ,,
,,	26	,, ,,	35–40	,, ,,
,,	24	,, ,,	27–35	,, ,,
Total	142	million tons with	46	per cent Fe.

Sweden yields ores with the highest iron content ; among the large deposits, those of the Lake Superior district come next ; while the oolitic ores of England, Germany, France, etc., are fairly low in the list, containing mostly 27–37 per cent.

TABLE OF THE TOTAL IRON ORE PRODUCTION

In Millions of Tons

	United States.	Great Britain.	Germany.	France.	Spain.	Russia.	Sweden.	Austria.	Hungary.	Belgium.	Italy.	Greece.	Algeria.
1911	53·27*	...	29·9	...	...	...	...	...	...	0·15	0·37	...	...
1901–1910	385·7*	146·6	234·2	85·0	87·6	some 47	40·9	21·4	17·0	2·0	4·1	6·1	7
1891–1900	164·0*	132·5	140·9	41·9	66·0	„ 43	19·1	14·4	some 13	2·5	2·1	3·8	4
1881–1890	91·0*	157·1	93·7	29·3	49·4	„ 15	9·0	some 9·4	„ 6	1·9	2·3		
1871–1880	43·8*	163·5	54·6	some 25	12·5	„ 11	7·5	„ 8	„ 4	4·3	2·2		
1861–1870		101·2	30·2	„ 23·5		„ 8	5·0	„ 7	„ 2·5	7·8	1·2		
1851–1860		some 99	some 12	„ 13·5		„ 6		„ 6	„ 2	7·1			
1841–1850	49·7*	„ 50	„ 6	„ 9		„ 5	41	„ 4	„ 0·8	3·5	some 10	some 20	some 20
1831–1840		„ 32	„ 4	„ 6	some 50	„ 5		„ 2·5	„ 0·6	some 15			
1821–1830		„ 17	„ 1·3	„ 20		„ 4		„ 2	„ 0·4				
1811–1820		„ 60	„ 25			„ 20		„ 20	„ 10				
Earlier	some 10												
Total to 1910	798	959	602	278	265	164	122	95	50	44	23	30	31

The total iron ore production in different countries up to the end of 1910 accordingly in round figures was :

Great Britain and Ireland	.	.	.	some	950	million tons.
United States	.	.	.	,,	800	,, ,,
Germany	.	.	.	,,	600	,, ,,
France	.	.	.	,,	280	,, ,,
Spain	.	.	.	,,	265	,, ,,
Russia	.	.	.	150 to 200		,, ,,
Sweden	.	.	.	some	125	,, ,,
Austria	.	.	.	,,	100	,, ,,
Hungary	.	.	.	,,	60	,, ,,
Belgium	.	.	.	,,	45	,, ,,
Algeria	.	.	.	,,	30	,, ,,
Greece	.	.	.	,,	30	., ,,
Italy	.	.	.	,,	25	,, ,,
Other countries	.	.	.	100 to 250		,, ,,

3560 to 3760 million tons.

or probably approximately 3750 million tons.   If the production of 1911 and 1912 be included the total iron ore production up to the end of the year 1912 would be approximately 4000 million tons.   This agrees very well with the figure which would be obtained on the basis of the pig-iron production up to the end of 1911, just given, in which the early production of lake- and bog ores is not included.

REMARKS ON THE ABOVE TABULATED STATEMENT

The estimates for the United States are given in long tons ; the others in metric tons.

In the case of the United States the starred figures (*) relate to the years 1910, 1900–1909, 1890–1899, 1880–1889, 1870–1879, and 1810–1869.[1]   In 1909, 1910, and 1911, the production amounted to 53·09, 53·27, and 42·15 million tons respectively.   In the year 1810 the production was at most 0·1 million tons, while in the still earlier period as far back as the year 1608, it was considerably less.[2]

The estimates for later years relate to the statistics of the iron ore production : for Great Britain and Ireland back to the year 1860 ; for Germany, including Luxemburg, back to 1862 ; for France to 1872 ;· for Spain to 1871 ; for Russia to 1901, together with some earlier years ; for Austria to 1869 ; for Hungary to 1895, together with some earlier years ; for Belgium to 1836 ; for Algeria to 1891 ; and for Greece, including manganiferous iron ore, to 1887.   In the case of Sweden complete statistics from 1301 are available.[3]

With the large industrial countries the estimates of the iron ore production for the still earlier years are based upon the production of pig-iron, the statistics of ore production being missing or not available.   It may be assumed that the iron ore in Great Britain, Germany, and France, yielded on an average 33·3 per cent of pig-iron, that of Austria and Hungary 35 per cent, and that of Russia 40 per cent.   Complete statistics for pig-iron date back to the year 1820 ; for the still earlier period however they are somewhat incomplete.   The figures of the iron ore production for the period before this date are consequently for most countries rather unreliable.   The production in the earlier centuries was however very low —to-day more iron is produced in a single year than formerly in a whole century—and any unreliability of the earlier estimates has therefore little effect on the approximate correctness of the total.

Mokta-el-Haddid from 1867 to 1877 produced 3,176,500 tons of iron ore.

Other countries: Norway previous to 1900 produced 3·7 million tons of iron ore, 1901–1910 some 730,000 tons, and in 1912 some 400,000 tons.   Bosnia and Herzegovina

---

[1] *Min. Res.* for 1908, p. 68.

[2] *Mineral Industry*, I., 1892, p. 270.   [3] *Ante*, p. 392.

in the thirteen years from 1899 to 1911 produced 1·7 million tons, to which must be added an earlier production. Tunis in late years has produced about 150,000 tons annually. India from 1897 to 1910 produced at least a total of 10 million tons. Japan from 1895 to 1902 produced 25,000–30,000 tons, and later 45,000–55,000 tons of pig-iron annually, *i.e.* in the last fifteen years more than 1,000,000 tons of ore, to which must be added an earlier production. South Australia from 1899 to 1909 produced some 1,800,000 tons of iron ore, while some ore was also produced by the other Australian colonies. Cuba in the year 1911 exported 1,147,879 tons of ore to the United States, in addition to several million tons earlier. Ontario in 1911 produced some 230,000 tons of iron ore ; some ore was also obtained from the other Canadian states, and a good deal from Newfoundland, namely, from 1895 to 1908 some 7 million tons. A minimum total of 100 million tons and a maximum of 250 million tons may therefore be reckoned for the ' other countries.'

Table of the Iron-Ore Districts with the Largest Production in Recent Years

1. The Lake Superior district : [1] production in 1910, 43·4 million tons of ore, equivalent to some 23 million tons of iron.

2. The German-French minette district : [2] Lorraine with 16·65 million tons in 1910 ; Luxemburg with 8·9 million tons in 1909, and Meurthe-et-Moselle with 10·67 million tons in 1909 ; or altogether 35·5 million tons of ore, equivalent to 12 million tons of iron.

3. The oolitic iron ores of Cleveland, Northampton, Lincoln, and North-East Yorkshire : [3] in recent years about 10 million tons of ore, equivalent to some 3 million tons of iron annually.

4. Bilbao : [4] yearly some 4·5 million tons of ore, equivalent to about 2·25 million tons of iron.

5. Norbotten in Sweden : [5] a present yearly production of some 3·5 million tons of ore, equivalent to about 2·2 million tons of iron.

6. The Clinton ores in the United States : [6] about 5 million tons of ore, equivalent to some 2 million tons of iron.

7. Kriwoj Rog in South Russia : [7] 1907, 3·75 million tons of ore, equivalent to some 2 million tons of iron.

8. Middle Sweden, including Grängesberg : [8] some 2·0 million tons of ore, equivalent to about 1·1 million tons of iron.

9. Siegerland : [9] 1910, 2·3 million tons of ore, equivalent to some 0·8 million tons of iron.

10. Cumberland and Lancashire : [10] yearly some 1·7 million tons of ore, equivalent to about 0·8 million tons of iron.

11. Eisenerz in Styria : [11] 1910, 1·7 million tons of ore, equivalent to some 0·7 million tons of iron.

12. The Urals : [12] 1906, 1·24 million tons of ore, equivalent to some 0·6 million tons of iron.

[1] *Ante*, p. 1062.   [2] *Ante*, p. 1003.   [3] *Ante*, p. 1022.
[4] *Ante*, p. 834.   [5] *Ante*, pp. 269-277.   [6] *Ante*, p. 1029.
[7] *Ante*, p. 1058.   [8] *Ante*, pp. 276, 378-394.   [9] *Ante*, p. 804.
[10] *Ante*, p. 826.   [11] *Ante*, p. 820.   [12] *Ante*, pp. 360-369.

13. Santander : 1907, 1·44 million tons of ore.

14. Murcia : 1907, 1·03 million tons of ore.

15. Almeria : 1907, 0·84 million tons of ore.

16. Algeria : 1910, 1·06 million tons of ore.

17. Upper Hungary : siderite ; [1] yearly some 1 million tons of ore, equivalent to about 0·4 million tons of iron.

18. Lahn-Dill district : [2] 1910, 1·0 million tons of ore, equivalent to 0·41 million tons of iron.

19. Peine : [3] yearly some 0·8 million tons of ore, equivalent to about 0·3 million tons of iron.

20. Scottish Blackband : [4] yearly almost 0·8 million tons of ore, equivalent to almost 0·3 million tons of iron.

21. Nučitz in Bohemia : [5] yearly some 0·65 million tons of ore, equivalent to about 0·2 million tons of iron.

The two economically most important iron-ore districts, that of Lake Superior in America, and that of Lorraine-Luxemburg including Meurthe-et-Moselle in Europe, at present yield together somewhat more than one-half, approximately 55 per cent, of the yearly iron ore and iron production. The first seven districts in the above table are responsible for at least three-quarters of the iron ore and iron production, while the twenty-one districts together produce nine-tenths of the world's total production. The remainder, some 10 per cent, is distributed over a considerable number of smaller districts.

The Iron Ore Production of the Separate Countries

In 1910 the iron ore production of the United States [6] was distributed among the various states as follows :

Lake District { Minnesota .	31·97	million long tons.
Michigan .	13·30	,, ,,
Wisconsin .	1·15	,, ,,
Alabama .	4·80	,, ,,
New York .	1·29	,, ,,
Virginia .	0·90	,, ,,
California, Colorado, New Mexico, Washington, and Wyoming .	0·86	,, ,,
Pennsylvania .	0·74	,, ,,
Tennessee .	0·73	,, ,,
New Jersey .	0·52	,, ,,
Georgia .	0·31	,, ,,
Other states .	0·32	,, ,,
	56·89	million long tons.

---

[1] *Ante*, p. 809.   [2] *Ante*, p. 1072.   [3] *Ante*, p. 1046.

[4] *Ante*, p. 1035.   [5] *Ante*, p. 1040.   [6] *Ante*, pp. 277, 373, 1028, 1062.

The production of the German Empire is contributed to by the different districts as follows :

IRON ORE PRODUCTION OF THE GERMAN EMPIRE DISTRIBUTED OVER THE DIFFERENT DISTRICTS

Districts.		Annual Output of Raw Iron Ore.	
		Amount including Moisture.	Average Iron Content after deducting Moisture.
		Tons.	Per cent.
Aachen Carboniferous-limestone district	1909	5,064	35·1
	1910	9,526	35·4
Berg limestone district	1909	22,917	32·6
	1910	17,466	34·2
Siegerland-Wied siderite district . .	1909	. 2,045,321	34·9
	1910	2,281,039	35·1
Nassau-Oberbesse (Lahn-Dill district) .	1909	907,461	41·0
	1910	1,004,263	40·9
Taunus district, including Lindener Mark	1909	262,106	23·5
	1910	278,055	24·0
Vogelsberg basalt district	1909	415,209	26·9
	1910	503,691	27·2
Waldeck-Sauerland district . . .	1909	30,024	27·6
	1910	28,194	25·9
Schafberg-Hüggel (Osnabrück) district .	1909	210,432	29·0
	1910	261,461	28·2
Wesergebirge district	1909	121,943	34·0
	1910	138,522	34·3
Sub-Hercynian (Peine, Salzgitter) district	1909	797,487	28·9
	1910	840,489	29·1
Harz district	1909	103,928	30·5
	1910	93,517	29·6
Bog ore district	1909	21,934	28·5
	1910	29,756	27·9
Silesian district	1909	268,213	29·5
	1910	272,579	29·6
Thüringia-Saxony district . . .	1909	185,791	38·5
	1910	237,870	38·4
Bavaria and Würtemberg-Baden district	1909	290,825	45·7
	1910	316,194	46·0
Lorraine minette district	1909	14,441,208	28·5
	1910	16,652,143	28·8
German Empire	1909	20,129,863	30·0
	1910	22,964,765	30·2

The production of Great Britain and Ireland was at its zenith during the period 1871–1884, when 16–17 million tons were produced annually. From this maximum it fell till in the early 'nineties it was 12–13 million tons, from which figure .t rose again till in the years 1906–1910 some 15–16 million tons were produced yearly. Latterly some 6–7, and in 1907 nearly 7–8 million tons have been imported yearly, this being in greater part Bessemer ore from Spain. The Cleveland and Northampton, etc., oolitic deposits [1] with an annual production in Cleveland of 6·07 million tons in 1908, and in Northampton of some 2·5 million tons in 1907, constitute the most important ore district. Up to 1909 roughly 250 million tons had been won in Cleveland alone. The Cumberland district comes next [2] with some 1·7 million tons yearly. In addition there are a number of smaller districts. The part played by the clay- and blackband ironstones is stated a little farther on. [3]

The production of France in the year 1909 was distributed among the different provinces as follows :

Meurthe-et-Moselle . . .	10,673,000 tons.
Pyrénées orientales . . .	277,900 ,,
Calvados	220,000 ,,
Orne	196,900 ,.
Haute-Marne	74,400 ..
Aveyron	49,000 ,,
Gard	40,000 ,,
Maine-et-Loire	29,700 ,,
Ariège	25,000 ,,
Ardèche	7,400 ,,

The minette accordingly yields more than nine-tenths of the whole production of this country.

The production of Russia in 1906 was distributed as follows :

South Russia	3,656,051 tons.
The Urals	1,242,000 ,,
Poland	300,905 ,,
Moscow basin	137,470 ,,
Northern district . . .	7,710 ,,
Siberia	6,940 ,,
The Caucasus	1,900 ,,

The most important districts in Spain, with their productions in 1907, are :

Biscay	4,736,193 tons.
Santander	1,437,707 ,,
Murcia	1,033,022 ,,
Almeria	844,676 ,,
Seville	302,957 ,,
Lugo	292,054 ,,
Teruel	215,845 ,,

Including a number of other districts the total production of Spain for that year was 9,196,178 tons. Bilbao in Biscay up to 1800 produced nearly 20 million tons, and from 1800 to 1860 some 3 million tons. The export of Bilbao in 1860 amounted to 70,000 tons, in 1865 to 102,000 tons, and in 1870 to 250,000 tons, while in the thirty-two years immediately before 1908 [4] 150 million tons were produced. The total production up to the end of 1912 was therefore approximately 200 million tons of iron ore. The production of Sweden has already been given. [5]

---

[1] *Ante*, p. 1022. [2] *Ante*, p. 826. [3] *Postea*, p. 1095.
[4] *Ante*, p. 834. [5] *Ante*, pp. 275, 391-394.

In Austria the most important district is Eisenerz in Styria [1] with a production up to the end of 1911 of 42–43 million tons.[2] In Belgium the production, which in the period 1836–1908 amounted to 30,846,000 tons, reached its zenith as early as the 'fifties and 'sixties when between 0·7 and 0·8 million tons were produced annually; the highest production was reached in 1865 with 1·02 million tons. In the last thirty years the production has been 0·2 million tons per year, more or less. The present important Belgian iron industry depends almost exclusively upon imported ore. The home iron deposits are not exhausted, but, with some exceptions, they cannot compete with imported ore.

THE LARGEST IRON DEPOSITS YET KNOWN AND THE TOTAL PRODUCTION OF SOME IRON DEPOSITS

The itabirite deposits of Brazil, and especially of Minas Geraes,[3] are regarded as the largest iron deposits yet known, though owing to the difficulties of transport they lie practically untouched. According to newspaper reports modern transport has been projected. Next in extent, so far as is at present known, come the Lake Superior deposits ; [4] the German-French minette ; [5] and the Norbotten deposits in Sweden, including Kiirunavaara.[6] Kiirunavaara contains the largest quantity of iron ore of exceptional richness within any limited district.

The Lake Superior district [7] from 1849 to the end of 1910 produced altogether 493 million tons of ore, equivalent to some 280 million tons of iron. The German-French minette district [8] up to the end of 1911 yielded altogether about 550 million tons of ore, equivalent to about 180 million tons of iron. Bilbao [9] has produced hitherto 190 to 200 million tons of ore, equivalent to some 100 million tons of iron. Cleveland [10] has yielded hitherto some 250 million tons of ore, equivalent to some 80 million tons of iron. The Scottish Blackband has yielded hitherto some 110 million tons of ore. While, finally, Eisenerz in Styria [11] has hitherto produced 42–43 million tons of ore, equivalent to some 17 million tons of iron.

THE IRON ORE-RESERVES OF THE WORLD

Upon the occasion of the XIth International Geological Congress at Stockholm, 1910, an attempt was made to estimate the iron ore-reserves

[1] *Ante*, p. 817.
[2] *Ante*, p. 820.
[3] *Ante*, p. 1060.
[4] *Ante*, p. 1062.
[5] *Ante*, p. 1003.
[6] *Ante*, pp. 269-275
[7] *Ante*, p. 1062.
[8] *Ante*, p. 1003.
[9] *Ante*, pp. 834, 1091.
[10] *Ante*, p. 1023.
[11] *Ante*, p. 820.

of the world. Although these reserves were divided into different classes according to the condition of their development, a study of the otherwise excellent work [1] shows that in the different countries the estimation was approached from very different standpoints, and that consequently the figures put forward cannot off-hand be compared with one another. This should be remembered when comparing the following tables. Further-more, the deposits in Australia, Asia, and Africa, have been so little investi-gated that the figures given for these continents are of little use. For Sweden, on the other hand, where the estimation was most carefully con-ducted, the figures given have, in the short time since the Congress, been proved to be too low. The tables accordingly are hardly suitable as the basis of any careful economic conclusions. They prove however with certainty that the iron ore-reserves, not only of the other continents but also of Europe, are gigantic. Even if the iron production in the future were doubled in successive periods of 20–25 years, the reserves known to-day would be sufficient to last for centuries. It may, however, be assumed that future advance in technology will render possible deeper mining than that being undertaken at present, while a more effective mechanical concentration of poorer ores may also be expected; further-more, smelting by electricity has already permitted the working of deposits situated within the reach of large water powers though far removed from coalfields; while, finally, the future discovery of other large deposits may be anticipated. Though therefore a dearth of iron ore is not to be feared in the next-coming centuries, several of the present important deposits will undoubtedly in a short time become exhausted, while a more distant future will undoubtedly produce considerable changes in the iron industrial centres.

[1] *The Iron Ore Resources of the World*, Stockholm, 1910.

[TABLE

IRON ORE-RESERVES OF EUROPE

	Actual Reserves.		Possible Reserves.	
	Ore.	Iron Content.	Ore.	Iron Content.
	Million tons.	Million tons.	Million tons.	Million tons.
France	3,300	1140	...	...
Luxemburg	270	90	...	...
Spain	711	349	considerable amounts	...
Portugal	...	...	75	39
Italy	6	3·3	2	1
Switzerland	1·6	0·8	2	0·8
Austria	251	90	323	97
Hungary	33	13	79	34
Bosnia and Herzegovina . .	...	...	22	11
Servia	...	...	...	...
Bulgaria	...	...	1·4	about 0·7
Greece	100	45	...	...
Turkey	...	...	...	...
Russia, European . . .	864	387	1,056	425
Finland	...	...	45	16
Sweden	1,158	740	178	105
Norway	367	124	1,545	525
Great Britain	1,300	455	37,700 [1]	10,830 [1]
Holland	...	...	...	...
Belgium	62	25	...	...
Germany	3,608	1270	considerable amounts	
Totals . . .	12,032	4733	41,029 + considerable amounts	12,085 + considerable amounts

[1] These figures are not on a basis for comparison with the others.

THE IRON ORE-RESERVES OF THE UNITED STATES

District.	Amount of Reserves.	
	Actual.	Possible.
	Million tons.	Million tons.
A. The Eastern Region—		
Archaean magnetite, solid ore . .	20	30
Archaean magnetite, concentrates .	40	10
Adriondack, red hæmatite . . .	2	2
Pennsylvania, soft hæmatite . . .	40	...
Cambrian-Ordovician brown hæmatite	65	181
Mesozoic and Tertiary brown hæmatite	10	15
Alabama, brown and red hæmatite .	27·5	27·5
Clinton, red hæmatite	505	1,368
Carbonate ores	...	308
B. Lake Superior District—		
Specularite and red hæmatite, brown hæmatite, etc.	3500	72,000
C. Mississippi Valley—		
Specularite and red hæmatite . .	15	5
Palæozoic brown hæmatite . . .	30	45
Tertiary brown hæmatite . . .	...	780
D. The Cordillera District—		
Magnetite and hæmatite	3	116
Titaniferous ores	...	218
Totals . . .	4258	75,105

THE WORLD'S IRON ORE-RESERVES

	Actual Reserves.		Possible Reserves.		
	Ore.	Iron Content.	Ore.	Iron Content.	
Europe . .	Million tons. 12,032	Million tons. 4,733	Million tons. 41,029	Million tons. 12,085	+ consider- able amounts
America . .	9,855	5,154	81,822	40,731	+ enormous amounts
Australia . .	136	74	69	37	+ consider- able amounts
Asia . . .	260	156	475	283	+ enormous amounts
Africa . . .	125	75	many thousands	many thousands	+ enormous amounts
Total . .	22,408	10,192	>123,377	>53,136	+ enormous amounts

THE DISTRIBUTION OF THE IRON ORE PRODUCTION AMONG THE VARIOUS CLASSES OF DEPOSIT

The greater part of the iron ore production is derived from sedimentary occurrences, among which the oolitic iron ores, especially of the Jurassic formation, play a very important part.   The well-known Jurassic minette deposits in Lorraine, Luxemburg, and Meurthe-et-Moselle, with a present annual production [1] of 36 million tons of ore, equivalent to some 12 million tons of iron, belong to this class ; as do also the Jurassic deposits in Cleveland, Northampton, Lincoln, and north-east Yorkshire, producing together some 10 million tons of ore annually, equivalent to some 3·5 million tons of iron.   The geologically fairly similar though Upper Silurian Clinton ores in the United States, yield some 5 million tons of ore per year, equivalent to some 2 million tons of iron.   Including these and a number of other occurrences, a total of some 55 million tons of oolitic ore containing about 20 million tons of iron is obtained as the production from such ores in 1910, a figure corresponding to nearly 40 per cent of the total iron ore production, and approximately 30 per cent of the total iron production.

The bean ores are of subordinate importance and altogether do not reach 1 per cent of the world's production.

The blackband- and clay-ironstones not many decades ago were of the utmost importance, especially those of England and Scotland, which about the middle of the nineteenth century were responsible for as much as some three-quarters of the production of those countries.   In Westphalia also and in many other coal-mining districts these ironstone beds were important.

[1] *Ante*, p. 1003.

Latterly, however, operations upon them, in consequence of greater cost, have rapidly declined and in some places have been completely suspended. The output of these ores in Westphalia in 1865 was not less than 1,154,000 tons,[1] while to-day operations have entirely ceased. In South Wales the production fell from 1,100,000 tons in 1872 to 40,000 tons in 1890 ; in South Staffordshire, from 715,000 tons in 1875 to 41,000 tons in 1890 ; in Derbyshire, from 493,000 tons in 1871 to 24,000 tons in 1890 ; several other counties show a similar decrease, so that England now produces but an inconsiderable quantity of such ores. In Scotland, where altogether 110 million tons of blackband [2] have been produced, and in the year 1881 some 1,402,700 tons of blackband- and 1,192,375 tons of clay-ironstone, or a total of 2,595,375 tons, operations have likewise declined, the production in 1894 having been 631,304 tons, and this almost entirely of clay-ironstone. A few years back Scotland again produced nearly 800,000 tons of blackband- and clay-ironstone. To this must be added a small quantity from other countries, so that, altogether, this class of iron ore may be said at the present time to be responsible for about one per cent of the world's output.

Chamosite and thuringite [3] together likewise yield about one per cent of the total production.

The detrital iron deposits [4] produce only one, or not quite one per cent of the total output, while the recent iron- or titaniferous-iron sands are to-day without, or with only quite subordinate importance.

The lake- and bog ores have of late years produced only some 35,000 tons per year, or 0·025 per cent of the total production.

On the other hand, some of the iron ore-beds in the fundamental rocks and in early Palæozoic crystalline schists are of the greatest importance. Thus, the Lake Superior occurrences alone, which are to be regarded as altered sediments, produced in 1910 no less than 43·4 million tons of ore, equivalent to some 23 million tons of iron, that is, approximately 30 per cent of the total ore production and 35 per cent that of the iron. Kriwoj Rog produces 3·5 to 4 million tons of ore per year, and in 1907 produced 3·7 million tons. To these figures must be added the production of a smaller mine in the northern Norwegian ferruginous mica- and magnetite-quartz schists as well as that of other occurrences belonging to this class, which altogether is therefore responsible for nearly 40 per cent of the world's iron production.

Of the metasomatic deposits, Bilbao produces yearly some 4·5 million tons of ore, equivalent to about 2·25 million tons of iron ; Cumberland-

---

[1] *Ante*, p. 1035.

[2] H. Louis, *The Iron Ore Resources of the World*, Stockholm, 1910, Vol. II. p. 638.

[3] *Ante*, p. 1039.   [4] *Ante*, p. 1044.

Lancashire some 1·7 million tons of ore, equivalent to 0·8 million tons of iron ; and Eisenerz in Styria some 1·7 million tons of ore, equivalent to 0·7 million tons of iron. Other deposits belonging to this class are : several occurrences in Algeria with a production of 1·06 million tons in 1910; others in Poland with 0·3 million tons in 1906 ; several important occurrences in Spain ; and several in Austria, Hungary, Germany, France, United States, Norway, etc. At the present day therefore some 12–15 million tons of ore, equivalent to some 5–7 million tons of iron, are derived from metasomatic occurrences.

Of the siderite lodes, Siegerland comes first with a production of some 2·3 million tons of ore per year, equivalent to about 0·8 million tons of iron ; then follows Upper Hungary with 1 million tons of ore, or 0·4 million tons of iron yearly ; so that altogether, including some smaller lodes, approximately 4 million tons of ore, equivalent to some 1·4 million tons of iron, are produced annually from this class of deposit. Lodes with preponderating hæmatite, specularite, or magnetite, are of subordinate importance.

Among the contact-deposits are reckoned :· Elba,[1] with a present annual production of some 0·45 million tons and with an ore-reserve of 5 to 6 million tons ; the Banat, with 0·12 to 0·15 million tons of ore yearly ; a number of smaller occurrences such as Traversella-Brosso, Schwarzenberg, Schmiedeberg, Pitkäranta, Christiania, etc. ; and several in North America of subordinate economic importance. Further, according to the authors' view, several of the principal occurrences in the Urals, with 1·24 million tons of ore in 1906; the occurrences in the fundamental rocks at Arendal with a yearly production of some 25,000 tons of iron ore ; and those of Dannemora-Persberg, etc., in Middle Sweden [2] with an annual production of some 0·75 million tons of ore, belong to this class. Contact-deposits produce then, altogether, some 3 million tons of ore, equivalent to about 1·5 million tons of iron, annually.

The magmatic segregations of titaniferous-iron ores in basic eruptive rocks are at the present time economically of quite subordinate importance. One Norwegian mine, Rödsand,[3] produces by magnetic separation some 15,000 tons of non-titaniferous concentrate containing some 64 per cent of iron. There are in addition a few other occurrences belonging to this class upon which mining operations are being prosecuted, though altogether these occurrences yield at most 0·1 per cent of the world's production. On the other hand, the iron- and apatite-iron deposits, which must be regarded as magmatic segregations in acid eruptives,[4] are of great importance. The Norbotten occurrences, for instance, produce to-day some 3·5 million tons of ore, equivalent to about 2·2 million tons of iron, a production

[1] Ante, pp. 369-372.   [2] Ante, pp. 383-389.
[3] Ante, p. 257.   [4] Ante, pp. 259-277.

expected shortly to be still larger ; while Grängesberg and other apatite-iron deposits in Middle Sweden produce some 0·8 million tons of ore, equivalent to 0·5 million tons of iron. To these must be added Sydvaranger in northern Norway, with 0·25 million tons of concentrate containing some 67 per cent of iron in 1912, and in some years probably some two-third million tons yearly ; several occurrences worked on a large scale in the Adirondacks, in the United States ; and perhaps the *Torrsten* or lean ores of the Striberg-Norberg type [1] in Middle Sweden. Altogether therefore, deposits of this class produce to-day some 6 million tons of ore, equivalent to about 3·5 million tons of iron per year.

The contribution of the different classes of deposit to the yearly production of iron is accordingly represented by the following percentages :

Magmatic ⎰ Titaniferous-iron ores .	.	.	at most 0·1 per cent.
segregations ⎱ Iron- and apatite-iron ores	.	.	about 5–6 ,,
Contact-deposits .	.	.	,, 2·5 ,,
Siderite lodes	.	.	2–2·5 ,,
Other lodes	.	.	little
Metasomatic deposits	.	.	about 8–10 ,,
⎧ Lake- and bog ores	.	.	. 0·025 ,,
⎪ Bean ore .	.	.	perhaps 1 ,,
Sedimentary ⎨ Oolitic ore .	.	.	roughly 30 ,,
deposits ⎪ Blackband- and clay-ironstone .	.	about 1 ,,
⎪ Chamosite and thuringite	.	.	,, 1 ,,
⎪ Detrital ore .	.	.	not quite 1 ,,
⎩ In fundamental rocks .	.	.	nearly 40 ,,
Roasted pyrite residues, old slags, etc. .	.	.	about 2 ,,
Undetermined .	.	.	,, 10 ,,

This table gives at all events an approximate representation of the present-day production of iron ore. Since however iron mining is a competitive industry, many important iron deposits for various reasons are not worked—as for instance the large occurrences in Brazil, owing to the difficulties of transport, and nearly all the titaniferous-iron occurrences, for metallurgical reasons—while the blackband- and clay-ironstone deposits, owing to high cost of production, are only worked on a small scale. The above summary is nevertheless in some respects very conclusive ; it shows that the sedimentary occurrences are far and away the most important of the iron deposits. As will be mentioned later, this is also the case with manganese. With both metals chemically so closely related, among these sedimentary occurrences the oolitic ores occur most extensively, these having been formed in greater part after the manner of the lake- and bog ores.

Many large deposits have also been formed by magmatic differentiation, contact phenomena, deposition as fissure-fillings, metasomatic processes, etc., as for instance, those of Kiirunavaara, Siegerland, Bilbao, etc. Taking however everything into consideration, the concentration of iron by these various means is subordinate when compared with that

[1] *Ante*, p. 382.

brought about on the surface by hydrochemical processes continued through long periods.

In so far that with all others the sedimentary origin is invariably of relatively subordinate importance, iron and manganese are unique among the heavy metals. With many metals, such for instance as chromium, nickel, tin, etc., no primary deposit formed by sedimentation is known. The reason of this unique position of iron and manganese is probably due in greater part to the fact that the other heavy metals occur considerably less extensively in the earth's crust; moreover, meteoric waters readily take iron and manganese into solution. With few exceptions the other heavy metals occur so sparingly in the crust that in solutions resulting from weathering they occur in very small amount. Small amounts of nickel, cobalt, zinc, copper, etc., it is true, are found in recent chemical deposits; nevertheless, apart from the garnierite- and asbolane deposits which were formed from rocks with relatively high nickel- and cobalt content, recent chemical deposits of these metals from meteoric waters appear to be non-existent, or are very subordinate. In order to form concentrations of copper, lead, zinc, tin, etc., large enough to constitute useful deposits, more energetic chemical agents are necessary than those at the disposal of meteoric waters. Those sedimentary occurrences of copper, lead, zinc, etc., which exceptionally do occur, were probably formed by the passing of these heavy metals into solution at great depth, such solution eventually bringing them to the surface; or they reached the surface as eruptive exhalations, etc.

THE MANGANESE ORE-BEDS

As already pointed out when discussing the lake- and bog ores,[1] a gradual passage occurs between the iron lake- and bog ores on the one side, and the manganese lake- and bog ores on the other. The rocks of the earth's crust contain on an average some 40–70 times as much iron as manganese.[2] Both metals upon weathering go equally readily into solution. Those solutions from which the manganese lake- and bog ores were deposited probably originally contained therefore in general more iron than manganese, even when such solutions came from rocks relatively rich in manganese and poor in iron, such for instance as sand, clay, etc. As already mentioned, at the oxidation of such solutions iron is precipitated earlier than manganese.[3] In harmony with this it can occasionally be demonstrated with manganiferous lake- and bog ores that the iron-ochre was precipitated earlier than the manganese-ochre.

Exceptionally, springs have been known to carry several times as

[1] *Ante*, pp. 982-988.   [2] *Ante*, p. 153.   [3] *Ante*, p. 980.

much manganese as iron.   Thus, M. Weibull [1] described spring water at
Alnarp near Lund, which throughout the different seasons generally contained between 4 and 69 mg. of MnO per litre, together with only traces
or occasionally as much as 7 mg. of FeO, a relation equivalent to 0·1–0·01
of iron, or still less, to 1 of manganese.   After having stood in a sealed bottle
for a period of two years a sample of this water was found to have precipitated a manganese-ochre with 43 per cent of manganese and 1·2 per
cent of Fe, or nearly forty times as much manganese as iron.   Weibull
cited other, certainly very exceptional data concerning water containing
relatively much manganese and little iron.

Small deposits of manganese-ochre poor in iron or almost free from
that metal, in glacial detritus and other gravel-deposits, are exceedingly
frequent and have often been described.   In this connection we would
refer to a paper by G. de Geer upon an occurrence of manganese-ochre
containing 73·19 per cent of Mn_3O_4, 16·27 per cent H_2O, 7·24 per cent of
insoluble residue, and only traces of iron, in the Upsala-Ås.[2]

Larger deposits of bog óre and occasionally also of lake ore with a
preponderating amount of Mn and very little Fe are also not uncommon.[3]

Beyschlag and Michael [4] describe manganese deposits from the
Alluvium of the Oder near Breslau.   F. Katzer [5] describes a somewhat
similar occurrence in the flats at the mouth of the river Amazon, where
manganese ore is found in numerous places within an area roughly 1000
km. long and 500 km. wide.   The ore in this occurrence, principally psilomelane with some pyrolusite, appears in peaty masses some 6 cm. and
more in thickness.   It is frequently accompanied by some limonite and
occurs in sand or sandstone—manganiferous sandstone with manganese
ore as the cementing material—in the present flood area of the Amazon
river, from the waters of which it was deposited.   The total quantity of
this ore is considerable, though in no place is it sufficiently concentrated
to be workable.   Some analyses of this psilomelane exhibited a strikingly
high barium content, 1 Mn to 0·19–0·27 Ba, and very little iron ; a high
BaO content is also present in one of the previously-mentioned [6] deposits
of manganese-ochre ; while the manganese ore of Kutais, Miguel Burnier in
Brazil, Chili, and many other places, likewise contains barium.[7]   Accordingly,
in the sedimentary occurrences manganese is occasionally accompanied by
a fairly large amount of barium.   More rarely some nickel, cobalt, copper,
etc., are found as oxides, and of these apparently more particularly cobalt.[8]

[1] 'Ein manganhaltiges Wasser und eine Bildung von Björnstorf in Schweden,' *Lunds
Universitets Årsskrift*, Afd. 2, Vol. II., 1907 ; and *Zeit. f. Untersuchung der Nahrungs- und
Genussmittel*, 1907, Vol. XIV. Part 6.   [2] *Geol. Fören. Förh.*, 1882, VI. pp. 42-44.
[3] *Ante*, p. 984.   [4] *Zeit. f. prakt. Geol.*, 1907, p. 153.
[5] 'Ein eigentümliches Manganerz des Amazonengebietes, *Österr. Zeit. f. Berg- u.
Hüttenwesen*, 1898, Vol. XLVI.
[6] *Ante*, p. 985.   [7] *Postea*, p. 1103.   [8] *Postea*, pp. 1103, 1107.

In the more recent geological formations a number of manganese ore-beds are known, which were undoubtedly formed by sedimentation and must be regarded as old manganese lake- and bog ores or shallow-water deposits. Of these, the Tertiary occurrences at Kutais containing sharks' teeth, and at Nicopol, both in Russia and described more fully later on, are noteworthy. From the tectonics at both places it is evident that the deposition of the manganese took place in shallow water, probably in large lagoons and in a shallow sea fairly near the coast. In regard to thickness and amount of ore per square metre of extent, these two Russian Tertiary occurrences are approximately comparable with the most extensive of recent manganese bog- or lake ores formed since the Glacial Period. The Kutais deposit, for instance, yields on an average 0·96 tons of ore with 50–51 per cent of manganese per square metre on the bed plane, equivalent to some 1·5 tons when the poorer material is included; while the small post-Glacial occurrence of manganese bog ore at Glitrevand near Drammen,[1] contains where thickest about 1 ton of manganese ore with 50 per cent of manganese, per square metre. A substantial difference however exists in the total areas of these two deposits; at Kutais the amount of manganese brought into solution must have been enormous.

At Nicopol gneiss or granite occurs in the immediate neighbourhood of the deposit, and at Kutais somewhat farther off. The solutions were probably derived from these crystalline rocks. It must be pointed out that manganese lodes and beds—not including the metasomatic occurrences associated with limestone and dolomite—are not particularly associated with basic rocks, the majority, though not all, occurring in genetic association with acid rocks, such as gneiss, granite, quartz-porphyry, etc.; we would refer in this connection to a table drawn up by Vogt.[2] The acid rocks on an average carry less iron and manganese than the basic. In the basic rocks also, iron is concentrated on the whole in relatively larger amount than manganese, with the result that the relation between iron and manganese in the acid rocks has, generally speaking, moved in favour of the manganese. The solutions from acid rocks probably therefore originally carried as a rule more manganese in relation to iron than similar solutions from basic rocks.

In the case of the sedimentary manganese occurrences it may in general be concluded that the solutions originally contained somewhat more iron than manganese, but that the bulk of the iron together with relatively much phosphoric- and silicic acid[3] was precipitated comparatively early. In harmony with this it is found that both the sedimentary manganese deposits of earlier formation, such as Kutais and Nicopol, and the present manganese bog ores, are frequently characterized by a high manganese

[1] *Ante*, p. 984.   [2] *Zeit. f. prakt. Geol.*, 1906, pp. 231-232.   [3] *Ante*, p. 980.

content with only relatively little silicic acid, etc., and by a fairly small amount of phosphoric acid. From the alternation of itabirite and manganese ore at Miguel Burnier in Brazil, it may be assumed that both ores were formed from the same solutions, and in such a manner that at first principally iron with little manganese, and later chiefly manganese

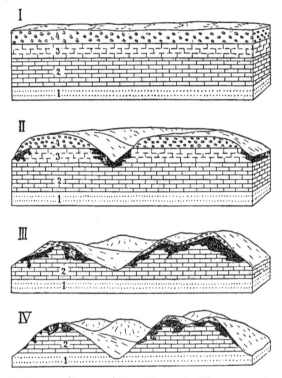

Fig. 442.—Diagrammatic representation of the genesis of the manganese deposits at Batesville, Arkansas. Penrose.

I, undisturbed bedding; *II, III,* and *IV*, different stages in surface disintegration and decomposition. 1, sandstone; 2, Lizard limestone; 3, St. Claire limestone; 4, Boone Chert; 5, manganiferous clay.

with little iron, were precipitated. This sequence naturally may have often been repeated.

Sedimentary manganese occurrences comparable with the recent deep-sea deposits of manganese nodules likewise occur; they appear however to be very rare. We would refer in this connection to the subsequent description of Čevljanović in Bosnia.[1]

[1] *Postea,* p. 1108.

A totally different origin, namely, by weathering in conjunction with an eluvial migration of some constituents and a corresponding enrichment of the manganese, has been investigated by R. A. Penrose in his work, *Manganese, its Uses, Ores, and Deposits*.[1] This authority explains the manganese occurrences at Batesville in Arkansas by the diagrams given in Fig. 442. There, the Upper Silurian St. Claire limestone possesses for a limestone a fairly high manganese content. Upon attack, especially by waters containing carbonic acid, carbonate of lime with some carbonate of magnesium, iron, and manganese, were removed in solution, while the remainder of the manganese remained behind as oxide, namely, as psilomelane and braunite, together with clayey material. By washing, this clayey constituent is removed and a saleable product recovered. The Batesville occurrence from 1850 to 1890 produced 30,000–35,000 tons of manganese ore, of which the greater quantity was produced in the 'eighties; the yearly production was at most only some 3000 tons. It may be taken for granted that similar deposits were also formed in earlier geological periods. Economically these deposits are only of subordinate importance.

Among the sedimentary manganese deposits, the manganese bog- or lake ores and the shallow-water formations probably predominate. The following analyses of practically clean manganese ore pertain to certain of the shallow-water deposits:

	Kutais.	Miguel Burnier.	
	No. 1.	No. 2a.	No. 3b.
SiO_2	3·85	0·53	1·27
Fe_2O_3	0·61	2·50	4·03
MnO_2	86·25	80·62	79·40
MnO	0·47	5·47	6·23
Al_2O_3	1·74	2·21	1·45
CaO	1·73	0·70	Trace
MgO	0·20	1·05	0·05
BaO	1·54	2·30	1·90
Na_2O+K_2O . . .	0·22	Trace	0·55
CuO	0·01	...	...
NiO	0·30	...	...
P_2O_5	0·32	0·07	0·05
S	0·23	...	...
SO_3	...	Trace	0·07
As_2O_3	...	...	0·03
CO_2	0·63	...	...
H_2O	1·85	4·95	4·74
Total . . .	...	100·40	99·77

No. 1, dried at 100°; cited after L. Demaret.—No 2a-b, from Miguel Burnier in Brazil; cited after H. K. Scott.

[1] *Geol. Survey of Arkansas, Ann. Rep.*, 1890, Vol. I.

The ore marketed is naturally, generally speaking, somewhat poorer
in manganese.   According to Holland and Fermor [1] average analyses of
ship-loads of manganese ore have given the following results :

	Kutais, Caucasus.		India.			Brazil.	
			Central Provinces, etc.	Vizaga-patam.			
Number of ship-loads.	77		26	22	4	25	
	With normal moisture.	Dried at 100°.	With normal moisture.			With normal moisture.	Dried at 100°.
Mn	45·3	49·6	50·5	51·3	46·0	44·6	50·3
Fe	0·76	0·83	6·3	5·5	10·3	3·4	3·8
SiO$_2$	9·3	10·2	5·7	6·1	3·1	1·8	2·0
P	0·147	0·161	0·126	0·096	0·291	0·046	0·052
Al$_2$O$_3$, CaO, etc. .	11·7	12·8	6·75	...	...	2·73	3·1
H$_2$O	8·7	...	0·72	0·71	0·76	11·4	...

The 22 ship-loads from the Central Provinces were obtained from Archaean, bed-like,
probably sedimentary occurrences.

Summaries of the manganese ore-occurrences of the world are found
in the afore-cited work by R. A. F. Penrose ; in a paper by Léon Demaret,
" Les Principaux Gisements des minerais de manganèse du monde," *Ann.
des Mines de Belgique*, 1905, X.; by Venator, in *Stahl und Eisen,* 1906; and
by P. Krusch, *Untersuchung und Bewertung von Erzlagerstätten,* 1st Ed.,
1907, Part 3.

THE MANGANESE ORE-BEDS IN EARLY, PRINCIPALLY
TERTIARY FORMATIONS

South-Russian Manganese Ore-Beds at Tschiatura in the Government Kutais, and at Nicopol in the Government Ekaterinoslav

LITERATURE

Upon Kutais : Abich, Akad. d. Wissensch., St. Petersburg, 1868, III.—A. Macco. Zeit.
f. prakt. Geol., 1898, pp. 203-205.—L. Demaret, 1905, *loc. cit.*—L. de Launay.   Les
Richesses minérales de l'Asie, Paris, 1911, pp. 265-269.
Upon Nicopol : N. Sokolow.   Mém. du Comité géol., St. Pétersbourg, 1901, XVIII. ;
Extract, Jahrb. f. Eisenhüttenwesen, 1903, II. pp. 213-216.

Of these two exceedingly important occurrences situated about 900
km. in a straight line from each other, Tschiatura lies to the south of the

[1] *Op. cit.*

Caucasus about 125 km. by rail from Poti, a port on the Black Sea, and Nicopol in the neighbourhood of the Dnieper about 200 km. from its mouth.

At the station Kviril a branch line, following the river of that name, leads from the main railway Poti-Tiflis to the manganese occurrence at Tschiatura. There a tableland of Senonian limestone and slate forms the base for Eocene sandstone, etc., and Olio-gocene and Miocene sandstone, slate, and limestone. The beds lie almost horizontal, dipping on an average 2½° north-east. The manganese ore-bed, lying near the surface of this plateau, is cut with extraordinary regularity by the valleys, upon the slopes of which it is recognizable from afar by its black colour. It belongs to the Lower Eocene and lies a little above the contact with the Upper Creta-ceous. Sandstone occurs both in the hanging-wall and foot-wall, while sharks' teeth and lizard remains occur in the ore itself. The whole deposit represents therefore a shallow-water formation. The thickness is generally 1·5–2·4, or on an average 2·1 metres. Within this thickness 5–12 manganiferous layers of varying thickness occur in alternation with marly sands, which, as illustrated in Fig. 443, in part are also impregnated with ore. These manganiferous layers consist of concentric oolites in a matrix of fine-grained ore. The hard layers contain on an average 56 per cent of manganese, while the softer layers, in consequence of their association with barren material, are poorer. The ore as delivered contains mostly 50–52 per cent of manganese—reckoned with ore dried at 100°—6–8 per cent of silica, only 1 or 2 per cent of iron, and 0·05–0·17 per cent of phosphorus. The present surface area of the whole ore-bed, of which about one-half has been removed by erosion, is estimated by different authorities at between 120 and 143 square kilometres, so that the area of the available deposit must be at least 60 square kilo-metres. Each square metre yields on an average 0·96 tons of ore, and the total reserves on the basis of the present development have been reckoned at 110 or 115 million tons. This estimate however is said to be too high, since the thickness of the ore

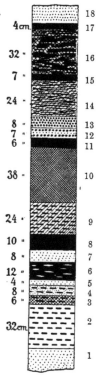

FIG. 443.—Section of the manganese deposit at Guemetti in the Caucasus. Demaret.

1, 5, 7, 18, Miocene sand-stone; 2, black coarsely granular ore in compact layers; 3, 4, brown man-ganiferous sandstone; 6, black ore in large grains; 8, 11, 15, 17, compact ore; 9, brown earthy ore; 10, black granular ore in compact layers; 12, granular ore with sandstone layers; 13, fine-grained ore; 14, brown earthy ore with sandstone layers; 16, brown earthy ore.

is subject to variation.   At other places in Transcaucasia small manganese ore-beds, and in some cases lodes, have been found.

The manganese ore-bed at Nicopol belongs to the Oligocene ; it is horizontally bedded, and has glauconitic clay and sand in hanging-wall and foot-wall.   Immediately beneath the Oligocene, granite or gneiss occurs, while at Horodizce, 18 km. north of Nicopol, a Tertiary manganese ore-bed lies almost immediately on granite.   The ore-bed at Nicopol, which is very similar to that at Tschiatura, is 0·3–1·8 or on an average 1–1·5 m. thick.   The superficies of the ore-bearing beds is estimated at 20 square kilometres, and the ore-reserves are stated to amount to 7·4 million tons.

Kutais from 1848 to 1897—work however has only been on any considerable scale since 1879—produced 1,682,000 tons of manganese ore, and up to the end of 1903 some 4,322,600 tons.   At Nicopol, from 1886 to and including 1903, some 753,000 tons were won.   Both deposits up to the end of 1907 had together produced 8·9 million tons of manganese ore, the greater portion of which was naturally obtained from Kutais.   The Russian manganese ore production in the year 1906 was 1,015,686 tons, from which high figure however it has of late years somewhat fallen. From the Urals, only one or a few thousand tons are produced yearly.

There are a large number of occurrences the geological conditions of which agree in the main with the Russian deposits just described. These however are all considerably smaller though the following are worthy of mention :

A bed-like occurrence in recent, practically horizontal beds, mostly of Diluvial and subordinately of Tertiary age, in the neighbourhood of Ciudad Real in Spain.[1]—A section of this occurrence from hanging-wall to foot-wall gave : 1 m. loam, 2 m. reddish-brown clay, 0·20 m. manganese ore, 0·60 m. red-brown clay traversed by manganese stringers, 0·30 m. grey clayey material with tuff fragments, 0·10 m. fragments of manganese ore embedded in clay, and 4 m. red manganiferous red clay, the manganese being in small concretions.   Basalts and basalt tuffs are found in the neighbourhood, the manganese ores being associated with the weathering- and decomposition products of these rocks.   Secondarily formed iron ore occurs immediately on the basalt, while the manganese ore appears somewhat farther removed.   Michael explains this occurrence as a fluviatile deposit formed from the decomposition products of basalt or basalt tuff.   His description however agrees very well with a primary deposition, during which at first chiefly iron and later chiefly manganese, were precipitated from solution.   The ore, which consists in greater part of psilomelane, contains on an average 43 per cent of manganese, while the phos-

---

[1] R. Michael, ' Die Manganerzvorkommen in der Nähe von Ciudad Real in Spanien,' *Zeit. f. prakt. Geol.*, 1908, pp. 129-130.

phorus varies between 0·098 and 0·272 per cent. The ore invariably contains some cobalt, namely, 0·14–0·37 per cent. The occurrence hitherto has been worked only in opencut.

Strullos in the neighbourhood of Larnaca on the island of Cyprus.[1] —At this place nodules of pyrolusite and psilomelane, and an earthy manganese-limonite formerly widely known as Cyprian umber, are found associated with Miocene marl, in the neighbourhood of a decomposed quartz-andesite or quartz-porphyry with attendant tuffs.

The San Pietro island off the south-west coast of Sardinia.[2]—Within a series of trachytes or rather trachyte tuffs, and red and white clays, occur two manganese ore-beds, 0·2 and 0·6 m. thick respectively, separated from each other by 2 m. of clay. The ore, which is chiefly pyrolusite with a little hausmannite, was probably formed by leaching from the tuffs. The annual production is estimated at 1000 tons of manganese ore.

The island of Milos off the coast of Greece.[3]—Here a 0·6–1·8 m. thick manganese ore-bed appears in Pliocene clays in the immediate neighbourhood of trachyte tuffs, from which the ore solutions were presumably derived. The production in the year 1902 was 15,000 tons of ore containing 45–52 per cent of Mn, 1 per cent of Fe, 8–12 per cent of SiO_2, and 0·09–0·10 per cent of phosphorus.

L. Demaret [4] describes several analogous occurrences in Liguria, in Eocene ; in Tuscany, in Devonian ; and in the Panama Republic, etc. At the last-mentioned place the deposits according to A. Schmidt [5] are probably of Tertiary age. R. A. F. Penrose [6] mentions some analogous Tertiary, though invariably fairly small deposits, at different places in North America.

In this connection we would also mention the deposits in the Coquimbo and Carrizal districts in Chili,[7] where manganese ores form fairly extensive beds in Jurassic-Cretaceous sandstone, slate, limestone, and gypsum beds. Immediately in the foot-wall of these occurrences an extensive eruptive occurs. Here also, the manganese ore, pyrolusite with hausmannite and some rhodonite, contains small quantities of barium ; the phosphorus content is low. In the years 1883–1894 Chili exported altogether 351,792 tons of manganese ore, and in 1894 alone 47,994 tons. Since that date however the industry has greatly declined.[8]

[1] Bergeat, *Die Erzlagerstätten*, 1904, I. p. 260 ; A. Gaudry, ' Géologie de l'île de Chypre,' *Mém. Soc. Géol. de France* (2), VII. pp. 191-192.
[2] L. Demaret, *loc. cit.* p. 68 ; F. Fuchs and L. de Launay, *Gîtes métallifères*, II. pp. 25-26 ; G. vom Rath, *Sitzungsber. d. Niederrhein. Gesellsch.*, 1883, XL. pp. 151.152.
[3] L. Demaret, *loc. cit.* p. 66, *Österr. Zeit. f. Berg- u. Hüttenwesen*, 1897, XLV. p. 514 ; Zenghelis, *Les Minerais et minéraux utiles de la Grèce*, 1903.   [4] *Loc. cit.*
[5] *Zeit. f. prakt. Geol.*, 1903, pp. 247-248 ; see also E. G. Williams, *Trans. Amer. Inst. Min. Eng.*, 1902, reviewed *Zeit. f. prakt. Geol.*, 1904, p. 369.
[6] *Loc. cit.*, 1891.
[7] Henry Louis, *A Treatise on Ore-Deposits*, 1896, p. 878.   [8] *Postea*, p. 1114.

The manganese ore-beds hitherto described, like the recent manganese bog ore, were probably all formed in shallow water.[1] The deposit at Čevljanović north of Sarajevo in Bosnia may, on the other hand, be regarded as a deep-sea formation. The geological position of this occurrence has been described by F. Katzer.[2] The manganese ore, which consists chiefly of psilomelane rich in barite—with 5–6·5 per cent of BaO—is associated with highly disturbed Jurassic siliceous radiolarite; it is also found, though only exceptionally and economically speaking in insignificant amount, in the Radiolarian limestone. The ore sometimes occurs as separate nodules and lenses, and sometimes in continuous bands, layers, or even thick beds. Owing to its conformity throughout an area 12 km. long and 6 km. wide, Katzer concludes a sedimentary origin for this deposit. He states 'the primary manganese ore nodules may be well explained as

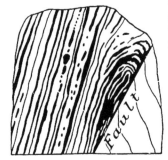

an analogous occurrence to the manganese nodules in the deep-sea clay of the present oceans. Of these latter it is stated however that they generally exhibit concretionary structure, which with the Čevljanovic hard manganese ore nodules, apart from the surface crust, appears to be exceedingly rarely the case.'

In Bosnia, small secondarily formed veins or leaders of manganese ore also occur, though these are of no economic importance. This country, in the thirty years 1881–1910, produced

FIG. 444.—Bands of manganese ore in radiolarite from the Grk mine in Bosnia. Katzer.

approximately 150,000 tons of manganese ore containing 45–47 per cent of manganese and but 0·03–0·07 per cent of phosphorus.[3]

MANGANESE ORE-BEDS IN CRYSTALLINE SCHISTS

THE MANGANESE DEPOSITS OF BRAZIL

LITERATURE

M. AR-ROJADA LISBOA. ' O Manganez no Brasil,' Jornal do Commercio. Rio de Janeiro, June 1898 and March 1899. Reviewed by E. HUSSAK. Zeit. f. prakt. Geol., 1899, pp. 256-257.—H. K. SCOTT. The Manganese Ores of Brazil, Iron and Steel Inst., London, 1900,

---

[1] P. Krusch, ' Über eine neue Systematik primärer Teufenunterschiede,' Zeit. f. prakt. Geol., 1911, p. 144.
[2] ' Die geologischen Verhältnisse des Manganerzgebietes von Čevljanović in Bosnia,' Berg- und Hüttenm. Jahrb. d. k. k. montanischen Hochschule zu Leoben, etc., 1906, LIV. Part 3. [3] Postea, p. 1114.

No. 1.—O. A. DERBY. ' On the Manganese Ore-Deposits of the Queluz (Lafayette) District, Minas Geraes, Brazil,' Amer. Journ. of Sc., 1901, XII.—J. C. BRANNER. ' The Manganese Deposits of Bahia and Minas Brazil,' Trans. Amer. Inst. Min. Eng., Sept. 1899.—P. CALOGERAS. As Minas do Brasil, Rio de Janeiro, 1905, Vol. II. pp. 281-349.—E. HUSSAK, ' Über Atopit von Miguel Burnier, Minas Geraes,' Zentralbl. f. Min. Geol., 1905, pp. 240-245 ; ' Über die Manganerzlager Brasiliens,' Zeit. f. prakt. Geol. pp. 237-239.—P. KRUSCH. Untersuchung und Bewertung von Erzlagerstätten, 1st Ed. p. 549, Stuttgart. See also literature on the Brazilian itabirite, cited on page 1060.

In the manganese ore-bearing formation stretching from Barbacana to Ouro Preto in the province of Minas Geraes, two districts may be distinguished, namely that of Miguel Burnier at Kilometre 498 on the central railway, and that of Lafayette, Queluz, at Km. 463. The most important deposits are marked on the map constituting Fig. 320.

In the first district the deposits occur in close association with the presumably Cambrian itabirite [1] which often includes manganiferous iron ore-beds. An alternation of limestone, itabirite, and manganese ore-beds,

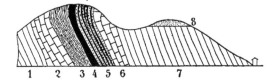

FIG. 445.—Section of the manganese deposit of Miguel Burnier at 502 km. in Minas Geraes, Brasil.   Kilburn Scott.

1, phyllite, greatly decomposed, micaceous and containing but little quartz ; 2, white limestone ; 3, manganiferous iron ore ; 4, clean manganese ore ; 5, itabirite and jacutinga ; 6, grey limestone ; 7, phyllite ; 8, Alluvial canga.

as illustrated in Fig. 445, is also frequently met. These manganese ore-beds were in all probability precipitated from carbonate solutions. The ore at Miguel Burnier, with an average manganese content of 50 per cent and a maximum of 55 per cent, is nearly always friable and therefore very hydrous, the water usually amounting to 14–20 per cent ; generally it is poor in phosphorus and quartz, containing only 0·05–0·07 and 1–3·5 per cent respectively.

In the district of Lafayette the itabirite is completely absent, and the manganese beds come in direct contact with granite-gneiss. The ore occurs here together with spessartine, the manganese-alumina garnet. Near the contact the spessartine occurs in alternating layers with others of preponderate rhodonite, and with distinct schistose structure.

O. A. Derby in 1901 regarded these manganese deposits as magmatic segregations ; E. Hussak in 1906 considered however that they were sedimentary beds which had been contact-metamorphosed by the eruptive

[1] *Ante*, p. 1060.

gneiss, spessartine being formed which later became decomposed, manganese oxide resulting; while from the descriptions it may be that they were formed similarly to the deposits at Långban in Sweden, namely, by increase of manganese during contact-metamorphism.

In Brazil, manganese mining began about the middle of the 'nineties. In 1894 the export was only 1430 tons, an amount which increased in 1900 to 127,343 tons, and in 1904 to 216,463 tons; while latterly the annual production has amounted to about 250,000 tons. Altogether, from the commencement of export in 1894 to the end of 1910 about 2,110,000 tons of manganese ore have been exported.

BRITISH INDIA

LITERATURE

Principal work by L. LEIGH-FERMOR. ' The Manganese Ore-Deposits of India,' 3 parts, Geol. Survey of India, Mem., 1909, XXXVII., also several other treatises by Fermor.— TH. H. HOLLAND and L. L. FERMOR. Geol. Survey of India, Records, 1910, XXXIX.— L. DE LAUNAY. La Géol. et les richesses minérales de l'Asie, Paris, 1911, p. 696.—P. KRUSCH. Untersuchung und Bewertung von Erzlagerstätten, 1st Ed. p. 461. Stuttgart, 1907.

The numerous and widely distributed manganese deposits in British India have of late years given development to a flourishing industry. These deposits, according to Holland and Fermor, belong to different geological formations, thus :

1. Occurrences in association with the so-called Kodurite series, this consisting of manganiferous eruptive intrusions in the fundamental rocks. Of these rocks, kodurite, consisting principally of potash felspar, spandite —garnet with a composition between spessartine and andradite—and some apatite, is especially typical, while, in addition, rhodonite and two or three other manganiferous pyroxenes also occur. Very large lenticular ore-bodies containing psilomelane, pyrolusite, braunite, manganese-magnetite, etc., occur in places.

The genesis of these ores is questionable ; possibly they are of metasomatic formation. Such deposits occur at Ganjam, Vizagapatam district, Madras. The Garbham mine with an ore-body 1600 ft. long and 100 ft. wide, from before 1896 to 1908 produced altogether 600,889 tons of ore, and the Kodur mine from 1892 to 1908, some 306,170 tons.[1]

2. Occurrences in the so-called Gondite series, these being conformable beds in the Dharwar horizon of the fundamental rocks, and regarded as metamorphosed manganese sediments. Deposits of this kind are found at numerous places in Bengal,[2] Bombay,[3] Central India,[4] and in the Central Provinces.[5] The original sediments were partly mechanical, namely, sand

---

[1] Ante, p. 1104.   [2] Gangpur.   [3] Panch Mahals.   [4] Jhabua.   [5] Balaghat, Nagpur.

and clay, these now being represented by quartzite, mica-schist, and phyllite; and partly chemical, namely, manganese oxides. There are two types of manganese ore-bed, namely, manganese ore-beds proper, and such as have been formed from a mixture of sand, clay, and manganese-ochre, and which to-day are represented by spessartine, rhodonite, etc. The spessartine-quartz rock, known as gondite, plays an important part and has given its name to the whole series. In addition, rocks consisting of spessartine and rhodonite, or rhodonite and quartz, etc., also occur. The ore-beds often reach a considerable length along the strike, 2·4, 2·8, and even 9·5–10 km., and in many cases also a great thickness, as for instance 30 m. of clean ore. Of the total manganese ore production of India, namely 579,231 tons, 916,770 tons, and 680,135 metric tons in the years 1906, 1907, and 1908 respectively, more than 400,000, some 500,000, and some 450,000 tons respectively, or somewhat more than one-half, were derived from this sedimentary series. The occurrences were formerly worked almost exclusively in opencut ; now however, deep mining has begun. The ore consists principally of braunite and psilomelane. As won it generally contains 50–55 per cent of manganese, 4–8 per cent of iron, 4–8 per cent of silica, this being contained principally in the braunite, and 0·07–0·14 per cent of phosphorus.[1]

Manganese ores are also found in the Archaean limestone of the Dharwar horizon. Such were probably formed by metamorphism from calcareous sediments and manganese-ochre ;[2] in regard to genesis they are therefore analogous to the manganese ores in the Gondite series. They are however only of subordinate importance.

3. Occurrences of the Laterite or Lateritoid Series.—These are found at the outcrop of different manganiferous rocks of the Dharwar horizon in various places in Bengal, Bombay, the Central Provinces, Goa, Madras, and Mysore. These ferruginous and manganiferous deposits are cavernous and show great similarity to the ordinary laterite. They are responsible for only a fraction of the manganese ore production of India.

Almost all the Indian deposits lie fairly distant from the coast, involving thereby high transport costs. From 1897 to and including 1909, British India produced altogether 4,052,000 tons of manganese ore, to which must be added a small amount for the years 1892–1896. Of late years the Indian production has at times surpassed that of Russia.[3]

Beds of manganese ore and manganese-iron ores in crystalline schists occur also in other places : for instance, a manganese iron ore-bed described by F. Kossmat and C. v. John [4] at Macskamezö in northern Transylvania, where different ferro-manganese silicates — knebelite = Mn-Fe olivine ;

---

[1] *Ante*, p. 1104.
[3] *Postea*, p. 1114.
[2] Holland and Fermor, *loc. cit.*, 1910, p. 164.
[4] *Zeit. f. prakt. Geol.*, 1905, pp. 305-326.

dannemorite = Mn - Fe hornblende ; and spessartine — masses of rhodo-chrosite, and in places manganese-magnetite, occur in Archaean mica-schists. The bed material is generally distinctly stratified, more particularly in those places where bands of the different constituents, magnetite or rhodochrosite, alternate with silicates. The occurrence is regarded as a regional-metamorphosed metalliferous sediment. At the out-crop the rhodochrosite especially, has given rise to an extensive secondary formation of manganite and limonite.

In the afore-cited paper an analogous deposit at Jakobeny in Buko-vina is also mentioned, where beds of rhodonite containing disseminated rhodochrosite and quartz are embedded in quartzose mica- or sericite-schists.[1] In alternation with the ore, mica-hornblende schists occur. At the outcrop, rhodonite and rhodochrosite are decomposed to manganite, upon which mining operations on a moderate scale are proceeding.

MANGANESE ORE-DEPOSITS OF QUESTIONABLE GENESIS

The Rhodochrosite-Rhodonite Deposit in the Huelva District

LITERATURE

Hoyer. ' Beiträge zur Kenntnis der Manganerzlagerstätten in der spanischen Provinz Huelva,' Zeit. f. prakt. Geol., 1911, pp. 407-432.

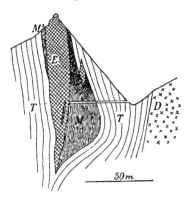

Fig. 446.—Section through the manganese deposit at Castillo de Palanco in the Huelva district. Hoyer.

M = primary ore ; M^2 = weathered ore ; E = ferru-ginous quartz ; T = clay-slate ; D = diabase.

In this district, illustrated in Fig. 217, are found, in addition to the famous pyrite deposits already de-scribed,[2] several hundred manganese deposits. These are conformably interbedded in clay-slates and por-phyroids, where they form lenticular bodies having frequently great thick-ness in proportion to length and depth. They consist of banded, compact, or irregular - coarse car-bonate-silicate manganese ores with jasper and hornstone. In the neigh-bourhood of the surface, the ore, as illustrated in Fig. 446, is oxidized. The lenses are seldom more than 150 m. long and have an average depth of 30 m., though this dimension

[1] Br. Walter, ' Die Erzlagerstätten der südlichen Bukowina,' *Jahrb. d. k. k. geol. Reichsanst.*, 1876, Vol. XXVI. pp. 372-376.   [2] *Ante*, pp. 315-327.

occasionally reaches more than 100 metres. They almost invariably occur in Culm beds and belong, at all events in part, to this formation. They are spacially closely connected with the Huelva eruptive area, and according to Hoyer are probably associated with effusive diabase. Subsequent to their formation they have suffered tectonic disturbance.

The most important primary minerals are rhodochrosite and rhodonite, with which a manganiferous garnet also often occurs. Quartz is always abundant, while light-coloured mica and in places chlorite, are likewise present. The ore is accompanied by jasper which often occurs in considerable amount. According to Hoyer, the occurrences may be explained syngenetically by sedimentation, or epigenetically by metamorphism and cavity filling. He considers the former as being more probable. Vogt, who in 1896 visited some of the manganese mines, and Krusch later, obtained the impression, on the other hand, that the deposits were epigenetic. The true genesis is therefore a very open question; deposition from springs may perhaps also be considered.

The occurrences were originally worked in the middle of the nineteenth cenury, though only on a moderate scale, for a manganite ore— pyrolusite with some psilomelane, manganite, wad, etc.—at the oxidized outcrop. At the beginning of the 'nineties the primary rhodochrosite-rhodonite ore, poor in iron and containing sulphur and phosphorus, began to be mined, and the production in some years, 1897–1900, reached 100,000 tons per year. Subsequently however, operations greatly declined, till now they have almost stopped. According to Spanish statistics published in *The Mineral Industry*, in the years 1881–1909 altogether 837,000 tons of manganese ore were mined in Spain, practically all of which came from the Huelva deposits.

[TABLE

TABLE OF THE MANGANESE ORE PRODUCTION AND ITS DISTRIBUTION
AMONG THE VARIOUS CLASSES OF DEPOSIT

The following statistics are taken from *The Mineral Industry* [1] and from Krusch's *Versorgung Deutschlands mit Erzen, u.s.w.*, Leipzig, 1913, p. 167.

THE WORLD'S MANGANESE ORE PRODUCTION IN METRIC TONS

	1882.	1890.	1900.	1905.	1909.
Germany . .		some hundred tons			
Sweden . .	1,673	10,698	2,691	1,992	5,212
Great Britain .	1,573	12,646	11,384	14,582	2,812
Belgium . .	345	14,255	10,820	Nil	6,270
France . .	7,538	15,984	28,992	6,751	9,378
Spain . . .	5,668	9,872	112,897	26,020	7,827
Portugal . .	17,336	...	1,971	Little	Little
Austria-Hungary	12,778	9,452	14,550	23,732	29,966
Bosnia . .	2,283	5,500	7,939	4,129	5,000
Italy . . .	6,978	2,147	6,014	5,384	4,700
Greece . .	...	13,547	8,050	8,171	5,374
Russia . .	14,431	182,468	802,236	508,635	574,938
Canada . .	...	1,205	34	22	Nil
United States .	4,605	26,098	11,771	4,118	1,547
Cuba . . .	...	22,161	21,973	8,096	2,976
Colombia . .	...	...	8,748	...	...
Brazil . . .	Nil	Nil	108,244	224,377	240,774
Chili . . .	...	48,759	25,715	1,323	...
British India .	Nil	Nil	129,865	250,788	652,958
Japan . . .	157	2,612	15,831	14,017	8,708
New Zealand .	2,216	490	166	55	6
Australia, etc. .	138	2,910	77	1,541	613
Totals, some .	0·1 million tons	0·4 million tons	1·2 million tons	1·1 million tons	1·5 million tons

In the case of the United States these statistics embrace only the true manganese ores poor in iron, and not the manganese-zinc ores of New Jersey, nor the manganiferous iron ores. In the case of some of the other countries however, and particularly Austria-Hungary, the last-mentioned ores are in part included. The three principal producers are Russia, British India, and Brazil.

Up to a few years ago, that is, before the large mining operations in Russia, British India, and Brazil began, the fairly moderate demand for manganese ore was met in greater part by lodes and metasomatic deposits.[2] Latterly however, by far the greater portion of the production has been obtained from sedimentary deposits. To this class belong nearly the whole of the Russian output of Tertiary manganese lake ores ; a considerable portion, possibly even one-half, of the total production of Brazil ; and more than one-half of the Indian production, together with smaller amounts from other countries. At present therefore it may be reckoned that at

[1] New York.　　　　　　　　[2] *Ante*, pp. 851-869.

least 75–80 per cent of the total manganese ore production comes from the sedimentary deposits. The remainder is distributed among lodes, metasomatic deposits, and contact-deposits.[1] Manganese ore is used now almost exclusively for the production of ferro-manganese, though small amounts are employed for the production of chlorine and for colouring glass, etc.

THE COPPER-SHALE BEDS

The German Zechstein formation, to which the copper-shale now only worked in the Mansfeld district, belongs, is known for the copper content of certain of its beds. Only the bituminous marl-shale bed, known as the Kupferschiefer, a few decimetres thick, is of economic importance, though the marl and clay above, and the Zechstein conglomerate below, are in some places cupriferous.

The Kupferschiefer is the second oldest member of the Zechstein formation, the Zechstein conglomerate constituting the actual foot-wall of that formation. This conglomerate however, being the basal conglomerate formed by the transgression of the Zechstein sea, is not continuous, but is found frequently only in depressions of the bed-rock, so that the Kupferschiefer is the lowest complete Zechstein bed. Where this rests immediately upon the Rotliegendes conglomerate this conglomerate is frequently bleached and then more aptly described as Grauliegendes or Weissliegendes. The hanging-wall of the Kupferschiefer is formed by Zechstein limestone, likewise belonging to the Lower Zechstein.

Strictly speaking, the bituminous shale should only be termed Kupferschiefer when the copper content is considerable. In ordinary use however the term is frequently applied to the geological horizon concerned, irrespective of the actual copper content. The distribution of the copper along this horizon is in no sense regular. While in the Mansfeld Syncline and in the Riechelsdorf Hills the copper content on an average is approximately 3 per cent, along the south and west borders of the Thuringian Forest it is only 1 per cent, and in Westphalia the same bituminous shale is practically without copper. The extensive though much broken Kupferschiefer of North Germany possesses therefore only in the two first-mentioned places a copper content high enough to warrant its description as copper ore. It must be pointed out that the fine state of division of the copper in the Kupferschiefer, unlike all other ores of similar content, does not permit the production of a concentrate by ordinary concentration processes. Kupferschiefer containing 3 per cent

---

[1] *Ante,* pp. 388, 394.

of copper cannot therefore be quite compared with another copper ore of the same content.

The form of the Kupferschiefer is as a rule that of a more or less inclined bed, the thickness of which, though never more than 50–60 cm., may vary considerably. The distribution of the ore in this bed has always been a point of the greatest interest to miners and geologists. It has been shown, for instance, that in section the copper content is everywhere greatest at the bottom, from whence it gradually decreases towards the hanging-wall. The extent of enrichment at the bottom is however just as variable as the rate of decrease towards the top ; these two factors determine the actual thickness mined. Since the bitumen content is likewise greatest towards the bottom, a certain parallelism exists between its distribution and that of the copper. The copper however is occasionally not confined to the Kupferschiefer, but extends also into the foot-wall, that is, into the Zechstein conglomerate or Weissliegendes ; the thickness of such impregnation in the foot-wall may reach as much as 10 centimetres.

The Zechstein formation is crossed by a number of fissures which when metalliferous constitute lodes. As stated when discussing the genesis of these lodes, the relation between the copper content of the Kupferschiefer and the proximity of the lodes is in so far important, that in many places a special concentration of ore along the lodes, and conversely a decrease in the copper content with distance from them, can be demonstrated. This fact is of importance in the determination of the average copper content, since samples taken from the immediate neighbourhood of the lodes generally exceed the average.

The copper in the Kupferschiefer exists principally as chalcopyrite, chalcocite, and bornite. The regional distribution of these minerals varies, in that with districts having a high average content, as in the neighbourhood of Mansfeld and Riechelsdorf, chalcocite and bornite predominate, while in districts poor in copper, chalcopyrite is the principal mineral. Similarly, the distribution in the bed itself is seen even with the naked eye to be irregular ; thin layers of mineral lie approximately along the bedding, veins cross the bedding, while finally here and there small nests occur. Under the microscope it is further seen that the fine shaly material also contains metalliferous grains more or less thickly distributed. Unfortunately the bottom and richest bed, in which the miner makes his undercut, is decomposed to such an extent that microscopic examination of thin slides is impossible.

Apart from the copper, the likewise very variable silver content is noteworthy. In the Mansfeld district this sometimes reaches 250 grm. per ton, or on an average some 0·55 parts of silver to 100 parts of copper ; in Thuringia, only a minimal silver content exists ; at Riechelsdorf the average is 40–50 grm. per ton; while, curiously enough, in Westphalia, where very little or no copper occurs, a silver content of some 10 grm. is present. Since in spite of great difference in silver content the copper content at Riechelsdorf and Mansfeld is approximately the same, while at places in Westphalia silver occurs without copper, the conclusion is justified that the silver is not necessarily associated with the copper but may also

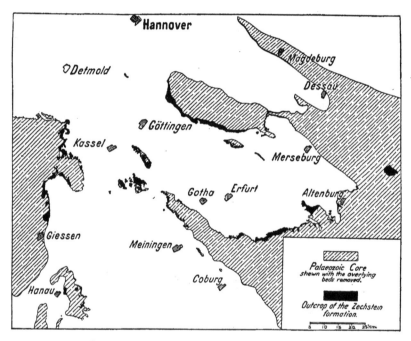

FIG. 447.—Extension of the Upper Zechstein including the Kupferschiefer, in Germany.
Geol. Landesanst. Berlin.

occur with other minerals, as for instance—according to investigation by Krusch in Westphalia—with pyrite.

Zinc is also characteristic of the Kupferschiefer. Taking into account not only the thickness mined but also the overlying thicknesses poorer in copper but frequently richer in zinc, these two metals occur in approximately equal amount.

The question of the genesis of the Kupferschiefer is one of the most difficult problems in ore-deposition. Two opposite views exist. While some authorities hold with the syngenetic origin, that is, simultaneous formation of the ore and shale, others regard the copper content as epigenetic, or younger than the shale. J. S. Freiesleben and A. v. Groddeck in 1879, and A. W. Stelzner and A. Bergeat in 1904, regarded the Kupferschiefer as one of the most typical examples of sedimentary deposits. Pošepný in 1894, Beyschlag in 1900, and Krusch, on the other hand, defended the epigenetic origin of the copper content.

In support of the older view of syngenetic sedimentation, the large presumably regular and continuous extent over districts lying far apart— the distance in a straight line between Goldberg in Silesia and Bieber on the Spessart is some 500 km.—was formerly quoted, and the presumed conformity thereby emphasized. Beyschlag however in 1900 showed that this property belongs only to the shale bed, and not to the copper mineralization. The investigations of Krusch in Westphalia have in recent years shown that the fluctuations in the copper content along the strike are more pronounced than was formerly realized, while the afore-mentioned differences in the silver content tend still further to shake the idea of conformity. In fact, the copper, silver, and zinc contents are not only fairly different in such different districts as the Harz, Thuringia, and Hesse, but, apart from such local fluctuations as occur along the previously mentioned fissures, considerable differences in content are found in one and the same district, as for instance at Mansfeld.

Beyschlag and Krusch hold the view therefore that the metalliferous content of the Kupferschiefer is later than the sedimentation of the shale, and that it accordingly represents a subsequent impregnation. According to them the circulation of the primary solutions took place along the intersecting fissures and along the contact between the Kupferschiefer and the conglomerate in the foot-wall. For this view the pronounced decrease of copper with distance from those fissures speaks, as does also the decomposed condition of the lowest portion of the ore-bed. The impregnation of the foot-wall is no hindrance to the acceptance of this view, since this impregnation could have been brought about just as well secondarily by solutions circulating above the immediate foot-wall, as primarily by a cupriferous Zechstein sea. Nor does the occurrence of cupriferous and native-silver fossils in any way necessitate the view of a syngenetic formation, since these are just as readily explained by epigenetic processes. On the other hand, the cross bedding or ramification of small mineral veins in the ore-bed, indicates an epigenetic genesis. When the metalliferous layers of the ore-bed are carefully studied it is often seen that even these in no sense exactly follow the bed in strike and dip, but that the agreement in these particulars is only general ; such layers often cross the bedding, though naturally at a very oblique angle.

By the acceptance of the epigenetic view of the genesis of the copper content all other essential phenomena of the occurrence may be explained.

That the form of the deposit is that of an ore-bed is explained by the well-known reducing effect which bituminous substances produce upon mineral solutions. The bitumen of the bed, so far as it existed, could have brought about the precipitation of the copper from solution.

The minerals, in character, are those usually found in the cementation zone of copper deposits. Krusch considers that chalcopyrite and cupriferous pyrite were the primary ores. The former occurs to-day as the dominant ore in the Thuringian Kupferschiefer at Schweina, while pyrite may be detected by the microscope in the bituminous shale of Westphalia. From ordinary pyrite poor in copper, cupriferous pyrite afterwards resulted from the play of the various cementation processes. These processes, which effected displacement of the original content by descending solutions, obscured the genetic conditions which governed the primary deposition of the ore from ascending solutions.

Earlier authorities who endorsed a sedimentary genesis quoted in its support, among other things, that the Permian time in Europe was characterized by considerable volcanic activity, and that thereby, at different places, certain metallic salts of copper, silver, and zinc, were added to the sea-water. They regard as a sediment not only the Kupferschiefer but also the copper-sandstone in the Perm government, Russia. Such a general distribution of sedimentary copper deposits in Permian time appears however very questionable. In such sedimentation it was considered that the precipitation of ore was brought about by means of sulphuretted hydrogen arising from the decomposition of the abundant remains of fish and other organisms. The numerous fossil fish—*Palæoniscus* and *Platysomus*—found in the Kupferschiefer are, as is well known, often distorted, a phenomenon considered to indicate that they were poisoned by the sudden entry of copper salts into the sea. This is however not convincing, since similar distorted fossil fish are frequently found outside of the Kupferschiefer, and in any case almost every dead fish becomes distorted by gases arising at its decomposition.

The adherents of the syngenetic theory find it impossible to refer the metalliferous content in copper, silver, zinc, etc., to the small amounts of these metals ordinarily contained in sea-water, but have to explain that content by local, ascending, concentrated increment of heavy-metal salts, probably in association with the Permian eruptions. In such a manner the difference of content at different places in the shallow Permian sea is explained.

It appears questionable however whether it is necessary at all to hold that precipitation was effected by sulphuretted hydrogen. It would be more simple, as suggested by E. Kohler,[1] to assume adsorption. The

---

[1] *Zeit. f. prakt. Geol.*, 1903, pp. 55-56; and *ante*, p. 974.

settling organic substances, clay particles, etc., would of themselves adsorb
the salts of copper, silver, zinc, etc., even from very weak solutions and bring
them to the bottom, where, by sulphuretted hydrogen arising from organic
decomposition or by the reduction of sulphates by organisms, they would
be converted to sulphides.   Such an adsorption would be a function of the
quantity of the settling material, and this may have varied at different
places.   There would accordingly, even upon the assumption of a sedimentary genesis, be no necessity to expect absolute conformity.   Generally
speaking, the copper content as well as that of the bitumen is greatest in
the bottom layers of the shale bed ; the zinc content, on the other hand,
is frequently greater in the upper layers.   Upon the assumption of a
syngenetic origin these facts however can readily be explained by precipitation by selective adsorption.[1]

The oft-recorded enrichment of the copper along the intersecting lodes
is however difficult to explain by the syngenetic theory.   These lodes
represent quartz-barite fissure-fillings containing cobalt-nickel ores, especially arsenides, but with little copper or zinc.   The relation between
copper, silver, and zinc, on the one side, and cobalt-nickel on the other, is
therefore in these lodes quite different from that in the Kupferschiefer.
Along these lodes in many places, especially in the Mansfeld district,
an enrichment has taken place, not however in the sense that the usual
copper-bearing bed has become richer, but that a somewhat higher-lying
bed contains bean- or kidney-shaped compact nodules consisting of chalcocite, bornite, chalcopyrite, and pyrite.[2]   In other places a diminution of
the copper content is experienced, while in others again no change whatever is recorded.   At Schweina in Thuringia it is usual to find that the
whole bed has been enriched ; the case of Riechelsdorf is mentioned later.[3]
Admittedly, as Bergeat has pointed out, under the assumption of a sedimentary origin a change in the copper content along the lodes can
also be explained by secondary processes ; solutions circulating through
the fissures could have forced their way into the copper-shale and have
effected a re-distribution of the metalliferous content.   It is difficult
however to explain in this manner the frequent enrichment.

Several objections, it is true, may be raised against the view that
the heavy metals of the Kupferschiefer were transported epigenetically
by infiltration from the lode fissures.   In the first place it appears difficult
to understand how in this manner, over districts far apart, only the thin
bituminous shale and none of the other beds cut by those fissures were
impregnated with ore.   And secondly, how the mineralization of the

[1] *Ante*, p. 975.
[2] ' Die Mansfeldische Kupferschiefer Gewerkschaft zu Eisleben,' Festschrift, 1907, p. 19.
[3] *Postea*, p. 1129.

Kupferschiefer should be so regular over large districts, and should for miles be characterized by copper with some zinc in the lower beds, and zinc with a little copper in beds a little higher. The relation of copper-zinc-silver to cobalt-nickel is also quite different in the Kupferschiefer from that in the cobalt lodes. While finally, Vogt points out that outside of Germany there are other occurrences comparable with the Permian Kupferschiefer, which have to be explained, though, it is true, most of them contain but little copper.

Thus, J. Kiär[1] describes, at Ringerike in the neighbourhood of Christiania, beds of a more or less sandy clay-slate, interbedded in sandstone and exceedingly rich in Upper Silurian fish and other fossils. He particularly emphasizes the fact that the thin bed containing most fish invariably also contained copper, though the amount was only 0·17–0·23 per cent. In this case Vogt considers that only a simultaneous deposition of material rich in copper and fish could be assumed. With fair probability the case here may be explained by adsorption.

The economic importance of the Kupferschiefer is based on both copper and silver, though in two districts only, namely, Mansfeld and Riechelsdorf, can the ore be regarded as in any way rich. The ore belongs to the class of self-fluxing ores, that is to say, after roasting it can be smelted without the addition of any flux. Experience has nevertheless shown that a profitable treatment, in the present state of metallurgy, is only possible with high copper prices.

MANSFELD

LITERATURE

J. C. FREIESLEBEN. 'Geognostischer Beitrag zur Kenntnis des Kupferschiefergebirges mit besonderer Hinsicht auf einen Teil der Grafschaft Mansfeld,' Freieslebens Geognostische Arbeiten, Vols. I.-IV. Freiberg, 1807–1815.—A. W. F. v. VELTHEIM. 'Über das Vorkommen der metallischen Fossilien in der alten Kalkformation im Mansfeldischen und im Saalkreise,' Karstens Archiv, 1827, I. Vol. XV. p. 89.—PLÜMICKE. 'Darstellung der Lagerungsverhältnisse des Kupferschieferflözes in der Zechsteinformation der Grafschaft Mansfeld,' Karstens Archiv, 1844, II. Vol. XVIII. p.. 139.—BAEUMLER. 'Über das Vorkommen von Nickelerzen im Mansfeldschen Kupferschiefergebirge,' Zeitschr. d. deutsch. geol. Gesellsch., 1857, IX. p. 25.—BRATHUHN. Generalkarte von den gesamten Mansfeldischen Kupferschieferrevieren, 1858. — J. HECKER. 'Erfahrungen über das Vorkommen der Sanderze in den Sangerhäuser und Mansfeldischen Revieren,' Zeitschr. f. d. ges. Naturw. Halle, 1859.—H. B. GEINITZ. Dyas oder die Zechsteinformation und das Rotliegende, 1861.—H. MENTZEL. 'Mansfelder Kupferschiefer-Bergbau,' Berg- u. Hüttenm. Zeitg., 1864, p. 213, and 1865, p. 65.—OTTILIAE. 'Notiz über die Metallführung des Mansfeldschen Kupferschieferflözes,' Abhandl. d. naturforsch. Gesellsch., Halle, 1866, Sitzungsber. IX. p. 18.—SCHRADER. 'Der Mansfelder Kupferschiefer-Bergbau,' Zeitschr. f. Berg-, Hütten- und Salinenwesen in Preussen, 1869, p. 251.—F. BEYSCHLAG. Geological map of the district of Halle on the Sieg ; Die Mansfelder Mulde und ihre Ränder, Berlin,

---

[1] 'A New Downtonian Fauna in the Sandstone Series of the Christiania Area,' Kris-tiania Ges. d. Wiss. math.-naturw. Kl., 1911.

1899 ; ' Beitrag zur Genesis des Kupferschiefers,' Zeitschr. f. prakt. Geol., 1900, p. 115.
W. BRUHNS in collaboration with H. BÜCKING. Die nutzbaren Mineralien und Gebirgsarten im Deutschen Reiche, Berlin, 1906, revised by v. Dechen.— Die Mansfeldische
kupferschieferbauende Gewerkschaft, IV., Allgemeinen Deutschen Bergmannstages zu
Halle a. d. S., 1889 ; X. ibid., Sept. 1907.

The centuries-old important copper industry in the two Mansfeld
districts commenced around the borders of the syncline where the Kupferschiefer came to surface. At present it is limited to the western portion
of the syncline at the foot of the Harz mountains.

The Mansfeld syncline, the birthplace of modern stratigraphy, owes
its name not only to the well-defined synclinal tectonics but to some extent
also to its orographic figure. It rests with its western border upon the
eastern Unterharz, and from the neighbourhood of Annarode sends through
Blankenrode and Bischofsrode, to Hornburg, an orographically and geologically equally well - defined outlier, the so - called Hornburg anticline,
which terminates the syncline to the south-east. At its northern boundary
it connects the extreme northern outliers of the Harz near Walbeck—by way
of Hettstedt, Gerbstedt, and Friedeburg, the so-called Hettstedt mountain
bridge—with the Carboniferous-Rotliegendes hills of Wettin and Halle,
on the other side of the Saale. The heights of the porphyry mountains,
which from Halle through Kröllwitz, Lettin, Brachwitz, and Wettin, flank
the Saale, form the north-east border of the syncline, which there also
is orographically well defined. The northern rim in the neighbourhood
of Gerbstedt and Hettstedt is least well defined. Towards the south-east
the syncline orographically and geologically is open. In addition to
erosion, the ages-continued, natural leaching of the salt- and gypsum beds
underground—such leaching being associated with the Tertiary folding
—has had its influence upon the orography of the district, in producing wide
areas of surface depression.

In the construction of the syncline, the Devonian of the Unterharz,
the Carboniferous, and the Lower Rotliegendes form the basement, while
filling the syncline are found Upper Rotliegendes, Zechstein, Bunter, and
Muschelkalk, all conformable and regularly bedded. Tertiary beds are
represented by small surface coverings economically important by reason
of the lignite they contain.

The basement beds also are in the form of a syncline, with synclinal
axis striking north-east, while the Mansfeld syncline strikes north-west.
The latter is so inserted in the basement that the beds of the Upper
Rotliegendes in very variable thickness first filled the irregularities, and
on the floor thus made even, the interesting marine beds were afterwards
deposited.

The Zechstein formation in the Mansfeld district is, from top to
bottom, divided into three divisions, namely, the Upper Zechstein consisting

from hanging-wall to foot-wall of Zechstein clay, the younger salt series with the main anhydrite, the grey saline clay, the older salt series with the potassium horizon, the older rock-salt, and the basal anhydrite; the Middle Zechstein, consisting of the oldest salt series with rock-salt, anhydrite and residual material, and dolomite; and finally, the Lower Zechstein, made up of Zechstein limestone, the Kupferschiefer, and Zechstein conglomerate.

The Zechstein conglomerate is frequently difficult to distinguish from the so-called Weissliegendes, the name given to bleached Rotliegendes. The uppermost layer of the Zechstein conglomerate is in places cemented

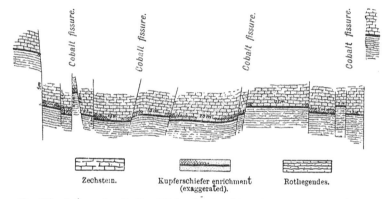

Zechstein. Kupferschiefer enrichment Rotliegendes.
(exaggerated).

Fig. 448.—Section across the Mansfeld Syncline at Eisleben, showing the enrichment along the cobalt fissures. *Festschrift*, 1907.

by a siliceous material, when it is known as the *Hornschale*. The Kupferschiefer consists of bituminous, blackish, marly shale, of compact, finely bedded character and of considerable solidity, fresh pieces ringing when struck; only at the outcrop is it friable and crumbly. Along the strike the bitumen content in the shale remains fairly constant, but, as stated already, it is higher in the lower layers than in the upper.

Within the Mansfeld syncline the miner differentiates the various layers of the Kupferschiefer as well of the Zechstein limestone in the immediate hanging-wall, in great detail. Along the north border of the syncline, in the neighbourhood of Hettstedt, the layers, and consequently the names, are somewhat different from those along the west border from Eisleben to Helbra. This difference is shown in the following table:

North border, Hettstedt.	West border, Eisleben.
Roof of Zechstein limestone. Occasional mineralization, particularly in the neighbourhood of the cobalt lodes, in the form of small bean-sized compact chalcocite nodules.	

North border, Hettstedt.	West border, Eisleben.
12-16 cm. *Dachberge* (*Oberberge*).	12-18 cm. *Graue Berge* (*Dachberge*).
Schwarze Berge (*Noberge*).	*Schwarze Berge* (*Noberge*).
10-12 „ *Lochberge*.	10-15 „ *Oberer und Unterer Schieferkopf*.
2-4 „ *Kammschale*.	2·5-4·0 „ *Kammschale*.
2-3 „ *Kopfschale*.	3·0-6·0 „ *Grobe Lette*.
6-8 „ *Oberer und Unterer Schieferkopf*.	
1·5-2·5 „ *Lochschale*.	2-4 „ *Feine Lette*.
2·0-3·5 „ *Lochen*.	
1·5-2·0 „ *Liegende Schale*.	

Weissliegendes, *i.e.* bleached Upper Rotliegendes, porphyry conglomerate.

The Mansfeld Kupferschiefer is palæontologically characterized by its many fish remains—*Palaeonicus Freieslebeni* and *Platysomus gibbosus*—and pine-wood needles—*Ullmannia*. The beds are marine, though the occurrence of vegetable remains, etc., indicates a deposition near the coast.

In the Mansfeld district the lower portion of the bed, with a maximum thickness of 3–5 cm., regularly carries copper- and silver minerals in particles so fine as only to be discernible to the naked eye when concentrated in fine layers or veins. The ore- and bitumen contents, as stated already, gradually diminish from foot-wall to hanging-wall, the bed itself being more constant than the mineralization, which mostly has rendered only the lowest 7–13 cm. payable. The assembly of fine metalliferous particles, constituting the so-called *Speise*, consists of bornite, chalcocite, chalcopyrite, seldom and to a less extent of galena and pyrite, and occasionally some tetrahedrite, cobalt ore, and stibnite. The presence of molybdenum, selenium, vanadium, and a fairly large amount of zinc, may also be demonstrated chemically.

In addition to the metalliferous *Speise* are met fine stringers of bornite and chalcopyrite, not infrequently running parallel to the bedding; coatings of chalcocite, bornite, chalcopyrite, and metallic silver, along the bedding-planes and transverse fractures; and finally, separate flakes, grains, and nodules. In amount, however, these other modes of occurrence are subordinate when compared with the *Speise*. The finer and denser this material, the higher is its copper content.

Although the whole thickness of the Kupferschiefer is metalliferous, generally the only portion worth smelting is that limited to the layers up to and including the *Kammschale*. When the latter and the next younger, the *Schieferkopf*, contain a strikingly large amount of copper, the foot-wall beds usually are poor. Accordingly, in the Hettstedt-Gerbstedt districts, as a rule, only the *Lochen*, the *Lochschale*, and the *Schieferkopf* are worked; in the Eisleben district, on the other hand, the *Lette*, in part with and in part without the *Kammschale*, seldom however the *Kopf*. The thickness mined varies in the first districts between 7 and 10 cm., in the last between 8 and 17 centimetres. In the neighbourhood of the cobalt fissures, especially in the Eisleben district, the *Schwarze Berge* and the roof material up to the *Fäule* are cupriferous, the copper ore occurring there in nodules.

The influence of the cobalt fissures and other faults upon the copper-shale at Mansfeld is to produce an increase or decrease in the metal content, not only immediately at the fissure but also at some distance therefrom. In the Eisleben district, as illustrated in Fig. 448, the fissures effect an enrichment of the ore-bed, while at Hettstedt the opposite is the case. On an average the Kupferschiefer of the Mansfeld districts proper, between Gerbstedt and Eisleben, contains 2–3 per cent of copper and 5·5 kg. of silver per ton of copper. The ore-bed along the whole of the northern rim as well as around the crest of the Hornburg anticline is poorer, and probably carries on an average hardly 1·5 per cent of copper, the amount of silver per ton of copper remaining as above.

In the neighbouring Sangerhausen district, adjoining the south border of the Harz and separated from the Mansfeld syncline by the Hornburg anticline, mining is at present practically stopped. In this second district the mineralization extends along the Weissliegendes, upon which, the Zechstein conglomerate being absent, the Kupferschiefer immediately rests. In the best case this mineralization consists of a 7 cm. wide impregnation of the Weissliegendes, with chalcopyrite and some chalcocite, and their decomposition products, malachite and azurite. The ore-bearing layer is known as ' yellow band ' ; it contains 5–10 per cent of copper. The lower-grade ore is described as ' sand ore.' In the Eisleben and Hettstedt districts sand ores are found isolated and sporadically. Apparently in those districts only occasional fissures were favourable for the formation of such sand ore.

The fissures associated with and intersecting the Kupferschiefer are either barren or they contain rich accumulations of ore, principally of copper and nickel. The ore as a rule is confined to the vertical space between the two faulted portions of the Kupferschiefer.

The following analyses taken from the previously cited Mansfeld *Festschrift* of 1907 complete this description :

[TABLE

	Average Ore.	Kammschale.	Kopf.	Schwarze Berge.	Dachberge.	Fäule.
	No. 1.	No. 2.	No. 3.	No. 4.	No. 5.	No. 6.
SiO_2	33·15	39·67	35·00	40·47	24·15	28·45
Al_2O_3	17·30	14·00	11·07	12·88	7·75	8·27
Fe_2O_3	...	...	...	...	0·69	0·68
FeO	3·34	2·56	2·02	2·43	1·50	1·33
CaO	10·40	5·94	12·50	10·71	22·16	24·90
MgO	1·00	4·83	7·49	6·69	9·36	4·98
K_2O	...	3·46	3·22	3·28	2·05	2·42
Na_2O	...	1·15	1·02	1·23	0·71	1·09
P_2O_5	...	0·23	0·15	0·21	0·13	0·13
S	2·31	2·30	1·64	1·34	0·61	0·56
SO_3	...	0·23	0·48	0·79	2·15	0·30
Cu	2·75	0·85	0·71	0·58	0·14	0·15
Ag	0·014	...	...	...	...	...
Zn	1·28	2·04	2·11	0·90	0·46	0·62
Pb	...	1·47	0·94	0·83	Trace	0·05
Ni	0·018	...	...	...	...	...
MnO	...	0·26	0·33	0·44	0·68	0·61
CO_2	9·24	7·56	16·26	14·02	26·16	24·39
C	9·06 [1]	9·96	3·70	1·61	0·32	0·22
H_2O	1·70	8·87	4·07	2·87	1·58	1·42

[1] Bitumen.

	Matte Slags.				Matte Slags.	
	No. 7.	No. 8.			No. 7.	No. 8.
SiO_2	49·90	48·56	Cu_2O		0·26	0·28
Al_2O_3	16·02	17·60	ZnO		1·76	0·85
FeO	5·58	2·46	PbO		0·09	0·09
CaO	15·23	21·81	NiO	}	0·013	0·009
MgO	6·61	3·00	CoO			
K_2O	4·32	4·18	MnO		0·37	0·32
Na_2O	0·44	0·69	C		0·13	0·06
S	0·18	0·19				

	Copper Matte.			Furnace-Refined Copper.	
	No. 9.	No. 10.	No. 11.	No. 12.	No. 13.
Cu	40·06	41·84	42·03	99·772	99·734
Ag	0·22	0·23	0·24	0·033	0·032
Fe	26·29	21·35	20·93	...	...
Pb	0·60	1·01	1·11	0·043	0·044
Zn	4·35	5·58	6·72	...	...
Mn	0·51	0·66	0·40	...	...
Ni	0·32	0·29	0·38	0·113	0·161
Co	0·32	0·26	0·34	...	...
S	24·96	25·03	25·24	...	...
As	0·02	0·02	0·04	0·014	0·019
Al_2O_3	1·15	1·06	1·19	...	...
Alkali salts [1]	0·51	1·99	1·02	...	...

[1] Alkali salts soluble in water.

No. 1, Composed of *Lette* and *Kammschale*. Average assay of a month's output from the Hoffnung's shaft, 1899.—Nos. 2–6, Some layers of the Kupferschiefer in the Otto shaft,

1891.—Nos. 7–8, Matte slags from the Krug and the Koch works, 1903. The analyses of matte slags give fairly exactly the composition of the shale bed after deducting the carbondioxide, the bitumen, and the metalliferous minerals.—Nos. 9, 10, and 11, average assays, 1906, from the Krug, Eckard, and Kupferkammer works.—Nos. 12 and 13, Furnace-refined copper.

From these and numerous other analyses it is reckoned that the ratio of copper to silver is 100 Cu : some 0·55 Ag ; that in the Kupferschiefer as a whole there is approximately the same amount of copper as zinc ; much less lead than zinc ; some 60–100 times as much copper as nickel and cobalt together ; and about the same amount of nickel as cobalt.

Mining operations at Mansfeld began in 1199 or 1200, and on June 12, 1900, the seven-hundred-years' jubilee was celebrated. At the beginning operations were on a fairly small scale. With time, however, they have so increased that Mansfeld is now the second largest copperproducing district in Europe, Rio Tinto in Spain being the first. In addition, Mansfeld is without doubt the largest silver producer in Germany. The copper production during the period [1] 1779–1877 amounted to 130,000 tons, that of 1878–1893 to 180,000 tons, and that of 1894–1907 to 280,000 tons. Including the earlier production and that of recent years, the total is some 800,000 tons. In the year 1906 the area exhausted reached a total of 1,509,008 square metres. From every square metre, on an average 0·458 tons of smelting ore yielding, as reckoned from the matte, 3·01 per cent of copper and 0·166 per cent of silver, was obtained, these figures being equivalent to 13·8 kg. of copper and 0·76 kg. of silver per square metre. The development of the industry in the last fifty years may be gathered from the following figures of production :

PRODUCTION OF MANSFELD MINES

	Copper.	Silver.		Copper.	Silver.
	Tons.	Tons.		Tons.	Tons.
1860	1,501	7·8	1890	16,391	88·1
1865	2,113	10·1	1895	15,079	75·9
1870	3,803	17·5	1900	18,676	97·5
1875	6,039	30·1	1905	19,878	101·3
1880	9,859	51·6	1911	20,851	113·3
1885	12,724	75·1	1912	20,503	112·7

In the year 1906 the total number of employees was 21,239, of which 16,386 were occupied underground. Including women and children, roughly 65,000 inhabitants are dependent upon this copper-mining industry.

[1] Ante, p. 198.

Other German Copper-Shale Deposits of the Mansfeld Type

LITERATURE

L. v. Ammon. 'Über eine Tiefbohrung durch den Buntsandstein und die Zechstein-schichten bei Mellrichstadt an der Rhön,' Bayr. geogn. Jahrb. XIII., 1900, pp. 149-193.— A. Heuser. 'Versuch einer geognostischen Beschreibung der im Richelsdorfer Gebirge aufsetzenden Gänge und sogenannten Veränderungen,' from Leonhards Taschenbuch, XIII., 1819, pp. 311-447.—Grassmann. 'Das Richelsdorfer Kupfer- und Kobaltwerk in Hessen,' Zeitschr. f. d. Berg-, Hütten- u. Salinenwesen, XXXIV., 1886, pp. 195-207.—Joh. Leb. Schmidt. 'Mineralogische Beschreibung des Biebergrundes,' from Leonhards Taschenbuch, II., 1808, pp. 45-70.—H. Bücking. 'Der nordwestliche Spessart,' Abhandl. d. pr. geol. Landesanstalt, 1892, N.F. XII., pp. 137-141.—G. Würtenberger. 'Über die Zechsteinformation, deren Erzführung und den Unteren Buntsandstein bei Frankenberg in Kurhessen,' Neues Jahrb., 1867, pp. 10-38 ; 'Zur Geschichte des Frankenberger Kupfer-werkes im Regierungsbezirk Cassel,' Zeitschr. f. d. Berg-, Hütten- u. Salinenwesen, XXXVI., 1888, pp. 192-209.—E. Holzapfel. 'Die Zechsteinformation am Ostrande des Rheinisch-Westphälischen Schiefergebirges,' Marburger Dissertation, 1879.—A. Denckmann. 'Die Frankenberger Permbildungen,' Jahrb. d. pr. geol. Landesanst., 1891, pp. 234-267.— A. Leppla. Über die Zechsteinformation und den unteren Buntsandstein im Waldecki-schen,' ibid., 1890, pp. 40-82.—F. Drevermann. 'Über ein Vorkommen von Franken-berger Kupferletten in der Nähe von Marburg,' Zentralbl. f. Min., 1901, pp. 427-429. Beschreibung der Bergreviere Arnsberg, Brilon und Olpe, sowie der Fürstentümer Waldeck und Pyrmont, published by the Kgl. Oberbergamt, Bonn, 1890, pp. 120-121, 139-143.— v. Festenberg-Packisch. Der metallische Bergbau Niederschlesiens, Vienna, 1881, pp. 75-77.—F. Pošepný. 'Über die Genesis der Erzlagerstätten,' Leobener Jahrb. XLIII. 1895.

The extent of the Kupferschiefer in Germany coincides with the former extent of the Zechstein sea, which stretched not only over the district of Mansfeld but also over the western parts of North and Middle Germany and far into Holland. Concerning the northern shore in Germany and the western in Holland, no evidence is forthcoming. The southern shore was in the neighbourhood of the present Odenwald and Spessart, from whence this sea extended eastwards, through the Frankenwald, Saxony, and the neighbourhood of Löwenberg in Silesia, to the Russian-Poland boundary and the north-east of East Prussia. All the districts within these limits, including the Mittelgebirge, the Thuringian Forest, the Harz in part, and the country around the present Mansfeld syncline, etc., were covered with Zechstein sediments. The present extent is con-ditioned by subsequent erosion. The division of the originally uniform marine Zechstein sediments into separate large synclines and basins took place principally during the Tertiary period, at the formation of the Mittelgebirge. At these orogenics also the Mansfeld syncline became separated, on the one side from the large sub-Hercynian Magdeburg-Halberstadt syncline, and on the other from the South Harz syncline, situated between the Harz and the Thuringian Forest.

In the Lower Zechstein near Riechelsdorf not only the Kupferschiefer, but also the foot-wall conglomerate or Grauliegendes carries copper, the several centimetres thick metalliferous bed so constituted extending over a wide area.

The copper ore occurs finely distributed as *Speise* in the Kupferschiefer, and as a compact impregnation in the conglomerate. While in the shale bornite and chalcocite predominate, in the conglomerate chalcopyrite chiefly occurs. Pyrite, galena, and sphalerite, are also found in places. As the old and very extensive workings prove, the copper content on the whole was fairly regularly distributed over this area; according to recent examination by Krusch, over large districts it was often more than 3 per cent. The thickness of the bed proper was 13 cm., with an average of 3·2 per cent ; above this came 4 cm., with 1·3 per cent copper, the material farther in the hanging-wall gradually merging into barren Zechstein limestone. Reckoning a thickness of 17 cm., an average copper content of 2·7 per cent is obtained ; or reckoning 15 cm., 2·9 per cent. The silver content varied between 20 and 30 grm. per ton, being therefore considerably lower than that at Mansfeld.

The amount of sand ore according to the last results was 31·4 per cent that of the shale ore, while its copper content, which on an average was 5–6 per cent, varied between 4·35 and 7·45 per cent.

The influence of the intersecting fissures, which are mostly developed as cobalt-barite lodes, upon the copper content of the bed, was such that ore-bearing fissures diminished the copper content, while those without ore increased it. In consequence of the fall in the price of copper, mining operations at Riechelsdorf were a few years ago suspended.

THURINGIAN FOREST

That part of the Kupferschiefer which forms a continuous border around the Thuringian Forest has in the last decades been explored principally in the neighbourhood of Schweina and Gumpelstadt. Here also the Kupferschiefer rests upon the Zechstein conglomerate, which in places is impregnated with copper for a thickness of 5–10 cm. The Kupferschiefer is 10–15 cm. thick. The mineralization, which is likewise finely distributed, differs from that of the previously-mentioned districts in that chalcopyrite predominates, while chalcocite and bornite recede. Here the relation between the copper content of the shale and the presence of the cobalt fissures is excellently expressed. While the content in the immediate neighbourhood of such fissures may reach 3 per cent, the average content is scarcely higher than 1 per cent. Silver plays no part whatever. The Schweina district, in addition to the Kupferschiefer, is known for its cobalt- and nickel lodes. From these also, an impregnation of the copper shale by cobalt took place, while the presence of zinc and arsenic can frequently be established. This occurrence to-day has unfortunately no economic importance.

With some other occurrences, such as Stadtberge and Goldberg, the section is essentially different from that obtaining in the Mansfeld district.

STADTBERGE, OR NIEDERMARSBERG

LITERATURE

Beschreibung der Bergreviere Arnsberg, Brilon und Olpe, sowie der Fürstentümer Waldeck und Pyrmont, published by the Mining Department, Bonn, 1890.—BUFF. Akten des Kgl. Oberbergamts zu Bonn.—W. BRUHNS, in collaboration with H. BÜCKING. Die nutzbaren Mineralien und Gebirgsarten im Deutschen Reiche, 1906.

At this place the Lower Zechstein consists of limestone beds 10–15 cm. thick, separated from one another by thin layers of marly shale. The copper is confined to these shale layers, which generally occur two to three times in any one section but may also occur in numerous narrow lenticular seams. The metalliferous minerals consist of malachite, and more rarely of azurite and chalcocite. The copper content varies between 1 and 6 per cent, increasing in the neighbourhood of the fault-fissures. Usually it is far below the limit of payability. This occurrence economically has never been important.

The fissures continue ore-bearing through the Zechstein into the footwall Culm silica-schist, where the second and more important occurrence is found, this being still worked. In the Oskar mine on the Juttenberg, the Minna mine on the Kohlhagen, and the Frederike mine on the Bilstein, the ore is found immediately beneath the overlying Zechstein.

The extremely fractured Culm silica-schist exhibits innumerable fractures coated with ore, and in that portion of the deposit being worked, chalcocite, bornite, and chalcopyrite are particularly frequent. A zone of this schist 15–20 m. thick must be regarded as constituting the metalliferous occurrence. In the neighbourhood of the surface the ore by atmospheric agencies has been altered to malachite, azurite, cuprite, and native copper. The metal content of the payable portions varies between 1·5 and 3·5 per cent, while the whole zone may be said to have an average of 1·6 per cent, with extremes of 0·5 and 5 per cent. The silica-schist is especially metalliferous in the immediate neighbourhood of the fissures.

Mining at Stadtberge occupied itself originally with the copper ore in the Zechstein and subsequently with the oxidized ores in the silica-schist, though latterly it has been necessary to pay attention to the sulphide ore. The production from the Oskar and Minna mines is some 50,000 tons of ore annually.

GOLDBERG IN SILESIA

LITERATURE

H. v. FESTENBERG-PACKISCH. Der metallische Bergbau Niederschlesiens. 1881.—
W. BRUHNS, in collaboration with H. BÜCKING. Die nutzbaren Mineralien und Gebirgsarten im Deutschen Reiche. Berlin, 1906.

On the northern slope of the Riesengebirge the Zechstein is cupriferous at the surface from Naumburg on the Queiss to near Goldberg, along which extent are found the old mining centres of Neukirch, Polish-Hundorf, Konradswaldau, Haasel, Prausnitz, and Goldberg.

At the once important centre, Haasel, the Lower Zechstein consists of alternating layers of limestone and marl-shale. The copper ores occur principally in the shale. Seven such shale layers make together a total thickness of 0·75–1·1 metres. Their copper content is on an average 1·6, and at a maximum 2·16 per cent, with 50 grm. of silver per ton. The limestone layers separating the marl-shales have an average thickness of 26 cm. and a copper content of 1·03–1·58 per cent, with 20 grm. silver. The ore consists of azurite and malachite. From 1866 to 1883 some 1100 tons of copper and 3437 kg. silver were won from about 85,000 tons ore.

OCCURRENCES PROBABLY ANALOGOUS TO THE KUPFERSCHIEFER

An Upper Silurian clay-slate rich in fish remains and with a low copper content has already been mentioned [1] as occurring in the Christiania district. In addition, oil-shales or other more or less bituminous shales have occasionally been found with so high a copper content as to justify mining operations, which at times have been upon a not inconsiderable scale. With the low price of copper during the last decades all such operations however appear to have been unprofitable. The primary ore is in many cases principally chalcocite.

Similar occurrences, also belonging to the Permian, have been found at several places in Bohemia ; noteworthy among these are Hohenelbe and Starckenbach in north-east Bohemia, where the deposits consist of copper impregnations in oil-shales and where the ore-bed previously worked was rich in vegetable remains. At Wernersdorf, likewise in Bohemia, the deposit consisted of sulphide ore, especially chalcocite, at different horizons in marly shales. According to Gürich, these are sedimentary deposits.[2]

---

[1] *Ante*, p. 1121.

[2] Katzer, *Geologie von Böhmen*, 1892, pp. 1188-1212, 1222-1225 ; Gürich, ' Die Kupfer-erzlagerstätte von Wernersdorf bei Radowenz in Böhmen,' *Zeit. f. prakt. Geol.*, 1893, pp. 370-371.

In Texas, over an extensive area between 98° and 100° west longitude and 33°-34° north latitude, a number of Permian copper deposits are found at different stratigraphical horizons. The ore here is associated more especially with bituminous shales or with marls rich in vegetable remains. It is not workable.[1]

In the neighbourhood of New Annan in Nova Scotia, occur Permian copper deposits, consisting principally of chalcocite in a thin micaceous sandstone containing vegetable remains.[2]

A so-called copper-shale occurring in crystalline schists at Stora Strand in Dalsand, Sweden, will be found described with the fahlbands.

It is questionable whether all the occurrences here briefly mentioned are comparable genetically with the German Kupferschiefer. The Permian copper-sandstone in Russia certainly deviates to such an extent from the Kupferschiefer that it must be reckoned with another class of deposit.

THE FAHLBANDS

The term 'Fahlband'[3] was derived originally from Kongsberg in Norway, to which place German miners were called in the seventeenth century. The designation *fahl* or 'faded' refers to the rust-coloured weathering of sulphide-bearing beds, which were termed 'bands.' By the term fahlband therefore is understood a crystalline schistose rock— gneiss, hornblende-schist, mica-schist, phyllite, quartz-schist, etc.—with a sparing, or at all events not particularly rich impregnation of metallic sulphides. Of these sulphides those which occur most extensively are pyrite, pyrrhotite, and chalcopyrite, with which frequently some arseno- pyrite, sphalerite, galena, etc., are associated. The cobalt fahlbands described below, occupy a place by themselves. It is not advisable to extend the term fahlband to impregnations of granular rocks, since these do not occur in bands. Moreover, from custom the term is confined to impregnations of metallic sulphides, and it is therefore not advisable to include those of metallic oxides.

Fahlband is a morphological and not a genetic term. Strictly speaking therefore, in any classification of ore-deposits based upon genesis a 'fahlband group' may not be formulated. There exists however so great uncertainty concerning the origin of many of these deposits that it is preferable for purposes of description to keep them together, though by so doing an odd

[1] Schmitz, 'Copper Ores in the Permian of Texas,' *Trans. Amer. Inst. Min. Eng.* XXVI., 1896, pp. 97-108.
[2] H. Louis, *Trans. Amer. Inst. Min. Eng.* XXVI., 1896, pp. 1051-1052.
[3] *Ante*, p. 660.

group including occurrences probably of exceedingly different genesis becomes formed.

Many mining engineers of the old school took it for granted that all gneisses, hornblende-schists, mica-schists, quartz-schists, etc., were of sedimentary origin, and that the sulphides in the fahlbands were also formed by sedimentation. It is known however that many, though not all gneisses, hornblende-schists, etc., of the fundamental rocks, are to be regarded as foliated eruptives, and the old conception that all fahlbands were sediments accordingly falls to the ground. Some fahlbands may indeed represent altered sediments, others however may have been formed by magmatic intrusion, and others again by deposition from aqueous or gaseous solution.

Economically the fahlbands, owing to their small sulphide content, as a rule are of subordinate importance. Some are, or were worked, principally for copper or cobalt.

THE COPPER DEPOSIT AT STORA STRAND IN DALSAND, SWEDEN

This occurrence is an example of a fahlband formed probably by sedimentation. As described by H. E. Johansson,[1] it is situated near the west shore of the Venern lake and occurs in the pre-Cambrian, perhaps Algonkian, Dalsand formation. The ore-bed, which with breaks is known for a length of 20 km., is found within an altered, fairly steep series of predominating calcareous clay-slates with conglomerates and quartz-sandstones; it is characterized everywhere by stratigraphical conformity. The clastic origin of the now more or less crystalline rocks may in several places be demonstrated.

The section immediately at the deposit is as follows : (1) at the bottom, altered calcareous clay-slate ; (2) green chloritic slate with cubical pyrite ; (3) brown siliceous bed with some chalcopyrite ; (4) brown so-called copper-shale consisting of mica, chlorite, quartz, albite, etc., with an invariable impregnation of chalcopyrite ; (5) grey-green copper-shale, the copper content diminishing towards the top ; (6) hornstone-like calcareous bed ; (7) calcareous clay-slate.

This copper-shale has been closely explored more particularly in its richest part 3·5 km. in length. With a thickness generally 0·5–1·3 m. it consists of a fine impregnation almost exclusively of chalcopyrite. Over a thickness of 0·5 m. the average copper content is 1·8 per cent, in addition to which comes 0·3 m. with 0·38 per cent. The quantity of copper per square metre on the bed plane is about 25 kg., losses in treatment, etc.,

[1] *Schwed. geol. Unters.*, Serie C., No. 214, 1909.

not being reckoned. To 100 parts of copper there are 0·19 parts of silver. In this respect, also, a similarity to the Mansfeld Kupferschiefer therefore exists. The copper-shale of Dalsand consists, according to various chemical analyses, of about 56 per cent of SiO_2, 19 per cent Al_2O_3, some Fe_2O_3, Fe, CaO, MgO, Na_2O, K_2O, with about 2 per cent sulphur and 1·3–1·7 per cent of copper. In composition therefore it differs from the marly copper-shale of Mansfeld principally by its low CaO content.

The total copper content in this long Stora Strand bed is very considerable. The ore however cannot be hand-sorted and is with difficulty concentrated.

In relation to genesis, Johansson compares this occurrence with the Mansfeld Kupferschiefer; it is however geologically older and somewhat metamorphosed.

As an example of intrusive fahlbands, the Bodenmais deposit in Bavaria has already been described.[1] This deposit consists partly of fairly clean pyrite and partly of gneissic material with a more or less sparing impregnation of sulphides, especially of pyrrhotite. The intrusive pyrite deposits exhibit similar phenomena. Thus, with many Norwegian pyrite occurrences,[2] fairly rich sulphide impregnations extending for considerable distances along the strike, occur as the continuations of the compact pyrite bodies, together with which, in greater part along the bedding-planes, they were intruded. In the Rio Tinto district the pyrite bodies proper are similarly accompanied by epigenetic impregnations of cupriferous pyrite, as for instance at Louzal in Portugal. In this connection the pronounced mobility of the fused sulphides must be remembered.

The magmatic nickel-pyrrhotite deposits also, which occur principally at the contact of gabbro rock with crystalline schists,[3] are often accompanied by fahlband-like impregnations of the neighbouring schists.[4] In different nickel mines at Ringerike it can, according to Vogt, be demonstrated that the pyritic sulphide within the relatively finely schistose gneiss in the immediate neighbourhood of the gabbro, occurs chiefly along the schist-planes, while in the more compact gneiss a network of intersecting sulphide veins is found.

The Kongsberg fahlbands, the extent of which within the central Kongsberg field is indicated in the rather antiquated map constituting Fig. 65, have been frequently described.[5] In this occurrence, more particularly mica-schists, garnet-mica schists, and chlorite-schists have been impregnated with metallic sulphides, these latter occurring chiefly along the schist-planes in these relatively finely banded rocks. The amount of these sulphides—pyrrhotite, pyrite, with chalcopyrite, some galena,

[1] *Ante*, pp. 337-340.   [2] *Ante*, pp. 304-313.   [3] *Ante*, pp. 280-299.
[4] *Ante*, p. 285.   [5] *Ante*, p. 660.

sphalerite, etc.—is generally only one or two per cent. In but few places is it higher, so that the earlier workings for pyrite containing some 40 per cent of sulphur and a little copper, were inextensive.

The amphibolites and amphibolite-schists also often carry some pyrite ; the same in places is the case with the so-called Kongsberg gneiss or gneiss-granite, which in the neighbourhood of the fahlband zones—consisting chiefly of mica- and chlorite-schist—is here and there pyritic. Within the foliated granite or gneiss-granite the pyrite partly follows the schist-planes and partly traverses the rock in fine zig-zag stringers.

According to investigation by C. Bugge, the amphibolites and amphibolite-schists occur in irregularly branched dykes—which in greater part are to be regarded as bedded dykes—in other Archaean rocks. The fahlbands, as illustrated in Fig. 330, mostly follow these intrusive amphibolites.

Kjerulf and Dahll in 1861 were of opinion that the sulphides of the fahlbands are of younger formation. On the other hand, A. Helland [1] believed in a contemporaneous formation of both the non-metalliferous material and the metalliferous. Among other things he pointed out that under the microscope pyrite is often seen in the centre of the garnet- and quartz individuals, without any apparent fractures along which it could have entered. Chr. A. Münster in 1894 regarded the rocks concerned, together with their sulphides, as altered sediments.

According to Vogt in 1899, the sulphides are found in quite different rocks, among which, some, such as the gneiss-granite, are of eruptive origin. In this gneiss-granite the sulphides form in part a network of zig-zag veins, and in this case therefore are certainly younger than the granite. Since then in its turn the granite is younger than the mica-schist, the sulphides also must be younger than the mica-schist. The epigenetic genesis of the sulphides of the Kongsberg fahlbands follows also from the above-mentioned more recent investigation by C. Bugge.

Cobalt Fahlbands

In the Scandinavian peninsula, four, or when similar impregnations in limestone are included, six cobalt fahlbands are known, these being distinguished from ordinary fahlbands by containing cobalt arsenides. As a rule cobaltite, $CoAsS$, is the most frequent ; in some mines in addition smaltite, $CoAs_2$, is found ; in others skutterudite, $CoAs_3$; and frequently also cobalt-arsenopyrite, glaucodote or danaite, $(Fe,Co)AsS$, etc. In these cobalt minerals, only very little cobalt is replaced by nickel. In some deposits nickel minerals such as gersdorffite, $NiAsS$, or

[1] *Archiv f. mathem. Naturw.*, Christiania, 1879, Vol. IV.

cobalt-nickel minerals such as linnaeite $(CoNi)_3S_4$, are also found, this last for instance playing an important part in the occurrence at Glad-hammer. Analyses of furnace products at Modum indicate the relation of cobalt to nickel in the ore to be on an average $1 : \frac{1}{15}$.[1]

Pyrite and pyrrhotite are also present in the cobalt fahlbands, though generally only in sparing amount. The same as a rule is also the case with chalcopyrite, though this at times occurs much more abundantly, some mines having formerly been worked for cobalt and copper at the same time. Bismuth minerals in some places are completely absent ; at other places they occur, though never in very large amount. Although nickel occurs in the earth's crust on an average much more extensively than cobalt,[2] nickel fahlbands corresponding to the cobalt fahlbands do not appear to exist ; at all events such deposits up to the present are not known. The relatively low nickel content of the cobalt fahlbands is also remarkable.

It may be questioned whether all these six fahlbands should in relation to genesis be included in one common group ; it is possible that the two deposits, Tunaberg and Håkonsboda, which occur in limestone, belong to the contact occurrences. Remarks concerning genesis will there-fore in greater part be confined to the best known cobalt fahlband, namely, that at Modum. This deposit, according to Vogt, is epigenetic. Further, its close association with large intrusions of amphibolite of gabbroidal character is striking. At Los in Helsingland the cobalt fahlband actually occurs in a foliated gabbro.

The twin elements nickel and cobalt are pronounced basic elements,[3] and accordingly are met chiefly in the basic eruptive rocks. In accordance with this the nickel-pyrrhotite deposits are notoriously found in connection with gabbro,[4] and the garnierite deposits in connection with peridotite and serpentine.[5] The nickel-pyrrhotite deposit at Erteli[6] and the cobalt fahlband at Skutterud in Modum are in a straight line only 15 km. distant from each other. In both places association with gabbroidal rocks is estab-lished. We may therefore assume, in any case for Modum, but also for Los and perhaps for other of the cobalt fahlbands, a genetic dependence upon gabbroidal rocks.

Cobalt, as is well known, is more readily dissolved than its twin element, and cobalt oxide is more easily precipitated from solutions than nickel oxide. In harmony with this, the cobaltiferous and nickeliferous lodes formed by aqueous deposition—the Mansfeld cobalt lodes and the lodes at Annaberg, Schneeberg, Temiskaming, etc.,[7]—contain on the whole more cobalt than nickel, though nickel occurs in the earth's crust much more

---

[1] J. H. L. Vogt, *Zeit. f. prakt. Geol.*, 1898, p. 386.

[2] *Ante*, p. 153.   [3] *Ante*, p. 158.   [4] *Ante*, p. 280.
[5] *Ante*, p. 952.   [6] *Ante*, p. 297.   [7] *Ante*, pp. 666, 677.

extensively than cobalt. The mineralogical and chemical analogy between the cobalt fahlbands on the one hand and the cobalt- and cobalt-bismuth lodes on the other, is also noteworthy.

The above considerations, according to Vogt, give ground for the view that the cobalt fahlband at Modum—and doubtless also those of Los and other districts—was deposited from solutions which in one way or another emanated from gabbroidal intrusions. These solutions followed the well-defined schist-planes, from which planes impregnation proceeded. The fahlbands at Modum are crossed by Archaean granite-pegmatite dykes, from which it follows that mineralization took place in the Archaean period. Probably the mineral solutions were an immediate after-effect of the gabbroidal intrusions.

In the first half of the nineteenth century the cobalt fahlbands satisfied a considerable portion, perhaps even more than one-half of the relatively small demand for cobalt preparations, smalt or cobalt-blue. By the fall in the price of smalt which followed the discovery and manufacture of artificial ultramarine in the middle of the forties, several of the companies exploiting cobalt fahlbands were ruined. At Modum work was soon resumed, though on a limited scale, only later to be suspended once again, and now the supply of cobalt preparations is derived in greater part from Temiskaming.[1] Owing therefore to the fact that the cobalt fahlbands are no longer worked, the study of these deposits is encompassed with considerable difficulty.

Modum in Norway

LITERATURE

J. F. L. Hausmann. Reise durch Skandinavien, Part 2, 1812, pp. 69-91.—K. F. Böbert. 'Ausführliche Monographie,' Karstens Archiv f. Mineralogie, Geognosie und Bergbau, 1847, Vol. XXI. ; ibid., 1832, Vol. IV. ; Nyt. Magazin f. Naturw., 1848, Vol. V. —M. Durocher. 'Observations sur les gîtes métallifères de la Suède, de la Norvège et de la Finlande,' Ann. des mines, Paris, 1849, 4, XV. pp. 319-328.—A. Helland. 'On the Occurrence of the Cobalt- and Nickel Ores in Norway (in Norwegian),' Archiv f. mathem. Naturw., 1879, Vol. IV.—Th. Kjerulf. Die Geologie des südlichen und mittleren Norwegens. Bonn, 1880.

At Modum, including Snarum—about 50 km. in a straight line west of Christiania and 40 km. north-east of Kongsberg—the so-called Bamle formation prevails, which is best developed at Bamle near Kragerö, and belongs to the upper, perhaps even to the uppermost portion of the fundamental rocks. This formation consists of thick quartzites with numerous, very large, and principally bedded intrusions of amphibolites and amphibole-schists ; to these must be added several gabbro massives, which at Snarum, as at Bamle, are accompanied by apatite lodes.[2] The

[1] Ante, p. 669.   [2] Ante, p. 452.

amphibolites, consisting of plagioclase, hornblende, some mica, garnet, etc., are, according to W. C. Brögger, altered gabbros. One or two kilometres from this cobalt deposit a number of serpentine- and ophi-magnesite occurrences are found.

The Modum cobalt fahlband on which the Skutterud mine is situated, extends, with many breaks, through the long-abandoned Snarum mines, for a length of about 10 kilometres. The strike is almost north-south, the dip 80° to the west. The most important cobalt mineral is cobaltite, which often appears in well-defined crystals ; with this is associated some skutterudite and cobalt-arsenopyrite, the latter containing generally 6–8 per cent,

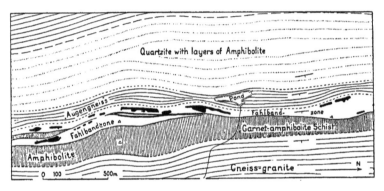

Fig. 449.—Map of the southern portion of the Modum fahlband zone including to the south the Skutterud mine. Many granitic dykes crossing the fahlbands are not shown.

and seldom as much as 18 per cent of cobalt ; some erythrite occurs as a secondary product. Pyrite, pyrrhotite, chalcopyrite, and some molybdenite, are also found, though only to a very small extent. Galena and sphalerite are absent from the fahlbands themselves, but occur, together with marcasite, in sparing amount in some quartz - calcite veins.

With the cobalt minerals the diopside mineral, malacolite or salite, occurs frequently, occasionally even in large amount ; also some anthophyllite, tremolite, epidote, brown tourmaline, plagioclase, quartz, yttrotitanite, etc., while among others, small crystals of apatite and rutile have also been encountered.

The fahlband zone, including the barren beds, reaches a thickness of 80 or 100 m., and, as indicated in Fig. 449, accompanies a thick band of amphibolite. Within the fahlband zone micaceous quartzites or even mica-schists, and a number of in greater part conformably intercalated intrusions of thin amphibolite seams, are principally met. In

addition, a peculiar augengneiss occurs containing quartz blebs rich in
sillimannite. This rock is identical with the intrusive orbicular granite
from the neighbourhood of Kragerö, investigated by W. C. Brögger, which
is characterized by segregations of quartz and sillimannite.

The cobalt minerals form impregnations chiefly within certain quartzite
layers rich in mica and tourmaline. They are absent however from the
thick purer quartzite somewhat farther to the west, but have been met in
the above-mentioned augengneiss. On the other hand, cobalt ore does
not occur in the compact amphibolite. Rich patches of this ore have
however often been found in the rock containing large orbicules of dark
biotite and garnet, occurring at the contact between the amphibolite and
the micaceous quartzites. In places where the impregnation is marked
malacolite occurs in preponderating amount, a feature which led to the
erroneous assumption that the malacolite might be used as a guide to
the discovery of ore-bodies.

The thickness of each individual cobalt fahlband within the wide fahlband zone is as a rule one or two, and at most some 5–8 metres. Frequently
however several parallel fahlbands occur close together but separated from
one another by barren rock and almost clean quartzite.

The Skutterud mine reached a depth of 170 m., reckoning from the
highest point at the outcrop. In this mine the cobalt ore, consisting chiefly
of cobaltite, occurs occasionally in rich streaks ; the rule however is for it
to occur as a scanty impregnation in pronouncedly schistose rock, and the
ore content accordingly is fairly low. Thus, statistics of mining operations
for the years 1878–1882 gave the following yearly average figures : rock
broken 2152 cbm., resulting in 1411 tons of picked ore, which by dressing
gave 96 tons of concentrate containing 10·2 tons of metallic cobalt.
Accordingly, on an average, after deducting the relatively large loss
in picking and dressing, 1 cbm. or 2·8 tons of rock yielded 4·75 kg.,
equivalent to 0·17 per cent of cobalt.

This fahlband was formerly frequently regarded as a sediment, a view
which Kjerulf successfully controverted. According to Vogt, the fact
that the mineralization of each individual fahlband is markedly irregular
both in strike and dip betokens an epigenetic formation ; in this direction also the association of the ore with such minerals as tourmaline
and yttrotitanite, malacolite, rutile, and zircon, tends to point. It must
further be emphasized that the fahlbands occur in the pronouncedly
schistose zones and not in the more compact quartzites and amphibolites, a
phenomenon probably due to the fact that the ore solutions followed the
well-defined schist-planes. Finally, conclusive evidence of the epigenetic
origin is afforded by the fact that the ore occurs not only in the sedimentary
quartzites but occasionally also in the intrusive augengneiss. The genetic

1140 ORE-DEPOSITS

association with the large amphibolite intrusions has already been mentioned.[1]

These mines were started in 1772, from which date up to 1822 they were worked by the State, passing over later into private possession. During the period of greatest success, from 1830 to the beginning of the 'forties, roughly 1000 hands were employed. Then, as the result of the discovery of artificial ultramarine, came the first great fall in the price of cobalt-blue. Operations however were continued, though on a limited scale, up to 1898, when further work was abandoned. During the period 1856–1898 cobalt products containing 257 tons of metallic cobalt were produced.[2]

SWEDISH COBALT FAHLBANDS

Gladhammer, in the neighbourhood of Westervik in Kalmar Län; Vena, not far from the sphalerite mine Åmmeberg,[3] north of Vettern; Los in Helsingland; Tunaberg, in the neighbourhood of Nykjöping in Södermanland; and Håkonsboda, south of Kafveltorp in Örebro Län.

LITERATURE

On Tunaberg detailed description by A. ERDMANN, Vet. Akad. Handl., 1848; on Gladhammer, Swedish Geological Exploration Publications, Ser. C. No. 64, 1884; on Vena, ibid. Ser. Aa, No. 84, section Askersund, 1889; on Håkonsboda, ibid. Ser. Bb, No. 4, 1889; on Los, ibid. Ser. C. No. 152, 1895.

The occurrences at Gladhammer and Vena are typical fahlbands, the former in mica- and quartzite-schists, the latter in fine-grained and micaceous gneiss. That at Los was also formerly described as an ordinary fahlband, as for instance by Durocher;[4] according to G. Löfstrand, the ore at this place occurs in the schistose portions of a foliated gabbro massive. At Håkonsboda it is found in slate and limestone; and at Tunaberg in a massive limestone. These two last occurrences differ more or less considerably from the ordinary cobalt fahlbands of the Modum, Gladhammer, and Vena type. At Tunaberg the ore is accompanied by contact minerals.

The cobalt ore with most of the Swedish occurrences is chiefly cobaltite, which at Los and Tunaberg is accompanied by smaltite; at Håkonsboda by cobalt-arsenopyrite; and at Gladhammer by linnaeite and gersdorffite. In some deposits bismuth occurs in small amount. In several cases the copper content was so considerable that the the mines were worked not only for cobalt but also for copper. Some of these deposits, especially Tunaberg but also Vena, were formerly in vigorous operation; the position now is that all have lain idle for years.

[1] Ante, p. 1136. [2] J. H. L. Vogt, in Statsökonomisk Tidsskrift, Christiania, 1900.
[3] Postea, p. 1171. [4] Annales des mines, Paris, 1849.

THE SILESIAN COBALT FAHLBANDS AT QUERBACH AND GIEHREN [1]

Querbach and Giehren lie south of Friedeberg. Here at the contact of mica-schist and gneiss a 1·5–5 m. zone of crystalline schist occurs, which in part contains abundant garnet and carries cobalt- and tin minerals finely distributed. Much of this zone consists of roundish garnets cemented by chlorite. The metalliferous minerals occur in this cementing material, the whole constituting the so-called garnet ore. No sharp demarcation of the ore-bearing beds is anywhere perceptible. The ore consists of pyrite, pyrrhotite, arsenopyrite, galena, sphalerite, smaltite, and cassiterite, the cobalt ore occurring chiefly at Querbach, the tin ore at Giehren.

The genesis of these deposits has hitherto not been satisfactorily established. Since the ore-bearing zone is profusely crossed by quartz- and calcite veins, the ore can hardly be syngenetic. The occurrence of tin ore with the cobalt ore distinguishes these Silesian fahlbands from the Scandinavian; it further indicates that pneumatolytic processes were active at the formation of the deposit. According to form however the deposits belong to the fahlbands. In the sixteenth and seventeenth centuries operations were not inconsiderable; in the year 1842 they were suspended.

THE PYRITE BEDS

The deposits of this type were formerly reckoned as belonging to the 'Pyrite group,' which included all pyrite- and marcasite occurrences, whatever their genesis. From this heterogeneous group the following genetic sub-groups have already been eliminated and described, namely, the intrusive pyrite deposits, the contact pyrite deposits, the pyrite lodes, and the metasomatic pyrite deposits. It now remains to discuss the occurrence of presumably sedimentary nature.

In relation to form, three kinds are to be differentiated, namely :

1. More or less continuous beds which have been subject to the same tectonic influences as the enclosing strata.

2. Concretionary deposits in clayey or slaty material, the concretions, sometimes more and sometimes less closely together, being found in all formations from the Diluvium to the Palæozoic. Fossils were frequently the cause of the formation of these concretions.

3. Impregnations in different rocks, particularly in slates, as for instance alum-slates; and in clays, as alum-shales.

---

[1] H. v. Festenberg-Packisch, *Der metallische Bergbau Niederschlesiens*, Vienna, 1881. W. Bruhns, in collaboration with H. Bücking, *Die nutzbaren Mineralien und Gebirgsarten im Deutschen Reiche*, Berlin, 1906.

In many cases the separation of the pyrite bed from the country-rock is not sharp ; in harmony with their bedded character it is rather the case that ore and country-rock replace each other indifferently. The pyrite bands on the Witwatersrand for instance, as illustrated in Fig. 4, merge frequently into quartzite ; where the quartzite becomes conglomeratic the pyrite confines itself exclusively to the cementing material. This pyrite consists either of more or less clean pyrite, or of pyrite intergrown with the sulphides of more valuable heavy metals.

The thickness of pyrite beds is very variable ; for payability great purity of the material and considerable thickness are essential. Since the pyrite which finds its way to the present-day market contains 42–49 per cent of sulphur, poorer pyrite beds can only be worked at a profit when the demand comes from the immediate neighbourhood. Alternating layers of pyrite with rock material occur frequently ; such deposits are only payable when the ore can be concentrated. The metalliferous minerals which as a rule make up the pyrite beds are, in addition to pyrite, the sulphides of copper, lead, and zinc ; with pyrite beds in ancient formations and in much foliated country, pyrrhotite also is frequently encountered. The finer the intergrowth, the more expensive is the production of saleable pyrite. Of the gangue-minerals, barite, which sometimes occurs intimately intergrown with the ore-minerals, comes first ; quartz and other minerals are rarer. Accessory precious metals such as gold and silver may be of importance. Both occur associated with the pyrite ; silver however also with the galena, which latter in its turn is intergrown with the pyrite. According to the extent of the precious-metal impregnation the pyrite deposits may merge into gold- or silver deposits.

Primary depth-zones owing to the small thickness of the bed are economically unimportant. Occasionally, it is true, a cleaner layer in the foot-wall can be differentiated from poorer ore in the hanging-wall. It has already been indicated that in the Rammelsberg deposit primary depth-zones in clean ore are occasionally also found.[1]

The secondary depth-zones are those ordinarily found with the pyritic sulphide deposits. Under favourable conditions a gossan is formed which contains relatively much heavy-metal sulphate but may be practically free from sulphur, leaving only the corroded nature of the country-rock to suggest the existence of a pyrite deposit in depth. In the cementation zone below this oxidation zone the precious-metal content may become concentrated, while the pyrite and any galena or sphalerite present will have only suffered alteration in so far as they effected a precipitation of the precious metals.

The primary zone below the zone of cementation can exhibit no

[1] *Ante,* p. 212.

primary depth-zones in the direction of the dip, since the tilting of the deposit is secondary. The foregoing occasional primary variations relate then exclusively to the original horizontal deposition of the ore.

The alum-ore occurrences may be regarded as exceedingly impure sedimentary pyrite deposits ; they consist of soft shale with pyrite finely disseminated throughout. To such shale the name of alum-shale was given because of its former application to the production of alum. Upon weathering, aluminium sulphate and alkali sulphates are formed, which can be leached. After the addition of any still lacking alum constituents to the resultant solution—generally more alkali sulphates are necessary— alum crystallizes out. The position of the alum-ores among the ores has already been discussed.[1]

Petrographically, it is noteworthy that the structure of alum-shale is very fine-grained and this shale is consequently characterized by earthy fracture. By weathering it loses its black colour and generally becomes brownish-grey. In respect to the usually considerable bitumen content, alum-shale resembles the Mansfeld copper-shale. Most alum-shales belong to the Palæozoic.

The Swedish alum-shales in Närke and Westergötland are especially well known. These are associated with loaf-shaped, peculiar accumulations of bitumen rich in carbon, known locally as *Kolm,* for which because of their radium content radium works were erected. The attempt was however abandoned as sufficient radium was not recovered, the application of alum-shale in this direction not being possible at present.

In the stratigraphy of the Palæozoic formation in Germany, the alum-shales of the Culm formation, which occur principally at the lower contact with Devonian and at the upper contact with Upper Carboniferous, are of importance. While the lower bed has only a very small thickness, the upper is sometimes more than 100 m. thick, and formerly was extensively mined.

In relation to genesis, it can be assumed that in all probability the pyrite content is primary, and that alum-shale accordingly is a true sedimentary deposit.

Pyrite concretions are found in the most varied geological formations. These at various places in Germany it was formerly attempted to exploit. Under present conditions however such deposits are invariably unpayable, since the pyrite masses available in other classes of deposit, especially the intrusive pyrite deposits, are so enormous that very favourable conditions must be fulfilled before the payability of a pyrite deposit is possible. Accordingly, to-day, only the compact pyrite beds, which, by the way, very seldom occur, are of economic importance. The number of these deposits according to present knowledge is, at most, but moderate.

[1] *Ante,* p. 73.

THE RAMMELSBERG ORE-BED NEAR GOSLAR

LITERATURE

H. CREDNER. Übersicht der geognostischen Verhältnisse Thüringens und des Harzes, 1843, p. 121.—B. v. COTTA. 'Über die Kieslagerstätte am Rammelsberg bei Goslar,' Berg- und Hüttenmännische Zeitg., 1864, p. 369.—SCHUSTER. 'Über die Kieslagerstätte am Rammelsberge bei Goslar,' Berg- und Hüttenmännische Zeitg., 1867, Vol. XXVI. p. 307.—FR. WIMMER. 'Vorkommen und Gewinnung der Rammelsberger Erze,' Zeit. f. Berg-, Hütten- und Salinenwesen im pr. Staate, 1877, Vol. XXV. p. 119.—A. W. STELZNER. 'Die Erzlagerstätte vom Rammelsberge bei Goslar,' Zeit. d. d. geol. Ges., 1880, p. 808.—G. KÖHLER. 'Die Störungen im Rammelsberger Erzlager bei Goslar,' Zeit. f. Berg- und Salinenwesen im pr. Staate, 1882, p. 31.—A. v. GRODDECK. Geognosie des Harzes, 1883, p. 118.—J. H. L. VOGT. 'Über die Genesis der Kieslagerstätten vom Typus Röros-Rammelsberg,' Zeit. f. prakt. Geol., 1894, p. 173.—F. KLOCKMANN. Berg- und Hüttenwesen des Oberharzes, 1895, p. 57.—U. SÖHLE. 'Beitrag zur Kenntnis der Erzlagerstätte des Rammelsberges,' Österr. Zeit. f. Berg- u. Hüttenwesen, 1899, p. 563. —L. BEUSHAUSEN. 'Das Devon des nördlichen Oberharzes u.s.w.' Abhandl. der pr. geol. Landesanst., 1800, N.F. Part 30.—A. BERGEAT. 'Über merkwürdige Einschlüsse im Kieslager des Rammelsberges bei Goslar,' Zeit. f. prakt. Geol., 1902, p. 289; Die Erzlagerstätten, 1st half, 1904, p. 329.—W. WIECHELT. 'Die Beziehungen des Rammelsberger Erzlagers zu seinen Nebengesteinen,' Berg- u. Hüttenm. Ztg., 1904, p. 285.—F. KLOCKMANN. 'Über den Einfluss der Metamorphose auf die mineralische Zusammensetzung der Kieslagerstätten,' Zeit. f. prakt. Geol., 1904, p. 153.—K. ANDRÉE. 'Über den Erhaltungszustand eines Goniatiten und einiger anderer Versteinerungen aus dem Banderze des Rammelsberger Kieslagers,' Zeit. f. prakt. Geol., 1908, p. 166.—B. BAUMGÄRTEL. 'Über Sphärosiderite in unmittelbarer Nachbarschaft des Rammelsberger Kieslagers,' Zentralbl. f. Mineralogie, Geologie, u. Paläontologie, 1909, p. 577.—W. LINDGREN and J. D. IRVING. 'The Origin of the Rammelsberg Ore-Deposit,' Econ. Geol., 1911, VI. p. 303.—SCHULZ. 'Beiträge zur Kenntnis der Kieslagerstätte des Rammelsberges,' MS. in Archiv der geol. Landesanstalt.

The pyrite deposit at Rammelsberg on the north-west border of the Harz is among the most interesting of known deposits, while among the pyrite deposits no other has been the subject of such a variety of explanation or so much study.

The general geological position has long been established. The Rammelsberg, as illustrated in Fig. 450, is an overturned air-anticline, the core of which consists of Lower Devonian Spirifer sandstone. The south-west overturned flank which encloses the bed is in the miner's foot-wall of this sandstone. It consists of Middle Devonian beds, first of the Calceola slate 3 m. in thickness, and then, geologically above this slate, of Wissenbach beds, the so-called Goslar slates. In these slates the ore-bed is intercalated. It is bounded chiefly by tectonic planes, of which the 'Wimmer Indicator' is the most characteristic. This indicator, which is an overthrust, is 0·5 m. thick and situated 2–3 m. from the deposit, in the miner's foot-wall—which owing to the overturning is the geological hanging-wall—following the deposit both in strike and dip.

The deposit in plan forms two ore-bodies, namely, the south-west or old bed and the north-east or new bed, these, representing the two transversely dislocated portions of one and the same deposit, being connected by scattered broken fragments.

The Goslar slates strike about north-east, are very poor in fossils, and dip at an angle of 45°–50° towards the south-east. Some years ago Goniatites were met in the ore-bed, while recently in the foot-wall slate a 0·5 m. bed containing *Pinacites Jugleri R.*, *Bactrites gracilis*, and *Orthoceras*, was found. Here and there quartzite beds are intercalated in the slates. One of these in the hanging-wall, exposed in the Julius-Fortunatus adit, carries sphaerosiderite concretions with frequent organic remains and veins of pyrite, quartz, and barite.

The separation between the ore-body and the country-rock is generally sharp. The old bed is often accompanied by a fine clay-parting, while

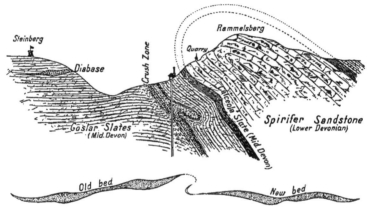

FIG. 450.—Section and plan showing the geological position of the Rammelsberg pyrite bed near Goslar. Klockmann.

the new bed, discovered only in 1859, exhibits in the hanging-wall a triturated slaty material of variable thickness, resembling lode-slate and known locally as *Anback*. This material is sharply separated from both ore-body and country-rock, and consists of deep-black slaty material with numerous pressure surfaces and slickensides. Since also it is traversed by many veins and fractures filled with gangue and ore-minerals, it bears all the evidences of being a disturbance zone. More complicated is the separation of the ore from the country-rock around the finger or tongue of pyrite which projects from the main body into the slates. In this branch body alternating layers of ore and slaty material, the so-called slate-band ore, occasionally occur, in which the thickness of the metalliferous layers varies. Sometimes these are microscopically fine, and sometimes several centimetres thick ; sometimes the ore predominates, and sometimes the slaty-material. This restriction of the banded ore to the branch body is striking; in this body it continues along the strike both in the foot-wall and hanging-wall, though it appears to be absent from the middle portion. In the old bed this banding in the branch is fairly regular; in the new, on the other hand, there is the most confused plication. Schulz from microscopical investigation considered that the metalliferous content of this banded ore was epigenetic and had in part replaced the slaty material.

The old bed reaches 20 metres or more in thickness, being thickest at the level of the Julius-Fortunatus adit. The new bed in the central portion is on an average 15 m. thick, though frequently it is much less. The deposit from end to end is known for 1200 m. along the strike. The old bed comes to the surface, while the new bed pinches out close under the surface, its thickness at places in the top levels being but small.

Of the numerous disturbances which have affected the deposit, one in the western part of the old bed, which in depth throws the ore-body into the foot-wall, is noteworthy. Numerous fractures within the ore-body are now filled with sulphides of copper, lead, zinc, antimony, and with pyrite. The ore-bed itself consists principally of pyrite, chalcopyrite, galena, sphalerite, and barite. The ore, always very fine-grained, often occurs highly contorted.

At Rammelsberg, as indicated in Fig. 451, the following ores are differentiated : pyritic ore, lead ore with a preponderating amount of pyrite ; mixed ore, lead ore with chalcopyrite and pyrite ; grey ore, lead ore with a preponderating amount of barite ; and brown ore, lead ore with a preponderating amount of sphalerite. The distribution of these ores has frequently been given upon the basis of the developments at particular dates. In the old bed it was believed that the following sequence from the present hanging-wall to the foot-wall was recognizable : copper ore, pyrite, mixed ore, lead ore, and grey ore. Schulz, who recently compiled a section on the basis of the latest developments, confirmed the occurrence of such a zonal arrangement in certain parts of the old bed ; while for the middle of that bed, at the horizon of the Julius-Fortunatus adit and the Bergesfahrt, he established from hanging-wall to foot-wall the sequence : pyrite, copper ore, pyrite, pyritic lead ore, mixed ore, and lead ore. Below the Julius-Fortunatus adit the pyritic layers intercalated between the mixed and copper ores pinch out, and the sequence from hanging-wall to foot-wall becomes, pyrite, copper ore, mixed ore, and lead ore. At the western end the old bed consists on the other hand exclusively of pyritic lead ore. At the eastern end also, the amount of pyrite considerably increases, while at the same time some barite found in the upper levels is in depth almost absent. The hanging-wall branch of the old bed, the occurrence of which is due to disturbance,

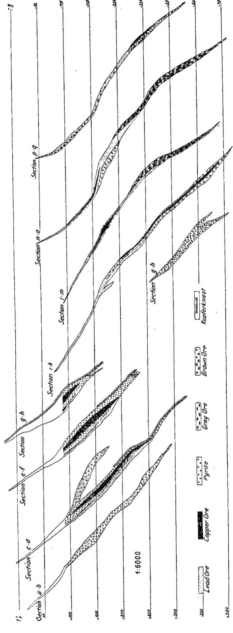

FIG. 451.—Plan and sections of the Rammelsberg ore-body. Schulz.

1147

consists in the upper western portion of pyritic lead ore, and in the east
of pyrite with traces of barite.

In the new bed no regularity of distribution can be established. The
filling of this much fractured and narrower ore-body consists chiefly of
lead-, copper-, and mixed ore. Pyrite recedes strikingly and occurs only
to the west, in small thickness. The occurrence of grey ore in the upper
portion and of brown ore in greater depth, is characteristic of this ore-bed.

The different varieties of ore are nowhere sharply separated, but
merge gradually into one another. They all consist of the same metalliferous minerals and only exist by reason of the varied proportions in
which these minerals occur.

COMPOSITION OF PRINCIPAL VARIETIES OF ORE

	No. I. Copper Ore.	No. II. Copper Ore.	Pyrite.	Mixed Ore.	Lead Ore.	Grey Ore.
	Per cent.	Per cent.	Per cent.	Per cent.	Per cent.	Per cent.
Au	0·00013	0·00015	0·00011	0·00015	0·00007	...
Ag	0·015	0·007	0·005	0·015	0·007	0·0017
Cu	18·32	10·10	2·24	4·45	0·65	0·51
Pb	4·73	2·31	2·10	10·69	11·94	15·30
Bi	0·12	0·11	0·12	Trace	Trace	...
As	0·10	0·11	0·15	0·06	0·04	...
Sb	0·12	0·12	0·08	0·13	0·10	...
Fe	24·56	33·55	35·94	13·65	16·24	4·87
Zn	9·75	4·50	4·50	20·25	18·00	2·16
Mn	1·34	1·22	1·42	1·59	2·61	0·39
Ni	0·06	0·07	0·07	0·06	0·05	...
Co	0·02	0·01	0·04	Trace	Trace	...
Al_2O_3	0·71	0·53	1·57	1·36	0·82	0·30
$CaCO_3$	1·79	4·22	5·90	4·09	6·79	0·65
$MgCO_3$	0·30	0·69	1·07	0·92	1·18	...
S	32·08	38·39	38·72	24·44	30·32	16·91
SiO_2	1·49	2·24	2·95	2·86	2·66	1·70
$BaSO_4$	2·93	0·62	0·52	13·82	6·74	55·04
Alkali	0·26	1·30	1·84	1·94	1·28	1·03
	98·70	100·08	99·22	100·33	99·50	98·86

The *Kupferkniest*, which occurs in the hanging-wall of the old bed, is
particularly interesting. This slate, according to Bergeat, is traversed
by a network of heavy-metal sulphides, including chalcopyrite, pyrite,
sphalerite, and galena, together with quartz, calcite, and barite, while
lenses and masses of pure chalcopyrite and pyrite are also often seen.
It occurs intercalated between the main body and the hanging-wall
branch, over an area about 200 m. long and of variable width, this at the
level of the Julius-Fortunatus adit reaching a maximum of 90 metres.
Along the dip this *Kniest* lies conformably with the bed ; at one place it
occurs actually in the bed. According to Krusch and other authorities
it has nothing to do with the primary formation of the bed.

The genesis of the Rammelsberg deposit is much disputed. Schuster in 1867 and others before him, ascribed a sedimentary origin to the deposit, a view which Wimmer, Stelzner, Klockmann, and Bergeat, later endorsed. The first to advocate an epigenetic origin were, Lossen in 1876 and Vogt in 1894, who emphasized the intrusive character, and maintained that the slates forming the country-rock of the deposit represented a facies formed in considerable depth, and that therefore a shallow-water deposition as assumed by Schuster and Klockmann was excluded. At the same time they pointed out that the hanging-wall and foot-wall boundaries of the deposit were formed by tectonic planes, and that in the place of a substantial conformity a gradual merging of the different varieties of ore might be observed along the strike. Finally, they considered that a genetic association between the deposit and the eruptive rocks occurring in its neighbourhood, might be assumed.

The adherents of the sedimentary theory, on the other hand, regard the hard arenaceous slates as well as the quartzites, as littoral formations, and explain the tectonic planes as having arisen after the formation of the bed ; they controvert also the relation between the neighbouring eruptives and the deposit. Klockmann, one of those best acquainted with the deposit, rejects any parallelism of the Rammelsberg bed with the Norwegian pyrite occurrences. In so far that barite, which occurs abundantly in the Rammelsberg ore, is absent from the Norwegian pyrite deposit, there is of course no parallelism. Klockmann regards the Rammelsberg deposit as contemporaneous with the surrounding country-rock, and as having been formed in a basin-like depression of the clay-slate. Evidence of this he sees in the fine slate bands intercalated in the ore like the growth rings in a tree. The heavy metals, according to him, were precipitated from solutions by one or other of the reduction processes.

The differences in the theories of genesis are in part due to the fact that formerly it was endeavoured to bring together into one class all the more important pyrite deposits. It having been demonstrated by Krusch that such a classification is untenable, pyrite can no longer be given an admittedly unjustifiable preference among metalliferous minerals, but the deposits which it forms, like deposits of all other minerals, must according to their genesis be divided into magmatic, contact-metamorphic, lode-like, metasomatic, and bed-like deposits.

Latterly, Bergeat and Bode have found Goniatites in the ore-bed, a discovery which affords considerable support to Klockmann's theory. In our opinion the Rammelsberg deposit, the original geological position of which owing to numerous disturbances is difficult to determine, can only be either a sediment or a replacement of Middle Devonian limestone. Since however for the latter view sufficient evidence is not forthcoming, it appears proper to Beyschlag and Krusch, in the present state of our knowledge, to regard the Rammelsberg ore-bed as a sedimentary deposit.[1]

Mining at Rammelsberg is very old, dating back at least to the year 972. According to Bergeat, the oldest workings of the upper levels and the old dumps date from the hoary past. The establishment of the town Goslar, a residence of the king of Saxony, coincided with the period of greatest activity. After numerous interruptions, some of which lasted for centuries, in 1635 and 1642 upon the establishment of the so-called communion—that is, the joint possession and management by Hanover and Brunswick, into which Prussia subsequently entered—the present industrial period began. Germany is poor in pyrite- and copper deposits, Rammelsberg being its most important pyrite occurrence. The economic importance of this deposit, the mines upon which have so far penetrated to a depth of roughly about 400 m., may be gathered from the following figures: in 1909 the production amounted to 22,467 tons of copper- and mixed ores, against 27,600 tons in the preceding year. Of this amount the Prussian portion was 12,838 tons—15,772 in the preceding year—worth £15,500. The total output had therefore a value of £27,125.

OTHER GERMAN PYRITE DEPOSITS, CONCRETION- AND ALUM-SLATE DEPOSITS, ETC.

LITERATURE

W. BRUHNS, in collaboration with H. BÜCKING. Die nutzbaren Mineralien und Gebirgsarten im Deutschen Reiche. Berlin, 1908.—A. SACHS. Die Bodenschätze Schlesiens. Leipzig, 1906.

At Misdroy small pyrite layers and veins in Senonian marls are found along the north of the Island of Wollin; these were formerly worked. The deposit at Rohnau in Silesia south of Kupferberg, where thick pyrite beds occur in crystalline schists and are won in opencut, is of some importance. From these beds are obtained sulphur, copperas, and iron oxide for red pigment. Mining operations began in the eighteenth century, were suspended in 1891, and revived again in 1904. In the Fichtelgebirge, pyrite, together with pyrrhotite and chalcopyrite, occurs as beds in gneiss, at Weirsberg in the neighbourhood of Kupferberg near Kulmbach. The ore is treated for copperas, mixed vitriol, and alum, while glass polishers' red is obtained from the residual material.

Closely associated with the pyrite beds are the alum-slate—vitriol slate—and alum-shale occurrences. Frequently the distribution of the pyrite in

[1] The evidence brought forward in support of an epigenetic genesis by Lindgren and Irving in their work above cited throws no new light upon the question, while according to recent investigation by Erdmannsdörfer, the results of which have not yet been published, the observations of these two authorities are disputable.

these rocks is so fine that a mechanical separation of the pyrite from the non-metalliferous material is not possible. The production of alum from such slates and shales reached its greatest development at the end of the seventeenth and in the first half of the eighteenth century. While the alum-slates, as already mentioned, occur most extensively in the Silurian and Culm, most alum-shales are found in the Brown Coal formation. Such slates have been worked in many places in the Thuringian Forest, as for instance at Sophienau, Garnsdorf, Wetzelstein, Arnsbach, Schmiedefeld, and Spechtsbrunn, while in Westphalia extensive operations were centred upon the alum-slate belt which to the south extends along the coal-bearing Upper Carboniferous in an easterly and north-easterly direction. Pyritic Upper Carboniferous and Permian slates also have been exploited at Dudweiler in Saarbrücken and at Kirn in Kreuznach. Alum- and vitriol slates are also known in the clayey coal of Lorraine, Würtemberg, and Bavaria.

The Brown Coal shales formerly were likewise largely used for the production of alum and copperas, and were in part even more important than the coal seams interbedded with them. Of these shales the following are worthy of mention : the brown coal with abundant pyrite at Buxweiler on the eastern slope of the Vosges, which was chiefly used for the production of alum and copperas and only to a small extent for fuel ; similar alum works formerly in operation at Grossalmerode in Hesse and at Riestedt in Thuringia ; shales containing large pyrite nodules formerly also worked for copperas at Rott. The number of these examples might be considerably increased. Among the youngest deposits, possibly still undergoing formation, belong the vitriol peats, which likewise may be used in the preparation of copperas. The peat is either impregnated by pyrite or it contains free sulphuric acid. The beds at Trossin in Torgau, Moschwig in Wittenberg, and those at Kamnig, Seifersdorf, Reichmannsdorf, Striegen-dorf in Grotkau, etc., have considerable extent. These deposits however no longer have any economic importance.

WITWATERSRAND GROUP

The Auriferous Conglomerates

To this group, named after the principal occurrence, belong the auriferous conglomerates which in fairly regular development may extend over large areas. On the Witwatersrand many such auriferous beds lie one above the other, separated from one another by barren material of variable thickness.

Only in exceptional cases do such conglomerates lie horizontally ;

generally they are folded mostly into anticlines and synclines, and disturbed by faults, so that in any district the determination of the tectonics is of great importance.

Genetically, the auriferous conglomerates of different districts are very different deposits. When the conglomerates have resulted from the disintegration of gold lodes the detritus of which after concentration by natural waters became subsequently cemented, such conglomerates represent fossil gravels. The gold occurs then principally in rounded nuggets and grains, while the larger fragments as well as the cementing material are mostly free from gold. In other cases the gold is mostly confined to the cementing material; such gold may have been precipitated during the formation of the conglomerate, or it may have subsequently reached the conglomerate through fissures.

The mechanical concentration of material subsequently cemented to conglomerate took place either in rivers—fluviatile conglomerates—or on the sea-coast—marine conglomerates. In the first case, long extending river terraces distinguishable from ordinary gravels more particularly by the regular presence of a cementing material were formed; while in the latter case terraces extending usually over large areas of mostly irregular form, resulted.

The distribution of the gold may vary in vertical section as well as in horizontal extent. In those cases, as with fossil gravels, where the precious metal occurs within pebbles or as small nuggets or grains, it is frequently concentrated in the lowest bed, so that a lower auriferous primary zone may be distinguished from an upper poorer zone. Should the gold have secondarily entered the conglomerate from fissures, such primary depth - zones are not observable, but an irregular horizontal distribution occurs, in so far that the gold content is frequently highest in the neighbourhood of such fissures, from whence it gradually diminishes on either side.

The gold conglomerates with syngenetic gold were formerly regarded as affording typical examples of uniformity in gold content. The possibility of poorer and richer portions alternating from place to place nevertheless exists even with these deposits. It is advisable in all cases therefore, be the origin of the gold what it may, not to sink new shafts and erect equipment on the results obtained from a few bore-holes, but before doing this, to test the conglomerate for its average gold content by a great number of such holes, etc.

With conglomerates carrying gold in the form of nuggets or grains the precious metal occurs as free- gold. With those, however, where impregnation took place from fissures it is found mostly in the form of auriferous pyrite and only quite subordinately as free gold. A similar phenomenon is sometimes exhibited when the gold has been deposited syngenetically with the cementing material. When such gold conglomerates come to the surface the gold associated with pyrite, just as happens with lodes, becomes dissolved by atmospheric agencies. In this manner, at surface an oxidation zone with but little free gold, and, not far below, a cementation zone with somewhat more free gold, may become formed. Extensive metal migration however does not as a rule take place, since the large amount of barren pebbles in the conglomerate, and the solidity of the cement, greatly hinder the circulation of water. The gold of these deposits is as a rule very pure ; in relation to the amount of silver accompanying it, it resembles that of the old gold lodes.

The economic importance of the Witwatersrand deposits, as will be gathered from the table of production already given,[1] is very considerable ; this district, to-day doubtless the most important gold producer of the world, was responsible in 1911 for about 35 per cent of the world's output.

Of this form of deposit, one important representative is known, namely :

The Conglomerates on the Witwatersrand in the Transvaal

LITERATURE

E. Cohen. 'Die goldführenden Konglomerate in Südafrika,' Mitt. des naturw. Vereins f. Neuvorpommern u. Rügen, 1887, Neues Jahrb. f. Min., 1887, B.-B. V.—W. H. Penning. Quart. Journ. Geol. Soc., London, 1885, XLI. ; and 1891, XLVII.—A. Schenk. 'Über das Vorkommen des Goldes in Transvaal,' Zeit. d. d. geol. Ges., 1889, Vol. XLI. p. 573.—W. Gibson. 'The Geology of the Gold-bearing and Associated Rocks of the Southern Transvaal,' Quart. Journ., 1892, XLVIII. p. 420.—L. de Launay. 'Les Mines d'or du Transvaal,' Ann. des mines. Paris, Jan. 1886; Jan.-Feb. 1891.—G. A. F. Molengraaff. 'Beitrag zur Geologie der Umgegend der Goldfelder auf dem Hoogeveld, u.s.w.,' Neues Jahrb. f. Min. u.s.w., 1894, B.-B. IX.—K. Schmeisser.—Über das Vorkommen und die Gewinnung der nutzbaren Mineralien in der südafrikanischen Republik. Berlin, 1894.—A. Pelikan. 'Über die goldführenden Quarzkonglomerate von Witwatersrand,' Verhandlungen d. k. k. geol. Reichsanst., Vienna, 1894, p. 421.—J. Kuntz. 'The Rand Conglomerates, how they are formed,' Trans. Geol. Soc. of South Africa, 1896, p. 118.—F. H. Hatch and J. A. Chalmers. The Gold Mines of the Rand. London, 1895. —G. F. Becker. 'The Witwatersrand Banket, with Notes on other Gold-bearing Pudding Stones,' U.S. Geol. Survey, XVIII., Ann. Rep., 1896-1897, Part 5.—G. A. F. Molengraaff. Ann. Rep. of the State Geologist of the South African Rep. for 1897. Johannesburg, 1898.— S. J. Truscott. The Witwatersrand Goldfields, Banket, and Mining Practice. London, 1898.—G. A. F. Molengraaff. 'Die Reihenfolge' und Korrelation der geologischen Formationen in Südafrika,' Neues Jahrb. f. Min. u.s.w., 1900, Vol. I. p. 113 ; Géologie de la République Sudafricaine du Transvaal, Paris, 1901 ; Geol. of Transvaal, 1904 ; numerous other papers by Molengraaff, especially in Trans. Geol. Soc. of South Africa up to the year 1905.—L. de Launay. Les Richesses minérales de l'Afrique, Paris, 1903, pp. 42-85.— J. W. Gregory. 'The Origin of the Gold in the Rand Banket,' Bull. Inst. of Min. and Met., London, 1907 ; 'The Origin of the Gold of the Rand Goldfield,' Econ. Geol., 1909, IV.— F. W. Voit. 'Übersicht über die nutzbaren Lagerstätten Südafrikas,' Zeit. f. prakt. Geol., 1908, pp. 137, 191 ; 'Der Ursprung des Goldes in den Randkonglomeraten,' Zeit.

---

[1] *Ante*, pp. 644-646.

d. d. geol. Ges., 1908, Vol. LX. Monthly report No. 5, p. 107, and No. 7, p. 181.—J. KUNTZ. ' Die Herkunft des Goldes in den Konglomeraten des Witwatersrandes,' Zeit. d. d. geol. Ges., 1908, Vol. LX. Monthly report No. 7, p. 172.—F. H. HATCH and G. S. CORSTORPHINE. The Geology of South Africa. London, 1909.—Also numerous works by CORSTORPHINE, especially in the later Trans. Geol. Soc. of South Africa, and by HATCH.—LEGGETT and HATCH. ' An Estimate of the Gold Production and Life of the Main Reef Series, Witwatersrand, down to 6000 feet,' Trans. Inst. Min. and Metallurgy, 1902–1903, XII.—HATCH. ' A Geological Survey of the Witwatersrand,' Quart. Journ., London, 1898, LIV. ; 'The Auriferous Conglomerates of the Witwatersrand,' Mining and Scientific Press, San Francisco and London, 1911 ; 'On the Past, Present, and Future of the Gold Mining Industry of the Witwatersrand,' The James Forrest Lecture, Inst. of Civil Engineers, London, 1911.

The oldest formation in South Africa is gneiss, upon which highly crystalline schists with intrusive granitic rocks are bedded. Immediately upon the crystalline schists, probably unconformably, follows the Witwatersrand formation which, particularly in the Rand district, is extensively developed and divided into a lower and an upper division. While in the lower division, in addition to quartzites, ferruginous schists occur abundantly, the upper division consists chiefly of quartzites and conglomerates, with only one single slate bed, known as the Kimberley Reef shale. These two divisions lie conformably to each other, the conglomerates of the Main Reef Series forming the base of the upper division. In general the thickness of the beds decreases towards the east.

While in the centre of the Rand the above-described normal section with the Main Reef conglomerates as the lowest obtains, in the west a number of other conglomerate beds occur still lower. In the Central Rand the horizontal thickness of the Witwatersrand formation is 24,000 feet ; to the west, at Klerksdorp for instance, it is substantially greater ; while to the east it is decidedly less. The beds, as illustrated in Fig. 452, form an extensive, approximately east-west syncline, the outline of which, particularly to the south, is disturbed by much detailed folding.

After the folding and tilting of the Witwatersrand formation followed a period of eruption, to which extensive occurrences of volcanic rock owe their existence. These constitute the Vaal River formation which, apart from relatively small thicknesses of quartzite, conglomerate, and coarse sandstone at its lower and upper limits, consists principally of eruptive rocks. Upon the Vaal River formation comes the Lydenburg formation, at the base of which is found the Black Reef ; and then the Cape formation, which in its turn is overlaid unconformably by the Karroo formation containing coal seams.

According to Voit, gold-bearing conglomerates and quartzites also occur in formations younger than the Witwatersrand beds, the best known of these being the Black Reef, and the Du Preez Series near Rietfontein.

Concerning geological age, that of the Witwatersrand formation has not yet been satisfactorily determined. While the crystalline rocks are Archaean, the Witwatersrand beds belong perhaps to the Algonkian, the Lydenburg formation possibly to the Silurian, and the Cape formation possibly to the Devonian.

With regard to gold content, only the conglomerates known as 'banket reefs' come into consideration. These in general strike east-west and at the surface dip at an angle of 60°–85° to the south, this dip with greater depth becoming flatter. The banket or conglomerate consists of well-rounded pebbles varying in size from a hazel-nut to a hen's egg, and united by a siliceous cement. The rock separating the different banket beds consists of quartzite or quartzitic sandstone. The gold occurs in the cement and is associated almost exclusively with pyrite. In the neighbourhood of the surface this pyrite is decomposed to limonite in which the precious metal occurs as free gold.

Almost the whole of the gold production is obtained from the bed known as the Main Reef Leader and another situated on an average about 30 m. in the hanging-wall and known as the South Reef. At Johannesburg, including the Black Reef conglomerate, which as stated above lies immediately at the base of the Lydenburg formation, seven such banket series are distinguished. Of these the most important is the Main Reef Series containing five thick conglomerates, namely, the South Reef, on an average 2½ feet thick; the Middle Reef, 4 feet; the Main Reef Leader, 1½ feet; the Main Reef, 10 feet; and the North Reef, about 2 feet thick. According to Schmeisser the following sequence obtains from the east to the centre of the Rand:

Du Preez Reef Series.	Kimberley Reef Series.
Main Reef Series.	Klippoortje Reef Series.
Livingstone Reef Series.	Elsburg Reef Series.
Bird Reef Series.	Black Reef Series.

The Main Reef Series is known for more than 80 km. along the strike. From the outcrop the mines become deeper as they go farther south, those at the outcrop being described as the outcrop mines, and those to the south as the deep-level mines. Among the latter, some in 1911 had already reached a depth of 1200 m., while others have been projected to a depth of 1800 metres.

The gold content of the Witwatersrand conglomerates varies considerably at different levels. In the neighbourhood of the surface, at a depth of about 30–70 m., it is sometimes very high, such richness being due to cementation; the Main Reef Leader where first worked yielded at times as much as 6 ounces per ton, while from the Jumpers, specimens from the neighbourhood of the outcrop were so encrusted with fine gold that assays of 600 ounces per ton were obtained. This cementation gold is completely crystalline, showing under the microscope well-defined

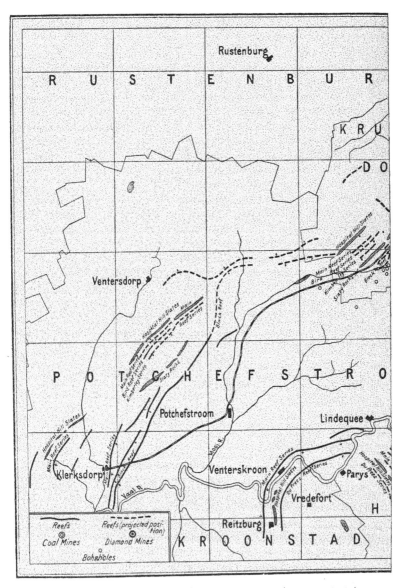

FIG. 452.—Map of the Witwatersrand showing the outcrop of the banket reefs and

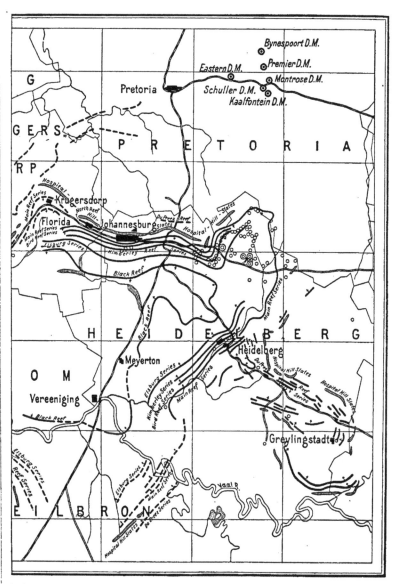

that of the principal slate bed, the Hospital Hill Slates. Mercer, Nicolaus & Co.

crystal faces and sharp edges. In the primary zone the gold content is exceedingly variable, not only in different mines but also on the different levels of one and the same mine. A number of average values are given later.

Almost the entire production comes to-day from the Main Reef Series, the Black Reef and other series contributing but little. One of the most productive mines is the Robinson, which in the year 1895 treated 140,655 tons, obtained from the different reefs in the following proportions :

From the Main Reef	.	.	.	.	36·92 per cent.
„ Main Reef Leader	.	.	.	30·86	„
„ South Reef	.	.	.	32·22	„

From this tonnage, according to the manager's report, 120,113 ounces of gold were won by amalgamation, 14,938 ounces by concentration, and 22,157 ounces by cyanidation, making a total of 157,208 ounces. This gives an average of somewhat more than 19 dwt. per ton by amalgamation and 3 dwt. by cyanidation, or together 22 dwt. 7 grm. The cost of mining and treatment at that time amounted to 19s. 2d. per ton.

In the Heidelberg and Nigel districts, banket reefs, the geological position of which has not yet been completely determined, are also worked. In 1894 the production of these districts amounted to about 52,500 oz. and in 1895 to 43,600 oz. of gold.

With increasing improvement in the mining and metallurgical equipments successively poorer ores have become worked, as the following statistics of average yield per ton of ore crushed indicate :

1890–1895	.	.	.	46·5 shillings	= 17	grm.
1896–1899	.	.	.	40·6 „	= 14·9	„
1902–1904	.	.	.	40·1 „	= 14·7	„
1905–1908	.	.	.	34·1 „	= 12·5	„
1909	.	.	.	29·1 „	= 10·7	„
1910	.	.	.	28·6 „	= 10·5	„

This great decrease is due chiefly, or perhaps exclusively, to the fact that the present large scale of operations and the modern equipments permit the mining of poorer ore than formerly was possible. Owing to the irregular distribution of the gold in the conglomerate it has not yet been possible to decide with certainty whether or not the deposits have on the whole become poorer in depth. The sum of numerous observations would however point to a small decrease of the average gold content in depth.

With the question of the genesis of the gold many authorities have concerned themselves. The most recent comprehensive works on this subject are by J. W. Gregory in 1907, and F. H. Hatch in 1911. There is no doubt about the sedimentary origin of the conglomerate. Although for a time there was divergence of opinion as to whether it was a fluviatile or littoral deposition, all authorities now regard the conglomerate as a littoral formation. It was natural therefore to regard the gold [content equally as sedimentary, which is the case for instance with the Cambrian gravel-deposits of Dakota. This was the view held by Cohen and Ballot in 1887 and later, though still in the early days of the field, by authorities such as Suess, Pošepný, Schenk, Fuchs, Halse, Pelikan, Gibson, Goldmann, Bleloch, and G. A. Denny, in addition to many practising mining engineers.

A multitude of facts however speak against this theory of sedimentation. Were it a mechanical sedimentation the petrographical character of the Rand conglomerate, excluding its condition of consolidation, should correspond with that of any other fluviatile or marine gravel, that is to say, the gold should be in the form of small, more or less indented water-worn grains, particularly in the lowest portion of the conglomerate, while auriferous quartz pebbles should also occasionally be met, and here and there larger nuggets. Such characteristics have hitherto however not been observed. A considerable difference also exists between the gold content of the Rand conglomerate and that of ordinary auriferous gravels, in that the Rand conglomerate is much richer than such gravels usually are. There is accordingly no justification for regarding this conglomerate as an ordinary fossil gravel.

Some authorities, as for instance E. Cohen, F. Pošepný, G. F. Becker, and J. W. Gregory, have endeavoured to remove the difficulties to the acceptance of these deposits as fossil gravels by assuming a subsequent chemical re-arrangement of the former gravel gold. According to their view this is an auriferous marine gravel the material of which was derived from gold-quartz lodes ; they consider that originally this gravel contained small grains of gold which subsequently by the action of solutions became dissolved, the gold being re-deposited in its present form. Gregory, who in 1907 concerned himself greatly with this question, considered the gold of the Witwatersrand banket to be marine gravel gold, and the pyrite to be altered magnetite. These modified gravel theories accordingly represent the Witwatersrand conglomerates as an analogous formation to the Cape Nome gravels.

Krusch, after examination of material collected by H. Weber from the last-mentioned gravels, gives the following substantial difference between the Witwatersrand auriferous conglomerate and ordinary auriferous gravel : ' In recent marine gravels, as already indicated by Beck in his *Lagerstättenlehre*, fine quartz grains predominate, while in the Witwatersrand conglomerate pebbles predominate. Although in both cases large nuggets are absent, the form of the gold in the Witwatersrand conglomerate is essentially different from that of Cape Nome. Observed with the

naked eye the gold particles of marine sediments are not unlike
crystal aggregates ; under the microscope however they prove to be
skeletons of former nuggets, highly corroded by the action of sea-water,
and scored in all directions by corrosion channels.   In auriferous gravels
therefore sea-water has principally dissolved and not deposited gold.'

If the gold in the Witwatersrand conglomerate be regarded as the
remnants of former nuggets, the relatively high gold content of the conglomerate is very striking.   Gregory's hypothesis assumes that between
the pebbles only the heavy material such as gold and magnetite remained,
while the lighter sandy material was washed away ; the assumption of the
alteration of magnetite to pyrite presents, however, great difficulties.   In
addition, there are other factors, mentioned below, which cannot be
brought into harmony with the alteration of a primary gravel.

A third school, including principally Penning, Hamilton, de Launay,
Stelzner, and F. W. Voit, in dealing with the origin of the gold in the Witwatersrand conglomerate,[1] advocates the theory of precipitation, according
to which the gold is a sedimentary chemical deposit.   Voit considers
that the gold in hot ascending solution reached the surface along the
littoral where the conglomerate was being formed, when together with the
pyrite it became precipitated.   He bases his view upon the occurrence of
the pyritic band illustrated in Fig. 4.   This precipitation theory also is
not free from objection.   Since the normal gold content of sea-water
is not sufficient in itself to explain the high content of the conglomerate,
auriferous springs must be assumed to have assisted.   Did such springs occur
however, it is remarkable that they were active only during the formation
of the conglomerates, and not—or only to quite a small extent—during the
sedimentation of the intercalated quartzites, etc.   If the co-operation of
springs be not assumed but the gold content of the sea-water be regarded
as having been sufficient, a further difficulty lies in the fact that in the
laboratory it is not possible with the agents present to precipitate gold
from such dilute solutions ; very energetic assistance is necessary to bring
this about.   Voit believes precipitation to have been promoted by surf
action and by comminuted organic substances, and points to the carbon
content of the beds.   This carbon however, though frequently observable, is in many mines completely absent.   Furthermore, with regard
to the pyrite, Gregory considers this as derived from magnetite, while
authorities generally insist upon its secondary character.

Most authorities, as for instance, E. Dorsey, A. R. Sawyer, Gardner
F. Williams, J. S. Curtis, J. Kuntz, P. Krause, J. Hays Hammond, F. H.
Hatch, Beck, and others, assume the subsequent impregnation of the
conglomerate.   According to this theory, the conglomerates were already

---

[1] Monthly Report of the *Deutsche geol. Ges.*, 1908, Vol. XL. No. 5.

consolidated at the entry of gold-bearing solutions representing the after-effects of the extrusion of the extensively occurring diabase.

The generally well - rounded quartz pebbles lie in a matrix which consisted originally almost exclusively of loose quartz sand, but which subsequently became cemented by secondarily infiltrated quartz. In addition to the quartz pebbles and sand, according to Hatch—the foremost authority on the geology of the Rand—only zircon in microscopic crystals, and as rareties, chromite and iridosmium in rounded grains, occur as primary constituents. Of secondary minerals, there are on the other hand a large number, namely, secondary quartz, chloritoid, chlorite, sericite, calcite, tourmaline, rutile, pyrite, marcasite, pyrrhotite, chalco-pyrite, sphalerite, galena, stibnite, cobalt- and nickel arsenides, graphite, gold telluride—admittedly only as a great mineralogical rarity—and gold.

In addition to silicification, a pyritization with an increment of about 3 per cent of pyrite principally in well-defined crystals, also took place. Pyrite concretions with radial and sometimes with concentric structure, also occur, while here and there quartz pebbles and quartz sand have been metasomatically replaced by pyrite. The secondary character of the pyrite is seen from the fact that the primary zircon crystals, and sometimes even the flakes of secondary chloritoid, are encrusted with pyrite. In other respects also, metasomatic processes have played an important part. For instance, pebbles and grains of quartz are found more or less altered to calcite, while pseudomorphs of quartz after chloritoid likewise occur.

Graphite occurs abundantly in the conglomerate of some mines, but is completely absent from that of others. Formerly, this graphite was explained by the alteration of organic substances deposited simultaneously with the conglomerate. According to C. R. Young[1] however, the graphite also is a younger formation, since it has occasionally replaced quartz.

Between the gold and the pyrite a very close association exists. Under the microscope, according to Hatch, it is seen that minute crystalline gold particles have grown upon pyrite crystals. Gold is also found along cracks between the aggregates of pyrite crystals, and as films around globular pyritic concretions. Where graphite occurs a close association between it and the precious metal may be observed in that gold occurs coating the graphite. Gold is also found in small veins of secondary quartz uniting broken quartz pebbles. Pyrite and graphite have accordingly occasionally precipitated gold. Further, a connection between the 30-40 m. wide diabase dykes which cut through the conglomerate in great number, and the average gold content of the conglomerate, may frequently be observed.

[1] *Trans. Geol. Soc. South Africa*, 1911, XIII. p. 65.

Often the gold content increases on both sides of such a dyke ; in other places a dyke forms the boundary between a rich and a poor patch, a fact indicating that the diabase intrusion cannot be older than the precipitation of the gold.

Since in many places in South Africa gold-quartz lodes occur in association with metasomatic deposits, as for instance in the De Kaap and Pilgrim's Rest goldfields, the possibility of the occurrence of mineral-bearing fissures is proved. Kuntz mentions similar quartz lodes on the West Rand, and even quotes diabase dykes as being auriferous.

This theory also explains the fact that the gold conglomerates occur at different horizons up to the Black Reef series, and are not confined to one horizon. The limitation of the gold content to the conglomerate beds—with the exception of the quartzite and quartzitic sandstone partings between and in immediate contact with the conglomerate—is explained by the greater interstitial space and permeability of the conglomerates.

In this connection also, the determination by Kuntz that the gold content of the conglomerate is particularly high along many diabase dykes, and that a connection exists between the degree of the inclination of the beds and their gold content, is significant. In many cases—according to Voit there are however quite a number of exceptions—the steep parts, that is, those more tilted and consequently disturbed, are richer in gold than those in greater depth and lying flatter.

Hatch gives the following summary of the different stages in formation :

1. Classification and sedimentation of the coarse and fine material, arising from the denudation of the surface of the old Swaziland formation—Archaean.

2. Consolidation of the quartz pebbles and sands under an increasing thickness of new sediments, to conglomerates and quartzites, through cementation by silica—quartz.

3. Subsidence of the earlier sediments to a considerable depth under younger formations. The formation of chloritoid from clayey material probably took place at this stage.

4. (a) Tilting of the beds with simultaneous folding and dismemberment by faults.

(b) Injection of the eruptive magma—diabase dykes—into fissures, together with the formation of pyrite, deposition of graphite, and precipitation of gold from thermal magmatic solutions given off at the consolidation of the eruptive intrusions.

5. Metasomatic replacement of allogenic as well as authigenic quartz by pyrite and calcite, and of chloritoid by quartz ; cementation of the small cracks by quartz ; these processes being accompanied by renewed precipitation of gold.

6. Decomposition of pyrite within the oxidation zone whereby the gold mechanically enclosed in the pyrite became liberated, and a simultaneous concentration of gold by the usual secondary enrichment processes.

History.—The Rand auriferous conglomerates in the southern Transvaal, 1800 m. above sea-level, have been known since 1884, in which year the brothers Struben—after Arnold in the previous year had discovered traces of gold in the district without finding a payable deposit—were successful in finding on the farm Wilge Spruit and soon afterwards also in other places south of the Witwatersrand, a number of auriferous beds.

In the year 1885 the first five-stamp battery was erected, from which in a trial crushing from the Bantjes mine, gold at the rate of 44 grm. per ton was recovered.   The conglomerate beds were then traced and followed with such energy that during the year 1896 nine farms were declared as public goldfields.   Then followed discovery after discovery, till in a short time the occurrence of gold-bearing conglomerates as far as Klerksdorp was established.   When at a depth of 100–300 feet the oxidation zone with its free-milling gold was bottomed and the primary zone entered, the disappointment of only being able to recover some 60 per cent of the gold by amalgamation was experienced, and work proceeded with great loss of gold.   This difficulty however was surmounted in the year 1890 by the introduction of the cyanide process.   This process, which for the pyritic or refractory ore is of vital importance, consists in treating the tailings after amalgamation with cyanide, whereby the gold is taken into solution and subsequently precipitated.   By this means generally 20–30 per cent of the total gold content, equivalent to 70–80 per cent of the content of the tailings, is won.   The total extraction by amalgamation and cyanidation is now 93–96 per cent, only about 4–7 per cent being unrecovered.

Another important event in the development of the industry was the connection of the Transvaal railways with those of Cape Colony on October 1, 1892, this connection taking place first over Kimberley, a distance of 480 km., and then over Vereenigung on the Vaal River, a distance of 67 kilometres.

A further favourable factor in the development was the availability of coal seams in the Witwatersrand syncline, which, though known long before the discovery of the gold, had, owing to the sparseness of the population, till then not been exploited.

Production.—The following table from Hatch gives the production of the Witwatersrand mines up to 1911 :

[TABLE

		Ore crushed.	Value of Gold recovered.	Value of Gold recovered per Ton.
		Tons.	£	Shillings.
	1887	No exact figures	81,045	
	1888	available, some	729,715	42·2 (?)
	1889	1 million	1,300,514	
	1890	702,828	1,735,491	49·4
	1891	1,175,465	2,556,328	43·4
	1892	1,921,260	4,297,610	44·7
	1893	2,215,413	5,187,206	46·8
	1894	2,827,365	6,963,100	49·2
	1895	3,456,575	7,840,779	45·2
	1896	4,011,697	7,864,341	39·2
	1897	5,325,355	10,853,616	39·7
	1898	7,331,446	15,141,376	41·3
War period	1899	6,639,355	14,093,363	42·4
	1900	about 1,000,000	2,484,241	49·7
	1901	412,006	1,014,687	49·2
	1902	3,416,813	7,179,074	42·0
	1903	6,105,016	12,146,307	39·8
	1904	8,058,295	15,539,219	38·5
	1905	11,160,422	19,991,658	35·8
	1906	13,571,554	23,615,400	34·8
	1907	15,523,229	26,421,837	34·0
	1908	18,196,589	28,810,393	31·6
	1909	20,543,759	29,900,359	29·1
	1910	21,432,541	30,703,912	28·6
Total	.	156,026,983	276,181,571	35·4
	1911	...	34,991,620	...

To this table must be added the production from the outside districts of the Transvaal, namely, from the gold-quartz lodes and the Lydenburg deposits,[1] which together in 1910 and 1911 were responsible for gold to the value of £1,297,823 and £1,448,141 respectively.

Of the total value of the Witwatersrand production up to the end of the year 1910, namely, some £276,000,000, £72,416,555, equivalent to 9s. 4d. per ton, was paid in dividends. In 1910, when the value of the gold recovered per ton of ore crushed was 28s. 6d., the average cost of production on fifty-six mines was 17s. 7d., leaving an average profit of 10s. 11d. per ton. In the case of the larger companies, as indicated in the following table after Hatch, work is carried on still more cheaply :

AVERAGE RECOVERY, COST, AND PROFIT PER TON CRUSHED OF SEVERAL WELL-KNOWN WITWATERSRAND MINES DURING THE FIRST QUARTER OF 1911

	Treated per Month.	Recovery per Ton.	Cost per Ton.	Profit per Ton.
	Tons.	Dwt.	Shillings.	Shillings.
Robinson	48,900	10·9	13·9	31·9
New Primrose	24,740	7·0	14·8	14·9
Simmer and Jack East . .	68,530	6·3	12·1	14·4
East Rand Prop. . . .	182,170	5·9	15·2	9·6
Witwatersrand	34,860	5·0	12·6	8·4
Knights Deep	57,820	4·7	11·7	8·4
Glencairn	18,520	3·7	13·7	1·8

[1] Ante, p. 638.

The differences in the value recovered per ton crushed in different mines, are noteworthy. In the case of some large mines the cost of production, including administration, has been reduced to as low as 13s. 8d. per ton, equivalent to $3\frac{1}{4}$ dwt or 5 grm., so that in such mines even with 5 dwt. or 7·8 grm. recovered, a profit of 6s. 6d. per ton is obtained.

In the year 1910, in thirty-six mines, altogether 81,000,000 tons of ore assaying 6·3 dwt. were developed.

The mines now generally work at a depth of 1000–2500 feet, though some have even reached 4000 feet. At a depth of 1000 feet the average rock temperature is 68·7° F. or 29·4° C., representing an increase of only about 1° F. per 208 feet of depth, or 1° C. per 374 feet, a rate of increase which is considerably lower than is generally the case elsewhere. At a depth of 7000 feet therefore a temperature of only 97·5° F., or 34·4° C., is to be expected. At this rate and with good ventilation, work may well be continued to a depth of 7000 or even 8000 feet.

Down to a depth of 6000 feet Hatch in 1911 estimated the amount of gold still to be won at £1,046,000,000, or sufficient for an annual production of £30,000,000 for a period of thirty-five years. In the next few years the production will undoubtedly exceed such an annual figure, but against this factor tending to diminish the life, must be set the probability of mining operations continuing deeper than 6000 feet. The Witwatersrand may accordingly play a dominating part for yet another generation, after which it will doubtless decline.

The gold industry of the Witwatersrand in April 1911 employed 25,000 Europeans and 194,000 natives, together therefore 220,000 employees. Chinese are no longer employed.

OTHER AFRICAN GOLD CONGLOMERATES

LITERATURE

A. R. SAWYER. 'The Tarkwa Goldfields,' Trans. Inst. Min. Eng. London, 1902.— S. J. TRUSCOTT. The Witwatersrand Goldfields, 2nd Ed., 1902, p. 487.—R. BECK. 'Einige Bemerkungen über afrikanische Erzlagerstätten,' Zeit. f. prakt. Geol., 1906, p. 208 ; Lehre von den Erzlagerstätten, Vol. II. Berlin, 1909.

1. *Mashonaland, Rhodesia*

According to Beck, the Ayrshire mine in the Lomagunda district, though working chiefly an auriferous hornblende-gneiss, discovered in the year 1904 upon a range some 15 km. farther south, a valuable conglomerate striking parallel with the Ayrshire bed and capable of being followed for a length of 4·5 kilometres.

This conglomerate consists of pebbles of granite, quartzite, and other crystalline rocks, in an abundant cement possessing the character of a quartz-biotite-amphibolite; where quartz predominates the tessellated structure typical of regional-metamorphosed sediments is exhibited. In addition to free gold, magnetite and pyrite are found.

The conglomerate bed in the Eldorado mine on the Hunyani is 1·8 m. thick, and in the upper levels contains on an average 24 grm. of gold per ton. Richness of this conglomerate immediately above the calcareous schists forming the foot-wall is characteristic. Beck, however, expressly points out that petrographically this conglomerate bed in no way resembles that of the Witwatersrand. From descriptions it is probably a fossil gold gravel.

2. *The Tarkwa Conglomerate in the Wassau District, Gold Coast*

This conglomerate, consisting chiefly of pebbles of quartz, quartzite, decomposed felspars, and fragments of phyllite, has been studied more particularly by Sawyer, Truscott, and Beck. The last-mentioned authority, beside much granular quartz and sericite, found in the cementing material, chlorite and a chlorite-mica closely related to chloritoid ; and in addition, ilmenite and magnetite, the latter in part with a hæmatite crust. As uncommon constituents he mentions also, zircon in rounded crystals, tourmaline, corundum, and free gold. The beds, which are traversed by east-west faults, strike approximately north-east and dip sometimes flatly, sometimes steeply, to the north-west. They are probably early Palæozoic. The precious metal occurs in small crystals or flakes and is characterized by great purity. As with the Witwatersrand conglomerate, the pebbles, apart from that which may have subsequently infiltrated, contain no gold. The gold content amounts to about 14 grm. per ton.

3. *The Ussungo-Sekenke Conglomerate, East Africa*

These deposits have been described in detail by Beck, to whom samples and descriptions had been sent by J. Kuntz.

In the Ussungo district they occur steeply inclined, in alternation with light-coloured quartzitic sandstones. Although this conglomerate at first glance appears to present much analogy to that on the Witwatersrand, Beck draws attention to a number of differences. Usually the pebbles are only as large as a hen's egg. They consist chiefly not only of quartz but also of crystalline rocks, such as iron-quartzite schists, fine-grained granite, and eruptive rocks rich in hornblende. The matrix represents a quartz aggregate with fine scaly sericite and abundant pyrite, this latter frequently occurring in bands or spherical aggregates. The gold occurs native or associated with pyrite, though it has also been found by Kuntz in pebbles of iron-quartzite schist. The gold content of the conglomerate is on an average only 3–4 grm., though in places it may increase to 14 grammes. Owing to this low content the deposit is of no economic importance.

According to Kuntz and Beck, at Sekenke, close upon the Wembere desert,[1] remnants of auriferous conglomerates are found in the neighbourhood of gold lodes, these conglomerates being thicker on the slope than higher up. They form a 20–60 cm. bed in a simple formation consisting in greater part of sandstones. The pebbles are almost exclusively of quartz, and subordinately of quartzite-schist. Since some of these are still angular, Kuntz's idea that the material was derived from disintegrated gold lodes is probably correct. The small amount of matrix consists, according to Kuntz, of opal and quartz. The gold content is low, reaching only 10 grm. per ton.

OTHER AURIFEROUS CONGLOMERATES

LITERATURE

W. LINDGREN. ' An Auriferous Conglomerate of Jurassic Age from the Sierra Nevada,' Amer. Journ. Sc. XLVIII., Oct. 1894.—G. F. BECKER. 'The Witwatersrand Banket, with Notes on other Gold-bearing Pudding Stones,' U.S. Geol. Surv., 8 Ann. Rep., 1896–1897, Part 5, pp. 153-184.—W. B. DEVEREUX. ' The Occurrence of Gold in the Potsdam Formation, Black Hills, Dakota,' Trans. Amer. Inst. Min. Eng., 1881–1882, Vol. X.—FRANKLIN R. CARPENTER. ' Ore-Deposits of the Black Hills of Dakota,' Trans. Amer. Inst. Min. Eng., 1888–1889, Vol. XVII. p. 570 ; ' Notes on the Geology of the Black Hills,' Preliminary Report of the Dakota School of Mines upon the Geology, Mineral Resources, and Mines of the Black Hills of Dakota, Rep. City, 1888.—J. F. KEMP. The Ore-Deposits of the United States and Canada. New York, 1906 —C. L. HENNING. Die Erzlagerstätten der Vereinigten Staaten von Nordamerika, Stuttgart, Ferdinand Enke, 1911, p. 165.—T. A. JAGGAR, jun., J. D. IRVING, and S. F. EMMONS. ' Economic Resources of the Northern Black Hills,' U.S. Geol. Survey, P.P. 26.

1. *The Homestake Conglomerate, Black Hills, South Dakota*

According to Henry Newton, the Black Hills, situated in greater part in South Dakota, consist of a roughly elliptical core—about 200 km. long in a north-west direction and 96 km. wide—of granite and old slates, overlaid unconformably by Cambrian—the so-called Potsdam sandstone —Carboniferous, Jurassic, Triassic, and Cretaceous. Tertiary volcanic rocks, principally phonolite, break through all these beds. The highest point of the Black Hills is the Terry Peak which reaches a height of 2155 metres. The folding which resulted in the present configuration took place in the Upper Cretaceous and in the Eocene.

According to Irving and Emmons, gold deposits in the form of old gravels occur in the pre-Cambrian, Cambrian, and Carboniferous. The conglomerates in the pre-Cambrian are said to be the earliest known gold

---

[1] *Ante*, p. 624.

deposits in the United States. In these the gold is found together with tourmaline, garnet, and cassiterite. Frequently the precious metal is intergrown with glassy quartz and calcareous schist. The deposits are to-day of no importance. The Carboniferous formation is represented by limestones which, chiefly west of Crown Hill and Portland, contain metasomatic gold- and silver deposits.

The Cambrian deposits however are those which deserve most mention. These conglomerates, up to 10 m. thick and having all the characteristics of basal conglomerate, occur immediately above the Algonkian crystalline schists. In places they merge into quartzite. Soft sandstones and quartzites, overlaid in turn by later Cambrian beds, form the hanging-wall. The conglomerates are auriferous, the gold occurring in particles which have resulted from the disintegration of auriferous lodes. As with many gravels, the deepest layer contains most gold ; in this case chemical re-distribution may have played a part. In places the conglomerate is traversed by rhyolite. According to Jaggar and Emmons this is an old Cambrian gravel-deposit. From the fact that the beds lying above the conglomerate contain marine fossils, in conjunction with the further fact that the con- glomerate is only of limited extent and merges on all sides into sandstone containing marine fossils, it is probably a littoral formation.

Becker describes a number of other auriferous conglomerates, some of which have been worked. Of these the following are worthy of mention : that in the Bald Mountain district in northern Wyoming, probably belonging to the Silurian ; the Triassic conglomerates containing a little gold, east of the gold district in the southern Appalachians ; three pre- Tertiary auriferous conglomerates in California, two of which belong to the Cretaceous and the third to the Jurassic ; a conglomerate probably of Carboniferous age at Corbetts Mills, Colchester Co., in Nova Scotia, where, as with recent gold gravels, the gold is found more particularly in the lower portions of the conglomerate, and subordinately in small cracks in the foot-wall slate ; conglomerates at several places in Australia, as for instance in the Peak Downs district in Queensland, at Tallawang and other places in New South Wales, South Australia, and in Tasmania, several of which are Carboniferous, and in which occasionally rounded nuggets, weighing as much as 5 oz., have been found ; that, probably belonging to the Upper Cretaceous, at Blue Spur in New Zealand at the contact of Tertiary and Cretaceous ; and finally, the Lower Carboniferous conglomerate at Bessèges in the department of Gard, France.

According to Becker, in all these cases it is a question of fossil gravels, in which the gold occurs more particularly in rounded grains and principally in the lower portion. Becker further points out that these old gravels— with the probable exception of those in New Zealand—are not of fluviatile, but of marine origin. This may be because fluviatile deposits upon sub-sidence to below sea-level would in general be destroyed by wave action, while coastal deposits, in part formed from such fluviatile gravels, in consequence of protection by later sediments, would survive.

THE LEAD-, ZINC-, AND COPPER BEDS

In relation to composition very different deposits are included in this group; it may be taken that in all probability from a genetic point of view also, there will be differences. In all cases however the form is that of a bed, that is, the mineralization is limited to certain geological horizons. The metalliferous bed however consists by no means entirely of payable ore, but very frequently of rich patches in a poorer and larger extent. The thickness is very variable; in many cases, and especially when there is only one bed, it is less than 1 m., while only exceptionally is it more.

The geological age of the beds here grouped together may be very different, as the following summary shows :

Tertiary : the Boleo and Alghero deposits.
Keuper : the Freihung, Alderley Edge, and Mottram deposits.
Bunter : the Commern, Mechernich, and St. Avold deposits.
Archaean : the Åmmeberg deposit.

Often, as at Commern, Mechernich, Boleo, St. Avold, and Freihung, the ore occurs in the form of roundish concretions or nodules, known in Germany as *Knotten*. In different concretions crystallization has acted differently; although in most cases the roundish irregular form predo-minates, with galena, for instance, more or less rounded cubes occur.

Compact ore-beds, though occurring, as for instance in places at Boleo and Åmmeberg, are uncommon; fairly regular aggregates of sand-stone grains cemented by ore are, on the other hand, as at Commern and Mechernich, more frequent.

With regard to composition, two main groups may be differentiated, namely, sulphide lead-zinc beds, as at Commern, St. Sébastien d'Aigre-feuille, and Freihung; and copper carbonate beds, as at Alghero, Boleo, St. Avold, and Alderley Edge.

The genesis of these deposits cannot always be satisfactorily established. Strictly speaking, only syngenetic deposits, that is, those in which the ore and country-rock are contemporaneous, should be reckoned to this group. Later investigation however, has shown that in some cases formerly regarded as syngenetic, the ore in fact was introduced subsequently.

Since therefore under existing circumstances it cannot always be determined whether such an occurrence is syngenetic or epigenetic, the authors consider it proper to include in this group all such as of which the epigenetic nature has not yet been finally established. In this connection, the sparseness or absence of gangue-minerals, which is a feature of these deposits, is an important factor ; in many epigenetic deposits, in addition to the ore, other minerals are found, the presence of which, in itself, is sufficient proof of a subsequent impregnation.

To a certain extent the Åmmeberg deposit occupies a special position in that it exhibits properties such as are generally observed with contact-deposits. It is true that this deposit is also exceptional in regard to its very great age, and in the course of geological time it may accordingly have suffered intense subsequent alteration.

Where primary sulphide lead-zinc ores occur, as for instance at Commern, Mechernich, Freihung, Saint Sébastien d'Aigrefeuille, Åmmeberg, these in the neighbourhood of the surface are altered to the ordinary oxidized ores. In the gossan are found, for instance, brown-coloured sandstones and conglomerates cemented by cerrusite, anglesite, or pyromorphite. The primary copper carbonate ores but seldom undergo alteration in the neighbourhood of the surface. When, on the other hand, chalcocite occurs as a primary ore—as for instance in the unimportant deposit at Senze de Itombe in Angola, in Cretaceous sandstone—the opportunity for the formation of secondary copper carbonates by atmospheric agencies is afforded. At Senze de Itombe accordingly, sandstone containing malachite and azurite, occurs in the neighbourhood of the surface, while chalcocite in sparing amount is found in greater depth.

Of the accessory precious metals with the sulphide lead deposits silver is the more notable, the amount of this in the occurrence at St. Sébastien d'Aigrefeuille being considerable.

The extent of these metalliferous beds may be considerable, sometimes over many kilometres. Taking the metal content into consideration however, it is found as a rule that the payable zones are but limited occurrences. Of these deposits three are economically important, namely, those of Boleo, Åmmeberg, and Commern-Mechernich, the last unfortunately being in greater part exhausted. The others are of small importance. On the whole, therefore, with this class of deposit extensive ore-deposits appear to be uncommon.

THE ÅMMEBERG SPHALERITE BED IN MIDDLE SWEDEN

LITERATURE

A. E. TÖRNEBOHM. Explanatory Text with the geological map, Mellersta Sveriges Berslag, Section No. 7, 1881, p. 31.—H. E. JOHANSSON. ' The Åmmeberg Zinc Ore Field,' Geol. Fören. Förh., 1910, pp. 1051-1078 ; also Pamphlet No. 35, ' Guides des excursions en Suède,' Inter. Geol. Congress. Stockholm, 1910.

The important metalliferous district of Åmmeberg, the position of which is indicated in Fig. 241, belongs to the Swedish granulite- or leptite formation, reckoned among the later fundamental rocks. In the immediate neighbourhood of the mining area different gneisses and granulites—or leptites—occur chiefly, and gabbro-diorite, amphibole-peridotite, eklogite, etc., and limestone, subordinately. To the south and west this intensely folded and crumpled area is cut off by an extensive granite occurrence, smaller masses of granite being also found in the immediate neighbourhood of the mines.

The ore occurs in the form of beds and fahlbands in a grey granulite rich in microcline. In addition to the principal sphalerite bed, which has been folded with the country-rock, several other ore-beds or fahlbands occur not far in the hanging-wall and foot-wall, among these being a poor sphalerite bed in the hanging-wall, and a 20–30 m. thick pyrrhotite bed containing 15–30 per cent of pyrrhotite, some sphalerite, etc., in the foot-wall. Immediately in the hanging-wall of this ore-bearing granulite zone several thick limestone beds, some 4 km. long, occur. The principal sphalerite bed may be followed almost without interruption for a length of 5 kilometres. Of this, however, only a number of lenticular swellings, which in places may reach 12–15 m. in thickness, are payable ; the average thickness of the payable ore is 4–5 metres.

The ore, occurring conformably to the country-rock, has a pronounced bedded structure. It is intergrown chiefly with the usual granulite minerals, namely, microcline, some quartz, subordinate plagioclase and biotite, and more rarely, pyroxene, hornblende, and garnet. A so-called wollastonite rock, which in addition to wollastonite contains some zoisite, garnet, and vesuvianite, occurs in the immediate foot-wall. In the close vicinity of the occurrence a pyroxene rock resembling the typical skarn of the Swedish iron deposits [1] is found. Granite- and pegmatite dykes frequently occur, sometimes conformably bedded and sometimes transgressive, in which latter case they cut across the ore-bodies.

In the principal bed, sphalerite with a low iron content forms the chief mineral, though galena, particularly in certain layers of the hanging-wall, also occurs. Chalcopyrite and pyrite are practically speaking completely

[1] *Ante*, p. 383.

absent. Pyrrhotite is hardly ever found in the zinc ore itself, though it occurs in certain layers or impregnated zones in the foot-wall. Native silver is found here and there in crevices, where it has probably been precipitated secondarily.

The ore-bodies have generally a steep dip to the north, along which up to the present they have been followed for a vertical depth of 350 metres. The mines since 1857 have belonged to the Vieille Montagne Company, and the operations have been extensive. About 10,000 tons of clean hand-picked ore containing 38–39 per cent of zinc, and some 45,000 tons of milling ore with 21 per cent of zinc, are won annually. From the latter about 21,000 tons of dressed ore with 37–38 per cent of zinc, 5000 tons of poor concentrate with 25 per cent of zinc, and 350 tons of lead ore containing 72 per cent of lead and 800 grm. of silver per ton, are obtained. The dressed ore ready for the market usually contains 38–38·5 per cent of zinc, 3–3·3 per cent of lead, 0·02 per cent of copper, and only 2·5–3 per cent of Fe ; while the insoluble residue amounts to 31 per cent, and the sulphur to 18·5–21 per cent. From 1857 up to and including 1909, altogether 1,968,729 tons of picked- and milling ore were produced.

Formerly this ore-bed was regarded by Swedish geologists and mining engineers preferably as a sediment. Johansson is inclined to consider it a magmatic segregation, to which class of deposit he reckons also the Swedish iron occurrences of the Persberg-Dannemora type.[1] The authors however regard these last as contact-metamorphic deposits, while they consider that the occurrence of wollastonite, vesuvianite, and skarn at Åmmeberg, justifies the assumption of a similar genesis for this deposit. Nevertheless, since the genesis of the deposit has not yet been agreed, the authors, following the example of many Swedish geologists, place it for the present among the ore-beds.

Some 8 km. from Åmmeberg is situated the previously-mentioned cobalt occurrence of Vena,[2] while in the immediate neighbourhood some iron deposits occur.

THE NODULAR LEAD DEPOSITS AT COMMERN AND MECHERNICH, EIFEL

LITERATURE

M. DARTIQUES. ' Sur les mines de plomb du Bleiberg,' Journ. d. Min. Vol. XXII. No. 131, pp. 341 - 360, 1807. — NÖGGERATH. ' Der Bleiberg im Roerdepartement, beschrieben in mineralogischer Hinsicht,' Ann. d. Wetterauisch. Ges. Hanau, Vol. III. pp. 29-40, 1814.—v. OEYNHAUSEN and V. DECHEN. ' Der Bleiberg bei Commern,' Karstens Archiv f. Bergbau, IX. pp. 60-133, Berlin, 1825.—D. A. GURLT. ' Erzvorkommen am Haubacher Bleiberge,' Verh. d. naturh. Vereins der pr. Rheinlande und Westfalens, Vol. XVIII., Sitzungsberichte, pp. 29-33, 1861.—C. DIESTERWEG. ' Die Beschreibung der Bleierzlagerstätten, des Bergbaues und der Aufbereitung am Bleiberge bei Commern,' Zeit.

---

[1] *Ante,* p. 390.   [2] *Ante,* p. 1140.

f. Berg- Hütten- und Salinenwesen im pr. Staate, XIV. pp. 159-179, 1866.—HABER.
'Genesis der Bleierze im Buntsandstein des Bleiberges bei Commern,' Berggeist, Vol. XI.,
1866, and Vol. XII., 1867.—F. W. HUPERTZ. Der Bergbau- und Hüttenbetrieb des Mecher-
nicher Bergwerkes-Aktien Vereins. Cologne, 1883. — M. BLANCKENHORN. 'Die Trias
am Nordrande der Eifel zwischen Commern, Zülpich und dem Roertal,' Abhandl. z. geol.
Spezialkarte von Preussen und den Thüringischen Staaten, Vol. VI. Part 2. Berlin, 1885.

The Triassic area on the north border of the Eifel forms an upland,
the heights of which to the west and south-east consist of Main Bunter,
while the flat undulating country in the centre owes its even surface to the
softer nature of the Upper Bunter.   On the north-east border later Triassic
beds occur, the Upper Muschelkalk always occupying the higher points.

The Bunter, economically the most important member of the Triassic,
is known for the many mining operations carried on principally in its lower
division.   It occupies the largest portion of this bay-like Triassic area, outside of which it also occurs in isolated patches.   Blanckenhorn divides the
Bunter into an upper and a lower stage.   While the former corresponds
to the Röth of Middle Germany and the Voltzia sandstone of West Germany,
the latter—the Main Bunter horizon—is equivalent to the Lower and
Middle Bunter of Middle Germany, or the Vosges sandstone of West
Germany.

In respect to mineralization, only the Main Bunter comes into consideration.   This consists of an alternation of coarse conglomerates and
coarse-grained sandstones, of which the former contain boulders of quartzite
grauwacke, sandstone, and white, opaque, though exceptionally transparent
quartz, up to 35 or 40 cm. in diameter.   Middle Devonian limestone
pebbles are more uncommon.   The cementing material is argillaceous and
more or less ferruginous.   In the neighbourhood of the metalliferous
sandstones—known generally as *Knotten* beds—the conglomerates are
generally white, and the cementing material consists of crystalline flaky
galena, cerussite, and copper carbonates.   Such conglomerates have been
worked more particularly at Call in the Caller Stollen, in the Meinertzhagen Bleiberg property, the Virginia mine, and in the Gottessegen mine on
the Griesberg.   The relation between the lead and copper is very variable.
With the ore, dolomite, calcite, pyrite, and barite, occur in druses.   Alternating with these conglomerates are coarse-grained sandstones, the grains
of which are often transparent, sometimes crystal-shaped, but generally
rounded.   The matrix is frequently siliceous-argillaceous, often ferruginous,
and occasionally somewhat calcareous.

These sandstones are characterized by the occurrence of ore nodules.
By these are understood roundish concretions of 1–5 mm. diameter, consisting of quartz grains, with galena, more rarely cerussite, azurite, and
malachite, as cement.   The barren or worthless beds, in the place of lead
ore, carry iron- or manganese hydrate.   It is an interesting fact that the

galena nodules frequently present the angularity characteristic of crystals,

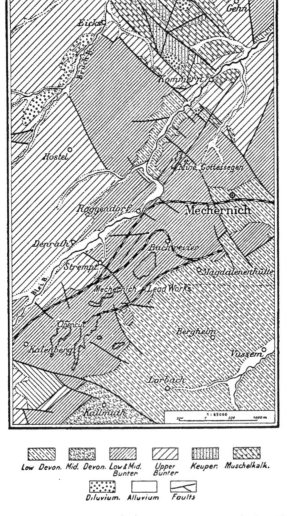

FIG. 453.—Geological map of the district around Mechernich.   Geol. Map of Prussia.

while occasionally even cube-shaped outlines may be recognized.   Cerussit occurs as a secondary alteration product at the outcrop.   The copper ore may also sometimes occur in pea-sized concretions of irregular shape, though generally they form a more regular impregnation. The relation of the nodules to the sandstone mass is very variable; while in some places these nodules occur close together, in others there are wide intervals where none are found.

Blanckenhorn mentions, in addition, both smaller and larger globular concretions, which differ from the enclosing sandstones in their greater compactness, and have a cement of crystalline dolomite, and exceptionally calcite. More rarely larger nodules are met with cerussite as the cementing material.

Unlike the white ore-bearing beds, the red sandstones in general are barren. Their cementing material is irregularly distributed, causing peculiar weathering phenomena, in that beds and irregular patches of more compact material have survived, while the looser material around them has succumbed to erosion. The seat of the Bunter upon the Devonian in

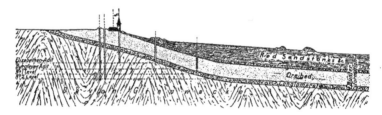

FIG. 454.—Diagrammatic section through the Bunter containing the nodular ore of Commern and Mechernich.

the foot-wall is but rarely to be observed. Usually the lowermost bed is a conglomerate, below which, occasionally, a thin layer of red clay containing many fragments of red-coloured Devonian rocks is found.

A definite sequence of the Main Bunter beds for the whole district, owing to the frequent petrographical changes of individual layers, cannot be formulated. The ore content of the same bed varies continually, each conglomerate bed may merge into sandstone and each sandstone bed into conglomerate. A considerable thickness may in one place be developed exclusively as sandstone, and in another as conglomerate. Nevertheless, in the Main Bunter on the right bank of the Blei in the south-east portion of the Triassic syncline at Bleiberg, a lower metalliferous portion with white nodular beds, and an upper, dark, non-metalliferous portion containing much iron, may be distinguished. These two portions are separated by a non-metalliferous conglomerate 2–46 m. thick.

The genetic conditions of the lead deposits at Commern are not simple. The Bunter, as illustrated in Fig. 161, is broken by a large number of trans-

verse faults into mutually displaced sections. Since these fissures in places
cut through and elongate individual nodules, it is evident either that, in
part at all events, they are younger than the nodules, or that movement
along them recurred at some later date. From this the too far-going
conclusion has been drawn that the formation of the nodules owes
nothing to the presence of the faults.

Concerning mineralization there are in general two views, namely,
those respectively of syngenetic and epigenetic formation. The Bunter
on the north border of the Eifel represents undoubtedly a littoral formation,
proof of this existing in the sudden change of sandstone beds into beds of
conglomerate, and *vice versa*. The possibility must therefore be considered
that plumbiferous mineral springs emerged through the sea floor and
saturated the accumulating sands, depositing their lead content in the
form of sulphide, in suitable places and around certain centres. Against

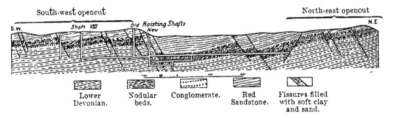

FIG. 455.—Diagrammatic section through the opencuts on the Griesberg near Commern,
belonging to the Gottessegen Company. Blanckenhorn.

this view of a syngenetic formation however, the sudden change in the
distribution of the ore nodules, as well as the occurrence of mineral druses
in the conglomerate, must be set.

Haber and Pošepný have pointed out that though the nodules generally
speaking lie parallel to the bedding, they also frequently form zones along
both sides of the steeply inclined fissures ; that is to say, they are likewise
arranged transversely to the bedding. In many cases a relation between
the nodules and narrow lead veins exists, in so far that the latter are continued as nodular zones. It is also a noteworthy fact that argillaceous
concretions are free from lead. In addition, not far from Commern, normal
lead lodes of considerable width occur in the Bunter and Devonian. The
Caller-Stollen mine for instance, according to P. G. Krause and Krusch,
works principally a lead lode, though certain Bunter beds in the neighbourhood of the lode fissure are altered to nodular ore-beds. For these reasons
the authors consider it more correct to regard the deposit at Commern as
epigenetic. Beyschlag has long recognized it as an impregnation connected
with lode fissures. While the fissure-filling in the Devonian took the form

of a normal lode, the ore solutions in the overlying permeable Bunter departed laterally from the fissure to saturate the rock, wherein their metal-liferous content eventually became deposited in the form of nodules.

The content of the deposit may be gathered from the following figures : according to Hupertz the ore *in situ* over a thickness of 22 m. contains on an average 0·5–3 per cent of lead in the form of galena. The ore mined, which comes from the richer parts, contains 2 per cent of lead. In dressing this ore 94 parts of the lead recovered is in the form of nodules containing 15–24 per cent of lead, 5 parts in smelting ore containing 55–60 per cent, and 1 part in low-grade ore containing 10–14 per cent. The nodules are subsequently further concentrated.

Lead mining on the north border of the Eifel was presumably begun by the Celts before the invasion of the Romans. At all events the Roman water ditches were carried over old dumps, while dump sand was used in the preparation of mortar for their construction. At that time, compact galena was probably won from fissures in the sandstone. The ore nodules of the Bunter became of importance only at a later date. The first to attract attention were the overlying coarse, more or less ore-cemented conglomerates which, mainly at the western boundary of the area, cover the ore-bearing sandstone in great thickness.

The present exploitation of these nodular deposits dates back to the year 1629, when the lords of Meinertzhagen obtained permission from the Duke of Arenberg to make and work an adit, as well as the sole right to the minerals in the districts of Peterheid, Bach, Schafsberg, and Kohlau. Up to the middle of the nineteenth century, apart from work in the con-glomerate, mining was prosecuted underground and on a small scale only, for azurite ore. The subsequent great expansion of work was consequent upon the introduction of opencut operations in the year 1852. While at first those portions were worked which were practically at the surface, since about the year 1868 some 40–50 m. of surface covering have had to be removed. In 1870 the working known as the 'new opencut' was begun, in which increasing compactness of the rock and decrease of galena in depth render mining each year more difficult. At present the deposit belongs to the Mechernich Company. Its economic importance may be gathered from the following table, kindly placed by the management at the authors' disposal :

[TABLE

ORE MINED

	Cubic Metres.[1]	Lead Content.
		Per cent.
1905	241,127·56	1·475
1906	218,624·23	1·435
1907	216,805·35	1·459
1908	152,672·17	1·574
1909	127,099·48	1·851
1910	150,951·75	1·839
1911	142,461·36	1·864

[1] 1 cubic metre = 2·545 metric tons.

DRESSED ORE PRODUCED

	Tons.	Lead Content.
		Per cent.
1905	21,901	41·28
1906	20,220	39·40
1907	19,962	40·24
1908	13,636	44·83
1909	12,250	48·89
1910	12,965	50·89
1911	13,332	50·70

THE LEAD DEPOSITS AT FREIHUNG, BAVARIA

LITERATURE

V. GÜMBEL. Geologie von Bayern, Vol. II. ; Geologische Beschreibung von Bayern, Cassel, 1894.—W. BRUHNS in collaboration with H. BÜCKING. Die nųtzbaren Mineralien und Gebirgsarten im Deutschen Reiche. Berlin, 1906.

At Freihung occurs the lead deposit in the Franconian Keuper upon which formerly extensive mining operations were prosecuted. This occurrence, which is connected with the Freihung-Kirchenthumbach fault, exhibits from hanging-wall to foot-wall the following section, the beds dipping 14°–16° south-west :

Red-brown and light-reddish sandstone	20 metres.
White medium- and coarse-grained sandstone with layers of shale and streaks of lead ore	10 ,,
Main ore-bed consisting of :	
White, friable sandstone, clayey, with 5–10 per cent of cerussite and galena	1·05–3 ,,
Red-brown clay, shale, and thin beds of sandstone and lead ore	0·05–2 ,,
White, clayey, ore-bearing sandstone	2 ,,
Red-brown shale and ore-bearing sandstone-slate . . .	1·05 ,,
Lower white ore-bearing sandstone	3 ,,
Red-brown and greenish shale	0·05 ,,
White, light red, and purple sandstone with nodules of cerussite and layers of red-brown clay	30 ,,

The thickness of the ore-bearing beds is very variable, frequently indeed they completely pinch out. The ore is chiefly cerussite, which in places, particularly in the main ore-bed, is replaced by irregular patches of galena. In depth the ore became poorer,. and the previously successful operations had consequently to be suspended in the year 1891.

With regard to genesis, it is significant that this deposit also is closely connected with a large fault, and in this case also the assumption of an impregnation zone is natural.

Similar, though less important lead deposits are recorded near Pressath, particularly on the Eichelberg, north of Freihung.

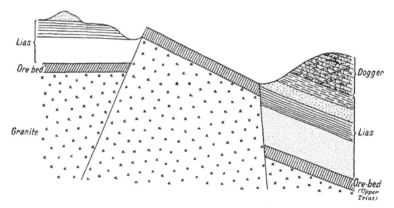

FIG. 456.—Diagrammatic section through the lead deposit at St. Sébastien d'Aigrefeuille in France, at right angles to the strike. Glockemeier.

THE LEAD DEPOSIT AT ST. SÉBASTIEN D'AIGREFEUILLE, DEPARTMENT GARD, FRANCE

LITERATURE

Written communications from Dr. E. NAUMANN and GLOCKEMEIER to KRUSCH

The lead deposit of Saint Sébastien d'Aigrefeuille in the Department Gard, occurs near the village of Générarques. The ore-bearing complex consisting of conglomerate and sandstone, is some 10 m. thick and belongs to the Upper Triassic. As indicated in Fig. 456, the Triassic lies immediately upon granite, and in a complete section is overlaid by Lias and Dogger. The ore-bed comes to the surface along an uplift, on either side of which it remains covered by Lias and Dogger.

The section of this uplift, given in Fig. 457, shows under a surface covering a bed 1·5–3 m. thick consisting of sandstone and marly shale, sometimes merging into conglomeratic sandstone. This lies upon a 1–2 m. conglomeratic sandstone which occasionally is ore-bearing and which at times may increase in thickness to include the sandstone bed above. The principal mineralization, however, occurs in the underlying coarse-grained conglomerate, which generally speaking is 4–6 m. thick though the thickness varies between 2 and 7 metres. In the foot-wall of this bed comes a lower conglomerate, 5–10 m. thick, which is frequently arenaceous and argillaceous, and lies immediately upon the granite.

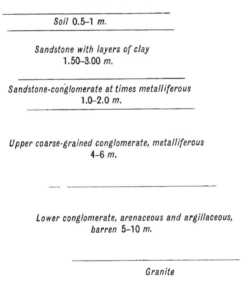

Soil 0.5–1 m.

Sandstone with layers of clay
1.50–3.00 m.

Sandstone-conglomerate at times metalliferous
1.0–2.0 m.

Upper coarse-grained conglomerate, metalliferous
4–6 m.

Lower conglomerate, arenaceous and argillaceous,
barren 5–10 m.

Granite

Fig. 457.—Diagrammatic section of the St. Sébastien d'Aigrefeuille deposit. Glockemeier.

The ore, consisting of galena and pyrite, fills the spaces between the pebbles, over large areas of the conglomerate. The payable masses have very irregular outlines ; their maximum lead content is 30 per cent, their average content 6–10 per cent, while the whole bed may contain 2–4 per cent. The silver found with the lead is very variable ; on an average it amounts to 1800 grm. per ton of lead. Sphalerite is very uncommon.

Fig. 458.—The extension of the Boleo copper ore-bed in Lower California.   F. Fuchs.

1181

THE COPPER DEPOSITS AT BOLEO IN LOWER CALIFORNIA

LITERATURE

F. Fuchs. 'Note sur les gisements de cuivre du Boléo,' Ass. fr. pour l'av. des Sc. Vol. XIV. p. 410. Grenoble, 1885.—M. Fuchs. Extract from Fuchs and de Launay, Traité des gîtes minéraux et métallifères, Vol. II. p. 349. Paris, 1893.—P. Krusch. 'Kupfererzlagerstätten in Niederkalifornien,' Zeit. f. prakt. Geol., 1899, p. 83.

The following description is in general taken from that of Fuchs. The district in which these deposits are found is situated on the west side of the Gulf of California, in 27° 25′ north latitude and 150° west longitude, 120 km. north-north-west of the small port of Muleje and directly opposite the port of Guaymas. It forms an approximately rectangular plateau 8 km. by 5 km. in extent, bounded to the north-east by the sea, and to the south-west by a large fault running, as illustrated in Fig. 458, almost parallel to the coast. This plateau is gently inclined towards the sea and cut by a number of deep valleys, while landwards it is backed by several volcanic peaks, including the Sombrepo Montado and the Juanita. In the deep-cut valleys complete sections of the many thin and regularly bedded sediments are exposed. Fuchs gives the following section from top to bottom:

Generally yellow, occasionally somewhat calcareous tuff . .	10–30	metres.
Conglomerate with calcareous cement, fossiliferous . . .	2–4	,,
Yellow or grey-mauve tuffs	15–20	,,
First copper bed, on an average	1	,,
Conglomerate	3–4	,,
Argillaceous tuffs, grey-mauve, exceptionally yellow or pink .	40–50	,,
Second copper bed	0·8–2·3	,,
Conglomerate of principally grey material . . .	4–5	,,
Slaty argillaceous tuffs, occasionally somewhat crystalline .	6–8	,,
Compact tuff of peculiar appearance, alcahueta . . .	1	,,
Pink tuffs, sometimes grey-mauve, becoming brown at the base	45	,,
Third copper bed	0·6–3	,,
Conglomerate with principally dacite-and labradorite fragments	3	,,
More or less crystalline tuffs, brown or greenish, only to be observed in the neighbourhood of trachyte . . .	50	,,

These sediments are intruded by a double series of eruptive rocks, the vents of which run parallel to the coast, and of which the western series cuts off the metalliferous beds. The rocks of this series consist of trachytes not markedly acid but closely related to the dacites. Fuchs supposes that the crystalline green tuffs at the base enclose a fourth copper bed. The whole system is overlaid by a thick basalt sheet. The tuffs are regarded by Fuchs as underground intrusions of volcanic mud. The youngest beds carry presumably Miocene or early Pliocene fossils; the copper-bearing beds are accordingly Miocene.

The country-rock of the copper bed is in no way distinguishable from the other tuffs. Above the ground-water level the ores are oxidized and accompanied by some iron-manganese ore and silica. Frequently black cupric oxide and cuprous oxide occur, with azurite, malachite, subordinate chrysocolla, and atacamite, this last as a mineralogical rarity. Manganese compounds resembling crednerite, $2Mn_2O_33CuO$, occur as associated minerals, a part of the manganese being replaced by iron. The grey-mauve, and when dry, fairly light-coloured tuffaceous country-rock contains 0·1–6 per cent of NaCl and variable amounts of carbonate and sulphate of lime. The ratio of ore to gangue varies considerably. The ore occurs in concretions, stringers, and oolites. An enrichment may always be remarked at the base, where frequently a compact ore-bed, 15–25 cm. thick, occurs.

In addition to these general characteristics each of the ore-beds has special peculiarities. The upper bed contains much cupric oxide, while carbonates and silicates occur only exceptionally. The second bed is somewhat more siliceous and carries copper silicates. The silica content is possibly connected with a large fault, along which siliceous geyser water probably ascended. In this bed the oolitic ore principally occurs, the grains of which, consisting of oxidized and carbonated ores known as *Boleos*, are several centimetres in diameter. The country-rock of this bed is practically free from copper ; the oolites contain 35–40 per cent. The dressing of the ore is accordingly very simple, and the concentrate obtained contains 25–30 per cent of copper. The third ore-bed consists of yellow clayey material which, though apparently but little mineralized, contains 10–15 per cent of copper ; the copper-manganese compounds related to crednerite contain 32–43 per cent of copper ; while black oxides with as much as 60 per cent of copper also occur.

Of great significance is the occurrence in greater depth of sulphides, including chalcocite, Cu_2S, with 75–80 per cent of copper, and covellite, CuS, with 60 per cent. From this it may be assumed that the oxidized ores and carbonates, etc., have been formed from sulphides. In the north and north-east of the district thick gypsum beds occur, these being sometimes coarsely-crystalline and sometimes fine-grained, in which latter case the material is an alabaster. At their base a mixture of gypsum and manganese ore with a low copper content and iron oxide, is found. The formation of these copper ores may be attributed to springs. The oxidation ores owe their existence in the first place to the influence of sea-water, but their subsequent accumulation to the action of meteoric waters.

The economic importance of the district may be gathered from the following table : [1]

---

[1] *Ante*, p. 890.

COPPER PRODUCTION IN METRIC TONS

1906	.	.	.	11,000 tons.
1907	.	.	.	11,200 ,,
1908	.	.	.	12,600 ,,
1909	.	.	.	12,400 ,,
1910	.	.	.	13,000 ,,

NODULAR COPPER DEPOSIT IN THE BUNTER

LITERATURE

W. BRUHNS in collaboration with H. BÜCKING.   Die nutzbaren Mineralien und Gebirgsarten im Deutschen Reiche.   Berlin, 1906.

At places malachite and azurite occur in deposits similar to the nodular lead deposits of Commern and Mechernich, when usually they are found over fairly extensive areas.

In Lorraine these minerals are known in the Upper Bunter at Sulzbach, Wasselnheim, and Pfalzburg, where they occur in the form of grains, nodules, patches, and fissure-fillings.   On the Grosser Zoll near Falk in the neighbourhood of Hargarten, they were formerly worked.   Other deposits lie on the Hochwald at Helleringen, between Oberhornburg and Helleringen, not far from St. Avold in Forbach, and at Herapel near Kocher.

Van Werwecke draws attention to the fact that on the Grosser Zoll four adits were made below the principal conglomerate of the Vosges sandstone, and that accordingly the Main Bunter also is not free from copper.   It is stated that still deeper adits occur, the lowest being about 20 m. below the principal conglomerate.   Of somewhat greater importance is the deposit occurring at Wallerfangen and St. Barbara, in Saarlouis, Rhenish Prussia.   There, in the uppermost division of the Bunter, the Voltzia Sandstone, on the Limberg, are found four ore-bearing beds consisting partly of sandstone and partly of clay.   At Santa Barbara, as an old Roman inscription testifies, mining operations were carried on by the Romans, while at Blauberg several shafts 25–48 m. deep were found in an area 1·3 km. long and 180 m. wide.   The azurite produced here during the period 1500–1538 was used as a pigment and exported largely to Italy.

Similar deposits are found at Berus and Felsberg on the right bank of the Saar near Beckingen in Merzig, and farther north, in the Bunter of Trier and Bitburg.   In the Bunter extending along the border of the Rhenish Schiefergebirge in the districts Düren, Schleiden, and Euskirchen, copper ores occur together with the previously described lead ores on the Griesberg near Commern.   Here were found not only blue azurite nodules, but also large patches of chalcocite and cuprite.   Between Bergheim and Bilstein two copper ore-beds, separated by a plumbiferous layer, are known.   The occurrence at Leversbach and Schlagstein, in which the copper ores occur in conglomerate beds and are closely associated with lead ores, has considerable extent.

At Berg and Floisdorf, between Glehn and Eicks, between Nöthen and Heistartburg, at Kufferath under the ironstone bed, and between Leversbach and Uedingen, copper ores occur in fine-grained sandstone. Since copper lodes occur in the neighbourhood of these occurrences, as for instance at Uedingen with chalcopyrite and ferruginous cuprite, the idea of a genetic connection between the ore-bearing beds and the fissure-fillings is natural, and accordingly the ore-beds presumably represent impregnation zones.

In Waldeck similar copper ore-beds are known. At Twiste, ore-bearing beds lying one above the other in the Bunter have been remarked at seven different places. Other places where similar deposits have been found are : Recklinghausen, Berndorf, Sachsenhausen, Schmilling-hausen, Herbsen, Rhoden, Wrexen, Huxmühle, Eilhausen, and Massen-hausen.

THE COPPER DEPOSITS AT ALDERLEY EDGE AND MOTTRAM ST. ANDREWS, ENGLAND

LITERATURE

J. A. PHILLIPS and H. LOUIS. A Treatise on Ore Deposits, p. 266. London, 1896.
—EDWARD HULL. ' On the Copper-bearing Rocks of Alderley Edge,' Cheshire Geol. Mag., 1864, p. 65.—E. HULL and A. H. GREEN. ' The Geology of the Country around Stockport, Macclesfield, Congleton, and Leek,' Mem. Geol. Survey, p. 39, 1866.

The copper deposits of Alderley Edge and Mottram St. Andrews are situated in Cheshire, about four miles from Macclesfield. The escarpment or ' Edge ' of Alderley shows a gradual rise from the Cheshire plain on the east, but forms a steep ridge to the north. It owes its origin to an east-west fault along which the Red Marl has subsided, while the soft sandstones of the Bunter, capped by Lower Keuper conglomerate, have been uplifted. The beds dip towards the plain on the east at an angle of from 5° to 10°. The general form of the Edge and its component beds is shown in Fig. 459, after Hull, who first determined the geological age of the copper-bearing sandstones. The geological section is as follows :

Red marl .	.	.	.	.	.	Red and grey laminated marl.
Waterstones	.	.	.	.	⎧ Lower Keuper	⎧ White and coloured sandstones,
Freestone	.	.	.	.	⎨ sandstone,	⎨ hard, quartzose conglomer-
Copper-bearing sandstone .	.	.	.	500 feet.	ate and marls.
Conglomerate with ore	.	.	.	⎩	⎩
Upper red and mottled sandstone	.	Bunter	⎧ Soft, fine-grained, yellow and ⎨ red sandstones, forming the ⎨ uppermost member of the ⎩ Bunter.

The metalliferous beds marked b^1 and b^2 in Fig. 459 lie at the base of the Keuper. Petrographically they are not without interest. The conglomerate is firmly cemented and consists principally of well-rounded quartz pebbles; it resembles in every respect the conglomerates of the Bunter. Copper, in the form of malachite and azurite, occurs at Alderley Edge only in small amount, while at Mottram St. Andrews, about a mile to the north-east, such copper is found in more considerable quantity.

The apparent position of the Mottram conglomerate below that of Alderley Edge is due to the afore-mentioned fault, which downthrows the beds in a northerly direction. This occurrence of copper was incidentally exposed in a quarry. A few years ago systematic mining was attempted, but without success.

The horizon of the principal mineralization at Alderley lies above the conglomerate. It consists in this case of a copper-impregnated zone involving a width of some 40–50 feet and three sandstone beds; on

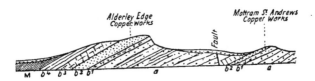

FIG. 459.—Diagrammatic section of the copper ore-bed at Alderley Edge and Mottram St. Andrews, England. Hull.

M = Red Marl; b^4 = Waterstones; b^3 = Freestone; b^2 = cupriferous sandstone; b^1 = conglomerate, all belonging to the Keuper; a = upper red and mottled sandstone belonging to the Bunter.

either side of this zone the mineralization quickly disappears. The colour of the sandstone may be green, blue, red, or brown. In addition to copper ore, barite and small amounts of lead-, manganese-, iron-, and cobalt ores, also occur. Lead is present chiefly in the form of cerussite, though galena, pyromorphite, and vanadinite, are also met. The thickness of the lowest of the three ore-beds sometimes reaches 66 feet, but varies very considerably. Above it lies a 1–6 feet argillaceous sandstone which forms the foot-wall of the second ore-bed, 18 feet thick. This bed in turn is overlaid by red argillaceous sandstones about 12 feet in thickness, forming the foot-wall of the top bed, 18 feet thick.

The average copper content of the workable portions of the sandstone amounted to 1·4 per cent. The ore contained traces of cobalt. The mines at Alderley Edge were worked for many years with considerable success; in 1877 however operations were suspended. An attempt was made to win the lead ore which occurs in certain bands of the sandstone, but so much of the cerussite was lost in process of dressing that only 55 per cent of the lead was recovered.

At Alghero in Sardinia,[1] south-east of Sassari, a bed of Miocene sandstone 0·1–1 m. thick, containing about 1–2 per cent of copper in the form of carbonates, occurs between conglomerate in the foot-wall and limestone in the hanging-wall.

THE ANTIMONY ORE-BEDS

The antimony ores belong to those heavy-metal compounds which are only exceptionally found in large amount; the occurrences mentioned when describing the antimony lodes and metasomatic antimony deposits were accordingly but few.

While formerly a bedded nature was assigned to a large number of antimony deposits, more recent genetic examination of some has proved incontestably that they were formed by epigenetic processes. The well-known deposit in the Caspari mine at Arnsberg, for instance, has in this work been described among the metasomatic occurrences. Furthermore, the deposits now to be briefly described are most probably not syngenetic, though some authorities may regard them as such; it may indeed be said that according to present knowledge not one single antimony deposit may be regarded as an ore-bed of indisputably syngenetic origin. The form is admittedly bed-like, the deposits consisting of thin layers along the bedding-planes, as well as lenticular masses a metre or two in thickness, but exceptionally thicker.

The material of the deposits when the thickness is small consists frequently of clean antimony ore, while when it is great other ores, rock-material, clay, and gangue-minerals, occupy much of the space. The most common and extensively occurring ore is stibnite, from which by oxidation stiblite, senarmontite, and valentinite are formed. In a few cases, as for instance at Djebel-Hamimat and Sidi Rgheiss in Algeria, the accumulation of these oxidized ores is so great that Coquand was inclined to regard them as primary sediments. Yet, from the occasional occurrence of stibnite more or less decomposed, as kernels in those deposits, a secondary formation may be concluded. Since with the Algerian deposits limestone and marl form the country-rock, the greater thickness of the occurrence may be attributed to oxidation-metasomatism, in process of which and at the decomposition of the stibnite the limestone of the country-rock became in part replaced.

With some deposits, as for instance those at Nuttlar and Brück on

---

[1] A. Stella, ' Rel. sulle ricerche min. nei giacim. cupr. di Alghero,' *Boll. R. Comm. geol. Italia*, 1908, pp. 1-34.

the Ahr, in addition to conformably intercalated ore-bodies, lode-fillings consisting chiefly of stibnite are found, and the idea is suggested that the so-called beds likewise represent cavity-fillings and associated metasomatic replacement of the country-rock, that is to say, bedded lodes and meta-somatic deposits.

The geological age of the formations in which antimony ore-beds occur is very various. Since at Gilham in the United States as well as at Nuttlar the antimony ores occur in Carboniferous beds, while the ore-bodies in the Caspari mine at Arnsberg [1] likewise have Carboniferous country-rock, the Carboniferous formation appears to be especially favoured. The ore at Brück belongs to older beds, namely, Devonian, while the country-rock at Djebel-Hamimat and Sidi Rgheiss, belonging to the Gault, and that at Coyote Creek, belonging to the Eocene, are younger. Petrologically antimony appears to favour no particular rock, it occurs in slate as well as in sandstone and limestone.

The size of the deposits is usually not great, the world's production of antimony ore being accordingly small and the price fairly high.

THE ANTIMONY ORE-BEDS OF THE RHENISH SCHIEFERGEBIRGE

LITERATURE

ERBREICH. 'Geognostische Beschreibung der Spiessglanzlagerstätte u.s.w. bei dem Dorfe Brück,' Karstens Archiv f. Min. VI. p. 44, 1827.—M. WEMMER. Die Erzlagerstätten der Eifel nebst Erzlagerstättentafel. Iserlohn, 1909.—Beschreibung der Bergreviere Arnsberg, Brilon, und Olpe, sowie der Fürstentümer Waldeck und Pyrmont, published by the Mining Department. Bonn, 1890.—W. BRUHNS, in collaboration with H. BÜCKING. Die nutzbaren Mineralien und Gebirgsarten im Deutschen Reiche. Berlin, 1906.

The antimony deposit of the Pass Auf mine at Nuttlar in the district of Brilon, lies on the south-west slope of the Wiemert near Vöckinghausen. The metalliferous beds, three of which are known, belong to the Culm. These three beds strike south and dip 80° south-west. The bed-filling consists of chert and black shales, in which stibnite occurs in the form of nests. In the departmental geological description [2] the beds are correlated with the Upper Carboniferous Millstone Grit, though from the occurrence of chert it is evident that they belong to the Culm, since hitherto no such chert has been known in the Upper Carboniferous.

The genesis of the deposit is not clear. Although the form coincides with that of beds, a secondary introduction of the antimony is not excluded.

At Brück on the Ahr in Adenau, a metalliferous zone, 24–32 m. wide and 160 m. long in a north-south direction, occurs in Palæozoic grauwacke-schist. The ore is in part interbedded conformably with the beds, though

---

[1] *Ante*, p. 784.   [2] *Loc, cit.*

numerous veins—a few centimetres wide, striking north-east and dipping 40°-50° south—indicate not a syngenetic, but an epigenetic occurrence. The ore-minerals, stibnite and pyrite, together with the gangue-minerals, quartz and dolomite, represent in part fissure-fillings, whence the material migrated into the bedding-planes and joints of the country-rock.

Antimony Ore-Beds of the United States

LITERATURE

F. L. Hess. The Arkansas Antimony Deposits, Bull. 340 of the Survey, pp. 241-252.— G. B. Richardson. 'Antimony in Southern Utah,' ibid. pp. 253-256 ; Mineral Resources of the Survey.—Charles L. Henning. Die Erzlagerstätten der Vereinigten Staaten von Nordamerika. Stuttgart, 1911.

With these, only stibnite comes into question. This ore occurs at Gilham in Sevier County, Arkansas, in thinly laminated Lower and Middle Carboniferous sandstone or arenaceous marls, where it forms lenticular masses of 1–7 m., and sometimes as much as 15 m. in thickness. The whole character of this occurrence indicates that the deposits are epigenetic and represent fissure-fillings. The stibnite is found associated with jamesonite, zinckenite, galena, tetrahedrite, pyrite, chalcopyrite, siderite, calcite, and quartz. At Coyote Creek in Garnfield County, Utah, stibnite also occurs in Eocene sandstone and conglomerate, though this occurrence likewise is of little value. The statistics of the United States for the last few years have included no production of antimony.

The Antimony Deposits at Djebel-Hamimat and Sidi Rgheiss in Algeria

LITERATURE

L. de Launay. Traité de Métallogénic, Gîtes minéraux et métallifères, Vol. I. p. 772. Paris, 1913.—Coquand. 'Sur les mines d'antimoine oxydé des environs de Sidi-Rgheiss au sud-est de Constantine,' Bull. de la Soc. géol. de France, 2°, Vol. IX., 1852, p. 342.

The deposits at these places situated 60 km. south-west of Guelma or 23 km. north 30° west of Ain Beida, are singular, being characterized by the occurrence of senarmontite, a mineral discovered by Sénarmont and described by Coquand.

At Djebel-Hamimat the ore occurs in an alternation of black limestones and slaty bituminous marls, these rocks according to Blayac belonging to the Gault. The deposit consists of senarmontite with little stibnite, and of antimoniferous zinc oxidized ore with some galena and cinnabar. The ore forms irregular masses which run fairly parallel with the beds, and occur principally at the contact of limestone and marl.   Fragments of limestone encrusted with ore are often found.   The deposit, operations on which have been started and stopped at different times, shows the antimony oxide either solid, crystalline, or even disseminated.   Occasionally crystals of senarmontite, 3 cm. in diameter, are found ; in the disseminated zones the crystals occur in shale.   In addition to the oxide, stibnite occurs in small amount ; this mineral has however in greater part been altered to oxide, though the senarmontite is frequently coloured black by the presence of finely distributed stibnite.   Coquand regards the deposit as a true senarmontite sediment ; de Launay however rightly remarks that many of its characteristics suggest a metasomatic formation.

Another antimony oxide deposit occurs 4 km. west of Hamimat at Aïn Bebbouch on the eastern slope of the mountain range of that name near Senza.   This deposit carries valentinite, probably as an alteration product of stibnite.

THE TIN-, GOLD-, AND PLATINUM GRAVELS

When discussing the formation of deposits [1] a differentiation was made between eluvial, fluviatile, marine, and glacial gravels.   In the case of tin, for instance, eluvial deposits occur on Mount Bischoff, fluviatile and marine deposits in the Malay States ; the most common are the fluviatile gravels, though these are usually so closely associated with the eluvial that where the former lie in the valleys, the latter are often found on the hills.   In the case of gold, glacial gravels occur in the Klondyke, fluviatile gravels in nearly all gold gravel-districts, and marine gravels in the Cape Nome district ; eluvial auriferous gravels are uncommon.   The platinum gravels of the Urals, the chief platinum district, are almost exclusively fluviatile, as likewise are those of Colombia.   As in the case of tin, these are frequently associated with eluvial gravels.

For all gravels, irrespective of the mineral-association, the following remarks apply : since they are generally found in river-courses which at the same time represent the important ways of communication to upland districts, the gravels of the heavy metals are usually discovered before the primary deposit, so that as a rule mining begins upon such gravels and extends only to the primary deposit when these have become exhausted.   All gravels have been formed by the disintegration of the hills, that is, by erosion, denudation, and abrasion.   In these processes only those minerals characterized by chemical resistance, hardness, and gravity, can be concentrated in the river-beds.   Minerals more readily

[1] *Ante*, p. 17.

disintegrated, such for instance as the sulphides, become oxidized and decomposed, while frequently they are to a considerable extent removed in solution.   Other minerals, such for instance as the felspars, become altered by atmospheric agencies to kaolin and clay, which in greater part are washed away.   Mica becomes chloritized and comminuted, while the other ferro-magnesian silicates are in greater part decomposed, comminuted, and carried away.   Quartz, the chief constituent of the most extensively occurring rock, granite, owing to its chemical resistance and hardness, plays an important part in gravels, in spite of its relatively low gravity. In addition, garnet, tourmaline, zircon, and other so-called ornamental or precious stones, are frequently found ; and finally, on account of their high specific gravity, magnetite, specularite, titaniferous iron, rutile, etc. With these are occasionally associated such metalliferous minerals as cassiterite, native gold, native platinum, monazite, and cinnabar.   Native silver and copper on the other hand do not occur in gravels, possessing insufficient power of resistance to survive a protracted natural concentration.   Finally, as further constituents of gravels, felspathic, micaceous, or chloritic clay, iron-ochre, etc., are frequently found.   The titaniferous-iron sands have already been described.[1]

In the case of very young gravels concentration is sometimes very incomplete.   For instance, those of Alaska consist principally of more or less rounded slaty material relatively little decomposed.   Upon these, only the first stages of disintegration, attrition, and concentration have been operative.

The material of gravels is brought to the rivers, etc., in greater part by the drag of the detritus down the slope.   It therefore reaches the river from the side, and, when the force of the drag is greater than that of the running water, it may be pushed across.   This uncommon case was described by Lungwitz[2] as obtaining at Anderson Creek,[3] British Guiana, where, surprisingly enough, the auriferous gravel extended across, and not along the stream as it almost invariably does.   The material of such transversely extending gravels is distinguished from that of normal fluviatile gravels in being angular, while that of the latter contains only pebbles.

Glacial gravels are characterized by rounded edges.   Of marine gravels, it is characteristic that in the surf action they have suffered the most intense attrition and assortment by sea-water, that is to say, by a dilute saline solution.   The chemically more readily attacked minerals, such as gold,[4] exhibit therefore in addition to attrition, corrosion phenomena.

A gravel, in addition to quartz which is usually the chief constituent, consists of rock fragments of the most varied size, and of clayey material.

[1] *Ante*, p. 1052.   [2] *Zeit. f. prakt. Geol.*, 1906, p. 212.
[3] *Ante*, p. 6.   [4] *Ante*, p. 634.

Here and there the sand or pebble individuals have subsequently been cemented together, and sandstone- or conglomerate beds have resulted. The vertical distribution of the valuable constituents of such a deposit varies. Only in exceptional cases is the whole thickness payable ; the ore or precious metal on account of its high specific gravity is generally concentrated towards the bottom. Gold is occasionally carried mechanically as well as chemically into fine crevices in the bed-rock, this rock in river-courses being always more or less fractured. A gravel of loose material is the most welcome, since cemented and very clayey beds are difficult. to work successfully. In most cases the ore content of gravels is considerably lower than that of primary deposits. With loose material, however, the working cost may be so small that under favourable conditions sands containing only a fraction of a shilling per cubic metre may be profitably treated. The chief conditions of payability are : firstly, the presence of such large quantities so bedded that a dredge or other simple apparatus can be applied ; and secondly, a relatively regular distribution of the valuable content.

Since in all cases water is necessary for the working of these gravels, only such as occur in the neighbourhood of water are, generally speaking, of economic importance.

THE TIN GRAVELS

Stanniferous deposits are particularly suited to the formation of gravels, cassiterite being one of the most resistant of minerals. Since also it possesses a high specific gravity, at disintegration of the primary deposit it remains behind either in the form of eluvial gravels or, when the material is carried to the river-courses where the cassiterite is comparatively quickly and completely separated from the considerably lighter materials, in the form of fluviatile gravels. The granitic areas which, strengthened by the tin veins they contain, resist erosion and project as hills from the country around, are accordingly frequently covered with eluvial gravels, while, as illustrated in Fig. 18, the fluviatile gravels occur in the river-courses. Finally, where the rivers carry tin ore to the sea, marine gravels become formed.

Since the eluvial gravels occur principally in districts of low rainfall, work usually suffers from a lack of water during a great part of the year. When the amount of water available on the spot is so small as to render concentration impossible, a rich tin gravel becomes unworkable unless water can be brought in.

Those of the gravel-deposits in the Malay Peninsula which occur in the form of large pocket- or funnel-fillings in limestone, are unique occurrences. Such deposits probably resulted from veins in the limestone, along which the limestone became decomposed, pockets becoming formed into which the vein material broke, and into which sand, clay, twigs, and secondary ores were brought by water. Such gravels represent therefore cavity-fillings with detrital deposits.

With eluvial and fluviatile tin gravels, cemented conglomerates are occasionally found so compact as to necessitate the use of rock-breakers.

In addition to cassiterite, the tin gravels contain those minerals of high specific gravity which occur together with tin ore in the primary deposit, and accordingly wolframite and tourmaline are frequently found in considerable amount.

Concerning the tin content, the payability of a tin gravel depends firstly upon the physical character of the material, and secondly upon the amount of water present. In general, and other conditions being favourable, payable gravels need have only a very low content, since the specific gravity of cassiterite is high and its concentration correspondingly simple. Furthermore, the larger the deposit and the greater the capacity of the concentration equipment, the lower may be the content. The price of cassiterite won from gravels is high, owing to its purity. The price of tin during the last thirteen years has shown a tendency to rise.

THE TIN GRAVELS OF THE STRAITS SETTLEMENTS, INCLUDING BANKA AND BILLITON

LITERATURE [1]

The tin gravel-deposits in this region are in many cases eluvial. The pocket-fillings belonging to this class have been briefly described above. The deposits are of Quaternary age. Their tremendous extent is in greater part due to the intense denudation experienced in the tropics. The most extensive of these deposits are those on the west coast of the Malay Peninsula, in Perak, Selangor, Sungei-Ujong, Jelebu, and Negri Sembilan; and on the east coast, in Pahang.

Under reference to the previous description of these occurrences [2] the authors here confine themselves to the following remarks. With the fluviatile gravels there lies on top a non-stanniferous or unpayable bed—the overburden—of sand and clay, which at Billiton is 4–6 m. thick, and at Banka 8–12 m., sometimes even 16 metres. At the bottom, and in many cases immediately on the bed-rock, comes the metalliferous bed, which is generally 0·1–0·25 m. thick though occasionally it reaches 1 metre. Apart from cassiterite the chief minerals are quartz, tourmaline, muscovite,

[1] *Ante,* p. 437 [2] *Ante,* p. 443.

hornblende, topaz, sapphire, wolframite, scheelite, magnetite, and some native gold. The richest deposits are found where the valleys debouch from the hills into the plain; such deposits are about 1 km. in length. With greater distance from the hills impoverishment rapidly sets in. In the neighbourhood of the primary deposit the gravels usually contain comparatively large but irregularly shaped ore-fragments; with distance the average content becomes lower, though the composition may be more regular. Mineralization varies exceedingly in different places, but is seldom so considerable as was formerly imagined; the economic value of these gravels lies rather in their great extent. At Banka and Billiton the tin content of the ore-bed itself is 2–4 per cent, while in the Malay Peninsula, including the overburden, it is 0·1–0·15 per cent. By washing these gravels a concentrate containing 68–73 per cent of metallic tin is easily obtained.

The gravels of the Malay Peninsula were already known in ancient times, when they produced tin sufficient to supply the demands of China and India. At Banka production began at the beginning of the eighteenth century, and at Billiton about 150 years later. The production of Banka has increased from 5000 tons annually during the period 1850–1880 to 10,000–12,000 tons of late years. At Billiton, on the other hand, production has sunk in recent years to 4000–5000 tons. Perak ranks first among the tin-producing Malay states, its production in 1911 having been 437,339 pikuls, equivalent to about 26,000 tons, out of a total production of 741,098 pikuls for the Peninsula. The production of earlier years has already been given; [1] the following figures relate to more recent years:

TIN PRODUCTION OF THE MALAY STATES
In Pikuls of 133 lb. or 60·5 kg.

	1909.	1910.	1911.
Perak	461,665	421,335	437,339
Selangor	266,007	240,192	231,175
Negri Sembilan . . .	48,072	34,697	29,230
Pahang	43,144	40,674	43,954
Totals	818,888	736,898	741,698
Metric tons	49,529	44,579	44,870

OTHER TIN GRAVELS

The Spanish-Portuguese tin district in the Orense and Pontevedra provinces resembles in more than one respect that of Cornwall. The

[1] *Ante*, p. 443.

lodes are associated with gravels. In Portugal such gravels reach to the Tagus. These gravels were already practically exhausted at the beginning of the Middle Ages. The chief deposit was at Miranda on the Douro.

In Cornwall, notoriously, the gravels played an important part in ancient time as well as in the Middle Ages, while even at the beginning of the nineteenth century they were of considerable importance.[1] To-day, however, they are almost completely exhausted. In 1911, for instance, they were responsible for only 70 tons of ore.[2]

In Queensland, tin gravels occur in the districts Herberton, Cooktown, and Stanthorpe. In the last-mentioned district in the year 1911 three dredging plants won tin ore to the value of £9130 from 279,300 cubic yards of gravel. Of the tin ore production of New South Wales in 1911, some 1742 tons or 68 per cent came from the Tingha and Emmaville gravels, the content of which may be gathered from the fact that 577,977 cubic yards of material yielded 147 long tons of tin ore worth £17,565, or on an average 0·57 lb. of cassiterite worth 7·29d. per cubic yard.[3] In another case 2,533,782 cubic yards of gravel yielded 1418 tons of tin ore worth £167,284, or on an average 1·25 lb. of cassiterite worth 15·48d. per cubic yard. The gravel-deposits of Tasmania have already been described under Mount Bischoff.[4] In Western Australia, the districts Pilbara and Greenbushes are known for their tin gravels; the former up to the present has produced altogether 5000 tons of cassiterite. At Greenbushes, a map of which is given in Fig. 18, tin lodes of the Saxon type occurring in granite have upon disintegration given rise to eluvial and alluvial gravels, though these, owing to the high cost of labour and the small amount of water available, were worked with but variable success. This occurrence is particularly interesting owing to the presence of tantalum- and niobium minerals;[5] a tantalo-niobate of iron and manganese—$(Fe, Mn) (Ta, Nb)_2O_6$—occurs in pebbles 5–6 inches in diameter. Since the specific gravity of this mineral is the same as that of cassiterite, it contaminates the cassiterite produced. Stibiotantalite, an antimony-tantalo-niobate—$Sb(Ta, Nb)O_4$—is a mineral confined exclusively to Greenbushes; it occurs in small pebbles with a dull surface, and to some extent resembles scheelite. Stibiotantalite has also the same specific gravity as cassiterite, so that the separation of these two minerals by water concentration is impossible; the antimony upon smelting mixes with, and deteriorates the tin. According to the Departmental Laboratory, a case occurred where a parcel of tin ore, apparently

---

[1] Ante, p. 436.   [2] The Mineral Industry, Vols. XIX. and XX.
[3] Eng. and Min. Journ., April 8, 1911.   [4] Ante, p. 444.
[5] P. Krusch, 'Die Zinnerzlagerstätten von Greenbushes,' Zeit. f. prakt. Geol., 1903, p. 383; E. S. Simpson, Notes from the Departmental Laboratory Geol. Survey of Western Australia, Bull. No. 6.

clean, contained only 53·14 per cent of SnO_2 and no less than 15·13 per cent Sb_2O_3, 19·85 per cent Ta_2O_5, and 3·56 per cent of Nb_2O_5.

In Alaska, gravel tin is won at Buck Creek, 14 miles north of York on the Seward peninsula ; the plant, which is quite new, in 1911 worked only thirty-four days, producing 92 tons of cassiterite assaying 64 per cent of tin.[1]

South Africa in 1911 produced 4795 long tons of cassiterite which were shipped almost exclusively to England ; as the production in 1909 was only 2858 tons, the increase was considerable. In Swaziland, in 1911, the tin gravels yielded 476 tons worth £42,250 ; the official production for that year is given as 1119 tons, part of which was probably derived from sources yet unpublished. In Rhodesia, tin gravels were discovered in 1909 in the Uniabi river, where the primary deposit consists of a cassiterite impregnation in chloritic schist in the neighbourhood of granite. At the Salisbury Enterprise, cassiterite is found in pegmatite dykes more or less altered to greisen. Still farther north, at Abercorn, cassiterite occurs in greisen and pegmatite. The conditions for the formation of gravels in this country are not favourable, though perhaps those in the Abercorn district are payable. In the Victoria district to the east, large cassiterite crystals weighing as much as 15 kg. have recently been found, such as might well have come from more promising deposits.

In the Congo state, the Bussanga district north of the copper region [2] is stated to be stanniferous.

The discovery of tin gravels in Nigeria has recently aroused great interest. These extend over an area of about 9000 square miles stretching from Ningo in the extreme east, to Duchin Wei about 40 miles east of Zaria in the extreme west, and from Liruen-Kano in the north, to Ninkada and Mada in the south. In the year 1910 about 11,800 tons of cassiterite were exported. The opinions concerning the importance of this district are various, and further prospecting may be necessary before a definite opinion is possible. The development of tin mining in this region is hindered principally by difficulties of transport.

In Siam, tin gravels are known on the island of Tonka somewhat south of the Isthmus of Kra, which in 1908 produced 3713 metric tons of cassiterite from which 2302 tons of metallic tin were recovered. The neighbouring province of Monthon Chumpon, somewhat farther into the Gulf of Siam, is also rich in tin. The principal districts are Lang-Suan and Lampun. It is intended to open up the former by a railway.

In China, the tin ore production is derived almost exclusively from Yunnan. Nearly 97 per cent is exported from the port of Mengtze. The

---

[1] *Eng. and Min. Journ.*, July 15 and Dec. 23, 1911.
[2] *Ante*, p. 918.

ore is smelted at Hongkong. In 1910 the export of Chinese tin amounted to about 6000 tons, and in 1911 to some 5000 tons. These figures embrace only the export to Great Britain and Europe, and not the home consumption nor that exported to other countries.

THE GOLD GRAVELS

The gold in gravel-deposits is found principally with quartz sand, quartz pebbles, clay, iron-ochre, iron oxides, garnet, and other heavy minerals. Where the gravels have resulted from the disintegration of country carrying not only gold but also platinum, monazite, precious stones, etc., the gold in such deposits is associated with these minerals. Platinum and monazite are frequently overlooked in the examination of gravel-deposits. It is a significant fact that quartz pebbles carrying gold in themselves occur but exceptionally, and only when the primary deposit exists in the immediate neighbourhood. The gold contained in gravels is invariably native gold, the other auriferous ores not being sufficiently resistant. The lower layer of the gravel is frequently the most auriferous. Fluviatile gravels usually carry most gold at those places where the rapidity of the stream was for some reason or other temporarily diminished, as for instance in pools along the river bed, at the mouths of tributaries, and at sudden changes in direction, while a fractured condition of the bed-rock is also favourable to its collection.

Concerning the processes by which the gold accrued in these gravel-deposits, there is great divergence of opinion. Some authorities advocate an exclusively mechanical concentration, while others assume a chemical concentration in which the gold was introduced in solution and then precipitated, probably by organic substances, etc. Much has been written on this question, general discussion being found in the following works :

A. G. Lock. Gold, its Occurrence and Extraction, pp. 746-800. London, 1882.—E. Cohen. ' Über die Entstehung des Seifengoldes,' Mitt. d. naturw. Vereins f. Neuvor-pommern und Rügen, XIX., 1887.—A. Liversidge. ' On the Origin of Gold Nuggets,' Journ. Roy. Soc. New South Wales, XXVII., 1893.—C. Hintze. Handbuch der Mineralogie, pp. 241-244, 1898.

F. A. Genth, United States, in 1859 ; and A. R. C. Selwyn, Victoria, in 1860, were undoubtedly the first to express the opinion that gravel gold was formed by precipitation from solution, or that existing gold nuggets might continue to grow by precipitation from solution. Since then this view has been maintained by : P. Laur, Paris, 1863 ; C. S. Wilkinson, New South Wales, 1866 ; R. Daintree, Victoria, 1866 ; J. A. Phillips, England,

1868; J. Cosmo Newberry, Victoria, 1868; W. Skey, New Zealand, 1870, 1872; E. Suess, Vienna, 1877; T. Egleston, United States, 1880, 1881, 1887; and many others. Mechanical concentration on the other hand has been adopted and defended by: Brough Smyth, Victoria, 1869; Rod. J. Murchison, England, 1872; J. D. Whitney, United States, 1880; J. S. Newberry, United States, 1881; W. B. Devereux, United States, 1882; F. Pošepný, Austria, 1887; E. Cohen, Germany, 1887; and A. Liversidge, New South Wales, 1893.

The adherents to the chemical theory advance the following points:

1. The gold in lodes does not occur in such large masses as it does in the gravels.
2. The general form of gravel gold and the nature of its surface speak against formation by mechanical concentration.
3. The gold content frequently increases towards the bottom of the deposit, as do also such organic substances as would effect the precipitation of the gold.
4. Gravel gold is poorer in silver than that from lodes.
5. If the gold had reached the gravels as the result of the mechanical disintegration of quartz lodes, the occurrence of quartz pebbles in the gravels would be more frequent.
6. Many gravels have a higher gold content than a derivation from lodes would render probable.
7. Gravels already worked have occasionally after a time been worked a second time with profit. This need not necessarily be due to renewed precipitation, but more probably to improved methods of washing or to the new material constantly being brought by the river.

Cohen has shown that in part the above points are not quite sound, and that in part they may equally well be explained by mechanical formation, for which, according to Newberry, the following factors speak:

1. Gold gravels and auriferous quartz lodes occur as a rule adjacent to each other.
2. The gold is coarsest in the neighbourhood of the quartz lodes, and becomes finer with greater distance from them.
3. The accumulation of gold in pockets is most easily explained by mechanical concentration.
4. The occurrence of isolated flakes and grains speaks against the chemical precipitation of the bulk of the gold. Crystallization and the formation of small auriferous veins within the gravel are as a rule not observed.
5. A rough and uneven surface of the gold is uncommon.
6. Newberry emphasizes the intense denudation suffered in gold-bearing districts;
7. Cohen, the fact that in the gravels, apart from iron oxide, no other mineral characteristic of secondary deposits is found;
8. Liversidge, that the quartz pebbles of the gravels come undoubtedly from quartz lodes, and it would be unnatural to explain the source of the gold differently from that of the quartz.
9. In several cases gold has been known to occur in lodes, particularly in the cementation zone, in just as large masses as in gravels.
10. Gravel gold always contains some silver. In addition, in California and in Australia it has been observed that the fine flakes in the gravels are poorer in silver than the large nuggets, a fact which points to a leaching of the silver content.

Taking all these points into consideration, it would appear that the explanation of the gold content in gravels by mechanical concentration is in general correct. It must be conceded however that, principally with the rich gravels, a small portion of the gold has sometimes been taken into solution and subsequently chemically precipitated. The most important gold solvent in such case would be water containing ferric sulphate, while the agents of reduction were principally pyrite, ferrous minerals, ferrous sulphate, and organic substances. Examples of chemically precipitated gold from Western Australian gravels rich in mechanical gold, were exhibited in the Paris Exhibition. According to Vogt these examples showed :

Gold along fine cracks in the iron-ochre of the gravels.
Traces of gold in the quartz pebbles.
Gold along fine cracks in the indurated clay.
Gold on tree roots found in the gravels.
Small crystals of gold on secondarily formed cobalt-manganese ore.
Gold in stalactites consisting of ochre and calcite.

Liversidge has described similar occurrences. It is furthermore noteworthy that the pyrite occurring in the Californian deep gravels as a secondary formation cementing the quartz pebbles, is auriferous.

Concerning geological age, the gravels as a rule are Alluvial and Diluvial, more rarely Tertiary. Examples of Tertiary gravels are known in the west of the United States and in Australia, where they lie in part under volcanic sheets. In pre-Tertiary periods also, doubtless many auriferous fluviatile gravels were formed, but only in particularly favourable cases have these survived.[1] At the subsidence of districts containing fluviatile gravels to below sea-level, and before new beds could be deposited as a protection, such gravels as a rule were destroyed by the sea. In the further concentration by sea-water, in which not only regional subsidence but above all river transport to the sea took part, marine gravels were formed. Such are known on the present west coast of America from California to Cape Nome in Alaska, in New Zealand, and elsewhere. In regard to the shape of the pebbles, etc., the marine gravels differ somewhat from the fluviatile. By far the majority of the Quaternary and Tertiary gold gravels are fluviatile or true gravels, in comparison with which marine gravels are subordinate.

Most, and practically all economically important gold gravels have been formed from the old gold-quartz lodes ; with the young gold-silver lodes denudation as a rule has been less extensive. Since the old gold lodes are widely distributed, small amounts of gold are found in most large rivers, as for instance in the Rhine, the Danube, the Garonne, etc., though seldom in amount worth mentioning. The largest gravel-deposits are situated in California, Alaska, Australia, Siberia, and British Columbia. Gravel gold can easily be recovered by simple washing, by hydraulic mining,

---

[1] G. F. Becker, 'The Witwatersrand Banket, with Notes on other Gold-bearing Pudding Stones,' *U.S. Geol. Soc.* 18th Ann. Rep. V., 1896–97 ; L. de Launay, *L'Or dans le monde*, Paris, 1907.

and under favourable circumstances by dredging. Gold-washing is as old as history. The Quaternary or recent gravels lying immediately at the surface are as a rule not very extensive, and therefore are comparatively quickly exhausted. For instance, the gravels of the Spanish rivers Douro and Tajo, famous in the time of the Phœnicians and Romans, were ages ago exhausted, this being also the case with the rivers of South and Middle Europe. Even the rich river gravels of California were almost completely exhausted in the course of twelve to fifteen years, that is, from 1848 to 1860. In Australia also, the Quaternary gravels did not last long. Richer and more productive were the Tertiary gravels in the west of the United States and in Australia, these in many cases having been protected by a volcanic covering.

The gold content of gravels is very variable. The precious metal occurs as fine as dust, in grains the size of a pin's head, and in nuggets some of which have reached more than 40 kg. in weight. Schmeisser mentions five nuggets from Australia, each weighing more than 40 kilogrammes. While the German rivers, as for instance the Rhine, contain only a very small amount of gold, many rivers of California, Alaska, Siberia, Australia, etc., are quite rich in gold. With the Australian rivers an average of 0·5–4 grm. of gold per ton is given, though in some cases even 15 grm. and more has been found.

INDIVIDUAL OCCURRENCES

THE AURIFEROUS GRAVELS OF THE UNITED STATES

LITERATURE

J. D. WHITNEY. 'The Auriferous Gravels of the Sierra Nevada of California,' Memoirs of the Museum Comp. Zool. VI., 1880.—W. LINDGREN. 'The Neocene Rivers of California,' Bull. Geol. Soc. Amer. IV., 1893.—W. LINDGREN and F. H. KNOWLTON. 'Age of the Auriferous Gravels of the Sierra Nevada,' Journ. of Geol. IV., 1896.—G. F. BECKER, A. J. BOWIE, R. E. BROWNE, A. C. LAWSON, J. S. DILLER, H. G. HAUKS, J. LE CONTE, B. SILLIMAN, H. W. TURNER, J. A. PHILLIS, and many others in J. F. KEMP, Ore-Deposits of the United States and Canada.—CHARLES L. HENNING. Die Erzlagerstätten der Vereinigten Staaten mit Einschluss von Alaska, etc. Stuttgart, 1911, Ferdinand Enke.

CALIFORNIA

In the west of the United States, on the western slope of the Sierra Nevada, the Quaternary river gravels or shallow placers are differentiated from the geologically somewhat older deep gravels, which are probably of Tertiary age. The latter are frequently covered with basalt flows or tuffs, while their material was derived from the Californian gold-quartz lodes, principally from the Mother Lode.[1] Lindgren in 1896 gave the following sequence, beginning with the youngest, of a typical section from California:

[1] *Ante*, p. 608.

Volcanic and inter-volcanic depositions. { Andesite tuffs deposited by large rivers.
Gravels resulting from inter-volcanic erosion.

Volcanic and inter-volcanic depositions. { Gravels from a rhyolitic period. Pliocene or Upper Miocene.
Rhyolite tuffs with a maximum thickness of some 60 metres. These beds are thicker higher on the Sierra Nevada where the eruption took place. Miocene.

Pre-volcanic depositions. { Bench gravels up to 90 m. thick, covering and often extending for a width of 1½–3 km. on either side of the deep gravels, frequently mixed with somewhat finer detritus than the deep gravels, and containing Miocene plants. Deep gravels which in general are hard and compact and contain large quartz pebbles ; these occupy the bottoms of the old valleys, and are as much as 60 m., though generally 30–45 m. thick. They carry no fossil plants and are older than the bench gravels. Possibly Eocene or Lower Miocene.

In depth the auriferous gravel is often cemented by secondary pyrite, when it is known as blue conglomerate ; above, owing to the weathering of the pyrite, it is brown.

The great volcanic denudation periods responsible for the Californian

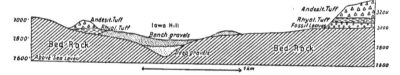

FIG. 460.—Section of the Californian gravel-deposits showing the relation between the bench gravels and deep gravels. Lindgren.

deep gravels belong, according to Whitney and Lawson, chiefly to the Pliocene. According to more recent examination by Lindgren and Knowlton, on the other hand, these deposits are of Eocene or Miocene age ; the old river-courses of these denudation periods have been extensively mapped by Lindgren. In any case it is apparent that the gold has experienced re-arrangement in different geological periods.

The shallow diggings at the surface were discovered in 1848, and exhausted in greater part as early as 1860, when the deep gravels were taken in hand and worked hydraulically. Since 1885 California has yearly produced gold to the value of 12-19 million dollars. The higher production of recent years may be attributed chiefly to the introduction of dredge mining in the place of hydraulic mining which, in 1887, owing to its disastrous effect on agriculture, was prohibited by law. Dredging has advanced of late years to such an extent that in 1907 gold to the value of more than 5 million dollars was recovered by this method of mining. Altogether, California up to 1900, from gravels and quartz lodes had produced about 1380 million dollars worth of gold.[1]

[1] *Ante*, p. 554.

Other gold gravels are found in Oregon, Washington, Wyoming, Idaho, Montana, and Colorado. In Oregon, along the coast, a belt cf auriferous sand occurs which often possesses a high metal content.

THE SUMPTER AND GRANITE DISTRICTS

LITERATURE

J. T. PARDREE. 'Placer Gravels of the Sumpter and Granite Districts,' U.S. Geol. Bull. 430a, pp. 51-57.—W. LINDGREN. 'The Gold Belt of the Blue Mountains of Oregon,' 22nd Ann. Rep. of the Survey, Part 2, p. 635.—CHARLES L. HENNING. Die Erzlager. statten der Vereinigten Staaten mit Einschluss von Alaska, u.s.w., Stuttgart, 1911, Ferdinand Enke.

In 1863 auriferous gravels were worked in the Powder, North Powder, and John Day river districts, in east Oregon. The geology of this region first became known in detail in the year 1909. Most of the gravels are now completely exhausted, and in but few places is gold still being won. According to Lindgren the following deposits are differentiated :

1. Pleistocene gravels. These form the present deposits and were the first to be worked. To-day they are considered exhausted.
2. Inter-volcanic gravels. Volcanic eruptions filled the deep valleys with lava and dammed up the upper Burnt, Powder, and John Day valleys, causing sand and pebbles to accumulate. These circumstances favoured the arrest of the gold, and auriferous gravels became formed at those places where rivers from gold-bearing districts emptied themselves into the impounded basins.
3. Pre-volcanic gravels. Eocene or later Miocene.

Such gravel-deposits could only retain their form through succeeding geological periods, when covered with volcanic lava. As however the lava flowed principally over the lower part of the hills and intense erosion has not taken place since, the gravels now generally lie deeper than the present river-courses, and search for them is consequently difficult. Only at Winterville and Parkville at the headwaters of the Burnt river have such deposits been worked.

The gravels beds of the Sumpter and Granite districts are bench- and deep gravels, that is, they belong to the first and second of the above-mentioned classes. They are probably genetically associated with glacier advance and the consequent accumulation of pebbles in the upper and deeper-cut valleys. Auriferous gravels in the United States belonging to the third class are of small importance. It is to be remarked however that the telluride gold lodes, though in general and as instanced by the occurrence at Cripple Creek little suited to the formation of gravels, have also occasionally given rise to such deposits.

THE GOLD GRAVELS OF ALASKA AND THE YUKON

LITERATURE

J. F. KEMP. The Ore Deposits of the United States and Canada. New York, 1900

Among the youngest gravels are those of the Yukon district. So far as these are yet known the richest lie in Canadian territory, though the earlier equipments were erected on the American side. The gold occurs in two different kinds of gravel. It occurs firstly on the bed-rock along the smaller stream beds and their tributaries, which latter are locally termed ' pups.' With these deposits the auriferous bed is overlaid by a barren, hard-frozen gravel of variable thickness, which in turn is frequently covered by peat. The auriferous bed is won in winter by thawing a shaft with fire and heated stones, and dumping the gravel until warmer weather allows it to be washed. Except for this frozen state these gravels do not in any way differ from normal gravels.

The second kind of auriferous gravel is the bench gravel [1] occurring on the valley side, above the present river level. These are regarded by Tyrrell as the moraines of small glaciers which disappeared after reaching but a short distance down the hillside. The precious metal of these gravels was probably derived from the quartz lodes of the Birch Creek, the Forty Mile district, and the Rampart Series.

South of the headwaters of the Yukon lies the Cassiar district which is reached from the coast through Wrangell and the Stickeen river. The most important discoveries here have been made in the Dease Lake district, where the precious metal is found in gravels which after their formation became covered by a considerable thickness of younger detritus. One of the roads to the Klondyke passes through this district.[2] Other Canadian gravels are known in the Chaudière river in east Quebec.

The auriferous gravels of the Cape Nome district are worthy of special mention.[3] This district is situated in Alaska, 64·4° north latitude and 165·1° west longitude, on Norton Sound on the north shore of the Behring Sea. It extends in a northerly direction from Mount Nome to Sledge Island, a distance of about 25 miles. The country here forms a gradually rising plain terminated by a low mountain chain. Through this plain the Snake and Nome rivers, with their many tributaries, meander. In these valleys the auriferous bed lies immediately under a peat covering 6–12 inches thick. The shore deposits, which in part have yielded good results, were discovered in 1899.

[1] *Ante*, p. 1201.
[2] G. M. Dawson, *Geol. Survey of Canada*, III., 1888, Rep. B ; E. D. Self, ' Cassiar District,' *Eng. and Min. Journ.*, Feb. 18, 1899, p. 205.
[3] H. Weber, ' Die Goldlagerstätten des Cape Nome Gebietes,' *Zeit. f. prakt. Geol.*, 1900, p. 133.

The precious metal of these marine gravels is distinguished from that of fluviatile gravels by the previously-mentioned[1] corroded appearance. Of recent years gravel-mining in this district has made considerable progress. In 1911 twenty-two plants were at work, either throughout the whole year or during the summer months. To these in the next year or so a further six will probably be added, the erection of these having already been commenced.

In 1910 the gold production of Alaska amounted to 11,985,000 dollars, and in 1911 to 12,700,000 dollars. This increase may be ascribed to the Iditarod-Innoko district, which, since the Fairbank and Seward districts in 1911 declined, is more important than these figures would indicate. In that year, according to Canadian statistics, the production from the Yukon gravels was about 4·58 million dollars, and accordingly most of these gravels lie outside Canada. British Columbia in the same year produced gold to the value of 468,000 dollars from gravel-deposits.

THE GOLD GRAVELS OF AUSTRALASIA

From the discovery of gravel gold in Victoria and New South Wales in 1851, or shortly after the sensational discoveries of gold in California in 1848, Australasia up to the present has produced almost as much gold as the United States. Thus, up to the end of 1910[2] Australasia had produced 4433 tons of gold, as against 4926 tons by the United States ; or according to other statistics and including 1911, Australasia 656 million sterling and the United States 716 million sterling.

Since the end of the 'nineties almost one-half of this Australian production has been derived from the telluride gold lodes of Western Australia ; [3] before this, Victoria with its exceedingly productive gravels was the most important producer. In the 'fifties, 'sixties, and 'seventies, the Australian gravels yielded even more than those of California. The following statistics covering the years in which the working of gravels was so important, give an idea of the productivity of these deposits.

GOLD PRODUCTION IN OUNCES

	To end of 1878.	1879–1895.
Victoria, since 1851	48,058,649	13,127,672
New South Wales, since 1851 . . .	8,811,346	2,583,246
Queensland, since 1861	2,901,092	7,623,352
South Australia	57,103	100,000
Western Australia, since 1886 . . .	...	686,361
Tasmania, since 1866	71,000	756,000
New Zealand, since 1857	8,959,482	3,797,240

[1] *Ante*, p. 1191.   [2] *Ante*, p. 644.   [3] *Ante*, p. 598.

In Victoria, which from 1852 to 1861 produced annually more than 2 million ounces—in 1856 even 3,053,744 ounces—and from 1862 to 1875 more than 1 million ounces, almost the whole of the gold was originally obtained from gravels. Lode-mining began fairly early, though even in the middle of the 'sixties the gravels produced double as much as the lodes. At the beginning of the 'seventies an approximately equal quantity came from the two classes of deposit; while only since the beginning of the 'nineties have the lodes produced more than twice as much as the gravels. In New South Wales also, the gravels were originally of great importance. In Tasmania, on the other hand, up to the end of 1895 only about one-quarter of the gold production was derived from gravels, and in Queensland only about one-tenth, the other three-quarters and nine-tenths respectively having been obtained from lodes. In Western Australia the gravels have always been of comparatively subordinate importance.

During the period when the Australasian gravels still played a prominent part in the production, Australasia—including that from lodes—produced during the quinquennial periods 1851–1855, 1856–1860, 1861–1865, 1866–1870, and 1871–1875 yearly averages of 69,573, 82,392, 77,634, 73,526, and 63,129 kg. of gold respectively. Owing to the great decline in gravel-mining, this average for 1876–1880 sank to 45,294 kg., and for 1881–1884 to only 43,186 kilogrammes. Later, gravel-mining suffered a still greater decline, while lode-mining on the other hand became more prominent, particularly in the Bendigo district, Victoria; [1] on Mount Morgan, Queensland; [2] and in the Waihi district, New Zealand. [3] In addition, towards the end of the 'nineties the exceedingly important telluride gold lodes of Western Australia [4] became large producers, considerably increasing the total production of Australasia, which in 1903 reached some 134,200 kilogrammes. Of recent years, however, it has fallen again, namely, to 107,200 kg. in 1909, and 92,800 kg. in 1911.

The gravel-deposits of Australia in section agree on the whole with those of America. The highest gold content is found in the immediate neighbourhood of the Silurian bed-rock. In Australia also, the early Pliocene deep gravels may be differentiated from the later overlying alluvials. Here also the old gravels are in part covered with basalt which, particularly in the Gippsland district, Victoria, reaches a thickness of 150 m., while the auriferous sands may exhibit a thickness of 10 metres. In addition, pre-Tertiary gravels of very early formation, since very old auriferous quartz lodes existed, are particularly interesting; gold is even found in the Upper Devonian and Lower Carboniferous conglomerates and sandstones.

[1] *Ante*, p. 611. [2] *Ante*, p. 640.
[3] *Ante*, p. 590. [4] *Ante*, p. 598.

The most important gravels are the Tertiary. The formation of these began in the Miocene, a period of intense erosion. In north Gippsland, between the Silurian rock and the auriferous sands, Maccoy found a bed containing fossil plants—*Cinnamomum polymorphoides*. Some of these old fluviatile gravels have been followed to their debouchment into Tertiary beds, and the assumption is justified that the deposition of gold continued in the marine beds at the river delta. Some Miocene gravels have been worked in Gippsland. Many others are covered with basalt, as for instance at Tangil in Gippsland. The Pliocene gravels are likewise fluviatile, more rarely marine. They owe their formation to the same period of erosion which, having commenced in the Miocene, reaches to the present.

Generally speaking, the gold of these deposits lies not far distant from the primary deposit. The more uncommon littoral gravels exhibit a fair regularity throughout their whole development. The fluviatile gravels are more irregular and are frequently covered with river deposits or basalt. When laid bare by subsequent erosion, they become 'shallow placers.' At Ballarat they occur in a Middle Pliocene valley. There the auriferous bed lies 50–150 m. deep and is 0·5–1·5 m. thick ; the bed-rock is likewise auriferous to a depth of 1·5–3 metres. The content of the gravels varies considerably ; the following weights per cubic metre hold good for Ballarat :

	In Loose Sands and Gravels.	In Cemented Beds.
1887	2·63 grm.	33·66 grm.
1890	2·15 ,,	3·63 ,,

In districts having sufficient water, gravels and sands lying at a depth of 100–150 m. and containing 1·5 grm. of gold per ton, can be worked profitably ; cemented beds however must contain at least 3 grm. in order to be payable. Some Australian gravels have yielded large nuggets : at Moliagul, for instance, a gold nugget weighing 71 kg. was found ; at Ballarat, several of 36–69 kg. at a depth of 45–50 m. ; and at Rheola in the Berlin district, some of about 30 kilogrammes.

The Pleistocene gravels are to-day of no importance.

In New Zealand [1] auriferous gravels are responsible for a considerable portion of the gold production, more particularly since the mountainous and well-watered character of the country permits hydraulic mining. Under these circumstances the cost of production is covered by 0·13 grm. of gold per cubic metre. As the recent gravels became exhausted

[1] L. de Launay, 'Les Richesses min. de la Nouvelle Zélande,' *Ann. des mines*, May 1894 ; *Traité de métallogénie*, 3rd Ed., 1913.

in a comparatively short time, to-day principally the old gravels are being worked. Here also glacial phenomena play an important part. The most important gravels are found in the south-west of the South Island, in the districts Westland, Inangahua, Grey, and Buller. At Westland south of the river Teremakau the sands are auriferous for a length of 150 km., along which length at Saltwater, Three Mile, and Five Mile, they have long been worked. Kumara in the centre, in the sixteen years up to 1895 produced gold to the value of 1·2 million sterling, from an area of 2,904,000 square yards. The Grey district in one year yielded 800 kilogrammes.

In Otago to the south of the island, gold washing is carried on more particularly at Blue Spur in the Tuapeka district and in the Clutha basin. At Blue Spur, work is almost exclusively confined to a conglomerate lying immediately on the bed-rock and consisting of pebbles of Ordovician quartzite-slate. This conglomerate appears to be of glacial formation and exhibits a stratification expressed by red ferruginous layers. It is particularly interesting owing to its great age. In its hanging-wall the lignites of the Oamaru formation occur ; these belong to the Eocene or perhaps even to the Upper Cretaceous. The gold deposit is therefore at least early Tertiary. In the Otago basin is situated one of the best-known auriferous districts of New Zealand, this district also extending to include the beds of the rivers Pomahaka and Waitahuna. Mining in this district is not confined to the young alluvials, but extends also, as in California, to bench or terrace gravels. The gold content is not high, but since the cost of production is low these gravels are payable.

THE GOLD GRAVELS OF SIBERIA

The auriferous gravels of this country, which play an important part in the Russian mining industry, are of considerable extent. Although these deposits represent a considerable portion of the natural wealth of Siberia, in consequence of the poor means of communication, generally speaking only those are worked in which the gold occurs in comparatively large grains. In the formation of these gravels glacial processes probably also played a part.

In the Urals the gravels to-day are almost exhausted. These are closely associated with gold-quartz lodes and are characterized by the irregular distribution of the gold, which is apparently arranged in strips following the lode-outcrops. In the gravel-deposit at Kotschkar, for instance, under about 10 m. of barren clay lies an auriferous sand 0·5–1·5 m. in thickness, containing 1–10 grm. of gold per ton. In the east of the Urals the auriferous beds reach 4 m., but on an average are 0·5–1 m. in thickness. They are usually 20–40 m. long, though exceptionally as much as 200 metres. In the Bogoslowsk district, on the other hand, deposits 4–6 km. in length and 20–40 m. wide, are known. In this district the auriferous sands fill pockets or irregularities, usually 1–4 m. deep, in the bed-rock. They are frequently covered by a peat layer. The richest portion is generally found at the base, immediately on the bed-rock,[1] though occasionally a second auriferous bed occurs at a higher level. The gravels occur in recent as well as in old valleys. The precious metal occurs as fine flakes which are sometimes very irregularly distributed in the sand. Large nuggets have been found in the Tzarevo-Alexandrowsky deposit in the Miass district. The gold is almost invariably associated with magnetite, hæmatite, ilmenite, and chromite, while platinum also occurs fairly frequently. Garnet, coloured precious stones, zircon, disthene, and now and then the diamond, are characteristic of the Ural deposits. The water-courses which disintegrated the gold lodes destroyed also many pegmatite dykes. These deposits are post-Tertiary and often very young; they contain remains of mammoth, rhinoceros, and occasionally artefacts. Almost all occur on the eastern slope, the western slope being poor in auriferous gravels.

The Siberian deposits are responsible for almost the whole of the Russian gold production, which since 1907, when the Yakutsk district became a producer, has risen considerably.

In the Yenissei basin the gravels occur on the right bank of the river, between Angara and Podkamennaia Tounyouska. They are also found on the Teya, the Enachimo, the Pit and its tributaries, as well as in the district of the Oudérei and Oudoronga.

In the Lena basin the gravels occur principally on the Potam plateau between the Vitim and the Lena, or on the Olekma. Of late years successful operations have been prosecuted on the Lenskoié in the neighbourhood of Bedaibo. In the Olekma district the section is as follows :

1. Recent fluviatile and eluvial formations.
2. Upper clay and loam up to 35 m. thick, as at Alexandrowsky.
3. Inter-Glacial sand and gravel.
4. Upper indurated clay, lower indurated clay.
5. Pre-Glacial sediments and Eluvium.

The gravels occur both in the lower portions of present valleys and along higher terraces, of which latter with some rivers there are several. The gold is in part so fine as to float in water, and in part coarse-grained; only in exceptional cases are large nuggets found, though such may occur weighing as much as 40 grammes. The auriferous bed is overlaid

---

[1] L. de Launay, *Traité de métallogénie*, 1913, Vol. III. p. 716.

by barren material sometimes 40 m. thick. The gravel itself usually lies immediately on the bed-rock. Exceptionally, a still higher auriferous bed is found having barren material for the foot-wall, as at Proroko-Iliinsky on the Nigri. Often at the base of the alluvials a thin bed consisting of clayey sand with rock fragments is met, the finer material of which may penetrate down along the bedding-planes of the up-tilted bed-rock for sometimes as much as one metre. In this bed gold nuggets such as may be picked out by hand occasionally occur. Upon it lies the main auriferous bed, which is usually 0·75–1 m. thick and only exceptionally more than 2 metres. This consists of clayey sand with numerous rock fragments, and contains pyrite, this mineral forming the bulk of the concentrate obtained by washing. The gold content is very variable; it may reach as much as 20 grm. per ton, though usually the material washed is poorer. In general the richest portions lie along the middle of the valley, from whence the content diminishes on both sides. In the Vitim basin the pre-Glacial gold deposits fill old and young valleys. They contain 15 grm., and exceptionally as much as 30 grm. of gold per ton. The alluvials consist of sandstone- and slate pebbles, these being frequently cemented by pyrite. Towards the top the sandy and clayey particles increase; in this portion, however, there is no gold.

In the inter-Glacial and post-Glacial sands also, the precious metal occurs at different horizons. In these the gold is rounded, while the pre-Glacial gold owing to the shorter distance it has been transported is usually angular. The average gold content of these more recent gravels is 4 grm. per ton.

In the Nijni deposit on the Bodaibo, the auriferous bed contains pyrite with 344 grm. of argentiferous gold per ton. At Konstantinovsky the pyrite has assayed as much as 1582 grm. of gold per ton. In these gravels gold and pyrite are accordingly most closely associated, a part of the gold having doubtless been deposited from solution. The total thickness of the alluvials reaches 150 metres. In the Kroutoi deposit, the most important in the Engagimo basin, the auriferous bed varies between 1 and 2 m. in thickness. It lies 25 m. below the surface, and consists of small pebbles with a clayey sand containing 7–8 grm. of gold per ton.

In the Amur basin most of the tributaries of the Amur are gold-bearing. The principal deposits are those of Djilinda and Djolon, etc. in the Zeya district. The gold occurs in comparatively large and angular pieces and appears to have suffered no great amount of transport. In the Zeya district the coarser beds with boulders of 30 cm. diameter usually contain much gold, while the gravels and sands containing finely comminuted gold are poor. In accordance with the varying nature of the bed-rock, the character of the auriferous gravels is very variable. Sometimes the bed-rock is covered with large boulders, while at other times, in the amphibolite-schist district, green sands containing much amphibolite detritus and pyrite are found. In the Leonovsky deposit—the most important belonging to the Djolon company, having in ten years produced gold to the value of 30 million francs—the average content was 10 grm. per ton, while in other gravels of the same district it has varied between 2 and 3 grammes. The gravels of the Kerbi, one of the tributaries of the Amgoun, appear in greater part to be eluvial ; they consist of altered slaty material. The gold content reaches exceptionally 6 grm. per ton, while the abundant pyrite, which remains behind in the sluice boxes, contains 3 grm. per ton. The Nagorny deposit on the Amur lies in the district of Klein-Khingan. Here, exceptionally, an upper older terrace, about 80 m. above the river Soutar, occurs. The gravel lies upon a fine-grained granite. The auriferous bed, consisting of clay, sand, and gravel, is 0·75–5 m. thick. It is overlaid by a 10–30 m. thick, more or less clayey or loamy bed. The gold content of the gravel is 2–23 grm. per ton. In the upper bed several other layers with approximately 1 grm. of gold per ton occur. Some authorities, including Jaczevsky who made a detailed examination of the Baikal district, expect that, analogous to the Californian deposits, auriferous gravels will be found under the basalt sheets.

The Gold Gravels elsewhere

In Manchuria, gravels formed by the disintegration of gold-quartz lodes occur, sometimes as older deposits on plateaus, sometimes as younger depositions in valleys, and finally as marine beds ; these last are the most important. The sea-bottom for about 100 m. from the coast is covered with pebbles, which lie on pre-Cambrian clay-slates. A bulk sample, washed by fifteen men in 2 hours, gave 166 grm. of gold, including one nugget which weighed 55 grammes.

In Korea [1] practically every one of the numerous river-courses contains some free gold. Among the best known deposits is that at Tangkogae, situated 160 km. north-east of Seoul between Kimsong and Hoyang, in the upper portion of the most northerly tributary of the Han river. These gravels have been worked for fifty years, and at times as many as 20,000 men have been employed. The country around consists of a cavernous limestone. By the collapse of this limestone a large depression 1 km. long and 500 m. wide became formed, into which three large and two small auriferous streams now flow. In these streams themselves gold gravels occur only to a small extent ; in the depression, on the other

[1] L. Bauer, 'Das Goldvorkommen von Tangkogae in Korea,' *Zeit. f. prakt. Geol.*, 1905, p. 69.

hand, extensive auriferous gravels are found. In thickness these vary between 3 and 15 metres. The contained boulders are large, blocks of 1 cubic metre being not uncommon, while sand beds occur subordinately. The gold content is rather irregular ; it increases towards the bottom, though no definite auriferous bed can be distinguished. Apparently, frequent floods have prevented the formation of a regular gravel. The Tangkogae gold is fairly coarse and of great purity ; small nuggets weighing as much as 16 grm. have been found, though usually the gold occurs as flat scales free from attached country-rock. Bauer states that native lead has been found in these gravels, such lead, according to Beck, being occasionally intimately intergrown with the gold. Since, however, at places in this district silver ores formerly were smelted, it is not impossible that the lead was introduced into the gravels by man.

In Russian Turkestan,[1] the gold gravels of Bokhara and of the Kaschgar district are to-day almost exhausted. These were among the first of all deposits to be exploited, some authorities even considering that they represent the Mount Ophir of King Solomon. The gravels were subsequently worked for a long time by the Mongolians and the Sarts. Modern mining operations began in 1894, in which year the deposits attracted the attention of the Russian government, following which they were examined, and in 1897 declared open. Prospecting work, however, has shown that these gravels are in greater part exhausted and that only the present river-courses carry gold to any extent. The primary gold deposits have not yet been definitely determined ; possibly they were Tertiary conglomerates with a low gold content.

Auriferous gravels have also been found in Mongolia, Chinese Turkestan, Tibet, etc. In eastern Tibet, an auriferous district lies east of Koukou-Nar. While formerly much gold was won from this district, since the invasion of the Dounganians (?) in 1863, further work has been abandoned. Workable deposits have also been found in the Altyn-Tag and Kuen-Lun mountains. According to Johnson, on the northern slope of these latter, 3000 men were employed in the Kerija mines, from which, according to Prjevalski, gold to the value of £110,000 was won annually. Farther west, on the northern slope of the Karakorum range, the upper portion of the Tarim or Yarkand-Daria river likewise carries gold. In western Tibet, auriferous gravels have long been worked east of the Upper Indus. This district, termed by the natives *Sarthol* or gold-land, is said to be the

¹ L. de Launay, *Richesses minérales de l'Asie*, 1911 ; A. v. Krafft, ' Mitteilungen über das Ost-Bokharische Goldgebiet,' *Zeit. f. prakt. Geol.*, 1899, p. 37 ; D. Levat, ' Notes géol. sur les richesses minérales de la Boukharie et du Turkestan,' *Bull. de la Soc. géol. de France* 4ᵉ série, Vol. II. p. 439, 1902 ; ' Turkestan et Boukharie,' *Mém. Soc. Ing. civils*, Sept. 1902, p. 42 ; ' Richesses minérales des possessions russes en Asie centrale,' *Ann. des mines*, 1903, p. 174 ; L. de Launay, *Traité de métallogénie*, Paris, 1913, Vol. III.

famous Gold Land of Herodotus. In it, systematic mining was carried on in the beginning of the nineteenth century. The gravels at Thok-dschaloung, like most of those in Tibet, lie very high, namely, 4980 m., and are worked principally in winter. Altogether, according to Reclus, the gold production of western Tibet amounts to about £8000 per year.

In India,[1] the principal gold deposits occur in a hornblende-schist, chlorite-schist, and clay-slate district, more or less associated with eruptive rocks; Foote embraces these beds under the term Dharwar series; they lie on gneiss. At the commencement of the nineteenth century gold was found west of the Nilgiri Hills in the courses of several rivers; in 1802 gold washing was discovered in progress near the village of Wurigam in eastern Mysore; in 1868 the gravels of Betmangla, which proved to be too poor to be worked, were discovered; and in 1870 those on the Hemagiri Hill. Early in the 'seventies small amounts of gold were won in the Bet-mangla and Kolar districts. By far the largest amount of the gold from the Mysore goldfield is however obtained from lodes. In the province of Orissa, alluvial gold occurs in the native states Dhenkanal, Konjhar, Pal Lahara, and Talchi; as with most other Indian auriferous gravels, the gold content is low, and these deposits are unimportant.

In Africa, auriferous gravels are known in Tunis, Senegal, in the Soudan, in French Guinea, on the Ivory Coast, in Abyssinia, in the Congo district, in Angola, the Transvaal, Rhodesia, and Madagascar. In Mada-gascar those at Ikopa, Betsiboka, Mahajamba and Bemarivo, and Manan-bulo, containing 1–2·5 grm. of gold per ton, deserve mention. In the Belgian Congo[2] during 1911 about £65,000 was spent by the state in exploration work undertaken largely in the auriferous district of Kilo to the west of Lake Albert. In 1910 the production from the state mines amounted to 876 kg. of gold worth £105,000, and in 1911 to 700 kg. worth £84,000, this decrease being consequent upon the exhaustion of some of the richer fluviatile gravels. The importance of the auriferous gravels at Nebula, about 110 miles north-north-east of Stanleyville, appears to have been exaggerated; the value of the gold produced in 1912 was anything between £4000 and £10,000. The deposit at Kanwa, 3 miles from Nebula, appears to be promising. In the year 1911 prospecting was undertaken at Moto, 325 miles north-east of Stanleyville. The total yearly gold production of the Belgian Congo has a value between £100,000 and £125,000.

The European auriferous gravels are to-day of no importance. In France, auriferous gravels are known in Brittany, on the Central Plateau,

---

[1] Phillips and Louis, A Treatise on Ore Deposits, 2nd Ed., London, 1896; V. Ball, 'Manual of the Geology of India,' Part 3, Economic Geology, p. 176, Calcutta, 1881.
[2] Sydney H. Ball, The Mineral Industry for 1911, p. 303.

in the Cevennes, in the basin of the Rhone, and in the Pyrenees, these having formerly delivered their gold to the Mint at Toulouse. In Germany, the Rhine is known to be auriferous, gold-washing in its bed dating back to the seventh century. Daubrée estimated the production in 1846 by 500 gold-washers at £1600.[1] In addition, he showed that not only the present river sands but also the older deposits, situated 10–20 km. from the present Rhine, carry gold. The gold content of the Rhine sand amounts to 0·014–1·011 grm. per cubic metre. In a nine hours' day one man has obtained gold to the value of 9 shillings. The gold in these gravels is associated with titaniferous iron, quartz, zircon, and, according to Doebereiner, some platinum. The primary deposit has not yet been located. In Spain, auriferous gravels occur along the Rio Sil and Rio Duerna in the Granada plain. These are unimportant, as also are others in northern Italy and in the Balkan Peninsula. In Transylvania, in the district of the young gold lodes, some gold was formerly washed from gravels. In north Finland, where some poor auriferous quartz lodes occur, gold has been washed from fluviatile gravels in the river Ivalo and its tributaries, that river itself being a tributary of the Enare. From the beginning of work in 1869, to 1899, gold containing only 5·5–6 per cent of silver and having a total value of about £55,440, was won; since then, production has been low. The richest gravels are stated to have yielded 2 grm. of gold per cubic metre. At Karasjok also, in the adjacent portion of Finmarken in Norway, gold has been obtained by washing, namely, from 1898 to 1901 some 7 kg. or including that obtained afterwards, some 20 kg., equivalent to £2500. According to H. Reusch [2] the gold occurs in the lower portion of the Åser, a bed consisting of material deposited from rivers under the ice sheet. These are accordingly fluvio-glacial gravels.

THE PLATINUM GRAVELS

Platinum was discovered in 1755 in the auriferous sand of the river Pinto in the province of Choco, Colombia. It was subsequently found in gravels in British Columbia, Minas Geraes, Borneo, etc. and particularly in the Urals, the gravels of this last place still yielding the largest part of the platinum production. This metal invariably occurs in an alloy of iron and platinum found associated with much chromite, magnetite, and the decomposition products of peridotite. The intergrowth of iron-platinum with other minerals, such as chromite, olivine, and pyroxene, of which

---

[1] Daubrée, 'Sur la distribution de l'or dans la plaine du Rhin,' *Ann. des mines*, 1846, 4e séries, Vol. X. p. 1 ; *Min. du Bas-Rhin*, 1872.

[2] *Norwegian Geol. Exploration*, No. 36, Report for 1903.

intergrowth excellent examples are found more particularly in Colombia, is especially interesting. The primary platinum deposits have already been described.[1]

The platinum gravels occur either in districts of fresh or serpentinized peridotite, or in rivers flowing from such districts. These gravels, in addition to iron-platinum frequently containing some iridium, osmium, ruthenium, palladium, rhodium, and copper, occasionally carry some osmiridium. The platinum content of the metal won varies between 70 and 96 per cent, and is usually 80–90 per cent ; the iron content may reach more than 10 per cent ; while rhodium, osmium, and iridium, may together amount to several per cent. The osmium-iridium content is manifested upon the dissolution of the platinum, when bright dark-grey flakes of osmium-iridium remain, these consisting in all probability of sisserskite, which mineral, according to Dana, contains 30 per cent of iridium. The determination of the amount of the other platinum metals present is therefore important. The platinum of different deposits is frequently of very dissimilar composition ; comparison of analyses from the Urals and Colombia would, for instance, show considerable differences. Although the average platinum content is approximately the same, the iron content of the Colombian platinum is substantially lower than that from the Urals. Another notable difference lies in the iridium-, osmium-, and osmium-iridium content. In the Ural platinum this amounts at most to a few per cent, while in that of Colombia it reaches nearly 7 per cent. In general the colour of the platinum from Colombia is somewhat lighter than that from the Urals.[2]

Owing to the high price of platinum, gravels may be worked when the metal content is but a fraction of a gramme per cubic metre, provided other conditions are favourable. These conditions are large quantities of gravel and sufficient water, so that dredge- or hydraulic mining may be practised. The payability of platinum deposits therefore is less burdened with conditions than is the case with gold.

The chief producer of platinum[3] is Russia, which country in 1909 produced according to official statistics 156,792 oz., but in reality about 250,000 ounces. This difference is explained in that the official figures do not take into account the considerable amounts stolen by employees. The production of recent years is given a few pages ahead.[4] Next in importance comes Colombia with a production of 8800 ounces in 1908. The United States produces only a fraction of the world's production, namely, in 1908 about 750 ounces ; then follow, New South Wales with

[1] Ante, p. 342.
[2] P. Krusch, Untersuchung und Bewertung von Erzlagerstätten, Stuttgart, 1911, 2nd Ed., Ferdinand Enke.
[3] P. Krusch, loc. cit.   [4] Postea, p. 1219, 1220.

about 530 ounces, and Sumatra and Borneo with 500 ounces, in 1908.   A
few kilogrammes of platinum and palladium are recovered annually from
the anode slime which falls as a by-product in the production of pure
nickel in the electrolytic nickel refinery at Kristianssand in Norway, which

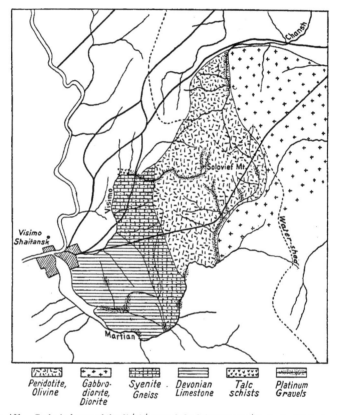

| Peridotite, Olivine | Gabbro-diorite, Diorite | Syenite . Gneiss | Devonian Limestone | Talc schists | Platinum Gravels |

FIG. 461.—Geological map of the district around the Solovief mountain in the Urals showing
the platinum gravels of the Martian basin.   Kemp.

treats furnace products from the Norwegian nickel-pyrrhotite.[1]   The
platinum - producing countries other than Russia together produce but
little more than 10,000 ounces annually, as against 250,000 ounces from
Russia.

In 1912 the price of platinum was about 6s. 3d. per gramme.   Since
native platinum as won from gravels contains only approximately 80 per

[1] *Ante*, pp. 155, 283.

cent of the pure metal, one gramme of such platinum is therefore worth
4s. 10d., while one gramme of gold is worth only 2s. 10d. To that of the
platinum must also be added the value of the other platinum metals
present, the prices of which in 1910 were :

Palladium	.	.	.	3s. 10d. per gramme.
Rhodium	.	.	.	16s. 0d. ,,
Iridium .	.	.	.	4s. 11d. ,,
Osmium .	.	.	.	3s. 10d. ,,
Ruthenium	.	.	.	7s. 11d. ,,

A few years ago the *Société Industrielle du Platine* was formed in
Paris, a company which is responsible for about 60 per cent of the Ural
production, and to-day has great influence upon the platinum market.
Russian producers are however endeavouring to render their deposits
independent of the foreign refineries and the price of native platinum
free from foreign influence. Although so far this has not been fully
achieved, a steady rise in the price has been obtained. In October 1908
one kilogramme of platinum fetched £140, while at the end of that year
£166 was obtained. Subsequently, for a time the price fell to £158, but
eventually rose to £187 in 1909, £216 in 1910, and £300 in 1911.

THE DEPOSITS

THE URALS.—The primary platinum deposits, discovered at the beginning of the 'nineties, have already been described,[1] together with some
features of the gravel-deposits. Three districts where platinum is won
are differentiated, namely, the most important, that of Nischni-Tagilsk
on the western slope of the Urals, and extending only a short distance on
to the eastern slope, as illustrated in Figs. 461, 462 ; that of Goroblagodatsk and Bissersk in the basin of the Iss and Dyja, illustrated in Figs.
462, 463 ; and that of Nikolai-Pawdinsk.[2] The two last-named districts
are situated on the eastern slope.

The gravels of Nischni-Tagilsk according to all accounts are eluvial.
The platinum occurs generally without gold. The superficial extent of
these gravels is 180 square versts ; they are grouped around the Solovief
mountain and are associated with the rivers Martian, Winzm, Tschausch,
Syssimka, etc. The deposits were discovered in 1825, and up to the year
1895 had produced 5514¼ poods of platinum.[3] The fluctuations in production are due to the irregularity of the platinum in the gravels and to the
varying price of the metal. The yearly output varies between 60 and

[1] *Ante*, pp. 342-345.
[2] *Ante*, p. 343 ; *The Mineral Industry for 1911*, p. 607.   [3] 1 pood = 16·4 kg.

130 poods. The gravels of Goroblagodatsk are centred around the Katsch-kanar and Sarannajy mountains; they were discovered in 1825 in the tributaries of the Iss, and active operations have been carried on since the 'sixties. The production varies between 100 and 180 poods per year. In the Bissersk district, where mining likewise began in the 'sixties, the production is smaller, only exceptionally reaching 50 poods.

The characteristics of the Ural gravels may be gathered from the following details.[1]

The distribution of the metal is everywhere irregular. As a rule the individual platinum grains are small, though nuggets weighing more than 1 kg. have been found in places. The platiniferous bed of the Tura, which may be taken as an example and the basin of which is illustrated in Fig. 463, consists of sand and large boulders, under which comes the platinum gravel 2–7 feet thick; on top, in places there is a covering of peat.

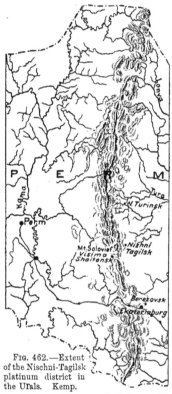

According to a report placed at the authors' disposal, the cost of production amounts to 4000 roubles per pood, equivalent to about 240 roubles per kilogramme, or about 6d. per gramme of platinum, a figure with which the value of the metal compares very favourably. Where hydraulic mining can be applied, that is, where sufficient water under the necessary pressure is available and the character of the deposit is suitable, work is cheaper than with dredging.

Fig. 462.—Extent of the Nischni-Tagilsk platinum district in the Urals. Kemp.

Exploitation is carried out either by contractors, the so-called *Starateli*, or by day-work. In the first case small plots are allotted to peasants who are paid for the metal they deliver. By this method no capital for plant is required and the work is done fairly cheaply; there is however the disadvantage that mining is not done systematically. When the beds are more extensive and thick peat layers occur, the whole district is divided up into plots, though large areas must be set apart on which to stack the peat; such areas are probably lost, even in respect to a possible subse-

[1] Krusch, *loc. cit.* p. 310.

quent working. From the nature of such an arrangement it follows that
the poor parts of the deposit are not worked, since the men are only
interested in obtaining as much metal as possible. In addition, the losses
incurred by the very primitive means of washing are said to amount to
one-third and sometimes even more. The price paid to the *Starateli*
varies between 1 rouble 50 copecks and 2 roubles 50 copecks per
zolotnik,[1] that is, from 1s. to 1s. 10d. per gramme.

With day-work the cost of production is greater, in spite of which
this method is better because the losses are less and work proceeds

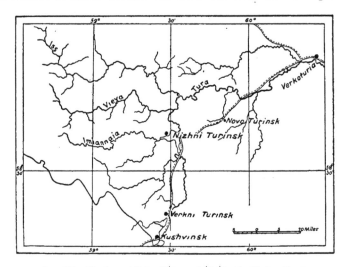

FIG. 463.—The Iss and Tura platinum district in the Urals.   Kemp.

systematically, that is, without leaving any unworked ground. By this
method the cost per cubic fathom of sand treated, including the cost of
lifting the peat, amounts to 5·50–5·70 roubles, equivalent to 1·90 roubles
per zolotnik, or 4000–7200 roubles per pood. By improvement in working,
this cost might be considerably reduced.

The payability of the platinum gravels depends firstly upon the
platinum content, but also to a considerable extent upon the means of
communication and transport and the thereupon dependent labour supply.
For the latter, the situation in regard to water and fuel, and the length of the
working hours, are important factors. The average wage of a workman in
summer is from 75 copecks to 1 rouble per day, and the maximum about
1 rouble 75 copecks. Water for gravel washing is everywhere abundant.

[1] 1 zolotnik = 4·26 grm.

Similarly, fuel and timber are generally available, since the forests, against payment of a small wood-cutting licence, are given over by the Forestry department to the requirements of the mines. One cubic fathom of birch timber, including felling and transport, costs 6 roubles. Seeing that

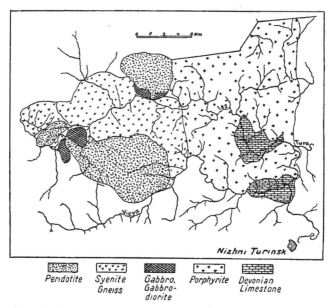

| Peridotite | Syenite Gneiss | Gabbro, Gabbro-diorite | Porphyrite | Devonian Limestone |

FIG. 464.—Geological map of the Iss platinum district in the Urals. Kemp.

snow falls thickly as early as October and remains on the ground until the beginning of April, the working period may be reckoned at only six months.

The platinum production of the Urals may be gathered from the following table :

	Official Statistics.	Actual Production.
	Ounces.	Ounces.
1900	163,060	212,500
1901	203,257	315,200
1902	197,024	380,806
1903	192,976	276,000
1904	161,950	290,120
1905	167,950	200,450
1906	185,492	210,318
1907	172,758	310,000
1908	156,792	250,000

The composition of the Ural platinum is given in the following analyses :

	Blagodad.	Nischni-Tagilsk.
	Per cent.	Per cent.
Platinum	81·5	88·87
Iridium	0·9	0·06
Osmium	0·06	Trace
Ruthenium . . .	...	...
Palladium . . .	0·05	1·30
Rhodium	2·9	4·44
Iron	Nil	Nil

According to Clark,[1] the iridium found in the Urals has the following composition :

Iridium .	.	.	.	76·80 per cent.
Platinum	.	.	.	19·64 ,,
Palladium	.	.	.	0·89 ,,
Copper .	.	.	.	1·78 ,,
Total	.	.		99·11 per cent.

The following are two analyses by Clark of the iridosmium of the Urals :

Iridium .	.	.	.	77·20 per cent.	43·94 per cent.
Platinum	.	.	.	1·10 ,,	0·14 ,,
Osmium .	.	.	.	21·00 ,,	48·85
Rhodium	.	.	.	0·50 ,,	1·65
Ruthenium	.	.	.	0·20 ,,	4·68
Copper .	.	.	.	Trace	0·11
Iron	.	.	.	—	0·63 ,,
Total	.	.		100·00 per cent.	100·00 per cent.

The export for the year 1911 is given as 221,201 ounces of native platinum, worth £1,580,000 ; and for 1910 some 272,815 ounces, worth £1,386,000. These figures, chiefly because of the amounts stolen by the miners employed, differ considerably from the figures of production, in which such amounts are not included.

COLOMBIA.—In Colombia, platinum occurs in Diluvial sediments, together with gold, zircon, magnetite, etc., in the river Cauca,[2] in the provinces Choco and Barbacaos. Platinum was indeed first discovered in the auriferous gravels of the Rio Pinto at Popoyan. The strip of country separating the headwaters of the Rio Atrato and Rio San Juan is particularly productive. A second platinum district is situated in Antioquia near Santa Rosa de Osos north-east of Medellin, concerning which however the available descriptions do not permit a clear conception of the circumstances in which the grains of platinum are found.

Krusch,[3] after Buttmann, gives more detailed information concerning

[1] 'The Data of Geochemistry,' *U.S. Geol. Survey*, Bull. 491, Washington, 1911.
[2] Hintze, *loc. cit.*   [3] *Loc. cit.*

the gravels on the west coast, in the area traversed by the rivers Opogodo, Condoto, and Iro. In this district also, the metal occurs in fluviatile and eluvial gravels. While in the former the platinum occurs as rounded flakes, in the latter it is present as smaller and larger, sharp-edged and but little rounded grains. The striking intergrowth of the metal with particles of more basic minerals, especially with a black mineral, presumably chromite, is significant. This mineral-association arose at the

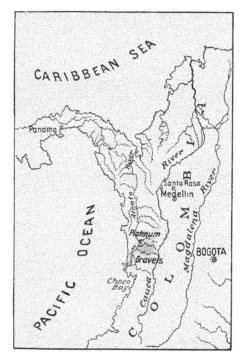

FIG. 465.—The platinum gravel-deposit of Colombia. Kemp.

consolidation of the original olivine eruptive magma, and its continued existence is proof that the platinum nuggets in which it is still maintained have suffered no great amount of transport.

The geological circumstances of the Colombian deposits may be gathered from the following remarks culled from a report by Buttmann. On the Guineo concession olivine- and serpentine boulders are found throughout the gravels, no matter whether these be fluviatile or eluvial. Where the fluviatile deposits lie immediately on the country-rock this latter weathers considerably, and between the fluviatile formation and the hard bed-rock, transition fluviatile and eluvial gravels become formed. The pebbles seldom reach a diameter of 20 centimetres. The gravels are in part laterized to a depth of 0·5–4 metres. Over large areas, however, they consist exclusively of soft material, such as in all probability would present no difficulty to hydraulic mining.

Concerning the platinum content of the Colombian gravels, the results obtained by the Colombian Syndicate have been placed at the authors' disposal. According to these, in the Guineo district the content varies between 0·115 grm. and 0·874 grm. per cubic metre. Assuming that the

FIG. 466.—Section through the platinum gravels of the Condoto district, Colombia. Kemp.

native platinum contains 80 per cent of pure metal, these figures correspond approximately to 4½d. and 3s. 10d.

The composition of Colombian platinum may be gathered from the two following analyses :

Platinum .	.	.	.	.	84·05 per cent.	85·50 per cent.
Osmium-iridium .	.	.	.		6·90 ,,	1·10 ,,
Soluble iridium .	.	.	.		1·72 ,,	1·05
Iron .	.	.	.	.	5·52 ,,	6·75
Copper .	.	.	.	.	1·76 ,,	1·40
Other platinum metals .	.	.			Trace	1·60 ,,
Total .	.	.			99·95 per cent.	97·40 per cent.

The following is an analysis by Clark of the iridosmium of Colombia :

Iridium	.	.	.	70·40 per cent.
Platinum	.	.	.	0·10 ,,
Osmium	.	.	.	17·20 ,,
Rhodium	.	.	.	12·30 ,,
				100·00 per cent.

UNITED STATES.[1]—The production of native platinum in the United States comes exclusively from the gravel-deposits of California and Oregon, where it is obtained as a by-product when working auriferous gravels. The Californian production probably amounts to 400 ounces, of which 85 per cent is derived from the auriferous gravels of Butte, Yuba, and Sacramento counties. The remainder is derived from the hydraulic mines of Trinity,

[1] Frederick W. Horton, *The Mineral Industry*, 1911, p. 597.

Shasta, and Humboldt counties, and Siskiyou, only a few ounces being won from the marine sands of del Norte and Humboldt counties.

The Oregon gravels are responsible for 70 ounces of native platinum yearly, two-thirds of this amount being obtained from the marine sands of Curry and Coos counties, and the remaining third from the river sands of Josephine and Jackson counties. The total production of the United States in 1911 amounted to only 470 ounces, the pure metal in which was about 70 per cent. To this must be added the amount recovered in refineries from bullion, etc., which may be taken as 600 ounces fine.

A nugget of American platinum gave the following analysis : [1]

Platinum	82·81 per cent.
Iridium	0·63 ,,
Palladium	3·10 ,,
Rhodium	0·29 ,,
Copper	0·40 ,,
Iron	11·04 ,,
Alumina	1·95 ,,
Lime	0·7 ,,
Magnesia	0·3 ,,
Total	100·32 per cent.

The platinum production of the United States may be gathered from the following table :

	Amount.	Value.
	Ounces.	Dollars.
1901	1408	27,526
1902	194	1,874
1903	110	2,080
1904	200	4,160
1905	318	5,320
1906	1439	45,189
1907	357	10,589
1908	750	14,250
1909	638	15,950
1910	773	25,277
1911	929	40,058

In 1911 the platinum consumed in the United States had a value of about 5,000,000 dollars.

CANADA.—In Canada, platinum is known in Quebec and British Columbia.[2] In Quebec, the auriferous gravels of the Du Loup and Plantes rivers in Beauce County carry some platinum, the primary deposit of which has not yet been located. Better known are the platinum gravels

---

[1] After Clark, loc. cit.

[2] J. F. Kemp, ' The Geological Relations and Distribution of Platinum and Associated Metals,' U.S. Geol. Survey, Bull. No. 193, Washington, 1902, p. 36 ; Hintze, Handbuch der Mineralogie, Vol. I. p. 142.

of British Columbia.[1] These occur together with auriferous gravels in
the south-west of the country, where the occurrence has proved to be by
far the most productive of such deposits in North America. Upon the
exhaustion of the Californian gravels the gold seekers penetrated north-
west into the hills, till in 1861 they found other deposits in British
Columbia. There, the discoveries in the Similkameen River, of which near

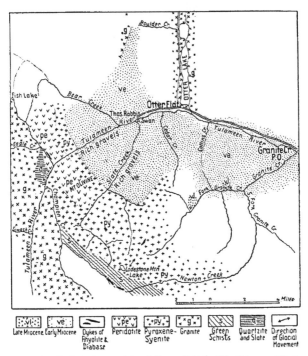

Fig. 467.—Geological map of the platinum district of the Tulameen River,
British Columbia. Kemp.

Princeton the Tulameen is a northern tributary, aroused great interest.
Gold mining however assumed no great importance, but from the beginning
the occurrence of platinum was noted. In 1885 this metal was found
together with gold in the sands of the Granite Greek, a tributary of the
Tulameen, about 12 miles above its junction with the Similkameen. The
districts richest in platinum occur on the Eagle Creek, a northern tributary

[1] M. Dawson, *Ann. Rep. Geol. Canada*, 1887, 3 R. ; ‘ Rep. on the Geology of the
Kamloops Map Sheet,’ *Geol. Survey of Canada*, Vol. VII. Part B. ; W. J. Watermann,
‘ Economic Geology of the Similkameen District, British Columbia,’ *Mining Rep. Vancouver*,
November 1900, p. 411.

of the Tulameen, and on the Slate Creek, a southern tributary of the same river, about 8 miles below the Granite Creek. A map of this region is given in Fig. 467.

The relation of platinum to gold in these occurrences is about 1 : 3 ; in the tributaries of the Tulameen above the Eagle Creek the platinum content gradually diminishes. With these occurrences also, the precious metal is derived from olivine rocks. The deposits have been fairly closely studied by the geologists mentioned.[1] The character of the metal may be gathered from the following analyses of 18·266 grm. of material, of which 17·894 grm. consisted of platinum, and the remainder of magnetite, some pyrite, and gold. The specific gravity of the platinum was 16·686 ; for purposes of examination it was separated into a magnetic portion equal to 37·88 per cent, and a non-magnetic portion to 66·12 per cent, the results being as follows :

ANALYSES OF PLATINUM FROM THE TULAMEEN RIVER,
BRITISH COLUMBIA

	Magnetic.	Non-magnetic.	Together.
Platinum	78·43	68·19	72·07
Palladium . . .	0·09	0·26	0·19
Rhodium	1·70	3·10	2·57
Iridium	1·04	1·21	1·14
Copper	3·89	3·09	3·39
Iron	9·78	7·87	8·59
Iridosmium . . .	3·77	14·62	10·51
Chromite	1·27	1·95	1·69
	99·97	100·29	100·15

BRAZIL.—Platinum is found in this country in auriferous sand at Corrego das Lages, Condado Serro on the Rio Abaete in Minas Geraes, and in Matto Grosso ;[2] at all these places diamonds are also found. The gold at Gongo Socco[3] in Minas Geraes contains palladium, and Brazilian gold in general, some platinum or iron. Iridium and iridosmium have been remarked in the auriferous gravels at Minas.[4]

[1] Ante, p. 1200.
[2] Hintze, Handbuch der Mineralogie, Vol. I. ; Hussak, ' Palladium und Platin in Brasilien,' Zeit. f. prakt. Geol., 1906.
[3] Ante, p. 623.
[4] J. F. Kemp, The Geological Relation and Distribution of Platinum and Associated Metals, Bull. No. 193, Washington, 1902.

GEOGRAPHICAL INDEX

Bindt, Hungary, 806
Bingerbrück, Taunus, 815, 862, 863, 864, 866
Bingerloch adit, Taunus, 866
Bingham Canon, Utah, 876, 877, 888, 943
Birkenberg, Bohemia, 705, 706, 707
Birmingham dis., United States, 1028, 1030
Birstal, Switzerland, 998
Birtavarre, Norway, 304
Bisbee, Arizona, 397, 877, 878, 880, 888
Bischofshofen, Austria, 904
Bislich, Germany, 1018
Bissade lode, France, 782
Bissersk, Urals, 344, 345, 1216, 1217
Bitburg, Rhenish Prussia, 1184
Bizen, Japan, 588
Björnevand, Norway, 264
Blaafjeld, Norway, 14, 257
Black Hills, Dakota, 447, 451, 525, 1167-1169
Black Hill mine, Victoria, 614, 616
Black Range dis., Arizona, 888
Black Reef, Witwatersrand, 1154, 1155, 1158
Blauberg, Germany, 1184
Bleiberg, Carinthia, 42, 44, 659, 722, 740-746
Bleischarley mine, Silesia, 727
Bliesenbach mine, Berg dis., 693
Blokken, Norway, 266
Blötberg, Sweden, 392
Blücher mine, Berg dis., 696
Blue mountains, Australia, 873
Blue Spur, New Zealand, 1168, 1207
Blue Tier mines, Tasmania, 446
Boccheggiano, Tuscany, 371, 873, 909-912
Bockau, Erzgebirge, 680
Bockswiese, Harz, 684, 685
Bodenmais, Bavaria, 173, 302, 337-340, 1134
Bodenstedt, Harz, 1047, 1048
Bodner fissure, Hohe Tauern, 631
Bogen, Norway, 1056, 1058
Bogoslowsk, Urals, 360, 485, 900, 1208
Bohemia, 204, 205, 601, 634-636
Boicza, Hungary, 518, 524, 547
Bokhara, Russian Turkestan, 1211
Boleo, Mexico, 168, 878, 890, 944, 1169, 1181-1184
Boleslaw, Russia, 725
Bolivia, South America, 168, 184, 185, 202, 205, 211, 215, 217, 218, 219, 423, 470, 515, 527, 529, 530, 578-585, 921
  statistics, 644, 647, 648, 940, 941, 942
Bömmelö, Norway, 188, 637
Bonenburg, Germany, 1013, 1019
Bonzel, Westphalia, 927
Bor, Serbia, 871, 875, 878

Börnecke, Harz, 1052
Borneo, East Indies, 343, 515
Borsa, Bukovina, 524, 1054
Bosnia, Austria, 249, 457, 657
  statistics, 1094, 1114
Bossmo, Norway, 304
Boulder Co., Colorado, 449, 525, 556, 557
Boundary dis., British Columbia, 398
Braastad, Norway, 378
Branch lode, Roudny, 635
Brand, Erzgebirge, 670, 674, 675
  Austria, 906
Bräunsdorf, Erzgebirge, 670
Brazil, South America, 259, 449, 470, 617-624, 851, 980, 1060-1062, 1092, 1108-1110, 1225
  statistics, 644, 1114
Breece Hill, Colorado, 763
Bresnay, France, 780
Briey, Lorraine, 1004
Brilon, Westphalia, 702
Brillador, Chili, 896
British Columbia, 342, 343, 398, 1224
  statistics, 554
British India, 1110-1112
  statistics, 644, 1114
Brixlegg, Tyrol, 657, 870, 872
Brocken, the, Harz, 684, 687
Brodenstein, Germany, 1052
Broken Hill, New South Wales, 121, 145, 180, 215, 219, 399-402
Brosso, Piedmont, 301, 373, 1097
Brown Face, Mount Bischoff, 445
Brownhill Extended mine, Western Australia, 593
Brück, Prussia, 1187
Bruseh, Malay Peninsula, 440
Brussa, Asia Minor, 783
Buchberg, Austria, 906
Büchelbach bed, Bieber, 845
Büchenberg, Harz, 1081
Buchholz, Erzgebirge, 677
Buck Creek, Alaska, 1196
Buena Fortuna mine, Spain, 831
Bukovina, Austria, 249
Bulgaria, statistics, 1094
Buller dis., New Zealand, 1207
Bullwhacker mine, Butte, 886
Bülten, Harz, 1046, 1047, 1048, 1049
Burdaly, Greece, 248
Burgberg, Hesse, 846
Burra-Burra, South Australia, 200, 216, 880, 898
Bussanga dis., Africa, 1196
Butte, Montana, 163, 165, 168, 199, 201, 214, 215, 216, 523, 558, 649, 871, 874, 877, 879, 880, 882, 883-889, 942
Buxweiler, Vosges, 1151

Pages 1-487 are contained in Vol. I., and pages 515-1225 in Vol. II.

Konjhar, India, 1212
Kons. Schlossberg and Dachsbau mine, Taunus, 866
Köppern, Taunus, 864, 866
Kordelio mine, Asia Minor, 783
Korea, Asia, 644, 1210
Kosaka, Japan, 897
Kosemitz, Silesia, 958
Koskulls Kulle, Sweden, 392
Kotschkar, Siberia, 1207
Kotterbach, Hungary, 806
Kragerö, Norway, 277, 278
Krappitz, Silesia, 723
Kraubat, Styria, 244, 249
Kremnitz, Hungary, 182, 524, 526, 527, 535, 536, 539-542
Kresevo, Bosnia, 872
Kressenberg, Bavaria, 1021
Kreuth, Carinthia, 743, 744
Křitz, Bohemia, 780
Kriwoj Rog, Russia, 1055, 1058-1060, 1088, 1096
Krompach, Hungary, 806
Kroutoi, Siberia, 1209
Krug-von-Nidda mine, Westphalia, 735
Kuen-Lun mountains, Tibet, 1211
Kufferath, Germany, 1185
Kumara, New Zealand, 1207
Kupferberg, Silesia, 37, 239, 350, 402-404, 922
Kupferplatte, the, Tyrol, 906, 907
Kuso, Sweden, 298
Kutais, Russia, 1100, 1101, 1104-1106
Kuttenberg, Bohemia, 659
Kwei Chan, China, 486

La Buena mine, Spain, 833
La Carolina, Spain, 709, 711
La Creu, Spain, 478, 479
La Cruz lode, Spain, 710
Lady Macdonald mine, Sudbury, 291
Lady Violet mine, Sudbury, 291
Lafayette, Brazil, 1109
La France, South Africa, 625
La Higuera mine, Chili, 892, 896
Lahn dis., Prussia, 1072-1078, 1089
Lake Superior, United States, 163, 170, 188, 198, 201, 346, 398, 877, 878, 928-937, 979, 1055, 1062-1071, 1088, 1092, 1096
Lake View mine, Western Australia, 203, 593
La Licoulne, France, 781
Lam, Bavaria, 340
Lampun, Siam, 1196
Lancashire, England, 161, 1088, 1097
Lancaster Gap mine, Pennsylvania (see also Gap mine), 946

Land of the Thousand Lakes, United States, 984
Långban, Sweden, 140, 382, 387, 389, 390, 966
Langeland, Prussia, 1013, 1019
Langerfeld, Westphalia, 736
Langesundfjord, Norway, 389
Langö, Norway, 378
Lang-Suan, Siam, 1196
Langvand, Norway, 901
La Peña, Spain, 316
Lapilla, Spain, 316
Las Caberas, Spain, 449
Las Prolongas lode, Spain, 711
La Touche, Spain, 659
Laubach, Hesse, 1000
Laurion, Greece, 145, 722, 728, 746-749
Lauterberg, Harz, 811, 870
Lauthenthal, Harz, 684, 686
La Villeder, France, 430
La Zarza, Spain, 316
Leadville, Colorado, 722, 760-765
Leavenworth Gulch, Colorado, 716
Lebong Soelit, Sumatra, 83, 589
Legrana, Greece, 746, 748
Lehrbach, Harz, 166
Lejana mine, Spain, 833
Lena dis., Siberia, 1208
Lengede, Hanover, 1046, 1047, 1048
Leo lode, Salzburg, 949
Leonovsky, Siberia, 1210
Leopoldine mine, Rhenish Schiefergebirge, 698
Letmathe, Westphalia, 735
Leubetha, Erzgebirge, 809
Leveäniemi, Sweden, 270, 274
Leversbach, Germany, 1184
Levins lode, Mississippi, 769
Lich, Hesse, 1000
Lidell, California, 465
Linares, Spain, 659, 709-711
Lincoln, England, 1027, 1088
Lindal Moor, England, 824, 825
Lindener Mark, Hesse, 161, 815, 862, 863, 867-869
Lintorf main lode, Westphalia, 704
Liruen-Kano, Nigeria, 1196
Little Bendigo, Western Australia, 614
Little Stobie mine, Sudbury, 291
Llallagua, Peru, 581
Llano County, Texas, 875
Loben, Carinthia, 821
Lobenstein, Erzgebirge, 809
Lochborn bed, Bieber, 845
Lodenblek, Harz, 1083
Lofoten Islands, 173, 253, 260, 266-269, 276
Logrosan, Portugal, 451

Michigan, United States, 556, 872
Michipicoten, Canada, 1062
Middle Reef, Witwatersrand, 1155
Miechowitz, Silesia, 723, 725
Mieres, Spain, 478
Miguel Burnier, Brazil, 1100, 1102, 1109
Mikultschütz, Silesia, 723
Mileschau, Bohemia, 780
Milluni, Bolivia, 581, 584
Milos, Greece, 1107
Mina Blanca, Atacama, 947
Minas Geraes, Brazil, 342, 617-624, 1045, 1055, 1060-1062, 1092, 1109, 1225
Mine Hill, New Jersey, 394
Mine la Motte, Missouri, 770, 946
Mineral Creek, Arizona, 888
Minna mine, Westphalia, 1130, 1131
Minne, Norway, 451
Mino, Japan, 588
Miranda, Portugal, 1195
Miravilla mine, Spain, 828
Misdroy, Germany, 1150
Mississippi, United States, 768-771
Missouri, United States, 140, 556, 768-771
Misvärtal, Norway, 294
Mitchell lode, Rhodesia, 627
Mittelberg, Thuringia, 854
Mitterberg, Salzburg, 788, 879, 903, 904-906, 945, 949
Mizusawa, Japan, 897
Modum, Norway, 162, 206, 947, 949, 1136, 1137-1140
Mohawk mine, Lake Superior, 932
Moldava, Hungary, 356, 359
Mommel, the, Thuringia, 817, 837-839
Moncanita, California, 461
Mönchenberg, Harz, 859
Montana, United States, 447, 523, 530, 558
  statistics, 554, 556
Montaro, Spain, 431
Monte Amiata, Tuscany, 458, 461, 471-474, 909
Monte Axpe, Spain, 828
Monte Blanco, Peru, 581, 584
Montebras, France, 430, 451
Montebuono, Tuscany, 472
Monte Catini, Tuscany, 300, 876, 877, 879, 880, 909
Monte Christo series, Victoria, 614
Monte Cruvin, Italy, 949
Monte-Cucchedu lode, Sardinia, 752
Monte Fumacchio, Tuscany, 410
Montel, France, 782
Monte Mulatto, Tyrol, 873
Monte Nova lode, Sardinia, 752
Monteponi, Sardinia, 728, 750, 753, 756
Monterey, Portugal, 430

Monte Rosa, Italy, 280, 281, 298
Monte Rotondo, Tuscany, 909
Montesinhos, Portugal, 430
Monte Somma, Italy, 350, 389
Montesund Avion hills, Portugal, 430
Monte Valerio, Tuscany, 410
Montevecchio, Sardinia, 750-752, 756, 910
Monte Vitozzo, Tuscany, 472
Monthon Chumpon, Siam, 1196
Montieri, Tuscany, 910
Montignat, France, 781
Moonta, Australia, 200, 871, 874, 877, 879, 880, 882, 898
Moravicza, Hungary, 357
Moresnet, Prussia, 731, 733
Morgen lode, Kupferberg, 403
Morgenrot lode, St. Andreasberg, 688
Morococha, Peru, 579, 581, 871
Morro Santa Anna mine, Brazil, 617
Morro Velho, Brazil, 618, 622
Moschellandsberg, Italy, 457, 458, 463, 483
Moschwig, Prussia, 1151
Moskedal mine, Norway, 309
Mosquito range, Colorado, 760
Moss, Norway, 189
Mother lode, California, 69, 601, 604, 608, 609, 610
Mottram St. Andrews, England, 1185
Mount Bischoff, Tasmania, 217, 444-446
Mount Blezard, Sudbury, 291
Mount Dere, New Caledonia, 249
Mount Dun, New Zealand, 244
Mount Lyell, Tasmania, 877, 899, 943
Mount Morgan, Queensland, 138, 640-643
Mount Tagora, Borneo, 487
Mücke, Harz, 1000
Muczari, Transylvania, 524, 546
Mug mine, Norway, 305, 310
Mühlberg, Silesia, 958
Mühlenweg, Harz, 1083
Münstergewand fault, Aachen, 732
Murcia, Spain, 452, 1089
Murray mine, Sudbury, 283, 291
Müsen, Siegerland, 7, 796
Mutsu, Japan, 860, 861
Mützhagen mine, Belgium, 733, 734
Myslowitz, Silesia, 725

Nagamatsu, Japan, 897
Nagorny, Siberia, 1210
Nagyag, Hungary, 168, 182, 219, 523, 524, 530, 539, 543-546
Nagybanya, Hungary, 165, 524, 536, 539, 542
Nakagawa, Japan, 782
Nakase, Japan, 782
Nammern-Klippen bed, Wesergebirge, 1021

SUBJECT INDEX

Those references marked with an asterisk denote Illustrations

Cementation zone—*continued*
  pyrite and arsenopyrite lodes, 921
  uranium-silver-gold lodes, 716
Cerargyrite, 85, 86, 145, 218, 219, 548, 561, 575, 679
Cerium, the metal, 152, 160
  the ore, 103
Cerussite, 86, 87, 541, 548, 715, 737, 740, 744, 753, 764, 1174, 1179
Cervantite, 101
Chalcanthite, 881, 894
Chalcedony, 104
Chalcocite, 89, 90, 674, 871, 872, 881, 885, 894, 915, 916, 917, 918, 919, 1116, 1124, 1130, 1131, 1170, 1183, 1184
  argentiferous, 85
Chalcopyrite, 89, 90, 91*, 104, 301-340, 397, 433, 608, 673, 674, 800, 871, 872, 881, 890, 894, 898, 905, 911, 1116, 1119, 1124, 1133, 1136, 1138
  argentiferous, 85
  auriferous, 76
Chamber deposits, 37-41, 720
Chambered veins, the term, 68
Chamosite, 92, 94
  beds, 1032, 1039-1044, 1096, 1098
Chirta, 831, 832
Chloantite, 95, 96, 667, 680, 905, 946
Chlorine, 150, 160, 172, 184, 219, 975
Chromite, 99, 347, 1161, 1213
  deposits, 173, 244-249, 344, 957
Chromium, the metal, 153, 156, 169, 172, 173, 206, 208, 347, 981
  the ore, 99, 207
Chrysocolla, 89, 90, 881, 886, 919, 934, 1183
Chrysoprase, 952, 961
Cinnabar, 84, 136, 457, 465, 468, 472, 476, 657
  deposits. *See* Quicksilver deposits
Classification of ore-deposits, in general, 12, 228-241
  Beck, 237
  Bergeat, 237
  Beyschlag, Krush, and Vogt, 238-241
  Burat, 228, 229
  Callon, 230
  von Cotta, 229
  Fuchs and de Launay, 234
  Grimm, 230
  von Groddeck, 230
  Gürich, 235
  Höfer, 235
  Kemp, 233
  Köhler, 230
  Lottner-Serlo, 230
  Naumann, 230
  Neve Foster, 230
  Phillips, 231

Classification of ore-deposits—*continued*
  Pošepný, 234
  Stelzner, 232
  Vogt, 236
  von Waldenstein, 228
  von Weissenbach, 228
  Werner, 228
  Whitney, 228, 231
  into sub-groups—
    copper deposits, metasomatic, 910
    copper deposits, native, 929
    copper lodes, 875, 878, 901
    copper, silver, and gold deposits, 892
    gold-silver lodes, young, 524, 525, 541, 545, 553, 555
    lead-silver-zinc lodes, 655
    oolitic iron-beds, 1002
    pyrite beds, 1141
Clay-ironstone, 92
  deposits, 45*, 1020, 1031-1039, 1095, 1098
Clay-partings (*see also* Gouge, and Preface, Vols. I. and II.), 637, 686, 695, 700, 705, 708*, 822, 915, 1145
Cobalt, the metal, 153, 156, 162, 168, 169, 172, 206, 207, 208, 347, 668, 873, 965, 981, 1099, 1100, 1136
  the ore, 97-99, 168, 206
  lodes (*see also* Asbolane deposits), 25, 27*, 206, 212, 793, 800, 801, 948, 1120, 1125, 1129, 1130, 1137
Cobalt-arsenopyrite, 97, 946, 1135, 1138
Cobalt-bismuth lodes. *See* Silver-cobalt-bismuth lodes
Cobalt fahlbands (*see also* Fahlbands), 162, 949, 1132, 1135-1141, 1172
Cobalt-nickel lodes. *See* Nickel-cobalt lodes
Cobalt-silver lodes. *See* Silver-cobalt lodes
Cobalt-uranium lodes, 655
Cobaltite, 97, 98*, 667, 801, 898, 946, 1135, 1138, 1140
Cockade ore (*see also* Concentric ore), 114, 652
Combed structure (*see also* Banded and crusted structures), 532, 675
Compact structure, 120, 399, 780, 853
Composite lodes (*see also* Preface, Vol. I.), 37*, 38-40, 540, 542, 576, 591, 593, 626, 650, 651, 686, 695, 884
Concentric crusted structure (*see also* Crusted structure), 113, 114, 115*, 120*, 652, 721, 802
Concentric ore, 114, 652, 686
Concretionary deposits, 45, 46, 1031
Concretionary structure (*see also* Oolitic structure), 982, 987, 994, 1002, 1017, 1108

Pages 1-487 are contained in Vol. I., and pages 515-1225 in Vol. II.

Iodine, 150, 160, 975
Iodyrite, 85, 86, 145, 218
Iridium, 83, 342, 1161, 1214, 1225
Iron, the metal, 137, 139, 152, 156, 160-162,
  167, 206, 207, 208, 342, 851, 981,
  1085, 1099
  the ore, 74, 92-94, 161, 196, 207, 1084-
  1099
  the deposits, 196, 1092, 1095-1099, 1172
  contact, 160, 179, 350, 353, 354-394,
    422, 1054, 1097, 1098
  gravels, 832, 847, 848, 993, 999, 1044-
    1054, 1061
  lodes, 116, 211, 212, 213, 655, 688, 706,
    786-811, 859, 949, 968, 1097, 1098
  magmatic, 14, 128, 160, 173, 194, 196,
    250-277, 1054, 1097, 1098
  metasomatic, 43*, 181, 197, 550, 812-
    850, 1054, 1096, 1098
  ore-beds, 979-1084
    bean ore, 990-998, 1001, 1002, 1095,
      1098
    blackband and clay-ironstone, 1031-
      1039, 1095, 1098
    bog and lake ore, 982-988, 1096, 1098
    chamosite and thuringite, 1032,
      1039-1044, 1096, 1098
    detrital (see also Gravels), 1044-
      1052, 1096, 1098
    oolitic, 1000-1031, 1095, 1098
    sands, 1052-1054, 1096
    sedimentary, in crystalline schists,
      1054-1071, 1096, 1098
    with metasomatic, 1072-1084
Iron cap. See Gossan
Iron-manganese ore (see also Manganese-
  iron ore), 161, 814
  deposits, 41, 169, 862-869, 980
  lodes, 674, 675
  ore-beds, 988-990, 1111
Irregular coarse structure, 109-111, 399,
  599, 607, 652, 686, 695, 700, 788, 799,
  802, 807, 813, 863, 908, 920, 925
Itabirite deposits (see also Ferruginous
  mica-schists), 113, 194, 626, 1054, 1060-
  1062, 1092, 1109

Jactuinga deposits (see also Itabirite deposits), 616, 619, 623, 1061
Jamesonite, 86, 87, 543, 707, 1189
Jasper, 861, 1113
Joints and jointing, 64, 65
  block, 64
  cylindrical, 65
  irregular polyhedral, 64
  prismatic or columnar, 65
  quadrangular, 65, 1048
  spheroidal, 65

Joints and jointing, tabular, 64
Juvenile springs, 135, 136, 935

Kaolinization, 134, 318, 362, 434, 526, 540,
  582, 918
Kidney ore (see also Botryoidal and Pencil
  ore), 119, 789, 811, 838, 859
Kinks in the lode, 49, 1021
Klump, 983
Knotten ore-bed. See Nodular lead deposits
Kolm, 1143
Korallenerz, 84, 460, 480
Krennerite, 76, 79, 80, 82, 543, 595
Kupferkniest, 1148
Kupferschiefer. See Copper-shale beds

Ladder lodes, 65, 616
Lake ore, 161, 192, 193, 977, 982-988,
  1096, 1098, 1099, 1100, 1101, 1103
Lateral displacement, 7*, 17, 26, 28*, 32-
  34, 48, 319, 434, 637, 690, 698, 787,
  798, 856, 905, 1081
Lateral secretion theory, 148, 157, 186, 189,
  190, 534, 736, 745, 770, 961, 963,
  967
Lead, the metal, 154, 155, 156, 164, 168, 172,
  206, 207, 208, 774
  the ore, 86, 87, 774-776
  native, 140, 1211
  the deposits—
    lodes, 675, 750, 823, 1176
    metasomatic, 738
    ore-beds (see also Nodular lead de-
      posits), 194, 195*, 1169-1180
Lead-copper lodes, 901
Lead-silver deposits—
  lodes, 164, 214, 422, 434, 515, 523, 526,
    529, 542, 550, 600, 649, 674, 675
  metasomatic, 760-765
Lead-silver-copper lodes, 901
Lead-silver-zinc deposits—
  contact, 121, 180, 399-402
  lodes, 186-188, 210, 428, 532, 542, 650-
    711, 944
  metasomatic, 168, 530, 649, 717-773
Lead-zinc deposits—
  contact, 398, 530
  lodes, 8, 190, 652, 693-696, 700-703, 770,
    793, 798, 799, 801, 870
  metasomatic, 41, 43, 109, 114, 218, 770,
    946
Lebererz, 84, 460, 480
Lievrite, 370, 410
Limonite, 92, 93, 674, 789, 808, 813, 814,
  831, 839, 841, 844, 847, 991, 998, 1007
  1016, 1021, 1028, 1046, 1052, 1069
  deposits. See Iron deposits
  lodes, 702

Prices of metals and ores—*continued*
rutile, 208
silver, 208
sulphur, 208
thorium, 208
tin, 208
tungstic acid, 208
zinc, 208, 774
Primary zone, 10, 75, 221
copper lodes, 883, 894
gold lodes, old, 602, 625, 637
gold-silver lodes, young, 594
lead-silver-zinc deposits, metasomatic, 218, 721
lead-silver-zinc lodes, 214, 652, 653, 679, 696
pyrite beds, 1142
uranium lodes, 714
Production of metals and ores, figures of—
aluminium, 207
antimony, 207, 778
arsenic, 207, 668
bismuth, 207, 681
cadmium, 207
chromium, 207
cobalt, 207, 668, 947
copper, 207, 327, 436, 887, 889, 890, 893, 895, 897, 899, 900, 911, 912, 932, 939-944, 1127, 1184
gold, 207, 538, 539, 544, 552, 554, 574, 585, 598, 643, 644-646, 893, 1164
iron, 207, 391, 392, 810, 834, 988, 1011, 1012, 1024, 1070, 1084-1099
lead, 207, 539, 574, 699, 730, 756, 771, 774-776, 1178
manganese, 207, 1114
molybdenum, 207
nickel, 207, 293, 668, 947, 956, 963-965
platinum, 207, 1219, 1223
pyrite, 303, 326, 327
quicksilver, 207, 463, 468, 470, 475, 478, 482, 485
silver, 207, 538, 539, 552, 556, 574, 583, 585, 647-649, 668, 675, 699, 730, 893, 1127
tin, 207, 424, 436, 443, 444, 585, 1194
uranium, 682
wolfram, 207
zinc, 207, 699, 730, 756, 771, 774-776
Propylite and propylitization, 134, 143, 185, 518-522, 531, 540, 571, 589, 590
Proustite, 85, 86, 525, 540, 575, 581, 662, 673, 885
Pseudo-brecciated structure. *See* Brecciated structure
Psilomelane, 94, 95, 789, 854, 859, 865, 1107, 1110

Pyrargyrite, 85, 86, 525, 540, 542, 575, 581, 673, 690, 885
Pyrite, 104, 215, 457, 651, 657, 673, 720, 727, 735, 763, 802, 920, 927, 975, 1136, 1138, 1141, 1148, 1161, 1166, 1209
argentiferous, 85, 673, 715
auriferous, 76, 602, 607, 610, 612, 622, 636, 715
cobaltiferous, 97
cupriferous, 89, 90, 119, 301-340, 889, 897, 899, 943
nickeliferous, 95
stanniferous, 99
deposits, 108, 160, 163, 165, 199, 213, 215, 631, 633, 878, 889, 899
intrusive, 173, 194, 301-340, 876, 878, 899, 920, 943, 978, 1134
lodes, 920-923
metasomatic, 923-928
ore-beds, 212, 1141-1151
Pyritization, 635, 1161
Pyrolusite, 94, 95, 790, 799, 853, 855, 859, 865, 867, 1048, 1107, 1110
Pyromorphite, 86, 87, 541, 702, 735, 764
Pyrosphere. *See* Barysphere
Pyrrhotite, 104, 301, 620, 623, 625, 627, 889, 923, 1136, 1138, 1171
cobaltiferous, 97
cupriferous, 89, 90
nickeliferous (*see also* Nickel-pyrrhotite deposits), 95

Quartz, 104, 105, 149, 413, 522, 531, 588, 629, 642, 651, 680, 686, 707, 780, 781, 800, 801, 968
auriferous, 601, 607, 611, 615, 618, 620, 623, 625, 628, 629, 638, 642
Quartz-copper lodes, 874, 875, 879, 891
Quartz-silver lodes, 655, 657, 673, 674
Quartz-tourmaline lodes, 637
Quartz-tourmaline-copper lodes, 874, 877, 879
Quicksilver, the metal, 84, 154, 156, 169, 206, 207, 208, 657
the ore, 84, 204
native, 456, 457, 470, 487
the deposits, 182, 185, 190, 204, 456-487, 530, 657, 909, 970
Quicksilver-tetrahedrite, 84, 482, 657

Radium, 102, 154, 711, 1143
Rare metals and earths (*see also* Platinum metals), 103, 151, 152, 154, 160, 162, 163, 675, 1195
Realgar, 101, 737, 777
Reefs, 602, 651
mullocky, 616
saddle, 63, 399, 401, 611

Structure of ores, 108-125, 652
  banded (*see also* Crusted and combed),
    111-117, 308, 350, 377, 380, 447, 453,
    686, 695, 799, 911, 919, 925, 959, 966,
    978, 1034, 1054, 1145
  bedded, 1171
  brecciated and pseudo-brecciated, 117,
    118*, 261, 273, 275, 284, 302, 377,
    433, 478, 545, 576, 608, 695, 721, 734,
    741, 754, 788, 813, 853, 856, 863,
    1046
  cellular, 721
  combed (*see also* Banded and crusted),
    532, 675
  compact, 120, 399, 780, 853
  concretionary (*see also* Oolitic), 982, 987,
    994, 998, 1002, 1017, 1108
  crusted and concentric crusted (*see also*
    Banded and combed), 111-117, 176*,
    417, 441, 453, 532, 589, 607, 652, 675,
    686, 700, 707, 720, 721, 735, 740, 780,
    788, 799, 802, 813, 853, 874, 885, 966
  drusy (*see also* Vuggy), 118-120, 122*,
    123*, 215, 788, 808, 831, 1176
  irregular coarse, 109-111, 399, 599, 607,
    652, 686, 695, 700, 788, 798, 802, 807,
    813, 863, 908, 920, 925
  microscopic, 123-125, 652
  net, 802
  oolitic (*see also* Concretionary), 982,
    1017, 1040, 1051
  reniform, 215, 788, 995
  stemmed, 622
  vuggy (*see also* Drusy), 690, 695, 853, 885
Subsidences (*see also* Preface Vol. II.), 29,
  796, 1008
Sulphide cobalt - nickel lodes (*see also*
  Arsenical cobalt-nickel lodes), 162,
  223, 965
Sulphide enrichment (*see also* Secondary
  enrichment), 216, 886, 894
Sulphide lead-copper lodes, 673, 674
Sulphide lead-zinc lodes, 190, 674, 680
Sulphide quartz-lead lodes, 655, 656, 657,
  659
Sulphur, the element, 136, 150, 166, 208,
  278-340, 667
  the ore, 73, 104
  native, 104, 208, 216
  deposits, 133, 190
Sweet springs, 135
Sylvanite, 76, 80, 82, 543, 545, 565, 595
Synclinal fissures, 62
Synclines, 17-20
Syngenetic deposits, 12, 13-34

Tectonic depressions. *See* Subsidences and
  Preface Vol. II.

Tectonic elevations. *See* Uplifts and Preface
  Vol. II.
Tectonic fissures and lodes, 61, 62, 433, 518,
  547, 798
Telluride gold, 79-83, 169, 188, 525, 543,
  593, 595, 603, 764
Telluride lodes, 40, 83, 169, 525, 563-566,
  1202, 1204, 1205
Tellurium, 151, 166, 167, 169, 184, 715
Temperature, 136, 278, 517, 562, 933, 966,
  1165
  critical, 132, 133, 147, 353, 419, 934, 967
Tenorite (*see also* Melaconite), 881, 894
Tension fissures, 63
Tetrahedrite, 89, 90, 525, 540, 581, 599,
  870, 872, 894, 907
  argentiferous, 85, 86
  quicksilver, 84, 457, 482, 905
Thorium, the metal, 152, 172, 208
  the ore, 103
Thuringite, 92, 94
  beds, 1032, 1039-1044, 1096, 1098
Tin, the metal, 154, 156, 167, 169, 206, 207,
  208, 417, 418, 657, 709, 873, 911, 921,
  981
  the ore, 99, 205, 423
  wood-tin, 99, 217, 548, 582, 921
  the deposits—
    contact, 405-411, 909
    gravels, 4*, 16*, 205, 423, 431, 436,
      437-446, 1190, 1192-1197
    lodes, 134, 143, 175, 186, 190, 204, 205,
      211, 217, 352, 412-431, 441, 444-448,
      516, 583, 657, 674, 870, 968, 970
Tin-copper lodes, 167, 209, 216, 421, 423,
  431-436, 870, 874, 878
Tin-rock, 416
Tin-silver lodes, 185, 205, 217, 423, 526,
  580-583
Tin-wolfram lodes, 421
Titaniferous-iron deposits, 160, 173, 250-
  259, 269, 1054, 1097, 1098
Titaniferous-iron sands, 1052-1054, 1096
Titanium, 151, 167, 172
  lodes, 135, 422
Topaz and topazification, 415
Torrsten, 276, 382, 390
Torsion fissures, 63
Tourmaline and tourmalinization, 184, 433,
  445, 603, 620, 808, 870, 873, 874, 892,
  893, 894, 902, 903, 947, 1166, 1193
Transverse fissures, 62, 63
Trough subsidences. *See* Subsidences and
  Preface Vol. II.
Tungsten. *See* Wolfram
Tungstic acid, 208
Types of deposit—
  Australian-Californian, 630

1262   ORE-DEPOSITS

Pages 1-487 are contained in Vol. I., and pages 515-1225 in Vol. II.

END OF VOL. II

THE DEPOSITS

OF THE

USEFUL MINERALS & ROCKS

THEIR ORIGIN, FORM, AND CONTENT

BY

PROF. DR. F. BEYSCHLAG, PROF. J. H. L. VOGT, AND
PROF. DR. P. KRUSCH.

TRANSLATED BY S. J. TRUSCOTT.

In 3 vols. Illustrated. 8vo.

Previously Published.

VOL. I. **18s.** NET.

ORE-DEPOSITS IN GENERAL.—MAGMATIC SEGREGATIONS
—CONTACT-DEPOSITS—TIN LODES—QUICKSILVER LODES.

SOME PRESS OPINIONS.

NATURE.—" A special feature of the work is the abundance and excellence of the illustrations. Each section is preceded by a bibliography, and the value of the book as a work of reference is thus greatly increased. Mr. Truscott has done his work so well that in reading the book one is apt to forget that it is a translation."

KNOWLEDGE.—" It can unhesitatingly be said that the authors have succeeded in their difficult task, and it is impossible to praise too highly the skill with which they have correlated the great accumulation of evidence. Furthermore, they have been eminently successful in their presentation of the genetic principles, and have shown that this branch of geology has a considerable claim to rank as a rational science. . . . The value of the book is enhanced by a large number of excellent illustrations indicative of the geological structure of the various deposits. . . . A most useful and readable book, which can be thoroughly recommended to all interested in the subject."

SCIENCE PROGRESS.—" The illustrations are numerous and excellent. Mr. Truscott is to be congratulated on this translation which makes available to English-speaking students the latest and finest Continental work on ore-deposits. We shall look forward with pleasure to the continuation of his work."

ATHENÆUM.—" Mr. Truscott's work as translator has been performed with conscientiousness and judgment."

MINING MAGAZINE.—" On the accomplishment of this self-imposed task Mr. Truscott deserves the congratulations and thanks of English-speaking geologists and engineers. . . . The work of translation has been admirably done. The sense of the original has been clearly brought out, and the rendering into English is excellent."

MACMILLAN AND CO., LTD., LONDON.

STANDARD WORKS ON
GEOLOGY AND MINERALOGY

THE WITWATERSRAND GOLDFIELDS:

BANKET AND MINING PRACTICE

With an Appendix on the Banket of the Tarkwa
Goldfield, West Africa

By S. J. TRUSCOTT

With numerous Illustrations. Third Edition.

Super Royal 8vo. 30s. net.

SOUTH AFRICA.—" The author has been at immense pains to bring together interesting data concerning the colossal work of crushing and extracting the gold from the Witwatersrand Banket. . . . The chapters on prospecting and mining will convey useful hints even to the oldest and most initiated heads, while the sections dealing with machinery, such as winding appliances, pumps and pumping, air compressors, rock drills, and other mechanical adjuncts, may be profitably perused by home manufacturers. . . . We wish it all the success it undoubtedly deserves."

MINING JOURNAL.—" It can be highly recommended to mining engineers and students all over the world, and more especially to those connected with mining on the Witwatersrand. It should be added that the get-up of the whole work in general, and the quality of the illustrations in particular, are in every way worthy of the reputation of the famous house that has published it."

GOLD MINES OF THE RAND. By Frederick H.
Hatch (Mining Engineer) and J. A. Chalmers (Mining Engineer). With Maps, Plans, and Illustrations. Super Royal 8vo. 17s. net.

GEOLOGY OF SOUTH AFRICA. By Dr. F. H.
Hatch and Dr. G. S. Corstorphine. Illustrated. Second Edition. 8vo. 21s. net.

GOLDFIELDS OF AUSTRALASIA. By K. Schmeisser.
Translated and Edited by Prof. Henry Louis. Illustrated. Medium 8vo. 30s. net.

CANADA'S METALS. A Lecture. By Sir W. C.
Roberts-Austen, F.R.S. 8vo. 2s. 6d. net.

MACMILLAN AND CO., Ltd., LONDON.

STANDARD WORKS ON
GEOLOGY AND MINERALOGY

MINERALOGY. An Introduction to the Scientific Study of Minerals. By Sir H. A. MIERS, F.R.S. 8vo. 25s. net.

MINERALOGY. By Prof. A. H. PHILLIPS, D.Sc. 8vo. 16s. net.

A TREATISE ON ORE DEPOSITS. By J. ARTHUR PHILLIPS, F.R.S. Illustrated. Second Edition. Revised by Prof. HENRY LOUIS. 8vo. 28s.

CRYSTALLOGRAPHY AND PRACTICAL CRYSTAL MEASUREMENT. By A. E. H. TUTTON, D.Sc., F.R.S. Illustrated. 8vo. 30s. net.

TEXT-BOOK OF GEOLOGY. By Sir ARCHIBALD GEIKIE, K.C.B. Illustrated. Fourth Edition. 2 vols. Medium 8vo. 30s. net.

GEOLOGY AND GEOGRAPHY OF NORTHERN NIGERIA. By J. D. FALCONER, D.Sc. With Notes by the late ARTHUR LONGBOTTOM, B.A., and an Appendix by HENRY WOODS, M.A. Illustrated. 8vo. 10s. net.

THE GEOLOGY OF NOVA SCOTIA, NEW BRUNS-WICK, AND PRINCE EDWARD ISLAND; OR, ACADIAN GEOLOGY. By Sir J. W. DAWSON, LL.D., F.R.S. Fourth Edition. 8vo. 21s.

A TREATISE ON ROCKS, ROCK-WEATHERING, AND SOILS. By Prof. GEORGE P. MERRILL. 8vo. 17s. net.

POPULAR LECTURES AND ADDRESSES. Vol. II. GEOLOGY AND GENERAL PHYSICS. By LORD KELVIN. Crown 8vo. 7s. 6d.

MACMILLAN AND CO., LTD., LONDON.

SOME STANDARD WORKS ON
METALLURGY AND CHEMISTRY

A HANDBOOK OF METALLURGY. By Dr. CARL SCHNABEL. Translated by Prof. H. LOUIS. . Second Edition. Medium 8vo. Vol. I. 25s. net. Vol. II. 21s. net.

A HANDBOOK OF GOLD MILLING. By Prof. HENRY LOUIS. Third Edition. Crown 8vo. 10s. net.

A COMPLETE TREATISE ON INORGANIC AND ORGANIC CHEMISTRY. By The Right Hon. Sir H. E. ROSCOE, F.R.S., and C. SCHORLEMMER, F.R.S. 8vo.

Vol. I. THE NON-METALLIC ELEMENTS. Fourth Edition. Revised by Sir H. E. ROSCOE, assisted by Dr. J. C. CAIN. 21s. net.

Vol. II. THE METALS. Fifth Edition. Revised by Sir H. E. ROSCOE and others. 30s. net.

INDUSTRIAL CHEMISTRY FOR ENGINEERING STUDENTS. By Prof. HENRY K. BENSON, Ph.D. Crown 8vo. 8s. net.

BLOWPIPE ANALYSIS. By FREDERICK HUTTON GETMAN. Globe 8vo. 2s. 6d. net.

BLOWPIPE ANALYSIS. By J. LANDAUER. Third Edition. Translated by J. TAYLOR. Globe 8vo. 4s. 6d.

A TEXT-BOOK OF ELECTRO-CHEMISTRY. By Prof. MAX LE BLANC. Translated from the 4th enlarged German Edition by WILLIS R. WHITNEY, Ph.D., and JOHN W. BROWN, Ph.D. 8vo. 11s. net.

ACHIEVEMENTS OF CHEMICAL SCIENCE. By JAMES C. PHILIP, M.A., D.Sc. Globe 8vo. 1s. 6d.

APPLIED ELECTRO-CHEMISTRY. By Prof. M. DE KAY THOMPSON. 8vo. 9s. net.

OUTLINES OF INDUSTRIAL CHEMISTRY. A Text-Book for Students. By FRANK HALL THORP, Ph.D. 8vo. 16s. net.

MACMILLAN AND CO., Ltd., LONDON.

Lightning Source UK Ltd.
Milton Keynes UK
UKHW010430150119
335567UK00011B/589/P